Lecture Notes on
Human Physiology

Lecture Notes on Human Physiology

EDITED BY

JOHN J. BRAY PhD

PATRICIA A. CRAGG PhD

ANTHONY D.C. MACKNIGHT PhD, MD, FRSNZ

ROLAND G. MILLS PhD

DOUGLASS W. TAYLOR MD, FRCPE

All of the Department of Physiology
University of Otago Medical School
Dunedin, New Zealand

THIRD EDITION

Blackwell
Science

© 1994 by
Blackwell Science Ltd
Editorial Offices:
Osney Mead, Oxford OX2 0EL
25 John Street, London WC1N 2BL
23 Ainslie Place, Edinburgh EH3 6AJ
238 Main Street, Cambridge
    Massachusetts 02142, USA
54 University Street, Carlton
    Victoria 3053, Australia

Other Editorial Offices:
Arnette Blackwell SA
1, rue de Lille
75007 Paris
France

Blackwell Wissenschafts-Verlag GmbH
Kurfürstendamm 57
10707 Berlin
Germany

Blackwell MZV
Feldgasse 13
A-1238 Wien
Austria

First published 1986
Reprinted 1987
Second edition 1989
Reprinted 1989, 1990, 1992, 1993
Third edition 1994
Reprinted 1995

Set by Semantic Graphics, Singapore
Printed and bound in Great Britain
at the University Press, Cambridge

DISTRIBUTORS

Marston Book Services Ltd
PO Box 87
Oxford OX2 0DT
(Orders: Tel: 01865 791155
        Fax: 01865 791927
        Telex: 837515)

North America
Blackwell Science, Inc.
238 Main Street
Cambridge, MA 02142
(Orders: Tel: 800 215-1000
            617 876-7000
        Fax: 617 492-5263)

Australia
Blackwell Science Pty Ltd
54 University Street
Carlton, Victoria 3053
(Orders: Tel: 03 347-5552)

A catalogue record for this title
is available from the British Library

ISBN 0-632-03644-3
ISBN 0-632-03722-9 (Four Dragons)

Library of Congress
Cataloging-in-Publication Data

Lecture notes on human physiology /
    edited by John J. Bray ... [et al.]. –
    3rd ed.
        p.   cm.
        Includes bibliographical references and
        index.
        ISBN 0-632-03644-3
        1. Human physiology.
        I. Bray, John J., PhD.
QP34.5.L44   1994
612—dc20

Contents

List of Contributors

This book was written by members of the Department of Physiology, University of Otago Medical School, Dunedin, New Zealand.

Dr J.V. Allison BMedSc, MB, ChB, PhD
Chapters 10–12
Dr D.P.G. Bolton MA, MSc, BM, BCh, PhD, MRCP
Chapters 20, 21
Dr J.M. Bowler MB, ChB, PhD
Chapter 17
Dr J.J. Bray BAgrSc, PhD
Chapters 2, 3, 10, 11
Dr P.A. Cragg BSc, PhD
Chapters 13–16
Dr E.R. Fawcett BSc, MB, ChB, FFARCS, FFARACS, DIC
Chapter 20
Associate Professor A.J. Harris BSc, PhD
Chapters 5, 7
Dr W.G. Hopkins BSc, BA, MSc, PhD
Chapter 21
Professor J.I. Hubbard BMedSc, MA, DM, PhD, FRACP, FRSNZ
Chapters 3, 8, 9
Dr B.I. Hyland BMedSc, MB, ChB, PhD
Chapters 8, 9
Dr E.M. Irvine BSc, MB, ChB, DA
Chapter 12
Associate Professor J.P. Leader BA, PhD
Chapters 18, 19
Professor A.D.C. Macknight BMedSc, PhD, MD, FRSNZ
Chapters 1, 2, 17–19
Dr R.G. Mills BSc, PhD
Chapters 3–5, 17
Emeritus Professor J.R. Robinson MA, MD, PhD, ScD, FRACP, FRSNZ, FNZIC
Chapters 1, 18, 19
Dr P.W. Sheard BSc, PhD
Chapters 3, 5
Dr P.M.E. Waite* BSc, PhD, MD, ChB
Chapters 4, 6, 7
Illustrators: Mr R.W. McPhee, Mr N.E. Still

* Present address: School of Anatomy, The University of New South Wales, Kensington, NSW, Australia.

Preface to the Third Edition

In writing a textbook of this nature one has to serve two masters — the student and the lecturer. In this edition we have attempted to please the former by improving the layout with short summaries to each section, and to accommodate the latter by the inclusion of what might be considered more advanced material in small print. Overly long chapters have been split into shorter ones and a new chapter, Cellular Communication, has been included to keep up-to-date with rapid advances in cellular physiology. All chapters have been revised and in a number of cases reorganized to make them more readable. Otherwise our intended readership remains the same as in the first edition; we have intentionally tried to cover somewhat more than is called for in a first-year course in physiology.

We are grateful to a number of people at the Otago Medical School for their contributions to this third edition, namely Dr C.P. Bolter for his comments on the chapters dealing with cardiovascular physiology, Professor B. Heslop and Dr M.A. Baird for writing the subsection 'The immune system' and Professor R.D.H. Stewart for help in revising Chapters 10 and 11.

Once again it has been a great pleasure to work with Blackwell Science Limited.

John J. Bray, Patricia A. Cragg, Anthony D.C. Macknight,
Roland G. Mills, Douglass W. Taylor

Preface to the First Edition

Writers of textbooks, like Dr Johnson's Lexicographer, are unwise if they aspire to praise and lucky indeed if they escape reproach. This is especially likely to be the case if the textbook attempts to cover in a rather short space, as befits the Lecture Notes series, a very big subject like physiology.

It has not been our aim to replace any of the large standard texts, but rather to produce a book that can be read through reasonably rapidly during, or in revising at the end of, the sort of course in human physiology that is commonly taught to a variety of students in pure science, medicine, dentistry, pharmacy and the like. To this end we have largely omitted detailed discussion of experimental evidence and have refrained from giving references to the original literature. We have also deliberately devoted but little space to applied physiology and to clinical illustration in the belief that these are commonly taught later in the course to students that require them.

The contents of the book largely represent what is taught in the Physiology Department of the University of Otago to students from five different faculties, most of whem study physiology for one year in their second year at university. While the individual chapters have naturally been written by specialists, all have been carefully read and fairly stringently edited by a small panel in an attempt to give a greater uniformity of style than is sometimes found in volumes of multiple authorship. The editors must therefore take the responsibility for the final text.

We are indebted to Professor J.R. Robinson, who began this book and continued to nurture it during his retirement. We also gratefully acknowledge the contributions of the other members of the Department of Physiology, University of Otago Medical School, and of our two illustrators. We wish to thank Mrs S.M. Keast for her patience and care in typing, proof-reading and indexing the manuscript, Mrs L.J. Donaldson for typing assistance and Drs J.G. Blackman and R.J. Harvey for their helpful comments. We thank those authors and publishers who generously gave us permission to reproduce figures or data, the sources of which are acknowledged in the legends. Finally, we are indebted to Mr Per Saugman and his colleagues at Blackwell Scientific Publications. Mr Saugman saw the need for such a book and gently nursed it to fruition.

John J. Bray, Patricia A. Cragg, Anthony D.C. Macknight,
Roland G. Mills, Douglass W. Taylor

Chapter 1
Cell Physiology

Physiology is the science of the functions and phenomena of living things. This book deals mainly with human physiology. Our bodies contain vast numbers of cells, which are the fundamental units of all living organisms.

1.1 CELLS

Physiology is about how cells are kept alive and about how their environment is maintained so that the integrity of the cells is preserved.

Mammalian cells are very small, of the order of 10^{-5} m in diameter. This is about midway between humans (of the order of 1 m) and atoms (10^{-10} m). A human has about as many cells—about 10^{14}—as a cell has molecules.

The detailed structure of cells is not apparent in the light microscope, with its limit of resolution of 0.2 µm. The electron microscope, with resolution down towards 1 nm (10^{-9} m), has revealed a complex ultrastructure. In life this is not static; it is maintained by continually assimilating matter from the environment. The cells, even of terrestrial animals, are tiny aquatic organisms—open systems that, like flames and waterfalls (and unlike crystals), need continuous supplies of matter and energy to maintain their form. The details of metabolism are the domain of biochemistry. Physiology deals rather with getting supplies to the cells at suitable rates and removing waste products. Each cell is surrounded by fluid (**extracellular fluid, ECF**) and the normal functioning of the cells requires that the composition of this bathing fluid is maintained constant.

This extracellular fluid is a middle-man fluid, a medium through which occurs all exchanges between cells and the external environment. The famous French physiologist Claude Bernard (1813–1878) called the ECF an internal environment in which the cells live a secluded life protected from the vicissitudes of a sometimes hostile external environment.

The constancy, necessary for the well-being of the cells, is maintained by **homeostatic mechanisms** that monitor and regulate temperature, osmotic pressure, pH and composition of the ECF. **Much of physiology is the story of homeostasis.**

1.2 HOMEOSTATIC MECHANISMS

The composition of the ECF, and therefore the environment of the cells, is maintained constant through homeostatic mechanisms that monitor and regulate the functions of the circulatory system, the alimentary system, the respiratory system and the renal system. This monitoring and regulation is coordinated through the nervous and endocrine systems and requires receptors and central integration.

A **homeostatic mechanism** is a regulating mechanism that is triggered by alteration in some physiological property or quantity and acts to produce a compensating change in the opposite direction. Minimum requirements are:

1 **receptors**, specialized to detect alterations in particular variables;
2 **effectors**, e.g. the circulatory system that carries nutrients and O_2 to the cells and removes metabolic waste products, including CO_2; the alimentary system that provides nutrients to the body; the respiratory system that carries out exchanges of gases with the external environment; the renal system that allows losses of unwanted solutes and water from the body; and the musculoskeletal system;
3 **coordinating and integrating mechanisms**, linking **1** to **2**. These are nervous and hormonal.

Nervous system

In essence this consists of **afferent nerve fibres** linking receptors to coordinating systems in the **brain** and **spinal cord** and **efferent nerve fibres** that carry information from the coordinating systems to the effector organs. There are two major subdivisions of the efferent system: the **somatic nervous system** uses striated, skeletal muscles as effectors for purposive behaviour and reflex actions and adjustments; and the **autonomic nervous system** sends its efferents to glands, the heart and smooth muscle in hollow organs and blood vessels.

Endocrine system

This comprises a group of ductless glands that secrete **hormones** into the blood which affect the function of **target cells** throughout the body. *The actions are generally slower and less sharply localized than those of the nervous system; they* are, however, under the control of the nervous system through hypothalamic and pituitary hormones which influence the other endocrine glands.

1.3 THE IMPORTANCE OF WATER IN THE BODY

About 60% of the body is water—its special properties are uniquely suited to life. Of this water, some 55% is in the cells and 45% is extracellular. The intracellular fluid (ICF) is high in K^+, the positive charge on which largely balances the negative charges on organic solutes. The ECF is high in Na^+ and Cl^-.

Properties of water

All substances exchanged between cells and their environment travel and react in solution in water, which is the main bulk constituent of all living systems. The biological importance of water depends upon a number of peculiar properties.

1 It is a liquid at ordinary temperatures.
2 It has a large heat of fusion.
3 It has a large heat capacity.
4 It has a large heat of vaporization.

The first of these is explained by hydrogen bonding; the others by the energy required to break bonds during melting, warming and vaporization.

5 It has a maximum density at 4°C.
6 It has a large dielectric constant.

This reduces electrostatic forces 80 times and makes water a superb solvent for ionic compounds. Moreover, water dipoles are strongly attracted to dissolved ions and to charged surfaces, coating these with relatively immobile layers of water molecules which greatly modify the properties of ions in solution.

Distribution of water in the body

In a man, about 60% of body weight is due to water. A 70-kg man therefore has about 42 litres. In a woman, the relatively greater percentage of the body weight contributed by fat cells (which contain little water) means that about 55% of body weight is water. A 50-kg woman therefore has about 27.5 litres of water.

This water is shared between the cells and the ECF which surrounds and supplies them. Difficulties in measuring the ECF volume has resulted in a variety of published values for cellular and extracellular water contents. Older figures, based on inadequate equilibration of markers throughout the extracellular volume, suggest that two-thirds of the water is cellular and one-third extracellular. A better approximation, which allows for the extracellular water in dense connective tissue and bone, is that ~55% of the total body water is intracellular and ~45% extracellular. Using these ratios, it can be calculated that a 70-kg man would have 23 litres of intracellular water and 19 litres of extracellular water. Three litres of the extracellular water is in the **blood plasma**; 15 litres (the **interstitial fluid** including lymph) provides an aquatic habitat for the cells and the remaining 1 litre represents the **transcellular fluids** (cerebrospinal fluid (CSF)), ocular, pleural,

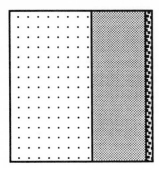

Fig. 1.1 The major subdivisions of the body water.
□ Cell water, ▦ interstitial fluid, ▦ plasma.

peritoneal and synovial fluids). Comparable approximate values for a 50-kg woman are intracellular water 15 litres, extracellular water 12.5 litres, plasma 2 litres and interstitial fluid 10 litres. The major subdivisions of body water are often illustrated as in Fig. 1.1, which is a two-dimensional representation with 55% cellular fluid and 45% ECF.

Clinically, the best guide to changes in body water is alteration in body weight. Total body water may be measured by desiccation. However, body water and water compartments are usually measured by methods that rely on the notion of volume of distribution.

Principle

A known amount of solute is injected intravenously. Then blood samples are obtained until a steady state concentration of the injected substance is reached. Values have to be corrected for any urinary losses.

$$\text{Volume of space} = \frac{\text{amount given} - \text{amount lost}}{\text{conc. in plasma } H_2O}$$

The volumes of the compartments can be estimated from the volumes of distribution of substances thought to equilibrate in different compartments. For example, total body water has been estimated from the volume of distribution of urea and isotopes of water (deuterium oxide, tritiated water). Similarly, inulin, sucrose, mannitol and isotopes of Na^+ and Cl^- have been used to estimate extracellular water, and isotopically labelled albumin to estimate plasma water. Interstitial water cannot be measured directly but is (ignoring the transcellular volume) the difference between extracellular volume and plasma volume.

The intracellular and the extracellular water have different solutes dissolved in them and so constitute two distinct fluids, the **ICF** and the **ECF**. Skeletal muscle cell fluid and plasma are typical of ICF and ECF, respectively. The **ICF** has mainly **potassium** with **organic anions**; ECF has mainly **sodium** and **chloride**, rather like diluted sea water. Their approximate compositions in mmol L^{-1} are shown in Table 1.1.

1.4 PASSIVE EXCHANGE OF WATER AND SOLUTES

The heart and blood vessels comprise a great transport system which carries water and solutes, including hormones and gases (O_2 and CO_2), throughout the body, and helps to keep the ECF evenly mixed. Circulating blood does not actually reach

Table 1.1 Blood plasma and muscle cell fluid as typical extracellular and intracellular fluids, respectively

	Plasma (mmol L^{-1})	Muscle (mmol L^{-1})
Na$^+$	150	10
K$^+$	5	150
Ca^{2+}	2	10^{-4}
Mg^{2+}	1	10
Cl$^-$	110	5
HCO$_3$$^-$	27	10
H$_2$PO$_4$$^-$	2	
Org P$^-$	5	130
Prot^{17-}	1	2
pH	7.4	7.1
Osmolarity	285 mosmol L^{-1}	285 mosmol L^{-1}

Plasma constituents are commonly expressed clinically per litre of plasma rather than per kilogram of water. Because protein occupies a finite volume, these values are somewhat lower than the values given in the table.

cells. In the tissues, **capillaries** bring blood within 5–10 µm of most cells; but to get from blood to the interior of a cell, solutes like glucose and O$_2$ must:

1 cross the capillary wall;
2 cross a layer of interstitial fluid between capillary and cell; and
3 cross the plasma membrane, which separates the ICF from the ECF.

The walls of most capillaries are relatively leaky and **diffusion** is the *dominant mechanism by which solutes and water cross the capillary wall* as well as being the mechanism by which they move through the interstitial fluid.

Diffusion

Diffusion is a consequence of the thermal energies of the individual molecules. It results in net flux of each species down its own gradient. For uncharged solutes, the rate of this net flux is given by Fick's law. Diffusion is very rapid over the distances within cells and between cells and capillaries.

Diffusion occurs as the spontaneous result of random thermal motions that tend to disperse accumulations of molecules and make concentration uniform. For example, water floating on copper sulphate solution shows diffusion—water diffusing into the solution, and copper sulphate diffusing into the water. Movements of molecules through the solution by diffusion are called **fluxes**, not flows, for the liquid as a whole does not move. Individual molecules move randomly and independently.

Consider movements of uncharged solute x in the imaginary situation illustrated in Fig. 1.2. At time zero assume that a solution of this solute can be placed only in one half of the container (region 1) and that water occupies the other half (region 2). Substance x will immediately begin to diffuse into the water in region 2, driven by its concentration in region 1. Some of the molecules of x arriving in region 2 will return to side 1 driven by the concentration of x in region 2. These

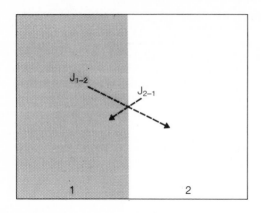

Fig. 1.2 Unidirectional fluxes. See text for explanation.

fluxes of x are termed unidirectional fluxes and can be expressed as:

$$J_{1-2} \propto x_1$$
$$J_{2-1} \propto x_2$$

where J_{1-2} represents the unidirectional flux from region 1 to region 2, J_{2-1} represents the unidirectional flux from region 2 to region 1 and x_1 and x_2 represent the concentrations of x in the two regions.

At any time, the net flux, J_{net}, can be represented as:

$$J_{net} = J_{1-2} - J_{2-1}, \text{ which is proportional to } x_1 - x_2)$$

In time, diffusional movements will result in the same concentration of x in regions 1 and 2. Now the two unidirectional fluxes will be equal but opposite and net flux will be zero.

More rigorously, for **uncharged molecules**, the **net flux** (i.e. the amount moving per unit time), is given by **Fick's law**:

$$\frac{dn}{dt} = -DA\frac{dc}{dx}$$

Here dn/dt is the number of molecules per unit time diffusing in the x direction across an area A at right angles to the x axis and dc/dx is the concentration gradient (the rate at which concentration increases with distance along the x axis). D, the diffusion coefficient for each substance, depends upon its relative molecular mass (M_r), the temperature and the viscosity of the solvent.

The minus sign is needed because net transport is down the gradient, from higher to lower concentration, and is therefore '**downhill' transport.** This is usually referred to as **passive transport** because no energy is needed beyond that inherent in the gradient. The driving force for diffusion is the gradient of concentration of the diffusing substance, and Fick's law states that the rate of transport is proportional to that driving force. This leads to automatic adjustment of the rate of transport to the demand. For example, if a cell uses O_2 or glucose faster, the gradient for diffusion from capillary to cell increases.

Rate of diffusion can also mean **distance covered in unit time**. The theory was developed by Einstein in 1905 for Brownian movement considered as a

Table 1.2 Time taken to reach 99% equilibrium by diffusion of solute from a plane boundary in water at room temperature (Jacobs 1935). (From Robinson, J.R. (1981) *A Prelude to Physiology*, p. 26. Blackwell Sicentific Publications, Oxford)

Distance from boundary	Time
10 cm	53 days
1 cm	128 h
1 mm	76 min
100 μm	456 s
10 μm	0.0456 s
1 μm	0.000456 s
0.1 μm	0.00000456 s

'random walk'. Distance is proportional to the square root of the number of steps, hence to the square root of time. This means it takes four times as long to go twice as far, but half the distance is covered in a quarter of the time. **Diffusion is therefore very rapid over short distances.** Jacobs in 1935 calculated the times taken for 99% equilibration at various distances from a plane where a constant concentration was maintained in water at room temperature (Table 1.2). Note the great speeds over distances as small as those within cells.

The relative sizes of the intracellular and interstitial fluid compartments are usually represented (as in Fig. 1.1) in two dimensions. In reality, because cells are three dimensional and the interstitial fluid surrounds them, the thickness of the interstitial layer is much less than would appear from Fig. 1.1. This is illustrated in Fig. 1.3, which represents a cross-section through a spherical cell and its thin surrounding interstitial fluid. With a cell radius of 10 μm, the thickness of the interstitial fluid layer will be 1.08 μm. Thus the diffusional distances through the interstitial fluid are very small.

Passive movement across the cell plasma membrane

Lipid-soluble solutes may permeate the plasma membrane by dissolving in the lipid bilayer. Ions and water-soluble solutes of low M_r may cross the membrane through channels, or though carriers that are specific membrane proteins. This

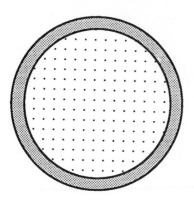

Fig. 1.3 The relationship between cell water and surrounding interstitial fluid for a spherical cell. ☐, Cell water; ▦, interstitial fluid.

permeation has the characteristics of specificity, saturation, competition, inhibition and a high temperature coefficient.

A membrane (about 7.5 nm in thickness) separates the cell interior from the surrounding interstitial fluid. This is properly referred to as the plasma membrane to distinguish it from the membranes within the cell (e.g. mitochondrial membrane). It is composed of a mosaic of globular proteins embedded in a lipid bilayer with the hydrophilic ends of the lipid molecules oriented towards the outside of the membrane (Fig. 1.4). Both the protein and lipid molecules are free to move in the lateral plane of the membrane.

Because of the nature of the lipid bilayer it is not surprising that the plasma membrane presents a formidable barrier to diffusing solutes.

Jacobs in 1952 calculated the times required for some diffusing solutes to reach 90% equilibrium in a region of water the same size as a red blood cell and compared these with actual times for red blood cells (Table 1.3a). Permeability of erythrocytes' membranes is therefore both low and extremely selective, even for solutes of low M_r.

Fick's law of diffusion can be modified for solute diffusion across plasma membranes. Assuming no significant concentration gradients within the interstitial

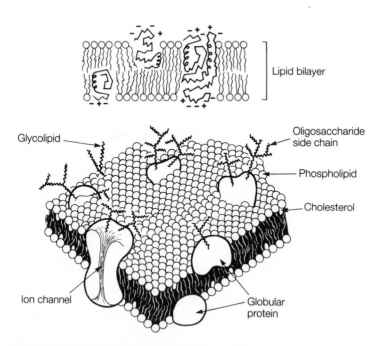

Fig. 1.4 The plasma membrane in cross-section (top) and in three dimensions (below). The globular proteins embedded in the lipid bilayer may be situated on one side of the membrane or extend through it. (Modified from Singer, S.J. & Nicolson, G.L. (1972) *Science* **175**, 720–731.)

Table 1.3 Times taken for diffusion of solutes into a red blood cell and a water droplet of similar size (Jacobs 1952). (From Robinson, J.R. (1981) *A Prelude to Physiology*, p. 53. Blackwell Scientific Publications, Oxford)

(a) Substance	M_r	Calculated time for water droplet(s)	Actual time for red blood cell	Slowing by membrane
Urea	60	0.00035	0.5 s	10^3
Glycerol	92	0.00049	67 s	10^5
Mannitol	182	0.00070	>2 days	10^8
Sucrose	342	0.00095	∞	∞

(b) Water	Permeability to gases, organic compounds and small anions	Permeability to cations
1	0–1 to 0.0001	0–0001 to 0.00000001

or cell fluids, the gradient which exists across the plasma membrane is:

$$\frac{\Delta c}{\Delta x} = \frac{c_{outside} - c_{inside}}{\text{membrane thickness } (\Delta x)}$$

Fick's law of diffusion can then be written as:

$$\frac{dn}{dt} = -D'A\frac{\Delta c}{\Delta x}$$

Here D' is a restricted diffusion coefficient for diffusion within the membrane. In practice, D', A and Δx are often unknown; they can, however, be incorporated into an empirical permeability coefficient $P' = -D'A/\Delta x$, i.e.

$$\frac{dn}{dt} = P'\,\Delta c$$

$$= P'(c_{outside} - c_{inside})$$

Although there are many exceptions, in general plasma membranes are more permeable to water than to solutes and, among solutes, far more permeable to gases, organic compounds and small anions than to cations. The orders of magnitude of permeabilities are related as shown in Table 1.3b.

Mechanisms of membrane permeation

These may be classified as follows.

1 Permeation by dissolving and diffusing in the membrane lipid, e.g. O_2 and CO_2, steroid hormones. In so far as plasma membranes offer a fairly complete layer of lipid between the ICF and ECF, this mechanism should favour lipid-soluble substances of low M_r.

For such substances there is a strong correlation between permeability and oil–water partition coefficients which measure the tendency of a solute to move out of an aqueous into a lipid environment. For a given driving force, permeation through the lipid bilayer should increase roughly in proportion to the membrane–water partition coefficient.

2 Permeation by temporary combination with a specific membrane component. This shows:

(a) **specificity**;

(b) **saturation kinetics**;

(c) **competition** between similar molecular species;

(d) **inhibition** and

(e) a **high temperature coefficient** (Q_{10}),

properties similar to reactions catalysed by enzymes. Movements of this type can be either through membrane *channels* or involve *carriers*.

Both channels and carriers are protein molecules that span the membrane lipid bilayer. Membrane carriers can be distinguished from channels in that a carrier mechanism implies a binding site that is exposed sequentially to one side or the other of the membrane (but never to both simultaneously), whereas a channel is accessible simultaneously from both sides of the membrane and contains one or more binding sites arranged in a transmembrane sequence. Movement mediated by a carrier is generally regarded as being slower by orders of magnitude than that through a channel. It is important to appreciate that these terms may simply describe two extreme states.

(a) **Membrane channels**, e.g. water, urea, ions. An individual channel can be highly selective for a particular ion or solute and exist in either an open or closed state. Channels open and close spontaneously. The fraction of the time that they remain open can be modified by interaction with specific molecules (ligands), by changes in membrane potential or by mechanical deformation. Thus we speak of ligand- or chemically gated channels (p. 37, e.g. the acetylcholine receptor channel involved in neuromuscular transmission) voltage-gated channels (e.g. the voltage-sensitive Na^+ channel involved in action potentials in excitable tissues) and mechanically gated channels (e.g. touch receptors).

The properties of channels are rapidly becoming known through a variety of powerful techniques, particularly patch-clamping (p. 38), which permits the measurement of current passing through a single open channel in a fragment of membrane. The rate of movement of a solute through a single open channel can be some 10^6–10^9 ions s^{-1}, a transport rate much greater than the highest known catalytic rates of enzymes or carriers, and comparable to the rates for free diffusion of ions in aqueous media. The permeability of the membrane may be regulated by activation or inhibition of single channels resident in the membrane and by the insertion into, or removal of, channels from the membrane.

(b) **Carrier-mediated transport**. This can be subdivided into **facilitated diffusion**, where the net movement is passive and driven by the electrochemical potential gradient (e.g. glucose exit from intestinal absorptive cell into the blood), and **primary** or **secondary active transport** (discussed below).

1.5 ENERGY-DEPENDENT TRANSPORT ACROSS PLASMA MEMBRANES

In active transport, movement occurs against the electrochemical potential gradient. There are two types. In primary active transport the consumption of metabolic energy is coupled directly to the movement, which is independent of the downhill movement of any other species. In secondary active transport the downhill movement of one species drives the uphill movement of another, and metabolic energy is not utilized directly. If the species move in the same direction across the membrane, this is termed co-transport; if the movements are in opposite directions, it is termed counter-transport.

Primary active transport

This has the following criteria:
1   it is **coupled directly to a continuous supply of energy**;
2   it is **independent** of the downhill movement of any other solute or of water;
3   it can be shown to occur **'uphill'**.

For **uncharged molecules**, uphill is from *lower to higher concentration*, i.e. against a chemical concentration gradient.

For **ions**, which are charged, uphill is *from lower to higher electrochemical potential* ($\tilde{\mu}$), for there is an electrical potential difference ($\Delta\psi$) across the plasma membranes as well as a chemical concentration gradient. The difference in electrochemical ptoential ($\Delta\tilde{\mu}$) for an ion of valency z between solutions is:

$$\Delta\tilde{\mu} = RT \ln \frac{c_1}{c_2} + zF\Delta\psi$$

chemical work + electrical work

where R is the gas constant ($8.314 \text{ J K}^{-1} \text{ mol}^{-1}$), T is the absolute temperature (K), and c_1 and c_2 are the concentrations in mol L^{-1} in solutions 1 and 2, respectively. Thus chemical work is expressed in J mol^{-1}. However, $\Delta\psi = \psi_2 - \psi_1$ is expressed in volts. As volts have the units J C^{-1}, Faraday's constant ($F = 96500 \text{ C mol}^{-1}$) is required to express the electrical potential term in the same units as the chemical potential term, so that the two can be summed.

Strictly speaking, ion activities should be used in these equations instead of concentrations. This allows for the effects of ion–ion interactions in the solutions. However, such interactions will be less the more dilute the solution and it is commonly assumed that the physiological solutions with which we are dealing are sufficiently dilute for concentration to be a reasonable approximation of activity.

Mechanism of active transport

At the molecular level primary active transport involves proteins which probably have their affinity altered by reacting with adenosine triphosphate (ATP). As well as the Na^+–K^+ pump, which is an Na^+–K^+ ATPase, there are plasma membrane Ca^{2+} and H^+ pumps which involve a Ca^{2+} ATPase and a H^+–K^+ ATPase, respectively.

As an example of active transport, consider the Na^+–K^+ pump. Cells accumulate K^+ and keep their intracellular concentration of Na^+ low, although they live in an ECF that is rich in Na^+ and poor in K^+. The membranes are permeable to both ions, and there is continuous exchange of both ions between ICF and ECF. To maintain their composition in a steady state, cells need energy from metabolism to expel Na^+, which diffuses into the cell, and to take up K^+, which diffuses out of the cell. Poorly metabolizing cells gain Na^+ and lose K^+ down their concentration gradients. If the cells recover, they take up K^+ and expel Na^+, both *against* gradients. These movements of Na^+ and K^+ are coupled directly to the consumption of metabolic energy in the form of ATP. The protein which moves the ions across the membrane is itself an ATPase. Its enzymatic function, which requires Mg^{2+}, is activated by cell Na^+ and extracellular K^+. Because it requires both ions for normal activation, it is often called the Na^+–K^+ ATPase. It is inhibited by cardiac glycosides, such as ouabain. It is generally assumed that 3 Na^+ are extruded from the cell and 2 K^+ taken up with each cycle.

Secondary active transport

The cellular Na^+ concentration is much lower than that of the surrounding interstitial fluid and the membrane potential is negative on the inside. There is, therefore, a favourable electrochemical potential gradient for Na^+ entry to the cells. The energy in this gradient is used to drive the net movements of other solutes against their potential gradients by coupling their movement across the membrane to that of the passive downhill movement of Na^+. *There is no direct coupling of metabolic energy to these movements.* Instead, some of the energy inherent in the Na^+ gradient across the plasma membrane generated by the Na^+–K^+ pump is dissipated by the coupled flow of Na^+ with accompanying solute across the membrane.

If both the Na^+ and the other solute move in the same direction, this is termed **co-transport**. Examples include the coupled entry of glucose with Na^+ (Fig. 1.5) and amino acids with Na^+ into small intestinal epithelial cells from the gut lumen, the movements of glucose and amino acids being against their own electrochemical potential gradients. If the Na^+ and the other solute move in opposite directions across the membrane, this is termed **counter-transport** (sometimes also referred to as **exchange**). Examples include the coupling of Na^+ entry to cells to Ca^{2+} extrusion (Fig. 1.5) and to H^+ extrusion, which plays a central role in the regulation of cell pH (see below). Note that secondary active transport can involve ions other than Na^+ (e.g. Cl^-–HCO_3^- counter-transport).

Why is active transport important?

1   Active transport is important for maintaining normal ionic concentrations, in particular of K^+, Ca^{2+} and H^+, which are essential for many intracellular activities.

2   Active transport is important for moving ions, water and other substances that accompany ions by sharing common carriers (i.e. co- and counter-transport) across plasma membranes, particularly in kidney, stomach and intestine.

3   Active transport maintains the gradients of ionic concentration which are the basis of **resting membrane potentials** and **action potentials** used for signalling in nerves and for the activation of muscles.

4   The active extrusion of Na^+ is important for regulating cellular volume.

Regulation of intracellular Ca^{2+} and pH

Although the concentrations of free ionized Ca^{2+} and H^+ are in the nanomolar range, regulation of both is essential for normal cell function. Regulation involves both membrane transporters and cellular processes.

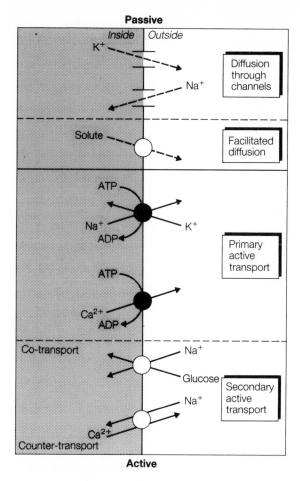

Fig. 1.5 Mechanisms of movement across plasma membranes. The solid horizontal line separates passive from active processes.

Although present in the cytosol at free concentrations (or, more correctly, activities) in the nmol range, Ca^{2+} and H^+ play important roles in the regulation of a variety of cell functions (p. 43) and so their concentrations need to be controlled precisely.

Intracellular Ca^{2+}

Cellular membranes generally have very low Ca^{2+} permeabilities and the concentration of free Ca^{2+} in the cytoplasm of inactive cells is regulated at around 10^7–10^{-8} mol L^{-1}. As shown in Fig. 1.6, Ca^{2+} diffusing into the cell from the interstitial fluid down its electrochemical gradient may be expelled either by primary active transport (Ca^{2+} ATPase) or by Na^+–Ca^{2+} counter-transport, an example of secondary active transport. Most Ca^{2+} in cells is compartmentalized or is bound to cellular constituents, such as membranes and binding proteins. Ca^{2+} may also be stored in the nucleus on binding proteins, or be transported

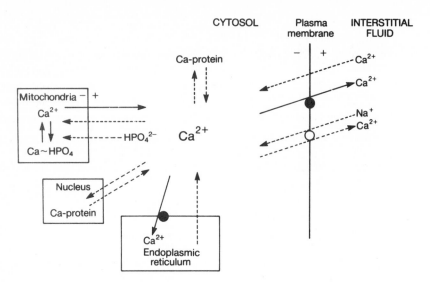

Fig. 1.6 Control of intracellular Ca^{2+}. The filled circles represent active transport; the open circle represents countertransport.

across the endoplasmic reticulum or mitochondrial inner membrane and stored in these organelles. If free ionized cytosolic Ca^{2+} is increased, both endoplasmic reticulum and mitochondria will rapidly accumulate the ion.

Mitochondria may accumulate calcium at the expense of ATP formation, causing calcium phosphate to precipitate. As a consequence, the mitochondria swell and their function may be irreversibly damaged. This may be of major importance in determining the extent to which ischaemic tissue will recover if perfusion can be restored.

Within cells, Ca^{2+} is an important regulator of many functions. It does this by influencing the activity of Ca^{2+}-binding proteins such as **troponin** and **calmodulin**. These proteins in turn control pathways that determine the activity of the cell (e.g. contraction, secretion). The importance of Ca^{2+} in controlling these pathways has led to its recognition as an important **intracellular messenger**. The level of free intracellular Ca^{2+} can rise as a result of an increase in the permeability of the plasma membrane to Ca^{2+} or its release from internal stores. Thus, action potentials in muscle and nerve terminals increase intracellular Ca^{2+} by opening voltage-dependent Ca^{2+} channels. Many neurotransmitters and hormones also act by opening ligand-dependent Ca^{2+} channels (either directly or indirectly via cyclic adenosine monophosphate (cAMP)) or by releasing Ca^{2+} from intracellular stores, such as the endoplasmic reticulum (via inositol trisphosphate; p. 43).

Intracellular pH

Many cell functions (e.g. enzyme activities, growth) are also influenced by cell pH, and changes in H^+ activity can affect both the activity of cytosolic Ca^{2+} and the levels of cAMP. These in turn influence a variety of specific cellular functions. Since H^+ ions are formed continuously as a consequence of cell metabolism, cells must be able to buffer H^+ ions and expel them to the interstitial fluid. In general, mammalian cell pH is approximately 7.0–7.2, i.e. slightly more acid than the plasma pH of 7.4. This is considerably more alkaline than the pH of about 6.5 predicted from the Nernst equation (p. 17) if H^+ ions were distributed across the plasma membrane in electrochemical equilibrium. A major contributor to the removal of H^+ from the cell is Na^+–H^+ counter-transport. Also of importance in cell pH regulation is HCO_3^- transport, for which a variety of pathways have been identified (e.g. Cl^-–HCO_3^- counter-transport and Na^+–HCO_3^- co-transport).

Endocytosis, exocytosis and transcytosis

Endocytosis, exocytosis and transcytosis represent energy-dependent processes by which molecules of larger M_r are encapsulated in membrane as they enter or before they leave the cells.

These mechanisms allow energy-dependent transport of material between cells and the interstitial fluid. They differ from those discussed above in that bulk movements occur by processes that involve portions of membrane (Fig. 1.7).

In **endocytosis**, invaginations of the plasma membrane first enclose extracellular material and then seal, forming vacuoles or vesicles within the cell that may fuse with lysosomes where they and their contents are degraded. Alternatively, the vesicular membrane may be recycled. Endocytosis is of particular importance for the uptake into cells of substances of large M_r, such as proteins, which do not otherwise cross plasma membranes.

For example, sensory and sympathetic neurones depend for their survival and development on the protein nerve growth factor, secreted by their target tissues and taken up by endocytosis at their axon terminals. This uptake, known as **receptor-mediated endocytosis**, involves specific high-affinity membrane receptors. There also exists a low-affinity non-specific mechanism of endocytosis at nerve terminals that permits uptake of exogenous proteins, e.g. horseradish peroxidase.

Pinocytosis and **phagocytosis** are forms of endocytosis. Pincocytosis ('cell drinking') refers to the endocytotic uptake of solutions into cells, whereas phagocytosis ('cell eating') refers to the engulfing of particulate matter, e.g. bacteria and viruses, by neutrophils and macrophages.

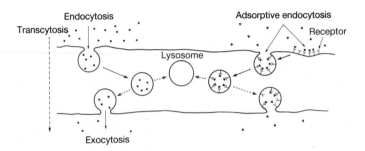

Fig. 1.7 Illustration of endocytosis, exocytosis and transcytosis.

Exocytosis is the reverse of endocytosis. Here substances are released from cells into the ECF by fusion of intracellular vesicles with the plasma membrane. This results in an increase in the area of the plasma membrane which is then reduced by endocytosis, giving rise to the so-called coated vesicles in the cytoplasm.

Important examples of exocytosis include the release of peptide hormones from endocrine glands, of enzyme precursors from exocrine glands in the gut, and of neurotransmitters from nerve terminals. (These processes appear to involve contractile proteins associated with the plasma membrane and are triggered by an influx of Ca^{2+} ions in response to specific stimuli.)

Transcytosis is a combination of endocytosis occurring at one membrane of a cell (endothelial or epithelial) and exocytosis occurring at the other membrane (Fig. 1.7). Thus a variety of macromolecules (e.g. plasma proteins) can be conveyed through the cell without degradation.

1.6 RESTING MEMBRANE POTENTIAL (RMP)

The RMP is largely a K^+ diffusion potential modified by the small finite permeability of the membrane to Na^+. Its magnitude can be estimated using the Goldman equation or modifications thereof. Although ultimately responsible for generating the ionic gradients on which the membrane potential depends, the Na^+–K^+ pump contributes only a few mV to the RMP as a consequence of its direct activity.

A difference of electrical potential is found across plasma membranes between ICF and ECF. This difference, called the **RMP**, has a magnitude of up to -80 to -90 mV with the cell interior being negative with respect to the interstitial fluid.

Diffusion potentials

The RMP is predominantly a **diffusion potential**. It arises because the plasma membrane is selectively permeable to ions.

Consider a simple membrane permeable only to K^+, separating two solutions of different KCl concentrations (Fig. 1.8). Initially, there will be net K^+ diffusion from the solution with the higher K^+ concentration driven by the chemical potential difference

$$RT \ln \frac{K_1}{K_2}$$

This will immediately create an electrical potential difference, with the side to which the K^+ is moving being positive. This electrical potential difference ($zF\Delta\psi$) will in turn increase the driving force for K^+ to move from side 2 to side 1. An equilibrium will be reached when the magnitudes of the two unidirectional fluxes of K^+ become equal (Fig. 1.8b). Now, the larger chemical gradient driving K^+ movement from side 1 to side 2 is balanced by the electrical gradient driving K^+ movement from side 2 to side 1. It is essential to appreciate that very few ions actually move during the establishment of the diffusion potential. For example, a

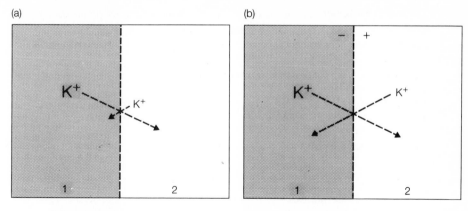

Fig. 1.8 A diffusion potential. The size of the K^+ symbols represents the relative magnitudes of the concentrations of the ion on sides 1 and 2.

change in potential difference of 100 mV would occur across 1 cm^2 of a membrane with properties similar to plasma membranes of cells if only 10^{-12} mol of K^+ were to move. In addition, this charge separation across the membrane would be confined to a distance of 10 nm.

Recall that, for any ion c, the electrochemical potential gradient across the membrane,

$$\Delta\tilde{\mu} = RT \ln \frac{c_1}{c_2} + zF\Delta\psi$$

(p. 11). At equilibrium there is no electrochemical potential gradient across the membrane, i.e. $\Delta\tilde{\mu} = 0$. Therefore,

$$-zF\Delta\psi = RT \ln \frac{c_1}{c_2}$$

Nernst equation

The above equation can be rearranged by dividing both sides by zF.

$$-\Delta\psi = \frac{RT}{zF} \ln \frac{c_1}{c_2}$$

The electrical potential difference ($\Delta\psi$) across the membrane at equilibrium is called the **equilibrium potential** and is given the symbol E. Therefore,

$$E = \frac{RT}{zF} \ln \frac{c_2}{c_1}$$

(Note the inversion of the concentration ratio to allow for the removal of the negative sign from the potential difference term, E.)

This is the **Nernst equation** introduced by the German theoretical chemist W. Nernst (1864–1941). In effect it states *either the maximum electromotive force* *that can be generated by a given ratio of concentrations of an ion, or the maximum ratio of concentration that can be sustained by a potential difference imposed from an external source.*

For use it is often convenient to put in actual values for the constants. This gives for monovalent ions at 37°C:

$$E = \pm 27 \ln \frac{c_2}{c_1} \quad \text{or} \quad E = \pm 61 \log_{10} \frac{c_2}{c_1}$$

where E has units of mV.

If the plasma membrane were permeable only to K^+ then the RMP would be the **equilibrium potential** for K^+ (E_K), with the magnitude given by the Nernst equation:

$$E(\text{inside}) = \pm 27 \ln \frac{c_{\text{outside}}}{c_{\text{inside}}}$$

where the sign is positive for cations and negative for anions. Therefore, using values for K^+ concentrations from Table 1.1, E_K is 27 ln 5/150 = 27 × – 3.33 = – 90 mV (inside negative). Similar calculations give for E_{Na}, 27 ln 150/10 = + 72 mV (inside positive) and for E_{Cl}, – 27 ln 110/5 = – 82 mV (inside negative).

The calculated equilibrium potential for K^+ (E_K) is somewhat more negative than the measured RMP of most cells, namely – 70 to – 80 mV. The reason is that the plasma membrane is not totally impermeable to Na^+, and Na^+ ions diffusing in down their gradient carry some positive charge into the cell and thus reduce the actual RMP to values less negative than E_K.

Equations for estimating RMP

The contribution of the diffusion of other ions to the RMP is expressed in the **Goldman equation**:

$$\Delta \psi = \frac{RT}{F} \ln \frac{P_K K_o + P_{Na} Na_o + P_{Cl} Cl_i}{P_K K_i + P_{Na} Na_i + P_{Cl} Cl_o}$$

where $\Delta \psi$ is the membrane potential, P is the membrane permeability to the ion denoted by its subscript, o indicates the outside concentration and i the inside concentration.

Note that because Cl^- is an anion, the Cl^- concentrations are inverted in the Goldman equation.

A simpler version of the above equation, which assumes that Cl^- is passively distributed across the membrane and that, therefore, there is no net Cl^- diffusion, is the **Hodgkin–Katz equation**:

$$\Delta \psi = \frac{RT}{F} \ln \frac{K_o + bNa_o}{K_i + bNa_i}$$

where b is the ratio P_{Na}/P_K and is taken to be 0.01 for the resting axon. The equation predicts $\Delta\psi = 27\ \ln(5 + 1.5)/(150 + 0.1) = 27 \times -3.07 = -83$ mV.

The good agreement of this predicted value with the measured RMP supports the idea that the RMP is largely a K^+ diffusion potential modified by the membrane's small permeability to Na^+ ions.

This is confirmed by the fact that the measured RMP in isolated nerve fibres varies as the Hodgkin–Katz equation predicts if internal or external K^+ concentration is changed experimentally. Also, if we calculate E_{Cl} using the Nernst equation, it is -82 mV, which is not significantly different from the measured RMP. Thus, as Cl^- ions are apparently in electrochemical equilibrium at least in nerve and skeletal muscle, the omission of Cl^- in the Hodgkin–Katz equation is justified for these tissues.

Rather than writing the Goldman equation in terms of ionic permeabilities, it is often expressed in terms of conductances (G, Ω or S), which can be related to membrane permeabilities (p. 78), but are more easily measured experimentally. If the total current across the membrane is zero, as it must be in the steady state, the membrane potential is given by:

$$\Delta\psi = \frac{E_K G_K + E_{Na} G_{Na} + E_{Cl} G_{Cl}}{G_K + G_{Na} + G_{Cl}}$$

where E represents the equilibrium potential and G the conductance of the membrane to ions indicated by the subscripts.

What direct effect does the Na^+–K^+ pump have on RMP?

Since the Na^+–K^+ pump creates the unequal distributions of K^+ and Na^+ between cells and ECF, its activity is indirectly responsible for the RMP. A pump which moved one K^+ in for each Na^+ out would carry no net charge and, therefore, no current across the membrane. However, the pump is not electrically neutral but transfers three Na^+ out for every two K^+ carried in. Thus current is generated directly by the pump, which is therefore termed **electrogenic**. However, provided that the membrane is much more permeable to K^+ than to Na^+, this electrogenic ion transport contributes relatively little to the resting membrane potential, which is dominated by the potassium diffusion potential.

To take account of this electrogenic transport, the Hodgkin–Katz equation can be modified by introducing the term n, where n is the coupling ratio for the pump (i.e. number of K^+ transported to Na^+ transported in each pump cycle):

$$\Delta\psi = \frac{RT}{F} \ln \frac{K_o + nbNa_o}{K_i + nbNa_i}$$

1.7 CELL VOLUME REGULATION

Cell volume reflects cell water content. Since plasma membranes are relatively permeable to water (p. 9), water will be distributed between cells and interstitial fluid to maintain constant the **activity of water** in each compartment.

Water activity is affected by solutes and by hydrostatic pressure. Water activity in a solution is decreased as the total concentration of solute particles is increased. It is increased when hydrostatic pressure is increased.

A difference in water activity between compartments will result in a net movement of water across the membrane down its activity gradient—a process termed **osmosis**.

Osmosis and osmotic pressure

Osmosis is the net movement of water across a membrane driven by a gradient of water activity. Osmotic pressure is the hydrostatic pressure required to prevent osmosis. Osmolarity expresses the total number of osmotically active particles per litre of water.

An **ideal semipermeable membrane** prevents diffusion of solutes but not solvent. Net movement of water across such a membrane (**osmosis**) will occur down the gradient of water activity and, therefore, from the more dilute to the more concentrated solution. It can be stopped by applying hydrostatic pressure to the more concentrated solution to raise its water activity. The hydrostatic pressure which just stops osmosis is the **osmotic pressure**. In other words, it is the pressure required to equalize water activities in the two compartments. A greater hydrostatic pressure will raise water activity in that compartment further and will cause water to leave the more concentrated solution. This is **ultrafiltration**.

The osmotic pressure ($\Delta\pi$) that will arise across an ideal semipermeable membrane can be estimated by the **van't Hoff equation**:

$$\Delta\pi = RT \, \Sigma(c_1 - c_2)$$

where R is the gas constant, T is the absolute temperature (K) and $\Sigma(c_1 - c_2)$ is the sum of the difference in molar concentrations (mol L^{-1}) of all osmotically active solutes on sides 1 and 2 of the membrane.

Traditionally, $\Delta\pi$ has been expressed in atmospheres; then $R = 0.082$ kg atm K^{-1} mol^{-1}. The van't Hoff equation predicts that an ideal 1 mol solution of a non-electrolyte, when separated by a semipermeable membrane from pure water at 0°C, should generate an osmotic pressure of 22.4 atm. Osmotic pressures generated using actual solutions usually deviate from those predicted from their molar concentrations because solute molecules interact with each other and with the solvent. Therefore, strictly speaking, in the van't Hoff equation $\Sigma(c_1 - c_2)$ refers to the **osmotic activities** of the solutes rather than to their concentrations but in dilute solutions these are approximately the same.

Since osmotic pressure exists only when a solution is separated from pure water by a semipermeable membrane in an experimental situation in which osmosis can occur, *it is wrong to speak of the osmotic pressure of a solution*. Instead, we use the term osmolarity (osmol L^{-1}) to describe the effects of the total number of osmotically active particles per litre of solution. The osmolarity of a solution can be estimated from the van't Hoff equation, by dividing the osmotic pressure in atmospheres by RT. An electrolyte like NaCl is dissociated in solution into two osmotically active ions, so its osmolar concentration will be approximately twice the molar concentration. The osmolarity of a solution can be measured in a commercial osmometer which relies on the fact that the osmotic pressure is related to the depression of the freezing point. Mammalian fluids have an osmolarity of about 0.3 osmol L^{-1} and thus would exert an osmotic pressure across an ideal semipermeable membrane of about $RT \times 0.3 = 7.6$ atm or 5800 mmHg or 770 kPa at body temperature.

It has become fashionable to speak of the **osmolality** of solutions (osmol kg solvent^{-1}). However, since in dilute aqueous solutions molal concentrations (mol kg^{-1} H_2O) closely approximate molar concentrations (mol L^{-1} solution), the terms **osmolality** (osmol kg$^{-1}H_2O$) and **osmolarity** (osmol L^{-1} solution) are often used interchangeably.

Effective osmotic pressure

Biological membranes are not ideal but in varying degrees leaky. The osmotic pressure difference across such a membrane ($\Delta\pi$) is less than the ideal and diminishes with time as some solutes diffuse through and eliminate their differences in concentration. The **effective osmotic pressure** will therefore be less than the pressure calculated above and will depend on the degree to which solutes cross the membrane. This is expressed by including the **reflexion coefficient**, σ, in the van't Hoff equation:

$$\Delta\pi = \sigma RT \, \Sigma(c_1 - c_2)$$

If none of the solute molecules colliding with the membrane cross it, $\sigma = 1$, i.e. all molecules are reflected. If all the solute molecules pass through the membrane $\sigma = 0$, none is reflected and there is no osmotic pressure. For example, the effective osmotic pressure of the ICF, often called its **colloid osmotic pressure**, is due to the large, reflected, non-penetrating solutes (e.g. proteins, organic phosphates). Some of these large solutes are negatively charged and attract small diffusible cations and repel small diffusible anions across the plasma membrane. This leads to an uneven distribution of small ions with a small excess inside the cell called the **Donnan excess**, which contributes further to the effective osmotic pressure of the ICF (see Box 1.1).

Net flow of water across membranes

The net flow of water across a membrane per unit time (J_v) is given by

$$J_v = ALp\,(\Delta P - \Delta\pi)$$

where A is the area of the membrane available for flow, Lp is the hydraulic conductivity which is a measure of the ease with which water flows through the membrane, ($\Delta P - \Delta\pi$) is the driving force, which is dependent on the differences in hydrostatic pressure (ΔP) and effective osmotic pressure ($\Delta\pi$) across the membrane.

Cell volume

Cell volume is determined by the cell content of osmoles and the extracellular osmolarity. Under physiological conditions, the colloid osmotic effect of impermeant cell solutes is offset by the Na^+–K^+ pump which effectively maintains Na^+ as an impermeant extracellular solute. Tonicity describes the behaviour of cells placed in a solution. If cells swell, the solution is hypotonic; if cells shrink, the solution is hypertonic; if cell volume is unchanged, the solution is isosmotic.

In the steady state, cell volume is constant and $J_v = 0$. Thus, $\Delta P = \Delta\pi$. Since animal plasma membranes cannot withstand appreciable gradients of hydrostatic pressure, both ΔP and $\Delta\pi$ between the ECF and ICF must be zero. This means that the

Box 1.1 Gibbs–Donnan distribution

Consider two compartments containing aqueous solutions of fixed volume separated by a membrane permeable to water and small M_r solutes but impermeable to protein (Fig. B1.1).

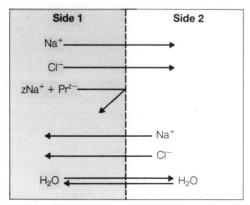

Fig. B1.1 Gibbs–Donnan distribution.

Suppose that the solution in side 1 contains protein (Pr) with a net negative charge (z^-) and that both solutions contain Na^+ and Cl^- ions. Because of the negative charge on the protein, diffusible cations, i.e. Na^+, will be attracted to side 1 and diffusible anions, i.e. Cl^-, repelled to side 2. As a consequence of the redistribution of ions, side 1 containing the protein will be slightly electrically negative with reference to side 2, that is, an electrical potential difference ($\Delta\psi$) will exist across the membrane. At equilibrium it will be equal to the equilibrium potential (E) for each ion, which is given by the Nernst equation:

$$E = \frac{RT}{F}\ln\frac{Na_1^+}{Na_2^+} = \frac{RT}{F}\ln\frac{Cl_2^-}{Cl_1^-}$$

Therefore

$$\frac{Na_1^+}{Na_2^+} = \frac{Cl_2^-}{Cl_1^-} = r$$

where r is the Donnan ratio. Moreover

$$Na_1^+ Cl_1^- = Na_2^+ Cl_2^-$$

The condition of bulk electroneutrality demands that the sum of all positive charges shall equal the sum of all negative charges in any solution, i.e.

$$Na_1^+ = Cl_1^- + Pr_1^{z-}$$

and

$$Na_2^+ = Cl_2^-$$

Box 1.1 *Continued*

Therefore

$$\text{Na}^+_1\text{Cl}^-_1 = [\text{Na}^+_2]^2 = [\text{Cl}^-_2]^2$$

Now $\text{Na}^+_1\,\text{Cl}^-_1$ is not a square and since the sum of the sides of a rectangle is greater than the sum of the sides of a square of equal area

$$\text{Na}^+_1 + \text{Cl}^-_1 > \text{Na}^+_2 + \text{Cl}^-_2$$

or

$$\text{Na}^+_1 + \text{Cl}^-_1 = \text{Na}^+_2 + \text{Cl}^-_2 + e$$

where e is called the Donnan excess. This excess concentration of diffusible ions, e, contributes to the difference in effective osmotic pressure between the two compartments so that:

$$\Delta\pi = RT\Sigma(\text{Pr}^{z-} + e)$$

The system can only reach a steady state when the volumes of the two compartments are fixed, for then $\Delta\pi$ is offset by an increased hydrostatic pressure in compartment 1. Without this restriction an osmotic flow of water and a diffusion of permeant solutes from compartment 2 to 1 will continue until all the solution is found in compartment 1.

It is important to recognize that the resting membrane potential of cells is *not* a consequence of the Gibbs–Donnan distribution as the intracellular and extracellular concentrations of ions do not comply with the Donnan ratio. In the case of the capillary wall where a Gibbs–Donnan distribution does exist, it can be calculated that the potential difference across the wall is only a few millivolts.

---

osmolarities of both fluids must be identical. Since the plasma membrane is freely permeable to water, water will move rapidly between ECF and ICF in response to any tendency for alterations in water activities between these two compartments. The determinants of cell volume, therefore, are:

1   the total number of osmotically active particles within the cell; and
2   the osmolarity of the ECF.

The cell has a considerable quantity of impermeant solutes (e.g. proteins, organic phosphates) whereas the interstitial fluid is relatively devoid of these. There is, therefore, a colloid osmotic gradient that would draw fluid into the cells. Why then do cells not swell under physiological conditions? It is believed that the Na^+–K^+ pump, by holding Na^+ extracellularly and preventing its accumulation in the cell, effectively offsets the colloid osmotic effect of cell macromolecules.

If this pump ceases to expel Na^+ (e.g. when cells are chilled), Na^+ enters passively down its concentration gradient, bringing Cl^- with it to preserve electroneutrality; to equalize its activity water then moves into the cells across the membrane and the cells swell.

Tonicity

The behaviour of cells in artificial bathing solutions cannot be predicted from the osmolarity of these solutions. It depends also on the permeability of the membrane to the solutes. For example, when cells are placed in 300 mmol L^{-1} urea, which is approximately isosmotic with mammalian fluids, the cells will swell because urea enters and water follows. A new term is required to define the strength of a solution as it affects the volume of cells. This is **tonicity**, which is defined operationally thus:

1 if cells **shrink** in a solution, the solution is **hypertonic**;
2 if cells **swell** in a solution, the solution is **hypotonic**; and
3 if the cell volume is **unchanged** the solution is **isotonic**.

Tonicity may also depend upon the functional activity and metabolism of the cells. Most cells swell grossly if their metabolism is inhibited, even in isosmotic NaCl solutions, because the cells gain NaCl. Consequently, isolated tissues need substrates and oxygen as well as the appropriate inorganic ions in bathing solutions designed to be isosmolal with normal ECF.

1.8 MOVEMENT OF WATER AND SOLUTES ACROSS EPITHELIA

Epithelia isolate the internal from the external environment. Net movement of solutes and water may occur via the epithelial cells (cellular route) and between these cells (paracellular route). Net movement from lumen to interstitial fluid is termed absorption; net movement from interstitial fluid to lumen is termed secretion. On the basis of the properties of the paracellular route, epithelia can be classified as tight or leaky. The major solute absorbed by tight absorptive epithelial is Na^+ and in the absence of hormonal stimulation, water permeability is low. Leaky absorptive epithelia transport a variety of solutes and have a relatively high water permeability. Epithelial cells in a variety of organs can secrete an isosmotic NaCl solution. In such secretory epithelia, this involves activation of apical membrane Cl^- conductance and movements of Na^+ and water from interstitial fluid to lumen via the paracellular route.

Epithelia separate the internal from the external environment and regulate movement of solutes and water to and from the body. Examples inlcude the surface layers of the skin, lungs, wall of the alimentary canal and lining of the urogenital systems. The epithelial cells, which may be one or more layers thick, are separated by their basement membranes from a variable amount of supporting connective tissue containing blood vessels, nerves and smooth muscle fibres. It is believed that the transport functions reside only in the epithelial cell layer.

All epithelial cells are separated from their neighbours by a space—the **lateral intercellular space**—whose size may vary (Fig. 1.9). They are held to each other at their luminal edges by junctions—**tight junctions** (Fig. 1.10).

Tight junctions were originally thought to be impermeable, but it is now appreciated that to a variable extent solutes and water may diffuse across them. There are thus two pathways by which solutes and water can cross epithelia, either through cells or between cells. Substances passing between cells must cross tight junctions and pass through the lateral intercellular spaces—the so-called **paracellular** or **shunt pathway**. The amount passing through this shunt pathway differs between epithelia, depending on the difference in permeability characteristics of

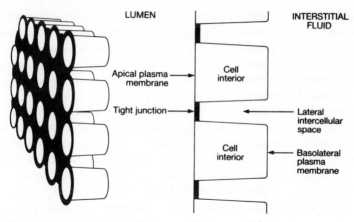

Fig. 1.9 Transporting epithelia.

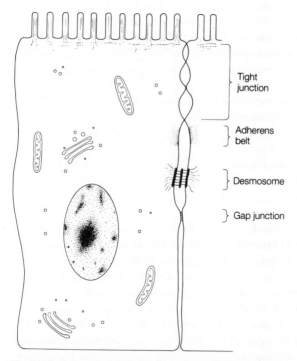

Fig. 1.10 Types of epithelial cell junctions. Tight junctions are formed by a series of connections between the outer surfaces of the plasma membranes of adjacent cells. They completely encircle the apical end of each cell, form an impermeable barrier for large molecules and are variably permeable to small solutes, ions and water. Cells are also held together by anchoring junctions, of which the adherens belt and the desmosomes are two examples. These junctions are connected to the cell cytoskeletal elements (adherens belt—actin filaments; desmosomes—intermediate filaments). Gap junctions (p. 34) provide a low-resistance pathway through which small solutes, ions and water can pass. Cells communicate with their neighbours through these junctions.

the tight junctions. Epithelia in which the paracellular pathway makes a significant contribution to total movements (e.g. proximal renal tubule, small intestine) are referred to as **leaky**, whereas those in which the cellular pathway predominates are referred to as **tight** (e.g. renal collecting duct, salivary gland ducts, amphibian skin and urinary bladder). The properties of the tight junctions are reflected in the electrical characteristics of the epithelium (Box 1.2).

The importance of the concept of leaky and tight epithelia lies in the underlying physiological significance. In general, leaky epithelia are specialized for the bulk handling of isosmotic solutions—either absorption or secretion—and lie more proximally, whereas tight epithelia lie more distally and are more selective in the manner in which they handle the load presented to them. Hormonal control of salt and water balance tends to be exerted on tight rather than on leaky epithelia.

Net flow of water across an epithelium per unit time (J_v) is determined by ALp ($\Delta P - \Delta \pi$) (p. 21) while net movements of solutes are determined by the following considerations. There are specific cellular transport pathways for a variety of solutes which allow unidirectional movement between lumen and interstitial fluid. In addition, solutes may diffuse in both directions through the paracellular pathway. The net solute flux (amount moved/unit time × unit area) is, therefore, the difference between two unidirectional fluxes. Net movement from lumen to interstitial fluid is termed **absorption**; net movement from interstitial fluid to lumen is called **secretion**. The net movement of any solute across an epithelium is determined by:

1   available surface area (since amount moved/unit time = area × flux);
2   time in contact with the available area;
3   electrochemical potential gradient, which will be influenced by blood and lymph flow;
4   properties of the epithelium, i.e. tight or leaky, together with the availability of specific transport mechanisms.

Absorptive epithelia

In general epithelial which actively absorb ions primarily move Na^+. This transepithelial Na^+ transport generates a potential difference across the epithelial layer. The transepithelial potential difference in tight epithelia is high (>20 mV) whereas in leaky epithelia it is low (0–5 mV). Since electrical neutrality must be preserved, Cl^- follows passively, as does water when water permeability of the epithelium is sufficient. Mechanisms by which Na^+ crosses epithelia are summarized for a typical tight epithelium in Fig. 1.11. Unlike muscle, nerve and blood cells whose plasma membranes are symmetrical, the plasma membranes of epithelial cells are asymmetrical. The apical portion, which faces the lumen, has unusual permeability and transport characteristics. It has a high Na^+ and low K^+ permeability and no Na^+–K^+ ATPase. In contrast, the basolateral portion, consisting of the membrane beneath the tight junctions and adjacent to the capillaries, is similar in its permeability and transport properties to other plasma membranes. In general, in tight epithelia only Na^+ readily crosses the apical membrane (passively via aqueous channels) and, in the absence of hormonal stimulation, water permeability is low.

Box 1.2 Electrical characteristics of epithelia

The differences in potential difference between tight and leaky epithelia reflect the characteristics of the junctions between cells—the **tight junctions**, which hold adjacent cells together at their apices. The greater their ion permeability, the smaller the transepithelial potential difference.

This is readily understood by considering the simplified electrical representation of an Na^+-transporting epithelium (Fig. B1.2).

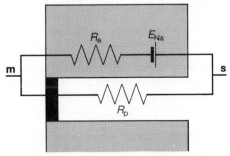

Fig. B1.2 Electrical representation of an idealized epithelium.

Sodium ions can only pass through the cellular pathway from mucosal (m) to serosal (s) solutions. In following this pathway they cross plasma membranes which offer resistance (R_a) to their flow and are actively transported across the basolateral membrane by the Na^+ pump (represented here by an electromotive force, E_{Na}). Univalent ions (Na^+, Cl^-, K^+) can all pass to varying extents between the cells through the paracellular pathway (tight junctions plus lateral intercellular spaces) which offers resistance R_p to their flow.

Since resistances R_a and R_p are arranged in parallel between mucosa and serosa,

$$\frac{1}{R_T} = \frac{1}{R_a} + \frac{1}{R_p} \quad \text{or} \quad R_T = \frac{R_a R_p}{(R_a + R_p)}$$

where R_T is the total resistance between mucosa and serosa, and transepithelial potential difference ($\Delta\psi$) is given by:

$$\Delta\psi = E_{Na}\frac{R_p}{(R_a + R_p)}$$

Therefore when R_p is high relative to R_a (as in a tight epithelium with a relatively impermeable paracellular pathway), $\Delta\psi$ is large. Conversely, when R_p is low relative to R_a (as in a leaky epithelium), $\Delta\psi$ is small.

Ussing introduced two important techniques for the study of epithelia. There can be no net passive movements of solutes across an epithelium in the absence of electrochemical potential gradients. Any net movement under such conditions must be due to active transport. To abolish chemical potential gradients, an isolated epithelium is mounted to separate two solutions of the same composition (e.g. sodium Ringer's). Any electrical potential difference arising across the epithelium is then nullified by passing a current across the epithelium. The magnitude of the

Box 1.2 *Continued*

current required to set the transepithelial potential difference to zero must equal the current actually flowing across the epithelium (but is of opposite sign). This current is called the short-circuit current (I_{sc}) and is equal to the net active movement of ions across the epithelium. Thus a simple electrical measurement allows determination of the magnitude of active transport across the epithelium.

Ussing also showed, using isotopes, that if a solute was simply moving passively across the epithelium the ratio of the two unidirectional fluxes ($J_{ms} : J_{sm}$) satisfied the following equation (the **Ussing flux ratio equation**):

$$\frac{J_{ms}}{J_{sm}} = \frac{c_m}{c_s} e^{-\frac{zF\psi_{ms}}{RT}}$$

where J_{ms} and J_{sm} are the unidirectional fluxes from mucosa to serosa and serosa to mucosa respectively, c_m and c_s are the concentrations of the solute in the two media, ψ_{ms} represents the transepithelial potential difference, and F, R and T have their usual meanings.

In leaky epithelia (Fig. 1.12), not only does Na$^+$ cross the apical membrane through channels, but entry of glucose, some amino acids and phosphate may be coupled to Na$^+$ entry. The electrochemical potential gradient for Na$^+$ entry is used to drive the accompanying solute into the cells from the lumen (i.e. co-transport). The active transport of Na$^+$ from cell to interstitial fluid is thought to

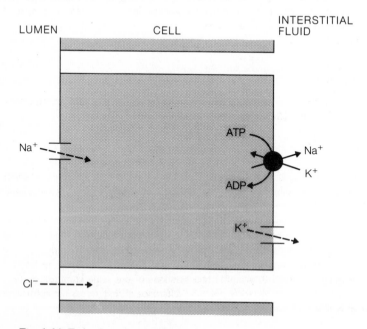

Fig. 1.11 Tight absorptive epithelia.

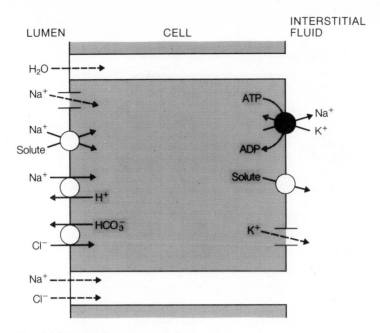

Fig. 1.12 Leaky absorptive epithelia.

involve the same Na^+–K^+ pump as already described (p. 11). The coupling of net water movement to net solute movement across leaky epithelia may involve the generation within the lateral intercellular spaces of a local region of Na^+ concentration, somewhat higher than that in interstitial fluid, reflecting the distribution and activity of the Na^+–K^+ pump. This provides an osmotic gradient moving water, either through the cells or tight junctions, from the lumen to the lateral intercellular space. This movement in turn creates a local hydrostatic pressure gradient so that fluid flows from the lateral intercellular spaces to the interstitium. However, the magnitude of any local osmotic gradient in the lateral intercellular spaces is now thought to be only a few mosmol kg^{-1} H_2O.

Counter-transport of Na^+ for cell H^+ is also often found. In addition, though often represented as coupled co-transport with Na^+ (Fig. 1.12), movement of Cl^- across the apical plasma membrane may be by counter-transport with cell HCO_3^- (or OH^-). In a cell with both Na^+–H^+ and Cl^-–HCO_3^- exchangers, the net effect will be Na^+ and Cl^- entry to the cell. H_2O and CO_2 are formed from the H^+ and HCO_3^- ions extruded from the cell.

Secretory epithelia

A variety of inorganic and organic ions are secreted by epithelia and specific mechanisms are discussed in the appropriate chapters. These secretions result in the transfer of solutes and water to the surfaces of the body (internal or external) and are produced by **exocrine** glands. (This contrasts with secretions from cells that are carried in the plasma (hormones) which are produced by endocrine glands.)

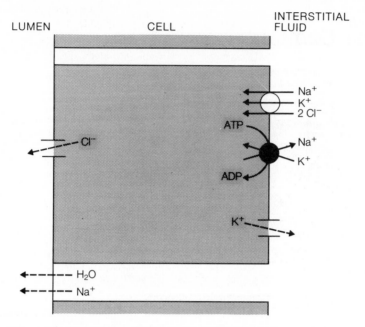

Fig. 1.13 Secretory epithelia. Symbols as in Fig. 1.11.

Of particular interest because of its contribution to the production of many secretions (sweat, tears, saliva, pancreatic, biliary and intestinal fluids) is the mechanism of isosmotic NaCl secretion, summarized in Fig. 1.13.

The secretory cells comprise a leaky epithelium. A Na^+–K^+–$2\ Cl^-$ co-transporter in the basolateral membrane is responsible for driving Cl^- into the cells against its electrochemical gradient. The accompanying Na^+ is recycled back to the interstitial fluid. In the quiescent state, the apical plasma membrane is impermeable to Cl^-. When secretion is stimulated a Cl^- conductance in this membrane is activated. The anion, therefore, moves from the cell to the lumen down its electrochemical gradient. As a consequence, Na^+ passes from the interstitial fluid to lumen through the leaky paracellular pathway to maintain electroneutrality and water moves through the same pathway down the osmotic gradient. Thus a primary isosmotic NaCl secretion is produced.

Chapter 2
Cellular Communication

Cells receive information from both outside and inside the body. Specialized sensory cells respond to chemical, electrical, light, mechanical or heat stimuli from external or internal sources, while cells usually communicate with one another by chemical signals. The cellular response to these signals may be simple and short-lived, such as depolarization, or it may be complex and long-lasting, such as the acquisition of memory. Nevertheless, the cellular mechanisms underlying these responses have much in common. Each of these signals is transduced into electrical or biochemical changes within the cell that lead to a characteristic response. For example, the stimuli that give rise to the sensations of taste, smell, hearing and vision, and also certain neurotransmitters and hormones, can all control the opening and closing of ion channels and depolarize or hyperpolarize cells.

In this chapter we discuss how information is transmitted between cells, and how signals from the external and internal environment are received and transduced into electrical and biochemical changes within the cell.

2.1 HOW SIGNALS ARE TRANSMITTED BETWEEN CELLS

Cells communicate with one another by chemical signals that either diffuse between cells (neurotransmitters, and paracrine and autocrine secretions) or are disseminated in the blood (hormones). These signals include small organic molecules (e.g. acetylcholine and adrenaline), and larger molecules such as proteins and steroids. Cells may also communicate with their immediate neighbours through gap junctions. These junctions transmit both electrical and chemical signals and their permeability may be regulated by changes in voltage, cytosolic pH and intracellular messengers.

Chemical signalling

The nervous and endocrine systems use chemical signals, neurotransmitters and hormones, respectively, for communication. **Neurotransmitters** are released from neurons, diffuse across a narrow synaptic space and act on adjacent neurons, muscle cells or secretory cells in glands; **hormones** are transmitted in the circulation and act on distant target organs. **Paracrine secretions** are local chemical mediators that are released by cells, diffuse a short distance in the extracellular space and act on cells nearby, while **autocrine secretions** act on the cells that

31

produced them. Therefore neurotransmitters, by definition, are paracrine secretions, although they are not usually classified as such, and some hormones, e.g. gastrointestinal hormones, may also be classified as paracrine secretions, because they may act locally as well as travelling in the blood stream to distant target organs. Secretions of neurons that are transported to distant target organs in the blood stream are referred to as **neurohormones.** Finally, certain cells secrete **growth factors** which are essential at critical stages in the development of their target cells.

Chemical signals may be ions or metabolites (e.g. Na^+, Ca^{2+}, glucose or amino acids) which stimulate certain endocrine cells; they may be simple organic molecules (e.g. the neurotransmitters acetylcholine, noradrenaline and glutamate) or they may be more complex molecules (e.g. protein and steroid hormones). Many substances that were first identified as hormones produced in endocrine glands are also synthesized in the nervous system where they act as neurotransmitters or neuromodulators (p. 100). Consequently, there is considerable overlap in the chemical messengers used by these systems.

Synthesis and secretion

Neurotransmitters and many hormones that do not readily cross plasma membranes are packaged into **secretory vesicles** within the cell prior to their release by **exocytosis** (p. 15). These include small polar organic molecules and large polypeptides. With appropriate stimulation, cytosolic Ca^{2+} increases. This increase triggers exocytosis; the vesicles fuse with the plasma membrane and release their contents to the extracellular fluid.

Small polar organic molecules may be synthesized by enzymes in the cytosol, and then are actively accumulated within the vesicles by membrane-bound transporters. Alternatively, they may be synthesized from their precursors by enzymes bound within the vesicles.

Polypeptide molecules, including proteins and glycoproteins which are destined for secretion, are first synthesized on the ribosomes of the rough (granular) endoplasmic reticulum as part of larger precursor polypeptides called **prepropolypeptides**. As they are synthesized, they are inserted into the intracisternal space of the endoplasmic reticulum and are subsequently modified by deletion of peptide sequences (Fig. 2.1). The 'pre' part of the prepropolypeptide is a **signal peptide**, which is involved in ribosomal attachment and transfer of the newly synthesized polypeptide across the membrane of the endoplasmic reticulum. After insertion the signal peptide is removed from the propolypeptide by a signal peptidase. Small vesicles containing the propolypeptide then bud off from the endoplasmic reticulum and transfer their contents to the Golgi apparatus by fusion with the Golgi membrane. In the Golgi apparatus the propolypeptide is packaged into secretory vesicles, and the 'pro' part is then split from the propolypeptide. Where the secretory product is a glycoprotein, carbohydrate groups are added by enzymes in the endoplasmic reticulum and these groups are further modified in the Golgi apparatus.

Lipophilic molecules are not packaged into vesicles for secretion. For instance, **steroid hormones** (e.g. cortisol, oestrogen and testosterone) are synthe-

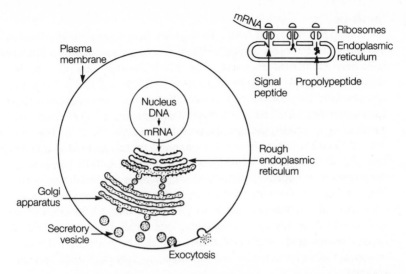

Fig. 2.1 Synthesis of polypeptides destined for secretion. *Inset*: attachment of the signal peptide and transfer of the nascent polypeptide across the membrane of the endoplasmic reticulum. During transfer the presequence is cleaved by a peptidase releasing the propolypeptide to the intracisternal space of the endoplasmic reticulum. (After Blobel, G. & Dobberstein, B. (1975) *J. Cell Biol.* **67**, 835–851.)

sized from cholesterol (Fig. 2.2) in steps that take place in the mitochondria, smooth endoplasmic reticulum and cytoplasm and which require acetate, O_2, NADPH and other cofactors. They are not stored to any extent in the cells that synthesize them and they readily cross the plasma membrane of the cell. Therefore, their rate of release is directly related to their rate of synthesis.

Other signals

Thus far we have been discussing soluble chemical signals that pass from one cell to another. There are also chemicals attached to cell surfaces or to the extracellular matrix that provide signals for cell migration during development and for the localization of receptors, ion channels and enzymes to particular areas of the cell surface that best serve their function (e.g. the clustering of acetylcholine receptors at the neuromuscular endplate).

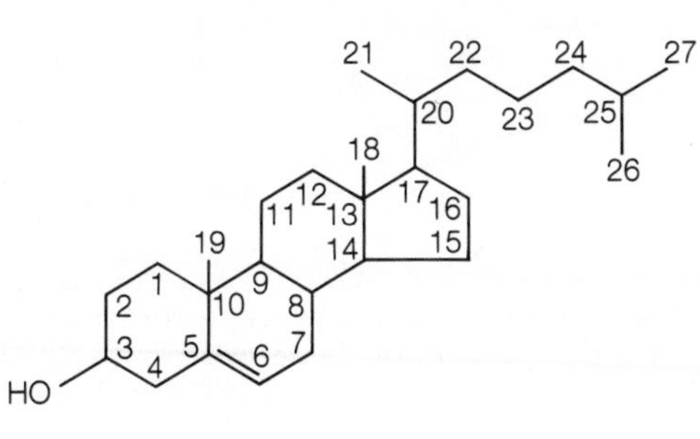

Fig. 2.2 Structure of cholesterol from which steroid hormones are synthesized.

Gap junctions

Membrane junctions that hold cells together (desmosomes and adherens belts) and those that impose a barrier to diffusion across layers of cells (tight junctions) have been mentioned previously (Fig. 1.10). Another type, the **gap junction**, allows small organic molecules and ions to pass from cell to cell. In the regions of gap junctions the plasma membranes of the two opposing cells come together to form a gap about 3 nm wide and small channels about 1.5 nm in diameter extend across the gap, linking the cytoplasm of the two cells (Fig. 2.3). The channels at gap junctions are formed from transmembrane protein assemblies called **connexons** (Fig. 2.3), of which there are several hundred at each gap junction. Two connexons from opposing membranes join together to form the aqueous pore or channel. Gap junctions are found in many tissues; in cardiac and smooth muscle cells they allow electrical signals to pass from one cell to another and thus synchronize their activity. In addition to ions, small molecules (<1500 M_r, e.g. sugars, amino acids and nucleotides) can also pass through the connecting channels. Cells connected by gap junctions are thus electrically and chemically coupled.

The opening and closing of these channels may be regulated by changes in voltage, cytosolic pH and intracellular messengers in much the same way that membrane ion channels are regulated (see below). For example, an increase in intracellular cyclic adenosine monophosphate (cAMP) or a decrease in intracellular Ca^{2+} can rapidly increase the permeability of gap junctions in certain tissues. This in turn will influence the rate of transfer of these messengers between adjacent cells.

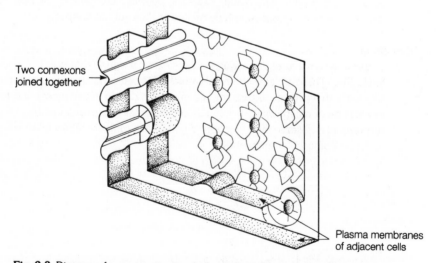

Two connexons joined together

Plasma membranes of adjacent cells

Fig. 2.3 Diagram of a gap junction in which proteins called connexons form channels between adjacent cells. Each connexon is composed of six subunits and joins with a connexon in the opposing plasma membrane to form a channel that links the cytoplasm of adjoining cells.

2.2 HOW CELLS RESPOND TO EXTERNAL STIMULI

Cells respond to a variety of external stimuli and transduce these signals into electrical and biochemical changes. In many instances this involves the interaction of a signal with a receptor in the plasma membrane. This may lead directly to activation of ion channels, but more commonly it induces the production of intracellular messengers that then regulate cell function. Guanosine triphosphate (GTP) regulatory proteins play a crucial role in coupling receptor occupancy to the generation of an intracellular biochemical cascade. Recognized intracellular pathways include the cyclic adenosine monophosphate (cAMP) system, and the polyphosphoinositide system. Other important intracellular messengers are cyclic guanosine monophosphate (GMP) and arachidonic acid derivatives (prostaglandins, thromboxanes and leukotrienes). Alterations in cytosolic Ca^{2+} activity may be linked to the biochemical pathways in mediating cell responses.

In some cases there is a direct link between a stimulus and activation of ion channels. Examples include stretch-activated channels in touch receptors and voltage-sensitive channels in excitable tissues. However, for chemical stimuli (e.g. gustatory, olfactory, neurotransmitters, hormones, growth factors, etc.) the first event leading to a cellular response is the interaction of the chemical signal (**ligand**) with its **receptor**.

Receptors

Receptors are molecular entities, either proteins or glycoproteins, that bind ligands with high affinity. (Note that cells, or components of cells, that respond to sensory stimuli are also referred to as 'receptors' but the two should not be confused.) Receptors for ligands that do not penetrate cells readily (e.g. neurotransmitters, peptide hormones) are located in the plasma membrane, but some lipophilic ligands (e.g. steroid and thyroid hormones) have their receptors in the nucleus.

The interaction of the ligand (L) and its receptor (R) can be represented as:

$$L + R \underset{k_{-1}}{\overset{k_1}{\rightleftharpoons}} LR$$

where k_1 and k_{-1} are the forward and reverse reaction rate constants. The law of mass action states that the velocities of the forward (v_1) and reverse reactions (v_{-1}) are proportional to the concentrations of the reacting species, i.e. $v_1 = k_1[L][R]$ and $v_{-1} = [LR]$. At equilibrium, $v_1 = v_{-1}$ and therefore the equilibrium association constant (K_a) is given by the equation:

$$K_a = \frac{k_1}{k_{-1}} = \frac{[LR]}{[L][R]}$$

This constant is referred to as the **affinity constant** and indicates the strength of binding between two molecules. Affinity constants may range from 10^6 L mol^{-1}, for the binding of some neurotransmitters to their receptors, to 10^{10} L mol^{-1} or higher for antigen–antibody reactions.

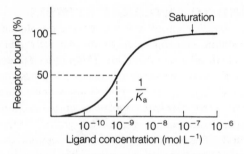

Fig. 2.4 Kinetics of binding of a radioactive ligand to its receptor. Specific binding is calculated by determining total binding of the radioactive ligand to its receptor and subtracting non-specific binding, which is measured in the presence of an excess of unlabelled ligand (agonist or antagonist). Note that binding becomes saturated as the concentration of radioactive ligand is increased and that the affinity constant (K_a) is given by the reciprocal of the ligand concentration at half maximal binding. Alternatively, it can be stated that the ligand concentration at half maximal binding is a measure of the equilibrium dissociation constant (K_d), which is the reciprocal of K_a.

In addition to their high affinity, receptors exhibit **specificity**. Only those cells that contain the appropriate receptors will recognize and respond to the ligand. The concentration of receptors in cells is very low and it is necessary to use radioactively labelled ligands to study the kinetics of binding (Fig. 2.4). As the number of receptors is limited, the binding also shows **saturation** as the concentration of the ligand is increased.

The response of cells to signal molecules may vary with the intensity of stimulation. High concentrations of signal molecules can temporarily inactivate receptors (e.g. by phosphorylation of the receptor)—a process known as **desensitization**. The number of receptors in a cell may also vary with stimulation. A change in activity may lead to a decrease in the number of cell receptors (**down-regulation**), or to an increase (**up-regulation**). The number of receptors in the plasma membrane at any time represents a balance between their rate of insertion into the membrane and their rate of removal from the membrane. Since a small fraction of ligand–receptor complexes are internalized by endocytosis, the ligand can enhance the rate of removal of receptor from the membrane.

The distribution of receptors on a cell is not necessarily uniform. For example, in adult skeletal muscle fibres the acetylcholine receptors are clustered at the neuromuscular junction but, if the nerve is cut, there is marked increase in the synthesis of receptors in the extrajunctional regions.

Ion channels

Ion channels are membrane proteins that contain small, highly selective aqueous pores. They allow specific ions, e.g. Na^+, K^+, Ca^{2+} or Cl^- to move down their electrochemical gradients across the membrane, and are often named after the ions for which they show the greatest selectivity. A channel is opened or closed by a molecular 'gate' which impedes ion flow and is regulated by a 'sensor' that forms part of the channel protein. In some channels the gate opens and closes spontaneously (**leak channels**) while in others the probability of their opening and closing is regulated by a number of factors.

The major types of regulated ion channel include:

1   voltage-gated channels , in which the sensor responds to the voltage across the membrane, e.g. the voltage-sensitive Na^+ channel involved in action potential generation;

2   chemically gated channels, in which the sensor is sensitive to a chemical signal such as the binding of a ligand to a receptor that forms part of the channel, e.g. the acetylcholine receptor channel involved in neuromuscular transmission; and

3   mechanically gated channels, in which the sensor is sensitive to mechanical deformation of the membrane, e.g. stretch-activated channels on sensory nerve terminals.

The properties of ion channels can be studied by the patch-clamp technique, which allows the measurement of current flow through a single channel in a very small patch of membrane (Box 2.1). All membrane channels so far described fluctuate between a fully opened or fully closed state in an all-or-none fashion; the processes that modulate 'gating' affect the probability of a channel being open or closed. Single-channel conductances range up to 200 pS or more, and members of the same type of channel have similar conductances.

Interaction of a ligand with its receptor (Fig. B.2.2) may lead to direct activation of an ion channel, i.e. the receptor is an integral part of the ion channel. We call this a receptor-operated channel which allows very rapid transfer of information. More commonly, the interaction of the ligand with its receptor leads to the activation of a membrane-bound enzyme and the production of an intracellular chemical, a slower process, which then regulates cell function.

Intracellular messengers

The first such intracellular chemical to be identified was **cyclic adenosine monophosphate (cAMP)** (Fig. 2.5). Sutherland, who discovered the role of cAMP in the 1950s while studying the action of adrenaline on glycogenolysis in liver, called it the **second messenger**. It is now known that many chemical stimuli result in an increase in concentration of cAMP in their target cells. Ca^{2+} is also an important intracellular messenger; stimuli may open channels allowing it to flow into the cell or be released from intracellular stores in the endoplasmic reticulum (p. 44). Other important intracellular messengers (e.g. cyclic GMP, inositol triphosphate and diacylglycerol) have since been identified (Fig. 2.4).

G proteins

In both the cAMP and the polyphosphoinositide systems that are described in detail later in this section, the step that follows the binding of the ligand to its receptor at the membrane surface, and that initiates the cell response, is the activation of a regulatory protein that binds GTP. Hence this protein is called a G protein. About 100 different ligand–receptor complexes are now known to activate G proteins, which have come to be known as **universal transducers**.

G proteins consist of α, β and γ subunits. In the inactive state, the α group of the G protein binds guanosine diphosphate (GDP) and the protein can be

Box 2.1 Patch-clamping

It has already been noted (p. 10) that passive ion movement across membranes is through water-filled pores or channels. These channels fluctuate between open and closed states. Therefore the number of ions crossing an area of membrane in a given time will also fluctuate. In macroscopic measurements of ionic conductance we are not aware of this, since the large number of channels smooths out these oscillations to give an average or steady value.

An elegant technique developed by Neher and his colleagues in the 1970s (for which they were recently awarded a Nobel prize) and known as 'patch-clamping', allows direct determination of the activity of single membrane channels. A specially prepared micropipette is brought up to a membrane surface and a tight seal formed between the tip of the pipette and the cell membrane. By various manipulations (Fig. B2.1) a small piece of membrane about 1–10 μm^2 can be isolated across the tip of the micropipette, in either an inside-out or an outside-out configuration. This may

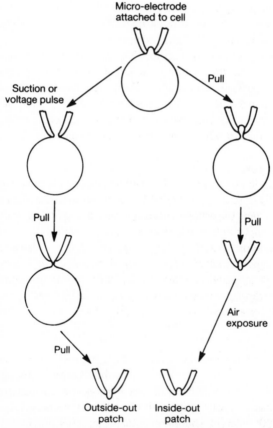

Fig. B2.1 Different membrane configurations used in patch-clamp analysis. (After Hamill, O.P., Marty, A., Neher, E., Sakmann, B. & Sigworth, F. (1981) *Pflügers Arch.* **391**, 85–100.)

Box 2.1 *Continued*

contain one or more channels and can be superfused with various ionic solutions on either its extracellular or cytoplasmic surface. Thus this technique allows current flow through individual channels to be recorded directly and affords the opportunity of studying the channels uninfluenced by other regions of the cell.

Experiments of this type have shown that a number of different ionic channels coexist within the membrane. The opening and closing (i.e. gating) of these channels may be voltage-dependent, ligand-dependent, stretch-dependent or spontaneous. For example, the acetylcholine receptor–ionophore complex at skeletal neuromuscular junctions is a 'ligand-gated' channel, the ligand being the neurotransmitter acetylcholine (Fig. B2.2a). The current generated across a patch of membrane containing an acetylcholine-gated channel exposed to a low concentration of acetylcholine is shown in Fig. B2.2b. It can be seen that the mean channel open time is about 1 ms and it can be calculated that on average approximately 4×10^7 ions pass through each channel per second.

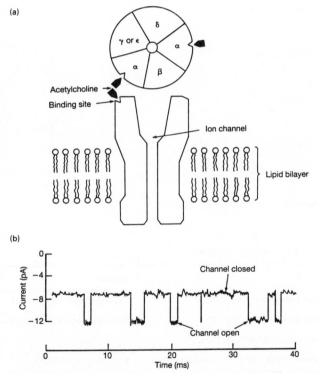

Fig. B2.2 Patch-clamp recording of current passing through an acetylcholine-gated channel of skeletal muscle.

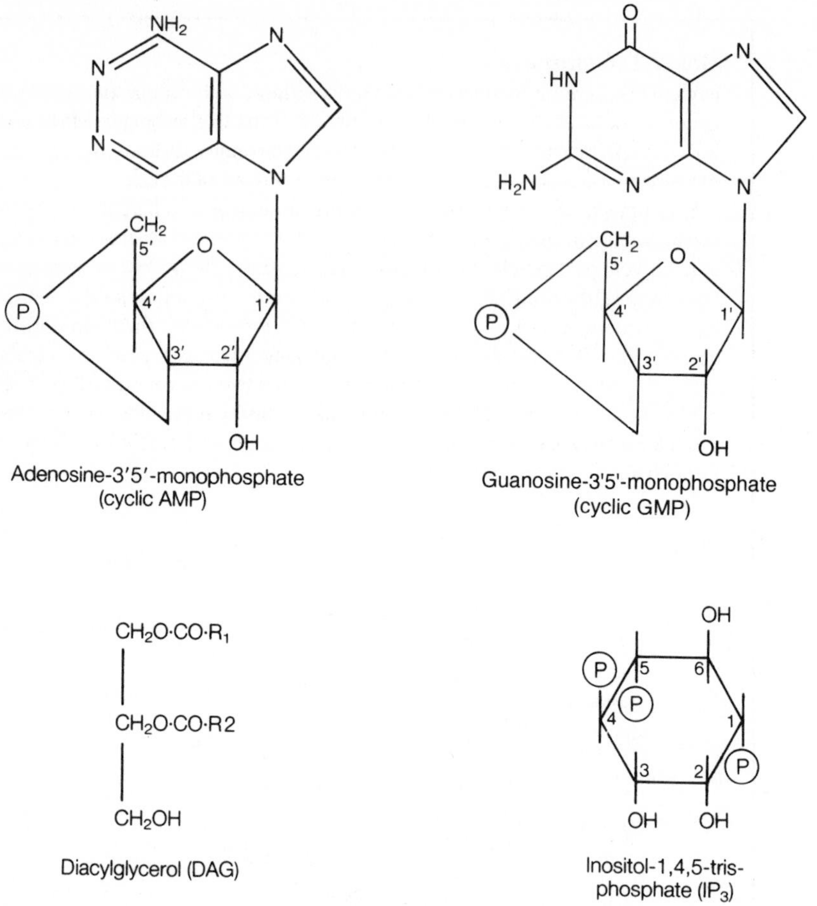

Adenosine-3'5'-monophosphate
(cyclic AMP)

Guanosine-3'5'-monophosphate
(cyclic GMP)

$CH_2O \cdot CO \cdot R_1$

$CH_2O \cdot CO \cdot R2$

CH_2OH

Diacylglycerol (DAG)

Inositol-1,4,5-tris-
phosphate (IP$_3$)

Fig. 2.5 Structure of intracellular messengers.

Table 2.1 G proteins and their actions

Type	Action	Comments
G_s	Stimulates adenylate cyclase	Activated by cholera toxin
G_i	Inhibits adenylate cyclase	Activation blocked by pertussis toxin
G_p	Stimulates phospholipase C	
G_t	Stimulates cGMP phosphodiesterase in photoreceptors	Known as transducin
G_o	Probably involved in gating of ion channels	Occurs in high concentration in brain (approximately 1% of total protein); activation blocked by pertussis toxin

Note that in addition to the G protein family there is another class of GTP-binding proteins, a low molecular mass group which includes the *ras* superfamily of proteins that play a role in the control of cell growth and which will not be considered further.

represented as βγα-GDP (Fig. 2.6). Following occupation of a surface receptor, a conformational change in the G protein results in the exchange of GDP for cytoplasmic GTP, giving βγα-GTP. The α-GTP subunit then dissociates from the βγ-subunits and binds to an effector molecule (e.g. an enzyme such as adenylate-cyclase, or a channel protein). As the α-subunit has GTPase activity, the GTP bound to the subunit is thereby converted to GDP. The α-GDP–effector complex then dissociates, the α-GDP subunit recombines with the βγ-subunits and the inactivated G protein is reconstituted.

A number of G proteins have been identified (Table 2.1). Some of these (G_s) stimulate an adenylate cyclase; others (G_i) inhibit this enzyme. Bacterial exotoxins have been particularly valuable in their identification: in the absence of hormones **cholera toxin** activates G_s by adenosine diphosphate (ADP)-ribosylation of specific amino acid residues in the α-subunit, while **pertussis toxin** blocks the receptor-mediated activation of G_i and G_o, also by ADP-ribosylation of the α-subunit.

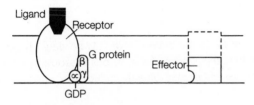

GDP - GTP exchange

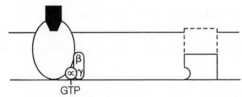

Effector stimulation

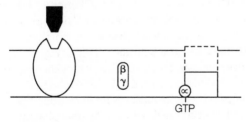

GTPase action

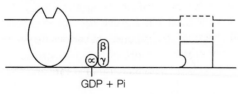

Fig. 2.6 Role of the G protein in transduction. The ligand binds to the receptor in the plasma membrane and activates the G protein, which consists of αβγ-subunits. This allows the α-subunit to exchange its bound GDP for GTP. The α-GTP then dissociates from the βγ-subunits and stimulates an effector (e.g. adenylate cyclase, phospholipase C, ion channel etc.). The GTPase on the α-subunit breaks down GTP to GDP, causing the α-GDP to dissociate from the effector and recombine with the βγ-subunits.

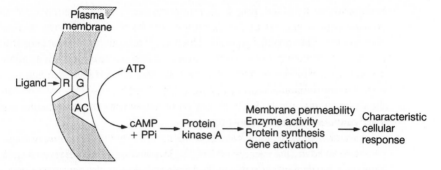

Fig. 2.7 Signals acting through cAMP. The ligand binds to its receptor (R) on the outer surface of the plasma membrane. At the inner surface are the GTP-regulatory protein (G) and the enzyme adenylate cyclase (AC), which catalyses the conversion of ATP to cAMP. Binding of the ligand to the R–G complex allows the G unit to bind GTP and stimulate adenylate cyclase activity. cAMP activates protein kinase A, which phosphorylates cellular proteins to produce the characteristic cellular response.

The high concentration of G_o in the brain (Table 2.1) no doubt reflects the important role of this protein in the actions of the various neurotransmitters and neuromodulators. In some cases G_o may directly induce the opening of ion channels.

cAMP system

Adenylate cyclase catalyses the synthesis of cAMP from adenosine triphosphate (ATP). Depending on the G protein (G_s or G_i) which is activated by the interaction of the signal and its receptor, adenylate cyclase activity may be either stimulated or inhibited.

Figure 2.7 summarizes the events that follow increased cAMP production. The raised level of cAMP within the cell stimulates **protein kinase A**, an enzyme that phosphorylates serine or threonine residues of other proteins. Protein kinase A is composed of two regulatory (R) units and two catalytic (C) units. cAMP binds to the R units, causing them to dissociate, and allows the free C units to phosphory-late cellular proteins. This may lead to specific changes in ion channel gating, enzyme activity, protein synthesis or gene activation, depending on the tissue involved. During this sequence of events, the initial signal is amplified many times. Eventually, the concentration of cAMP is restored to its basal level by degradation to AMP, this step being catalysed by the enzyme **cyclic nucleotide phosphodi-esterase**. The phosphorylated proteins are deactivated by **phosphatases**.

In view of their specific nature of action it may appear odd that the actions of so many hormone and neurotransmitters are mediated by cAMP. First, only certain tissues contain receptors which will react with a particular ligand to produce an increase in intracellular cAMP. Second, the responses of various tissues to raised levels of cAMP are different. For example, in response to an increase in cAMP, liver cells break down glycogen but heart cells increase their rate and strength of contraction.

cGMP system

The only other cyclic nucleotide found in animal tissues is **cGMP** whose formation from GTP is catalysed by guanylate cyclase. Some neurotransmitters and hormones operate by stimulating a membrane-bound guanylate cyclase and increasing cGMP concentrations in their target cells (e.g. atrial natriuretic factor). There is also a cytoplasmic form of guanylate cyclase. Nitric oxide, formed in cells from the amino acid L-arginine, appears to be an important physiological activator of cytoplasmic guanylate cyclase. Nitric oxide has important roles as a paracrine secretion and as a neurotransmitter.

In some cases a decreased level of cGMP mediates the actions of a signal. For instance, when light energy is absorbed by the photoreceptor pigment (rhodopsin) in the retina, rhodopsin interacts with a G protein transducin (G_t). This in turn stimulates a phosphodiesterase, leading to decreased cGMP concentrations and closure of cationic channels (p. 167). In this instance the channels are directly gated by binding of cGMP.

Polyphosphoinositide system

Other chemical stimuli operate through the polyphosphoinositide system. Phosphoinositides are a family of plasma membrane phospholipids containing inositol. Binding of a hormone to its receptor activates a different G protein or proteins (G_p) to those involved in the adenylate cyclase system. G_p then stimulates a phosphodiesterase (**phospholipase C**) in the plasma membrane (Fig. 2.8). This enzyme, on the inner face of the plasma membrane, cleaves phosphatidylinositol 4,5-bisphosphate (PIP_2) in the membrane to form **inositol trisphosphate** (IP_3) and **diacylglycerol** (DAG). IP_3 migrates to the endoplasmic reticulum where it stimulates the release of Ca^{2+}, while DAG remains in the membrane and, in the presence of Ca^{2+} and phosphatidylserine, activates **protein kinase C**. This enzyme phosphorylates many intracellular proteins, including some that are also phosphorylated by protein kinase A. Phorbol esters, a group of chemicals that can induce tumours, are structurally related to DAG and also activate protein kinase C. This action may explain their cancer-inducing effects.

Finally, IP_3 and DAG are metabolized. The end-products may be recycled to form phosphatidylinositol, which is used to resynthesize PIP_2 as depicted in Fig. 2.8 or, in the case of DAG, a lipase may release arachidonic acid, the precursor of the eicosanoids, including leukotrienes and prostaglandins (Fig. 2.9). Ca^{2+} is reaccumulated by the endoplasmic reticulum or transported out of the cells.

Thus, in this system, Ca^{2+} acts as a third messenger and events in both branches determine the appropriate cellular response. Increased cytosolic Ca^{2+} may stimulate the activity of protein kinases or other regulatory enzymes in the cell either directly or through a Ca^{2+}-binding protein called **calmodulin**. Calmodulin, which binds up to four Ca^{2+} ions per molecule with high affinity, regulates the activity of many cellular enzymes, e.g. Ca^{2+}/calmodulin-dependent kinase, phosphorylase kinase, adenylate cyclase, cyclic nucleotide phosphodiesterase, Ca^{2+} ATPase and myosin light-chain kinase. Not only does Ca^{2+} (via calmodulin) regulate enzymes involved in the synthesis and degradation of cAMP but the two intracellular messengers nearly always function together, although not necessarily in harmony.

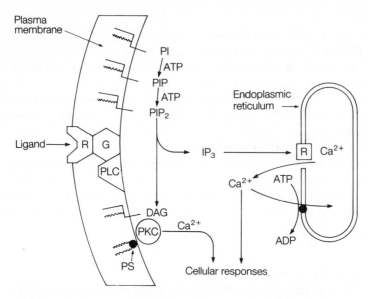

Fig. 2.8 Signals operating through the polyphosphoinositide system. The binding of the ligand to its receptor (R) in the plasma membrane activates the enzyme phospholipase C (PLC) via the GTP-regulatory protein (G). This enzyme cleaves phosphatidylinositol 4, 5-bisphosphate (PIP_2) to form the second messengers inositol 1,4,5-trisphosphate (IP_3) and diacylglycerol (DAG). IP_3 mobilizes Ca^{2+} from the endoplasmic reticulum while DAG, in the presence of phosphatidylserine (PS) and Ca^{2+}, activates the enzyme protein kinase C (PKC), and these actions lead to specific cellular responses. IP_3 and DAG are rapidly metabolized and their end-products recycled to form phosphatidylinositol (PI), which is phosphorylated to form PIP and finally PIP_2.

Arachidonic acid derivatives

Arachidonic acid, a 20-carbon moiety, can be formed from DAG produced by the polyphosphoinositide system or be released from membrane phospholipids through the action of the family of enzymes known as **phospholipase A_2** that require Ca^{2+} for their activity. Receptor-coupled G proteins may be involved in the regulation of phospholipase A_2 but their role is poorly understood. Metabolism of arachidonic acid leads to the formation of eicosanoids: prostaglandins and thromboxanes produced by the cyclo-oxygenase pathway and leukotrienes by the lipoxygenase pathway (Fig. 2.9). Not every cell produces the full range of eicosanoids and not all of the eicosanoids are second messengers, as defined above. Many leave the cell in which they are synthesized and diffuse locally to exert their effects on other cells by interacting with specific surface receptors coupled to G_p.

Arachidonic acid arises from the metabolism of phospholipids containing ω-6 (or n-6) fatty acids, so called because the first double bond is found following the sixth carbon from the methyl end of the carbon chain. Such fatty acids are obtained from diets containing lipids from vegetables and land-based animals. Metabolism of phospholipids containing ω-3 (or n-3) fatty acids (e.g. eicosapen-

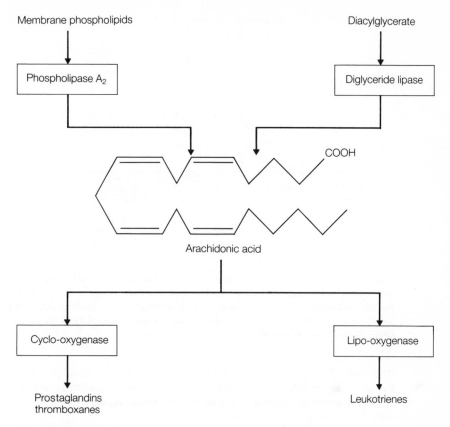

Fig. 2.9 Synthesis of prostaglandins, thromboxanes and leukotrienes.

taenoic acid) found in lipids from fish and marine mammals yields a different group of eicosanoids. There is presently much interest in the relationships between eisosanoids derived from the ω-6 and ω-3 fatty acids, for there is some evidence that the intensity of the effects of ω-6 derivatives are moderated by the presence of the ω-3 derivatives and that this may be of significance in a variety of diseases, including ischaemic heart disease.

Tyrosine kinase activation

Finally, there are a number of substances which do not appear to act through the recognized intracellular messenger systems discussed above. Some of these (e.g. insulin, epidermal growth factor) are known to interact with membrane-bound receptors that have intrinsic tyrosine kinase activity and can autophosphorylate their own tyrosine residues. This is in contrast to the protein kinases mentioned previously which phosphorylate serine or threonine residues. The **insulin receptor** is composed to two α- and two β-subunits (Fig. 2.10). Insulin binds to the α-subunit, which is in contact with the extracellular fluid, and activates the tyrosine kinase of the β-subunit, which is located on the cytoplasmic side. This results in autophosphorylation of tyrosine residues on the β-subunit. It also causes phospho-

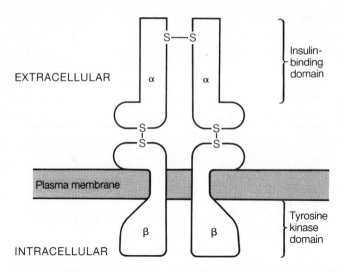

Fig. 2.10 Model of the insulin receptor. It is a glycoprotein composed to two α- and two β-subunits (M_r 130 000 and 90 000, respectively) which are linked by disulphide bonds, the precise location of which has not yet been established. Insulin binds to the α-subunit on the extracellular surface and activates the tyrosine kinase domain of the β-subunit which is located on the intracellular surface.

rylation of other protein substrates but the significance of these and subsequent events is not understood. At high ligand concentrations, binding to the second α-subunit may suppress the response of the tyrosine kinase.

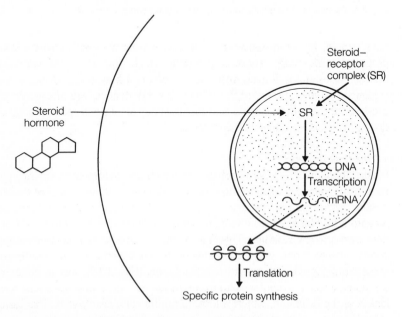

Fig. 2.11 Steroid hormone action at the cellular level.

Activation of nuclear receptors by lipophilic signals

In contrast to signals involving hydrophilic ligands discussed above, those involving lipophilic ligands (e.g. steroid hormones), that readily cross the plasma membrane, require intracellular receptors. Originally it was thought that steroid hormones bind to specific cytoplasmic receptors and are then translocated to the nucleus, but now it seems more likely that their receptors are located in the nucleus (Fig. 2.11). The hormone–receptor complex by binding to DNA activates (or suppresses) the transcription of specific genes and, after splicing of the exon sequences, leads to the production of specific messenger RNA species. Consequently, the action of steroid hormones leads to changes in the concentrations of specific cellular proteins. Thyroid hormones also bind to nuclear receptors that belong to the same family as steroid hormone receptors. Unlike steroid hormones, however, thyroid hormones appear to react with receptors that are already bound to DNA, and subsequently trigger the synthesis of specific RNA and protein species.

Chapter 3
Introduction to the Nervous System

The nervous system is a network of cells specialized for the reception, integration and transmission of information. It comprises the brain and spinal cord (the **central nervous system**; CNS), and sensory and motor nerve fibres that enter and leave the CNS or are wholly outside the CNS (the **peripheral nervous system**; PNS). The fundamental unit of the nervous system is the neuron. The cell bodies of neurons tend to be segregated into compact groups (nuclei, ganglia) or into sheets (laminae) that lie within the grey matter of the CNS or are located in specialized ganglia in the PNS. Their nerve fibres course in the white matter of the CNS or along peripheral nerves. Groups of nerve fibres running in a common direction usually form a compact bundle (nerve, tract, peduncle, brachium, pathway). Many of these nerve fibres are surrounded by sheaths of lipid material called myelin which gives rise to the characteristic appearance of the white matter. In addition to neurons there are glial cells which play a supporting role. There are about five times more glial cells than neurons and they occupy approximately half the volume of the brain.

3.1 CELLS OF THE NERVOUS SYSTEM

Neurons are specialized for the transmission of signals. Each neuron consists of a cell body, dendrites, an axon and synaptic terminals. The dendrites receive signals from other cells, the axon conducts signals from the cell body to distant targets, and the synaptic terminals transmit signals to other cells. Neurons are surrounded by glial cells, which maintain the structure of the nervous system, form myelin sheaths around axons, repair damage to nervous tissue, help to maintain the surrounding ionic environment, and assist in the uptake of certain neurotransmitters.

Neurons

Neurons are more diverse in size and shape (Fig. 3.1) than cells in any other tissue of the body. Nevertheless they have certain features in common: they usually possess dendrites, a cell body, an axon and synaptic terminals (Fig. 3.2a).

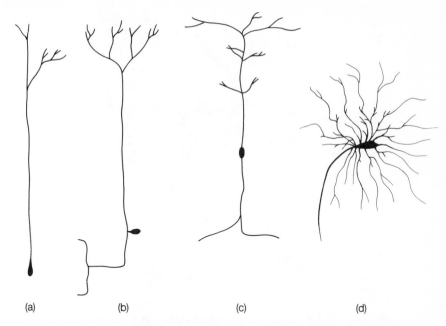

(a)   (b)   (c)   (d)

Fig. 3.1 Various types of neurons: (a) unipolar; (b) pseudounipolar; (c) bipolar; and (d) multipolar.

Dendrites are branches which leave the cell body and share with it the function of receiving information from synaptic connections with adjoining neurons. They are usually tapered and possess various protuberances, such as dendritic spines (Fig. 3.3) which are sites of contact with other cells.

The **cell body**, also known as the soma or perikaryon, contains the nucleus. In the cytoplasm surrounding the nucleus is an extensive array of rough endoplasmic reticulum, previously known as the Nissl substance, which synthesizes proteins for export to the axon. Large neurons replace as much as one-third of their protein content every day.

The **axon** is a slender process ranging in length from a few hundred micrometres to more than a metre. The diameter of axons usually varies from 0.1 to 20 μm, although in some invertebrates it may be as much as a millimetre. The axon contains mitochondria, microtubules, neurofilaments, microfilaments and vesicles (Fig. 3.2b). At its junction with the cell body, also known as the **axon hillock**, there is some rough endoplasmic reticulum but beyond the axon hillock there are no ribosomes. (In contrast, dendrites do contain ribosomes.) Axons are usually longer than dendrites and at their ends they branch to form terminals which make contact with other cells at **synapses**.

Synaptic terminals are the sites of release of chemicals (**neurotransmitters**) and are usually in close proximity to another neuron or effector cell. A synapse consists of a presynaptic element containing synaptic vesicles, a synaptic cleft and a postsynaptic element (Fig. 3.3). The synaptic vesicles, which contain the neurotransmitter, vary in size in different cells from 25 to 250 nm. The synaptic

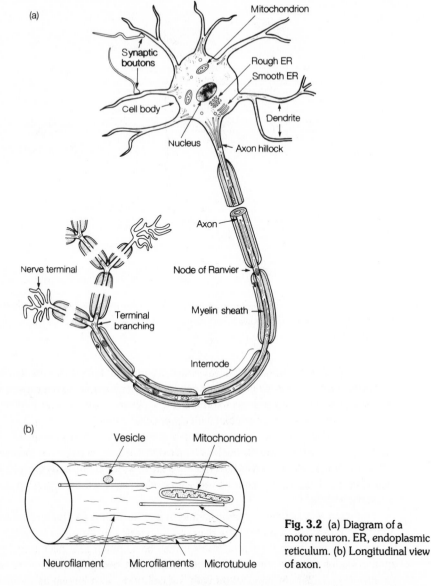

Fig. 3.2 (a) Diagram of a motor neuron. ER, endoplasmic reticulum. (b) Longitudinal view of axon.

cleft between neurons is 20–30 nm wide and at skeletal neuromuscular junctions it is 50–100 nm wide. In electron micrographs, electron-dense areas on the cytoplasmic faces of pre- and postsynaptic membranes indicate sites of specialization for exocytosis of vesicle contents and for localization of postsynaptic receptors, respectively. While most neurotransmitters are released close to effector cells, some neurons release substances which act at some distance from their site of release to modulate the activity of surrounding neurons. Such substances are referred to as *neuromodulators*.

Rapid transmission of information by neurons relies on their ability to conduct action potentials, coupled with neurosecretion at synapses. These processes are

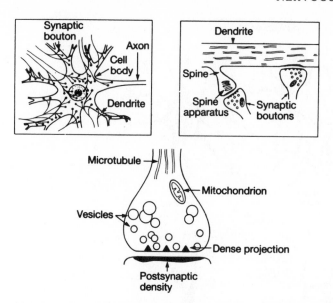

Fig. 3.3 A typical synapse found in the central nervous system. Left inset: numerous branches of axons from distant sites form synaptic terminals (boutons) on a cell body and dendrites of a neuron. Right inset: typical synapses on a dendrite.

discussed in more detail in the next chapter, as too is the ability of neurons to process and integrate information. Neurons also have a trophic ('nourishing') influence on the differentiation of sensory receptor cells (e.g. taste cells) and effector cells (e.g. skeletal muscle fibres). Their effect on taste cells is most likely due to the release of trophic factors but their action on skeletal muscle fibres is also due to the electrical activity they induce in those cells.

In summary, neurons are specialized:

1 to receive information from the internal and external environment;
2 to transmit signals to other neurons and to effector organs;
3 to process information (integration); and
4 to determine or modify the differentiation of sensory receptor cells and effector cells (trophic functions).

Neuroglia

There are three types of neuroglial or glial cells in the CNS—astrocytes, oligodendrocytes and microglia (Fig. 3.4)—and one type in the PNS—Schwann cells. Glial cells fill the spaces between neurons and accordingly have a role in maintaining neural structure.

Astrocytes are star-shaped cells with processes extending into the surrounding tangled network of unmyelinated nerve fibres (neuropil). These processes contain bundles of fibrils and usually run at right-angles to the direction of the nerve fibres. Some processes expand into end-feet which are applied to the surfaces of blood vessels. They are also in contact with one another through gap junctions and have a high permeability to K^+ (and resting membrane potentials close to the potassium equilibrium potential (E_K), which help to dampen in-

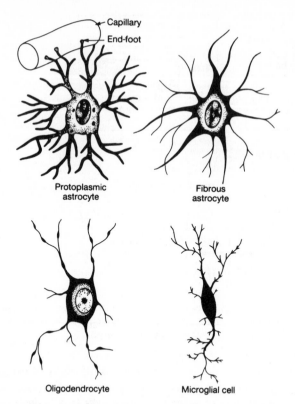

Protoplasmic astrocyte

Fibrous astrocyte

Oligodendrocyte

Microglial cell

Fig. 3.4 Neuroglial cells.

creases in extracellular K^+ concentration that might arise in local areas of the brain where there are high levels of activity.

Because of their high permeability to K^+, astrocytes adjacent to active neurons quickly become depolarized by the increase in extracellular K^+. Current then flows through gap junctions between depolarized astrocytes and non-depolarized astrocytes surrounding them. This electrical current causes K^+ to flow into the surrounding depolarized astrocytes, decreasing the extracellular K^+.

Another function of astrocytes is to assist in the removal of certain neurotransmitters (e.g. glutamate) after their release from neurons.

Oligodendrocytes can be distinguished from astrocytes by having fewer and thinner processes; they do not possess gap junctions. They form *myelin* sheaths around axons in the CNS, as do **Schwann cells** in peripheral nerves, by enveloping them with concentric layers of plasma membrane (Fig. 3.5). Myelin forms an insulating sheath around an axon with small areas of axonal membrane exposed between successive myelin segments known as the *nodes of Ranvier*. Schwann cells also encircle unmyelinated axons in the PNS without forming concentric layers of myelin.

Microglia are smaller glial-type cells of uncertain origin that appear to act as scavenger cells. Injury to the brain activates microglia which undergo proliferation and migrate to the point of injury where they turn into macrophages and remove

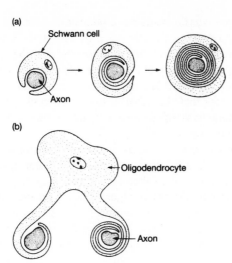

Fig. 3.5 Myelination of axons. (a) A Schwann cell in the peripheral nervous system forms myelin around a portion of a single axon. (b) An oligodendrocyte in the central nervous system can form myelin around several axons.

the debris. Astrocytes then proliferate and wall off the damaged area, forming a glial scar.

3.2 DEVELOPMENT, GROWTH AND REGENERATION OF THE NERVOUS SYSTEM

Cells of the nervous system are derived from embryonic ectoderm. Neurons are generated in the ventricular layer of the neural tube; they then migrate to their final position and send processes to their target regions. Structural components, which are synthesized in the cell body and carried by axonal transport, are incorporated at the growing tip of the axon, called the growth cone. The growth cone responds to environmental cues that help guide axons to their final destination. During development many neurons undergo a period of natural cell death which reduces their numbers. Once development has occurred, neurons can no longer divide; regeneration of their axons can occur in the PNS but is restricted in the CNS.

Early development of the nervous system

Cells of the very young embryo comprise three primordial types—ectoderm, endoderm and mesoderm. The nervous system develops from cells of ectodermal origin. At the earliest stage of development, these cells form a flat plate called the neural plate. The neural plate then rolls up to form the **neural tube**, the progenitor of the brain and spinal cord. On top of this tube is found **neural crest** tissue which migrates down either side to give rise to dorsal root ganglion cells, autonomic ganglion cells and a wide variety of non-neural cells including Schwann cells and endocrine cells. The innermost cells in the neural tube line the future ventricles of the CNS, and are termed the ventricular layer. Neurons are generated by mitosis in the ventricular layer. Young postmitotic neurons migrate away from the ventricular layer, guided by radial glial cells, to settle in their final positions in the brain and spinal cord.

During early development the neuron sends out a number of neuronal processes, one of which develops into the axon; the others form the dendrites. Prior to differentiation these neuronal processes are referred to as *neurites*, which possess growth cones at their elongating tips. The growth cones react to various environmental cues, such as the extracellular matrix and cell adhesion molecules, it is by specific interaction with these cues that the growing axons are guided to the correct target region. These interactions enable neural circuits to be established accurately without major trial and error.

Once initial contact has been established with the target region, the target cells may become *hyperinnervated*. Such is the case with the early motor input to developing skeletal muscles. This hyperinnervation may represent an important trigger in the development and maturation of the target. Hyperinnervation of the target is not maintained to maturity, and there is a normal period during development, usually after the neural centre and target have established a functional relationship, when this hyperinnervation is lost by retraction of terminals from multiply innervated sites.

Neurotrophic factors

Substances derived from glial cells and target tissues can promote the development and survival of neurons. The best characterized of such substances is **nerve growth factor** (NGF), a protein of M_r 135 000, which was first shown to stimulate the growth of neurites of dorsal root ganglia *in vitro*. A limited number of other neurotrophic factors have subsequently been isolated. Sympathetic and sensory neurons have an absolute requirement for NGF during the early stages of their development and disappear following the administration of antibodies to NGF at this time. Sympathetic neurons also require NGF for optimal maintenance at later stages. NGF is produced by target tissues and is specifically taken up into the nerve terminals of sympathetic and sensory neurons by receptor-mediated endocytosis (p. 15) and then transported to their cell bodies.

Extracellular matrix and cell adhesion molecules

Axons tend to grow along pre-established pathways and the extracellular matrix is an important source of this guidance. Extracellular matrix molecules, such as laminin, fibronectin and collagen, promote neurite outgrowth apparently by interaction with receptors called integrins in the growth cones of the neurites. The growth cone is motile and makes and breaks contact with adjoining structures, apparently 'tasting' its surroundings to make sure its direction of growth is appropriate. Axons also tend to grow along neighbouring axons to form bundles and axon–axon association is promoted by cell adhesion molecules which bind to one another at their apposing surfaces.

Cell death

Many more neurons are produced during development than are actually required by the mature nervous system. There is a brief period of normal development when many neurons die. The initial overproduction and subsequent death of neurons are thought to provide a mechanism which allows for optimal functional and numerical matching between the neural centre and its target cell population. The interactions between neuron and target are critical throughout the life of the neuron, since removal of the target tissue usually leads to the death of the neuron.

Role of activity

There are certain critical periods in the development of young animals during which particular groups of neurons must be used if their synaptic connections are to be functionally maintained. For example, if one eye of a young kitten is covered during the second to the fourth month of age the animal remains functionally blind in that eye for the rest of its life and there is permanent deficit in the number of cortical synaptic connections relaying information from the deprived eye to the visual cortex. Strabismus (squint or 'lazy eye') in children has a similar result; the critical period here is from 2 to 5 years of age.

Growth, maintenance and axonal transport

In a large neuron the axon can occupy more than 99% of the volume of the cell. Since there are no ribosomes in axons and nerve terminals, proteins required for the growth and maintenance of the axon must be synthesized in the cell body and carried towards the nerve terminals by **axonal transport**. In mature axons substances carried by axonal transport exchange with existing structures, while in growing axons they are incorporated into the growth cones of their elongating tips.

Axonal transport away from the cell body is referred to as 'anterograde' while that in the opposite direction is called 'retrograde'. Rates of anterograde transport have been determined by measuring the movement of radioactively labelled proteins after injection of labelled amino acids into regions containing cell bodies. This technique has revealed that there are several distinct rates of transport in axons—a *fast rate* of up to 400 mm per day, and *two slower rates* of 2–8 mm per day and approximately 1 mm per day.

Fast axonal transport involves the movement of organelles, such as synaptic vesicles and mitochondria, along the axon. This movement can be best visualized in extruded squid axoplasm by video-enhanced contrast microscopy. Such studies show that the organelles travel along the outside of microtubules and that movement is powered by a protein motor, **kinesin**, and adenosine triphosphate. Recently, it has been shown that organelles also move along actin filaments (microfilaments) and that such movement is probably powered by a myosin-like protein.

Cytoskeletal proteins move at the slow rates, although whether they move as monomers or filaments is not known. Tubulin and the neurofilament triplet proteins, which form the microtubules and neurofilaments respectively, move in the slowest component of slow transport. Other structural proteins, notably actin which forms the microfilaments, and also a number of soluble proteins move in the somewhat faster component of slow transport.

Retrograde transport is elegantly demonstrated in histochemical studies in which a tracer substance, e.g. horseradish peroxidase, is injected into a region containing nerve terminals. The tracer is taken up into the nerve terminals by endocytosis, transported along the axon in vesicles, and subsequently appears in the cell body. This technique has proved to be of considerable value for tracing neural pathways in the brain. The rate of retrograde transport approaches that of fast anterograde transport; its mechanism is similar, but requires a different protein motor, **cytoplasmic dynein**. Substances of physiological and pathological importance that are taken up at the nerve terminal and transported retrogradely include NGF, tetanus toxin and poliomyelitis virus.

Neural regeneration

PNS

If a peripheral nerve fibre is severed, the distal portion of the axon dies. There-upon, Schwann cells surrounding the axon dedifferentiate and undergo mitosis, filling the nerve sheath distal to the cut. Cell bodies of cut neurons undergo chromatolysis, a process involving dispersion of Nissl substance and synthesis of new RNA. Thus RNA coding for proteins involved in synaptic functions is lost and replaced by newly synthesized RNA, which codes for proteins necessary for regrowth of the axon. The proximal ends of cut axons sprout and will grow within the old nerve sheath, provided that they can find their way to that site. This applies to both motor and sensory neurons. If, however, the more central portion (dorsal root) of a sensory nerve is damaged, regeneration into the spinal cord usually does not occur. There is little specificity in the regrowth of the peripheral axons, so that normal relations between particular motor neurons and muscles, and particular sensory neurons and their peripheral fields of innervation, are not accurately restored. The success of reinnervation of the original target area depends on the re-establishment of the original pathway. Hence, in attempting to restore function, it is helpful to suture the cut ends of the corresponding bundles (fascicles) together again.

Although there is little specificity in the regrowth of mammalian peripheral axons, it is noteworthy that motor axons grow back to form terminals at the original endplate sites on skeletal muscle fibres. It appears that they are directed by cues in the form of marker molecules present on the synaptic basal lamina that has been left behind. The marker molecules have been isolated and identified as proteins called **agrins**.

CNS

Damaged neurons in the CNS of higher vertebrates have a very limited capacity for regeneration. Axotomy often results in the death of the injured neuron, possibly because cells must be nourished by the retrograde transport of trophic factors. Cells surviving axotomy do produce axonal sprouts, but these fail to grow along their original pathway and form a tangled mass (neuroma) near the point of section. Recently it has been demonstrated that two proteins in the myelin produced by oligodendrocytes (but not Schwann cells) inhibit neurite regenera-tion. While this may appear to be a disadvantage, these neurite growth inhibitors could play an important role in restricting the growth of nerve fibres to specific pathways during CNS development.

3.3 DIVISIONS OF THE PNS

The PNS is that portion of the nervous system that lies outside the spinal cord and brain; it comprises both the somatic and the autonomic divisions. The somatic division contains all the peripheral pathways responsible for communication with the environment and the control of skeletal muscle. The autonomic nervous system comprises all the efferent pathways from controlling centres in the brain and spinal cord to effector organs other than skeletal muscle.

Somatic nervous system

This division includes all sensory (**primary afferent**) fibres from tissues such as eyes, ears, skin, joints and skeletal muscles and it includes the motor (**efferent**) fibres to skeletal muscles. The sensory fibres have their cell bodies in the dorsal root ganglia (Fig. 3.6a) or brain; the motor fibres have their cell bodies in the ventral horn of the spinal cord or in the brain. Fibres enter or leave the spinal cord via pairs of spinal nerve roots (dorsal and ventral) at more or less regular intervals along each side. Fibres from the cell bodies of sensory neurons in the dorsal root ganglia pass via the dorsal roots to the periphery and to the spinal cord. In the spinal cord they branch and usually make synaptic contact with interneurons in the grey matter. The neurotransmitters released at these synapses include *glutamate, substance P* and other unidentified substances. Fibres from the cell bodies of motor neurons in the ventral horn of the spinal cord leave via the ventral roots.

The sensory fibres and motor fibres are collected together to form **peripheral nerves** which are covered by a connective tissue sheath (**epineurium**; Fig. 3.7). Within the nerve the fibres are arranged into bundles (*fascicles*); each fascicle is surrounded by a **perineurium** and each fibre by an **endoneurium**. The somatic sensory fibres range in size from small unmyelinated (class C) to large myelinated (class A) axons; the motor nerves are all large and myelinated (class A). The most distal portions of the somatic nerves are usually branched; the terminal regions of any one sensory fibre go to adjacent receptors of the same type and the intramuscular portions of motor nerves branch to innervate a number of similar muscle fibres. Somatic motor neurons release the neurotransmitter *acetylcholine*, which has an excitatory effect on skeletal muscle; there are no peripheral inhibitory actions exerted on skeletal muscle.

Autonomic nervous system

The actions of the peripheral autonomic nervous system are normally involuntary and are directed to the control of individual organ function and to homeostasis. Classically the autonomic nervous system is regarded as being *solely motor* in function, its fibres going to cardiac muscle, to smooth muscle and to glands. Sensory information comes from visceral and somatic afferent inputs. The peripheral autonomic nervous system is usually divided anatomically into the **sympathetic** and **parasympathetic systems**, and many tissues are innervated by both systems. When this occurs the two systems usually have opposing effects. In addition to these nerves there is a network of nerves that can act independently of the CNS. This network is often considered as another division of the autonomic

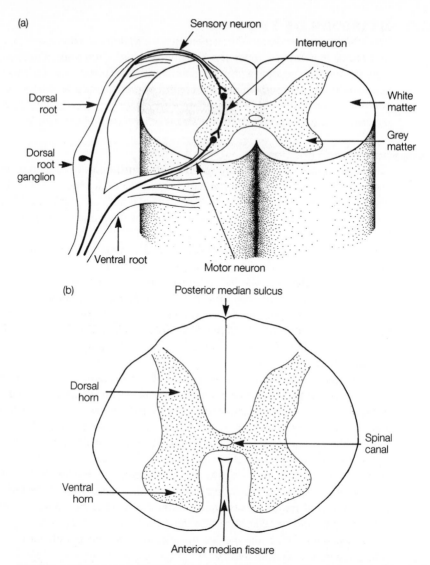

Fig. 3.6 (a) Diagram showing the relationship of the dorsal and ventral roots to the spinal cord. (b) Cross-section of the spinal cord.

nervous system and is referred to as the **enteric system**. Both the sympathetic and parasympathetic systems modulate the activity of this system.

The organization of the autonomic nervous system differs from the somatic division as all final motor neurons lie completely outside the CNS. The cell bodies of the peripheral neurons are grouped together to form **ganglia**. The efferent fibres passing from the CNS to the ganglia, the **preganglionic fibres**, are slow-conducting (class B and C) fibres which release the neurotransmitter *acetylcholine*. The final motor neurons from the ganglia to the tissues, the **postganglionic fibres**, are mainly slow-conducting, unmyelinated class C fibres. These fibres and those from visceral sensory receptors run together as **visceral nerves**.

(a)

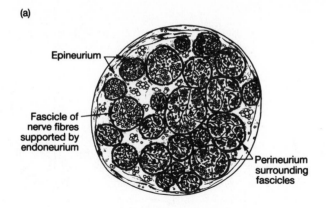

(b)

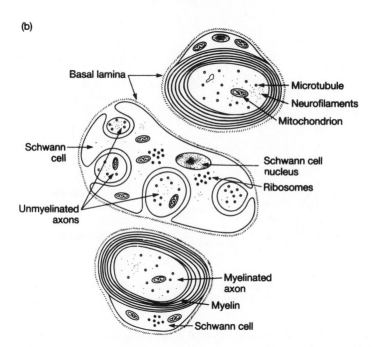

Fig. 3.7 (a) Cross-section of a peripheral nerve. (b) Enlargement of part of a fascicle showing myelinated and unmyelinated axons. Note that a number of unmyelinated axons are enclosed by a Schwann cell while each myelinated axon is enveloped by layers of myelin from a single Schwann cell.

Sympathetic system

All preganglionic sympathetic nerves have their cell bodies in the thoracic and upper lumbar segments (T_1–L_3). The axons, along with somatic motor fibres, pass out of the spinal cord in the ventral roots (Fig. 3.8). On leaving the spinal column, the preganglionic fibres separate from the somatic nerves to form the **white rami communicantes** and join a distinct group of sympathetic ganglia, the **vertebral** (paravertebral or chain) **ganglia**. These ganglia are segmentally arranged and lie along each side of the spinal column. The preganglionic fibres either:

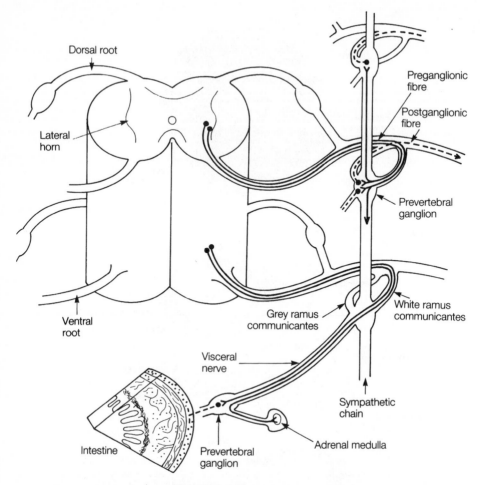

Fig. 3.8 Outflow of sympathetic fibres from the thoracolumbar spinal cord. Solid lines represent preganglionic fibres; broken lines, postganglionic fibres.

1 synapse with postganglionic neurons in one or more of the vertebral ganglia; or

2 leave the vertebral ganglia in visceral nerves and pass to **prevertebral ganglia** in the abdomen or to the **adrenal medullae**.

Many of the postganglionic fibres from vertebral ganglia leave the ganglia as the **grey rami communicantes** and join the spinal nerves that pass to peripheral tissue. Others, and also postganglionic fibres from prevertebral ganglia, join visceral nerves to particular organs.

The distribution of the postganglionic sympathetic nerves is not necessarily the same as the somatic motor nerves from the same segment and there is a lot of overlap between adjacent outflows. In brief, and very approximately, the T1 outflow passes to the head, T2 to the neck, T3–6 to the thorax, T7–11 to the abdomen and T12–L3 to the legs (Fig. 3.9). Most of these postganglionic fibres

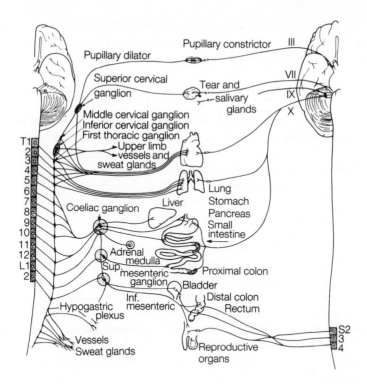

Fig. 3.9 Distribution of sympathetic (left) and parasympathetic (right) fibres.

release the neurotransmitter *noradrenaline* (norepinephrine) and are referred to as adrenergic fibres. The adrenal medullae can be thought of as modified ganglia that release the hormone *adrenaline* (epinephrine), and also some noradrenaline, into the blood stream rather than directly on to effector cells. Some sympathetic nerves that innervate the blood vessels of muscles, sweat glands and hair follicles in the skin release acetylcholine instead of noradrenaline.

Parasympathetic system

The preganglionic nerves of the parasympathetic system come from both the brainstem and sacral spinal cord (S2–4); the axons from the brainstem leave in cranial nerves III, VII, IX and X and the sacral axons leave in the ventral roots (Fig. 3.9). There are no vertebral or paravertebral ganglia in this system; instead all ganglia are found adjacent to or within the effector organ. The postganglionic fibres from parasympathetic ganglia are relatively short and nearly all release *acetylcholine* as a neurotransmitter. In most cases the distribution of the parasympathetic outflow is more restricted than that of the sympathetic system. Thus, cranial nerve III supplies the smooth muscle of the eye (ciliary muscle and pupillary constrictor muscle), VII the lacrimal and submaxillary glands, and IX the parotid gland. The sacral parasympathetic outflow supplies the lower colon, rectum, bladder, the lower part of the ureters and the external genitalia. The major and most widely distributed parasympathetic outflow travels in cranial nerve X, the vagus. Approximately 70% of all parasympathetic preganglionic fibres leave the CNS in this nerve and supply all of the viscera in the thorax and most of the viscera in the abdomen. However, note that the vagus also carries many sensory fibres and a number of somatic motor nerves, e.g. to laryngeal and pharyngeal muscle.

Enteric system

The autonomic nerves in the gastrointestinal tract differ from those in the other divisions as they form an extensive network in which many cells are not influenced by the CNS. The network is composed of ganglia and interconnecting bundles that lie in the wall of the intestinal tract (from the oesophagus to the rectum) and form the myenteric and submucosal plexuses (see Fig. 17.3). A number of different neuronal types have been identified in the plexuses but a detailed knowledge of the relationships between ganglia and between cells is lacking. However, it is recognized that this system contains sensory neurons, interneurons (integrative) and excitatory and inhibitory motor neurons. Many of the excitatory neurons (interneurons and motor neurons) release *acetylcholine* as a neurotransmitter, but others, particularly interneurons, may release *5-hydroxytryptamine* (5-HT, serotonin) or other compounds. The inhibitory neurons to smooth muscle do not release acetylcholine or noradrenaline as a transmitter and hence are referred to as **non-adrenergic, non-cholinergic nerves**. It has been suggested that they may be purinergic and release *adenosine triphosphate*, but more recent evidence suggests that a peptide (possibly *vasoactive intestinal polypeptide*; VIP) or nitric oxide (NO) may be involved.

It should be noted that there is good evidence indicating that non-adrenergic, non-cholinergic nerves are also active in the regulation of other tissues, such as the smooth muscle of the respiratory tract and urinary system. The identity of the transmitter released by these fibres is not proven but it may be VIP, *substance P* or NO.

3.4 BASIC DESIGN OF THE CNS

The CNS comprises the brain lying within the skull and the spinal cord lying within the vertebral column. The brain consists of three major subdivisions—the brainstem, the cerebellum and the diencephalon and cerebrum. The brainstem, which links the spinal cord and the cerebrum, is composed of the medulla oblongata, the pons and the midbrain. The cerebellum is attached to the brainstem. The diencephalon comprises the thalamus, the subthalamus and the hypothalamus. The cerebrum consists of the right and left cerebral hemispheres, which are connected by the corpus callosum. The outer grey matter of the cerebral hemisphere constitutes the cortex. Within the white matter of each hemisphere are several collections of neuronal cell bodies, such as the basal ganglia.

During embryonic development the neural tube grows to form three longitudinal swellings at its anterior end. These swellings develop into the **fore-, mid- and hindbrain**. Their cavities form the **ventricles** and remain in communication with the canal, in the **spinal cord**, which is formed from the remainder of the tube. The forebrain is divided into two regions, rostrally the **telencephalon** and caudally the

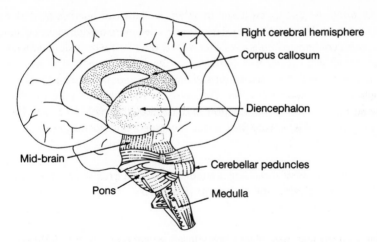

Fig. 3.10 A lateral view of the brainstem after removal of the left cerebral hemisphere and the cerebellum. (Adapted from Curtis, B.A., Jacobsen, S. & Marcus, E.M. (1972) *An Introduction to the Neurosciences*, p. 13. W.B. Saunders, Philadelphia.)

diencephalon. The telencephalon is formed by a mid-portion and two lateral outpouchings, the primitive **cerebral hemispheres** or **cerebrum**, which grow first forwards and then backwards to cover most of the mid- and hindbrain (Fig. 3.10). The diencephalon gives rise to the **thalamus and hypothalamus**. The hindbrain gives rise to the **pons** and **medulla**, which together with the midbrain form the **brainstem**. Further smaller raised swellings develop in the hindbrain to form the **cerebellum**. The cerebellum and the cerebral hemispheres have two unique features:

1 their surfaces are extensively folded, forming depressions called **fissures** or **sulci** and raised portions called **gyri**; and

2 unlike the brainstem and spinal cord, they have their grey matter on the outside and white matter on the inside.

Spinal cord

The spinal cord has a segmental structure with pairs of nerve roots, one pair on each side, arising at more or less regular intervals (Fig. 3.6a). Each pair comprises the **dorsal root** carrying information into the spinal cord from peripheral receptors and the **ventral root** carrying signals out, which control affector organs such as muscles. In cross-section, each segment of the spinal cord shows a central butterfly-shaped area of grey matter and a peripheral zone of white matter (Fig. 3.6b). The grey matter contains cell bodies of neurons. In the ventral horn of the grey matter lie the cell bodies of motor neurons whose axons leave in the ventral roots to innervate muscles; in the dorsal horn of the grey matter lie cell bodies of interneurons concerned with the processing of signals entering in the dorsal root axons.

The white matter consists of the axons of cells in the brain descending to spinal cord cells and the axons of cells in the spinal cord ascending to synapse with cells in the brain. As like axons travel together, well-formed bundles (*tracts*) can be distinguished. Thus, in the dorsal quadrant of the white matter we can distinguish the dorsal columns composed of axons ascending to the brainstem with informatiuon from skin, joints and muscles; in the lateral quadrant there are both ascending tracts to the cerebellum and brainstem and descending tracts from the cerebral cortex and brainstem. The spinal cord segments perform the initial processing of afferent information and contain the neural circuits for many reflexes, some of which are the basis of movement and posture.

Brain

The brain consists of three major subdivisions—the **brainstem**, the **cerebellum** and the **diencephalon and cerebrum** (Fig. 3.10).

Brainstem

This is an ancient part of the brain in an evolutionary sense. Although much smaller than the cerebellum, the brainstem is essential to life while the cerebellum may, with considerable residual but not life-threatening disability, be dispensed with. The brainstem links spinal cord and cerebrum and is composed of three regions—the medulla oblongata, pons and midbrain (Fig. 3.10).

The **medulla oblongata** is continuous with the spinal cord and contains the same fibre tracts. The grey matter, however, is not organized in a continuous column but is broken into discrete nuclei, including motor and sensory nuclei for the throat, mouth and neck, and nuclei involved in the control of the respiratory and cardiovascular systems and of movement and posture.

The **pons** may be recognized on the ventral surface (Fig. 3.11a) by the bulge of the brainstem formed by axons descending from the cerebrum and turning up into the cerebellum which, in the intact brain, conceals the dorsal surface of the pons. The pons is continuous with the medulla and contains the same ascending and descending tracts in the reticular grey matter (**reticular formation**). It also contains nuclei that are involved in the control of the respiratory system and in motor and sensory functions of the face.

The **midbrain**, the smallest part of the brainstem, is continuous with the pons below and the diencephalon above. The ventral surface is characterized by the large **cerebral peduncles** (Fig. 3.11a) lying laterally and carrying axons from the cerebrum to the brainstem and to the cerebellum. The dorsal surace may be recognized by two pairs of protuberances, the **superior** and **inferior colliculi** (Fig. 3.11b). The superior collicular neurons are concerned with the processing of visual information and the inferior collicular neurons with auditory signals. The midbrain also contains nuclei concerned with the state of wakefulness of the brain and ascending tracts from the spinal cord on their way to the diencephalon.

From the brainstem the various **cranial nerves** have their superficial origins at different levels (Fig. 3.11a). There are 12 pairs of these and they innervate the skin and muscles of the face, structures in the head and neck and, in the case of the vagus (X) nerves, the thoracic and many of the abdominal viscera. Strictly speaking, the olfactory (I) nerves are connected to the forebrain, and not the brainstem. The optic (II) nerves are joined to the diencephalon via the optic

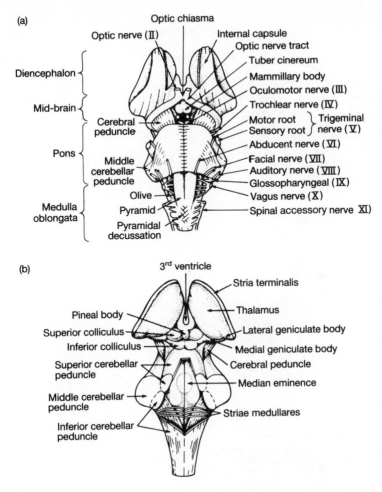

(a)

Optic chiasma

Optic nerve (II)

Internal capsule

Optic nerve tract

Tuber cinereum

Diencephalon

Mammillary body

Mid-brain

Oculomotor nerve (III)

Trochlear nerve (IV)

Cerebral peduncle

Motor root ⎱ Trigeminal
Sensory root ⎰ nerve (V)

Abducent nerve (VI)

Pons

Middle cerebellar peduncle

Facial nerve (VII)

Auditory nerve (VIII)

Glossopharyngeal (IX)

Olive

Vagus nerve (X)

Medulla oblongata

Pyramid

Spinal accessory nerve XI)

Pyramidal decussation

(b)

3rd ventricle

Stria terminalis

Thalamus

Pineal body

Superior colliculus

Lateral geniculate body

Inferior colliculus

Medial geniculate body

Superior cerebellar peduncle

Cerebral peduncle

Median eminence

Middle cerebellar peduncle

Inferior cerebellar peduncle

Striae medullares

Fig. 3.11 The brainstem: (a) ventral view; (b) dorsal view. (Adapted from Truex, R.C. & Carpenter, M.B. (1969) *Human Neuroanatomy*, 6th edn, p. 31. Williams & Wilkins, Baltimore.)

chiasma. Cranial nerve pairs III and IV emerge from the midbrain, V–VIII from the pons and the remaining four pairs from the medulla.

The nuclei from which these cranial nerves have their origins may extend through a considerable length of the brainstem. Note that many of the nerves contain a mixture of fibres, motor and sensory, somatic and autonomic.

Cerebellum

This is a large organ overhanging the dorsal aspect of the brainstem to which it is attached to each side by three large bundles of white matter—the **superior, middle** and **inferior cerebellar peduncles** which carry neural signals in and out of the cerebellum (Fig. 3.12a). Three cerebellar lobes can be recognized. On the superior aspect (Fig. 3.12b) the anterior lobe is separated from the larger middle lobe by

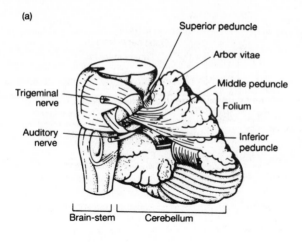

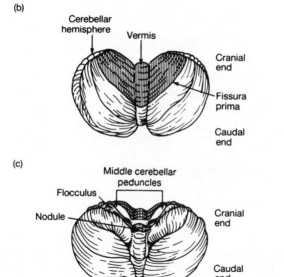

Fig. 3.12 The cerebellum: (a) lateral aspect; (b) superior aspect; (c) inferior aspect. (Redrawn from (1949) *Gray's Anatomy*, Johnson, T.B. & Willis, J. (eds) 30th edn, pp. 949, 954. Longmans, Green, London.)

the primary fissure. The anterior and middle lobes are further divided into a narrow median strip—the vermis—and two lateral **cerebellar hemispheres**. Rostrally, and seen only on the inferior aspect (Fig. 3.12c), are the **flocculus** and **nodule**, together forming the **posterior lobe**. The anterior and middle lobes are transversely folded, the folds being termed **folia**. This folding vastly increases the surface area of the cerebellum.

All parts of the cerebellum have the same structure—a thin superficial layer of grey matter, the **cerebellar cortex**, covering a much larger expanse of white matter formed by axons entering and leaving the cortex. Deep in the white matter, close to the brainstem, lie the **intercerebellar nuclei**. A section at right angles to the long axis of a folium shows that it is much wider laterally than medially and that subsidiary foldings of the folium, each with cortical grey matter and white matter, give the section the appearance of a tree, hence the name **arbor vitae** (Fig. 3.12a). the cerebellum has its major role in the control of the rate, range and direction of movement. The small, phylogenetically old, posterior lobe is connected to the organ of balance (labyrinth) and is concerned with the reflexes which ensure an upright posture.

Diencephalon and cerebrum

The **diencephalon**, nearest to the cerebrum, is covered by it in the intact brain and can be seen only if a brain is sectioned transversely or if one hemisphere is removed (Fig. 3.10). Developmentally it represents that part of the forebrain from which the rudimentary cerebral hemispheres have budded off. Its walls have been thickened by the growth on either side of the mass of grey matter that constitutes the **thalamus**, and ventrally by the structures that make up the **hypothalamus**, so that the original cavity has become the narrow cleft of the **third ventricle** (Fig. 3.13b). The *thalamic nuclei* function in at least three ways. One group relays signals concerned with all types of afferent information, except olfaction, to the cerebral cortex on the same side. Another group relays signals to the motor cortex on the same side and receives information from the cerebellum and basal ganglia. A third group is concerned with sleep and wakefulness. The *hypothalamic nuclei* serve as regulating centres for autonomic functions (e.g. body temperature, heart rate, blood pressure). They control the release of hormones by the pituitary gland, and are involved in the expression of the emotions and the regulation of food and water intake.

The **cerebrum** consists of the right and left **cerebral hemispheres** (Fig. 3.13a). Connecting them in the midline is a large thick band of white matter known as the **corpus callosum** (Fig. 3.13b) made up of axons passing between the hemispheres. Each hemisphere comprises an outer layer, about 1 cm thick, of grey matter which is the **cerebral cortex**, covering a dense thick inner layer of white matter. The grey matter has a rich blood supply; the white matter is less well-endowed. The white matter is composed of axons connecting different regions of the cortex with each other and connecting the cortex with the rest of the brain.

The cerebral cortex is folded into gyri and sulci, thereby increasing the surface area. The deepest sulci are called fissures. On the lateral surface of each hemisphere (Fig. 3.13b) the *lateral fissure* partially separates off the temporal lobe from the rest. This lobe contains the **primary auditory cortex** which receives signals from the auditory receptors in the inner ear.

The *central sulcus* runs from the medial surface to the lateral fissure (Fig. 3.13a). It is not as prominent as the lateral fissure but there are usually well-formed and continuous gyri on either side of the sulcus. The central sulcus forms the posterior boundary of the frontal lobe. The gyrus forming this posterior boundary and the anterior wall of the central sulcus is the *precentral gyrus*. This is the **primary motor cortex** containing neurons whose axons run down through the brainstem and spinal cord and synapse with motor neurons.

The **parietal lobe** lies behind the central sulcus. The posterior border of the sulcus is the *postcentral gyrus* of the parietal lobe—the **primary somatosensory**

(a)

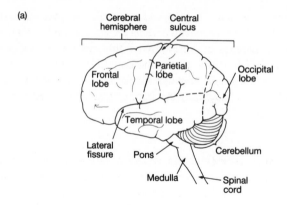

(b)

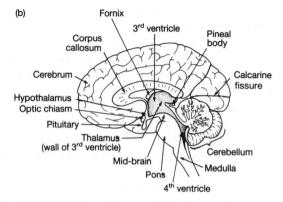

Fig. 3.13 The human brain: (a) lateral view; (b) view of a mid-sagittal section.

cortex which receives information from receptors in skin, joint and muscle.

The **occipital lobe** is most posterior and little of it can be seen on the lateral surface. Examination of the medial surface (Fig. 3.13b), however, shows the *calcarine fissure* of the occipital lobe where the **primary visual cortex** is located. The occipital lobe is entirely concerned with the processing of visual information.

The **corpus callosum** connects the two hemispheres and is surrounded by the *cingulate gyrus*, which is involved in the control of emotional behaviour.

Within the white matter of each hemisphere is a large collection of neuronal cell bodies, the **basal ganglia**. These are a group of nuclei which are concerned with the initiation and control of movement. They lie between the thalamus and cerebral cortex in each hemisphere and comprise the **claustrum**, the **putamen** and the **caudate nucleus**, and the **globus pallidus** (Fig. 3.14). Other smaller nuclei are for functional reasons usually considered along with the basal ganglia. The largest of these are the **subthalamic nucleus** and the **substantia nigra**, a darkly pigmented nucleus lying in the midbrain. The caudate nucleus and putamen are often referred to together as the *neostriatum*, while the putamen and globus pallidus are sometimes referred to together as the *lentiform nucleus* because of

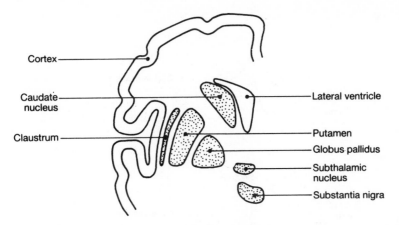

Cortex

Caudate nucleus

Claustrum

Lateral ventricle

Putamen

Globus pallidus

Subthalamic nucleus

Substantia nigra

Fig. 3.14 Coronal section of the right cerebral hemisphere showing the location of the basal ganglia (claustrum, caudate nucleus, putamen and globus pallidus) and associated structures (subthalamic nucleus and substantia nigra).

their combined shape. The term *corpus striatum* includes all of these structures but not the claustrum.

Limbic system

This is a functionally related group of structures forming a fringe on the medial side of each hemisphere (Fig. 3.15). It is important in the regulation of behaviour and in memory (p. 260). The structures concerned are many synapses away from the primary sensory or motor pathways and receive information from the overlying cortex which, after processing, is directed back to the cortex. The limbic system comprises parts of the frontal lobe on either side, namely the **cingulate gyrus**, the **hippocampus**, the **septum**, the **amygdaloid nucleus** and the **anterior thalamic nucleus**. Some authors also include the hypothalamus. The cingulate gyrus lies above and around the corpus callosum (Fig. 3.15a). The hippocampus lies on the medial wall of the temporal lobe bordering the inferior horn of the lateral ventricle (Fig. 3.15b). It is a layered structure like the cerebral cortex but of simpler cellular pattern. The amygdaloid nucleus lies in the same region just in front of the hippocampus. The septal nuclei lie in the midline anterior to the hypothalamus between the lateral ventricles and ventral to the corpus callosum, the genu (knee) of which forms their anterior boundary on either side (Fig. 3.15a). A large fibre bundle known as the fornix takes origin from the hippocampus and overlaying cortex on each side (Fig. 3.15a,b). This bundle arches forward underneath the corpus callosum accompanied by a similar but much smaller bundle of axons (stria terminalis) from the amygdaloid nucleus. The fornix and stria terminate in the septum and hypothalamus.

Meninges, ventricles and cerebrospinal fluid

The CNS is covered by three membranes or **meninges** (Fig. 3.16) comprising a tough outer layer, the *dura mater*, a middle layer, the *arachnoid mater*, in which

(a)

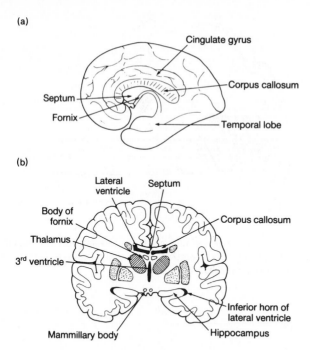

(b)

Lateral
ventricle   Septum

Body of
fornix

Corpus callosum

Thalamus

3rd ventricle

Inferior horn of
lateral ventricle

Mammillary body   Hippocampus

Fig. 3.15 The limbic system of the brain shown in (a) the medial surface of the right cerebral hemisphere and (b) a frontal section through the junction of the third and lateral ventricles. See text for further information.

lie the blood vessels, and an inner layer, the *pia mater*. Between the arachnoid mater and the pia mater is the *subarachnoid space* which is in communication with the ventricles and contains **cerebrospinal fluid** (CSF), a specialized extracellular fluid. At the base of the brain, the subarachnoid space becomes enlarged to form cisterns, the largest being the *cisterna magna* (Fig. 3.17a).

The **ventricles** comprise the *lateral ventricles* in the cerebral hemispheres, the *third ventricle* in the diencephalon and the *fourth ventricle* in the hindbrain. The third and fourth ventricles are connected in the midbrain by a canal called the *cerebral aqueduct* (Fig. 3.17a).

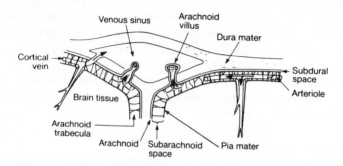

Fig. 3.16 Meninges of the brain.

(a)

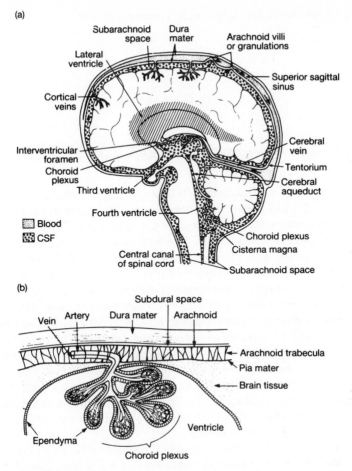

(b)

Fig. 3.17 (a) Flow of cerebrospinal fluid. (After Rasmussen, A.T. (1937) *The Principal Nervous Pathways*, p. 4. Macmillan, New York.) (b) Choroid plexus.

Blood–brain barrier

Slow diffusion of many substances of low M_r between the blood and CSF and between the blood and brain suggests the existence of a **blood–CSF barrier** and a **blood–brain barrier**. These barriers are permeable to respiratory gases, to glucose and to fat-soluble drugs like volatile anaesthetics. The endothelial cells of the capillaries within the CNS are held together by **tight junctions** and it is these which limit diffusional exchanges of water and water-soluble solutes. However, these cells are involved in transport of solutes between blood and brain interstitial fluid. Note that the blood–brain barrier is absent in certain structures called **circumventricular organs** which abut on the third and fourth ventricles, e.g. the subfornical and pineal organs, the area postrema and the median eminence.

CSF

The CSF is formed predominantly by the **choroid plexuses**—rich networks of

Table 3.1 Composition of CSF compared with protein-free plasma*

CSF : plasma water	
Na$^+$	1.0
K$^+$	0.7
Ca^{2+}	0.5
Cl$^-$	1.1
HCO$_3^-$	0.9
Glucose	0.6

*Note that the protein content of CSF is 0.3 g L^{-1} compared with 70 g L^{-1} for plasma.

blood vessels covered with epithelial cells (*ependyma*) projecting into the ventricles (Fig. 3.17b). Fluid formed in the lateral ventricles passes to the third and fourth ventricles and to the central canal of the spinal cord. This movement is aided by the cilia on ependymal cells. The fluid escapes into the subarachnoid space through foramina in the ependymal lining of the fourth ventricle and circulates around the brain and spinal cord (Fig. 3.17a). Finally it is reabsorbed through the arachnoid villi into the sinuses of the venous system. In most regions of the brain, substances are free to diffuse between the ependymal cells and so there is a ready exchange of solutes between the CSF and the extracellular spaces of brain and spinal cord. The local environment of the neurons is further controlled by the activity of glial cells which can adjust both K$^+$ and H$^+$ ion concentrations. Of the 700 ml of CSF formed per day, about 70% is derived from the choroid plexuses and the other 30% comes from endothelial cells lining the brain capillaries. In adult humans the volume of CSF is about 140 ml (compared with about 250 ml for brain interstitial fluid) and when lying supine its pressure is about 10 mmHg, i.e. a little less than local venous pressure.

The composition of CSF differs from what one would expect if it were simply an ultrafiltrate of plasma (Table 3.1), indicating that it is actively secreted. One model for the formation of CSF by the choroid plexuses is illustrated in Fig. 3.18. Note the unusual feature, that the Na$^+$–K$^+$ ATPase is located in the apical plasma membrane. Note also that though CO$_2$ crosses the blood–brain barrier readily, HCO$_3^-$ does not. The HCO$_3^-$ in CSF is synthesized within the epithelial cells, in the reaction

$$CO_2 + H_2O \rightleftharpoons H_2CO_3 \rightleftharpoons H^+ + HCO_3^-$$

The initial reaction is catalysed by carbonic anhydrase which is present in the epithelial cells. The primary secretion has more HCO$_3^-$ than plasma (about 45 mmol L^{-1}) and it buffers H$^+$ ions produced by neuronal and glial metabolism; this reduces [HCO$_3^-$] to about 23 mmol L^{-1}. The decrease in [HCO$_3^-$] may also reflect the mixing of choroidal CSF with fluid secreted by the capillary endothelial cells, which is thought to have a higher Cl$^-$ to HCO$_3^-$ ratio.

The functions of the CSF are as follows:

1 Conferring of buoyancy. The brain has little mechanical strength or rigidity. It weighs 50 g in CSF, compared with 1500 g in air and so flotation in CSF protects it against deformation and damage from the innumerable accelerations imposed

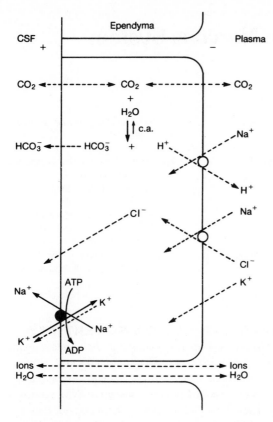

Fig. 3.18 Model for the formation of CSF by the choroid plexuses, ca, Carbonic anhydrase.

by movements of the head. Counterpressure of CSF surrounding the blood vessels within the cranium and the spinal cord compensates for gravitational effects of alterations in posture (p. 453) or external acceleration.

2 Maintenance of a constant ionic environment in brain. This is important because the activity of neurons is highly sensitive to ionic changes. The CSF also provides another route for removal of substances which have limited lipid solubility or are too large to move easily across capillary walls. In other tissues the lymphatic system, which is not found in the CNS, serves this function.

Chapter 4
Signalling in the Nervous System

4.1 LOCAL POTENTIALS AND ACTION POTENTIALS

Electrical signals in neurons are derived from changes in their resting membrane potential and can be divided into two types—local potentials and action potentials. In general, local potentials occur at points of discontinuity (e.g. environment–sensory receptor, central nervous system (CNS) synapses and nerve–muscle junctions). They have their maximum amplitude at their point of origin and away from this (0.2–2 mm) the amplitude falls rapidly. In contrast, action potentials are transient changes in membrane potential that are conducted over longer distances, in some instances a metre or more, with little change in amplitude. Each of these signals can be accounted for by the properties of the cell membrane and its response to changes in membrane potential.

The role of these signals in the nervous system can best be illustrated by outlining a simple common reflex, the **withdrawal reflex**. This reflex results in the rapid withdrawal (latency about 30 ms) of a limb when a part of it is subjected to a noxious stimulus. For example, a foot placed on a sharp object may be withdrawn initially without conscious direction. Underlying this reflex withdrawal is a specific sequence of neural events. In brief, the pathway (Fig. 4.1) comprises a sensory (afferent) neuron, an interneuron, a motor (efferent) neuron and skeletal muscle fibres. However, when the reflex is initiated many pathways with this structure will operate in parallel.

The noxious stimulus causes a local change of potential (**receptor potential**) in the sensory nerve terminal, resulting in the generation of action potentials which are conducted along the sensory fibre to the CNS. Their arrival at the central terminal of the sensory fibre is followed by release of a neurotransmitter, which produces local potentials in the dendrites and cell body of the interneuron. The

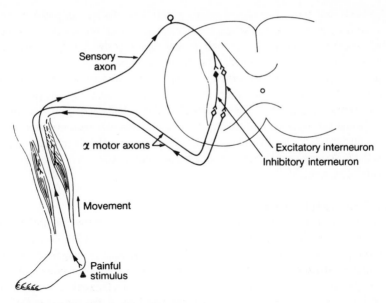

Fig. 4.1 Withdrawal reflex pathway.

local potentials again initiate action potentials that are conducted along the axon of the interneuron to its point of synaptic contact with the motor neuron. Again, a transmitter substance is released, causing a local excitatory potential followed by action potentials in the motor neuron. These action potentials are then conducted down the peripheral axon to the skeletal muscle fibres. At the nerve–muscle junctions, a similar sequence of events occurs, i.e. transmitter release, generation of a local excitatory potential in the muscle fibre followed by action potentials, conduction of the action potentials along the muscle fibre, and finally contraction.

It should be noted that this is a very simple model and that withdrawal reflexes may spread across the spinal cord and to other spinal segments to coordinate muscle activity. the nerve supplies to some skeletal muscles may be excited while others may be inhibited. In the case of withdrawal from a noxious stimulus the total response may also involve activity in autonomic nerves and the tissues that they innervate. For example, there may be an increase in heart rate and an increase in blood pressure. Although the pathways involved in these responses are more complex than the withdrawal response they also are the result of a series of local potentials and action potentials.

Membrane potentials

Neurons, like other cells, are electrically polarized so that their interior is negatively charged with respect to the outside of the cell (p. 16). Typical resting values for neurons are about – 70 mV, for glial cells – 90 mV, for skeletal muscle cells – 80 mV and for smooth muscle cells – 70 mV.

The value of the membrane potential can be estimated from the equilibrium potentials of the various ions and their conductances across the plasma membrane

(p. 19). Ions move through channels which have the special property of being selectively permeable to particular ion species. At rest, the conductance through K^+-specific channels (G_K) is much greater than that through Na^+-specific channels (G_{Na}) and so the membrane potential approaches the K^+ equilibrium potential $(E_K \sim -90$ mV). When a tissue is stimulated and specific channels are activated (gated), the current flow through them makes a greater contribution to the membrane potential. The potential developed will depend on the equilibrium potential for each ion and on the relative contribution each makes to the total conductance (p. 19). The major types of gated channels in excitable membranes include voltage-gated channels, chemically gated channels, and mechanically gated channels. (p. 36).

Local (electronic) potentials

A net movement of charge across the membrane changes the potential difference. A decrease in the magnitude of the potential difference with the inside becoming more positive is called a **depolarization**; an increase in magnitude is called a **hyperpolarization**. A depolarization is caused by the movement of positive charge into (or negative charge out of) the cell and a hyperpolarization is caused by movement of charge in the opposite direction. If the movement of charge (current) responsible for the change in potential is limited to a particular site on the membrane of an elongated cell, the change is greatest at that point and falls off exponentially with distance. As a result of the new potential gradients established along the cell and across the membrane, currents will flow between adjoining regions and through the membrane. As the potential gradients inside the cell and across the membrane decline with distance (Fig. 4.2), so too do the currents. Such local potentials, sometimes referred to as **electrotonic potentials**, are typically restricted to within 1 or 2 mm of their point of origin. In addition to their being thus **decremental**, such local potentials are **graded**, i.e. they increase in proportion to the increase in current. Because the membrane behaves as a capacitor (it stores electrical charge), the changes in potential are **distorted**, i.e. the rate of change in potential is slowed and lasts longer than the stimulus (Box 4.1). Local potentials also **summate**.

These local changes in potential are important as receptor potentials in sensory receptors, as synaptic potentials and as the basis for the initiation and propagation of action potentials.

Action potentials

Nerve and muscle plasma membranes have the capacity to generate action potentials. While small currents produce the local potentials discussed above, large inward currents (such as the influx of Na^+ at a sensory nerve terminal) depolarize the membrane to a point, termed **threshold**, beyond which an all-or-nothing response, termed an **action potential**, ensues.

Studies using intracellular microelectrodes show that during an action potential the polarity of the membrane potential temporarily reverses (i.e. the cell interior becomes positive with respect to the interstitial fluid) before returning to its original value. For example, the membrane potential of a neuron may move from a resting potential of -70 mV to a peak of $+40$ mV and back again in less

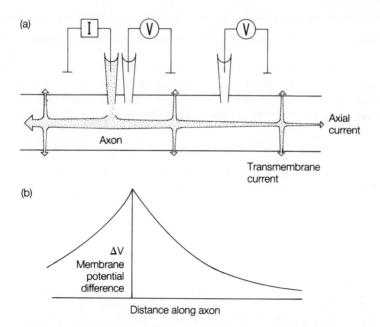

Fig. 4.2 Electrotonic spread of current along an axon and the resulting change in membrane potential. (a) The injection of current (I) and the recording of potential (V); (b) the decrease in the potential away from the site of current injection.

than 2 ms (Fig. 4.3). The magnitude of the action potential is not dependent on the magnitude of the signal initiating it. Moreover its duration is fixed and does not vary with the duration of the stimulus.

Immediately after each potential there is a period called the **absolute refractory period** during which a second action potential cannot be initiated. Hence, no matter what the stimulus, summation of action potentials is impossible. The absolute refractory period is followed by the **relative refractory period** during which the cell membrane can be brought to threshold only with a larger than normal stimulus. The durations of the absolute refractory periods of fibres of different sizes are shown in Table 4.1.

Ionic basis of the action potential

This was first described by Hodgkin and Huxley in the early 1950s on the basis of voltage-clamp experiments (Box 4.2). When an excitable cell is depolarized slightly (but is still below threshold), Na^+ conductance (G_{Na}) is increased by the opening of a small number of rapidly activated, voltage-dependent Na^+ channels. Under normal circumstances the probability of a voltage-sensitive channel opening is very low but as the membrane potential moves toward zero the probability increases. Once a channel is opened, it automatically closes again within a millisecond or so. Despite the brevity of each opening, the increase in Na^+ influx results in Na^+ making a greater contribution to the membrane potential and the cell is further depolarized. At the same time there is a net movement of K^+ out and Cl^- into the cell because of the shift away from the equilibrium potentialsof

Box 4.1 Basic electrical properties of biological membranes

Current, voltage, conductance and capacitance
Small ions can flow across biological membranes, and this current can result in the separation of electrical charge. Some of the properties of such membranes can be expressed in electrical terms. The **potential difference** (V) across biological membranes is normally some 60–100 mV, the cytoplasm being negative with respect to the interstitial fluid. The potential difference is a measure of energy separated across the membrane per unit charge (Joules/coulomb, $J\ C^{-1}$) and is therefore independent of area. In contrast, current (coulomb/second, $C\ s^{-1}$), which is directly proportional to the potential difference, increases with the available area. For any V, the actual current flow (I) is determined by the electrical conductance (G) of the membrane. Conductance expresses the ease with which current flows. Its reciprocal, the membrane resistance (R), is more familiar from elementary physics. Ohm's law states that:

$$I = V/R \quad \text{or} \quad I = GV$$

Conductance (siemens, S) and resistance (ohms, Ω) are also determined by the area involved. When area is taken into account, the resistance is often referred to as the **specific membrane resistance** and has the units $\Omega \times$ area. In practice, it is sometimes possible to determine the resistance of a plasma membrane by inserting microelectrodes into a cell, passing a known current across the membrane, and measuring the change in potential difference which results. (This is permissible since the cytoplasm offers comparatively little resistance to current flow.) In some cells, membrane resistance so measured is found to be influenced by the direction of the induced change in potential, thus there is not a linear relationship between current and voltage—Ohm's law being obeyed strictly over only a narrow range. Consequently the resistance calculated from the change in potential following a hyperpolarizing current (one which makes the interior of the cell more negative) may differ from that calculated when potential is decreased by the same amount following a depolarizing current. This asymmetrical behaviour, in which current moves across the membrane more easily in one direction than the other, is referred to as **rectification**.

The conductance (reciprocal of the resistance) of a membrane can be thought of as reflecting the number and selectivity of water-filled channels spanning the lipid bilayer, and the concentration and charge of the ion species moving through those channels. Conductance (S) measured by electrical techniques, should not be confused with permeability ($cm\ s^{-1}$) measured by isotope exchange or other chemically based methods, even though the two have been used synonymously. If an ion moves only by simple passive diffusion across a membrane, then it is possible to relate its measured conductance and measured ion permeability by using appropriate conversion factors and transference numbers which express the fraction of total current carried across the membrane by that ion.

Plasma membranes do not instantaneously change to a new steady potential difference when a current is passed across them. Initally, some of the current passed changes the amount of charge (Q) stored on the membrane itself. The membrane, in separating charge across it, acts like a capacitor and the charge separated per unit driving force ($C\ V^{-1}$) is the **capacitance**, C, equal to Q/V, and its unit is the farad (F). The capacitance per unit area or **specific membrane capacitance** is similar for most biological membranes—about $1\ \mu F\ cm^{-2}$.

Box 4.1 *Continued*

Because use is made of electrical analogues in understanding ion movements across single membranes and epithelia and because such concepts as resistance and capacitance are employed in discussing circulatory and respiratory physiology as well, the following should be noted for the model in Fig B4.1.

1 The total resistance (R_T) in a circuit containing resistors in series is $R_1 + R_2 + R_3$. Since $G = 1/R$, the total conductance (G_T) for this circuit is given by $1/G_T = 1/G_1 + 1/G_2 + 1/G_3$ or $G_T = G_1G_2G_3/(G_1G_2 + G_1G_3 + G_2G_3)$.

2 The total resistance (R_T) in a circuit containing resistors in parallel is given by $1/R_T = 1/R_1 + 1/R_2 + 1/R_3$ or $R_T = R_1R_2R_3/(R_1R_2 + R_1R_3 + R_2R_3)$. Here the total conductance is given by $G_T = G_1 + G_2 + G_3$.

3 The total capacitance (C_T) in a circuit with capacitors in series is given by $1/C_T = 1/C_1 + 1/C_2 + 1/C_3$ and with capacitors in parallel it is $C_1 + C_2 + C_3$.

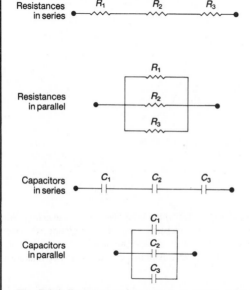

Resistances in series

Resistances in parallel

Capacitors in series

Capacitors in parallel

Fig. B4.1 Resistors and capacitors in series and in parallel.

Membrane space and time constants

Nerve and muscle fibres are elongated, like submarine cables, and are endowed with so-called cable electrical properties. A simple electrical model for the passive properties of an axon is shown in Fig. B4.2a. Because the membrane is not a perfect insulator, when current is injected into a fibre it flows longitudinally in both directions and progessively leaks out across the membrane. By inserting glass microelectrodes into large fibres at points of increasing distance from the site of current injection (Fig. 4.2), it is possible to record electrotonic potentials along the fibres. In such experiments, it can be seen that when a small, square current pulse is passed across the membrane of a cell the membrane capacitance will slow the time-course of the consequent voltage change. Indeed, the electrotonic potential will rise exponentially before it reaches a plateau (Fig. B4.2b). The rate of the exponential rise or fall of potential is expressed by the **membrane time constant** (τ) which is defined as the *time* required for the voltage to reach $1/e$ of its final value. For cells in which the

Box 4.1 *Continued*

applied current is distributed homogeneously, as in the case of a spherical cell, it can be shown that τ equals the product $R_m C_m$ where C_m is the specific membrane capacitance (F cm^{-2}). Thus, C_m can be calculated, if resistance is known, from the time-course of the change in the voltage following the passage of current across the membrane. This relationship is valid only when current is applied uniformly across the whole surface of a cell or fibre. If, however, current is applied focally by injection into an elongated cell, as in the experiments considered above, then the apparent time constant becomes longer the further away from the point of current injection.

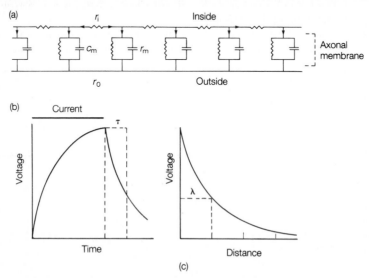

Fig. B4.2 Electronic potentials recorded with a brief injection of current. (a) A circuit diagram representing the cytoplasmic resistance and the membrane resistance and capacitance: (b) The time-course of voltage change produced by a current pulse when measured at the site of injection. (c) The attenuation of the maximum potential away from the injection site.

In addition to there being a slowing of the electrotonic potential, the amplitude is seen to decline exponentially with distance from the current electrode (Fig. B4.2c). This exponential process is expressed by the **membrane space constant** (λ) which is defined as the distance in which a voltage falls to $1/e$ (about 37%) of its original value. It is independent of the size of the event. It can be shown that

$$\lambda = \frac{1}{2} \sqrt{\frac{DR_m}{R_1}}$$

where D is the diameter (cm) of the fibre, R_m is the specific membrane resistance (Ω cm^2) and R_i is the specific intracellular resistance (Ω cm). Note that R_m refers to the specific transverse resistance of 1 cm^2 of membrane, i.e. $R_m = r_m \times \pi D$ where r_m is the membrane resistance of 1 cm length of fibre (Ω cm), and R_i refers to the specific resistance through 1 cm^3 of axoplasm, i.e. $R_i = r_i \times \pi D^2/4$ where r_i is the intracellular resistance per 1 cm length of fibre (Ω cm^{-1}).

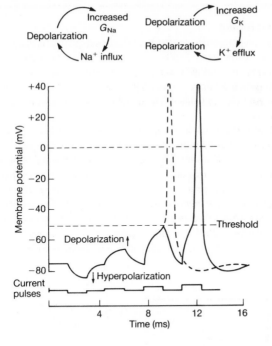

Fig. 4.3 Generation of an action potential. Top: diagrams showing the effects of changing the membrane potential on Na$^+$ and K$^+$ conductances during the depolarizing and repolarizing phases respectively of the action potential. Bottom: the responses of a neuron to localized current pulses beginning with a hyperpolarizing current and followed by a series of depolarizing currents of increasing strength. The stimulus that produces an action potential on 50% of occasions is a threshold stimulus.

these ions ($E_K \sim -90$ mV; E_{Cl} = resting membrane potential). As long as the depolarizing influence remains, the movement of K$^+$ and Cl$^-$ tends to counter the increased contribution of Na$^+$ and limits the change in membrane potential. If the stimulus increases, the resulting depolarization opens more and more Na$^+$ channels until with a depolarization of about 15 mV the current generated by the influx of Na$^+$ is balanced by the movement of K$^+$ and Cl$^-$. This potential is the threshold for the initiation of action potentials. Beyond this point the change in Na$^+$ conductance becomes a self-regenerating, positive-feedback process (Fig. 4.3); the increase in Na$^+$ influx leads to greater depolarization and the opening of many more channels which then allow Na$^+$ to enter. This explosive increase in G_{Na} (Fig. 4.4) continues until the membrane potential reaches a value that approaches E_{Na} ($\sim +70$ mV) and the potential would stay at this value but for the automatic closure of the Na$^+$ channels.

Table 4.1 Properties of mammalian nerve fibres (data from Ruch T.C. & Fulton J.F. (1960) *Medical Physiology and Biophysics*, 18th edn. p. 77. W.B. Saunders, Philadelphia)

	Nerve fibre type		
	A	B	C
Fibre diameter (μm)	1–20	<4	0.3–1.5
Conduction speed (m s^{-1})	5–120	3–15	0.6–2.5
Spike duration (ms)	0.3–0.5	1.2	2.0
Absolute refractory period (ms)	0.4–1.0	1.2	2.0

Box 4.2 Voltage clamping

The ionic basis of the action potential which has been discussed above was described by Hodgkin and Huxley in the early 1950s. Their interpretation was based on voltage-clamp experiments on squid giant axons. In these experiments a feedback apparatus was used to deliver a current across the membrane in order to clamp membrane potential at any selected voltage. The current passed in order to clamp this voltage was monitored.

If the command voltage was set to the resting potential, no current flowed. When the potential was changed rapidly to more positive values and an action potential invoked, there was a brief inward current followed by a strong prolonged outward current (Fig. B4.3). Replacing external Na^+ with the impermeant cation choline abolished the inward current, indicating that it was carried by Na^+ ions. This was confirmed by clamping the cell to a voltage more positive than E_{Na}. This reversed the inward current, so there were now two outward currents, with different time-courses, the first carried by Na^+, the second by K^+. From such experiments it was possible to derive the ionic conductance changes underlying the action potential, as shown in Fig. 4.4.

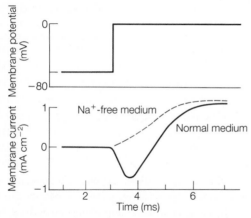

Fig. B4.3 Voltage clamping of a squid giant axon showing the membrane potential set by the command voltage and the current flow in Na^+-containing and Na^+-free media. (After Hodgkin, A.L. & Huxley, A.F. (1952) *J. Physiol* **166**, 449–472.)

This automatic shutting of the voltage-sensitive Na^+ channels is known as **sodium inactivation** and inactivated Na^+ channels cannot be reactivated until the membrane potential has returned to a value that is near the resting membrane potential (Fig. 4.5). The self-regenerating nature of the increase in G_{Na} and its inactivation ensure that the action potential is **all-or-none** and transient, and that it is followed by a brief period of inexcitability—the **absolute refractory period**.

The structure of the voltage-sensitive Na^+ channel (Fig. 4.5) is not the same in all tissues but in mammals may be heterodimeric (α, β), as in peripheral nerve and muscle, or heterotrimeric (α, β_1, β_2), as found in brain. The α-subunit is a large transmembrane polypeptide which exhibits voltage-sensitive conductance and undergoes inactivation; its characteristics are modified by the presence of β_1-type subunit.

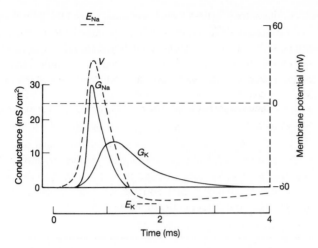

Fig. 4.4 The time-course of the computed ionic conductances durin an action potential in a squid giant axon. (Adapted from Hodgkin, A.L. & Huxley, A.F. (1952) *J. Physiol.* **117**, 500–544.

Repolarization following an action potential also depends on the outward movement of K^+ ions to bring the cell back to a potential near E_K. This is brought about by the opening of voltage-dependent K^+ channels in the membrane. These channels appear to be similar in structure to the voltage-sensitive Na^+ channels except that they are selective for K^+, they react more slowly than Na^+ channels to depolarization and they remain open longer than Na^+ channels. As a

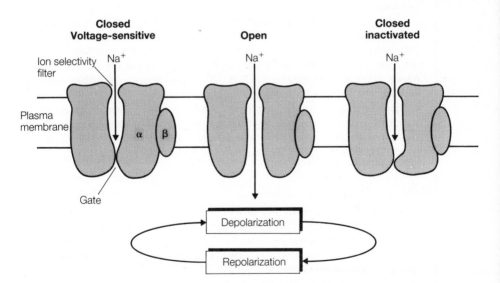

Fig. 4.5 Diagrammatic representation of the three configurations adopted by the voltage-sensitive Na^+ channel during the depolarizing and repolarizing phases of an action potential.

consequence K^+ conductance (G_K) reaches a maximum some time after the peak in G_{Na} (Fig. 4.4). The increased G_K remains until the potential has returned to the resting membrane potential. This process, called **delayed rectification**, is responsible for the rapid repolarization of the membrane during termination of the action potential and together with residual Na^+ inactivation is the cause of the relative refractory period.

During each action potential the cell gains some Na^+ ions and loses some K^+ ions. It can be calculated (Box 4.3) that only about 1/3000 of the K^+ ions in a nerve fibre of 20 μm diameter is actually exchanged for Na^+ ions during the course of an action potential and this is insignificant. However, after several hundred impulses have passed, the total amount of K^+ ions lost becomes significant (particularly in small-diameter fibres) and the ionic gradients must be reestablished by the Na^+–K^+ pump using metabolic energy.

Box 4.3 Number of ions crossing the membrane

The number of ions that must be separated to give rise to the membrane potential can be calculated from the relationship $C = Q/V$, where Q is the amount of charge in coulombs. With a membrane potential of 100 mV (10^{-1} V) and a specific capacitance of 1 μF, the charge separated across each square centimetre of membrane is therefore $10^{-1} \times 10^{-6}$ coulombs. Since 1 mol of univalent ion carries 96 500 coulombs, or approximately 10^5 coulombs, 10^{-7} coulombs represents about 10^{-12} mol of univalent ion. Since the number of ions in a mol (Avogadro's number) is 6.0×10^{23}, this represents 6×10^{11} ions. Therefore, the number of ions separated across the membrane to give a membrane potential of the magnitude measured across plasma membranes is infinitesimal when compared with the molar concentrations of ions in biological fluids, and the principle of bulk electroneutrality is not violated.

During the course of an action potential, Na^+ ions move into the nerve fibre during the rising phase and K^+ ions move out to repolarize the membrane during the falling phase. The net gain of Na^+ ions and net loss of K^+ ions have been determined in squid giant axons using radioactive isotopes of Na^+ and K^+ as tracers. These experiments show that $3–4 \times 10^{-12}$ mol of each ion actually crosses each cm^2 of membrane during the passage of one impulse.

The number of K^+ ions crossing the membrane can be calculated in relation to their intracellular concentration by equating a nerve fibre to a hollow cylinder. Its surface area is given by $2\pi rl$ and its volume by $\pi r^2 l$, where r is the radius and L the length. Therefore the volume enclosed by a cylinder of 1 cm^2 surface area is:

$$\frac{\pi r^2 l}{2\pi r} = \frac{r}{2}$$

In the case of a 20 μm axon, this equals 5×10^{-4} cm^3. Since the axon contains about 140 mmol L^{-1} K^+, there are 7×10^{-8} mol of K^+ ions in each segment enclosed by 1 cm^2 of membrane. Therefore, in comparison, the 3×10^{-12} mol of K^+ ions per cm^2 of membrane that actually leave the interior of the nerve fibre during the passage of one impulse represent only about 1/3000 of the internal K^+ pool and in the short term this is insignificant. In the smallest axons, however, it may become significant and such fibres can conduct only a few action potentials before their gradients must be recharged by the Na^+–K^+ pump.

The existence of specific membrane channels and their role in the generation of action potentials is supported by the action of specific pharmacological blocking agents. For example, **tetrodotoxin** blocks voltage-dependent Na^+ channels and hence the increase in G_{Na}; **scorpion venom** prevents Na^+ inactivation and so, after initiation of an action potential, the cell remains in a depolarized condition; and **tetraethylammonium** blocks K^+ channels and leads to a delayed recovery of the action potential. **Local anaesthetics** prevent the generation of the action potential by inhibiting the voltage-dependent opening of the Na^+ channels.

Propagation of the action potenital

A propagated action potential can be considered as a wave of excitation passing down an axon. At any one time only a portion of the axon is depolarized. This local depolarization gives rise to a passive spread of current up and down the axon (Fig. 4.6) and this current in its turn depolarizes adjacent regions ahead of the action potential, bringing them to threshold and so generating an action potential in a new region. As regions which have just been excited are refractory, propagation of the action potential is **unidirectional**.

Because the action potential mechanism is relatively slow, the further local currents can spread, the faster the action potential can travel. The **conduction velocity** can, in principle, be enhanced by increasing the **membrane space constant**, because current spreads further along the fibre, and by minimizing the **membrane time constant**, because the time taken to reach threshold decreases. The intracellular resistance is dependent on fibre diameter because in larger fibres more ions are available to carry the current, whereas the extracellular resistance is not normally a limiting factor because of the large volume of the extracellular fluid. In considering biological adaptations that can increase conduction velocity, one solution is to increase fibre diameter (thus decreasing intracellular resistance and increasing the space constant), as for example in squid whose giant nerve fibres

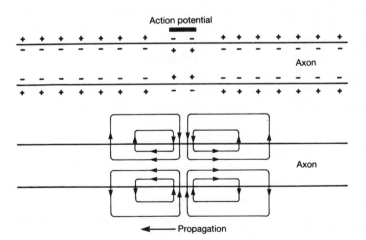

Fig. 4.6 Flow of current associated with the propagation of an action potential in an unmyelinated axon.

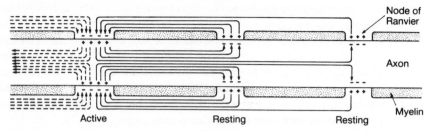

Fig. 4.7 Current flow during saltatory conduction of an action potential in a myelinated axon.

have diameters of up to 1 mm. The conduction velocity of such a squid axon is about 25 m s^{-1}, compared with about 1 m s^{-1} for small (about 1 μm) unmyelinated fibres. An alternative development has been to increase the membrane resistance and to decrease the membrane capacitance by enveloping axons with myelin sheaths. Since resistances in series sum and capacitances in series sum inversely (Box 4.1), by surrounding an axon with a spiral layer of myelin, the total membrane resistance is increased and the total membrane capacitance decreased. This greatly increases the space constant without increasing the time constant and the conduction velocity of small nerve fibres is enhanced. The conduction velocity of a myelinated nerve fibre of 20 μm diameter is about 100 m s^{-1} at 37°C. Most mammalian fibres greater than 1 μm diameter are myelinated.

In addition to the effects on membrane resistance and capacitance, myelination restricts membrane current and the generation of action potentials to the **nodes of Ranvier** (Fig. 4.7); the membrane in between is not excitable. The passing of an action potential from node to node is called **saltatory conduction**. However, it should be realized that at any instant an action potential travelling along a myelinated axon will occupy a large number of nodes and not simply a single node. For example, an action potential lasting 0.5 ms travelling at 100 m s^{-1} will occupy a 50 mm length of the axon and so may cover some 40 nodes. In summary, the benefits of myelination are:

1   higher conduction velocities for rapid signalling;
2   small diameters for conserving space; and
3   higher metabolic efficiency because of the reduced flux of ions and hence the reduced expenditure of energy required to restore ionic concentrations.

Extracellular recording of nervous activity

Peripheral nerves contain both sensory and motor axons ranging from large-diameter, myelinated fibres with high conduction velocities to small, unmyelinated, slowly conducting fibres. During the conduction of an action potential along these axons, current is caused to flow in the tissue or medium surrounding the nerve. As the extracellular resistance is low (compared with the axonal membrane resistance and the intracellular longitudinal resistance), the field potentials generated in the surrounding medium will be small. However, if the whole nerve or an isolated axon is immersed in mineral oil to limit current spread, the increase in extracellular resistance by minimizing current dissipation results in the recording of greater potentials when two electrodes are placed on the surface of the nerve or axon.

Action potentials recorded by *extracellular electrodes* and a differential voltage amplifier are seen to be waves of *negativity* moving along the axon. Before an action potential reaches the recording electrodes there is no potential difference between them (Fig. 4.8). When an action potential reaches the first electrode, the underlying membrane reverses its polarity and becomes negative with respect to the membrane under the second electrode. By convention this is shown graphically as an upward deflection. When the action potential is affecting the membrane under both electrodes simultaneously, the potential difference falls to zero between the electrodes. Later as the electrical activity moves away from the first electrode to affect predominantly the area of the membrane under the second electrode, the potential difference is reversed. As this recording shows both negative and positive components, it is referred to as a *di-* or *biphasic* action potential. The recording can be simplified if the action potential fails to reach the second electrode. This can be accomplished if a small segment of the axon

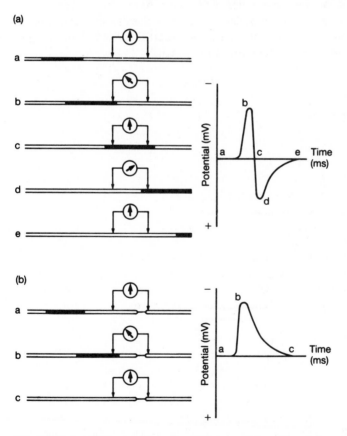

Fig. 4.8 Extracellular recording of an action potential conducted along an axon. The potentials corresponding to the positions of the conducted wave are indicated by a–e. (a) Biphasic record. (b) Monophasic record after crushing a portion of the axon between the recording electrodes.

between the electrodes is crushed or a local anaesthetic applied. Then the action potential is recorded as a single negative deflection—a **monophasic** action potential (Fig. 4.8b).

These mono- or biphasic action potentials, when recorded from a single axon, show the all-or-none and refractory characteristics of the intracellularly recorded action potential. In contrast, when the activity in a whole nerve is recorded, the record is the summed activity of action potentials being conducted along all the fibres. This is called the **compound action potential**. It is not all-or-none because of the different thresholds of various fibres. It increases in amplitude as the stimulus is increased until all fibres are excited. Its absolute refractory period is the absolute refractory period of the largest axons (Table 4.1).

The compound action potential may also display a number of peaks (Fig. 4.9). The separation of the peaks will depend on the conduction velocities of the fibres in the nerve and on the distance of the recording electrodes from the point of stimulation; the greater this distance, the greater the separation of the peaks (Fig. 4.9). In addition, the amplitude of these peaks and their relationship to each other depend on the fibre composition of the nerve. As the contribution of each fibre to the compound action potential is proportional to its diameter, the large fast-conducting fibres, which also have the lowest threshold, dominate the record, even though the small fibres are more numerous.

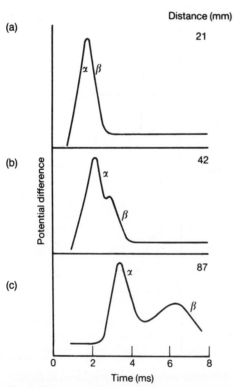

Fig. 4.9 Components of a compound action potential. (After Erlanger, J. & Gasser, H.S., (1937) *Electrical Signs of Nervous Activity*, p. 13. University of Pennsylvania Press, Philadelphia.)

For convenience, Gasser and his colleagues in the 1930s divided the axons in peripheral nerves into three broad groups (A, B and C) according to their conduction velocities. The A fibres include all the peripheral myelinated fibres from 1 to 20 μm in diameter; this group is further subdivided into groups of decreasing size and conduction velocity, the α, β, γ and δ fibres. The B group comprise the small myelinated fibres in visceral nerves (preganglionic autonomic) and the C group all small unmyelinated, afferent and efferent fibres. The conduction velocities and other properties of the fibres in the various groups can be seen in Table 4.1. A different grouping (I–IV) is often used for sensory fibres (p. 143). As mentioned earlier, the conduction velocity of a myelinated fibre is closely related to its fibre diameter since this determines the cross-sectional area and the intracellular resistance (p. 86). However the relationship between diameter and velocity is not linear; in the case of large fibres the conduction velocity (in metres per second) is approximately the diameter (in micrometres) multiplied by 6, for small myelinated fibres the factor is 4.5 and for unmyelinated fibres it is 1.7.

Conduction along fibres may be blocked by cold, anoxia, compression and drugs. The conduction velocity decreases by approximately 3% of the maximal velocity for each 1 °C fall. This is important in limbs or superficial tissues where the temperature may be well below the core temperature. Conduction block occurs in the large myelinated fibres at about 7 °C and in the unmyelinated fibres at about 3 °C. Compression and anoxia preferentially block large myelinated fibres . Thus painful stimuli, which are transmitted in small unmyelinated fibres, can still be felt when other modes of sensation are lost. In contrast, local anaesthetics tend preferentially to block small fibres, the surface area to volume ratio of which is high.

As changes in the conduction velocity of peripheral nerves can also occur as a result of accident or disease, studies of the compound action potential can be useful clinically. Thus neural degeneration and subsequent regrowth lead to the formation of thinner fibres with lower conduction velocities. Similarly, demyelinating diseases, e.g. *multiple sclerosis* and *diphtheria*, may result in a reduction in thickness or a localized loss of myelin. Upon remyelination there is also a reduction in the distance between the nodes of Ranvier (internodal length). Under these conditions there is a decrease in the length constant and an increase in the membrane capacitance, both of which diminish the longitudinal spread of current and hence the conduction velocity. Similarly with an extensive localized loss of myelin there may also be a conduction block.

Spontaneous action potentials

Some neurons spontaneously undergo fluctuations in their resting membrane potentials. These may generate action potentials even in the absence of external excitation. Such fluctuations in membrane potential have recently been shown to be generated by specific changes in membrane conductances which may involve Na^+, Ca^{2+} and Ca^{2+}-dependent K^+ channels. Cells with these properties are important in the generation of rhythmic activity reflected in respiration and the electroencephalogram (EEG). They have also been shown to underlie the expression of some behavioural patterns in invertebrates.

4.2 SENSORY RECEPTORS

Receptors form the interface between their surroundings and the nervous system. They usually detect a particular change in the environment, known as the adequate stimulus, and transduce that into electrical activity; most receptors respond to a stimulus with a depolarizing receptor potential which, if it exceeds threshold, will initiate action potentials in the sensory neuron. Some receptors respond to a continuous stimulus with a prolonged depolarization but others may adapt and respond only briefly. In most instances the frequency of action potentials in the sensory axons and the number of axons recruited are a reflection of the amplitude of the stimulus.

Sensory receptors are the link by which all information about the internal and external environment enters the nervous system. They are structures which convert different forms of stimulus energy into nerve impulses. However, they are limited in the range of stimuli that they can receive and information about our environment is thus limited to these stimuli. Sensory receptors may be endings of sensory neurons or adjacent specialized cells that initiate activity in an afferent neuron. The action potentials generated then travel along the afferent nerve fibres to the CNS. Within this section we will deal mainly with somatic and visceral receptors and not the receptors of the special senses such as taste, smell, vision and sound (Table 4.2 and see Chapter 7).

Specificity

Receptors are, to a large extent, specific or selective in their response, being sensitive primarily to one particular kind of energy. This energy, or change in energy, forms the **adequate stimulus** and a receptor transforms or **transduces** this particular kind of stimulus into a change in membrane potential. It may also be able to transduce other forms of energy but its **threshold** will then be higher, i.e. the stimulus will need to be of higher intensity. Receptors are usually classified

Table 4.2 Receptors grouped according to their location in the body

Receptors	Location
Somatic receptors	
Superficial receptors	On the body surface and accessible mucous membranes
Deep receptors	In muscles, tendons and joints
Visceral receptors	In the walls of blood vessels, gastrointestinal tract, bladder and in membranes and linings of body cavities
CNS receptors	In the brain and spinal cord
Receptors of special senses (see Chapter 6)	
Visual receptors	In the eye
Auditory receptors	In the ear
Orientation receptors	In the vestibular apparatus
Olfactory receptors	In the nose
Gustatory receptors	In the tongue and oral mucosa

according to the kind of energy for which they are most specific, i.e. as mechano-receptors, thermoreceptors, chemoreceptors, photoreceptors or nociceptors (Table 4.3). Within this broad-ranging classification receptors may show further specialization; thus, for example, thermoreceptors may be most sensitive to either an increase or decrease in temperature (but not both) and mechanoreceptors may respond selectively to pressure or vibration. This high degree of specificity may be associated with a particular structure (e.g. vibration and the Pacinian corpuscle) but often receptors with different specificities have no obvious structural differences (e.g. thermoreceptors).

Table 4.3 Classification of receptors

Receptor type	Adequate stimulus	Location	Examples of effective stimulus
Mechanoreceptors	Mechanical deformation	Skin	Touch, pressure, vibration
		Muscles and tendons	Changes in muscle length and tension
		Joints	Joint position and movement
		Viscera (e.g. blood vessels, lung, stomach, bladder)	Distension
		Cochlea	Sound vibrations, about 16–20 kHz
		Vestibule (e.g. semicircular canals, utricle)	Linear acceleration, angular acceleration
Thermoreceptors	Heat changes	Skin	Warming or cooling
		Hypothalamus	Warming or cooling
Chemoreceptors	Certain chemicals	Carotid and aortic bodies	Changes in plasma P_{O_2}, P_{CO_2}, pH
		Medulla oblongata	Local changes in pH
		Tongue and gut	Acids, salts, sugars
		Nose	Odorous chemicals
		Hypothalamus	Changes in plasma osmolality
Photoreceptors	Electromagnetic radiation of particular wavelengths (400–700 nm)	Eye	Light
Nociceptors	Mechanical, thermal or chemical but only at an intensity which threatens or causes tissue damage	Skin	Pinch, crush, sting, heat above 45–50°C
		Deep structures (muscle, joint and viscera)	Excessive stretch

Transduction mechanisms

A stimulus may act directly on the membrane of the nerve ending, e.g. free nerve endings in the skin. Alternatively, as shown in Fig. 4.10, receptors may be highly specialized and the stimulus may act indirectly either via an **accessory structure**, e.g. a capsule or hair shaft, or via a **receptor cell**, e.g. rod and cone cells in the eye, taste buds in the tongue, hair cells in the ear. But even where the stimulus acts directly on the nerve terminal, the relationship of the ending to the surrounding tissue will still affect the transmission of the stimulus to the ending. Where an accessory structure is present, this usually plays an important role in transmission of the stimulus to the receptor.

A stimulus to a receptor cell causes a change in the conductance of the receptor membrane and this in turn causes a local change in the membrane potential, the **receptor potential**. The receptor potential evokes a local **generator potential** in the sensory nerve terminal, either directly by electrotonic spread or indirectly by release of a chemical transmitter (Fig. 4.11). The generator potential initiates action potentials which are propagated along the sensory nerve to the CNS. Where the stimulus acts directly on a sensory nerve ending, the receptor potential and the generator potential are one and the same thing. There are, however, advantages in having separate receptor cells. For example, synthesis of cellular material can be regulated locally by the receptor cell and regeneration of a damaged receptor cell is quicker than regeneration of a sensory fibre.

The **receptor potential** usually arises from the opening of specific channels.

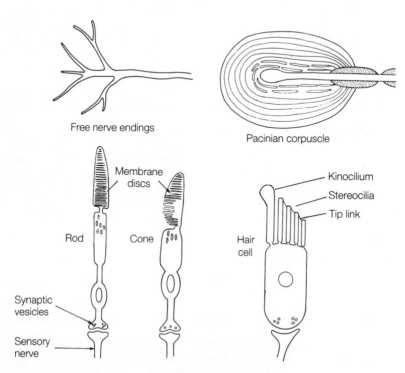

Fig. 4.10 Structure of various sensory receptors.

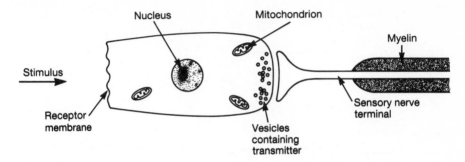

Fig. 4.11 Synaptic contact between a receptor cell and sensory fibre.

Except in the case of electroceptors (e.g. of fish), these channels are not voltage-sensitive but may be mechanically gated, chemically gated, or temperature-gated. The potential change is commonly one of depolarization, although there are exceptions. For example, vertebrate photoreceptors are hyperpolarized by light while vestibular hair cells are either depolarized or hyperpolarized according to the direction of hair displacement. The receptor potential is non-propagating, has no refractory period and its amplitude is graded and dependent on the strength of the stimulus. Repeated stimuli can therefore summate. The relationship between stimulus intensity and receptor potential (transfer function) is as a rule logarithmic.

Where there is a separate receptor cell, the generator potential is a synaptic potential probably due to an increase in both G_{Na} and G_K. Its properties are similar to those of the receptor potential and it may provide amplification of the receptor potential signal. Both receptor and generator potentials are more resistant to local anaesthetics than are action potentials. The relationship between generator potential amplitude and the frequency of action potentials in the sensory fibre is approximately linear. The final result of this combination of transfer functions is different in different systems. In some cases, the frequency of action potentials is proportional to the logarithm of stimulus intensity (Weber–Fechner law); in others, the relationship has been shown to be linear.

Adaptation

This is the term applied to the decline in the receptor potential shown by most receptors during the application of a constant or maintained stimulus. Receptor potentials in **slowly adapting** receptors, e.g. muscle spindles and Golgi tendon organs, are prolonged and decay slowly while those in **rapidly adapting** receptors, e.g. hair receptors and Pacinian corpuscles, quickly repolarize to values below threshold. The mechanisms of adaptation are not all known but in some cases the repolarization is a direct consequence of the increased influx of Na^+ and the activation of the electrogenic Na^+–K^+ pump, while in other cases adaptation is influenced by the nature of the accessory structure. For instance, the outer viscoelastic lamellae of the Pacinian corpuscle will remain deformed by a constant pressure stimulus while the inner lammellae quickly recover. In this case the failure to maintain transmission of the stimulus to the terminal membrane of the nerve results in a decrease in the amplitude of the receptor potential.

Changes in receptor or generator potential are reflected as changes in the

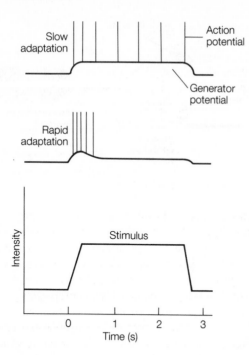

Fig. 4.12 Adaptation of generator and action potential discharge in sensory fibres during a stimulus.

discharge frequency of action potentials in the afferent fibres. During a constant stimulus, the impulse frequency in sensory neurons with slowly adapting receptors may remain at a relatively constant level but in sensory neurons with rapidly adapting receptors it may decline rapidly and cease altogether within seconds (Fig. 4.12).

Coding of sensory information

A stimulus is transformed within a **sensory unit** into one or a number of stereotyped action potentials. The term sensory unit applies to the afferent nerve fibre of a single sensory neuron, all its peripheral branches and central terminals, and any non-neural transducer cells associated with it (Fig. 4.13). The impulses so generated in the unit provide information to the CNS about the intensity and duration of the stimulus.

Information about the nature or **modality** of sensation (i.e. the type of sensation experienced, vision, touch, etc.) is derived from the arrival of the action potentials at particular sensory regions within the CNS. The selective nature of sensory mechanisms is a result of both the selectivity of the receptors and the specificity of afferent nerve connections in sensory regions of the CNS. If a receptor is activated by other types of stimuli, necessarily at higher intensities, the brain still interprets the sensation as that of the adequate stimulus. Thus, activity in optic nerve fibres is interpreted as the sensation of vision, whether the activity was produced by light or by sharp pressure to the eye.

Information on **stimulus position and distribution** is conveyed to specific parts of the brain in a way that depends on the particular sensory unit or groups of

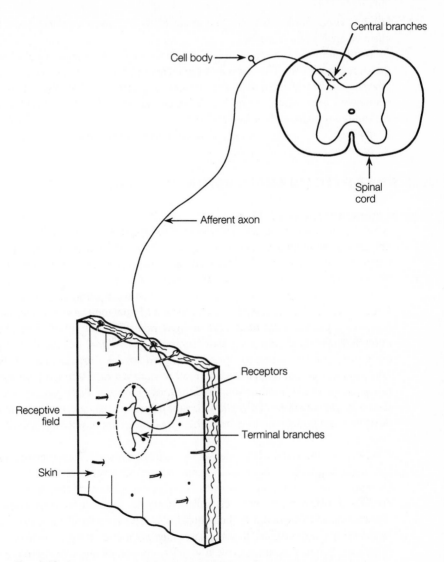

Fig. 4.13 A sensory unit and the receptive field supplied by its terminal branches.

sensory units that are activated. For example, each cutaneous sensory unit responds only to appropriate stimuli applied over a localized region of the body surface. This region forms the **receptive field** (Fig. 4.13) for that sensory unit. Its size reflects the threshold of the receptors and the degree of branching of the neuron. Certain regions of the body, e.g. fingertips and lips, receive a denser innervation with individual units having smaller receptive fields than in other regions, e.g. trunk and thigh. These densely innervated areas are therefore capable of providing more precise information about the position, size and shape of a stimulus. Receptive fields of individual sensory units overlap, so the stimulus is usually effective in exciting a number of sensory units in the skin. In an analogous fashion, visual fibres have effective visual fields and some joint receptors have effective angles of movement.

A suprathreshold stimulus usually produces a burst or train of impulses rather than a single impulse. The response from tonic or phasic receptors increases as the intensity or rate of change of the stimulus increases. The number of neurons responding also increases and this is known as **recruitment** of sensory units. **Stimulus intensity** is therefore coded by:

1   the frequency of impulses in individual sensory units, and
2   the number of active sensory units.

4.3 SYNAPTIC TRANSMISSION

Neural synaptic transmission

Neurons communicate with one another at synapses. At the majority of synapses the arrival of activity in the presynaptic region is followed by the release of chemical transmitters which diffuse across the synaptic cleft. Transmitters inter-act transiently with receptors on the postsynaptic membrane and produce a change in membrane conductance and, consequently, membrane potential. Depending on the conductance change the membrane may be either depolarized (excitatory postsynaptic potential) or hyperpolarized (inhibitory postsynaptic potential). Synaptic potentials can summate: if the final potential change is excitatory, action potentials may be initiated in the postsynaptic cell but if inhibitory, action potential generation will be suppressed. Electrical synapses are rarely found in the mammalian CNS; at these synapses the adjacent cells are joined by gap junctions which allow current from the presynaptic neuron to pass directly to the postsynaptic neuron where it changes the membrane potential.

A typical neuron in the CNS receives inputs from many neurons (**convergence**) and makes synaptic contact with many others (**divergence**; Fig. 4.14). For example, a motor neuron in the ventral horn of the spinal cord may receive some 20 000–50 000 synaptic contacts called **synaptic boutons** (p. 49). These are located mainly on the dendrites (axodendritic synapses) and cell body (axosomatic synapses) of the neuron but in some cases they are found on the axon (axoaxonal synapses). Synaptic boutons contain vesicles, the contents of which are released into the intercellular space (**neurosecretion**). They contain mitochondria and the enzymes necessary to synthesize neurotransmitters (other than peptide neuro-transmitters).

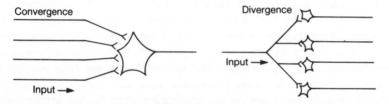

Fig. 4.14 Convergence and divergence of neural inputs.

When an action potential invades a nerve terminal it causes an **influx of Ca^{2+}** ions which trigger the release of neurotransmitter by **exocytosis** (p. 15). This neurotransmitter diffuses across the cleft and binds to receptor molecules linked to ion-selective channel (**ionophores**) in the postsynaptic membrane. The link between the receptor and the ionophore may be direct and very rapidly activated, or it may be indirect through second messengers. A conductance change, the nature of which is characteristic of the receptor–ionophore pair, not of the transmitter. The **synaptic delay**, which is the time between excitation of the nerve terminal and the permeability change in the postsynaptic membrane, ranges from less than 0.5 ms in directly linked systems to more than 1 s in systems using second messengers. Diffusion across the narrow cleft separating the pre- and postsynaptic membranes is very rapid and contributes only a small fraction to this delay; in direct-linked systems most of the delay can be attributed to the time for presynaptic release of the transmitter but in indirectly linked systems the delay is largely a result of the postsynaptic metabolic processes.

Vesicles within the synaptic bouton are normally bound to a protein called synapsin I. Following depolarization of the terminal and the influx of Ca^{2+}, calmodulin-dependent protein kinase I is activated. This phosphorylates synapsin I, causing dissociation of the protein–vesicle complexes and the vesicles move to the presynaptic membrane for release. More recent experiments indicate that fusion of the vesicle to the presynaptic release sites involves a number of proteins that are restricted to the vesicular and presynaptic membrane and that one of these also exhibits Ca^{2+}-dependent binding.

The change in conductance induced by a transmitter in the postsynaptic membrane may produce a small localized change in membrane potential (**synaptic potential**) which is either a depolarization (**excitatory postsynaptic potential, EPSP**) or a hyperpolarization (**inhibitory postsynaptic potential; IPSP**; Fig. 4.15a, b). The action of the transmitter is terminated by diffusion away from the cleft, by enzymatic breakdown or by reuptake into either the terminal or surrounding cells. These small synaptic potentials last longer than an action potential and when they succeed each other by a sufficiently short interval, their effect is additive (**temporal summation**). Synaptic potentials elicited simultaneously at different sites on a neuron can also summate (**spatial summation**). If the net result is a depolarization of sufficient magnitude, an action potential is triggered (Fig. 4.15c). In contrast inhibitory synaptic potentials tend to prevent the cell from reaching threshold. The coexistence of excitatory and inhibitory synaptic inputs and their capacity for temporal and spatial summation allow the neuron to **integrate** signals from a variety of other neurons.

One particular part of the neuron, the **initial segment**, which is situated close to the axon hillock, has a **lower threshold** than that of the dendrites and cell body, and so the action potential is usually initiated in this region. Action potentials are not normally fired in the dendrites or the cell body. The low threshold of the initial segment is a result of the unusually high density of voltage-sensitive Na^+ channels in this region, which results in more voltage-sensitive Na^+ channels being open for a given level of depolarization. As the amplitudes of synaptic potentials decay rapidly with distance, it is generally believed that synapses located close to the

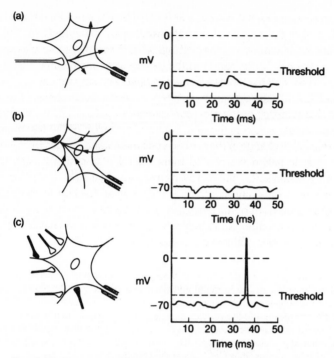

Fig. 4.15 Synaptic transmission in the CNS. (a) Excitatory postsynaptic potentials. (b) Inhibitory postsynaptic potentials. (c) Summation to threshold and generation of an action potential.

initial segment are more effective than those synapses located further away. However, in some situations, e.g. in motor neurons, it appears that particular distal synapses may compensate by generating larger synaptic potentials.

Ionic basis of synaptic potentials

By altering the relative conductances to Na^+, K^+ and Cl^- ions, the membrane potential can be set to any value between E_K and E_{Na}. Synaptic transmission is mainly effected by changing one or two of these conductances, the relative contributions of which can be estimated from the **reversal potential** (Box 4.4). An EPSP is a depolarization of a few millivolts (Fig. 4.15a) resulting from an increased conductance to Na^+, or both Na^+ and K^+ ions. The movement of the positively charged Na^+ into the cell may be in part negated by the outward movement of K^+, but the net influx of positive charge partially depolarizes the postsynaptic region. This brings the membrane closer to threshold and makes it more likely that an action potential will be triggered.

Some inhibitory transmitters activate only G_K so that a slight hyperpolarization, i.e. an IPSP, occurs (Fig. 4.15b), moving the membrane further away from threshold. More commonly, there is a simultaneous increase in both G_K and G_{Cl}. As E_{Cl} is usually close to the membrane potential, an increase in G_{Cl} alone would not change the membrane potential, but it means that more excitatory current is

Box 4.4 Reversal potentials

Synaptic potentials result from current flow through ion-selective channels. This flow tends to change the membrane potential towards a particular value at which there is no net flow of ions across the membrane. The membrane potential at which this occurs is called the reversal potential (E_R). Its value depends on the ratio of the ionic conductances and may be used to calculate this ratio. The reversal potential is determined experimentally by altering the membrane potential of the postsynaptic region until the evoked synaptic potential is reduced to zero. For example, at the skeletal neuromuscular junction, activation of acetylcholine receptors increases Na^+ and K^+ conductances in the postsynaptic membrane. The ratio of the Na^+ and K^+ conductances can be calculated if the postsynaptic membrane potential is set to a value (-5 mV) at which there is no change when the nerve is simulated.

Assume $E_{Na} = +60$ mV, $E_K = -90$ mV and ignore G_{Cl}. At the reversal potential, the acetylcholine activated channels open but there is no change in membrane potential as the fluxes of Na^+ and K^+ are equal but opposite, i.e.

$$I_K = I_{Na}$$

Using Ohm's law ($I = GV$), where the driving force (V) on a particular ion is the difference between the reversal and equilibrium potentials,

$$G_K(E_R - E_K) = -G_{Na}(E_R - E_{Na})$$

$$\frac{G_{Na}}{G_K} = \frac{(E_R - E_K)}{(E_{Na} - E_R)}$$

$$= \frac{85}{65}$$

$$= 1.3$$

That is, the channels opening in response to acetylcholine have a higher conductance to Na^+ than K^+ and when opened at the resting membrane potential the result is depolarization. Note, however, that the reversal potential is not actually reached as the current flow is localized and is opposed by current drawn from neighbouring regions of the cell. The reversal potentials and conductances associated with the generation of both excitatory and inhibitory synaptic potentials can be determined in this way if the membrane potential in the postsynaptic region can be controlled accurately. This is not possible with thin, branching, irregularly shaped cells and so channel conductances are determined using patch-clamp procedures (p. 38).

required to depolarize the cell as the membrane resistance is lowered and in addition any depolarization is opposed by Cl^- moving into the cell. The combined effect of increasing G_K and G_{Cl} is to move the membrane potential to a value between E_K and E_{Cl}. Cells already depolarized by excitatory activity will produce larger IPSPs but cells in a hyperpolarized state will generate smaller or no IPSPs while still being effectively inhibited.

Synaptic potentials are local potentials that spread passively along the membrane. Both EPSPs and IPSPs can arise in the same neuron because of the coexistence at separate sites of different conductance mechanisms in the same cell. As previously mentioned these conductance mechanisms in the membrane are ion-selective channels coupled to receptors which are activated by specific transmitters. Receptors may be clustered in the postsynaptic membrane or in some cases scattered over the surface of the cell. In some cases the receptor may not be coupled directly to the opening of an ion channel but may instead result in an increase in the concentration of a specific intracellular messenger (e.g. cyclic adenosine monophosphate, Ca^{2+} and derivatives of inositol lipids, p. 41). Such messengers then start a cascade of events that modulate the activity of specific ion channels and through these modify the membrane potential.

The records of small postsynaptic currents from isolated neurons and membrane patches indicate that fast EPSPs and IPSPs resulting from an action potential in a single presynaptic bouton are generated by the neurotransmitter released from 1–3 vesicles and the opening of 20–40 postsynaptic channels. Similar studies indicate that the concentration of transmitter in the synaptic cleft is high for only a brief period (approximately 1 ms) and that the time-course of synaptic potentials (the time taken for depolarization and repolarization) is determined by the number and kinetics of the channels activated by transmitter.

Many transmitter substances have been discovered in brain and peripheral tissues, including acetylcholine, monoamines, amino acids, purine derivatives, peptides and, more recently and unexpectedly, nitric oxide. A list of established and putative neurotransmitters and neuromodulators (p. 50) is given in Table 4.4. Some are excitatory, some are inhibitory, and some have both actions. In many cases the dual action of a single transmitter is associated with activation of subgroups of receptors which can be classified according to their effect and their affinity for particular pharmacological agents. Nitric oxide appears to be unusual as a neurotransmitter. Whereas 'conventional' neurotransmitters are synthesized, stored in vesicles and then released to act on surface receptors, nitric oxide is synthesized when action potentials cause an elevation in the intracellular concentration of Ca^{2+} in the prejunctional terminal, and it then diffuses across pre- and postjunctional membranes to act directly on the guanylate cyclase of the postsynaptic cell (p. 42).

Table 4.4 Some of the substances that act, or are thought to act, as neurotransmitters or neuromodulators in the brain, spinal cord and peripheral nervous system

Acetylcholine
Nitric oxide
Monoamines: adrenaline, noradrenaline, dopamine, histamine, serotonin (5-hydroxytryptamine)
Amino acids: aspartic acid, glutamic acid, glycine, γ-aminobutyric acid
Purine derivatives: adenosine, adenosine triphosphate
Peptides: adrenocorticotrophic hormone, angiotensin II, antidiuretic hormone, calcitonin gene-related peptide, cholecystokinin, β-endorphin, leu-enkephalin, met-enkephalin, gastrin, glucagon, gonadotrophin-releasing hormone, neuropeptide Y, neurotensin, oxytocin, somatostatin, substance P, thyrotrophin-releasing hormone, vasoactive intestinal polypeptide

Of the compounds in Table 4.4, glutamic acid, γ-aminobutyric acid (GABA) and glycine appear to play major roles in synaptic transmission in the CNS of vertebrates and may be used to illustrate the complexity of synaptic transmission. Glutamate is a powerful excitatory agent that appears to operate as the principal excitatory transmitter in some regions of the CNS, e.g. the cerebellar cortex. Most of the actions of glutamate appear to be transmitted through fast, direct-coupled receptors which have been divided into two types according to the pharmacological agents (e.g. N-methyl-D-aspartic acid (NMDA), α-amino-3-hydroxy-5-methylisoxazolepropionic acid (AMPA) and kainate) that best activate ionophore. Thus there are NMDA and AMPA/kainate direct-coupled receptors that respond to the release of glutamate. The ionophore opened by the activation of the AMPA/kainate receptors show a preference for monovalent ions, in particular Na^+ (and K^+) and the result of receptor activation is depolarization. In the case of the NMDA receptor/ionophore the selectivity is for divalent cations and there is a large influx of Ca^{2+} that can have both an immediate depolarizing and a long-term effect on synaptic transmission. A third type of glutamate receptor, a so-called metabotropic receptor, is linked by a G protein and phospholipase C to the synthesis of inositol trisphosphate (IP_3).

In contrast to glutamate, GABA and glycine have largely inhibitory functions within the vertebrate CNS; GABA predominates in higher centres and glycine predominates in the brainstem and spinal cord. When released into the synaptic cleft, GABA can activate two types of receptors. $GABA_A$ and $GABA_B$, both of which inhibit the initiation of action potentials in neurons; $GABA_A$ receptors through direct-coupled ionophores that have a high conductance to Cl^- and $GABA_B$ receptos through G proteins and channels with a high conductance to to either K^+ or Ca^{2+}. Glycine, like GABA, stabilizes neural membrane potentials by increasing the conductance to Cl^-.

From the above it would seem that synaptic transmission is a relatively straightforward sequence of action potential invasion of the terminal bouton, transmitter release, diffusion across the synaptic cleft, interaction with postsynaptic receptors and conductance change. But in addition to these events there are subtle, but effective, interactions occurring within synapses. For instance, many of the transmitter substances mentioned above appear to provide a rapid, self-limiting feedback that inhibits transmitter release. This autoinhibition is the result of transmitter in the synaptic cleft acting on presynaptic receptors (autoreceptors) that, either directly or indirectly (via second messengers), modify the rate of release. Some of the transmitters listed above also exert slow modulatory influences on synaptic transmission. Such actions are more aptly called **neuromodulation** and may have their action at either pre- or postsynaptic sites.

Synaptic modulation

Presynaptic inhibition

In this type of inhibition the release of excitatory transmitter is suppressed. As it controls the input to a neuron and not its excitability, presynaptic inhibition enables selective control of the inputs converging on a neuron. It occurs where the terminal of one neuron makes synaptic contact with the presynaptic ending of a second neuron which in turn excites a third neuron (Fig. 4.16). The first neuron controls the amount of excitatory transmitter released from the second neuron and hence the level of stimulation of the third neuron. It appears that more than one process may be involved in the effect of the first neuron on the second. Control may be exerted by the release of a transmitter which causes depolarization and partial inactivation of the voltage-dependent Na^+ mechanism. This in turn reduces the amplitude of any action potential that arrives at the axon terminal and presumably the influx of Ca^{2+} ions which determine the amount of transmitter released. Alternatively, the transmitter may modulate the conductance of Ca^{2+} channels via G proteins and second messengers.

An example of presynaptic inhibition can be found on spinal motor neurons where the release of excitatory transmitter from muscle spindle afferents is inhibited by activity from other muscle sensory receptors. It appears that the inhibition of transmitter release results from the activation of a GABA-releasing interneuron. The GABA apparently acts on $GABA_A$ receptors to increase the conductance to Cl^-, but as the boutons actively accumulate Cl^- the result is an efflux of Cl^- and depolarization of the terminal; the GABA may also act via presynaptic $GABA_B$ receptors to reduce the influx of Ca^{2+} during the action potential.

Postsynaptic modulation

In addition to controlling the release of transmitter from presynaptic terminals, some of the transmitters and co-transmitters released into synaptic clefts can modify the postsynaptic response. For example, neurally released peptides (some released as co-transmitters) have been reported to potentiate the effect of smaller 'classical' transmitters, and the repetitive stimulation of glutamate pathways to a number of localized regions in the CNS have resulted in a *long-term potentiation* of synaptic transmission that can last for hours or days (p. 261). The precise

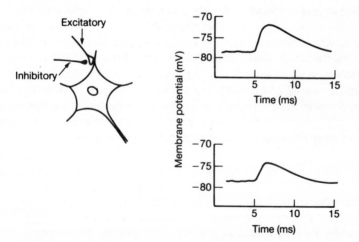

Fig. 4.16 An excitatory postsynaptic potential (a) before and (b) during presynaptic inhibition.

mechanisms underlying long-term changes in synaptic efficiency of this kind have not yet been established but might be the result of an increase in transmitter release or synaptic surface area, a decrease in transsynaptic distance, an increase in the number of postsynaptic receptors or ion channels, or an increase in production of second messengers.

Neuromuscular transmission at skeletal muscle

Synaptic transmission at skeletal neuromuscular junctions has a high safety margin. The invasion of the nerve terminal by an action potential is followed by the release of acetylcholine from several hundred vesicles. Acetylcholine diffusing to the postsynaptic membrane activates 100 000 or so nicotinic receptors and their cationic channels. The depolarization (endplate potential) induced by this change in conductance in turn activates voltage-sensitive channels, generating action potentials which then transmit excitation to the remainder of the muscle fibre. Rapid degradation of the acetylcholine ensures that uncontrolled repetitive activation is avoided and that high frequencies of neural activation are followed.

Myelinated motor fibres give off unmyelinated terminal branches which form synapses (**neuromuscular junctions**) with skeletal muscle fibres. In adult mammalian skeletal muscles there is only one neuromuscular junction in the middle of each skeletal muscle fibre. At the junction the terminal branch of the axon lies in a shallow depression of the muscle fibre surface. The pre- and postjunctional membranes are separated by the junctional cleft which is about 100 nm wide and contains a basement membrane (**basal lamina**). The postjunctional membrane is thrown into a series of folds, which, together with the prejunctional membrane and junctional cleft, is called the **endplate** (Fig. 4.17). The neurotransmitter released at the neuromuscular junction is **acetylcholine**. It is synthesized in the cytoplasm in the reaction:

CH_3COCoA + $HOCH_2CH_2N^+(CH_3)_3$
acetyl coenzyme A   choline

$\rightarrow CH_3COOCH_2CH_2N^+(CH_3)_3$ + $HCoA$
acetylcholine   coenzyme A

which is catalysed by the enzyme **choline acetyltransferase**. Acetylcholine is then packaged, together with adenosine triphosphate (ATP) and protein into vesicles of about 50 nm diameter.

The mechanism of release of acetylcholine is similar to that of other transmitters (p. 97). Briefly, when a nerve fibre is stimulated, an action potential invades the nerve terminal, causing the influx of Ca^{2+}. As a consequence, acetylcholine is released into the junctional cleft by exocytosis (Fig. 4.17), together with ATP and, probably, soluble proteins contained within the vesicles. The role of the ATP and protein is not known. Exocytosis of acetylcholine is blocked by **botulinus toxin** produced by the bacterium *Clostridium botulinum* and by high Mg^{2+} (*in vitro*). The vesicular membrane which fused with the prejunctional membrane during exocytosis is subsequently removed by endocytosis and reutilized. In contrast to

(a)

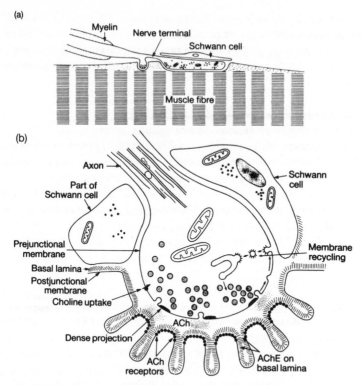

(b)

Fig. 4.17 Fine structure of a mammalian neuromuscular junction (a) in longitudinal section and (b) in cross-section.

synapses in the CNS where a few vesicles are released by each nerve impulse, the invasion of the motor neuron terminal by an action potential is followed by the release of several hundred *quanta*, each quantum containing approximately 10 000 molecules of acetylcholine. A quantum represents the number of acetylcholine molecules contained within each vesicle. More than sufficient numbers of acetylcholine molecules are released from the nerve terminals during each nerve impulse to ensure the generation of an action potential in the skeletal muscle fibre that it innervates. For this reason neuromuscular transmission is said to have a high safety margin.

The acetylcholine diffuses across the cleft and binds to receptor sites on channels that are localized on the crests of the postjunctional folds (Fig. 4.17b). The receptor and ionophore are directly coupled and the binding of acetylcholine to the two receptor sites (p. 39) on a closed channel induces a rapid conformational change that opens the channel and increases G_{Na} and G_K. Patch-clamp experiments indicate that the channel may open (for about 1 ms) and close several times before the acetylcholine dissociates. The conductance change that results from the opening of a large number (10^4–10^5) of these channels results in a net influx of positive ions and a localized depolarization of the endplate called the **endplate potential** (Fig. 4.18a). It has been calculated from the reversal potential (Box 4.4) that opening of these channels by acetylcholine increases G_{Na} and G_K in the ratio of 1.3 : 1. Under normal circumstances, the endplate potential exceeds

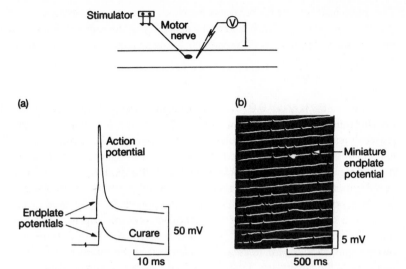

Fig. 4.18 Potentials recorded with a microelectrode inserted into a skeletal muscle fibre adjacent to the endplate. (a) Evoked endplate potentials in the absence and presence of curare. (b) Spontaneous miniature endplate potentials.

threshold and an action potential is generated but in the presence of receptor-blocking compounds, e.g. curare, subthreshold endplate potentials can be observed. During transmission there is a synaptic delay of 0.2–0.3 ms, much of which is attributed to the release of acetylcholine rather than to its subsequent diffusion across the cleft and the conductance changes that it provokes. There is a spontaneous release of acetylcholine from the nerve terminal; this is either *non-quantal*, presumably the result of continuous leakage, or *quantal*, due to the exocytosis of individual vesicles. The quantal release gives rise to small transient fluctuations in membrane potential, **miniature endplate potentials** in the postsynaptic region (Fig. 4.18b).

The postsynaptic receptors at the skeletal neuromuscular junction are referred to as **nicotinic** because nicotine mimics the effects of acetylcholine at these junctions. Interaction of acethylcholine with its receptors is blocked reversibly by **tubocurarine** (curare) and almost irreversibly by α-**neurotoxins**. The neurotoxins (e.g. α-bungarotoxin, α-cobrotoxin) are small proteins (M_r 8000) obtained from snake venoms and, when labelled with radioactive iodine, have been used to estimate the number of acetylcholine receptors (10^7) at the neuromuscular junction.

The acetylcholine released by each impulse is rapidly hydrolysed to choline and acetate by the enzyme **acetylcholinesterase** which is located in the basal lamina of the postjunctional membrane. Some 60% of the acetylcholine released by an action potential is hydrolysed before it crosses the synpatic cleft and the remainder is degraded within a few milliseconds. The choline produced by hydrolysis is taken up by a specific transport mechanism in the nerve terminal and reutilized in the synthesis of acetylcholine. Any that diffuses away from the endplate into the bloodstream is destroyed by **pseudocholinesterase**.

The hydrolysis of acetylcholine is blocked by **organophosphates**, such as diisofluorophosphate, which bind to the active site of the acetylcholinesterase and result in the accumulation of transmitter in the synaptic cleft and the prolonged activation of receptors.

Myasthenia gravis

This disease affects about 0.01% of the population and is characterized by a failure in neuromuscular transmission with repetitive nerve stimulation. The symptoms are weakness and fatigue, particularly of extraocular muscles, difficulty in swallowing and speech and, in advanced stages, respiratory failure. Hyperplasia of the thymus gland is frequently associated with this disease. Biopsy samples of intercostal muscles of myasthenic patients show a reduction in the amplitude of the endplate potentials and miniature endplate potentials. Consistent with these findings, a decrease in endplate acetylcholine receptors (up to 90%) has been found in binding studies with ^{125}I-labelled α-neurotoxin. The symptoms of this disease therefore appear to reflect a deficiency in the number of acetylcholine receptors at the neuromuscular junction. Removal of the thymus gland often has a beneficial effect and it has long been suspected that myasthenia gravis is an autoimmune disease. This has been confirmed by the demonstration of antibodies of acetylcholine receptors in these patients.

4.4 TRANSMISSION IN THE AUTONOMIC NERVOUS SYSTEM

Tissues innervated by the sympathetic, parasympathetic and enteric supplies of the autonomic nervous system receive either an excitatory, or both an excitatory and inhibitory neural input. The neural activity to these tissues first passes through ganglia which have both a distributing and integrating function. The postganglionic nerves are largely unmyelinated fibres which terminate in a densely branching network of varicose fibres in the effector tissue. Vesicle-bound transmitter substances such as acetylcholine, noradrenaline and other compounds are released from the varicosities into the neuroeffector junction and either excite or inhibit the effector cells by interacting with specific postjunctional receptors. Many of the autonomically innervated tissues are also influenced by the secretion of adrenaline and noradrenaline from the adrenal medullae.

The pattern of activity in the preganglionic autonomic nerves is controlled by the **hypothalamus**, and by centres in the **brainstem** and **spinal cord** (Section 3.4). The output to some tissues is phasic; to others it is continuous (tonic) and of low frequency (e.g. 1–2 Hz to blood vessels). The activity in all these pathways is either initiated or modified by sensory input from visceral or somatic receptors, or by changes in emotional state. When the change in activity occurs as a result of visceral input it is often referred to as an **autonomic reflex**.

Synaptic transmission in autonomic ganglia

The activity in many postganglionic fibres (p. 58) is wholly dependent on preganglionic activity, and the ganglia act mainly as distribution centres for this activity.

The extent of divergence can be estimated from the ratio of preganglionic to postganglionic axons. It varies enormously but a ratio of 1 : 190 has been found in the human superior cervical ganglion and 1 : 30 in sympathetic-chain ganglia. However, distribution is not the only function of autonomic ganglia. They can also show integrative activity, namely:

1 postganglionic summation of preganglionic activity;
2 complex, excitatory–inhibitory synaptic potentials; and
3 inhibitory input from visceral receptors.

The integrative capacity of ganglia appears to be most highly developed in the enteric system as coordinated patterns of contractions are maintained even when all connections to the CNS are severed. As local anaesthetics block this activity it is thought to be due entirely to peripheral reflex activity.

Within autonomic ganglia there are both divergence of the presynaptic fibres and convergence on to postsynaptic neurons. In some cases synaptic transmission from individual fibres is always successful and the ganglion cells act as relays; in others the summed activity of converging neurons is required for successful transmission. In each case the pre- to postganglionic transmission is **cholinergic**, acetylcholine being released from the preganglionic terminal to act on postganglionic **nicotinic** acetylcholine receptors. The resultant changes in permeability lead to a fast EPSP (Fig. 4.19); if it is large enough or if several smaller EPSPs sum to reach threshold, an action potential will be initiated and propagated to the effector tissue. Although this aspect of transmission is similar to that in skeletal muscle (p. 104), ganglionic transmission does differ in that, first, the structure and pharmacology of the nicotinic cholinoceptors is somewhat different and, second, the acetylcholinesterase in the synapses does not appear to be important in the termination of transmitter action; rather the transmitter diffuses away.

In addition to the fast excitation, preganglionic activity can lead to slow IPSPs (Fig. 4.19) in the cell bodies of ganglionic neurones due to small catecholamine- or peptide-containing interneurons in the ganglia; some neurons also produce slow EPSPs due to a late cholinergic effect on **muscarinic** cholinoceptors (p. 110).

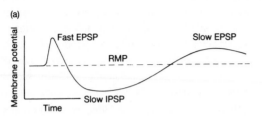

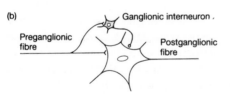

Fig. 4.19 Synaptic transmission in autonomic ganglia. (a) Synaptic potentials in a postganglionic neuron. (b) The neural pathways thought to be responsible for the fast EPSP and slow IPSP.

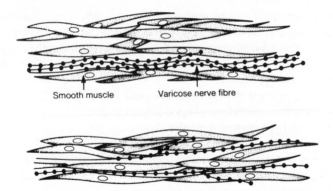

Fig. 4.20 Terminal branches of autonomic neurons. In some effector tissues they remain grouped together in bundles (top) but in others single fibres branch off from the bundle and the terminal branches make close contact with the effector cells (bottom).

Autonomic neuroeffector innervation

Of the sympathetic and parasympathetic systems, the former has been studied more extensively. The axons of the these postganglionic adrenergic neurons are long and unmyelinated, but the terminal regions branch extensively. Within effector tissues the extent and arrangement of the terminal branches vary. In some tissues they remain in bundles with only the occasional single terminal branch evident, whilst in others, single terminals predominate (Fig. 4.20). The neuro-effector junctions formed by the terminals also vary from narrow (25 nm) to wide (100 nm) junctional clefts. Irrespective of their structure the transmitting regions in these terminals are small swellings (approximately 1 μm in cross-section) called **varicosities**. Each neuron may have thousands of varicosities and each is responsible for the storage (except of nitric oxide; NO) and release of transmitter.

Autonomic neuroeffector transmission

In parasympathetic nerves acetylcholine is synthesized as it is in somatic motor neurons. In sympathetic neurons noradrenaline is synthesized from tyrosine in reactions catalysed by the enzymes outlined in Fig. 4.21. All the reactions but the last take place in the cytoplasm. The final step, from dopamine to noradrenaline, occurs in the storage vesicles. In addition to containing the transmitter and the enzyme dopamine β-oxidase, vesicles also contain chromogranins (proteins) and ATP, the functions of which are unknown. The rate-limiting step in this pathway is the conversion of tyrosine to dihydroxyphenylalanine, and the synthesis of noradrenaline is regulated by a small quantity of noradrenaline that leaks from the vesicles and inhibits the activity of the enzyme tyrosine hydroxylase (i.e. end-product inhibition).

In addition to the 'classical' transmitter substances, acetylcholine or noradrenaline, neurons of the autonomic nervous system are known to release neuropeptides (e.g. neuropeptide Y, somatostatin, vasoactive intestinal polypeptide or enkephalin) and ATP. At present the physiological significance of these agents is not fully understood but there is increasing evidence that in particular tissues they

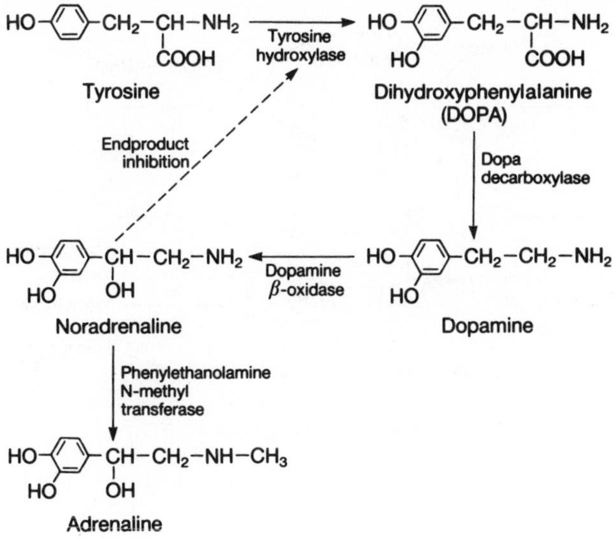

Fig. 4.21 Synthesis of noradrenaline.

may have a role as **co-transmitters**, while in other tissues they may act as neuromodulators.

ATP is a common constituent of vesicles and may act as a co-transmitter. For instance, in vesicles derived from sympathetic neurons, there is a 4 : 1 molar ratio of noradrenaline and ATP. In the vas deferens of some animals the membrane potential of the smooth muscle responds to nerve stimulation with a biphasic, fast and then slow wave of deplorization. In this tissue it appears that the transmitter responsible for the fast component is ATP, which acts on direct-coupled purinergic receptors, while the slow component is due to noradrenaline which acts through α-adrenoceptors and G proteins.

Like chemical transmission at most other sites, the release of vesicle-bound transmitters like acetylcholine and noradrenaline from the **prejunctional** terminal follows invasion of the terminal by an action potential and an influx of Ca^{2+}. The contents of a few (1–3) vesicles are then released by exocytosis from most of, but not all, the varicosities into the junctional regions where they are free to interact with receptors (Fig. 4.22a). Unlike skeletal muscle where receptors are limited to the neuromuscular junctions, the postjunctional receptors are not all localized beneath the nerve and many are spread over the surface of the effector cell. In addition to these receptors on the effector cell there are receptors on the prejunctional membrane which are thought to modulate the release of transmitter.

The noradrenaline released from sympathetic nerves (and adrenaline from the adrenal medullae) produce their effects by interacting with α- or β-adrenoceptors, classified by Ahlquist in 1948. These receptors are now classified using pharmacological and molecular biological techniques (Box 4.5).

In general, interaction of catecholamines with α_1- and β_1-receptors leads to excitation and interaction with β_2-receptors leads to inhibition. Thus stimulation of arteriolar smooth muscle occurs through α_1-receptors, stimulation of the heart

(a)

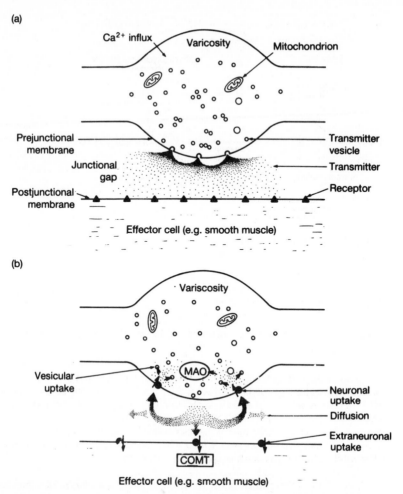

Fig. 4.22 (a) Chemical transmission at the sympathetic neuroeffector junction.
(b) Inactivation of noradrenaline at the sympathetic neuroeffector junction.

through β_1-receptors, and inhibition of bronchial smooth muscle through β_2-receptors. However, it must be remembered that some tissues have mixed populations of receptors and the response depends on the agent, the receptors and their location. The α_2-receptors are found on both pre- and postsynaptic membranes; activation of presynaptic receptors appears to inhibit transmitter release.

The acetylcholine released from the postganglionic parasympathetic fibres interacts with **muscarinic** cholinoceptors on the postjunctional membrane. These receptors differ from the nicotinic receptors, in that they are stimulated by muscarine, not nicotine, and are blocked by atropine and not tubocurarine. Like adrenoceptors the muscarinic receptors have been subdivided by pharmacological and molecular biology techniques into several groups, some of which are found predominantly in particular types of tissues (Box 4.5).

In most tissues the excitatory effects of transmitters appear to arise through an increase in membrane conductance, mainly to Na^+ but in some also to Ca^{2+} and

Box 4.5 Subdivision of adrenoceptors and muscarinic receptors

Many of the receptors involved in neuroeffector transmission in the automonic nervous system have been characterized by the use of compounds that specifically activate (agonists) or prevent the activation (antagonists) of the receptors on the effector tissue or the nerves that control it. This technique and the successful isolation and purification of receptors and their monomeric units have allowed the classification of receptors and the intracellular pathways they control to exceed the known physiological functions. The following receptors and the second messengers that they modify have been proposed but their relative roles in many functions have not been fully determined. For simplicity the list (Table B4.1) has been limited to the receptors activated by noradrenaline and acetylcholine but receptors for many other active substances, monoamines, peptides, kinins etc., have also been divided into subgroups.

Table B4.1 Classification of receptors

Adrenoceptors

Type	α_{1A}	α_{1B}	α_{1C}	α_{2A}	α_{2B}
Agonist potency	Nor>Adr	Adr = Nor	Adr = Nor	Adr>Nor	Adr>Nor
Second messenger	IP_3/DG	IP_3/DG	IP_3/DG	⇓cAMP	⇓cAMP

Type	β_1	β_2	β_3
Agonist potency	Nor>Adr	Adr>Nor	Nor>Adr
Second messenger	⇑cAMP	⇑cAMP	⇑cAMP

Muscarinic receptors

Type (locality)	M_1 neural	M_2 cardiac	M_3 smooth muscle glandular	M_4(?)	M_5(?)
Agonist	ACh oxotremorine	ACh	ACh	ACh	ACh
Second messenger	IP_3/DG	⇓cAMP	IP_3/DG	⇓cAMP	IP_3/DG

ACh, acetylcholine; Adr, adrenalin; cAMP, cyclic adenosine monophospate; DG, diacylglycerol; IP_3, inositol trisphosphate; Nor, noradrenalin.

K^+. The subsequent depolarization of the effector cell, usually accompanied by an action potential, initiates the response of the effector cell by causing an influx, or internal release, or Ca^{2+}. The actions of neurotransmitters may also be mediated by other intracellular messengers (Box 4.5). For example, cyclic adenosine monophosphate promotes the opening of Ca^{2+} channels in the plasma membrane of cardiac muscle and so increases the force of contraction. Other tissues are activated by the intracellular formation of inositol triphosphate and diacylglycerol which release intracellular stores of Ca^{2+} and stimulate protein kinase C, respectively. Inhibitory actions of the transmitters can be accounted for in many tissues by an increase in conductance, mainly to K^+, and subsequent hyperpolarization of the membrane; in some tissues this is accompanied by an elevation in the concentration of cyclic guanosine monophosphate.

There is good evidence to suggest that transmission from some powerful inhibitory autonomic nerves found in viscera, the so-called non-adrenergic, non-cholinergic nerves, is the result of neurally released NO. The NO, or its precursor, appears to be synthesized in the nerve terminals from L-arginine by a Ca^{2+}-activated NO synthase; from there the NO diffuses to the effector cell where it stimulates the activity of the intracellular guanylate cyclase and thus elevates the concentration of cyclic guanosine monophosphate.

The termination of the action of acetylcholine appears to be essentially the same as in skeletal muscle (p. 105), although diffusion out of the junction may also be important. In contrast, once released from the neuron the action of noradrenaline is *not* terminated by degradation but it is **inactivated** either by **uptake** or by **diffusion** away from the junctional region. The uptake processes transport the intact amine across plasma membranes, either into the neuron (in which case it is called reuptake, or **neuronal uptake**) or into the effector cell (**extraneuronal uptake**; Fig. 4.22b). Estimates of the relative activities of these two processes vary greatly but it would appear that in densely innervated tissues some 70% of the amine released is removed by neuronal uptake and 25% by extraneuronal uptake. Extraneuronal uptake and diffusion may be more important in diffusely innervated tissues. The major portion of the amine accumulated by the neuron is transported into vesicles for reuse as a transmitter, with obvious benefit to the neuron.

That portion of the transmitter taken up by the nerve and not recycled is degraded by **monoamine oxidase** whilst the amine taken up by the extraneuronal tissue is degraded largely by **catechol O-methyltransferase**. As the effects of catecholamines are not potentiated immediately by inhibition of either jor both of these enzymes, it is believed that they are not involved in the termination of transmitter action. Rather, enhanced responses may be seen in the presence of inhibitors of neuronal uptake (e.g. cocaine and desipramine) or of inhibitors of extraneuronal uptake (e.g. corticosteroids).

Opposing actions of the parasympathetic and sympathetic systems

Many tissues have a dual autonomic innervation and stimulation of one component usually results in effects opposite to those produced by stimulation of the other (Table 4.5). Further, it is clear that the effect of stimulating any one component may vary from tissue to tissue, being excitatory in some tissues and inhibitory in others. Nevertheless, a closer consideration of Table 4.5 reveals that the **parasympathetic** effects are largely directed towards **maintenance** and **conservation** of bodily function. Thus, responses to parasympathetic stimulation include slowing of the heart, constriction of the pupils, contraction of the bladder (detrusor muscle) and increased secretion and motility in the digestive tract.

In contrast, the effects of **sympathetic** stimulation, if examined all together, are directed towards coping with stress and comprise the '**fight or flight**' response described by Cannon in 1939. These changes include increased heart rate and contractility, bronchodilatation, pupillary dilatation, inhibition of intestinal motility, constriction of the splanchnic vascular bed, decreased muscle fatigue, and elevated blood glucose and free fatty acids. However, this is not to say that the sympathetic system always act *en masse* or that it is active only in stress. Indeed, most of the activity in the sympathetic nervous system is associated with normal

Table 4.5 Effects of sympathetic and parasympathetic stimulation

Effector	Parasympathetic stimulation	Sympathetic stimulation	Adreno-ceptors
Muscle			
Gastrointestinal			
Longitudinal and circular	Increase motility	Decrease motility	α, β
Sphincters	Relax (enteric NO)	Contract	α
Urinary bladder			
Detrusor	Contract	Relax	β
Trigone (internal sphincter)	Contract		α
Cardiac	Decrease rate	Increase rate	β_1
	Decrease force (atria only)	Increase force	β_1
Blood vessles			
Arterioles			
Skin and mucosa		Vasoconstrict	α_1
Abdominal, mesenteric		Vasoconstrict	α_1
Skeletal		Vasoconstrict	α_1
		Vasodilate (circulating adrenaline)	β_2
		Vasodilate (cholinergic)	
Coronary	Vasodilate	Vasodilate	β_2
		Vasoconstrict	α_1
Pulmonary	Vasodilate	Vasoconstrict	α_1
Brain	Vasodilate	Vasoconstrict	α_1
Penis, clitoris	Vasolidate	?Vasoconstrict	α
Veins		Vasoconstrict	α
Splenic capsule		Contract	α
Tracheal–bronchial	Contract	Relax	β_2
Eye			
Dilator of pupil		Contract	α
Sphincter of pupil	Contract		
Ciliary	Contract	Slightly relaxed	β_2
Genital organs			
Seminal vesicle		Contract	α
Vas deferens		Contract	α
Uterus		Relax (depends upon hormonal status)	β_2
Piloerector		Contract	α
Skeletal		Increased glycolysis	β
		Increased contractile force	β
Glands			
Salivary	Voluminous serous secretion	Decrease mucous secretion (submaxillary)	α
Pancreas and liver	Secrete	Decrease secretion or no effect	α
Nasopharyngeal	Secrete	Decrease secretion or no effect	
Bronchial	Secrete	Decrease secretion or no effect	
Sweat		Secrete (cholinergic)	
Tear	Secrete		
Metabolism			
Liver		Glycogenolysis	β_2
		Gluconeogenesis	
Adipose cells		Lipolysis	β_1
Insulin		Reduced	α

homeostatic activity and its actions are localized (e.g. pupillary dilatation and regional changes in blood flow), and occur without substantial changes in other tissues.

The activity of tissues having a dual innervation depends on the balance between parasympathetic and sympathetic discharge. Often one of the systems is dominant, for example, the diameter of the pupils and the resting heart rate are largely determined by the level of activity (tone) in their parasympathetic nerve supplies. Changes in effector activity are usually the result of reciprocal changes in both parasympathetic and sympathetic activity. Mutual antagonism in the periphery can occur.

1 as a result of dual reciprocal innervation of effector cells, such as occurs in heart, bronchial smooth muscle and detrusor smooth muscle; and

2 from suppression of synaptic transmission in ganglia.

The latter has been found in parasympathetic ganglia supplying detrusor muscle and in the ganglia that form the myenteric plexus of the intestinal tract. In fact most of the sympathetic supply to the intestinal muscle layer, excluding that accompanying blood vessels, ends in the myenteric plexus and appears to act by presynaptic inhibition of transmitter release.

Circulating catecholamines

The activity of some autonomically innervated tissues is influenced by both noradrenaline released from nerves and by catecholamines released into the blood stream from the adrenal medullae. In humans, adrenaline comprises some 80% of the catecholamines released from the gland, the remainder being noradrenaline. The adrenaline and noradrenaline are synthesized and stored in different cells. An additional step in the synthetic chain (Fig. 4.21) produces adrenaline by *N*-methylation of noradrenaline. The activity of the enzyme **phenylethanolamine-*N*-methyltransferase** responsible for this step is increased by steroids from the adrenal cortex.

The free plasma concentrations of these amines are low at rest, being 1–2 nmol L^{-1} for noradrenaline and 0.2–0.8 nmol L^{-1} for adrenaline. All of the adrenaline comes from the adrenal medullae and the noradrenaline from the adrenal medullae and sympathetic nerves. Low levels are maintained at rest because the rate of release from the adrenal medullae is normally low and the half-life of the circulatory catecholamines is relatively short (less than 1 min in experimental animals). During periods of stress, physical or emotional, the rate of release of adrenaline can increase eight- to 10-fold.

Adrenaline has many actions similar to noradrenaline but, because it is a more potent β_2-agonist, it has more pronounced metabolic actions. These include elevation of blood glucose and free fatty acids and an increase in metabolic rate. Adrenaline also produces vasodilatation in those vascular beds in which the β- to α-receptor ratio is high, e.g. in skeletal muscle and heart muscle.

Chapter 5
Muscle

There are three kinds of muscle in the body, classified according to their structure and function. **Skeletal musc**les are characterized by the presence of thin light and dark bands (striations) that are seen to lie across fibres when viewed through a microscope. These muscles form some 40% of the fat-free body weight, they are under voluntary control, and they are the only tissue through which humans can directly influence their environment. In contrast to skeletal muscles, **smooth muscles** (which are also called involuntary or visceral muscles) lack transverse striations and are not under conscious control; smooth muscles are found in viscera and blood vessels. **Cardiac muscle**, like skeletal muscle, is striated but unlike skeletal muscle it is spontaneously active; it generates the pressures required to drive blood around the vascular system and is described in Chapter 13.

5.1 SKELETAL MUSCLE

Skeletal muscle cells are long, multinucleate cells characterized by transverse striations and the ability to contract rapidly. Usually the contraction of a skeletal muscle maintains or moves one component of the skeleton relative to its neighbour. The force necessary to do this is generated by a regular array of protein-containing filaments, the actin and myosin filaments, and is accompanied by a large increase in the consumption of adenosine triphosphate (ATP) and the generation of heat. The contraction of each muscle fibre is preceded by the generation, near the endplate, of an action potential which travels along the sarcolemmal membrane and down the transverse (T) tubules. The action potential synchronizes and initiates each contraction by promoting the release of Ca^{2+} from the sarcoplasmic reticulum. The elevation of intracellular Ca^{2+} stimulates the binding of myosin cross-bridges to the actin; the subsequent flexing of the cross-bridges while they are attached to the actin generates force. The force generated by a muscle is dependent on both the frequency of the action potentials in the muscle fibres and the number of muscle fibres that are active. The latter is determined by the number of active motor neurons, each of which innervates a number of muscle fibres. The motor neuron and the muscle fibres

(a)   (b)

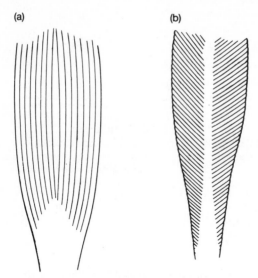

Fig. 5.1 Parallel (a) and bipennate (b) arrangements of muscle fibres.

that it innervates form the basic functional contractile unit which is called a motor unit. The needs of the body for slow, sustained contractions and fast, brief contractions are satisfied by the presence of muscles having the appropriate structure and fibre composition.

Skeletal muscles have two major functions: to develop tension and to produce heat. The tension developed by muscles is used to move or resist the movement of limbs, to close sphincters that control the emptying of hollow organs, to move the tongue and to regulate the vocal cords, and to perform other specialized functions. The heat produced by muscles is used to maintain body temperature (p. 685). The level of non-shivering heat production is regulated by hormones whilst shivering and the heat it produces are under direct neural control.

Muscles vary enormously in their capacity to generate force (tension) and in the rate at which this force can be developed. As the maximal force that muscles develop is proportional to their cross-sectional area (up to 40 N cm^{-2}), so the 'strength' of a muscle is dependent on the number of muscle fibres and on their diameters. In general, a pennate arrangement of fibres gives rise to contractions that are more powerful than a simple parallel arrangement (Fig. 5.1). As the contractions of muscles depend on the shortening of a large number of subcellular units (sarcomeres; see below) arranged in series, the speed with which a muscle changes length depends on the number of units in the series, on the rate of their change in length, and on the magnitude of any external applied force opposing the shortening of the muscle. Muscles also contain varying amounts of fibrous and connective tissue which contribute to their mechanical properties.

Cellular structure of skeletal muscle

Muscle fibres are long multinucleate cells 50–60 μm in diameter and ranging in length from a few millimetres to 3 or 4 cm. In small animals each fibre usually runs from tendon to tendon and terminates at each end in the connective tissue of the tendons. In large animals, however (including humans), muscles are often too long for the efficient development of tension via very long fibres, so in these cases two or more fibres may be arranged in series, each short fibre in the series being directly activated by its own nerve terminal. The force developed by the muscle fibres is generated by intracellular contractile proteins which are arranged into **myofilaments** (Fig. 5.2). The myofilaments are in bundles, called **myofibrils**, which run the whole length of the fibre. Each myofibril is surrounded by the **sarcoplasmic reticulum**, and between the lateral cisternae of the sarcoplasmic reticulum are fine **T tubules** opening out on the surface membrane (the *sarcolemma*; Fig. 5.3). The complex of a T tubule and two adjacent portions of cisterna is known as a *triad* (Fig. 5.3) and in human muscles these are located at the junction of the A and I bands.

The myofilaments are arranged into **sarcomeres** which are the contractile units of muscle fibres. Each sarcomere is approximately 2 μm in length and its limits are defined by a Z disc at each end. From the Z disc *thin* (approximately 5 nm wide and 1000 nm long) **actin** myofilaments project towards the middle of each sarcomere (Fig. 5.2) and in the central region of each sarcomere the

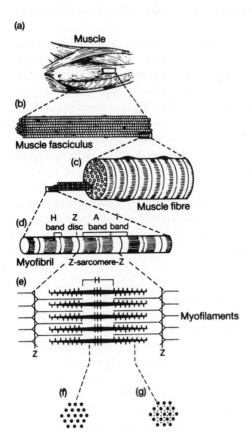

Fig. 5.2 Organization of muscle structure from whole muscle to myofilament (a–c) and transverse sections (f, g) showing the pattern of myofilaments. (Adapted from Bloom, W. & Fawcett, D.W. (1975) *A Textbook of Histology*, 10th edn, p. 306. W.B. Sanders, Philadelphia.)

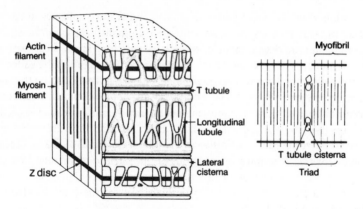

Fig. 5.3 Left: diagram illustrating the sarcoplasmic reticulum and T tubules in mammalian skeletal muscle. Right: a transverse section through sarcoplasmic reticulum and T tubules illustrating the relationship between a T tubule and two adjacent lateral cisternae (a triad).

filaments interdigitate with *thick* (approximately 12 nm wide and 1600 nm long) **myosin** filaments. Each thick filament is surrounded by a hexagonal array of thin filaments (Fig. 5.2) and from each thick filament helically arranged cross-bridges extend towards the thin filaments.

The striated appearance of skeletal (and cardiac) muscle fibres is a result of the serial and parallel repetition of the myofilaments and the differing abilities of the actin- and myosin-containing regions to transmit light. As polarized light is not transmitted through the myosin-containing region (i.e. it is anisotropic), this region is called the A band (Fig. 5.2). But light is transmitted through the actin-containing region (i.e. it is isotropic) and so it is referred to as the I band. In the middle of the A band where the myosin and actin filaments do not overlap, there is a lighter H band which marks the region devoid of cross-bridges and in the middle of this a finer dark M line. The Z disc lies in the middle of each I band. The regular structure of skeletal muscles is maintained at rest and during changes in length by a network of stable filaments formed from such proteins as *titin*, a very large, elastic molecule which links myosin filaments to their adjacent Z disc; *nebulin*, a more rigid molecule linking Z discs to actin filaments; and *desmin*, a strong inelastic molecule which connects adjoining Z discs in a myofibril and adjacent myofibrils.

Resting membrane potential and action potentials

The resting membrane potential of many skeletal muscle fibres is in the range of -75 to -85 mV. The basis of this membrane potential is similar to that found in other excitable cells, that is, a high intracellular concentration of K^+ and a selective permeability that favours potassium (p. 16). The membrane potential of skeletal muscles is stable and spontaneous activity of myogenic origin is only found in diseased or damaged muscles, or muscles deprived of their innervation (p. 133). Under normal circumstances contractions are initiated and maintained by the transmission of action potentials from the motor nerves to the muscle. The local depolarization (endplate potential) generated by the interaction of neurally released acetylcholine with the endplate nicotinic receptors (p. 103) is more than sufficient to take the sarcolemma surrounding the endplate to the threshold for action potentials. The ionic basis of the action potential is the same as that described in nerves, i.e. a rapid increase in the conductance to Na^+ and then K^+ (p. 77). However, each of these processes is slightly slower than those found in axons and so the total duration of each muscle action potential may be several milliseconds longer than an axonal action potential. Once initiated in the middle of each muscle fibre, the action potentials are conducted at about $4–5 \text{ m s}^{-1}$ towards both ends of the fibre; the mechanism of conduction, by local current flow, is the same as that seen in unmyelinated axons (p. 85).

Contractile process

If a muscle changes its length, the sarcomeres also change in length. However, the thin and thick filaments remain the same length and the change in length is achieved as a result of the filaments sliding over each other—the **sliding filament theory** developed by Hansen and Huxley in the 1950s. The forces generated during contractile activity arise in the regions where the actin filaments overlap the cross-bridges. A rod-like light meromyosin component of myosin molecules forms the backbone of the thick filament, while a heavy meromyosin component forms the cross-bridges. During contractions the myosin cross-bridges attach to adjacent actin filaments and flex towards the centre of the sarcomere (Fig. 5.4), thereby generating a tension. Only if the active tension generated by the interaction between actin and myosin exceeds any passive tension applied to the muscle externally will the muscle shorten. As activation occurs at both ends of the myosin filament, the opposing actin filaments are drawn in towards the centre, the Z discs are pulled closer and the muscle fibre shortens. Force is transmitted from Z discs to the extracellular matrix via **costameres,** subsarcolemmal structures that contain the protein dystrophin.

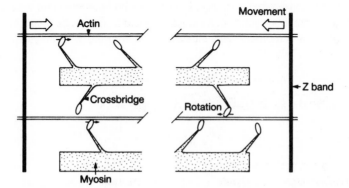

Fig. 5.4 Relative movement of actin and myosin filaments. This is accomplished by the rotation of the cross-bridge head which contains the myosin ATPase. A second flexible point appears to exist where the cross-bridge joins the backbone of the filament.

This process has at least three requirements:

1 that the actin and myosin must interact;

2 that the myosin cross-bridges must flex; and

3 that the system must be able to convert chemical energy to mechanical energy.

The actin filaments are composed mainly of globular actin molecules in a helical arrangement (Fig. 5.5) and it has been shown that purified actin molecules combined with the cross-bridge component of myosin molecules. This interaction is inhibited in normal resting cells by the presence of a troponin–tropomyosin complex. Early X-ray diffraction studies of muscles at rest and in **rigor mortis** (see below) showed that the cross-bridges could have two stable positions—the resting position and the flexed position. But pulsed irradiation–diffraction (synchrotion) studies of muscles have demonstrated movements of the cross-bridges during contractions.

Unravelling the mechanism involved in the transformation of chemical energy to mechanical energy has proved elusive. There is overwhelming evidence that ATP is the fuel used to operate the contractile machinery and that the other sources of energy, e.g. creatine phosphate, carbohydrates and fatty acids, are used to produce ATP. Myosin cross-bridges contain an ATPase and the rate of contraction of the sarcomeres appears to depend on the speed with which this enzyme can hydrolyse ATP. But at which stage of the cross-bridge cycle the hydrolysis occurs and how it affects the structural change is still problematic. A common schema suggests that after a myosin cross-bridge has flexed, ATP combines with the cross-bridge and causes it to dissociate rapidly from actin. After this it appears that the myosin head is 'recocked'; the recocking appears to consume little energy and the myosin remains in an intermediate but stable configuration (myosin–ADP–P_i). When the cross-bridge next combines with actin the reaction products from the hydrolysis of ATP dissociate and there is a considerable release of energy as the cross-bridge flexes. ATP is then required to initiate dissociation and a new cross-bridge cycle. When muscles are depleted of ATP the actin and myosin can no longer dissociate and rigor mortis occurs, that is, the muscles become stiff and inextensible.

It should be realized that a single cycle of cross-bridge movement would not cause a large contraction as the actin filaments would be moved by only some 1% of the length of a sarcomere. Significant muscle shortening requires that the

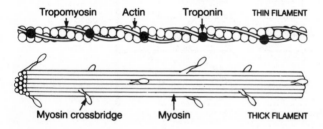

Fig. 5.5 Each actin filament is composed of two chains of actin monomers. In the grooves between these chains lie strands of tropomyosin and at regular intervals of about 40 nm are troponin molecules. Each myosin filament is composed of many myosin molecules, each of which comprises a rod-like component and neck and head regions.

myosin cross-bridges undergo repetitive cycling. In each cycle the cross-bridge attaches to the actin, flexes, and then dissociates before returning to its initial configuration and a new binding site on the actin filament. The repetitive asynchronous activity of the cross-bridges from surrounding myosin filaments ensures that the force exerted on an actin filament is maintained during a contraction. This activity can result in the actin filament being pulled between the myosin filaments. When the muscle is unable to shorten, the elastic properties of the muscle fibres allow the cross-bridge mechanism to operate and force to be generated.

Excitation–contraction coupling

Myosin ATPase activity and the contraction of muscle depend on both cytoplasmic Mg^{2+} and Ca^{2+}, but these ions have completely different roles. Whereas the Mg^{2+} is necessary for myosin ATPase activity and is in adequate supply, Ca^{2+} controls the interaction between the actin and myosin filaments and its supply is regulated. At rest the free cytoplasmic Ca^{2+} concentration is so low (approximately 10^{-8} mol L^{-1}) that little interaction occurs. However, during activity the concentration rises sharply, thus the development of tension within a muscle fibre is regulated by the free cytoplasmic Ca^{2+}.

The interaction between actin and myosin (and the ATPase activity) is inhibited by the **troponin–tropomyosin** complex (Fig. 5.5). The troponin component is a complex molecule spaced regularly along the actin filament and has specific tropomyosin-binding (TnT), calcium-binding (TnC) and inhibitory (TnI) subunits. The inhibitory effect of the complex on actin–myosin binding is removed when Ca^{2+} binds to the C subunit. The change is thought to be associated with movement of the tropomyosin strands that lie in the grooves between the strands of actin molecules. The movement of the tropomyosin and possibly the elevated intracellular Ca^{2+} concentration appear to cause subtle changes in the conformation of actin which enhance its interaction with myosin. Ca^{2+} concentrations of 10^{-5} mol L^{-1} cause maximal inhibition of the troponin–tropomyosin influence; the lowering of the Ca^{2+} concentration removes its permissive influence and thus inhibits the cycling of myosin cross-bridges.

In resting muscle, the free cytoplasmic Ca^{2+} concentration is low because the sarcoplasmic reticulum contains a membrane-bound pump (ATPase) that avidly collects Ca^{2+} and then transports it to the lateral cisternae. However, this Ca^{2+} can be released by depolarization of T-tubule membranes that come into close apposition with the lateral cisternae. An action potential propagating along the muscle and down the T tubules causes Ca^{2+} release and an elevation in the free intracellular Ca^{2+} concentration (Fig. 5.6); the propagation of the action potential down the T tubule ensures that the contractile activity in adjacent myofibrils is synchronized.

Energy balance

As mentioned above, ATP is the immediate source of energy for muscle contraction. Very little ATP is stored in a muscle, and it must constantly be renewed because of its consumption in both the contractile process and the Ca^{2+}-transporting mechanism of the sarcoplasmic reticulum. The ability of muscles to accomplish this restoration is shown by our inability to demonstrate changes in the

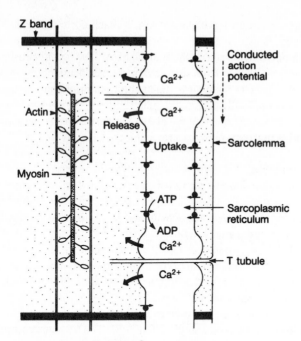

Fig. 5.6 The release of Ca^{2+} from the sarcoplasmic reticulum and its reaccumulation by an active transport process.

level of ATP unless its synthesis is blocked. The short-term reserve for replacement is creatine phosphate (PC), which forms a dynamic balance with free ATP, and the enzyme creatine phosphokinase (CPK) ensures that this equilibrium is reached rapidly. Thus the hydrolysis of ATP results in

$$ATP \rightarrow ADP + P_i$$

but the level of ATP is rapidly restored by the reaction

$$ADP + PC \overset{CPK}{\rightleftharpoons} ATP + C$$

As the last reaction is reversible, the PC is restored by the production of new ATP. This may be derived from the metabolism of glucose and free fatty acids from blood, or from reserves of glycogen and lipid droplets in muscle fibres. Depending on the type of muscle (p. 128) the major portion of the ATP may derive from either anaerobic or aerobic metabolism.

Under anaerobic conditions the breakdown of muscle glycogen proceeds via the glycolytic pathway to lactic acid. The end-product is lactic acid rather than pyruvic acid and the oxidized nicotinamide dinucleotide (NAD^+) generated by the conversion of pyruvic acid to lactic acid is used in an earlier step. During aerobic conditions both fatty acids and pyruvate can enter the citric acid cycle via acetylcoenzyme A and thus a far greater amount of adenosine diphosphate (ADP) is converted to ATP. For example, 3 mol of ATP is generated per mol of glucose-6-phosphate formed from glycogen during an aerobic metabolism (i.e. 2 mol of ATP is generated per mol of glucose). In contrast, aerobic catabolism through the citric acid cycle of the 2 mol of pyruvate produced per mol of a

6-carbon sugar generates 36 mol of ATP, and oxidation of a mol of a 6-carbon fatty acid generates 44 mol of ATP.

The whole process, that is contraction and relaxation, operates with an *efficiency* of conversion of metabolic energy into external work of the order of 10–20%; the remainder is dissipated as heat.

Heat production

Our everyday experiences reveal that muscular work is accompanied by the liberation of heat and that the amount of heat generated by muscles is proportional to the effort. The rate at which muscles produce heat may rise during maximal contractile activity to 20–50-fold the resting level. To a large extent the amount of heat (mW g^{-1} tissue) produced by a muscle depends on the physiological characteristics of the muscle; the steady-state heat production of fast-contracting muscles is some six times greater than that of slow-contracting muscles. As these two types of muscles have similar abilities to develop force (N cm^{-2} cross-sectional area), it is clear that fast-contracting muscles are less efficient than slow-contracting muscles. Shivering when exposed to cold may also result in an increase in the production of heat by muscles; with intense shivering it may rise to some eight times the resting level.

Precise measurements of the heat released from an isolated muscle which is contracting at a fixed length reveal that heat production is maximal at the *in vivo* length and diminishes with either increases or decreases in length. These studies also reveal that if a muscle is stimulated continuously for a brief period (e.g. 80 Hz for 1 s) the heat is produced in three phases. The most rapid phase, which is referred to as *activation heat*, is generated prior to flexing of the cross-bridges and appears to be associated with excitation–contraction coupling. As the contractile force develops and is maintained, there is a sustained release of heat, called *maintenance heat*, which is dependent on both cross-bridge activity and cross-bridge-independent activity (e.g. Ca^{2+} accumulation). Finally, *recovery heat* is released for several minutes after the contraction is complete and is approximately equal to the activation heat; unlike activation heat, recovery heat is dependent on an adequate supply of O_2. If, in addition to developing force, a muscle is permitted to shorten during a contraction, there is an additional production of *shortening heat* that appears to be related to the increased rate of cross-bridge cycling.

Contraction of muscle

Muscles are said to be contracting when the contractile machinery is active and energy is being consumed in this process. This term applies whether the muscle is shortening, remaining at constant length or lengthening. In the latter case, the contractile process may be activated, the muscle may actively develop tension, but the muscle as a whole may be forced to lengthen by the imposition of external force. Such **eccentric** contractions are a feature of normal muscle function, and are essential to our ability to move ourselves about while opposing the force of gravity. Whilst in everyday use the muscles operate in all these ways, it is convenient to study muscle contraction when either the length of the muscle or its load is constant. When the length remains constant (**isometric** contractions), we measure the force (tension) generated by the contractile machinery. When the load remains constant (**isotonic** contractions), we measure the rate of shortening of the muscle. These forms of contraction are, of course, also used in everyday activities, e.g. isometric contractions are involved in the maintenance of posture and isotonic contractions in the lifting of limbs.

Isometric studies of muscle show that shortly after a muscle is stimulated by a *single* stimulus there is an increase in muscle tension which then decays. The time-course of an action potential and a contraction is shown in Fig. 5.7. The time taken for the development of peak tension varies from 10 to 100 ms; its rate of decline also varies and both depend on the type of muscle being studied. A single contraction of this type is called a **twitch**. If the muscle is stimulated a second time, before it has had time to relax completely, the second response may add to the first and a greater-peak tension is developed. This is referred to as **mechanical summation**. If the stimulation of the muscle continues it fails to relax completely and during the period of stimulation the tension fluctuates (Fig. 5.7). With increasing frequency of stimulation the maximum tension is increased, the oscillations become smaller and, eventually, at **fusion frequency** a smooth **tetanic contraction** is produced. The tension produced in a tetanus may be two to three times as great as that produced in a twitch. Note that while the twitch provides a useful experimental measure of the properties of the muscle fibres, it is not a behaviourally useful action since the time-course of the twitch is usually too short for a behavioural response to

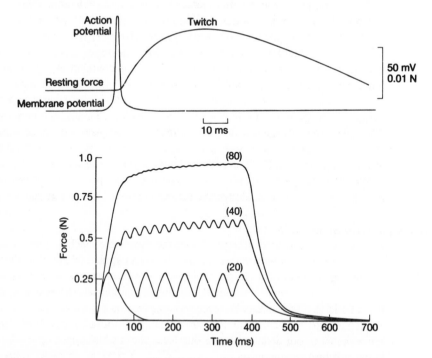

Fig. 5.7 Above: the time-course of an action potential recorded intracellularly from a single fibre and the accompanying isometric twitch recorded from many fibres. Below: isometric contractions from a rat extensor digitorum longus muscle showing the response to a single stimulus and to bursts of increasing frequencies (Hz), indicated in parentheses.

occur. Skeletal muscles, therefore, are activated in normal behaviour by volleys of action potentials which produce fused contractions.

The substantial difference between the maximal tensions reached in a twitch and in a tetanus has been attributed to the physical properties of the muscle and to changes in the cytoplasmic Ca^{2+} concentration. First, muscles are not rigid and the forces generated by the contractile machinery are transferred to limbs by elastic structures (the tendons and myofilaments). These are embedded in a viscoelastic medium (the cytoplasm, sarcolemma, sarcolemmal connective tissue and the connective tissue around fibre bundles) and many of these elements are arranged parallel to the contractile machinery. Thus much of the energy consumed in a twitch is used in overcoming the damping action of these elements. With continued activation as in a tetanus, the elastic elements are stretched, the movement of filaments is minimal and the maximum muscle tension is attained. Second, there is evidence from invertebrate striated muscles that a higher level of cytoplasmic Ca^{2+}, and hence muscle activation, is reached during a tetanus, but this has not been shown to occur in mammalian muscle.

Length–tension relationship

The tension generated by a muscle contracting isometrically depends on the initial length. In humans, maximal muscle tension can be generated when the muscle is approximately at its maximal natural length in the body. The relationship between the length of a muscle and the contractile (active) force that it develops can be examined by measuring the forces generated by a muscle at different lengths. Two forces can be measured: the *passive* force and the *total* force.

When a relaxed (unstimulated) muscle held between a movable clamp and a force transducer (Fig. 5.8a) is progressively stretched, an increasing force (tension), derived from an increasing resistance to stretch, can be measured (Fig. 5.8b). As the contractile machinery is not active this force is passive and is due to the resistance exerted by elastic elements in the muscle. Not all of this resistance to stretch is a result of extracellular components of the muscle but may also be due to the elongation of myofilaments. The filaments most likely to contribute in this way are the titin filaments which appear to span the gap between opposing actin filaments and are highly elastic. It should be noted that the force generated by stretch is not directly proportional to the increase in length (the elastic modulus increases with length), and that the elasticity varies greatly amongst muscles. Some muscles, for example the back muscles of humans and the hind-leg muscles of kangaroos, which contain large amounts of elastic extracellular matrix material, powerfully resist extension by purely passive mechanisms.

The curve of *total* tension is constructed by stimulating the muscle at various lengths and measuring the forces generated (Fig. 5.8b).Each of these forces will be the sum of both the passive force and the *active* force developed by the contractile machinery. The amplitude of the active force at the various lengths is obtained by arithmetically subtracting the passive force from the total force. The maximal active force is seen to occur near the natural resting length and to decrease with changes in length.

The most elegant way to relate these changes to the contractile machinery is to do this experiment with a single living muscle fibre. Then the force generated is

(a)

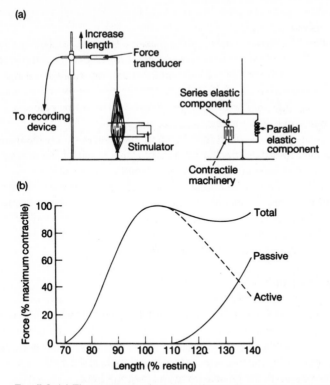

(b)

Fig. 5.8 (a) The experimental set-up used to study isometric contraction at different lengths (left) and a model of the contractile and elastic elements in muscle (right). (b) The relation between force and muscle length. Note that the total force generated at each length is the sum of the active force generated by the contractile elements and the passive force due to extension of the elastic elements.

seen to be related to the degree of overlap of the actin and myosin filaments (Fig. 5.9). It can also be seen that at long lengths, when there is no overlap of the actin and myosin filaments, the fibre is incapable of generating a force; in the intermediate range, when overlap of filaments is optimal, the force generated is maximal; at shorter lengths, the actin filaments overlap and interfere with each other and the force decreases. Eventually, at very short lengths (60–70% of the maximal natural length) the Z discs will be pulled against the myosin filaments and the external force will again fall to zero. At this point the contractile machinery may still be active but the energy is used to distort the myosin filaments.

Force–velocity relationship

The length–tension curve has described the ability of muscles to develop tension when the muscle is held at fixed lengths (isometric contractions). But, as mentioned earlier, the movement of limbs may be associated with the shortening of muscles under a constant load (isotonic contractions). It is an everyday experience that the lighter the load, the more rapidly it can be lifted. In fact, both the rate and the degree of muscle shortening depend on the load. The relationship between the

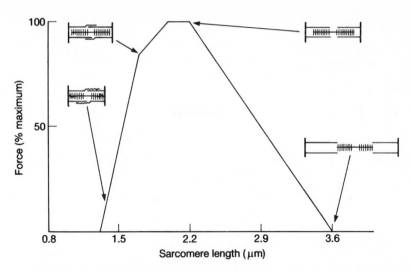

Fig. 5.9 The relationship between the contractile force and sarcomere length in a single muscle fibre. The insets illustrate the degree of overlap of the myofilaments at the sarcomere lengths indicated. (After Gordon, A.M., Huxley, A.F. & Julian, F.J. (1966) *J. Physiol.* **184**, 170–192.

rate of shortening and the load carried by a muscle is illustrated by the force (load)–velocity curve.

This relationship is determined by measuring the rate of shortening of a muscle as it lifts a variety of loads. The muscle is not initially subject to each load as this would alter the starting length of the muscle. However, before it can shorten the muscle must obviously first lift each load. Such an event is called an **after-loaded** contraction. When stimulated tetanically an after-loaded muscle starts to contract. Initially and until the tension exceeds the load, the contraction is isometric. After this, the muscle shortens isotonically and continues to shorten until it reaches the length at which (according to the length–tension curve) the maximal force it can develop is equal to the load. It is clear that with zero load the time required initially to shorten (the latency) will be minimal and the velocity of the contraction maximal (Fig. 5.10); as the load is increased the latency is increased and the velocity decreases. Finally, when the load is too heavy, the velocity of shortening is zero and the muscle is contracting isometrically. It can be seen from the force–velocity curve that the *power* (force × velocity) which a muscle develops is not constant. The power output of a muscle is in fact optimal when both the load and the velocity are moderate—hence the advantage of multiply geared bicycles.

The reasons for the shape of the force–velocity curve are not known. One suggestion is that the myosin cross-bridges move continually as a result of thermal agitation and that there is only a limited space within which a cross-bridge and an actin site can interact. If this is correct and the actin filament is moving, the probability of successful union will decrease as the velocity of movement increases. Thus at high velocities few cross-bridges will be formed and, as the force generated is dependent on the number of cross-bridges, it will be low at high velocities.

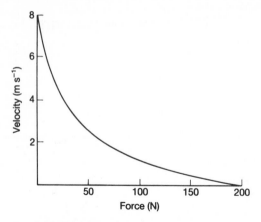

Fig. 5.10 The effect of force (load) on the velocity shortening of human muscle. (Adapted from Wilkie, D.R. (1950) *J. Physiol.* **110**, 249–280.)

Accordingly, the velocity of shortening will increase until the force generated by the muscle equals the load. If the force is either greater or less than the load, the velocity will either increase or decrease, respectively, which will in turn decrease or increase the number of bridges formed and the force generated. This idea is supported by the observation that the velocity of shortening in isotonic contractions is relatively constant.

Muscle fibre types

The position of our body and the complex pattern of movements it performs are the results of the pulling actions of muscles. However, the diversity of these actions requires that the muscles used to perform them have different properties. Thus, some muscles are called upon to maintain a high level of tension for long periods without fatigue while others are required to produce intermittent rapid movements. These two extremes of activity are illustrated by the postural soleus muscle that reaches a peak tension in 80–200 ms and the extraocular eye muscles that develop their peak tension in 7–8 ms (Fig. 5.11a). The soleus muscle contains predominantly slow-contracting muscle fibres and the extraocular muscles predominantly fast-contracting muscle fibres. Muscles that have to perform both endurance and rapid actions have a more even mixture of these fibre types. When the properties of the slow (**type I**) and fast (**type II**) muscle fibres are compared, pronounced differences are evident. The slow fibres have a low myosin ATPase activity and a high capacity to produce ATP by oxidative phosphorylation, which is aided by a well-developed blood capillary network and high levels of intracellular **myoglobin**. The latter is an O_2-binding protein (like haemoglobin) which both facilitates the diffusion of O_2 into these muscle cells and stores a small quantity of O_2 in the cells. The combined effect of these characteristics is slow, fatigue-resistant contractions as the rate of production of ATP is sufficiently rapid to replace that split by the myosin ATPase. The high concentration of myoglobin and high capillary density in these muscles have led to the use of the term 'red muscle'.

There are two distinct groups of fast-contracting fibres. Both have a greater

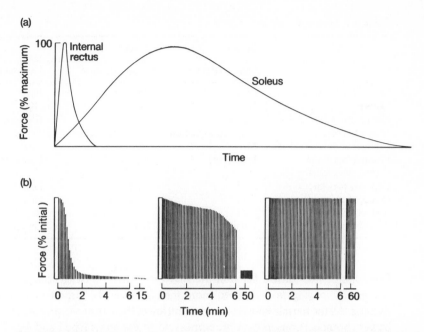

Fig. 5.11 (a) Isometric twitch contractions of cat internal rectus and soleus muscles scaled to the same peak height. (Adapted from Cooper, S. & Eccles, J.C. (1930) *J. Physiol*, **69**, 377.) (b) Fatigue of fast (left), intermediate (middle) and slow (right) muscle fibres that were stimulated through their nerve supply at 40 Hz for 330 ms once each second. (From Burke, R.E., Levine, D.W., Tsairis, P. & Zojac, F.E. (1973) *J. Physiol*, **234**, 723).

diameter and a higher myosin ATPase activity than the slow fibres but their resistances to fatigue differ (Fig. 5.11b); the resistance to fatigue is correlated with a high oxidative capacity and those fibres with a high resistance are often referred to as **intermediate** fibres. The largest and fastest contracting type II fibres (the so-called *fast* fibres) have a poorly developed oxidative metabolism and depend largely on glycolysis for the production of ATP. Although ATP production by glycolysis is rapid, the high rate of consumption of these fibres during powerful contractions results in their rapid fatigue. A summary of these and other properties of the different fibre types is given in Table 5.1.

Regulation of contraction

Because muscles are composed of many fibres and the activity of individual fibres can be graded, the total force generated by a muscle depends on the number of active fibres and the level of activity in each fibre. Each motor axon entering a muscle makes contact with a number of muscle fibres; each of these fibres is innervated by a single terminal branch of that axon. Thus, groups of muscle fibres are activated synchronously.

Motor units

A **motor unit** comprises a motor neuron and the group of muscle fibres innervate by the branches of its axon (Fig. 5.12). Motor units vary greatly in size, ranging

Table 5.1 Muscle properties

	Type I	Type II	
	High oxidative	High oxidative	Low oxidative
Rate of contraction	Slow	Fast	Fast
Myosin ATPase activity	Low	High	High
Main pathway for ATP production	Oxidative phosphorylation	Oxidative phosphorylation	Glycolysis
Number of mitochondria	Many	Many	Few
Myoglobin content (muscle colour)	High (red)	High (red)	Low (white)
Capillary density	High	High	Low
Glycogen reserves	Low	Intermediate	High
Rate of fatigue	Slow	Intermediate	Rapid
Fibre diameter	Small	Intermediate	Large

from one or two muscle fibres in the smallest units of muscles controlling fine movements of fingers or eyes to more than 2000 in the largest units in limb muscles. All the muscle fibres in a motor unit are of the same type, and they tend to be very homogeneous in their properties and so the terms type I and type II are used for both motor units and muscle fibres. In general, the type I units of slow muscles are rather similar in size, although they are not particularly large; type II units, in contrast, supply fast muscles and range from very small to very large. The larger a motor unit is, the larger the axon and the nerve cell body of the motor neuron supplying it. This probably reflects the need for production by the cell of all the materials needed to keep every one of its nerve terminals functioning.

Gradation of tension

Increments in tension can result from an increase in the force generated by individual motor units or by bringing into action (**recruitment**) additional units. Extracellular recordings of the electrical activity of muscle fibres (electromyography) have shown that both these events occur, but not at the same rate. Thus, the initial development of muscle tension is thought to be largely due to recruitment of units. As explained below there is, in addition, an increase in firing

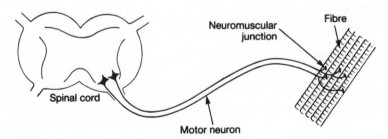

Fig. 5.12 Two motor units, each consisting of a motor neuron and the muscle fibres that it innervates.

frequency but the contribution of this to increments in tension is thought to be important mainly in the generation of larger forces.

The recruitment of motor units is not random but occurs in an orderly fashion from small to large. Low tensions are produced and precisely controlled by the selective mobilization of a number of small units. In fact, under most circumstances, a small proportion—paradoxically, the smallest ones—do most of the work. The largest units are activated only when a maximal effort is required and even then their activity is often brief.

Recruitment of motor neurons

The ordered recruitment from the *pool* of neurons supplying a muscle arises because the smallest cells are the most easily excited. The smaller surface area of the small motor neurons results in these cells having a higher input resistance. When similar excitatory synaptic currents are generated in the small and larger motor neurons the small ones reach threshold first. In contrast, inhibitory synaptic currents generated in neurons within an activated pool appear to be more effective on the larger neurons as these are the cells closest to their threshold. As the intensity of excitatory synaptic activity in a motor neuronal pool increases, larger and larger motor units are recruited and at the same time the frequency of discharges increases. However there are also neural mechanisms that limit the discharge frequency of individual motor neurons to a frequency appropriate to the type of muscle fibres they innervate.

At present the 'size principle' outlined above is thought to be the basic process governing the order of recruitment of motor neurons. It does not however imply that the order of recruitment will always be precisely the same. Indeed, it is well-documented that small changes in the required movement or the imposition of various stimuli will change the order. Given that the somata of the smaller motor neurons within the pool that control a muscle will be very similar in size and that there will be considerable background synaptic activity, this variation is not surprising.

It should be noted that the contractions of skeletal muscles that move objects, maintain posture and adapt to changes in load and fatigue are not regulated solely by the motor units. These activities also make use of sensory information, including that from the muscles and limbs involved. The role of the muscle receptors (the muscle spindles and Golgi tendon organs) in motor control is discussed in Chapter 8.

Development and maintenance of skeletal muscles

The speed with which muscles can contract and their ability to do work are not constant throughout life, but change as a person grows; their performance is also influenced by exercise. The development, growth and maintenance of muscles are all dependent on the presence of an intact motor nerve supply.

Development of muscles

Skeletal muscle fibres are derived from cells of embryonic mesodermal origin. These myogenic precursors have their origin in the **somites**, the tissue blocks which are adajcent to the developing brain and spinal cord. Myogenic precursors (myoblasts) migrate from the somites to the appropriate position in the body where, under the influence of unknown environmental signals, they may exit the mitotic cycle and fuse with one another to produce multinucleate embryonic muscle fibres. This process occurs in several waves from several different generations of myoblast, and gives rise to several (two in small animals, three or more in large animals) generations of muscle fibre. The number of fibres in skeletal muscles appears to be genetically determined but the expression of the full genetic capacity is dependent on the normal development of the nerve supply to the muscles. If during early development the motor nerves fail to maintain contact, for example, as a result of physical or drug-induced damage, the muscles will be smaller than normal due to a decrease in the number of their fibres.

As well as influencing the number of fibres in a muscle, some property of the neural input also appears to influence fibre type. This has been demonstrated in a number of ways, but most obviously in that the muscle fibres within a motor unit are homogeneous with respect to such properties as contraction time, resistance to fatigue, enzymes of anaerobic and aerobic metabolism and myosin ATPase.

These properties are determined early in development but they are not irreversible and changes can be seen in both developing and adult muscles, for example after denervation (see below). The ability of nerves to regulate the properties of muscles is referred to as a **neurotrophic** influence but it is not known precisely how this influence is exerted. There is good evidence that nerve-induced muscle activity at the appropriate frequency (tonic low frequency for slow muscles and phasic high frequency for fast muscles) is important. There is also evidence suggesting that specific messengers, *neurotrophic factors*, are released by motor nerves to influence the muscle fibres they innervate.

Effects of training

Type I fibres make up about 30–40% of the cells in human muscles and they are approximately the same size in men and women (the mean diameter being approximately 60 μm). Type II fibres are larger in men (average diameter 69 μm) than in women (50 μm). Two distinct responses to regularly performed strenuous exercise can be seen in muscle: hypertrophy of the fibres with an increase in strength (e.g. weight-lifters) and an increased capacity for aerobic metabolism (e.g. long-distance runners, cross-country skiers, swimmers).

Endurance exercise training gives rise to an increased capacity for oxidation of pyruvate and long-chain fatty acids. This is due to an increase in the density of mitochondria and hence in the absolute amount of enzymes, for example, those of the tricarboxylic acid cycle and those involved in the activation, transport and oxidation of long-chain fatty acids. There is an increase in capillary density and myoglobin, which speeds the rate of diffusion of O_2 from cell membrane to mitochondria. Trained individuals have increased intramuscular stores of triglyceride and lowered concentrations of serum triglycerides and their muscles can utilize lipids directly from blood.

The consequences of these changes are that during submaximal exercise trained individuals derive more energy from fat and less from carbohydrate than do untrained individuals. Furthermore, in the trained individual, liver and muscle glycogen stores are better maintained during exercise and a greater proportion of oxygen is extracted from the blood supply to muscles. **Fatigue** has two major causes, an inability to maintain an adequate motor drive from the CNS and a failure of excitation–contraction coupling. Fatigue resulting from voluntary exercise is normally evident before significant depletion of muscle energy reserves has occurred and, no matter how severe, never deplete muscle energy supplies to the point of inducing rigor. The process of fatigue is always reversible and both its onset is delayed and its intensity reduced by training.

Effects of ageing

The total number of fibres in a muscle decreases with age, and it has been suggested that the average number of fibres per motor unit gets larger. It is a generally held opinion that the units lost are the smaller ones and that the progressive loss of muscle fibres is due to a loss of the nerve cells that supplied them. The potency of synaptic transmission also declines with age, possibly as a result of the observed decrease in the synthesizing ability of aged nerve cells.

Effects of damage to nerve or muscle

After the nerve to a muscle is sectioned, there are changes in both the muscle and the axons. The loss of the neural influence results in pronounced changes in the muscle fibres. The earliest changes, such as the partial decrease in resting membrane potential and the increase in sensitivity to applied acetylcholine (which results from the addition of acetylcholine receptors to all of the sarcolemma) and the generation of spontaneous activity (**fibrillation**) can be seen in a few days. But other changes, such as a pronounced decrease in the ability to develop tension, the change in enzymic composition and the decrease in fibre diameter (**atrophy**), may take longer to develop. In humans, the fibres may shrink down to some 10 μm and remain so for months or until they are reinnervated; reinnervated fibres grow and develop according to the characteristics of the motor neuron providing their input. If muscle fibres remain denervated for prolonged periods (months to years), they will gradually be replaced by connective tissue and fat.

When the nerve to a muscle is sectioned, some of the motor neurons die, but others regenerate their axons. However, in higher vertebrates there is little or no specificity in the re-establishment of nerve–muscle connections. Regrowth of the axons is aided and directed by the presence of the old nerve sheaths (hence the accurate suturing together of the cut ends of a nerve is important). Normal muscles have fibres within individual motor units well-scattered across the muscle, but, following regeneration of a cut nerve, muscle fibres of a single motor unit occur as a clump of cells (Fig. 5.13), as if the ingrowing nerve made connections with all the muscle fibres in its immediate vicinity. However, reinnervation may not always be successful and when a whole limb is denervated there is very little evidence of orderliness in nerve regeneration to muscles and normal coordination of movement is never fully restored.

Damaged muscle fibres do not possess an intrinsic capacity for regeneration. However, damage is repaired very efficiently by the activation of small mononu-

(a) Axons

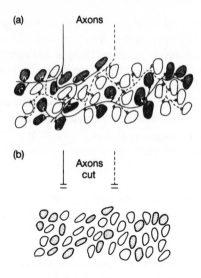

(b) Axons cut

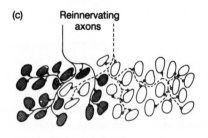

(c) Reinnervating axons

Fig. 5.13 Reinnervation of muscle fibres. (a) Prior to denervation the fibres innervated by each motor neuron are intermixed. (b) The muscle fibres atrophy following section of the nerve. (c) After reinnervation the muscle fibres of each motor unit tend to be grouped together.

cleate cells (**satellite cells**) that normally lie beneath the basal lamina of mature muscle fibres. These cells seem to be a special generation of myoblasts which migrate into the muscle region from the somite during development, but which do not then contribute immediately to the formation or growth of a muscle fibre. Provision of the appropriate stimulus triggers them to undergo mitosis, increase in numbers and ultimately fuse to repair the damaged fibres or to make new multinucleate muscle cells. In animals, it is actually possible to remove a muscle, mince it, pour the mince back into the appropriate place in the animal, sew up the skin, and produce a new but smaller functional muscle. Regeneration of the muscle in such cases is critically dependent on the presence of the nerve.

5.2 SMOOTH MUSCLE

Smooth muscle has a wide range of functions and all muscular forces not generated by striated or cardiac muscle are generated by smooth muscle. Smooth muscle tissues contain bundles or sheets of smooth muscle cells which are thin elongated cells which may be connected to their neighbours electrically by low-resistance gap junctions which help coordinate contractile activity. the contractile activity is initiated by an increase in the concentration of intracellular Ca^{2+} which acts through calmodulin and phosphorylation of myosin light chain.

The contractions of smooth muscle are slower than that of skeletal muscle but more efficient.

Smooth muscles vary in their level of activity, from those that show more or less continuous activity to those that are quiescent for prolonged periods. In some, only localized contractions occur (e.g. intestinal sphincters), whilst in others the whole organ may be involved (e.g. bladder). Activity in smooth muscles may depend on a number of factors, including the character of the smooth muscle cells, their environment, neural input and hormones. All neural influences are exerted by the autonomic nervous system; some tissues are innervated by only one division, while others are innervated by both the parasympathetic and sympathetic divisions.

Smooth muscles control the movement of material through most hollow organs; for example, they propel material in the gastrointestinal tract, they restrict flow in arteriolar blood vessels and bronchi, and they expel material from the bladder and vas deferens. Smooth muscles also control piloerection and influence the input of sensory information into receptors as when the dilator and constrictor muscles of the iris affect the amount of light reaching the retina.

Smooth muscle structure

A connective tissue sheath, the **epimysium**, surrounds the smooth muscle of each organ. Thin septa extend inwards from the epimysium to form the **perimysium**, which contains fibroblasts, capillaries, nerves and collagenous elastic fibres; the collagen fibres are synethesized by the muscle cell and form the major component of the extracellular space. The perimysium divides smooth muscle into discrete *bundles* of fibres. These bundles range from 20 to 200 μm in width, and anastomose with one another; these anastomoses can be seen at roughly 1-mm intervals along a given fibre bundle (Fig. 5.14). An exception is found in arteriolar walls which may be only a few cell diameters in thickness and the smooth muscle of which is not organized into bundles.

The individual smooth muscle cells within a bundle are 2–10 μm in diameter, and vary in length from about 50 μm in arterioles to 400 μm in most other organs. The smooth muscle cells within each bundle are fusiform, or irregular elongate cells that interweave and overlap with each other (Fig. 5.14) to form a network interlaced with collagen.

Individual smooth muscle cells come into close contact with 10 or so neighbouring cells; at these points they may be connected by specialized intercellular junctions of relatively low electrical resistance called **gap junctions**. At these junctions the sarcolemma of the cells is separated by 3–5 nm but the gap is bridged by structures which allow small ions to pass from cell to cell. The relatively low electrical resistance of these junctions allows current which may have either an excitatory or an inhibitory effect, to pass from cell to cell. Where bundles exist the direct coupling of cells within each bundle may result in the bundles being the functional (contractile) unit.

Pronounced differences between the structure of smooth muscles and striated muscles are seen at the ultrastructural level. For instance, smooth muscle cells

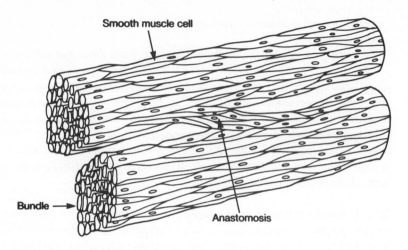

Fig. 5.14 Smooth muscle cells arranged in bundles that are interconnected by an anastomosis.

possess few mitochondria, the sarcoplasmic reticulum is poorly developed and located close to the sarcolemma, there is no postjunctional thickening or specialization at the neuromuscular junction and the myofilaments of actin and myosin are irregularly arranged. The actin filaments appear to be inserted into specialized structures in the sarcolemma, so-called dense bodies, and radiate out in a longitudinal direction from these. A detailed knowledge of the actin and myosin filaments in smooth muscle is lacking but relative to skeletal muscle the actin filaments are very long, there is a much higher ratio of actin to myosin and there is very little troponin. There is instead a high concentration of a specific Ca^{2+}-binding protein, **calmodulin** (p. 44) associated with the contractile proteins.

Contractile activity of smooth muscle

The contractions of smooth muscles are in general much slower than those of skeletal muscles. When excited by a single stimulus, there is often a long latency, a slow rise to peak tension (>1 s) and then a slow decline to the resting state (Fig. 5.15). In many tissues, this single contraction may take several seconds; with repetitive stimulation the forces generated by smooth muscles increase and can

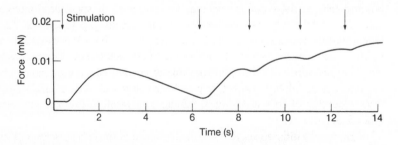

Fig. 5.15 The contractile response of a smooth muscle to a single stimulus and the response to repetitive stimuli.

reach levels similar to those found in skeletal muscles (30–40 N cm^{-2}). However, unlike many skeletal muscles, smooth muscles can maintain their tension at a high level for long periods and over a wide range of muscle lengths. It seems probable that the low activity of the myosin ATPase may account for both the slow development of force and the relatively low O_2 consumption during contractions (<1/100 that of skeletal muscle). The ability to contract over a wide range of lengths (up to four times the resting length) may be a result of the irregular arrangement of the myofilaments.

As with skeletal muscle, the force generated by smooth muscle is controlled by the level of intracellular free Ca^{2+}. In smooth muscle during stimulation, this Ca^{2+} may come from the interstitial fluid as a result of a change in membrane permeability to Ca^{2+} or it may be released internally from bound stores such as the sarcoplasmic reticulum. The incoming Ca^{2+} appears to contribute substantially to the rising phase of the smooth muscle action potential (see below). As a result of these changes the cytoplasmic concentration of Ca^{2+} may rise from a resting level of 10^{-8} mol L^{-1} to 10^{-6} mol L^{-1} or higher. With this rise in concentration, more Ca^{2+} combines with the regulatory protein calmodulin to activate a protein kinase that phosphorylates myosin. As phosphorylation of the myosin is a prerequisite for the activation of the smooth-muscle actin–myosin complex, it is the level of free Ca^{2+} that regulates the contractile activity of smooth muscles. Inactivation of the contractile mechanism is accomplished by the lowering of the intracellular concentration of Ca^{2+} and the activity of a phosphatase that dephosphorylates myosin light chain. The Ca^{2+} is transported to the extracellular fluid or sequestered internally. Expulsion of Ca^{2+} from the cell is energy-dependent and may be the consequence of the activity of a Na^+–Ca^{2+} exchange mechanism or a Ca^{2+}-dependent ATPase.

While the above may be the dominant controlling mechanism in many smooth muscles, there are other regulatory pathways influencing the contractile activity of smooth muscles. For instance, studies have shown that the level of contraction is not always well-correlated with changes in membrane potential. Indeed, some smooth muscles depolarized to the reversal potential of an excitatory transmitter will contract when exposed to the transmitter, thus indicating that there are membrane potential-independent mechanisms for excitation–contraction coupling, e.g. through receptor activation of G proteins and the production of inositol trisphosphate. Similarly, the force generated by some smooth muscles fails to correlate well with the level of free intracellular Ca^{2+}, which suggests that Ca^{2+} independent mechanisms exist for the initiation of cross-bridge activity. Finally, in muscles that contract for long periods, there may be a decrease in the level of phosphorylation of myosin light chain while tension is maintained. It appears that dephosphorylation of the cross-bridge while it is attached to actin may slow its dissociation. This 'latching' of cross-bridges may provide an energetically efficient means of maintaining tension.

Resting membrane potential and action potentials

The resting membrane potential of many smooth muscles is in the range of − 60 to − 70 mV and the basis of this membrane potential is similar to that found in other excitable cells. However, the resting membrane potential is some 20 mV below the equilibrium potential for K$^+$ and this may be the result of a relatively large resting permeability to Na$^+$.

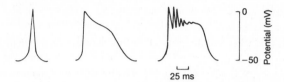

25 ms

Fig. 5.16 Action potentials of smooth muscle may be spike-like (left), plateau-like (middle), or a mixture of these (right).

Not all smooth muscles exhibit action potentials, but in those that do they are usually spike-like, but somewhat slower than in skeletal muscle; plateau-type action potentials are seen in some tissues (e.g. in the ureter; Fig. 5.16). A depolarization of some 20 mV is required to reach threshold and initiate an action potential. The action potentials reach a peak potential of some + 10 mV and if the stimulus is maintained, repetitive firing may occur, the frequency depending on the degree of depolarization.

The inward current responsible for the action potential in nerves and skeletal muscles is carried by Na^+ ions. This can be shown by the fact that the magnitude of the overshoot of the action potential is directly proportional to E_{Na}, and that action potentials fail in solutions depleted of Na^+. This is not true for some smooth muscles, where removing Na^+ from the bathing fluid actually makes the action potential larger. On the other hand, removing Ca^{2+} ions from the bathing fluid does abolish action potentials, while increasing Ca^{2+} produces larger action potentials. Mn^{2+}, which blocks Ca^{2+}-mediated action potentials in other tissues, blocks the smooth-muscle action potential.

Types of smooth muscle

As mentioned earlier, the activity of smooth muscles may be rhythmical and in these tissues dependent on spontaneous myogenic mechanisms; other tissues are quiescent until stimulated by an incoming signal. The former have often been referred to as unitary (cells acting together) and the latter as multiunit (cells acting independently) smooth muscles, respectively, but these divisions are now of little use as they represent extremes. Smooth muscles are now conveniently divided into three groups according to their membrane properties.

Spontaneously active smooth muscle

Many organs containing smooth muscle contract rhythmically (e.g. stomach, small intestine, ureter, parturient uterus). As this coordinated activity is seen in the presence of neurotoxins or local anaesthetics, it is concluded that it is initiated and coordinated by the smooth muscle cells, i.e. it is **myogenic** (as in the heart). Such activity usually depends on the spontaneous generation of action potentials, and the presence of a conducting system (the gap junctions). Two types of mechanism are responsible for the spontaneous generation of action potentials, *pacemaker* potentials and *slow waves*.

In some smooth muscles (e.g. uterus and *Taenia coli*), it appears that focal pacemaker regions slowly depolarize a group of cells to threshold; the subsequent action potentials are then conducted through the tissue. The pacemaker regions are not constant in location, and it is thought that all regions within these tissues have the capacity to assume the role of pacemaker.

The rhythmic activity of the stomach and intestine results from depolarizations called slow waves (see Fig. 17.8). These are discrete, plateau-type depolarizations, lasting 2–8 s, that can be recorded from all regions of the external muscle layer. As slow waves can be initiated by depolarizing currents and are propagated, their frequency over a relatively short distance (cm) is determined by the cells having the highest rate. In the intestine, where they are about 20 mV in amplitude, the slow waves initiate action potentials or spikes and the amplitude of contractions depends on the number of action potentials. The situation is less certain in the fundus and body of the stomach where the slow waves are larger but the spikes are smaller and only at the beginning of the slow wave. In this area the slow wave can exceed the membrane potential for the initiation of contraction and the amplitude of the contraction appears to depend on the degree and time for which the membrane potential exceeds this threshold. The ionic mechanism responsible for the generation of the slow waves is not fully understood at present but it has been attributed to changes in membrane conductance and the activity of a rheogenic pump.

The spontaneous contractile activity can be altered by nervous activity which may be either excitatory or inhibitory. For example, in the intestine, acetylcholine, the transmitter released from parasympathetic nerves, causes the smooth muscle to depolarize. As a consequence, the number and frequency of action potentials on each slow wave are increased and the contractions are more forceful. In contrast, the inhibitory action of noradrenaline, the transmitter released from sympathetic neurons to the detrusor muscle, and the inhibitory actions of non-adrenergic, non-cholinergic autonomic neurons to the gut (p. 62) are due to hyperpolarization and movement of the membrane potential away from threshold. This may result in complete cessation of contractile activity while the spontaneous fluctuations in membrane potentials continue at a subthreshold level.

Electrically inexcitable smooth muscle

This term applies to an extreme, but not unimportant, group of smooth muscles that do not generate action potentials (e.g. bronchial, tracheal and specific arterial smooth muscles of some species). In these tissues, the membrane potential remains stable until the tissue is stimulated. Stimulation may be the result of neurotransmitter release or the activity of local or bloodborne agents (e.g. histamine, bradykinin). Stimulation is accompanied by depolarization and subsequent contractions. In tissues with a sparse innervation, excitation can spread because of the presence of gap junctions. The physiological advantage of these muscles may reside in their generally slow and sustained response to nerve stimulation.

Intermediate smooth muscle

This category is the most widely distributed (e.g. iris, piloerector, blood vessels, vas deferens, seminal vesicles). These have a stable resting membrane potential and when stimulated they exhibit spike-like action potentials. The cells are linked by gap junctions but conduction is decremental and so the contractions fail to spread throughout the tissue. The force of contraction is proportional to the frequency of the action potentials and is usually under neural control.

Activation of smooth muscle

Contraction of all smooth muscles is dependent on changes in the intracellular Ca^{2+} level. This can occur as a result of inherent myogenic mechanisms which regularly depolarize the muscle fibres or as a result of neural or hormonal action. Contractions may also be induced by other means. For instance, some smooth muscles are relatively plastic when slowly stretched but rapid stretching results in a stretch-induced depolarization and contraction. Such behaviour may be important in **autoregulation** of blood vessels (p. 422). In other tissues, local agents modify the force of contraction (e.g. the actions of O_2 and CO_2 on blood vessels of the lungs, and histamine on bronchial smooth muscle).

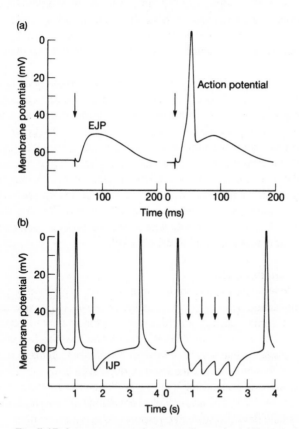

Fig. 5.17 Junction potentials in smooth muscles (a) Excitatory junction potentials (EJP) recorded intracellularly from the vas deferens following stimulation (arrows) of its sympathetic nerve supply (left, subthreshold EJP; right, suprathreshold EJP leading to an action potential). (b) Inhibitory junction potentials (IJP) recorded intracellularly from guinea-pig *Taenia coli* following stimulation (arrows) of the intramural nerves (left, single stimulus; right, repetitive stimuli).

In many tissues, the dominant influence is exerted by the nerves. Irrespective of whether they are sympathetic, parasympathetic or enteric, excitation is usually the result of a depolarization (an **excitatory junction potential**; Fig. 5.17) which is caused mainly by an increased conductance to Na^+ and, to a lesser extent, K^+. In the case of inhibition, the hyperpolarization of the smooth muscle membrane (the **inhibitory junction potential**; Fig. 5.17) may be due to an increased conductance to K^+ (and possibly Cl^-). In nearly all cases, the increase in conductance arises as a result of the neurotransmitter interacting with specific surface receptors on the muscle fibres (p. 35); one notable exception appears to be nitric oxide which is released as a neurotransmitter but acts directly on smooth-muscle guanylate cyclase (p. 42). If the neuromuscular junction has a relatively small junctional cleft (20 nm), then the junctional potential is distinct, with a fast rate of rise and may last 0.5 s; it seems that wider (400–500 nm) neuromuscular junctions may not exhibit the same rapid junctional potentials and the response to nerve stimulation is relatively slow. In fact in some tissues (e.g. the tunica media of blood vessels), many of the muscle fibres may not be directly innervated; they may, however, be under some neural influence, as current will spread from neighbouring innervated regions (through gap junctions).

The contractile activity of many smooth muscles is also influenced by circulatory hormones and paracrines (p. 31) and discussions of these are presented in the relevant chapters. From these it can be seen that the actions of many hormones are limited to particular smooth muscles. Thus, the stimulatory actions of the gastrointestinal hormone, gastrin, on smooth muscle, are largely limited to the stomach and gallbladder; and the stimulatory actions of oxytocin are limited to the uterus. The specificity of these interactions is demonstrated by the inability of hormones with similar structures (e.g. gastrin and cholecystokinin, and oxytocin and vasopressin) to have the same effects; this presumably reflects the highly specific nature of the receptors on the plasma membrane.

Other hormones have widespread activites; thus, adrenaline can change the contractile activity of many tissues (e.g. blood vessels, bronchioles, the intestines). Similarly, the highly potent prostaglandins and thromboxanes have pronounced effects on a number of smooth muscles. Thus prostaglandin F_2 is a potent stimulator of uterine contractility (and is used to induce labour at term), of intestinal smooth muscle and of bronchial smooth muscle. However, the physiological roles of these compounds are uncertain.

Chapter 6
Sensory Systems

The term sensory systems applies to those parts of the nervous system concerned with the detection, transmission and analysis of information about stimuli from the internal and external environments. Sensation is the conscious perception of such stimuli; much of the sensory information that is received by the body is outside the realm of consciousness. Sensory systems include the **receptors**, the **afferent nerve fibres** of sensory neurons and the **central pathways** activated by appropriate stimuli. This chapter considers the sensory systems associated with the skin, muscles and joints, which are generally referred to as the somatic senses or somatosensory pathways, and those associated with the viscera. The following chapter deals with the pathways from the special senses which arise in the eye, ear, nose and mouth.

6.1 SENSORY NEURONS

Sensory neurons comprise the peripheral endings, afferent nerve fibres, cell bodies and their central processes through which changes in the environment are transmitted to the central nervous system (CNS). For the somatic senses, the cell bodies lie in the trigeminal and dorsal root ganglia. Sensory neurons vary in their sensitivity to stimuli, in their axon diameter and in their electrical and chemical properties.

Like receptors, sensory neurons may be classified as somatic or visceral (Table 4.2). Their cell bodies are located outside the CNS in various ganglia. These cells are pseudounipolar (Fig. 3.1) and their cell bodies do not receive synaptic inputs.
   Somatic sensory neurons have their cell bodies either in the trigeminal or dorsal root ganglia. Their peripheral axons (somatic afferent fibres) travel in peripheral nerves carrying information from the skin, skeletal muscles, tendons, joints and bone. Their central processes travel in trigeminal or dorsal roots to the CNS. **Visceral sensory neurons** have their cell bodies in the various cranial nerve ganglia or in the dorsal root ganglia of the thoracic, lumbar and sacral regions. Their peripheral axons (visceral afferent fibres) travel with sympathetic and parasympathetic nerves and carry information from receptors in blood vessels and

internal viscera, particularly the heart, lungs, bladder, rectum and genital organs. Their central axons travel in the cranial VII, IX and X nerves to the nucleus of the solitary tract (tractus solitarius) in the brainstem or, in the thoracic, lumbar and sacral regions, join the dorsal roots to enter the dorsal horn of the spinal cord with the somatic afferent fibres. The receptors for the somatic and visceral afferent fibres include mechanoreceptors which are sensitive, for instance, to skin indentation, muscle length or visceral stretch; thermoreceptors which detect changes in skin temperature; and nociceptors which detect potentially damaging stimuli (Table 6.1).

Within sensory ganglia, cell bodies differ in size, staining properties and immunoreactivity. The central processes of some ganglion cells may release glutamate as a transmitter. In some C fibres, however, substance P is the likely transmitter, though other peptides such as somatostatin, vasoactive intestinal polypeptide and cholecystokinin are also candidates.

As already discussed (p. 88), peripheral nerve fibres are classified according to their diameters. In general, the greater their diameter the greater their conduction velocities. Unfortunately, there are two overlapping classification systems for afferent fibres. However, it is common to use the alphabetical A–C system for cutaneous afferents and the numerical I–IV system for muscle afferents.

Associated with the differences in diameter, peripheral nerve fibres also show differences in the following functional properties.

1 *Modality.* Although there are no rigid correlations between diameter and modality, some general trends are clear, as shown in Table 6.1.

2 *Conduction velocity.* This increases with increasing fibre diameter and with myelination (p. 86).

3 *Electrical threshold.* When a nerve is stimulated by electrical pulses, the strength of the stimulus required decreases with increasing fibre diameter. Consequently, unmyelinated fibres require the greatest stimulation. This is a property of the axons and does not reflect thresholds to physiological stimuli.

4 *Refractory period.* This decreases with increasing diameter; action potentials

Table 6.1 Cutaneous sensory nerve fibres and their receptors

Fibre type	Numerical equivalent	Diameter (μm)	Receptors	Modalities represented
Aβ	Smaller fibres of I and all II	4–16	Merkel cells, Pacinian corpuscles, Ruffini endings, Meissner's corpuscles, larger hairs	Touch, pressure, vibration, position sense
Aδ	III	1–4	Small hairs Specialized nerve endings Specialized nerve endings	Touch, pressure Cold Pricking pain
C	IV	0.5–2	Unknown mechanoreceptors* Unknown thermoreceptors Unknown nociceptors	Touch, pressure Cold, warmth Dull aching pain

*Not found in humans.

are of shorter duration and recovery is faster in large fibres. Thus, the frequency of firing in unmyelinated fibres seldom exceeds 10 Hz while short bursts of 500 Hz or more are occasionally seen in large myelinated fibres.

5 *Sensitivity to local anaesthetics.* Local anaesthetics, such as procaine and Xylocaine, act on excitable membranes to reduce the voltage-dependent influx of Na^+ associated with the action potential and hence block activity. Small fibres are the most susceptible because of their high surface area to volume ratio. Thus, after application of local anaesthetic to a cutaneous nerve, sensations of temperature and pain are blocked before touch and pressure associated with large fibres.

6 *Sensitivity to ischaemia.* A reduction in blood supply to a nerve, for example, as a result of an embolism, a thrombosis or an occlusion due to a pressure cuff, blocks large fibres first, probably because of their greater volume and hence higher metabolic demands. Thus, in this type of block, vibration sensitivity and position sense are lost before the ability to appreciate pain or changes in skin temperature.

6.2 SOMATOSENSORY PATHWAYS

Inputs giving rise to sensation follow one of the two routes within the CNS—the dorsal column pathway or the anterolateral pathway. The dorsal column pathway is more discriminative and precisely organized, while the anterolateral pathway is the only route for thermal or nociceptive inputs. Relay nuclei are synaptic stations along both pathways where information can be modified by segmental and descending controls.

Afferent nerve fibres, except for those of the special sense organs, enter the CNS mainly via the spinal dorsal roots, and the trigeminal and vagal roots. The innervation areas of peripheral nerves to the skin show little overlap and hence section of a nerve gives a characteristic region of sensory loss. However, fibres from one peripheral nerve enter the spinal cord over several dorsal roots, so that a particular innervation area is represented over a number of segments. The innervation area of the skin supplied by a single dorsal root is known as a **dermatome**. Dermatomes from adjacent dorsal roots overlap considerably so that section of a single root does not produce a region of complete anaesthesia.

After entering the spinal cord, fibres usually bifurcate giving ascending and descending branches which mostly end in adjacent grey matter. Inputs giving rise to sensations usually follow one of two ascending pathways within the spinal cord— the **dorsal column pathway** and the **anterolateral pathway.**

Fibres destined for the dorsal column pathway are mainly large myelinated (Aβ) fibres which enter the spinal cord via the more medially positioned dorsal rootlets. They bifurcate, giving one branch which enters the **dorsal horn** and a second branch which enters the ipsilateral **dorsal column** (Fig. 6.1).

The branches in the dorsal horn have a variety of destinations. Branches of the largest fibres from the annulospiral endings of the muscle spindles make direct synaptic contact with motor neurons in the ventral horn. Branches from muscle, joint and skin mechanoreceptors synapse in the dorsal horn with the cells of origin of the **spinocerebellar tracts** which supply the cerebellum with information used in the control of posture and movement. Other branches contact dorsal horn cells

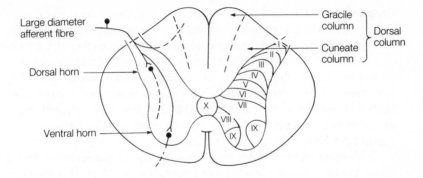

Fig. 6.1 Diagram showing the entry of a large-diameter (Aβ) afferent nerve fibre into the spinal cord and the passage of its branches into the dorsal horn, ventral horn and dorsal column. The Roman numerals indicate the laminae of the spinal cord—10 anatomically defined layers. (From Rexed, B. (1964) *Prog. Brain Res.* **11**, 58–90.)

in different laminae (Fig. 6.1) depending on their modality and are involved in local segmental activity.

Dorsal column pathway

The branches of the primary afferent fibres in the dorsal column ascend without synapsing to the medulla. There they synapse with neurons in the ipsilateral **dorsal column nucleus** (Fig. 6.2). These neurons send their axons to the opposite side of the medulla, forming a well-marked histological feature—the **sensory**

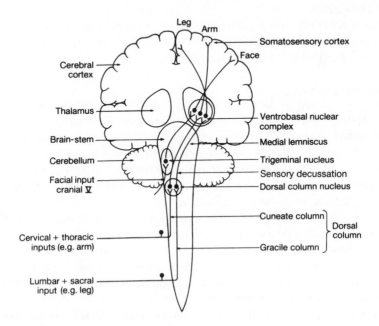

Fig. 6.2 Dorsal column pathway.

decussation—where the fibres from the two sides cross. After crossing, the fibres ascend in a tract known as the **medial lemniscus,** which passes through the midbrain and terminates in the **ventrobasal nuclear complex** of the thalamus. The thalamic neurons send their axons to layer IV of the six layers in the postcentral gyrus of the cerebral cortex (**somatosensory cortex**). Somatic inputs from the head enter the brainstem via the trigeminal root and first synapse in the trigeminal nucleus in the pons and medulla. From there, second-order fibres ascend to join the medial lemniscus, cross the midline and synapse in the medial part of the ventral basal complex (Fig. 6.2).

The dorsal column pathway carries information necessary for fine tactile discrimination, vibration sensitivity and position sense. Throughout the pathway the sensory modalities are preserved. Because the dorsal column pathway provides for these rather refined sensibilities, it is usually called the **discriminative** (specific) pathway. Since the medial lemniscus is a major component of this pathway it is also referred to as the **lemniscal system.**

Somatotopy

Receptive fields in this pathway remain fairly small and localized and, throughout the whole pathway from dorsal columns to the somatosensory cortex, adjacent fibres or cells correspond to adjacent areas in the periphery. Thus, the body surface is represented in an orderly, topographic fashion known as **somatotopic organization,** which is more fully discussed on p. 150. Moreover, certain peripheral areas have higher innervation densities peripherally and therefore disproportionately larger central representations throughout the whole pathway.

Conduction properties

Activity in this pathway is relatively stable, showing little alteration with arousal and attention and a lower susceptibility to blockage by general anaesthetics and hypoxia than that in the anterolateral pathway.

Anterolateral pathway

Fibres contributing to this pathway are the smaller myelinated (Aδ) and unmyelinated (C) fibres which enter the spinal cord via the more lateral rootlets, together with some branches of larger fibres. Many bifurcate, sending branches rostrally or caudally in the dorsolateral tract (tract of Lissauer) before synapsing in the dorsal horn with second-order neurons most commonly found in laminae I and V. The axons of second- or third-order neurons cross to the opposite side of the cord and ascend in the anterolateral tract (Fig. 6.3).

Axons in the anterolateral tract terminate in a number of regions (Fig. 6.4). Some axons travel directly to the thalamus (**spinothalamic tract**), terminating either laterally in the ventrobasal complex or in more medially located regions such as the intralaminar nuclei and nucleus submedius. These thalamic nuclei in turn provide inputs to various cortical regions, including the somatosensory cortex. Other axons of the anterolateral tract terminate in the reticular formation in the medulla, pons and midbrain (**spinoreticular and spinomesencephalic tracts**). These regions then relay to areas such as the thalamus and hypothala-

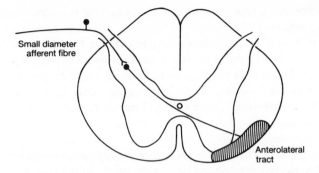

Fig. 6.3 Diagram showing the entry of a small-diameter (Aδ or C) afferent nerve fibre into the spinal cord and its first relay in the dorsal horn; the fibre of the second-order neuron then crosses to the anterolateral segment.

mus. The route taken by trigeminal inputs is not well-understood, although it probably involves the more caudal regions of the trigeminal nucleus in the medulla.

Fibres entering the anterolateral pathway carry information about touch, pressure, cold, warm and noxious stimuli. While some cells remain modality-specific, others respond to both noxious and innocuous stimuli. At the thalamic level some cells can respond to somatic, auditory and visual inputs. Receptive fields in the anterolateral pathway also tend to be larger than in the dorsal column pathway and may include over half the body surface. Somatotopy in this pathway is therefore less precise than in the dorsal column pathway with more overlap and less clear topography. In addition, responses can show marked alterations with attention, can habituate to repeated stimuli, and are less resistant to anaesthesia.

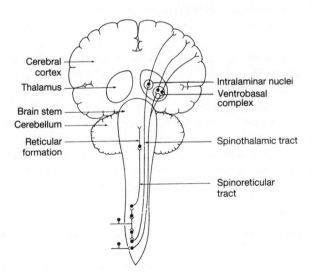

Fig. 6.4 Anterolateral pathway.

For these reasons this pathway is often referred to as the **non-discriminative** (non-specific) pathway.

Because this is the principal pathway carrying impulses from nociceptive and thermal inputs, damage to the anterolateral cord can produce analgesia in the opposite side of the body below the lesion. Such damage can give rise to a dissociated sensory loss: touch to a particular region can still be felt (via the dorsal column pathway) but pain and temperature changes are not perceived.

Function of relay nuclei

On any sensory pathway, there are always several interruptions at synaptic regions known as **relay nuclei**; information can never travel via a single fibre from receptor to cortex. For the somatosensory pathways, the nuclei consist of cells in the dorsal horns of the spinal cord, the dorsal column nuclei, the corresponding nuclei in the trigeminal system and the ventrobasal nuclear complex of the thalamus. These nuclei are important in the functioning of the sensory pathways as they provide for interaction and modification of the input by excitatory and inhibitory mechanisms. These mechanisms are activated by ascending and descending pathways.

The mechanisms of excitation and inhibition at synapses have been dealt with in Chapter 4. Inhibition can be either presynaptic or postsynaptic. One situation in which both pre- and postsynaptic inhibition are thought to contribute to the functioning of sensory pathways is surround inhibition.

Surround inhibition

This is also called lateral or afferent inhibition. It is important in all sensory systems, particularly the visual and auditory systems, in the sharpening of contrasts, in the localization of stimuli and in spatial discriminative ability.

A sensory unit commonly has an excitatory receptive field surrounded by a region, the inhibitory surround, which when stimulated reduces the output of the unit. The inhibitory surround of a unit is located in the excitatory fields of other adjacent sensory units which, when stimulated, exert an inhibitory effect on that unit by way of interneurons (Fig. 6.5). These interneurons may act either by pre- or postsynaptic inhibition. Thus, the unit with the greatest afferent input imposes the greatest inhibition on its neighbours.

At each central synapse, diverging connections may result in activity becoming widespread. Surround inhibition is important because it reduces the spread of activity. It occurs at synaptic relays at all levels of the sensory system, thus ensuring that the focus of activity is kept sharply localized, with weakly excited surround connections being inhibited by the strongly active centre. An example of how this works in two-point discrimination (p. 153) is shown in Fig. 6.6.

Descending control

Descending pathways originating in various regions of the brain exert control over the sensory as well as motor functions of the spinal cord. They act at every synaptic level to reduce irrelevant activity and to sharpen contrasts and improve discrimination. Pathways originating in the sensory cortex can influence transmission at all

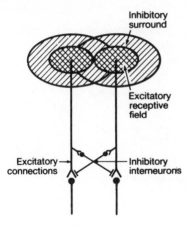

Fig. 6.5 Surround inhibition.

levels. For example, the removal of descending influences by the application of a cold block to the spinal cord alters the activity of dorsal horn cells caudal to the block. Conversely, electrical stimulation in the midbrain around the periaqueductal grey matter produces analgesia in humans and animals. Such stimulation is thought to activate a descending inhibitory pathway from the raphe nuclei in the medulla. This pathway is serotonergic and exerts a powerful inhibitory effect on dorsal horn nociceptive cells.

These descending control pathways can be influenced by incoming activity in sensory pathways. In the control of pain, for example, activity in spinoreticular pathways has been shown to influence descending inhibition. Nociceptive specific cells in lamina I of the spinal cord also project via a dorsolateral pathway to areas in the midbrain and medial thalamus and may modify the emotive aspects of pain and its control.

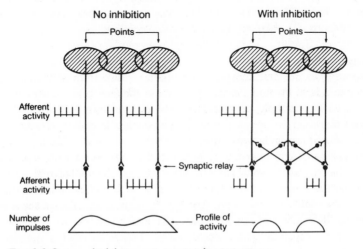

Fig. 6.6 Surround inhibition in two-point discrimination.

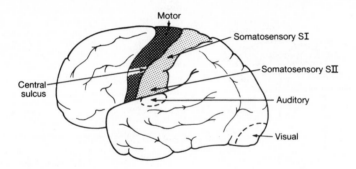

Fig. 6.7 Somatosensory cortex—primary (SI) and secondary (SII) areas.

6.3 SOMATOSENSORY CORTEX

Inputs from the somatosensory pathways end in a region of cerebral cortex known as the somatosensory cortex. This is organized somatotopically with inputs from the legs ending medially and those from the face laterally. Mass activity in the cortex in response to sensory stimulation can be recorded as evoked potentials.

The primary or *first* somatosensory cortex (SI) lies along the central sulcus (Fig. 6.7). It is classically considered to lie behind the sulcus in the postcentral gyrus, though many sensory responses occur precentrally. A *second* and smaller somatosensory area (SII) lies along the lip of the lateral fissure. Layer IV of the somatosensory cortex receives inputs from the thalamus. From there, activity spreads to more superficial and deeper layers before passing to other cortical regions. Recordings in this area have shown that neurons in columns perpendicular to the surface have similar receptive fields and respond to inputs from similar receptors. Responses can be obtained from cutaneous as well as deep mechanoreceptors (mainly in muscle). Thus, the somatosensory cortex, like the visual and auditory cortex, is organized functionally into columns.

Somatotopy

Responses in the first somatosensory cortex result from stimulation on the opposite side of the body, except for parts of the face, and are somatotopically organized. For example, responses from the foot and leg are represented medially in the somatosensory cortex, responses from the face laterally and responses from the hand and arm in between (Fig. 6.8). As in the dorsal column pathway, the density of peripheral innervation determines the size of the corresponding cortical area. The particular area which is enlarged varies in different animals and reflects the behavioural use of different peripheral areas in discrimination. For instance, in the pig the snout representation is particularly large; in rats, the whiskers; while in humans the hands and lips are disproportionately represented. SII also shows somatotopic organization but responses are frequently bilateral and mainly to cutaneous mechanoreceptors.

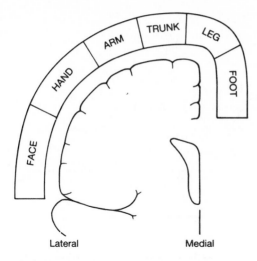

Fig. 6.8 Somatotopic organization. The sensory representations are indicated on a cross-section through the somatosensory cortex. (Adapted from Penfield, W. & Rasmussen, T. (1968) *The Cerebral Cortex of Man: A Clinical Study of Localization of Function*, p. 44. Hafner, New York.)

Lesions to the somatosensory cortex cause severe impairment of fine tactile discrimination in the corresponding peripheral area. Electrical stimulation of the cortex gives rise to a tingling sensation in the corresponding part of the periphery, though normal tactile sensations are not elicited.

In primates, somatosensory neurons which respond to nociceptive stimuli have been found together with neurons that respond with a slowly adapting discharge to pressure on the skin. These neurons are found in the sulcus between the pre- and postcentral gyri. Two populations that respond to nociceptive stimuli have been described. One has restricted contralateral fields and could be involved in the sensory–discriminatory aspects of pain. The other has large, whole-body receptive fields and is thought to be involved in the motivational and arousing aspects of pain. Two lines of evidence suggest there is a similar representation in humans. Epilepsy generated in this region has as its aura pricking pain, and superficial lesions of the region produce a major impairment of sensitivity to cutaneous pain.

Evoked potentials

On stimulation at the surface of the body, a mass response consisting of the inhibitory and excitatory postsynaptic potentials of many activated cells can be recorded from the surface of the somatosensory cortex, or even through the scalp if averaging techniques are used to improve the signal to noise ratio. This mass response is known as an **evoked potential** (Fig. 6.9). Evoked potentials have also been studied in the visual and auditory systems. They most commonly consist of a series of alternating positive and negative waves, the first wave being positive.

The *latency* of an evoked potential varies in different sensory systems and in different animals but is very consistent under similar stimulating and recording conditions. This has proved valuable in the early diagnosis of certain diseases like **multiple sclerosis**, in which the latency of the evoked potential is increased

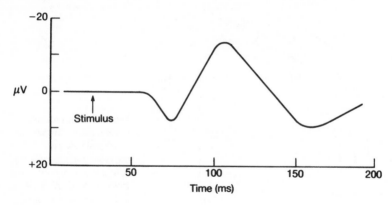

Fig. 6.9 Evoked potential recorded from the cortical surface.

because of loss of myelin along the pathway. Evoked potentials can also be used to assess the integrity of a sensory path in infants or comatose patients.

6.4 MODALITIES OF SENSATION

Stimulation of sensory receptors gives rise to a number of perceived sensations. Thus mechanoreception results in such sensations as touch, visceral distension and kinaesthesia or limb position. Thermoreception produces sensations of warmth or cold, and nociception may result in pain. Various anomalies in pain sensations, such as referred pains, are common.

Now that the nature of the receptors, the sensory neurons and the central pathways involved in sensation have been discussed, we can consider in more detail the various modalities of sensation.

Mechanoreception

The sensations usually associated with gentle mechanical stimulation of the skin are **touch** and **pressure**, although more complex sensations such as **vibration** and **tickle** are also recognized. Figure 6.10 illustrates the kinds of mechanoreceptors to be found in hairy and glabrous skin of mammals. Muscles and joints contain receptors which contribute to **position sense** as well as responding to pressure and stretch. Viscera have only a few mechanoreceptors.

Mechanoreceptors are specialized to detect particular properties of the stimulus such as its position, its intensity, its duration, its velocity or its acceleration. Receptors that respond when the stimulus is stationary are **position** or **length** detectors. As the stimulus is applied, the frequency of nerve impulses increases with the amplitude of deformation and then remains steady or declines gradually when the displacement is static. Such slowly adapting receptors can also gauge the **intensity** and **duration** of mechanical deformation of the skin or muscle. During a change in position or length they may also provide information about the velocity of deformation. Examples of such receptors are Merkel cells and Ruffini endings. Both are probably involved in the sensation of pressure on the skin.

Receptors detecting primarily **velocity** respond mainly during movement and

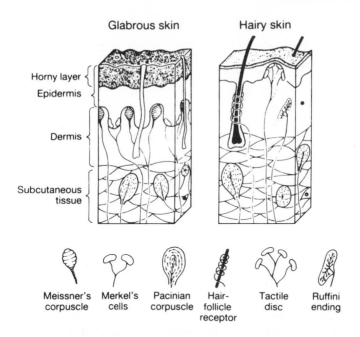

Glabrous skin   Hairy skin

Horny layer
Epidermis

Dermis

Subcutaneous
tissue

Meissner's   Merkel's   Pacinian   Hair-   Tactile   Ruffini
corpuscle   cells   corpuscle   follicle   disc   ending
                                    receptor

Fig. 6.10 Mechanoreceptors in glabrous and hairy skin. (From Schmidt, R.F. (1981) *Fundamentals of Sensory Physiology*, p. 86. Springer-Verlag, New York.)

cease to fire after movement has stopped. Such rapidly adapting receptors are called **phasic** receptors, in contrast to **tonic** receptors that continue to respond to a constant stimulus. Phasic receptors are involved in sensations such as touch and tickle. Examples of this type of receptor are Meissner's corpuscles in glabrous skin and hair follicle receptors in hairy skin.

A third type of receptor responds to rapid trasients in skin displacement, i.e. to **acceleration**. They are very rapidly adapting and may discharge only one impulse for each stimulus. Two types are known—Pacinian corpuscles and certain hair receptors. Rapidly repeated stimulation of these receptors evokes a sensation of vibration. Pacinian corpuscles are extremely sensitive to vibration and detect very small amplitudes of the order of 1 µm, particularly between 150 and 300 Hz.

Touch

Tactile sensations can be produced by indenting the skin on the fingertips by as little as 10 µm. Localized regions or **touch points** occur which are more sensitive to touch or pressure than surrounding areas. In the fingertips the thresholds of the receptors are signifciantly lower than surrounding areas. The capacity for **spatial discrimination** also varies over different parts of the body. For example, two-point discrimination assessed with the pointed arms of a pair of calipers is 1–3 mm over the tips of the fingers and tongue; in contrast, the minimum spatial discrimination is 20–50 mm over the forearm and back. Usually touch involves an active movement of the skin over an object. This movement greatly improves our ability to discriminate surface texture, as well as the size, shape and consistency of an object.

Kinaesthesia

In muscle and tendons there are mechanoreceptors providing information about muscle length and tension. These are usually associated with **muscle spindles** (p. 214) and **Golgi tendon organs** (p. 217). In joints there are mechanoreceptors that respond to bending of the limbs. These may be Ruffini endings in joint capsules, Golgi endings in ligaments or Pacinian corpusules, often found associated with joints and along bones and ligaments. All of these receptors (and mechanoreceptors in the skin) may be involved in the subconscious control of posture and movement, and also in the conscious awareness of position and movement, which is called **kinaesthesia**. All receptors signalling information about the position of the limbs and body are called **proprioceptors**. Our present understanding of the roles of the various proprioceptors in position sense is summarized below.

Recordings from joint receptors indicate that they respond mainly at the extremes of movement of a joint and may respond at both extremes or to movements in more than one axis of rotation (e.g. flexion–extension and abduction–adduction). They are therefore unlikely to provide accurate information on mid-range positions. However, stimulation of single-joint afferents can be perceived as movement of the appropriate joints.

Muscle spindles and Golgi tendon organs also give information about the state of joints and can provide accurate information about position. Indeed, various lines of evidence suggest an important role for these receptors in kinaesthesia. Thus, vibration of muscles (a powerful stimulus for muscle spindles) can give rise to illusions of movement and inaccurate estimation of joint position. These illusions remain even if the joints and skin are anaesthetized. One organ, the tongue, has muscle spindles but no joints or joint receptors. If the tongue is anaesthetized so that information from mucous membranes is lost, its position can still be distinguished. However, activity of *single* muscle spindle afferents around a joint is not perceived as joint movement, so it is presumably the activity in the population of fibres that is important in kinaesthesia.

The contribution of cutaneous afferents to joint position in still uncertain. These afferents respond mainly at the limits of joint movement. They can facilitate detection of passive finger movement, but skin anaesthesia has little effect on kinaesthetic performance.

In conclusion then, it appears that summed activity from muscle receptors may be the principal input for normal position sense. However, joint and skin receptors may play a role in situations where input from muscles is impaired, and at the extremes of movement.

Thermoreception

There are two types of cutaneous receptors associated with thermal sensation— **cold receptors** that respond to a decrease in skin temperature and **warm receptors** that respond to an increase. These receptors are active in detecting localized changes in temperature (e.g. when a cold object is touched) as well as in detecting changes in ambient temperature. These receptors, together with receptors in the hypothalamus and spinal cord, are also involved in the regulation of body temperature (p. 684). The structure of cutaneous thermoreceptors is not well-defined but they are probably free nerve endings. They are supplied by small myelinated (Aδ) and unmyelinated (C) fibres.

The afferent fibres associated with temperature receptors usually show a basic discharge at the normal physiological skin temperature (32°C). As the skin temperature falls, activity in the afferent fibres from cold receptors increases and may peak at a skin temperature of 15–20°C; in those from warm receptors the activity increases with temperature until it reaches a peak of 40–45°C. At this temperature there may also be a transient paradoxical response from cold receptors, although they are normally silent above 40°C.

Our ability to detect actual skin temperature and temperature changes depends on factors like the area of skin affected, the initial skin temperature and the rate of temperature change. For steady temperatures, the body has a neutral zone from about 20 to 36°C (the actual range being variable) in which there is no persisting sensation of body temperature. For temperatures above or below this range, persisting sensations of cold or warmth are experienced. Above about 45°C, pain rather than warmth is usually reported. Changes in temperature of the order of 0.1–0.3°C can be detected if rapid; changes of more than 5°C may be required if they occur slowly.

Pain

Pain can be described as the sensation resulting from stimuli which are intense enough to threaten or to cause tissue injury. Such stimuli may be mechanical (e.g. scratch), chemical (e.g. acid) or thermal (e.g. burn); neurologists commonly test pain sensitivity by a pinprick. However, pain sensations may show no simple correlation with the intensity or extent of tissue injury. Areas of apparently normal skin may show increased sensitivity (hyperpathia) while massive tissue destruction may occur without pain. Thus, the location and type of injury are important. Pain sensitivity also varies in different individuals, races and cultures. Even the same injury in the same individual can cause different degrees of pain depending on the situation. For example, injuries in accident victims are often not remembered as painful at the time of the accident. It is clear then that the sensation of pain is complex and that the body can alter the threshold in different situations.

It is usual to consider pain as having two components—the sensation *per se* and the emotional overtones of suffering and distress associated with it. While these two components usually occur together, certain procedures may leave one without the other. Thus, sensation may occur without distress after frontal lobotomy operations, while abnormal activity in certain thalamic areas may lead to a sensation of pain that is particularly distressing and intractable. Certain drugs may also have different effects on the two components; morphine, for instance, is especially effective on the emotional component. These considerations make studies on pain mechanisms difficult.

Somatic pain

The receptors in skin, muscle and joint that respond to painful stimuli— nociceptors—are probably free nerve endings. They are supplied by either small myelinated (Aδ) fibres or unmyelinated (C) fibres. The endings of Aδ fibres signal high-intensity mechanical stimuli, while the endings of C fibres signal high-intensity mechanical or heat stimuli or are less selective, responding to high-intensity mechanical, thermal and noxious chemical stimuli (polymodal nociceptors). The response is usually a vigorous burst of activity. Sometimes the two groups of nerve fibres give rise to a double sensation, the pain differing in latency and quality. There is a sharp initial pain signal which is caused by faster conducting A fibres, then later a long-lasting, aching pain due to activity in C fibres.

Many nociceptors show **sensitization**, that is, an increase in activity in response to repeated stimulation. Such sensitization probably underlies the hyperalgesia found in an injured area. Various factors such as K^+, acetylcholine, kinins and prostaglandins released by tissue damage have been implicated as mediators activating nociceptors but no single factor has yet been identified. There is also evidence that factors such as substance P may be released from peripheral nociceptor terminals and may contribute to vasodilatation and the inflammatory process. **Cramping pains** may occur with excessive use of skeletal muscles as a result of stimulation of nociceptors by unknown factors. **Itch** is probably a distinct sensation with its own receptors; it also requires the liberation of a chemical substance, probably histamine.

Visceral pain

Responses from visceral receptors, including nociceptors, have been studied in the vagus and other visceral nerves. As in the skin, visceral nociceptors are thought to be free nerve endings and occur in the walls of most hollow viscera, mesenteries and blood vessels. Like somatic nociceptors they are supplied by small myelinated and unmyelinated afferent fibres.

It is unclear to what extent viscera contain specific nociceptors. Many afferents from the heart, lungs, intestine and bladder respond to innocuous mechanical stimuli as well as to damaging stimuli. However, specific responses have been reported from heart, ureter and gallbladder. Effective stimuli are either excessive stretch or ischaemia. For example, in the heart afferents have been described which respond only to interruption of coronary blood flow, while in the ureter there are afferents responding specifically to overdistention. Nociceptors are not present at all in some tissues, for example, brain, so brain tissue can be cut without giving rise to pain. The dura, on the other hand, is extremely sensitive.

The sensations most commonly detected from viscera are either fullness or pain. Thus, stretch in the rectum or bladder gives rise to the sensation of fullness and triggers emptying reflexes. Excessive stretch or distension of many viscera gives pain which is often colicky or intermittent, for example, biliary colic or ureteric colic. Alternatively, severe visceral pain can also occur with ischaemia, as for instance in angina pectoris. Visceral pains are commonly poorly localized and may be referred to other parts of the body.

Referred pain

Damage to an internal organ is commonly associated with pain or tenderness not in the organ but in some skin region sharing the same segmental innervation. A classic example of this is the referral of cardiac pain to the left shoulder and upper arm. The most likely explanation for referred pain is that some central cells receive both cutaneous and visceral inputs. Such cells have been described in the lamina V of the dorsal horn. In the normal situation, lamina V cells are activated by skin inputs. When the visceral organ is injured, it activates the same cell in the dorsal horn and this is interpreted as injury to the skin. The skin, though uninjured, feels tender and local anaesthetics applied to the skin help to relieve pain by reducing the total sensory input to central cells.

Phantom-limb pain

Amputation of a limb is often followed by the feeling that the limb is still present. The phantom area gradually shrinks and may completely disappear, but a small percentage (5–10%) of amputees are left with a persistent and severe pain apparently from the absent region. The pain is thought to arise centrally because sectioning of the contralateral anterolateral tracts may give only temporary analgesia, and the pain may return and be felt at the same site as preoperatively. However, in some cases the pain may be relieved for long periods by a prolonged or severe stimulus to the stump, although in others the stump may be reamputated or anaesthetized without any relief.

Opioids

Opiates, such as morphine, heroin and codeine, are amongst the most powerful analgesics known. They act centrally by combining with receptors in many regions, including those associated with pain perception, such as the periaqueductal grey matter of the brainstem, parts of the limbic system and the substantia gelatinosa (lamina II) of the spinal cord.

An increased understanding of the way in which they are active has occurred with the discovery of the endogenous **opioid peptides**, comprising enkephalins, endorphins and dynorphins, which are powerful analgesics when administered intracranially. The **enkephalins** are pentapeptides (tyrosine–glycine–glycine–phenylalanine, then methionine or leucine) produced within the brain in regions like the limbic system and thalamus and in the substantia gelatinosa of the spinal cord. The **endorphins** and **dynorphins** are larger molecules containing a core of amino acids similar to the enkephalins (p. 278). They were first found in the pituitary and later shown to be present in several regions of the brain. In the CNS, opioid peptides are localized in nerve terminals and are thought to act as neurotransmitters or neuromodulators. They have pre- and postsynaptic inhibitory actions when iontophoresed on to neurons.

Although quite different in structure from the opiates, opioid peptides act by binding to the same type of receptors. The opioid peptides are as potent, or even more potent, in relieving pain than opiates when injected into the brain. They are not effective when injected intravenously, however, because they do not cross the blood–brain barrier and are rapidly degraded. It is claimed that their levels are decreased in chronic pain states and increased by electrical stimulation of the periaqueductal region. Such stimulation produces powerful analgesia, presumably by activating descending inhibitory controls. Opioid peptides may play a role in **acupuncture** as some of the effects of acupuncture can be blocked by the opiate antagonist naloxone.

As well as acting as higher levels of the CNS, the opioid peptides may also act at the level of the spinal cord and in peripheral tissue. **Substance P**, a peptide present in the terminals of afferent fibres, has been suggested as a transmitter for nociceptive afferents in the dorsal horn. Opioid peptides may suppress pain by acting presynaptically to block release of substance P from these afferent fibres. Substance P may also be released from peripheral nerve terminals and play a role in vasodilatation and inflammation. There is recent evidence that, besides their central action, opioids may act peripherally to modify this release.

Chapter 7
Special Senses

7.1 VISION

Our brain derives knowledge of the world around us from visual stimuli. Vision is the process in which the brain uses information from light-sensitive receptors in the retina to create a representation of the external world; at least one quarter of the human cortex is devoted to this task, marking its importance in human perception. The eye is the receptor organ for vision. Light enters the eye through the pupil and is focused by the cornea and lens on to the retina at the back of the eyeball. When focusing on a close object (near response), the lens rounds up, the pupil constricts and the eyes converge. Focusing defects arise from too short (hypermetropia) or long (myopia) an eyeball, incorrect curvature of the cornea and lens (astigmatism), and loss of lens elasticity (presbyopia). Six different muscles move each eyeball in its socket, allowing us to follow an object (smooth-pursuit movement) or switch our gaze from one spot to another (saccadic movement).

Functional anatomy of the eye

The eye is a hollow sphere, with three peripheral layers: a tough outer fibrous layer (the **sclera** and **cornea**); a middle layer comprising the vascular **choroid** and the muscular **ciliary body** and **iris**; and an inner neural layer, the **retina** (Fig. 7.1). Nerve fibres leave the eye at the **optic disc**. Blood vessels supplying the retina also

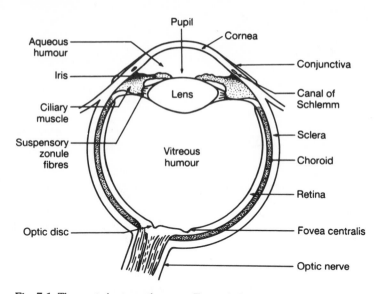

Fig. 7.1 The eye in horizontal section. To reach the photoreceptors, light must traverse the cornea, and aqueous humour, the lens, the vitreous humour and the inner layers of the retina.

enter and leave the eye at this site. Many blood vessels run over the inner surface of the retina and can be examined with an ophthalmoscope. This is useful clinically in the examination of diseases, such as hypertension and diabetes. The two inner chambers of the eye contain the **aqueous** and **vitreous humours**; suspended between these is the **lens**. Other parts of the eye are as shown in Fig. 7.1. The shape of the eye is maintained by an intraocular pressure of around 15 mmHg. Light (electromagnetic radiation with wavelengths in the range 400–700 nm) enters the eye through the cornea, passes through the aqueous humour and the iris (which forms an aperture), then through the lens and vitreous humour and, finally, through the neural elements of the retina to reach the photoreceptors (**rods** and **cones**) on the posterior surface of the retina. Highest-acuity (central) vision depends on focusing light rays on to the **fovea centralis** (Fig. 7.1). Behind the neural layer of the retina lies the **pigment epithelium**. This absorbs stray light, preventing blurring of the image, and is involved in the metabolism of visual pigments (see later).

Physiological optics

When focused at infinity the eye has 60 **diopters** (D) of refraction (power in diopters = 1/focal length in metres). Most refraction (45 D) occurs at the air–cornea interface while the lens provides 15 D, which can be increased for focusing on objects close to the eye. Changing the point of fixation from an object at infinity to one nearby involves the **near response**, which has three components.

1 Accommodation. The refractive power of the lens is made greater by increasing its surface curvature, particularly its anterior surface. In childhood, the power of accommodation is about 10 D, but this steadily decreases with age (**presbyo-**

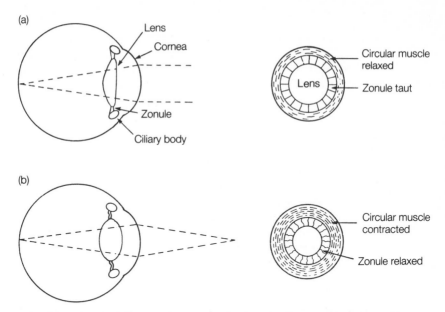

Fig. 7.2 Accommodation for near vision. In distant vision (a) the circular muscles of the ciliary body relax and the fibres of the zonule become taut, causing the lens to flatten; in near vision (b) the circular muscles constrict and the fibres of the zonule relax, allowing an increase in curvature of the lens, particularly its anterior surface.

pia). When focused at infinity, the lens is pulled flat by tension in the **zonule** (lens ligament). Accommodation is achieved by contracting the ciliary muscles, under the control of parasympathetic nerve fibres in cranial nerve III (oculomotor nerve). This causes the zonule to move forwards and to slacken, releasing the lens from tension and allowing it to round up due to its own elasticity (Fig. 7.2).

2   **Constriction of the pupil**. This is referred to as **miosis** and results in a better depth of focus, since with constriction light rays pass only through the central part of the lens. Pupil size is controlled by both sympathetic and parasympathetic nerve fibres to the muscles of the iris. Sympathetic fibres (cervical) contract the radial muscle and thereby dilate the pupil (**mydriasis**); parasympathetic fibres (cranial nerve III) contract the circular muscle and thereby constrict the pupil.

Constriction of the pupil reduces **spherical aberration** (light entering near the periphery of the lens being refracted more strongly than near the centre). **Chromatic aberration**, caused by the differing degrees of refraction of short (blue) and long (red) wavelengths of light, is also reduced, although never fully corrected. The perceptual consequences of this are minimized by the fovea having very few blue cones, and containing the yellow pigment, macular yellow, which strongly absorbs blue and violet light. Thus short-wavelength light does not affect the cones and the fovea responds only to longer wavelengths of light.

3   **Convergence of the eyes**. This ensures that the light rays from an object fall on corresponding parts of each retina to give a fused image. It is achieved by contraction of the left and right medial recti muscles (see Fig. 7.4) controlled by cranial nerve III.

Light reflex

The light reflex induces miosis in response to light or when looking at near objects. This reflex has two components. When light is shone into one eye the **direct** light reflex results in the pupil constricting. At the same time, the **consensual** light reflex causes the other pupil also to constrict. The afferent pathway for the light reflex is via the optic nerve to the pretectal region of the midbrain and the efferent pathway is along parasympathetic fibres of cranial nerve III.

Defects in focusing

Three common refraction errors in eyes are hypermetropia, myopia and astigmatism. In addition, all normal eyes eventually suffer from presbyopia.

1   Hypermetropia (long-sightedness). In this condition, the horizontal axis of the eye is too short and even with full accommodation images of close objects are focused behind the retina (Fig. 7.3a). This can be corrected with a *convex lens*.

2   Myopia (short-sightedness). The axis of the eye is too long, and even with full relaxation images of objects at infinity are focused in front of the retina (Fig. 7.3b). This can be corrected with a *concave lens*. The causes of myopia are considered on p. 186.

3   Astigmatism. Here, the curvature of the cornea (or occasionally the lens) is different in one plane from that in another, giving different degrees of refraction in different planes. This can be corrected with a *cylindrical lens* to give additional refraction in the required meridian, or by contact lenses with accurately spherical surfaces.

4   Presbyopia. The near point of vision is the distance between the eye and the closest object which can be brought to a clear focus. It recedes with age, being approximately 10 cm from the eye at 20 years and receding to about 80 cm when all accommodative power is lost by about age 50–60. With ageing, the inelastic

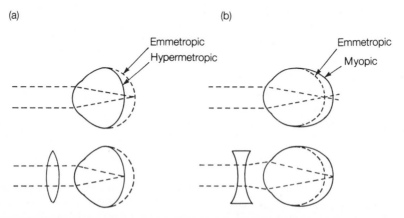

Fig. 7.3 Defects in focusing. The horizontal axis of a hypermetropic eye (a) is too short, and of a myopic eye (b) too long for accurate focusing. These defects may be corrected with concave or convex lenses, respectively. The position of the retina of the normal (emmetropic) eye is shown by the broken line.

crystalline core (lens nucleus) increases in size, and the surrounding lens capsule loses its elasticity.

Eye movements

Each eyeball is moved by four rectus muscles (superior, lateral, inferior and medial) and two oblique muscles (superior and inferior; Fig. 7.4a). The lateral rectus is controlled by cranial nerve VI (abducens nerve), the superior oblique by cranial neve IV (trochlear nerve) and all the others by cranial nerve III (oculomotor nerve). When looking straight ahead, these skeletal muscles have the actions shown in Fig. 7.4b. The two eyes may be moved in the same direction (conjugate movements or versions) or in opposite directions (disjunctive movements or vergences), as occurs for instance during accommodation. When altering fixation from one object to another, the eyes move in rapid jumps called **saccades**. Even at rest the eyes are never still, but move in constant microsaccades, an activity which is essential for vision. **Smooth-pursuit** movements occur when the eyes follow a moving object. A **squint** (strabismus) occurs when the visual axis of one eye is not fixed on the object being viewed by the other.

Transduction in photoreceptors

Light focused on to the retina is absorbed by visual pigments in the photo-receptors—rods for monochromatic and night vision, cones for colour and day vision. Rod photoreceptors contain the pigment rhodopsin; cones contain three colour pigments (red-, green- and blue-sensitive opsins). Each pigment also contains retinal which is isomerized by photons and split from opsin in a process

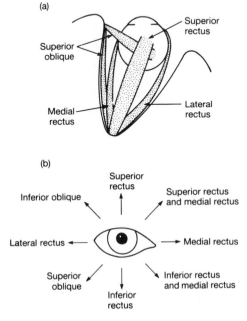

Fig. 7.4 (a) Extraocular muscles of the right eye as seen from above (inferior oblique is situated below). (b) Directions of movement of the right eye, mediated by its extraocular muscles.

Table 7.1 Ways in which rods differ from cones

Rods	Cones
120×10^6 per eye	7×10^6 per eye
Not found in the fovea, but uniformly distributed throughout the rest of the eye	Red- and green-sensitive cones highly concentrated in fovea, and blue-sensitive cones near fovea. Less concentrated peripherally
High sensitivity to light; longer outer segments, relatively more photopigment	Lower sensitivity; shorter outer segments, relatively less photopigment
Used for night vision; pigment is fully bleached in bright light so that rods are inoperative in daytime	Used for daytime vision
Many rods linked to each retinal ganglion cell so that receptive fields are larger, hence relatively lower acuity	Smaller receptive fields, especially in fovea where limit is diameter of one cone (≈ 2 μm)
Do not give colour sensation	Give colour sensation
Pigment (rhodopsin) is in membranous discs inside the rod outer segment	Pigment (red-, green- or blue-sensitive) is incorporated in folds of the cone outer segment membrane
Sensitive to light rays with wide angle of incidence, including those traversing the periphery of the lens	Greater directional sensitivity, less sensitive to light rays traversing the periphery of the lens

known as bleaching. Light induces a hyperpolarizing potential in the photorecep-
tors. On entering the dark pigment is regenerated, allowing dark adaption.

The photoreceptors, **cones** (used for daylight vision) and **rods** (used for nocturnal vision) are arrayed along the outer surface of the retina next to the pigment epithelium. The photoreceptors are supplied by blood vessels from the choroid, not by those on the inner retinal surface. This is why detachment of the retina from the choroid (*retinal detachment*) leads to extensive damage to rods and cones. Some of the differences between rods and cones are summarized in Table 7.1.

Photopigments

The transduction of light energy into a receptor membrane potential begins when photons are absorbed by photopigments located in the outer segments of the receptors. Each photopigment consists of a protein **opsin**, and a chromophore **retinal** (the aldehyde of vitamin A). There are four forms of opsin giving four photopigments—the rod photopigment **rhodopsin**, and the three cone photopigments: **erythrolabe** (red-sensitive), **chlorolabe** (green-sensitive) and **cyanolabe** (blue-sensitive). Their absorption spectra are shown in Fig. 7.5. Perception of detect different colours depends on the different degrees of excitation of the three cone types by light of a particular wavelength.

The threshold sensitivity of the eye to dim light depends on the wavelength

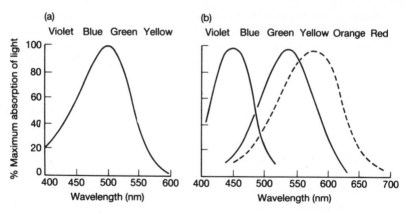

Fig. 7.5 Absorption spectra of (a) the rod photopigment and (b) the three cone photopigments.

and on whether the eye is adapted for night vision (**scotopia**) or day vision (**photopia**). The scotopic spectral luminosity curve (Fig. 7.6) shows maximum sensitivity to light of 507 nm wavelength and is characteristic of the absorption spectrum of rhodopsin, while the photopic spectral luminosity curve (Fig. 7.6) has a peak at 555 nm and is characteristic of the combined absorption spectra of the three cone pigments. Thus, in scotopic vision, the eye is most sensitive to green while in photopic vision it is most sensitive to yellow. The shift from scotopic vision to photopic is called the **Purkinje shift**.

The outer segments of rods and cones are long narrow thread-like structures 2 μm in diameter. They act as waveguides, so that once a photon has entered the outer segment it will be internally reflected, maximizing its chances of being absorbed by photopigment. Absorption of a photon results in isomerization of the

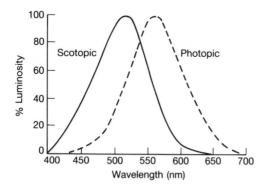

Fig. 7.6 Relative sensitivity to different wavelengths of light in dark-adapted (scotopic vision) and light-adapted (photopic vision) eyes. The scotopic sensitivity is measured by adjusting the luminosity of light of different colours until the intensity matches that of a reference light. Photopic sensitivity is measured in a similar manner using a **flicker photometer** where the reference and test lights alternate at a frequency of 20 Hz so that differences in colours are not noticed.

chromophore retinal from 11-*cis* (which binds to opsin) to all-*trans*, which does not bind to opsin. Isomerization starts a series of reactions ending in separation of retinal from opsin, a process known as bleaching. The all-*trans* retinal is transported to the pigment epithelium where it is regenerated to the 11-*cis* isomer.

Adaptation

The sensitivity of the eye depends on the ambient light intensity. On going from a light environment into a darker one, there is a gradual increase in sensitivity allowing dimmer lights to be seen, a mechanism known as **dark adaptation** (Fig. 7.7). This has three principal components:

1   **pupil dilatation**, which can account for up to a 16-fold increase in sensitivity;

2   **neural changes** mediated by amacrine cells in the retina so that retinal neurons become more excitable and the inhibitory components of receptive fields are suppressed; and

3   an increase in **receptor sensitivity** due to regeneration of photopigments from their bleached forms, a process requiring less than 10 min in cones and about 30 min in rods.

Sensitivity is limited by the amount of pigment available; rods are fully bleached in daylight. On entering a dark environment, regeneration of photopigments starts but since rods contain more photopigment than cones they take longer to regenerate pigment. Therefore dark adaptation has two components, with an inflection in the curve at the time when rods become more sensitive than cones (Fig. 7.7). The overall change in sensitivity is by a factor of about 10^5 occurring over 30 min, of which a smaller fraction is due to dilatation of the pupil and neural mechanisms within the retina, and most to regeneration of photopigment. A much faster (3–5 min) period of **light adaptation** occurs when going from a dark environment into a brighter one, resulting from pupil constriction and bleaching of photopigments.

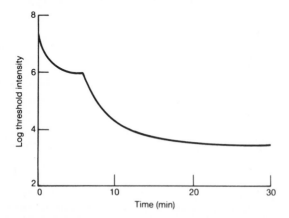

Fig. 7.7 Dark adaptation. The absolute threshold for perception of light, measured with time after entering a dark room. The inflection in the biphasic curve reflects the time at which rod vision becomes more sensitive than cone vision. Note the logarithmic scale of luminance; overall sensitivity increases by 5 orders of magnitude.

After-images

Visual sensations outlast the stimulus, giving rise to both **positive** and **negative after-images**, which reflect bleaching of photopigments. Intense stimulation causes a lasting change in the level of bleached photopigment so that sensation of after-images persists for seconds to minutes. A very brief flash of light is followed first by positive and then by negative after-images in which the light source appears dark against a brighter background. If the light is coloured, the negative after-image may be tinged by the complementary colour. A stimulus lasting several seconds is followed by:

1 *persistence of vision*, the apparent continuation of the stimulus for about 1 s, at rapidly diminishing subjective brightness;

2 *after-colours*, which may last seconds or minutes, frequently changing colour (these require very bright stimuli).

The **critical fusion frequency**, the frequency at which a flickering stimulus appears to be continuous, depends on the frequency of action potentials fired by retinal ganglion cells. In dim light, when action potential frequency is relatively low, it may be less than 10 Hz, but in bright light it may range up to more than 60 Hz. This enables us to perceive the intermittently presented images in films and television as continuous.

The receptor potential

Photoreceptors are maximally depolarized in the dark, and the bleaching of photopigment by light induces a **hyperpolarizing** receptor potential. The outer membrane of the outer segment contains cyclic guanosine monophosphate (cGMP)-gated ion channels (selective for Na^+, Ca^{2+} and Mg^{2+}) responsible for the receptor potential. In the dark, the cGMP-gated channels are open, allowing a steady current to enter. This inward current keeps the cell depolarized, near -40 mV, and transmitter (probably glutamate) is continually released from the synaptic terminal. Light closes the outer segment channels, the membrane hyperpolarizes and transmitter release decreases.

During normal photopic vision, light induces a hyperpolarizing receptor potential and dark a depolarizing receptor potential (a dark object on a lighter background can be considered an excitatory visual stimulus) by way of a complex sequence of reactions. In the dark, when eyes are at their maximum sensitivity, these reactions permit a very large degree of amplification, so much that the reception of a single photon evokes a receptor potential large enough to induce action potentials in the optic nerve, and psychophysical experiments show that single photons result in the perception of an event (but not a visual image!).

Light-triggered cGMP cascade

The membranous discs in the outer segments of rods, and the equivalent membrane folds in cones, contain three major proteins: visual pigment (retinal plus **opsin**); a G protein called **transducin** (p. 40); and **cGMP phosphodiesterase**. Upon absorbing a photon, 11-*cis* retinal isomerizes and triggers a series of conformational changes in the protein part of the visual pigment molecule. The isomerized molecule is free to diffuse laterally in the membrane and can activate a number of molecules of transducin. This catalyses the exchange of guanosine triphosphate (GTP) for guanosine diphosphate (GDP) bound to the α-subunit of transducin (p. 41). The binding of GTP causes liberation of an active α-GTP subunit, which in turn disinhibits cGMP phosphodiesterase, leading to rapid hydrolysis of cGMP in a local region of the outer segment. A single photoisomerized rhodopsin molecule can trigger liberation of about 500 α-GTP subunits, and hence activate many phosphodiesterase molecules, each in turn hydrolysing about 500 molecules of cGMP, with the result that a single photon closes about 300 outer-segment ion channels. The mechanisms by which the light response is terminated are not fully understood. They include phosphorylation of isomerized rhodopsin to inactivate it, and hydrolysis of the α-GTP subunit to α-GDP.

In the dark, Na^+, Ca^{2+} and Mg^{2+} ions enter the outer segment through the light-regulated conductance channels. Na^+ accumulating in the outer segment diffuses into the inner segment, where it is pumped out by a Na^+–K^+ pump. Ca^{2+} ions are removed by the Na^+–Ca^{2+} exchange carrier, while the exit path for Mg^{2+} is not known. In the dark, elevation of the concentration of cGMP increases the influx of Ca^{2+}. Ca^{2+} ions inhibit guanylate cyclase, the enzyme responsible for synthesizing cGMP, so an increased concentration of Ca^{2+} lowers cGMP, providing negative feedback control of cGMP levels. Conversely, in the light, as the cGMP-regulated channels close, Ca^{2+} continues to be removed and its concentration becomes reduced, allowing cGMP levels to increase. (Light → hydrolysis of cGMP → Ca^{2+} declines → disinhibition of guanylate cyclase → raised cGMP.) This system is involved in adaptation to light, and it stabilizes levels of cGMP in the dark so that only 1–2% of the cGMP-regulated channels are open, reducing the load on the photoreceptor cell metabolism.

The cGMP-regulated channel openings in the outer segment are unitary events with a mean duration of about 1 ms and a unit conductance of about 0.1 pS. This small effective conductance ensures that the amplitude of dark current noise (spontaneous channel openings due to random thermal motion) is small compared to the electrical signal resulting from reception of a photon, thereby allowing detection of single photons.

Electro-oculogram and electroretinogram

There is a steady potential difference of about 6 mV between the cornea (positive) and the posterior portion of the interior of the eye, the fundus (negative), generated predominantly across the pigment epithelium. As a result of this potential difference, electrodes placed on the skin at each side of the eye detect a change in potential when the eye is moved in a horizontal direction. The record so obtained during eye movements is called an **electro-oculogram**.

Illumination of the eye causes a series of smaller changes detected between an electrode on the cornea and an indifferent electrode. This is known as an **electroretinogram** and its early part reflects electrical activity within the receptors.

Retina and visual pathways

Light has to pass through three layers of retinal neurons to reach the photoreceptors. A change in photoreceptor potential is transmitted to the ganglion cells, either directly via bipolar cells or indirectly via horizontal and amacrine cells. Action potentials are not generated until the ganglion cells are reached. Axons from the ganglion cells converge at the optic disc (blind spot) and travel in the *optic nerve to the lateral geniculate nucleus; those from the nasal half of each retina cross over (decussate) at the optic chiasm, while those from the temporal halves stay on the same side. They synapse in the magnocellular and parvocellular layers of the lateral geniculate nucleus before being relayed to the visual cortex to give rise to conscious sensation of vision, or to other nuclei (e.g. superior colliculus) as part of various visual reflexes (e.g. for eye movements, pupillary dilatation and lens accommodation). The magnocellular system processes information pertaining to movement, texture and stereoscopic vision; the parvocellular system processes information about colour and high-acuity perception of form and contours.*

The retina

The retina is an outgrowth of the brain and, in addition to the photoreceptors (rods and cones), contains four other types of neuron—**bipolar** cells, **horizontal** cells, **amacrine** cells and **ganglion** cells. It is composed of three cellular layers with two synaptic layers (plexiform layers) interposed between them (Fig. 7.8). Light passes through the retina to reach the photoreceptors (rods and cones) at its outer

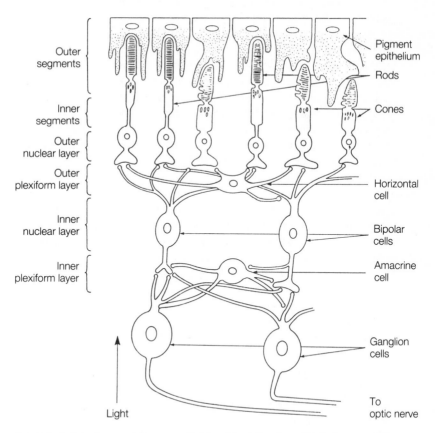

Fig. 7.8 Cellular layers of the retina. (Adapted from Dowling, J.E. & Boycott, B.B. (1966) *Proc. R. Soc. (B)* **166**, 80–111.)

surface; these are the only light-sensitive cells in the retina. There are no photo-receptors overlying the optic disc and light falling on it cannot be seen—hence the term **blind spot**. Ganglion cells are the only output cells from the retina; their axons (about 10^6) leave the eye at the optic disc to form the optic nerve. Ganglion cell axons are unmyelinated and hence transparent until they reach the optic nerve. At the fovea, where visual acuity is greatest, the orderly layering of the retina is modified with neural elements, other than cones, displaced to the sides and retinal blood vessels absent.

All signals originating in photoreceptors must pass to ganglion cells via bipolar cells, which are part of both **direct** (photoreceptor → bipolar → ganglion cell) and **indirect** (photoreceptor → horizontal or bipolar → amacrine → ganglion cell) pathways (Fig. 7.8). Horizontal cells contact photoreceptors over a relatively wide area, and are electrically coupled to one another via gap junctions. Every photoreceptor is contacted by a horizontal cell. In primates, outputs from horizontal cells are not well-known, and may be directed back to photoreceptors, or to bipolar cells, or to both. There are at least 20 types of amacrine cells, which link bipolar and ganglion cells, providing alternative, indirect routes between them. Amacrine cells are involved in a number of complex circuits which give individual ganglion cells particular receptive field properties.

Bipolar cell receptive fields are circular, detecting differences in luminance between an inner circle and a surrounding annulus (p. 175; see Fig. 7.13). Those excited by light in the centre of their receptive field are named **ON-bipolar cells** and those excited by dark in the centre of their receptive field as **OFF-bipolar cells**. Direct connections from bipolars to ganglion cells are all excitatory, so ON-centre bipolars supply ON-centre ganglion cells, and OFF-centre bipolars supply OFF-centre ganglion cells. Photoreceptors, bipolars and horizontal cells do not fire action potentials (some amacrine cells may); the first action potential in the direct pathway occurs at the ganglion cell.

Ganglion cells

Ganglion cells, the output cells of the retina, respond to a variety of different properties of the stimulus. Some respond to objects brighter than the background, others to darker; some give sustained responses, some transient; some are colour-coded, some are not, and so on.

Ganglion cells can be classified into three main groups—A, B and C. The A cells (Fig. 7.10a) are relatively large, with wide dendritic fields and fast-conducting myelinated axons in the optic nerve. They respond to low spatial frequencies, have high contrast sensitivity, are very sensitive to movement and do not respond to colour contrasts. The B cells are smaller with relatively small dendritic fields and slower-conducting myelinated axons; they respond to high spatial frequencies, have poor contrast sensitivity and show colour opponency. The C cells are small and their axons are unmyelinated in both the retina and optic nerve; their properties are less well-characterized.

Retinal ganglion fibres project in the optic nerve to two major subcortical visual centres, the **superior colliculus** and the **lateral geniculate nucleus** (LGN). The pathway to the superior colliculus is concerned with eye movements and orientation to visual stimuli; that to the LGN projects to the cortex and is concerned with the sensation of vision. There is partial cross-over of the optic nerve fibres. Fibres from the nasal halves of the retinas cross while those from the temporal halves remain ipsilateral, so that visual stimuli in the left halves of the visual fields of both eyes excite the right visual cortex and vice versa (Fig. 7.9).

The A ganglion cells project to the **magnocellular** layers of the LGN; the B ganglion cells project to the **parvocellular** layers of the LGN (Fig. 7.10). C cells project to a variety of nuclei, including the superior colliculus (reflexes involving attention to movement), the suprachiasmatic nuclei (involved in diurnal rhythms), ventral LGN (part of the reflex pathway regulating the size of the pupil), pretectal complex (concerned with the near response, i.e. vergence, miosis and accommodation), and the accessory optic system (projecting to the dorsal, lateral and medial terminal nuclei in the walls of the midbrain and mediating eye movement

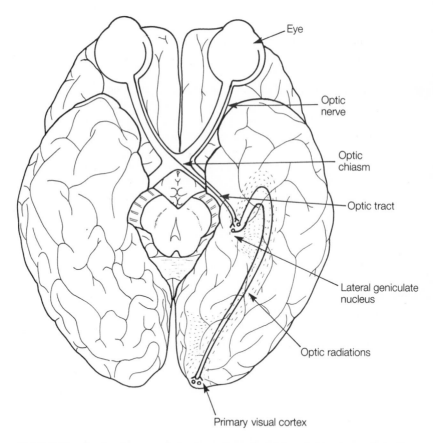

Fig. 7.9 The visual pathways and visual area 1, as viewed from below. Images from the right field of view fall on the left hemiretinas of the two eyes, and these project, via the optic nerve, optic chiasm (where axons from the nasal half of each eye cross over and those from the temporal half stay on the same side), LGN and optic radiation, to visual area 1 in the left cortical hemisphere.

(a)

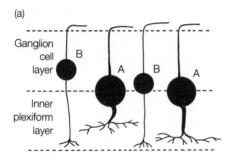

(b)

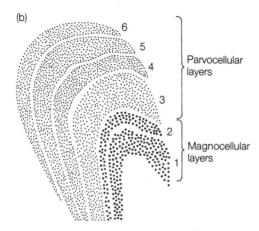

Fig. 7.10 (a) A and B ganglion cells in the retina and (b) the magnocellular and parvocellular layers of the left LGN cut parallel to the face. (Adapted from Perry, V.H., Silveira, L. & Cowey, A. (1990) *Ciba Symp.* **155**, 5–14.)

reflexes). Some of these nuclei, particularly the superior colliculus, also receive inputs from A and B ganglion cells.

Magnocellular and parvocellular divisions of the visual system

The six-layered LGN has two major subdivisions comprising two ventral magnocellular layers (1 and 2) and four dorsal parvocellular layers (3–6; Fig. 7.10b), which receive inputs from A and B ganglion cells, respectively. These layers receive inputs from either the ipsilateral (layers 2, 3 and 5) or contralateral (layers 1, 4 and 6) eye.

The magnocellular and parvocellular divisions perform different tasks in analysing visual information from the retina. The magnocellular system processes information pertaining to movement, texture and stereoscopic vision. It is distinguished by its high temporal resolution and is probably colour-blind. Cells from the magnocellular layers project to layer 4Cα of the primary visual cortex (Fig. 7.12).

Cells from the parvocellular layers project to layer 4cβ of the visual cortex (Fig. 7.12). The parvocellular system has two major subdivisions. One projects to the so-called blob regions (containing high levels of cytochrome oxidase (p. 172)) of the visual cortex and the other to interblob regions. The blob component is concerned specifically with colour; the interblob component has high spatial resolution, orientation selectivity and end-stopping, indicating involvement in high-resolution perception of form (shape). (End-stopping is a receptive field property such that cells respond to discontinuities in contour, such as corners, curves and ends of lines.)

Regions of cerebral cortex involved in vision

Over one-quarter of the human cortex is thought to be involved in processing visual information, reflecting the profound influence of visual sensation in human perception. The **primary visual cortex** (V1, also known as the striate cortex), which is located in the occipital cortex on the walls of the calcarine fissure, receives inputs from LGN (Fig. 7.9). The entire contralateral visual field is represented, with a small extension over the vertical meridian. Other visual areas, V2–5, are located in the peristriate and more anterior cortex and are concerned with different aspects of visual perception. Thus V2 is specialized for stereoscopy, colour and high-acuity vision; V3 for form perception; V4 for colour; and V5 for movement.

Cytochrome oxidase staining

Histochemical studies of the visual cortex to demonstrate the distribution of cytochrome oxidase (an enzyme that characterizes tissues with high metabolic activity; Fig. 7.11) show a remarkable correlation with areas of the visual system involved in the analysis of colour. In V1, layers 4A and 4C stain uniformly and densely for cytochrome oxidase, while **blobs** of cytochrome oxidase activity extend pillar-like through the full thickness of layers 2 and 3, and can be seen more faintly in layer 1 and in layers 4B, 5 and 6 where they lie in precise register with the overlying upper layer blobs. V2 has thick and thin cytochrome oxidase stripes separated by lightly staining interstripes. Cells in the thin stripes show colour-opponency and are not orientation-sensitive. Cells in the thick stripes are concerned with stereoscopic depth; they are orientation-sensitive and selective for binocular disparity. Cells in the pale interstripes are concerned with high-acuity perception of form; they are orientation-sensitive and over 50% are end-stopped.

Processing visual information

Retinal and lateral geniculate cells have centre-surround receptive fields, while most neurons in the visual cortex respond to more complex stimuli. Uniform illumination of the whole receptive field gives, on average, little excitation; neurons respond best to contrasts in illumination. Some neurons in the visual cortex are stimulated best by bars or edges of a particular orientation or spatial frequency, or by specific colours, or by movement. Information about the disparity of images on the two retinas is used to give stereoscopic perception of depth and distance. Neurons of the visual cortex with closely related functions are arranged anatomically in radial columns with cells responding to the same region of the retina, dominated by the same eye and sharing the same orientation or colour preference. Thus a small region of the visual cortex will receive inputs from all possible receptive fields belonging to a small region of visual space.

Parallel processing

Information processing is done in both a hierarchical and parallel manner. A hierarchically ordered series of ascending connections, from lower to higher centres, explains the responses of many cells to specific stimuli. At each stage the

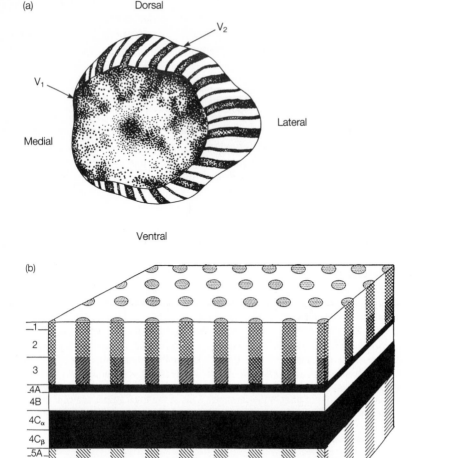

Fig. 7.11 Distribution of cytochrome oxidase-positive regions in visual areas 1 and 2 of the monkey visual cortex. (a) Sections of monkey cortex cut parallel to the surface, showing blobs in V1 and thick and thin stripes and pale interstripes in V2. (b) Schematic representation of the three-dimensional distribution of cytochrome oxidase rich cells in V1. The blobs are found in all layers except layer 4; they are most obvious in layer 3, as indicated by gradations in the degree of stippling. (From Tootell, R.B.H. *et al.* (1988). *J. Neurosci.* **8**, 1500–1530.)

relatively simple properties of individual cells combine to form a picture of increasing complexity. In addition, information concerned with different aspects of visual perception, e.g. colour, high-acuity vision, movement, stereoscopic depth and periodic pattern detection, is carried along separate parallel paths.

Separation of different components starts in the retina (e.g. with different receptors and ganglion cells), is maintained in the LGN (magnocellular and parvocellular divisions) and continued in the cerebral cortex (Fig. 7.12). The

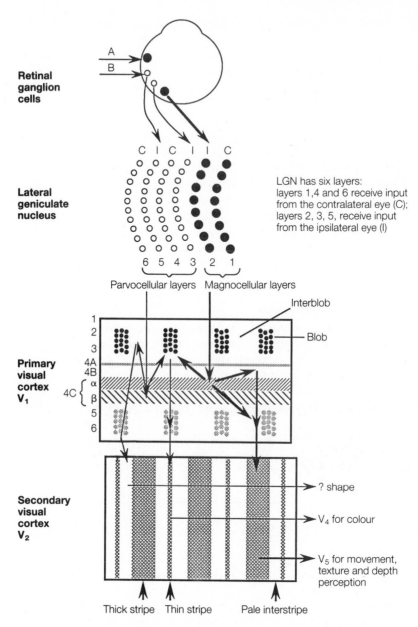

Fig. 7.12 Diagrammatic illustration of the magnocellular and parvocellular pathways from retina to visual cortex (*see* text for explanation). (Adapted from Livingstone, M.S. (1988) *Sci. Am.* **258**, 68–75.)

magnocellular pathway projects through layer 4Cα of V1 to the thick stripes of V2 (via layer 4B) and then to V5, and carries information on movement, texture and depth perception. In addition it carries information on luminance (brightness) which is conveyed to the blob region to be combined with colour in the parvocellular system. The parvocellular pathway projects to layer 4Cβ and is split into two streams. One projects via the interblob regions of layers 2 and 3 of V1 to the interstripe region of V2 and gives information on shape (form) and contour. The other goes via the blob regions of layers 2 and 3 of V1 to the thin stripes in V2 and then to V4 and gives information on colour. There are intracortical connections between the two systems and between different cortical layers.

Receptive fields

A cell is defined as being part of the visual system if it responds to visual stimuli presented to the eye. 'Receptive field' is an abstract concept, being more than just an *area* of visual space within which a stimulus will affect the cell. 'Receptive field' in this context includes the pattern of the stimulus which will alter the response of the cell: that is, the shape, colour, orientation, movement, direction and binocularity of the effective stimulus. A receptive field may be a simple ON-centre, OFF-surround (as for ganglion cells, see below). Other cells may have complex fields, such as a cell responding only to a minimum of 6 lines orientated 10–15° to the horizontal, moving from left to right and presented binocularly; and to no other stimuli!

Centre-surround receptive fields

Retinal ganglion cells and LGN neurons have centre-surround receptive fields. Many cells have action potentials even in the dark; small light or dark spots presented within the receptive area may increase or decrease the frequency of action potentials. Excitatory and inhibitory regions of the receptive field are arranged in the form of an annulus surrounding a central circle (Fig. 7.13). If a bright spot stimulates the central region (ON-centre unit), then light will inhibit the

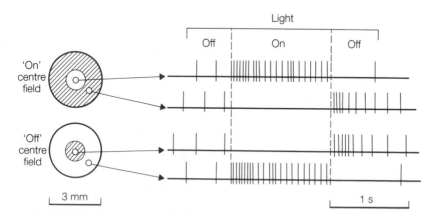

Fig. 7.13 Receptive fields of retinal ganglion cells and of neurons in the LGN. Most receptive fields at these levels have a centre-surround arrangement. Most cells, when impaled with a microelectrode, are observed to fire irregularly, even in complete darkness. ON-centre units are excited by a light spot in the central region of their receptive field and are inhibited when the light spot is moved to the surrounding annulus. When the light spot in the surrounding annulus is turned off, there is a brief period of excitation. OFF-centre units show the opposite pattern of responses. The centre and surround regions interact in an antagonistic way and so uniform illumination gives little or no response.

surround. Conversely, if a dark spot stimulates the central region (an OFF-centre unit), then dark will inhibit the surround. Units with centre-surround receptive fields give little or no response to uniform stimulation throughout their field; they respond only to *contrasts* in illumination. In consequence, signals in the optic nerves are primarily responses to *differences* in illumination within the visual fields, while constancies must be inferred by the visual centres in the brain. This allows a high degree of economy in the amount of information needed to form a visual perception. It also limits the size of a visual object within which differences in luminance can be detected.

The retina and LGN also possess colour-sensitive units, with centre-surround receptive fields to spectrally opponent colours, for example, red centre, green surround and vice versa.

Orientation sensitivity

Some neurons in layer 4 of V1 have centre-surround receptive fields, but the majority of cells in other layers respond to contours, or to contrasts in either colour or luminance which are *linear*. A contour must lie within a narrow range of orientation ($\pm 15°C$) if it is to be effective. The luminance contrast across the contour must be at least 1% to evoke a response in a neuron in a magnocellular pathway, and at least 10% for a parvocellular neuron. Receptive fields of characteristic neurons in the visual cortex are shown in Fig. 7.14.

Simple cells respond to bars or edges with a particular orientation within a fixed portion of visual space (Fig. 7.14a).

Complex cells respond to bars or edges with particular orientations and with a particular optimum width, but placed anywhere within a larger region of visual space (Fig. 7.14b).

End-stopped cells may be simple or complex. They have the particular property that they are inhibited if the contour to which they respond extends outside a certain limited region of their receptive field (Fig. 7.14c). This property means that these cells are stimulated by contours only if they end within their receptive field, or if they change in direction. End-stopped cells respond, therefore, to corners, curves and discontinuities.

Periodic pattern-sensitive cells respond to visual stimuli such as gratings, or scenes of leaves or pebbles, with a very small range of variation in their spatial frequency (Fig. 7.14d). They also have precise requirements for orientation.

Colour sensitivity

Colour-sensitive cells are not sensitive to orientation. Single opponent cells, as seen at the level of retinal B ganglion cell or LGN, are excited by one colour in their centre and by the opponent colour in their surround (e.g. red centre/green surround or blue centre/yellow surround). Double-opponent cells, first seen in the blob regions of layers 2 and 3 of V1, are excited by one colour and inhibited by its opponent in their receptive field centres, and inhibited by the colour and excited by its opponent in their surrounds. Some blob cells, of magnocellular origin, are not colour-sensitive but respond to luminance anywhere within their receptive field; they provide information necessary for perception of colour brightness.

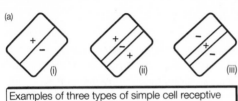

(a)

(i)   (ii)   (iii)

Examples of three types of simple cell receptive fields. + on zone, − off zone

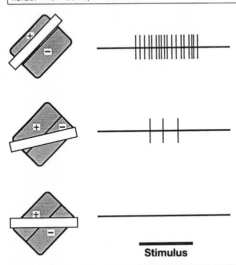

Stimulus

Simple cell (type i above): a narrow slit of light evokes a maximal response only when it is placed at a particular orientation within a fixed portion of the receptive field (rectangle)

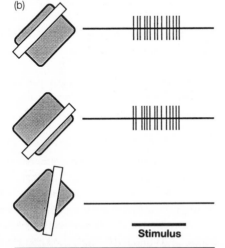

(b)

Stimulus

Complex cell: a narrow slit of light evokes a response wherever it is placed within the receptive field (rectangle) of a complex cell, provided the orientation is correct

Fig. 7.14 Receptive fields of neurons in the visual cortex. (a) Simple cell responses (p. 178) to a bar of light. (b) Complex cell responses. (c) Comparison of complex and end-stopped complex cell responses (p. 178). (d) Example of responses from one of the most selective period pattern-sensitive cells in monkey visual cortex to small changes in spatial frequency and orientation of gratings of light (p. 179). The optimum patterns (centre) and patterns (on either side) giving half-maximal responses are shown above the curves. (From von der Heydt, R., Peterhans, E. & Dürsteler, M.R. (1992) *J. Neurosci.* **12**, 1416–1434.)

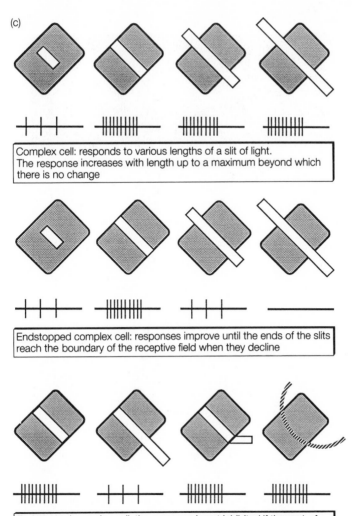

Complex cell: responds to various lengths of a slit of light. The response increases with length up to a maximum beyond which there is no change

Endstopped complex cell: responses improve until the ends of the slits reach the boundary of the receptive field when they decline

Endstopped complex cell: the response is not inhibited if the part of the stimulus extending from the excitatory zone has a different orientation. This means that endstopped cells can detect curvature

Fig. 7.14 *Continued.*

Disparity

Disparity-sensitive neurons, which include about 70% of neurons in the magnocellular layers of V1, give their optimum response to stimuli presented at slightly disparate points on the two retinas, and are thought to have a role in stereoscopic perception of depth and distance. *Tuned excitatory* neurons are disparity-sensitive over a narrow range, and account for over 50% of all disparity-sensitive neurons. These neurons fire at a maximal rate when the stimulus projects an image on two nearly corresponding points on the two retinas (usually within 6′ arc), and the response falls off rapidly for any divergence from this optimum. *Tuned inhibitory* neurons have binocular responses which are suppressed over the same narrow

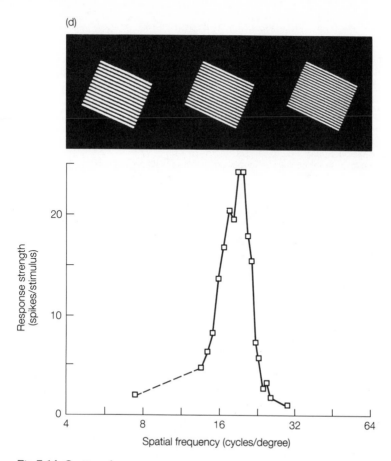

Fig.7.14 *Continued.*

range of disparities. *Reciprocal* neurons have reciprocal selectivity for stimuli closer to or further from the fixation point. *Far* neurons are excited by disparate points past the fixation point and inhibited by those nearer, whereas *near* neurons behave in the opposite fashion. *Flat* neurons are not disparity-selective, and give similar responses to stimuli at varying distances within the disparity range of binocular interaction.

Columnar arrangement of visual cortex

Neurons in the visual cortex are arranged in radial columns, in which all cells have closely related functions.

Ocular dominance columns

The left half of each visual field projects to the right half of each retina, divided down the vertical meridian. This in turn projects to V1 in the right hemisphere. The opposite is true for the right half of each visual field, which projects to the left hemisphere (Fig. 7.9). Corresponding regions of each retina project to closely

(a)   (b)

Dorsal

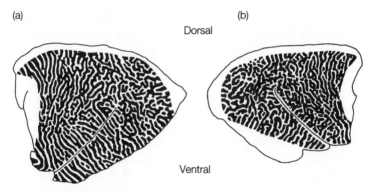

Ventral

Fig. 7.15 Ocular dominance columns in the striate cortex of a monkey as revealed by autoradiography of sections tangential to the exposed surface of the left (a) and right (b) occipital lobes. The dark stripes are metabolically active areas, as revealed by the radioactive 2-deoxyglucose technique, and correspond to one eye while the other eye is masked. Note the differences between hemispheres. (From Tootell, R.B.H. *et al.* (1988). *J. Neurosci.* **8**, 1500–1530.)

adjacent neurons in appropriate regions of V1, and these neurons are arranged in ocular dominance columns (Fig. 7.15), that is, cells in one column respond preferentially to inputs from one eye. A microelectrode passing radially through the cortex might successively record signals from 50 different neurons, but all these neurons would have a dominant input from one eye. If the microelectrode was moved sideways by a millimetre, and the recording repeated, it might now register a series of neurons dominated by the other eye, all with receptive fields in the same small region of visual space. Columns with the same ocular dominance are arranged in short lines or stripes approximately 400–500 μm wide (Fig. 7.15). Each individual, and each hemisphere, has a different pattern of ocular dominance stripes showing that these form dynamically during early childhood development rather than being rigidly genetically determined.

Orientation columns

Neurons within a cortical orientation column (approximately 50–100 μm wide) share the same orientation preference, although they may be simple, complex or end-stopped. If the recording electrode is moved sideways a few tens of microns, the orientation preference would change by 15–30°, in a progressive and orderly manner, coming full circle within a millimetre (Fig. 7.16). Both ocular dominance and orientation columns are arranged in orderly rows, but these are not congruent with one another, so that if you recorded along a row of cells with a common orientation preference, ocular dominance would shift alternatively from one eye to the next, and receptive fields would appear in adjacent regions of visual space.

Cytochrome oxidase blobs

Cells in the blobs (Fig. 7.11) form columns concerned with processing of colour;

orientation columns lie in the interblob regions. Both blob and interblob cells form part of ocular dominance stripes.

Hypercolumns

The orderly arrangement of cortical columns into orientation or ocular dominance stripes or colour-sensitive blobs has the consequence that a small region of cortex, a square millimetre or so in area, will receive inputs from all possible receptive fields within a small region of visual space. This region is a hypercolumn, responsible for the primary processing of visual information from this space (Fig. 7.16). A hypercolumn contains cells with all possible orientation preferences from both eyes, and colour-sensitive neurons, receiving projections from a restricted small region of visual space. An adjacent part on the retina is represented within an adjacent block of cortex. Thus the whole visual field is analysed in an orderly topographic arrangement.

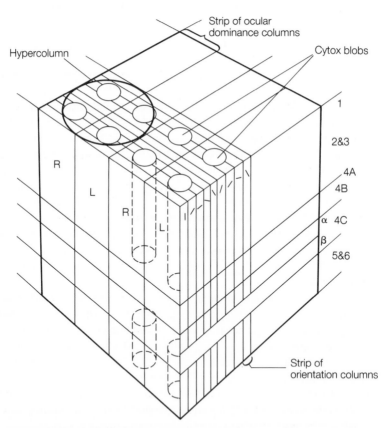

Fig. 7.16 Modular arrangement of cortex in visual area 1. A small region of cortex, constituting a hypercolumn, contains columns of cells of all possible orientation preferences, with left or right eye dominance, and cytochrome oxidase blobs containing cells devoted to analysis of colour. (From Livingston, M.S. & Hubel D.H. (1984) *J. Neurosci.* **4**, 309–356.)

Recent evidence indicates that this columnar organization is a feature of higher visual areas as well as V1.

High-acuity vision

High-acuity analysis of contour and form is a function of the parvocellular system. Acuity is not limited by optics, but by the resolving power of the retina. In the fovea the limit to visual acuity is set by the distance between two cones—about 2.5 μm or 0.5' (~0.01°) of arc. Away from the fovea, receptive fields get larger, involving groups of cones, and acuity is correspondingly lower.

For routine clinical purposes, one measures acuity using a *Snellen chart* in which letters of the alphabet are arranged in lines and the size of the letters in each line is such that the details of the letters each subtend 1' of arc at defined distances from the eye. The test is usually conducted at 6 m from the chart to exclude the effects of accommodation, and the line of smallest letters which the subject can correctly read is determined. Visual acuity is defined as the ratio of the distance of the individual from the chart to the distance at which the details of the correctly read line subtend 1' of arc. For a normal individual this is 6/6. A myopic person could detect only larger letters, for example those subtending an angle of 1' at a distance of 3 m, giving a measure of acuity of 6/3.

Stereoscopic vision

With binocular vision, an image falls on corresponding but slightly disparate points on the two retinas. Images at different distances from the observer produce different horizontal disparities. Stereoscopy involves the use of this horizontal retinal disparity to give rise to three-dimensional perception. With knowledge of disparity at the two retinas, an estimate of distance can be made from the geometry of the situation. Stereopsis is a basic function of vision, and does not require recognition of objects. Other cues for depth perception include relative size of objects, occlusion/superposition, parallax, perspective, shadow casting (e.g. to distinguish pits and bumps), texture gradients, rotation and the relative movement of parts of an object, and awareness of the need to accommodate.

Stereoacuity

The threshold for depth perception is a few seconds of arc difference in projection between the two retinas, which is less than the size of individual photoreceptors. Any defocusing (blurring) of the image reduces stereoacuity much more than visual acuity.

Perception of shape

A sphere, viewed from any angle, appears as a circle. It is because a sphere is normally unevenly illuminated that aspects presented to the two eyes are different; the fusion of these two images places the light and dark shades in their true relationships in three dimensions and so creates the appearance of solidity.

Colour vision

'Light' is defined as the spectrum of electromagnetic energy that we can see (wavelengths 400–700 nm). Light with a uniform mixture of wavelengths is seen as white (we may also see a mixture of two monochromatic lights as white). A pigment absorbs some light and reflects the rest and the pigment appears the colour of the reflected light. Green leaves, for example, absorb long- and short-wavelength light, and reflect the middle wavelengths.

Perception of colour adds an extra dimension to visual perception; in animals it helps in camouflage-breaking and in the recognition of food. Humans can discriminate about 100 variations of wavelength and 30–80 variations of spectral purity of each wavelength, even when stimuli have been equated for luminance. In other words, several thousand colour combinations can be discriminated at a constant level of luminance, compared with about 30 discriminable variations in intensity contrast. We are colour-blind in dim light (scotopic vision).

Visual receptors

We have three types of cones, with peak absorptions at 430 ('blue', actually violet) 530 ('green', actually green) and 560 nm('red', actually yellow-red; Fig. 7.5). The rod pigment, rhodopsin, reflects blue and red and therefore looks purple, with peak absorption in the green. The three cone types have broad sensitivity curves with much overlap and the sensation of colour depends on the extent to which each is excited. Three types of cones is the minimum number possible such that no monochromatic light, at any wavelength, looks white. We perceive illumination as white when all types of cone are stimulated equally.

Adding white to any spectral colour *desaturates* it. To match two colours, both hues and saturation must be made the same, and the intensities matched.

Inherited variation in human colour vision

Individuals with normal colour vision can match any spectral colour by using a mixture of the three primary colours, and are known as **trichromats**. About 8% of males and 0.5% of females differ from the majority in the proportions of the three primary colours needed to match a given colour. These fall into two classes: **dichromats** and **anomalous trichromats**. Dichromats have lost one of the three systems and therefore require only two primary colours to match any spectral colour. Anomalous trichromats have a reduction in the spectral sensitivity of one of the systems. Dichromats and anomalous trichromats are then divided into three types depending on whether the blue, green or red cone systems are defective. Loss or reduction in the red or green systems is most common and causes difficulty in distinguishing between these two colours; defects in the blue system are rare.

Colour constancy

Colour is a perception, not a measure of the wavelength of light, and the trichromat theory refers to the receptors for colour vision, not to processes in the visual cortex. The perceived 'colour' of an object is changed little when the colour or intensity of illumination is changed, even though a physical instrument would detect a change in the wavelength composition of the light reflected from the object. Conversely, in a monochrome environment, for example an experimental room with uniform illumination at some particular wavelength, it is not possible to give an accurate description of its colour; in fact, it is likely to be described as white or light grey regardless of its illumination. Colour constancy appears to derive by comparing the relative activities of red, green and blue cones stimulated by light emitted by the object of regard with the wavelength composition of light from the surrounding visual field. Land's retinex theory of colour vision proposes that the brain constructs lightness records of reflectance in short, middle and long wavebands, and by comparing these derives the perception of colour. This property of colour vision is essential for object recognition.

Visual perception of movement

Perception of moving objects depends on specific processing within the visual system, and is the function of the magnocellular system. This system has about half the spatial resolution of the parvocellular system at a given eccentricity, and is not sensitive to colour. The threshold for movement detection in humans depends on factors such as the angular velocity, the size, contrast and position of the moving stimulus, and on the duration of the movement.

The 'movement centre'

In the V5 cortical region of the monkey, most cells respond to a movement in a certain direction, and preferred directions change systematically through the cortex. Other cells respond to rotating stimuli in a clockwise or anticlockwise direction. Complex movement-sensitive neurons may have very large receptive fields, responding to the same preferred direction and velocity anywhere within the visual field. Lesions in V5 lead to a specific loss of perception of movement, without other defects in vision. One patient, for example, said she was unable to pour a cup of tea as she could not see the moving fluid, and while she could walk between parked cars to cross the road, she could not see moving vehicles.

Perception of texture

A class of neurons in V1, known as periodic-pattern detectors, respond specifically to patterns of fine lines. They are specific in their requirements, responding to spatial frequencies within a very narrow range of variation, and they also are orientation-specific. Perception of very detailed patterns, such as leaves of trees, blades of grass or areas of sand or pebbles would require transmission of enormous amounts of data, and the function of texture-sensitive neurons is thought to be the summarization of large amounts of information into a single channel.

Development of vision

Ocular dominance columns and stereoacuity

Ocular dominance columns are not present during initial formation of projections from LGN to V1, but appear in a dynamic fashion near the time of birth (in monkeys, and presumably human infants). Their formation and maintenance require synchronous use of both eyes, and they are fully formed in children by about 5 months of age. Similarly, depth perception is present in children by 4 months of age, and appears to be fully developed, to adult levels, by 5 months of age. This may reflect the segregation of ocular dominance columns. By comparison, visual acuity improves rapidly during the first 6 months, but continues to improve for several years.

Critical periods in development

There is a period, beginning at birth, when deprivation of form vision has profound and permanent effects of visual function. In monkeys, this sensitivity reaches a peak at 2 weeks of age, and then tapers off over a period of about a year. Even a few days of deprivation at 2 weeks of age may have permanent effects. Children born with congenital cataracts which are not immediately corrected have profoundly defective vision.

In experimental animals (cats or monkeys) raised with one eye closed, lateral geniculate projections from the deprived eye form very few synaptic connections in V1. The damage is caused by form deprivation, not light deprivation, and the animal is behaviourally blind in the deprived eye. After a week or more of eye closure (at the peak critical period), opening the closed eye does not give rise to any recovery. If, at the same time, the previously open eye is closed and the monkey is still in its critical period, recovery does occur (but projections from the newly closed eye are suppressed).

Similarly, disturbances of binocular vision can occur, for example, with a squint (**amblyopia**) and have profound effects on the development of disparity-sensitive cells and hence on depth perception. Commonly in this situation, one eye becomes dominant and develops excess cortical connections while inputs from the other 'lazy' eye are suppressed.

In general the development of the visual system is tuned to visual experience. Animals raised in an environment with vertical black and white stripes, and deprived of other forms of visual experience, show a dramatic decline in numbers of cells responding to other orientations. Animals brought up in an environment illuminated by flashes of a strobe light lack movement-sensitive neurons.

Focusing defects and eyeball size

Myopia and hypermetropia reflect anatomical defects in the size of the eye, the posterior chamber being too long or too short, respectively (p. 162). At birth there is a large scatter between individuals in their ability to achieve accurate focus, with rapid growth towards emmetropia (normal eye length). This requires great accuracy in the anatomy of the eye, since depth of focus at the retina for a 5-mm pupil is only a few microns. Studies of the statistical distribution of eye focus within large groups of people show that the range is much narrower than would be expected, providing evidence for feedback control of growth of the eye.

There is much anecdotal evidence for eye growth being affected by environmental circumstances as well as the genetic background of the individual, for example the Bedouin or American Indian with remarkable distance vision, or the extremely myopic Japanese or Talmudic scholar. There is also epidemiological evidence, such as a sudden increase in incidence of myopia in Eskimos with the coming of television, or the current 80% incidence of myopia in Taiwanese high-school pupils.

Animal experiments have demonstrated the existence of two mechanisms of developmental regulation of eye growth. One is local with a blurred image on a region of retina resulting in excessive growth in the adjacent region of the sclera. The other depends on visual feedback or a signal from brain to eye. In human eye development, both local (defocusing) and brain-dependent (over-accommodation) signals may be important.

7.2 TASTE

The sense of taste (gustation) is important in the selection and enjoyment of food, but full appreciation requires olfactory, tactile and visual inputs as well. Taste cells occur in taste buds on the tongue and oral mucosa. They are activated by a variety of chemicals such as sugars, acids, salts and organic compounds such as quinine. Such activation provides the basis for the sensations of sweet, sour, salty and bitter.

Gustatory receptors

Taste cells are modified epithelial cells, which are organized together with support-ing cells into **taste buds**. Most taste buds are located on nipple-like protuberances, called **papillae**, on the tongue surface (Fig. 7.17), with a few on the epiglottis, soft palate and pharynx. They are 50–70 μm in diameter and open on to the surface of the tongue at the taste pore (Fig. 7.18). In humans, there are several thousand taste buds, although the number has been shown to be highly variable with no significant age or gender differences. Most taste buds are located on the sides of the *vallate* (circumvallate) papillae, which are found at the back of the tongue (Fig. 7.17). Smaller *fungiform* and *foliate* papillae, on the sides and tip of the tongue, may also carry a few taste buds. Taste buds are not found on the numerous *filiform* papillae.

Taste buds contain several cell types (Fig. 7.18). Besides the sensory taste

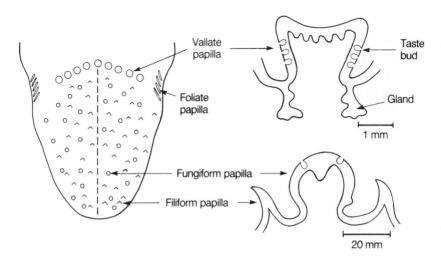

Fig. 7.17 Distribution of taste buds on the tongue and appearances of different papillae.

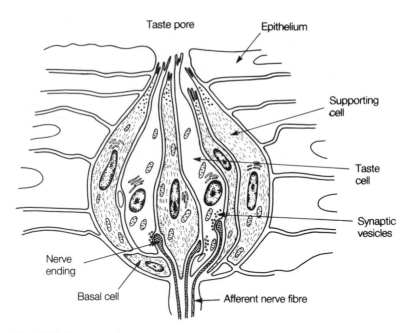

Taste pore

Epithelium

Supporting cell

Taste cell

Synaptic vesicles

Nerve ending

Basal cell

Afferent nerve fibre

Fig. 7.18 Structure of the taste bud.

cells, there are basal cells and supporting cells. Basal cells give rise to taste cells; supporting cells can be divided into type 1 (dark) cells and type 2 (light) cells. While some consider type 1 and type 2 cells to be supporting cells, others suggest that they represent different stages in the cycle of the taste cell. Taste cells are constantly shed, have a half-life of 10–14 days, and are renewed by division of the underlying basal cells. Taste cells make synaptic contact with the afferent nerve fibres (Fig.7.18) upon which they depend on their continual existence; denervated taste buds degenerate and new buds reappear when the nerve regenerates.

Classically, four basic tastes are recognized: **sweet, sour, salty** and **bitter**. A fifth taste, *umami*, produced by compounds like monosodium glutamate, has also been described. All complex tastes are thought to be accounted for by combinations of these basic tastes; most of what the layperson calls 'taste' is a combination of taste and smell.

Most sugars taste sweet, for example, glucose, sucrose, lactose produce the sensation of sweetness, but so too do unrelated molecules like glycine, alanine, saccharine and even certain proteins. The sensation of sourness is produced by acids and is related to the H^+ concentration, but not all acids are equally sour at equivalent pH. Salty sensations are elicited mainly by cations such as Na^+ and Li^+, but different anions (Cl^-, SO_4^{2-}, NO_3^-) alter the taste quality. Many different organic compounds are described as bitter, e.g. quinine, caffeine, nicotine, morphine and strychnine.

Taste cells are chemoreceptors. Chemicals enter the pores of the taste buds and react with the microvilli at the apical surface of the receptor cells. The transduction process is poorly understood but there is evidence for different mechanisms for different chemicals. For example, Na^+ salts depolarize taste cells by Na^+ influx through Na^+, channels in the apical membrane, some bitter compounds may cause depolarization by direct interaction with ion channels, while sugars may interact with specific receptors on the microvilli and gate ion channels via cyclic nucleotides or inositol trisphosphate/Ca^{2+} (p. 35). There is also evidence that ligand-gated channels mediate transduction for some amino acids. Depolarization of taste cells leads to action potentials and the release of neurotransmitter (possibly serotonin or vasoactive intestinal peptide). The transmitter elicits generator potentials and hence action potentials in the afferent nerve fibres (p. 92).

Afferent nerves and central pathways in gustation

Taste buds are innervated by the chorda tympani, lingual and vagal nerves. These afferent fibres synapse in the medulla in the nucleus of the solitary tract before projecting to the thalamus and cortex. Most central taste cells respond to a variety of chemical stimuli.

Each taste bud receives branches from many afferent fibres. Taste buds on the anterior two-thirds of the tongue are supplied by the chorda tympani branch of the facial nerve (VII), those on the posterior one-third by the lingual branch of the glossopharyngeal nerve (IX). The vagus nerve (X) supplies taste buds on the epiglottis. Recordings from single afferent fibres in the chorda tympani show that over half respond selectively to either salt (NaCl) or sucrose, while the remainder show a broad sensitivity to combinations of salt, sweet, acid and bitter stimuli. In the glossopharyngeal nerve, approximately 30% of the fibres are selective for bitter stimuli.

Taste fibres from the chorda tympani, glossopharyngeal and vagal nerves travel to the medulla and synapse in the nucleus of the solitary tract (tractus solitarius). Here some neurons show selective sensitivity, for example, to salt or sugar. Others, however, show a broader sensitivity, responding, for example, to salt and acid, or to salt, acid and bitter stimuli. There is also convergence here between gustatory and thermal inputs, many cells responding to cooling of the tongue as well as to chemical stimuli. The second-order fibres project to the most medial part of the ventroposterior medial nucleus of the thalamus. Thalamic neurons then project to a small area of the insular cortex and operculum near the somatosensory area for the tongue. There is also evidence for a secondary area in the orbitofrontal cortex. Gustatory information is also conveyed to other areas such as the hypothalamus and the limbic system.

7.3 SMELL

In many animals the sense of smell (olfaction) is important for the selection of food, in sexual activities and in the recognition of other animals and of territories. Olfactory cells lie in a specialized region in the roof of the nasal cavity. They are modified neurons with a superficial receptor area and a fine axon passing to the olfactory bulb. Odours combine with these cells to produce depolarization and impulse activity.

Olfactory receptors

Olfactory cells lie in the olfactory epithelium, which in humans covers an area of about 5 cm^2 in the roof of the nasal cavity. The olfactory cells are specialized bipolar neurons (Fig. 7.19). At the apical region of the cell is a modified dendrite which ends on the surface in a knob bearing 15–30 long, non-motile cilia, embedded in a layer of mucus. The ciliary plasma membrane is thought to contain the odorant receptors. The basal region gives rise to a fine (0.1–0.4 µm diameter) unmyelinated axon. The epithelium also contains supporting cells, which have apical microvilli and secrete mucus, and basal cells which lie on the basement membrane. Recently, a fourth type of cell of unknown function, the microvillar cell, has been described. The olfactory cells are the only nerve cells in adult mammals to show continual degeneration and replacement; they are replaced by division and differentiation of the basal cells.

Several thousand chemicals can be smelled. Odoriferous compounds are mainly organic, containing 3–20 carbon atoms, but there is no simple relationship between chemical structure and odour. Threshold concentrations vary, being extremely low for certain substances; many odorants can be detected at picomolar $(10^{-12}$ mol L$^{-1})$ concentrations. Odoriferous compounds reach the olfactory epithelium by diffusion and this can be aided by sniffing to increase air flow. The molecules must dissolve in the mucus layer before they can interact with receptors on the cilia of the sensory neurons. An odorant-binding protein may bind the odorant and carry it to the receptor.

Recent evidence indicates the existence of hundreds of different membrane receptors. Binding of the odorant complex to the receptor leads, probably via a guanosine triphosphate-binding protein (G_{olf}) and cyclic adenosine monophosphate (cAMP, p. 37), to an increased conductance to Na$^+$ and hence to a depolarizing generator potential. In this instance cAMP directly gates a cation-selective channel. Although cAMP appears to be the major second messenger in olfactory transduction, there is increasing evidence for the involvement of phosphoinositide-derived second messengers.

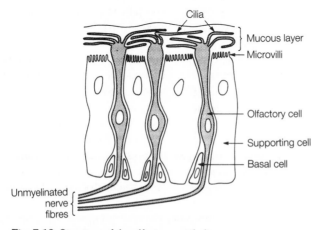

Fig. 7.19 Structure of the olfactory epithelium.

Recordings from single neurons have shown that depolarization and hence impulse activity increase with increasing concentration of odorant, and that single neurons respond to many odours. Olfactory neurons have very high specific membrane resistances and consequently the opening of a single channel can initiate an action potential. Nevertheless, it appears that a large number of neurons respond to a particular odorant generating a diffuse pattern of neuronal firing. The summated generator potentials of the olfactory neurons can be recorded as an **electro-olfactorogram**.

The sense of smell shows marked adaptation and is subject to masking of one smell by another—the basis of air fresheners. Olfactory acuity has been shown to be best in the third to fifth decades of life and is also better in women than in men.

Afferent nerves and central pathways in olfaction

Olfactory axons terminate in synaptic regions of the olfactory bulb, termed glomeruli. From there second-order cells project to the olfactory cortex and to other regions, including the thalamus and limbic system.

The unmyelinated axons of the receptor cells form the first cranial nerve; the axons pass in bundles (olfactory fila) through the cribriform plate and enter the olfactory bulb where they terminate in spherical glomeruli ($100–200$ μm in diameter). Glomeruli are regions of synaptic contact between the afferent axons and the dendrites of mitral and tufted cells. There is marked convergence in the glomeruli as $10^4–10^5$ primary axons synapse with about 200 secondary cells. The bulb also contains many interneurons, such as periglomerular and granule cells involved in the complex processing of the sensory input. Mitral cells project via the olfactory tract and lateral olfactory stria to the primary olfactory, prepiriform and peri-amygdaloid cortices. Fibres in the medial olfactory stria travel, via the anterior commissure, to the contralateral bulb. Tufted cells probably project to areas such as the olfactory tubercle. Other areas involved in olfaction include the mediodorsal nucleus of the thalamus, the orbitofrontal neocortex and various regions of the limbic system, such as the septum, amygdala and hypothalamus. Little is known about the role of these centres in olfactory sensations.

7.4 HEARING

The auditory system has evolved as a survival system for warning and hunting; it also functions as the afferent part of spoken communication. To carry out these functions of recognition, location and communication, it must detect sounds of particular interest, such as human speech, and ignore irrelevant background sounds. Each ear accepts sound energy from a limited range of directions and then performs a frequency analysis on it so that the auditory nerve carries information on the intensities of the frequency components of a sound. Sound vibrations are transduced into nerve action potentials by the hair cells in the cochlea. The brain has centres specialized for analysing the relative intensity and phase of individual sounds so as to compute their direction of origin. It also has higher centres dedicated to recognizing the source and nature of sounds and the content of speech.

Aspects of a sound stimulus

The sensation of sound detected by the ear is caused by variation in air pressure within a specific range of frequencies and intensities. Outside this range the sensations evoked are variously described as vibration, flutter, tickle and pain.

A source of sound such as a tuning fork causes the surrounding air molecules to oscillate. The disturbance spreads out from the source in sinusoidal waves which consist of areas of compression of air (high pressure) alternating with areas of rarefaction (low pressure). Sound can thus be described, like all waves, in terms of both frequency and amplitude (pressure difference). The **loudness** of the sound is related to the amplitude of the sound wave and the **pitch** to the frequency.

Frequency

A normal young person can hear sounds ranging in frequency from 20 to 16 000 Hz. Higher frequencies are called ultrasound and lower frequencies infrasound. Frequency discrimination by the human ear is best at about 1 kHz, where changes of as little as 3 Hz can be detected.

Sound of only a single frequency, a **pure tone**, is uncommon in daily experience. Most sounds are composed of numerous, simultaneously occurring frequencies. Fourier (1768–1830) first suggested that all complex waveforms could be analysed by breaking them down into simpler constituent waves of single frequencies. When subjected to Fourier analysis, musical sounds are found to be composed of a fundamental frequency plus a number of harmonically related frequencies, i.e. the higher ones (the harmonics of upper partials) are integral multiples of the lowest (the fundamental). The number and intensity of each determine the tone or timbre of the sound, whilst the fundamental determines the pitch. Thus the range and strength of the violin's upper partials are greater than for the tuba and give the violin a brighter tone.

Amplitude

This is the objective aspect of a sound stimulus, measurable with instruments, whilst loudness is the subjective, conscious aspect, which is harder to quantify. The amplitude of sound waves is measured using the decibel scale (the unit of sound amplitude, the **bel**, is named after Alexander Graham Bell). This is a logarithmic scale defined as sound pressure level in decibels.

Decibel (dB) = $20 \log_{10} P_t/P_r$

where P_t is the test pressure and P_r a reference pressure of 2×10^{-5} N m^{-2}, just loud enough in the range 1–3 kHz to be audible to an average listener. Alternatively, the decibel may be expressed in terms of intensity (W m^{-2}), in which case it equals 10 times the logarithm of the ratio of the intensity of the test sound to the intensity of the reference sound.

The threshold of hearing varies with the stimulus frequency and is least over the range 1–3 kHz (Fig. 7.20). The resulting displacement of the tympanum is about the diameter of a hydrogen atom. The threshold rises to 60 dB (i.e. 10^3 times this value) at 50 Hz and 30 dB at 10 kHz. Threshold for pain is about 120 dB, although sounds above 100 dB may damage the organ of Corti in the cochlea. Loudness discrimination is about 1 dB under the most favourable experimental

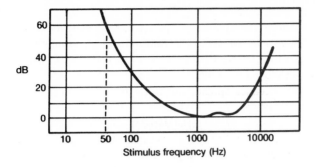

Fig. 7.20 Threshold of audibility curve.

conditions (but more usually 2–3 dB), so that at 1–3 kHz there are no more than about 120 steps of loudness from the threshold of hearing to the threshold of pain.

Duration

To establish a pitch requires a minimum duration of 15–20 ms, which, surprisingly enough, hardly varies with frequency.

Direction of sound source

At lower frequencies the direction of origin of a sound can be determined only by sensing the difference in time of arrival of the wavefront, whereas for higher frequencies the shielding effect of the head and pinnae of the ears results in differences in amplitude of the sound between the two ears. As a result, discrimination is best at higher frequencies where both these mechanisms can be employed. The direction of origin of a pure tone of frequency less than 1 kHZ is difficult to determine; real sounds generally contain harmonics at frequencies greater than this, allowing their position to be sensed.

Functional anatomy of the ear

The ear is divided into outer, middle and inner ear compartments (Fig. 7.21).

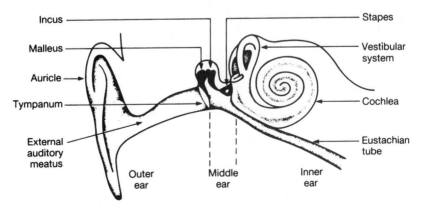

Fig. 7.21 Diagram of the outer, middle and inner ear compartments.

Outer ear

The **external auditory meatus**, about 27 mm long, conducts air pressure variations from the **auricle** (pinna or ear flap) to the **tympanum** (eardrum). The air column has a resonant frequency of about 3–4 kHz and conducts energy to the eardrum most effectively in this range.

Middle ear

The outer ear and the middle ear are separated by the tympanum (Fig. 7.21). The cavity of the middle ear is air-filled and is connected to the pharynx by a narrow passage called the **Eustachian tube**, which allows the pressure of the air in the middle ear to equilibrate with the outside pressure. Vibrations of the tympanum are conveyed through the middle ear to the **oval window**, which sits between the middle ear and the fluid-filled inner ear, via a series of small bones called **ossicles**. These are the **malleus** (hammer), which is attached to the tympanum, the **incus** (anvil) and the **stapes** (stirrup), which is attached to the oval window. The ossicle chain, together with the smaller area of the oval window relative to the tympanum, operates to promote the efficient transfer of energy from air to the liquid environment of the inner ear, i.e. it acts as an impedance-matching device. The resonant frequencies of the middle-ear mechanism are around 800–1200 Hz and, with the broadly tuned resonance of the external auditory meatus, are mostly responsible for the low threshold of hearing in this frequency range.

The effectiveness of the ossicle chain can be modified by the activity of the middle-ear muscles. The **tensor tympani muscle** pulls the tympanum inwards and this results in the stapes being pushed into the oval windows. The **stapedius muscle** pulls the stapes out of the oval window and pushes the tympanum outwards. When they contract together, they stiffen the middle-ear mechanism to reduce the transfer of energy into the inner ear. They do so in response to high-intensity sounds, with a latency of about 10 ms, but soon relax again during constant stimulation.

Inner ear

This has a bony coiled tube of two and a half turns, the **cochlea**, divided by membranes along its length into three canals or scalae, called the **scala vestibuli**, the **scala media** or **cochlear duct** and the **scala tympani** (Fig. 7.22). The scala

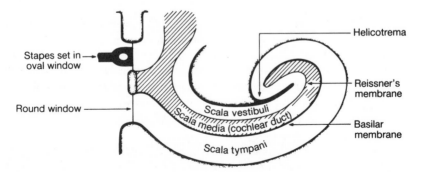

Fig. 7.22 Diagram of the inner ear. (The coiling of the cochlea is reduced for simplicity.)

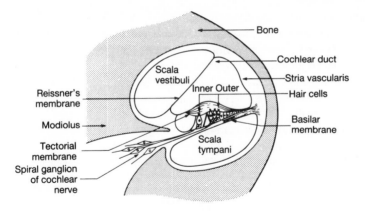

Fig. 7.23 A cross-section through the three canals of the cochlea.

vestibuli and scala tympani are filled with **perilymph** and join at the **helicotrema**; the cochlear duct is filled with **endolymph**. (Perilymph is very close to extracellular fluid in composition while endolymph, which is continuously secreted by a plexus of blood vessels called the **stria vascularis**, is very close to intracellular fluid in composition.) The cochlear duct is separated from the upper scala vestibuli by **Reissner's membrane** and from the lower scala tympani by the **basilar membrane**. When the stapes is pushed into the oval window, a wave is conducted along the scala vestibuli and through the flexible Reissner's membrane to the cochlear duct. The basilar membrane is depressed and the round window bulges out into the middle ear where the sound energy is dissipated into the air.

The sound receptors of the ear are found in the **organ of Corti** which lies on the basilar membrane within the cochlear duct (Fig. 7.23). It is composed of an epithelium of **hair cells** (Corti cells) and supporting cells. The hair cells are arranged into inner and outer groups. Each hair cell is anchored on the basilar membrane and has a bundle of hairs (**stereocilia**, but no kinocilium; Fig. 4.10) projecting from its tip and embedded in the shelf-like **tectorial membrane**. The hair cells make contact at their basal surface with afferent cochlear fibres, which pass into the bony **modiolus** where their cell bodies form the **spiral ganglion**. About 90% of the spiral ganglion cells in each ear innervate inner hair cells, about 10 axons converging on each hair cell; the remainder diverge to innervate many outer hair cells.

In addition to afferent cochlear fibres which run with the vestibular fibres in the auditory nerve (cranial nerve VIII), there is a small tract of efferent fibres ending on or close to the Corti hair cells, the olivocochlear bundle, which has an inhibitory action, reducing the afferent discharge at a given stimulus intensity. Furthermore, outer hair cells may be induced to vibrate and so actively dampen local regions of the basilar membrane.

Stimulation of Corti hair cells

The basilar membrane is a fibrous structure running from the inner core or modiolus to the outer side of the bony tube (Fig. 7.23). When the basilar membrane moves up and down, it tends to move rather rigidly, bending about an axis near the modiolus (Fig. 7.24). The tectorial membrane is pushed up and pulled down and there is a shearing motion between the tectorial membrane and the reticular lamina, bending the stereocilia to one side and the other. Bending of the stereocilia triggers the opening of mechanically gated ion channels in the hair-cell membrane. This leads to a depolarization of the hair cell, causing the release of a chemical transmitter from the cell's basal surface which in turn initiates action potentials in the cochlear fibres.

Hair cells are amazingly sensitive and, as already mentioned, can detect mechanical displacements as small as the diameter of a hydrogen atom. The tips of the stereocilia are connected by thin threads (Fig. 4.10), which are physically linked to ion channels. This arrangement permits very small movements to be transduced into changes in membrane potential.

In order to get a fuller picture it is necessary to consider not only radial movements of the basilar membrane, but also how movements along its length from base to apex are determined by the frequency of the sound. Sound of a particular frequency causes a wave of displacement to travel along the basilar membrane. The position and properties of the basilar membrane, which at the window end is narrower and stiffer than at the apex or helicotrema, are such that the region nearest the window resonates best with high-frequency sound while more distant regions resonate best with low-frequency sound. As the membrane resonates, energy is transferred across it and is dissipated by being conducted back to the round window, so if sound containing a mixture of frequencies is passed through the oval window, its components are progressively removed with distance along the basilar membrane, in order from high to low frequency. Figure 7.25a shows the displacement caused by a 200-Hz wave instantaneously (solid line) and the envelope of all displacements over a period of many cycles (dashed line). It can be seen that there is a region of maximum displacement over a period of many cycles (dashed line) and that there is a region of maximum displacement quite far from the oval window. Although all the basilar membrane from the oval window

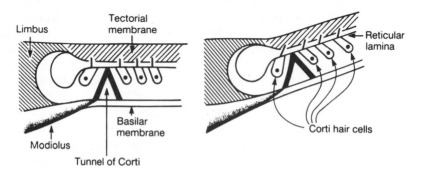

Fig. 7.24 Movement between the tectorial membrane and the reticular lamina causing the hairs of the Corti cells to bend. (From Davis, H. (1957) In: Bullock, T.H. (ed.) *Physiological Triggers and Discontinuous Rate Processes*, pp. 10–71. American Physiological Society, Washington.)

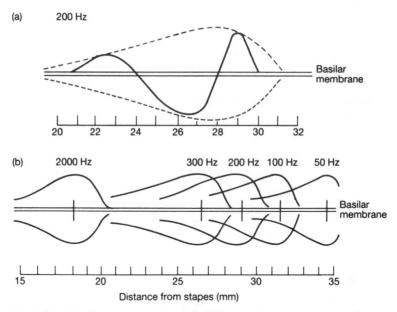

Fig. 7.25 (a) Deflections at one instant (solid line) and an envelope of all deflections over a period (dashed line) caused in a basilar membrane by a 200-Hz sound wave. (b) Envelopes of basilar membrane deflection for various frequencies of sound stimuli. (Adapted from von Bekesy, G. (1974) *J. Acoust. Soc. Am.* **19**, 452–460.)

up to this point moves to a lesser degree, the amplitude falls off quickly beyond the maximum due to damping. Figure 7.25b shows the envelopes of deflection for various frequencies of sound stimuli; the peak displacement position of the basilar membrane is closer to the stapes the higher the frequency.

From the above discussion, it can be seen that the hair cells are mechanically stimulated at particular frequencies depending on their position in the basilar membrane. Furthermore, individual hair cells vary in the length and stiffness of their stereocilia; those with short stiff cilia are mechanically more sensitive to high-frequency stimuli than those with longer and more flexible stereocilia. Finally, at least in some lower vertebrates, hair cells are electrically tuned due to sponta-neous oscillations in their membrane potential which show a characteristic fre-quency for each cell. In summary, each hair cell is tuned over a very narrow range by its position along the basilar membrane, by its individual mechanical properties, by active damping of local regions of basilar membrane by outer hair cells, and possibly by its electrical resonant properties.

Action potential discharges in the auditory nerve

Many single afferent fibres are spontaneously active in the absence of a stimulus. At sine-wave stimulus frequencies up to 1 kHz, the firing of single fibres can follow the sine wave in the sense of occurring only in corresponding halves of successive cycles (Fig. 7.26). Since nerve fibres cannot carry more than about 1000 action potentials per second due to their refactory periods, when the stimulus frequency is increased beyond this, a fibre which has previously given one spike in each cycle

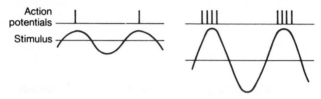

Fig. 7.26 Action potential discharges in a single afferent fibre produced by two different intensities of a fixed frequency.

will begin to drop out once every few cycles and so on. At these higher stimulus frequencies, a *group* of fibres, in which individual fibres misfire in an independently assorted fashion, can give discharges which when recorded from the group rather than from a single fibre can keep in step up to about 3 kHz. Thus one fibre responds to cycles 1, 4, 7, and so on; a second fibre to cycles 2, 5, 8; and a third fibre to cycles 3, 6, 9. Above 3 kHz, no clear visible relation can be made out in the discharge of a single fibre to the stimulus waveform.

If the stimulus thresholds are determined for a single fibre at various frequencies and intensities (Fig. 7.27), it is found that they rise rapidly on both sides of one small range of frequencies (the 'best' frequency) but more steeply on the higher frequency side, so that usually some high frequencies cannot provoke discharge, however intense they are. Note that as the stimulus intensity is raised the fibre responds over a wider frequency range.

We are now in a position to describe how the auditory information is encoded in auditory nerve fibre discharges with reference to the four properties of a sound stimulus, discussed earlier.

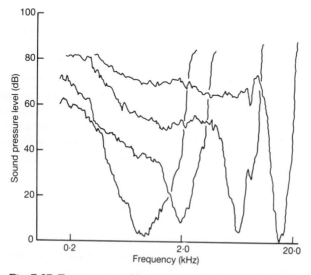

Fig. 7.27 Tuning curves of four sensory units from the cat auditory nerve. (Adapted from Kiang, N.Y.S. (1984) In: Brookhart, J.M. & Mountcastle, V.B. (eds) *Handbook of Physiology, Section I. The Nervous System*, Vol. III, pp. 639–674. American Physiology Society, Washington.)

1   Pitch discrimination is mainly determined by the area of basilar membrane occupied by firing receptors for a given stimulus frequency (the 'place theory' of hearing). High notes stimulate receptors closer to the base of the basilar membrane and low notes stimulate receptors closer to the apex (Fig. 7.25b).

2   Intensity sensation is derived from the total number of impulses per second in the auditory nerve fibres. At a given stimulus frequency only certain fibres will respond to a low-intensity stimulus. As the stimulus intensity is raised, the discharge frequency in these firing fibres is increased (Fig. 7.26) and previously silent fibres with higher thresholds at that stimulus frequency are recruited.

3   Duration of stimulus is signalled by the total duration of the afferent spike discharge caused by the stimulus.

4   Direction of the sound source is signalled by preservation in the auditory pathway of (part of) the initial time difference in receptor activation on the two sides of the head, as well as continuing intensity differences.

This information is now delivered to the cochlear nuclei in the medulla, where all the auditory fibres in cranial nerve VIII terminate. The auditory system as a whole is characterized by orderly segregation of nerve fibres and their cell bodies concerned with particular frequencies (tonotopic organization). A series of relay nuclei between the cochlear nuclei and the auditory cortex processes this information further, allowing, for example, analysis of sound location (superior olivary nucleus) and correlations with visual location of objects (inferior colliculus).

Central auditory pathways

Auditory inputs pass through several synaptic relays before reaching the primary auditory cortex. In the cortex responses are tonotopic and bilateral. Descending controls can modify activity at all levels of the pathway.

The auditory pathway from the spiral ganglion to the primary auditory cortex has several major relay stations, including the cochlear nucleus, the superior olivary nucleus (many axons bypass this nucleus), the inferior colliculus and the medial geniculate nucleus (Fig. 7.28). Axons from the cochlear nucleus ascend to the medial geniculate nuclei in either the ipsilateral or the contralateral lateral lemniscus, often after synapsing in the superior olivary nucleus, the first binaural relay station. The main cortical area is deep in the lateral or Sylvian fissure, emerging at the surface near the middle of the dorsal border of the superior temporal gyrus. Both ears are represented in both right and left cortices. Collateral branches from the pathway travel to spinal levels, the cerebellum and reticular formation of the brainstem and to the superior colliculi. All these are for coordination with other systems.

There are also inhibitory endings derived from higher levels of the central nervous system at each afferent synaptic level so that some modification of the afferent discharge must occur in these integrating centres.

Deafness

Hearing impairment in the sense of a raised threshold to sound stimuli may be due to impaired sound transmission in the outer or middle ear (conduction deafness) or to damage to the receptors or to the neural pathways (sensorineural deafness).

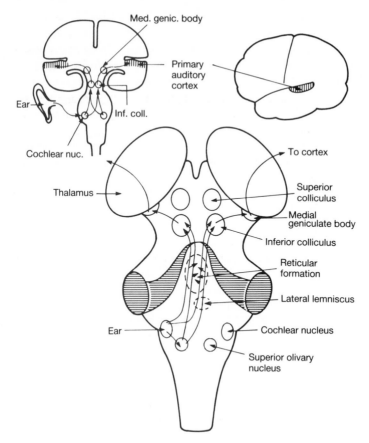

Fig. 7.28 A dorsal view of the brainstem (cerebral cortex and cerebellum removed) showing the central auditory pathways. Top left inset: simplified representation of the pathways in a roughly frontal section of the brain. Top right inset: a lateral view of the left cerebral hemisphere showing the location of the primary auditory cortex.

Conduction deafness may be caused by:
1 narrowing or blockage of the external meatus as by pus, wax, coal dust etc.;
2 thickening of the tympanum following scarring or repeated middle-ear infections;
3 exudate in the middle ear (otitis media);
4 dislocation or fixation (ankylosis) of the ossicle chain; and
5 otosclerosis, which is a narrowing of the gap between the footplate of the stapes and the surrounding bone, progressing to fixation of the footplate when the gap is bridged.
Some restoration of hearing can be achieved in patients with otosclerosis by an operation in which the window is unblocked by breaking the bony bridge or by removal of the whole stapes and footplate and replacement with a prosthesis, commonly of polythene tubing.
Sensorineural deafness caused by damage to the organ of Corti results from high-intensity stimulation of long duration, predominantly of high frequencies, as

in such trades as boiler-making. Ear protection is considered desirable where the ambient noise level is more than 85 dB above threshold in the range 300–2400 Hz. Other sorts of nerve deafness can be caused by some antibiotics (streptomycin), by mechanical damage to cranial nerve VIII (auditory nerve), by a tumour, or by diseases such as meningitis, rheumatism, malaria and syphilis.

One can differentiate diagnostically between conduction and sensorineural deafness with apparatus no more complex than a tuning fork. In bilateral conduction deafness, the tuning fork will be more audible when applied to the vertex of the skull, causing energy to be conducted through the skull bones, than when it is held close to either ear. But in the case of sensorineural deafness, the bone conduction route does not give greater audibility of the tuning fork. An objective measure of the degree of deafness is commonly carried out with an audiometer.

7.5 VESTIBULAR FUNCTION

The vestibular system provides information on the spatial orientation and movement of the head and plays an essential role in regulating movement of the body and limbs as well as the maintenance of body posture. In addition, afferent discharges from the vestibular organs influence reflex centres responsible for maintenance of a stable retinal image by controlling neck muscles and extraocular eye muscles. Rotatory movement of the head is detected by hair cells in the semicircular canals, while linear acceleration and the direction of gravity are detected by hair cells in the otolith organs.

There are two parts to the vestibular apparatus; the **semicircular canals** and the two otolith organs, the **utricle** and the **saccule** (Fig. 7.29). The **ampullae** of the semicircular canals contain sense organs which respond to rotatory acceleration of

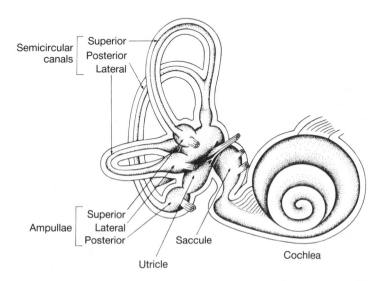

Fig. 7.29 Vestibular apparatus of the right side in three dimensions. (Adapted from Melloni, B.J. (1957) *The Internal Ear, An Atlas of Some Pathological Conditions of the Eye, Ear and Throat*, pp. 26–31. Abbott Laboratories, Chicago.)

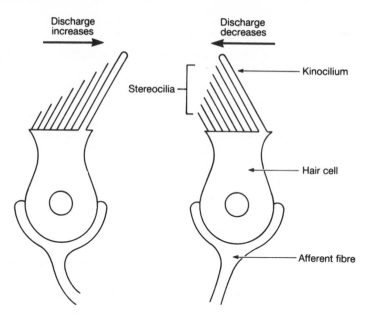

Fig. 7.30 Diagram of the vestibular hair cell bending the cilia towards the side of the kinocilium increases the discharge frequency; bending the cilia away from that side decreases it.

the head. The otolith organs are sensitive to the direction of the force of gravity and to linear accelerations of the head. All these tissues constitute the **membraneous labyrinth**, a system of tubes filled with endolymph and surrounded by perilymph. This in turn is encapsulated within the **bony labyrinth**, a series of bony tubes containing both the auditory and vestibular sense organs and which lies in the cavities in the temporal bone.

Stimulation of vestibular hair cells

Vestibular sensation results from stimulation of hair cells in the ampullae and otolith organs. Bundles of 40–70 **stereocilia** project from the free surface of each cell. These are packed hexagonally and increase progressively in length towards the side of the bundle from which the **kinocilium**, a specially long cilium with a different structure, projects (Fig. 7.30). Hair cells are polarized, so that movement of their cilia towards one side of the kinocilium causes depolarization and to the other side hyperpolarization. As they are depolarized, the hair cells release progressively more transmitter, resulting in an increased rate of firing of the afferent axons with which they are in contact. Hair cells in the resting state are typically in the mid-range of polarization so that there is a steady release of transmitter and tonic firing of sensory nerve impulses. Any slight movement of their hairs results in an increase or a decrease in the rate of firing.

The semicircular canals

The three semicircular canals lie in nearly orthogonal planes, i.e. each is at right

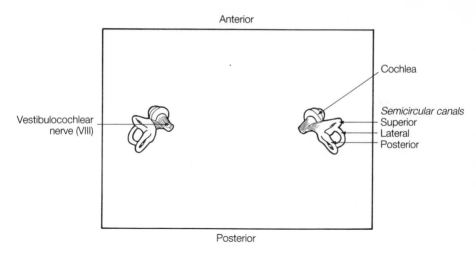

Anterior

Cochlea

Semicircular canals
Superior
Lateral
Posterior

Vestibulocochlear
nerve (VIII)

Posterior

Fig. 7.31 The orientation of the semicircular canals as seen from above. Note that, while the lateral canals on the left and right side lie in the same plane, the superior canal on each side lies in almost the same plane as the posterior canal on the opposite side.

angles to the other two (Fig. 7.31). With the head erect the lateral (horizontal) canal is raised anteriorly at about 30° to the horizontal while the superior (anterior) and posterior canals are at 45–55° to both sagittal and frontal planes, respectively. The two lateral canals on the left and right side of the body are in a single plane, and the posterior canal of one side is in a plane nearly parallel to the superior canal of the other side, so that these respective pairs are stimulated similarly and at least one pair is affected by any given angular acceleration. The canals of either side alone can generate afferent impulses signalling movement in any direction.

The labyrinths are filled with endolymph, a watery fluid with ion concentrations similar to that of intracellular fluid, with $K^+ \sim 150$ mmol L^{-1} and $Na^+ \sim 20$ mmol L^{-1}. The vestibular epithelium contains **dark cells**, which are responsible for the ion transport which maintains these concentrations. Each canal is connected at both ends to the utricle, so that the endolymph is free to circulate. At one end of each canal is an **ampulla** (Fig. 7.29) containing receptors which transduce circulation of endolymph into nervous impulses. The receptor apparatus, the **crista ampullaris** (Fig. 7.32) consists of the **cupula** (or cupola), a gelatinous wedge-shaped structure running fully across the cross-section of the ampulla to form a diaphragm that effectively blocks any bulk flow of endolymph. As it has nearly the same specific gravity as the endolymph, it neither floats nor sinks and thus is almost unaffected by the direction of gravity. The cupula is mounted on a ridge carrying many hair cells that have their cilia embedded in the base of the cupula. The hair cells in each ampulla all have the same orientation; those in the ampullae of the horizontal canals have their kinocilium facing towards the utricle, while in the vertical canals they face away from the utricle. Thus, movement of fluid towards the utricle depolarizes (stimulates) the hair cells of the lateral canals, but hyperpolarizes (inhibits) those of the vertical canals.

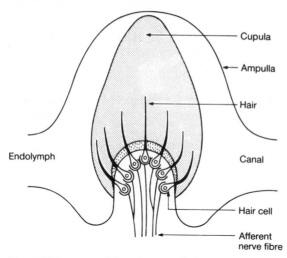

Fig. 7.32 Diagram of the crista ampullaris.

During any angular acceleration of the head, the endolymph tends to circulate through the semicircular canals as a consequence of its inertia, and the consequent fluid pressure causes deflection of the cupula and bends the cilia projecting from the hair cells. Movement of the fluid is soon damped and during steady motion the cupula rapidly returns to its resting position and produces no further sensation.

Figure 7.33 illustrates the changes in frequency of impulses recorded from

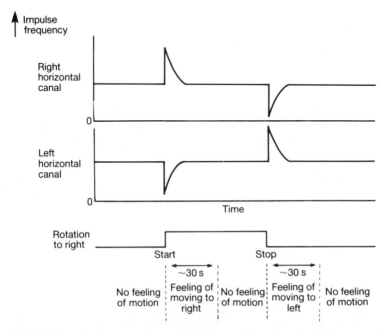

Fig. 7.33 Impulse frequency in vestibular afferent fibres when the head is rotated to the right at a constant angular velocity and then stopped.

axons in the vestibular nerve when the head is rotated to the right at a constant speed. The sensation of rotation persists beyond the period of firing in vestibular axons, demonstrating that this information is integrated and stored, probably in the brainstem. There is a spontaneous resting discharge which changes with any movement of the head. When the head accelerates to reach constant velocity, there is a sudden increase in frequency; the discharge then returns to its frequency at rest before the sensation of movement ceases in 25–30 s. When the rotation is stopped quickly, the opposite sequence takes place. The discharge frequency of the receptors changes in the opposite sense to that at the beginning of rotation and there is a feeling of rotating in the opposite direction until the cupula again returns to its resting position.

The discharge patterns from the ampullae of corresponding semicircular canals on the two sides of the body are mirror images, so that the brain receives both positive and negative sense information on any movement.

The utricle and saccule

Both these organs have **maculae**, thickened regions containing hair cells, and vestibular nerve terminals. The hair-cell cilia are embedded in a gelatinous mass called the **otolith** membrane which contains a mass of calcium carbonate crystals called **otoconia** (Fig. 7.34). the specific gravity of the otoconia is about 2.9, much higher than that of the endolymph. If the head is still, the otolith membrane tends to fall to the lowest possible point, giving rise to sensation of the direction of the force of gravity. During movement, its equilibrium position is determined by the direction of the vector sum of all accelerations of the head—linear, centrifugal and gravitational. It should be noted that conscious awareness of an accelerating force depends on cutaneous receptors and muscle and joint proprioceptors. Sensation of 'up' includes visual awareness of the origin of brightest light and of the horizon, as well as proprioceptive and cutaneous information.

Hair cells are arranged in each macula in a complex pattern of orientation (Fig. 7.35). There is a line of reversal in the orientation of the hair cells in the maculae

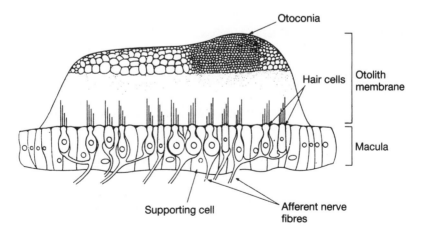

Fig. 7.34 Diagram of the otolith organ found in the utricle and saccule. (After Lindeman, H.H. (1969) *Ergeb. Anat. Entwicklungsgesch.* **42**, 1–113.)

Macula of utricle

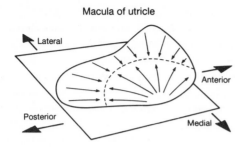

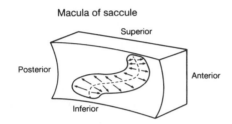

Macula of saccule

Fig. 7.35 Directions of orientation of hair cells in the maculae of the left utricle and saccule. The arrows indicate the direction of bending of the hairs for excitation. (From Spoendlin, H.H. (1966) In: Wolfson, R.J. (ed.). *The Vestibular System and its Diseases*, pp. 39–68. University of Pennsylvania Press, Philadelphia.)

of both utricle and saccule. In the macula of the saccule, the kinocilia are oriented away from each other on either side at the line of reversal rather than towards each other, as in the utricle. As a consequence of these patterns of orientation of the hair cells, each different movement of the otolith causes a different spatially organized pattern of discharge in the afferent fibres. These highly organized afferent patterns generated in different parts of the vestibular system probably account for the direction of the compensatory eye movements which occur.

There is a small tract of efferent fibres to the vestibular apparatus whose function is probably inhibitory, as is the corresponding supply to the cochlea. These efferent fibres may play a role in **habituation** to repeated patterns of acceleration or during constant rotation.

Functions of the vestibular organs

In experimental animals, spontaneous rates of firing in different vestibular fibres vary from a few per second up to about 200 per second, averaging 90. The frequency of this discharge is altered by bending the hairs of the hair cells. To a first approximation, the ampullae of the semicircular canals are sensitive to angular acceleration and the maculae of the utricle and saccule to the direction of gravity and to linear acceleration. However, they can all respond to other modes of stimulation to some extent.

The vestibular apparatus evokes two interrelated categories of reflex response, **dynamic** and **static**. Oculomotor reflexes, important in stabilization of the visual image on the retina, depend primarily on the dynamic function, which is mediated principally by the semicircular canals. To maintain a stable retinal image during head movements, two quite different reflexes are involved in causing compensatory eye movements. If an object is being looked at, movements of the object or movements of the head are compensated by the **fixation reflex** which, by smooth-pursuit movements of the eyes, keeps the image of the object on the part of the retina giving the clearest vision, the fovea centralis. Experimentally, this reflex can be evoked by non-geniculate afferents to the superior colliculus, but it normally involves participation of the visual parts of the cerebral cortex to distinguish objects of interest. There is a limit to the amount by which an eye can be turned in the head to follow an object and to extend this it is necessary for the head to move, or for the eye to return quickly to roughly the straight-ahead position and to find another object of interest to follow as, for example, when looking out from a moving vehicle. This type of movement is called **optokinetic nystagmus**. The word 'nystagmus' refers to any oscillatory movements of the eyes, whether normal or pathological, and especially to this slow turn–quick return sequence just described. The direction of the nystagmus is conventionally named according to the quick phase. The other mechanism by which the eyes are moved so as to keep an image as far as possible steady on the retina is driven by the vestibular apparatus and takes place to some extent whenever the head is moved, even with the eyes closed or in total darkness, and is called **vestibular nystagmus**. Here the slow drift is in the opposite direction to the rotation of the head and the quick phase is in the same direction. Even in the dark, this reflex may provide full compensation for head movements (between 60 and 100% compensation, depending on the methods of testing). In the light, this reflex collaborates with the fixation and optokinetic reflexes. The vestibular nuclei are connected with the oculomotor neurons via the medial longitudinal bundle.

The conscious sense of movement depends also on the dynamic sensory components of the vestibular apparatus, and involves a part of the cerebral cortex close to the main auditory sensory area in the temporal lobe, tucked into the lateral or Sylvian fissure.

The static reflexes, mediated primarily by the utricle and saccule, are important for the maintenance of the upright position of the head and for body posture. The maintenance of balance depends on the distribution of tone in the body muscles, which is regulated via the vestibular nuclei in the medulla and their connections with the vestibulospinal tract and the cerebellum.

Central vestibular pathways

Axons from the vestibular apparatus project to the vestibular nuclei in the brainstem. Vestibular inputs are involved in a variety of reflexes concerned with stabilizing the eyes (via inputs to the oculomotor nuclei), stabilizing the head (via inputs to neck motoneurons) and maintaining balance (via pathways to the cerebellum and spinal cord).

Vestibular axons on either side have their cell bodies in the vestibular ganglion (Scarpa's ganglion) which lies near the internal auditory meatus (Fig. 7.36). They project to the vestibular nuclei, which lie in the pons and the dorsorostral part of the medulla. They are components of a number of reflex arcs responsive to movement of the head and to apparent changes in the direction of the force of gravity. Such reflexes are responsible for stabilization of the eyes, holding the

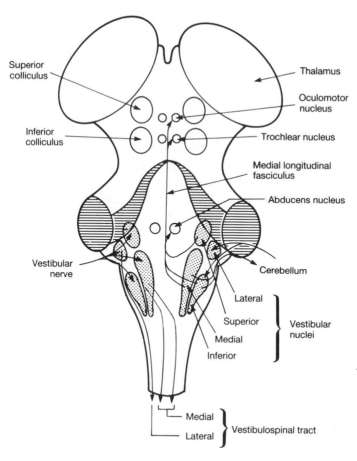

Fig. 7.36 A dorsal view of the brainstem (cerebral cortex and cerebellum removed) showing the central vestibular pathways.

head erect and maintenance of body stability. They are often not consciously perceived, and awareness of head movement or the direction of gravitational force is often secondary to the automatic reflex responses to acceleratory forces on the head.

Each vestibular nucleus has a particular pattern of projections to the oculomotor system, to the spinal cord and to the cerebellum. The **lateral vestibular nucleus** (Deiter's nucleus) receives axons from the macula of the utricle and from the cerebellum and spinal cord. Its neurons send their axons into the lateral vestibulospinal tract that terminates ipsilaterally in the ventral horn region along the length of the spinal cord. These axons powerfully facilitate α and γ motor neurons, innervating antigravity muscles. Neurons in the lateral vestibular nucleus respond selectively to tilting of the head. Their resting discharge is increased by tilting in one direction and decreased by tilting in the other. The input from the cerebellum is inhibitory, and removal of this tonic inhibition gives rise to decerebrate rigidity. The **medial** and **superior vestibular nuclei** receive inputs from the ampullae of the semicircular canals. Outputs running in the medial vestibular tract terminate bilaterally in the ventral horn of the cervical region of the cord, making monosynaptic connections with neck-muscle motor neurons. These provide for the reflex control of neck movements to maintain the position of the head and provide a stable base for eye movements. Both nuclei also participate in vestibulo-oculomotor reflexes, having outputs which run in the medial longitudinal fasciculus to the oculomotor nuclei, and give rise to rotatory nystagmus movements. The **inferior vestibular nucleus** receives excitatory inputs from the semicircular canals and from both the utricle and the saccule, and inhibitory inputs from the cerebellum. It has vestibulospinal and vestibuloreticular outputs and also has powerful effects on the cerebellum. Although its specific functions remains controversial, this nucleus integrates input from the whole vestibular apparatus and affects higher brain centres.

Disorders of vestibular function

Semicircular canal function can be tested by rotating the subject in a special chair whilst the head is held in such a posture as to stimulate one functionally associated pair of canals selectively. One can examine the consequences of stopping rotation on either eye movements or on body muscle tone. The latter is manifested by a tendency to deviate from a straight line when walking with the eyes closed. In these procedures, both right and left canals are inevitably stimulated together.

Unilateral stimulation can be produced by running water at above or below body temperature into the external meatus of the ear—the **caloric test**. The temperature difference gives rise to convection currents in the endolymph which excite the hair cells. This produces nystagmus which can be compared with a known normal response.

Diseases affecting the vestibule or its afferent fibres can cause abnormal discharges producing a sensation of **vertigo**, or giddiness, in which the external world may seem to move, the body may be felt to be moving, or the posture of the limbs, especially the legs, may be felt to be unsteady. Along with this there may be nystagmus, double vision, actual falling, and autonomic signs such as pallor, sweating, pulse rate and blood pressure changes, nausea and vomiting. **Ménière's disease** is characterized by attacks of vertigo and progressive impairment of hearing. The cause is unknown but on postmortem examination the endolymph-filled chambers of the inner ear are found to be enlarged.

When an often-repeated cyclical pattern of strong stimulation of the labyrinth occurs, as in sea travel and land travel in a car along a winding road, similar physiological consequences occur and are known generally as **motion sickness**. This is most marked when there is a lack of coordination between visual and vestibular information, and is often ameliorated or halted if it is possible to visualize the true horizon.

Chapter 8
Motor System

Movements are produced and postures maintained by the contraction of skeletal muscle fibres. The timing and force of contraction of muscle fibres in turn depend on the pattern of action potentials discharged by α motor neurons innervating the muscle. The problem of understanding motor control therefore reduces to the problem of the control of activity in α motor neurons. There are two main inputs regulating α motor neuron activity—sensory signals from the periphery and descending signals from higher in the neuraxis. The former are responsible for spinal reflexes and the latter for integrated postural reflexes and voluntary movement. Many of these influences, however, are directed through networks of local spinal interneurons.

8.1 SPINAL CORD MOTOR SYSTEM

The spinal motor apparatus includes α and γ motor neurons which project out to the muscles, and networks of interneurons. These elements together are responsible for the final patterning of muscle activities and are capable of generating some basic patterns of movement.

Motor neurons

There are two classes of motor neurons—large-diameter (up to 70 μm) α cells with axons of 12–20 μm diameter that innervate skeletal **extrafusal fibres** in skeletal muscles, and small γ cells with axons of 1–8 μm diameter that innervate the **intrafusal fibres** of the muscle spindles. The cell bodies of both classes of neurons are located in the ventral horns of the spinal cord, and in analogous positions in the brainstem.

Motor neurons are multipolar cells (many dendrites, one axon). Their dendrites spread from the ventral horn to the dorsal horn and even across the midline of the cord, and also up and down a particular cord segment, thus giving many opportunities for synaptic contact. Electron microscopy has shown that the vast dendritic tree of any motor neuron is covered with tens of thousands of synaptic terminals originating from multiple sources within and above the spinal cord and brainstem. This extreme convergence reflects the fact that motor neurons are the final common pathway from the central nervous system to muscle.

γ **motor neurons** are scattered amongst the α motor neurons of each motor nucleus in the medial and lateral columns of the ventral horn. Their axons pass out of the ventral roots together with the larger-diameter axons of the α motor neurons.

Distribution in spinal cord and brainstem

The cell bodies of motor neurons of trunk and proximal limb muscles lie in groups (**motor nuclei**) in the medial part of the ventral horn (Fig. 8.1a), forming a long column running almost the entire length of the spinal cord (C1–L4). In the brainstem, the column becomes discontinuous but the cell bodies of the motor neurons of the cranial nerves III, IV, VI and XII can be found in an analogous position. In the cervical and lumbar enlargements of the cord, the cell bodies of motor neurons of all distal limb muscles lie together in nuclei in the lateral part of the ventral horn while those of the proximal limb muscles lie medially (Fig. 8.1a).

The anatomical separation of medial and lateral columns of motor neurons reflects a functional distinction between the muscle groups they control. The trunk muscles, together with the proximal limb muscles, are involved in functions such as balance, posture and walking which require body or whole-limb movement. The

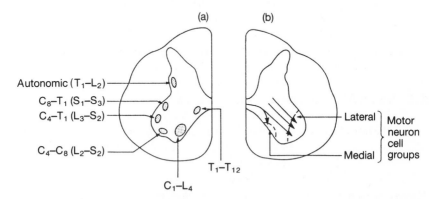

Fig. 8.1 Transverse section of the human spinal cord illustrating: (a) the cellular columns of the motor neuron groups. The longitudinal extent of each column of cells is indicated in figures and letters by the cord segments that it spans (adapted from Brodal, A. (1981) *Neurological Anatomy*, 3rd edn. Oxford University Press, Oxford); (b) the lateral and medial groups of interneurons (arrows) projecting mainly to the more lateral and more medial motor neuron cell groups (enclosed in broken lines), respectively (adapted from Sterling, P. & Kuypers, H.G.J.M. (1968) *Brain Res.* **7**, 419–443.

muscles moving the fingers and wrists do not have postural functions but are involved in manipulatory movements. The descending inputs to the ventral horn similarly divide into those primarily influencing the medial neurons and hence posture, and those influencing the lateral neurons and hence manipulatory activity.

Role in regulation of muscle force

The motor neurons innervating a given muscle form a group which share much the same afferent inputs. The excitability of all members of the group therefore tends to change in much the same way at the same time. In general, however, the smallest-diameter motor neurons have the highest input resistance and so reach threshold first (p. 131). Since the smaller-diameter motor neurons form smaller motor units, their discharge generates a small increment of tension. Initially, as each small motor neuron discharges, there is a small increment of tension, larger increments being added later in the contraction when larger-diameter cells are excited. In addition, the firing rate of each active motor neuron increases as the level of excitation increases, leading to mechanical summation of muscle activity (p. 124). Recruitment and changes in firing rate together enable muscular tension to be varied over a wide range with a relatively small number of motor units.

Interneurons

Most inputs to the spinal motor neurons arise from interneurons which lie dorsally at the base of the ventral horn. There are **excitatory** and **inhibitory** interneurons, utilizing a wide range of transmitter substances. Their topography is similar to that of motor neurons, the more medial interneurons projecting mainly to the medial motor nuclei and the more lateral interneurons to the lateral motor nuclei (Fig. 8.1b). The interneurons receive inputs from sensory nerves and from other motor centres. Many act locally within the same spinal segment, but some are specialized for sending information to other spinal levels (**propriospinal** interneurons) or across to the other side of the cord.

Functions of spinal circuits in motor control

The neural circuits formed by the interneurons are responsible for the final shaping of motor neuron activity resulting from sensory or descending inputs. Most pathways to the spinal motor apparatus act by synapsing on elements of this network, rather than by acting directly on motor neurons (certain notable exceptions to this include monosynaptic connections to motor neurons from muscle spindle afferents and some axons from cortical and motor brainstem structures).

In addition to processing extrinsic control signals before passing them on to α motor neurons, spinal networks exert intrinsic control over motor neuron activity. An example of this is **recurrent inhibition**, in which motor neurons are inhibited by the activity of interneurons which they themselves activate. These inhibitory interneurons, known as **Renshaw cells**, are excited by transmitter released from collateral branches of the motor neuron axon (Fig. 8.2). Recurrent inhibition appears to play an important role in dampening the activity of motor neurons; in particular, it appears to limit the discharge frequency of tonically active α motor neurons.

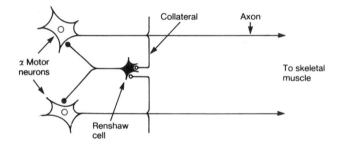

Fig. 8.2 Circuit diagram for recurrent inhibition of a motor neuron by a Renshaw cell.

Spinal networks are also involved in controlling more complex aspects of motor coordination. Interneurons carry information:

1 between motor nuclei within the ventral cord to coordinate muscle activities within a limb;

2 across the cord to coordinate activity of limb pairs; and

3 up and down the cord to other spinal segments via propriospinal interneurons, enabling coordination between limb girdles. Certain basic movement patterns such as walking, which requires complex integration of activity in diverse motor neuron pools, may be organized within the cord by such intrinsic interneuronal networks. The role of descending signals from the brain and of sensory signals from peripheral structures is to initiate and modulate the movement patterns produced by these networks. For example, when walking, the pattern is adapted to cope with surface irregularities—reflexly by sensory inputs, and predictively by descending inputs from the brain. An important source of descending inputs which trigger the spinal circuits into action is the **pedunculopontine nucleus** (see Fig. 8.9). Stimulation of this nucleus in decerebrate animals will induce locomotion on a treadmill. The pedunculopontine nucleus is connected to neurons of the reticular formation, which send signals inducing locomotion to the interneurons and motor neurons in the spinal cord. The pedunculopontine nucleus is regulated in turn by the subthalamic nucleus of the basal ganglia (Fig. 8.9), which receives input from the motor cortex.

8.2 REGULATION OF MOTOR NEURON ACTIVITY BY SENSORY SIGNALS

The spinal motor systems receive sensory inputs conveying information from muscle spindles about muscle length and rate of change in length, and from Golgi tendon organs about tension developed by the muscle. This information is used to regulate α motor neuron activity, including the generation of various spinal reflexes.

The activity of α motor neurons is influenced by sensory information arising from receptors in skin (e.g. nociceptors), muscles and tendons. Muscle length and tension are detected by mechanoreceptors within specialized structures known as **muscle spindles** and **Golgi tendon organs**, respectively. As in the skin, the different types of endings are supplied by nerve fibres of different diameter but they are usually referred to by the I–IV classification (p. 143). Information from these sensory pathways is utilized directly by the spinal circuits to produce rapid and relatively stereotyped responses to perturbations—the **spinal reflexes**. It is important to note that the same sensory information is also distributed to other regions of the central nervous system (CNS) where it contributes to the subconscious control of posture and movement (p. 221) and to the control and conscious awareness of limb position (proprioception; p. 154).

Muscle spindles

Muscle spindles (Fig. 8.3) are made up of two to 12 small-diameter striated muscle fibres termed **intrafusal** muscle fibres. The whole bundle of intrafusal fibres is enveloped in a capsule which lies in parallel with the main (extrafusal) muscle fibres and is attached to them at its ends. The middle third of the intrafusal fibre, known as the equatorial region, is non-striated and largely non-contractile. The outer thirds, known as the polar regions, are striated and contractile. Each muscle contains many muscle spindles.

There are two types of intrafusal muscle fibres:
1 nuclear bag fibres, commonly two per spindle, which are longer and thicker and have a group of nuclei clustered around the centre of the fibre; and

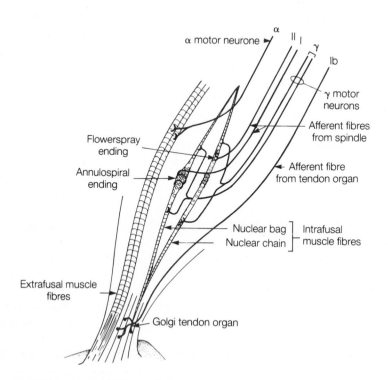

Fig. 8.3 The innervation of extrafusal muscle fibres, the muscle spindle and the Golgi tendon organ.

2 nuclear chain fibres, commonly four to five per spindle, which are shorter and thinner and have fewer nuclei arranged in a chain along the centre of the fibre (Fig. 8.3).

Group Ia sensory axons coil around the equatorial region of the nuclear bag and chain fibres to form **primary** receptor endings (**annulospiral endings**). (Group Ib sensory axons of similar diameter serve Golgi tendon organs.) Group II sensory axons (smaller and slower-conducting) end mainly on nuclear chain fibres to form **secondary** receptor endings (**flowerspray endings**).

The receptor endings of both Ia and group II afferents are mechanoreceptors and they are stimulated when the equatorial regions of the intrafusal fibres are put under tension, as happens when the whole muscle is stretched. Conversely, activity in the spindle endings is reduced or silenced when tension is removed from the equatorial region of the spindle, as might occur when the whole muscle shortens following contraction of the extrafusal muscle fibres. This pattern of responses arises because the spindles lie adjacent and parallel to the extrafusal muscle fibres.

Responses to different degrees of stretch on the muscle spindle provide information about the current length of the muscle. They are known as the **static** or length-sensitive responses and are in some ways analogous to the position response of cutaneous mechanoreceptors. Both primary (Ia) and secondary (II) endings give a static response. During static firing, the frequency of impulses is proportional to the muscle length. The static response is probably associated with stretch of both bag and chain fibres.

Spindles also respond to the rate of change of length of the muscle. This is known as the **dynamic** or velocity-sensitive response. Only the primary (Ia) endings give a large dynamic response. The faster the stretch, the higher the impulse frequency. During release from muscle stretch both primary and secondary endings commonly show a decrease in impulse frequency.

Role of γ motor neurons

In addition to the two types of sensory ending, muscle spindles also have their own motor innervation. This innervation is supplied by fusimotor or γ motor neurons. Stimulation of the γ motor neurons does not alter tension in the whole muscle but it produces contraction of the polar regions of the intrafusal fibres. This, in turn, stretches the equatorial region of the intrafusal fibres, which alters the sensitivity of the spindle to stretching of the whole muscle.

The γ motor neurons have two main roles in regulating spindle sensitivity: one role is the maintenance of sensitivity during muscle shortening and the other role is to adjust the sensitivity as required in different situations. First, shortening of a muscle due to active contraction of extrafusal muscle fibres would slacken the spindles, rendering them less sensitive to imposed stretches at the new muscle length. This is overcome by concurrent activation of γ motor neurons along with the α motor neurons (α–γ co-activation). This has the effect of maintaining tension on the sensory regions of the intrafusal muscle fibres during muscle shortening, and thus maintains their sensitivity. Second, the motor control system requires that spindle sensitivity to stretch be adjusted in certain situations. For example, during the step cycle there are phases in which lengthening of muscles is required and a strong stretch reflex would interfere, while in other situations a heightened stretch reflex is required. To cope with these situations, the γ motor neurons are able to adjust spindle sensitivity at any one muscle length by variation in the level of their activity. For this, independent control of α and γ motor neurons is required.

Static and dynamic responses of spindles can be adjusted independently via two types of γ efferent fibres. These are associated with two types of ending on the intrafusal muscle fibres: **plate endings**, which are found on both nuclear bag and nuclear chain fibres at the ends or poles of the fibres and may be associated primarily with dynamic γ efferent fibres; and **trail endings**, which are found mainly on nuclear chain fibres just next to the equatorial region of the fibres and are associated primarily with static γ efferent fibres. Stimulation of static fibres increases the static response of both group Ia and group II afferents, while stimulation of the dynamic fibres increases both the dynamic and static responses of group Ia fibres.

Another, quite different role for γ motor neurons has been proposed in the **initiation** of movements. Contractions could be initiated through this pathway if activity in γ motor neurons caused muscle spindles to increase their rate of discharge, leading to activation of α motor neurons and muscle shortening. In principle this could offer greater sensitivity of control, and might allow automatic compensation for different initial muscle lengths. In most cases, however, α–γ coactivation occurs in movement.

Stretch (myotactic) reflex

Muscle spindles are the starting point for the simplest of all reflexes found in mammals—the **stretch reflex**. When a muscle is stretched, the elongation leads to a discharge in the muscle spindle afferents that is conducted into the spinal cord via the dorsal roots. Within the cord the sensory fibres branch and some terminals make **monosynaptic** excitatory contacts with α motor neurons to the extrafusal muscle fibres of the *same* muscle (Fig. 8.4). Impulses generated in these motor

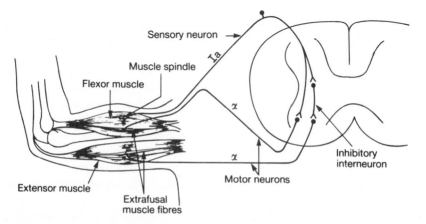

Fig. 8.4 Circuit diagram of a monosynaptic stretch reflex showing reciprocal inhibition.

neurons leave the cord in the ventral roots and conduct impulses back to the muscle, causing it to contract in opposition to the change in length. Thus the stretch reflex tends to restore the initial muscle length.

Such reflex contraction may be elicited by changes in muscle length of around 1–2 mm. Such imposed length changes are normally caused by changes in the orientation of limbs in relation to gravity. Stretch reflexes have both phasic and tonic components, are of short latency, and are present to some extent in all muscles but are best seen in the antigravity muscles (e.g. leg extensors). Movements involving the stretch reflex are facilitated by the pathways that produce **reciprocal inhibition**. For example, stretching the biceps (flexor) muscle spindles produces inhibition (via an inhibitory interneuron) of motor neurons to the triceps (extensor) muscle (Fig. 8.4) and, similarly, stretching the triceps causes inhibition of the biceps.

In medical practice, simple phasic stretch reflexes are evoked by tapping muscle tendons (**tendon jerks**); this produces a brief muscle stretch which excites the dynamically sensitive spindle afferents. Changes in responsiveness can be indicative of neurological disease: for instance, the response may be reduced by a peripheral nerve lesion and exacerbated with lesions of descending pathways to α motor neurons.

The same reflex pathway can be activated by electrical stimulation of the muscle nerve. At low stimulus strengths, the sensory Ia fibres coming from the spindles can be selectively activated; these then activate the α motor neurons. The reflex muscle response thus induced (known as the **H-reflex** after Hoffmann, who first described it) can be recorded electromyographically as a distinct wave (H-wave) with a latency of around 35 ms. Recordings of H-reflexes in humans are useful for examining excitability changes occurring within the spinal cord, since standard repeatable stimuli can easily be applied to the nerve and changes in muscle properties or spindle sensitivity do not affect the response.

Role of spindle reflexes in normal muscle tone

When a joint is passively moved a certain resistance is normally felt. This resistance or **tone** is in fact reflex and not due to any elasticity of the muscles. Such a movement produces a series of reflexes in the stretched muscles, which can be seen as a series of waves in electromyographic recordings. The first, of short latency, is generated by the phasic action already described. The origin of later waves which provide the tonic-maintained aspects of this effect is not well-understood. On the one hand, it is claimed that the later responses are due to primary spindle activity being directed to motor neurons by way of the motor cortex (cortical loop). On the other hand, it is claimed that they result from a slow spinal reflex pathway which perhaps originates in the spindle secondary endings.

Golgi tendon organs

These are found on tendons usually near the muscle–tendon junction (Fig. 8.3). Their afferents are group Ib axons which branch over several tendon fascicles. They are covered by a capsule which, at its ends, becomes continuous with the connective tissue of the muscle or tendon.

A Golgi tendon organ responds to changes in **tension** in its muscle fascicle (and hence in the tendon); an increase in tension increases the frequency of impulses in the Ib afferents. This is due to its position on the tendon which is *in series* with the muscle; the tendons do not change length but rather are exposed to the force or tension developed by the muscle. Note that in comparison with muscle spindles, Golgi tendon organs provide information about muscle performance during isometric contraction and also when a muscle actively shortens, since in both cases tension is developed in the tendon. Furthermore, this structural arrangement means that the Golgi tendon organ is much more sensitive to active muscle tension than to passive stretching of the muscle, since in the latter case much of the applied force is dissipated in lengthening the muscle. Thus, Golgi tendon organs respond not only at extreme tension but also during normal tension development.

Golgi tendon organ reflex

Information from Golgi tendon organs is utilized for reflex regulation of muscle tension, somewhat analogously to the spindle reflex control of muscle length. However, in contrast to the stretch reflex, the Golgi tendon reflex pathway operates via an inhibitory interneuron, which has synapses on α motor neurons of the muscle. It is thus an example of a negative feedback loop. The role of this loop can be most easily appreciated in maintained isometric contractions, such as holding a pen for a long period of time. The 'desired' amount of force is initially generated by a certain level of activity in α motor neurons. An inhibitory influence from the Golgi tendon organs would be exerted on the α motor neurons, but this would be matched by excitatory influences from elsewhere in the CNS. Now, if the muscle force is reduced by fatigue, then activity in the Golgi tendon organs would decline, inhibition of the motor neurons would decrease and the muscle would be excited more strongly. If, on the other hand, the force exerted is too great, the inhibitory effect will be increased and the force correspondingly reduced.

Role of spindle and tendon organ reflexes in ongoing motor control

It would be a mistake to think that these sensory systems act only in response to external perturbations. They are in fact of vital importance for the precise regulation of α motor neuron activity during the performance of all voluntary movements. The mechanical properties of the musculoskeletal system pose many problems for motor control systems. For example, during a movement, development of force by a muscle for any degree of α motor neuron activation will vary as the muscle changes length, due to the force–length relationship of muscles and to changes in mechanical leverage (p. 126). Similarly, the response of a muscle will differ at different times owing to fatigue. Thus a constant signal will not always produce the same increase in tension. The spinal cord circuits utilizing information from muscle spindles and Golgi tendon organs are able automatically to compensate for these variations from moment to moment. These control processes are best called *servomechanisms*, that is, mechanisms that modify output during performance to adapt to varying conditions. Such systems have distinct advantages as they can quickly and accurately follow command signals without the need for intervention by higher centres.

Flexor (nociceptive) reflexes

A typical flexor reflex is the withdrawal of a limb following painful or potentially painful stimulation (p. 75). The receptors are skin nociceptors, the central connections are polysynaptic, and excitation is evoked in motor nuclei of most of the limb flexor muscles while antagonistic muscles are inhibited. At the same time, contralateral motor neurons of extensor muscles are excited via interneurons which cross to the other side of the cord (**crossed extensor reflex**), facilitating support of the body while the hurt limb is withdrawn.

8.3 SUPRASPINAL CONTROL OF MOVEMENTS

Descending commands to motor neurons arise from various sites in the brainstem, and from regions of the cerebral cortex. The brainstem structures have diverse functions in motor control, many relating to less voluntary aspects such as automatic regulation of muscle tone and upright posture. Cerebral cortical involvement relates more to the control of discrete voluntary movements. In turn, cerebral cortical motor functions are modulated by important contributions from the basal ganglia and the cerebellum.

Pathways to spinal motor circuits

Inputs to spinal motor circuits arising from higher in the neuraxis are necessary for the instigation of coordinated movements and their postural concomitants. Most of these inputs act via synapses on spinal cord interneurons, but some synapse directly upon the motor neurons. There are six main pathways (Fig. 8.5). The corticospinal tract originates in the cerebral cortex; the rubro-, lateral vestibulo- and reticulospinal tracts originate in the brainstem and extend the length of the spinal cord; the tecto- and medial vestibulospinal tract also originate in the brainstem but extend only through the cervical region and are primarily concerned in regulating the motor nuclei of neck muscles.

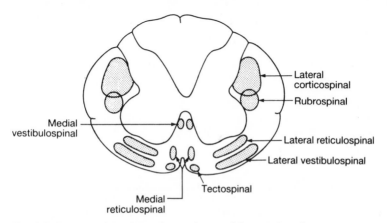

Fig. 8.5 Diagram of the descending pathways of the spinal cord in transverse section.

Brainstem motor systems

Role of individual nuclei in the brainstem

Regions of the brainstem that are important in controlling movements are the red nucleus, the lateral vestibular nucleus, the medial vestibular nucleus, the reticular formation, the superior colliculus, the locus coeruleus and the raphe nuclei (Fig. 8.6).

The **red nucleus** (nucleus ruber) receives inputs from the cerebral cortex and from the dentate and intermediate nuclei of the cerebellum. In their relationship to movement, neurons of this nucleus are similar to those found in motor cortical regions (p. 225); however, they may be involved in the regulation of more automatic, well-learnt movements. Cells of the posterior part of the nucleus send axons in the **rubrospinal tract** (Fig. 8.5). Fibres cross immediately and run down the contralateral side of the brainstem and spinal cord. Their terminals are distributed similarly to those of the corticospinal tract, some synapsing directly with α motor neurons.

The **lateral vestibular nucleus** (Fig. 8.6) receives input from the otolith organs of the vestibular apparatus, i.e. dynamic information about linear accelerations and static information about head tilt. It also gets a major input from the cerebellum and the pontine reticular formation. Neurons of this nucleus project ipsilaterally into the **lateral vestibulospinal tract** (Fig. 8.5), which terminates in the medial part of the base of the ventral horn. A few fibres terminate directly on

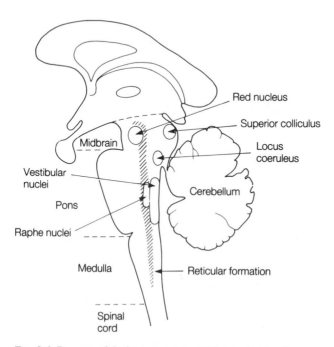

Fig. 8.6 Diagram of the brainstem in sagittal view showing the approximate positions of structures involved in motor control.

medial motor neurons. This system seems to be involved in defending the upright stance against perturbations likely to be associated with head movements.

The **medial vestibular nucleus** receives information about dynamic angular accelerations of the head from the semicircular canals; it projects via the uncrossed **medial vestibulospinal tact** (Fig. 8.5), the caudal equivalent of the medial longitudinal fasciculus, and influences nuclei in the brainstem and spinal cord which control eye and neck muscles. This system is involved in the **vestibulo-ocular reflexes** (p. 206) that stabilizes the eyes in the head; it also activates neck muscles in response to vestibular stimulation and so keeps the head stable during body movements.

The **reticular formation** is a diffuse aggregation of cells in the core of the brainstem, separated by a tangle of fibres running in all directions. The medial nuclei of this system are concerned with regulating eye, neck, trunk and limb movements. They receive somatic and proprioceptive sensory signals, as well as descending inputs from the cerebral cortex and limbic system. They are also interconnected with the fastigial and intermediate nuclei of the cerebellum. The reticular system can be divided into pontine and medullary parts: the pontine part projects ipsilaterally down the spinal cord, while the medullary part sends axons down both sides of the cord. The **reticulospinal fibres** (Fig. 8.5) terminate on the medial group of interneurons and to a small extent on medial motor neurons. The pontine part is important for automatic maintenance of the upright stance, through facilitation of antigravity reflexes; the medullary part seems to suppress spinal reflexes during sleep and also to allow descending controls to override them during voluntary movement.

The **superior colliculus** (tectum) is a region that receives inputs conveying visual information; it seems to be important in generating head movements that help direct one's gaze at a particular point. It may also be important in mediating very fast reflexive movements in response to visual stimuli. Fibres from this region join the **tectospinal tract** (Fig. 8.5) which is a crossed pathway terminating on interneurons in the cervical cord.

The **locus coeruleus** is located near the floor of the rostral part of the fourth ventricle. It projects widely in the CNS, including the spinal cord. Neurons in the locus coeruleus contain the neurotransmitter noradrenaline. Neurons from the **raphe nuclei** also project in the cord; they are a narrow continuous collection of cells in the middle of the brainstem which contain the neurotransmitter 5-hydroxytryptamine (serotonin). The locus coeruleus and raphe nuclei are activated in a non-specific way by external events and internal states, and appear to modulate the responsiveness of motor neurons. It seems likely that these monoamine systems set the background level of excitability of motor neurons in relation to demands likely to be made of them.

Role of brainstem motor systems in postural control

By **posture** we mean the maintenance for a period of time of position (of head, limbs, trunk) in space, as a prelude or background to movement. Fundamentally, posture implies maintenance of a particular distribution of muscle tone, to achieve a particular disposition of the body. Postural control includes a wide variety of motor acts and is a function of all levels of the motor system. For instance, postural activities occur with or even in advance of voluntary movements to maintain balance; these **anticipatory postural adjustments** are organized by the cerebral motor cortex. At the other extreme, the spinal cord networks and the various spinal reflexes are responsible for more simple, immediate reflex responses to peripheral perturbations of individual limbs. In between, there are more complex responses, some of which are controlled by motor structures of the brainstem.

The vestibular nuclei contribute to postural reflexes in response to changes in head position. For instance, when the head is dorsiflexed, the otolith organs (p. 204) are stimulated and pathways from the vestibular nuclei activate appropriate motor neurons to promote extension of all four limbs (assisting postural support), and the eyes 'counter-roll' downwards. The eye movement responses are most clearly seen (**doll's head phenomenon**) in severely brain-damaged patients with an intact brainstem. However, a major additional component of limb and eye responses to changes in head position arises from activation of neck proprioceptors (e.g. muscle spindles of neck muscles), which produces tonic neck reflexes. Normally, the vestibular and neck muscle reflexes interact with each other and with ongoing motor activities; they are most clearly demonstrated in isolation in preparations which are not producing spontaneous movements, and in which only the neck or vestibular systems are intact.

Linear acceleration also stimulates the otolith organs, and the reflexes which follow are organized by the lateral vestibular nuclei which give rise to the vestibulospinal tracts. An example of this type of reflex is the **vestibular placing reaction**. This can be evoked by suddenly tipping a blindfolded cat head-down, whereupon its forelegs extend. If the blindfold is removed, a cat will still extend its legs even if its utricles are destroyed—a **visual placing reaction**. In general, visual stimuli can substitute for labyrinthine stimuli in postural reflexes. Thus, animals and humans with bilateral labyrinthine damage have normal posture if they can use their eyes.

Angular acceleration stimulates the semicircular canals (p. 200) and the ensuing reflexes involve the medial (and other) vestibular nuclei which give rise to the medial longitudinal fasciculi and the medial vestibulospinal tracts. The eye movements are called **vestibular nystagmus** (p. 206); limb muscle responses are subtle, but involve a biasing of limb muscle tone that would tend to counter any imposed rotation.

The role of the brainstem in postural control is most clearly demonstrated in quadrupeds where the brainstem alone is capable of organizing entire sequences of postural responses into coherent adaptive behaviour. This was revealed in experiments in which the brain and spinal cord of quadrupeds were transected at various levels. Where the spinal cord is isolated from higher centres (a spinal preparation), the animal cannot remain upright for any length of time, although it will exhibit stepping motions. If the cut is made at the level of the upper medulla, such an animal can stand for long periods but cannot right itself if laid on its side, while a cut through the midbrain will permit such righting responses. Thus the complex integrated response of standing up can be organized in the brainstem in such species. In primates, however, descending inputs from higher centres are required for full expression of these behaviours.

Motor cortex

The motor cortex consists of several regions in the cerebral cortex which are directly involved in the regulation of brainstem and spinal cord motor systems. In general terms, the motor cortex is important in the control of voluntary movements, particularly those accurately performed. The cortical regions (Fig. 8.7) involved include:

1 the **primary motor cortex**, which lies in the precentral gyrus immediately rostral to the primary sensory areas;

2 the **pre-motor** and **supplementary motor** areas, which lie laterally and medially, respectively, in front of the primary motor cortex; and

3 the **frontal eye field**, which lies further rostrally again, and possibly additional areas in the cingulate gyrus close to the supplementary motor area.

These areas are identified as being important in motor control on several grounds: neuronal activity in these areas is found to be related to movements, electrical stimulation of these regions can evoke movements, damage to them may disturb movement control, and they contain neurons which project from the cortex to the brainstem and spinal motor systems in the **corticobulbar** (to brainstem) and **corticospinal (pyramidal) tracts**, sometimes synapsing directly with motor neurons.

The origin of the names for some of the regions requires explanation. The *primary* motor cortex was the first recognized, and is most clearly implicated in the

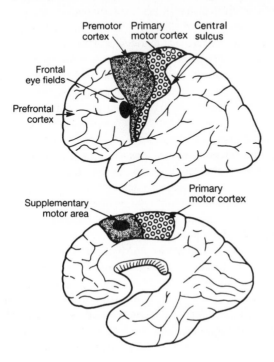

Fig. 8.7 Motor areas of the brain. Top: lateral surface of the left cerebral hemisphere showing the primary motor cortex (precentral gyrus), the premotor cortex and the frontal eye field. Bottom: medial view of the surface of the right cerebral hemisphere (with the left hemisphere removed), showing the medial extension of the primary motor cortex and also the supplementary motor area in front of the primary motor cortex.

immediate control of discrete movements. Lesions here produce a localized paralysis of the contralateral body; electrical stimulation is effective at lower intensities and reveals a more detailed somatotopic map (see below) than in other areas; and neuronal activity is more closely coupled to the execution of movements. Stimulation of the *supplementary* motor area reveals a smaller, less ordered somatotopic map; the cells are somewhat less tightly coupled to movement, and lesions produce only subtle deficits. The *premotor* area was at first not considered to have any direct influence on the spinal motor apparatus, but limited projections have now been found. Electrical stimulation produces some movements, at a rather high intensity of stimulation.

Corticospinal (pyramidal) tract

Axons projecting from the cortex to the brainstem or spinal cord arise in layers III and V of both motor and sensory regions of cortex, in almost equal porportions. The sensory areas project into the dorsal cord, presumably modulate sensory systems predominantly, and are not considered further here. The axons of the motor cortex descend through the posterior limb of the internal capsule and the cerebral peduncles to pass through the pons and medulla. In the midbrain and medulla, some axons (corticobulbar fibres) leave the tract and pass to the motor nuclei of cranial nerves. These are crossed pathways but no definite tracts can be seen as the axons pass across the brainstem in small separate bundles. Axons also pass to the red nuclei and to the reticular nuclei, all of which are themselves the source of spinal pathways.

In the caudal medulla corticospinal fibres cross the cord, forming the pyramidal-shaped structures which give the tract its alternative name. The crossed fibres thereupon enter the spinal cord, forming the lateral corticospinal tract (Fig. 8.5). A few fibres (15%) do not cross at this level and pass down ipsilaterally.

The fibres of the lateral corticospinal tract terminate on the lateral group of interneurons in the dorsal part of the ventral horn. In the lower cervical cord, about 10% of the fibres also terminate directly on motor neurons, mostly in motor nuclei which serve more distal muscles, such as of the forearm and hand.

Somatotopic maps in the motor cortex

Electrical stimulation of the cerebral hemispheres of humans and experimental animals evokes movements on the opposite side of the body. Experiments in which the cortex is systematically stimulated millimetre by millimetre reveal at least two representations or maps of the body in terms of movements evoked, one within the precentral gyrus (primary motor cortex), and another on the medial surface of the brain (supplementary motor area).

In the primary motor cortex, also known as the MI, the movements evoked are always only brief twitches rather than complex series of movements. Leg movements are elicited from regions extending on to the medial surface of each hemisphere, while movements of the trunk, arm, hand and head muscles are elicited in a fairly ordered (**somatotopic**) sequence across the mediolateral extent of the hemisphere. The somatotopic representation of the body is similar to that in the postcentral gyrus (see Fig. 6.8) except that the size of the representation in the precentral gyrus is proportional not to the density of receptors in the body region but to the accuracy with which movements are made. The basic plan of the somatotopic representation within the primary motor cortex arises because each region is connected via the corticofugal tracts with the appropriate interneurons and motor neurons to produce muscle contraction of that part of the body. However, the details, especially boundaries between body parts, are highly variable between individuals and even flexible over time within a single individual. The boundaries seem to be controlled by local cortical inhibitory connections, and by patterns of thalamic inputs relaying sensory information from the periphery.

The map within the supplementary motor cortex extends rostrally from and is continuous with the primary motor cortex leg representation. Although also broadly somatotopic, in that arm and head movements tend to be evoked from more rostral regions, the map is much smaller, extending for only about 2–3 cm, and the areas are less well-defined. It was identified first in humans and, when stimulated, complex synergistic movements of the contralateral limbs and body, often with vocalization, were produced. These responses were quite unlike the localized contralateral contractions evoked by stimulation of MI. Subsequently, however, responses similar to those seen in MI have been found, and it may be that the region of the supplementary motor area contains two subfields: a caudal one containing an MI-like map of the body with isolated twitch responses, and a more rostral region from which complex, tonic movement sequences can be obtained with higher stimulus intensities.

Functions of motor cortex subregions

Primary motor cortex

As noted above, stimulation of this area of brain reveals a fairly detailed map or representation of the movement of the body. However, each joint, and in fact each muscle, is represented several times at different sites within the map, perhaps reflecting organization at the cortical level in terms of combinations of muscles required for different movements, rather than of muscles *per se*.

Neurons within each region are found to be active in relation to movement of the same joint or muscle; a proportion of these project to the spinal cord (**corticospinal cells**). Functionally, there appear to be two types of corticospinal cells in MI. There are larger cells with fast-conducting axons which discharge mostly in relation to movement (phasic) and smaller cells with low conduction velocity which discharge continually (tonic), though there is also a change during movement. Many cells change their activity just before muscle activation, and this is believed to be the signal that evokes voluntarily controlled muscle action. The activity of some cells also specifies the force produced by the muscles during the ensuing contraction. Although the majority of corticospinal cells act by way of connections with spinal interneurons, some of them synapse directly on α motor neurons and often control several muscles which work together (**synergists**). These special nuerons are called **corticomotoneuronal cells** and are thought to be especially important for control of accurate movements of distal muscles (e.g.

of the fingers during precision grip). This function is the most severely affected by damage to the primary motor cortex or the pyramidal tract.

In addition to a role in control of individual muscle action, large numbers of MI neurons act together to specify movement direction, especially at the shoulder. In fact, during a shoulder movement, most cells in the shoulder region of the cortex change their activity in some way. This coding of movement direction by large number of cells (**population coding**) means that an infinite variety of movement directions can be specified by changing the mix of activity in a large but finite number of cells. It also means that no individual cell has to specify the movement accurately—the accuracy arises out of the average activity of many cells. In addition, no one cell is vital: individual cells can die, but the population will still be able to specify the movement.

Primary motor cortex neurons in turn are regulated by inputs, which arise from other regions of the motor cortex, from the sensory cortex of the postcentral gyrus, and from the parietal cortex. In addition, inputs are relayed through the thalamus from the cerebellum and basal ganglia. Traditionally, the primary motor cortex has been seen as the final common path through which most higher motor control is routed; now, however, it is realized that many regions of the wider motor cortex have direct access to the spinal cord, and thus act to some extent in parallel to the primary motor cortex.

Frontal eye field

This region is in front of the premotor cortex on the lateral surface of the brain (Fig. 8.7). Stimulation in the frontal eye field evokes movements of the eyes and sometimes turning of the head in the direction of movement. The commonest movement elicited is a conjugate deviation of the eyes to the contralateral side. The frontal eye fields are probably the primary motor cortex for the eyes, since no eye movements are evoked by stimulation within the precentral gyrus. Neuronal activity in this region is enhanced during voluntary eye movements and destruction of the frontal eye fields prevents voluntary movements of the eyes, though reflex movements are unaffected.

Supplementary motor area (SMA; also known as MII)

As noted above, the SMA is found on the medial surface of each hemisphere above the cingulate gyrus and in front of the leg area of the primary motor cortex (Fig. 8.7). There is debate over the function of the SMA. On the one hand, it may have higher-order functions, such as in planning or programming of movements generated internally, rather than in response to external cues. Patients with lesions here are reported to show a lack of spontaneous movement and speech, and recent studies of regional blood flow in humans have shown that when only *thinking* about a motor task there is an increased blood flow in the SMA of both sides, while an increase in the contralateral MI only occurs if the movement takes place. Electrical signs of cortical activity before movement recorded from human subjects (Fig. 8.8) are also suggestive of a preparatory role for the SMA. In these experiments, subjects perform the same movement many times, and the tiny

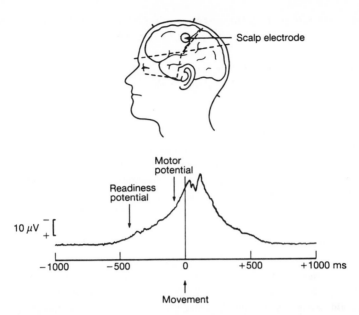

Fig. 8.8 Averaged cortical potentials recorded on the scalp over the precentral gyrus during voluntary flexion of an index finger. The premovement potential can be divided into an initial readiness potential and a later rapidly rising motor potential. (From Deeke, L., Sheid, P. & Kornhuber, H.H. (1969) *Exp. Brain Res.* **7**, 158–168.)

electroencephalogram signals recorded from the subject's scalp are averaged before and during the movement. Prior to movement a very early negative potential, presumably indicative of excitation of cortical neurons, is recorded which begins more than 0.5 s before the movement and comes to a peak just after movement. This initial surface potential is termed a **readiness potential** and can be recorded widely over the skull but is maximal over the SMA and is assumed to originate there. It is found to be biggest in internally driven rather than in externally cued movements. It is also bigger when a complex, rather than simple, series of repetitive movements is performed.

Neurons in the SMA are more likely to be active very early prior to movement than are MI neurons. Recently, neurons have been found in the SMA that seem to be active only when an animal is performing a remembered sequence of movements, rather than the same movements in response to visual signals.

On the other hand, a role for the SMA in more direct motor control is also presumed, because of the projections from the SMA to the spinal cord, the presence in the SMA of MI-like responses to cortical stimulation and of MI-like neurons active in relation to simple, triggered movements. The actual direct motor control function of the SMA is unknown. It may be important in suppressing unwanted movements, since lesions of the area in humans do affect the ability to inhibit movements, such as the grasp reflex in response to objects put into the hand, or the SMA may have a special role in facilitating coordinated movement of the two sides of the body. Evidence for this includes the fact that ablation of SMA in monkeys brings about a disturbance of bimanual coordination. Also, neurons are found in the SMA that discharge in relation to whether the hands are used together or singly, rather than in relation to activity of a particular muscle.

Premotor cortex

This lies between the precentral gyrus and the prefrontal cortex on the lateral surface of the brain. It shows a limited response to brain stimulation at rather high stimulus levels. It is possible that several subregions exist. Overall, it is postulated to be important in planning for movements, particularly those performed in response to external cues.

Other areas involved in movement control

Other areas, including parts of the cingulate gyrus and parietal cortex, have recently been included within the motor cortex on the basis of connections with the primary motor cortex and spinal cord, and the relation of their neuronal activity to movement. Functions of the cingulate areas are at present obscure, but seem similar to those ascribed to the SMA and premotor cortex. The parietal cortex has long been known to be important in movement performance; a person who suffers a left-sided parietal lobe lesion may be unable to carry out actions requiring a series of sequential movements on command, though quite capable of the individual movements. This condition is termed **apraxia**. However, it eventually becomes partly a subjective decision where motor cortex begins and ends; for instance, lesions of visual processing areas in the parieto-occipital cortex can produce specific deficits of movement normally elicited by visual stimuli (**visuomotor apraxia**). These movement abnormalities are not seen in acutely blind patients, suggesting that some aspect of movement control must already be processed in these regions. In the end, it can be argued that all of the brain is involved to some extent in movement control!

Role of the ventrolateral thalamic nuclei

The ventral division of each thalamus consists of anterior, lateral and posterior nuclei. It is the lateral group which is concerned with motor function. It is well-established that the neurons of the ventrolateral nucelus project to and receive projections from the ipsilateral motor cortex. The thalamic neurons are in turn excited by axons from the contralateral deep cerebellar nuclei (Fig. 8.12) and also receive input from the basal ganglia (p. 232). This input is topographically distributed.

There is thus the potential for complex integration of signals from the basal ganglia and cerebellum with signals from the motor cortex representing recent motor commands. Some indication of the importance of this integration is given by the clinical finding that the tremor and involuntary movements of Parkinson's disease can be relieved by lesions of this region.

Effects of damage to the descending motor pathways

Effects on control of posture and tone

The projections of the brainstem and cortex systems to the spinal motor apparatus are very important in regulating the excitability of the α motor neurons which in turn is an important determinant of the reflex resistance of a muscle to stretching (muscle tone). Immediately following a spinal transection, for example in the cervical region below the level of the phrenic motor neurons, there ensues a state of profound torpor: the muscles lie inert, the strongest stimulation of a sensory nerve evokes no response, reflexes are unobtainable, the blood vessels dilate, the blood pressure falls, thermal sweating is absent, the bladder distends with urinary overflow and the viscera in general are quiescent. This condition is known as **spinal shock**. It is not due to the trauma nor to the low blood pressure but to interruption in the normal flow of impulses down the long descending tracts of the cord that impinge on interneurons and motor neurons so that their excitability is lowered. The lateral vestibular and pontine reticulospinal tracts are most important in this respect and spinal shock follows any section caudal to the lateral vestibular nucleus. The course of recovery varies with the species from a few minutes in frogs to some weeks in humans. There is a gradual increase in excitability; first, flexor withdrawal reflexes return with extensor inhibition, then bladder and rectum empty reflexly, the blood pressure rises, and after some months (in humans) extensor activity returns and tendon jerks can be elicited.

The recovery of excitability of the motor neurons which underlies the reflex recovery may be due to sprouting of the terminals of their remaining inputs. This may enable the input fibres to excite the motor neurons more readily. There may also be an increase in sensitivity of the motor neurons to the transmitter released by the remaining inputs, so that the same quantity of transmitter is now more effective. This would be akin to the post-denervation supersensitivity of skeletal muscle.

Complete brainstem section just rostral to the vestibular nuclei causes the tone to be exaggerated instead of lost. The exaggeration of tone is so extreme in antigravity muscles that a particular posture is maintained—**decerebrate rigidity**. Humans in this condition have arms and legs extended, the back arched and the head dorsiflexed. The feet are ventroflexed and the arms pointed. The wrists are flexed and the fingers little affected. The rigidity is due, essentially, to the removal of inhibitory influences with the maintenance of facilitatory influences, and can be made more extreme by making a more rostral transection, in the midbrain, leaving the pontine reticular nuclei intact. Neurons in this nucleus, like the neurons of the lateral vestibular nucleus, excite the motor neurons of antigravity muscles.

Decorticate rigidity occurs when the motor systems above the brainstem are severely damaged, e.g. removal of most cerebral cortex. It differs from decerebrate rigidity in that it can be modified by reflex means and shows great variation in severity in different species. Cats, for instance, can walk about in a decorticate condition, but a human being in this condition is unconscious, with legs extended and arms in a position determined by the head position (through neck reflexes). Thus, if the head is turned to the right, the right arm is extended and the left arm flexed and if the rotation is to the left the opposite is seen.

Effects on voluntary movement

A complete transection of the spinal cord will cause permanent loss of all voluntary movement controlled by motor neurons caudal to the plane of section. People with a spinal-cord transection above the lumbar but below the cervical enlargement are paralysed in both legs and are said to be **paraplegic**. A section above or in the cervical enlargement below C4 will give rise to paralysis of all four limbs. Such people are called **quadriplegics**. A transection above C4 leads immediately to death as a consequence of respiratory paralysis.

Paralysis or weakness of one side of the body (**hemiplegia**) commonly results from a sudden interruption of the blood supply to an area of the nervous system, causing disruption of the cortical motor pathways to the spinal cord. This is commonly referred to as a **stroke**. Because of the crossed nature of this motor innervation, a hemiplegia signals a lesion of pathways on the opposite side of the brain.

Although reflexes are regained 2–4 weeks after interruption at any level of the pathways from the motor cortex in humans, recovery of voluntary movements takes much longer and is rarely complete. In particular, finger and hand movements are seldom completely regained.

The signs commonly displayed by a patient with hemiplegia, on examination, include the following.

1 Muscle weakness without wasting, particularly of distal muscle groups, e.g. finger muscles are more affected than shoulder. Also the arm muscles are commonly more affected than the leg muscles. Because of their bilateral innervation, the muscles which move the eyes, the muscles of the upper third of the face and the trunk muscles are spared in unilateral lesions.

2 Changes in muscle tone. For weeks or months there is commonly reduced resistance to movement—**hypotonia**. In many patients, this is then replaced by **spastic tone** in which the resistance to movement typically increases as the movement continues.

3 Changes in reflexes. For weeks or months reflexes are lost or diminished in amplitude, but eventually in many patients the amplitude of reflexes such as the jaw, biceps, knee and ankle jerks is increased. **Clonus**, i.e. repeated responses to a single stimulus, may be present.

4 Absence of superficial reflexes. The abdominal reflexes—a brisk contraction of the abdominal muscles following stroking of the overlying skin—are lost.

5 Extensor plantar response (**Babinski's sign**). Normally, the big toe and all the other toes turn down when the lateral border of the foot is stroked. With interruption of the corticospinal pathway the big toe dorsiflexes instead. This is said to be the most important sign in clinical neurology.

This constellation of signs is known to neurologists as the **upper motor neuron syndrome**. In addition, the hemiplegic will display other signs which will indicate the level of the lesion. For example, difficulties with speech with a right hemiplegia suggest a left-side cortical lesion, while the presence of field defects and disorders of sensation may indicate a lesion in the internal capsule, and the presence of a third-nerve palsy, a midbrain lesion.

The neurological basis of the signs of hemiplegia is to some extent obvious. For instance, the loss of movement and its distribution and the initial loss of tone and reflexes reflect the interruption of the cortical output to motor neurons. The spastic tone and brisk tendon jerks indicate an increase in excitability of α and γ motor neurons. This may arise as the result of changes in the spinal cord. These changes include a slowly increasing effectiveness in the excitation of motor neurons by the primary endings of muscle spindles and a slowly decreasing diminution of the effectiveness of the reciprocal inhibition of antagonist motor neurons.

The abnormal reflexes can also be explained by the loss of cortical input. In babies of a few months old, Babinski's sign is present, but as myelination of nervous pathways proceeds and the corticospinal input takes control, the reflex disappears. Its return in hemiplegia is thus no surprise.

8.4 BASAL GANGLIA

The basal ganglia are a group of interconnected nuclei deep to the cerebral cortex in each hemisphere (see Fig. 3.14). They comprise the **neostriatum**, made up of the caudate nucleus and putamen, the **globus pallidus**, the **claustrum**, the **subthalamic nucleus** and the **substantia nigra**, a darkly pigmented nucleus lying in the midbrain

The basal ganglia seem to be important in regulating the function of the cerebral cortex. All regions of cortex project to the basal ganglia, which, via the thalamus, eventually project back upon the cortex. The net effect of this loop is excitatory to the cortex, i.e. a kind of *positive feedback*. However, activity through the loop is modulated by circuitry within the basal ganglia, so that only specific pathways are active. Thus, rather than simply amplifying cortical activity as a whole, the basal ganglia may be important in selecting certain patterns of cortical activity to be promoted, at the expense of others. In circuits involving motor control regions of cortex, this may play a role in the selection and reinforcement of **motor programs**— patterns of cortical activity responsible for generating movements.

Circuits within the basal ganglia

Afferent pathways

The neostriatum is the input structure of the basal ganglia. It receives projections from all regions of the cortex (which are known to be excitatory); there are also inputs from the intralaminar nuclei of the thalamus and the substantia nigra.

Afferents to the neostriatum make direct synaptic contact with the cells which project out of this structure. However, the responses of these cells are modulated by circuits within the basal ganglia. First, the output cells influence each other via recurrent axon collaterals. These are inhibitory to neighbouring output cells, since they use γ-aminobutyric acid (GABA) as the transmitter. Second, local interneurons are present in low numbers, an important group of which use acetylcholine as their transmitter. Third, the **nigrostriatal pathway** from the substantia nigra to the neostriatum (Fig. 8.9a) releases *dopamine* into the striatum; this modulates the activity of output cells and interneurons, and may affect transmission at other synapses. This dopamine system is disordered in **Parkinson's disease.**

(a)   (b)

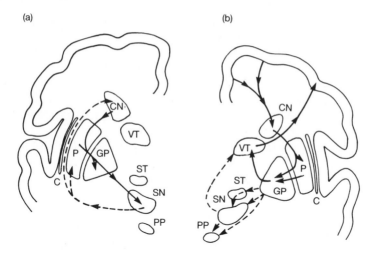

Fig. 8.9 Coronal sections of the cerebral hemispheres illustrating pathways thought to be important in the function of the basal ganglia. (a) Pathways thought to be disordered in Parkinson's disease. Note the striatonigral pathway (unbroken line) from the caudate nucleus and putamen through the globus pallidus to the substantia nigra; the nigrostriatal pathway (broken line) from the substantia nigra to the caudate nucleus and putamen. (b) Pathways by which the basal ganglia influence posture and movement. Note the circuit (unbroken lines) comprising the pathway, cortex, caudate nucleus and putamen–globus pallidus–thalamus–motor cortex, which is thought to play an important role in the supraspinal control of movement and posture and the pathway (broken lines) projecting to the pedunculopontine nucleus which influences locomotion. CN, caudate nucleus; (C), claustrum; GP, globus pallidus; PP, pedunculopontine nucleus; P, putamen; ST, subthalamic nucleus; SN, substantia nigra; VT, ventral thalamic nucleus. (Adapted from Brodal, A. (1981) *Neurological Anatomy in Relation to Clinical Medicine*, 3rd edn, p. 216. Oxford University Press, Oxford.)

Efferent pathways

Output neurons of the neostriatum pass to and inhibit neurons in the globus pallidus and the substantial nigra. Neurons in these target organs are tonically active; they project to and inhibit cells in the ventral group of thalamic nuclei, which in turn project upon and excite neurons in the cortex. Note that around the loop there is first excitation of neostriatal cells; these inhibit globus pallidus and substantia nigra cells, which normally inhibit thalamic cells. Thus activation of the neostriatum will inhibit cells that are inhibiting the thalamic cells, i.e. the thalamic cells will be released from inhibition, resulting in activation of their cortical targets.

In addition to the cortex–basal ganglia–cortex loop, which may be of particular importance for control of limb movements, there are some projections from the basal ganglia to brainstem motor structures, for instance to the pedunculopontine nucleus (Fig. 8.9b) in the midbrain which is involved in the control of locomotion, and from the substantia nigra to neurons in the superior colliculus which control neck and eye movements. Output to the pedunculopontine nucleus is both inhibitory, from the globus pallidus and substantia nigra, and excitatory, from the subthalamic nucleus. The subthalamic nucleus seems to be particularly important in locomotor control; it receives direct excitatory cortical inputs, as well as inhibitory inputs from the globus pallidus.

Functions of the basal ganglia

Neurons in basal ganglia of experimental animals discharge in relation to movement of the **contralateral** limb, as is the case for the majority of motor cortical neurons. In fact, basal ganglia neurons have similar properties to those discovered in the motor cortex, apart from their activities being poorly correlated to force production. There are cells which are activated by sensory instruction signals, during delay periods, in advance of voluntary movements and in association with movement. Recently it has been suggested that each of the cortical–basal ganglia–cortical loops from different cortical regions acts in parallel, structuring activity in each cortical region rather than funnelling information from non-motor cortex to motor cortex, as was once thought. This structuring is thought to be mediated both by the recurrent inhibitory collaterals in the striatum, and by the dopamine input from the substantia nigra. This latter function reflects the postulated role of the dopamine system in **reward**. Dopamine cells are selectively activated by stimuli and actions which are potentially rewarding to the animal, and their activity may be necessary to allow rapid selection of appropriate motor programs likely to result in achieving the reward.

Diseases affecting the basal ganglia

The association of the basal ganglia with the control of movement was first proposed as the result of studies on patients with diseases of, or injuries to, this region. Lesions of the neostriatum or subthalamic nucleus produce disorders characterized by excessive, abnormal involuntary movements such as **athetosis** (slow writhing movements), **chorea** (involuntary jerky movements of the extremities and facial muscles) and **ballismus** (violent flailing movements usually one limb, involving the proximal muscles). Ballismus is a characteristic sign of damage to the subthalamic nuclei, while **Huntington's chorea** is a genetic disease associated with degeneration of the neostriatum.

On the other hand, destruction of the dopamine cells of the substantia nigra which project to the neostriatum produces **Parkinson's disease**, a syndrome which includes **bradykinesia** (a slowness in initiating and changing movement), **resting tremor**, which is characteristically involuntary, occurs at rest, and disappears during movement, and **rigidity** due to abnormally high tone in both agonists and antagonists acting about joints. The tone, however, may wax and wane with the tremor so that the rigidity also fluctuates, giving rise to a characteristic **cogwheel rigidity**. The symptoms of Parkinson's disease can be alleviated by treatment with a dopamine precursor, L-dopa, which crosses the blood–brain barrier and is able to increase levels of dopamine in the basal ganglia. It may be possible in future to treat the condition by implanting dopamine-synthesizing tissue into the brain.

8.5 CEREBELLUM

The gross anatomy of the cerebellum has already been discussed in Chapter 3

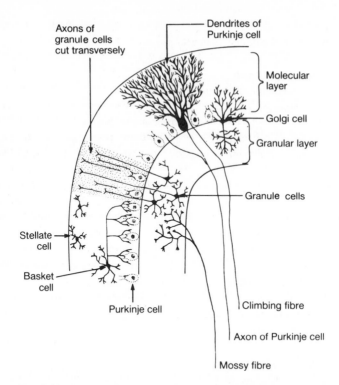

Fig. 8.10 Transverse section of a folium of the cerebellum. (Redrawn from Johnson, T.B. & Willis, J. (eds) (1949) *Gray's Anatomy*, 30th edn, p. 956. Longmans, Green, London.

(p. 65 and Fig. 3.12); its structure at the cellular level is fairly well-understood, as are its neural circuits. It is thus particularly frustrating than an equivalent level of understanding of its function has so far not been achieved. All the lobes of the cerebellum have a three-layered cortex covering white matter in which are embedded the deep nuclei. Histologically, the cerebellar cortex displays a remarkable degree of uniformity (Fig. 8.10).

Cerebellar cortex

The three layers comprise:

1 a molecular layer containing a large number of axons which run along each folium parallel to its long axis, and which intersect Purkinje dendritic trees at right angles;

2 a thin Purkinje cell layer containing the cell bodies of the large Purkinje cells; and

3 a granular layer containing 20 billion granule cells, afferent mossy and climbing fibres, and the elaborate synaptic complexes between incoming mossy fibre axons and granule cell dendrites.

In addition to Purkinje and granule cells, there are Golgi cells, basket cells and stellate cells.

Purkinje cells are about 15 million in number. Their dendritic trees pass up into the molecular layer to intersect with parallel fibres (Fig. 8.11). Spines on their dendrites are the sites of excitatory synapses between granule cell axons and Purkinje cell dendrites. A second type of excitatory synapse is made by climbing fibres which wind round the Purkinje-cell dendrites. The axons of the Purkinje cells pass to other cell types and out of the cortex to the deep cerebellar nuclei. Everywhere the discharge of the Purkinje cells has an **inhibitory** effect.

Granule cells are **excitatory** neurons which receive excitation from incoming mossy fibres and send their axons up to the molecular layer to form the parallel fibres (Fig. 8.11).

Golgi cells, basket cells and stellate cells are all **inhibitory** interneurons (Fig. 8.11b). **Golgi cells** have their cell bodies near the Purkinje cell layer. They inhibit granule cells and their dendrites receive excitation in both the molecular and granular layers. **Basket cells** have their cell bodies in the molecular layer close to the Purkinje cells. Their dendritic trees intersect the parallel fibres which excite them and their axons form a dense meshwork around the initial segment of each Purkinje cell. **Stellate cells** resemble basket cells but they lie more superficially in the molecular layer and their axons terminate on Purkinje-cell dendrites.

Deep intracerebellar nuclei

These are four masses of grey matter embedded in the white matter of each half of the cerebellum. They lie in line from lateral to medial. The largest, the **dentate nucleus,** lies most laterally and the **fastigial nucleus** lies most medially under the vermis, forming part of the roof of the fourth ventricle. Between the fastigial and

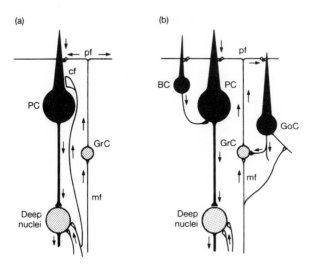

Fig. 8.11 Synaptic relationships within the cerebellum. (a) Input–output circuits (arrows). (b) Inhibitory circuits (inhibitory cells are shown in black). BC, Basket cell; cf, climbing fibre; GoC, Golgi cell; GrC, granule cell; mf, mossy fibre; PC, Purkinje cell; pf, parallel fibre. (Adapted from Eccles, J.C. (1966) In: Granit, R. (ed.) *Muscular Afferents and Motor Control,* pp. 19–36. John Wiley, New York.)

dentate nuclei on each side are smaller masses of grey matter—the **globose** and **emboliform nuclei** in humans, but which in experimental animals form the one **nucleus interpositus**.

The computations performed in the cerebellar cortex influence movement and posture via these nuclei. Their neurons receive excitatory impulses continuously from branches of all types of incoming afferent fibres on their way to the cerebellar cortex (Fig. 8.11). Superimposed on this activity are periods of inhibition of varying length produced by the inhibitory output of the Purkinje cells related to the

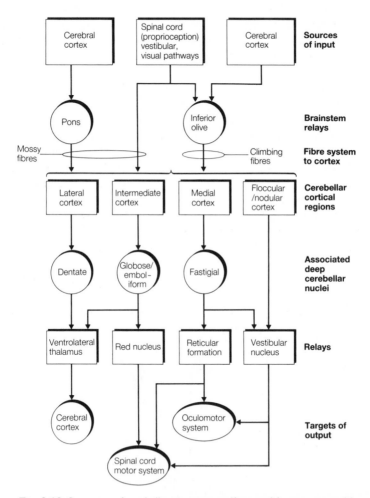

Fig. 8.12 Summary of cerebellar connections (for simplification, several features are not represented). In particular, note that: (1) different inputs to the cerebellum are differentially distributed, to some extent, within the cortical regions; (2) the mossy and climbing fibre systems have side branches that synapse within the deep nuclei; and (3) some of the cerebellar outputs reach structures which are sources of input, forming information-processing loops through the cerebellum.

particular nucleus. The functioning unit in the cerebellum comprises a group of neurons in a deep intracerebellar nucleus and the neurons in a related region of the cerebellar cortex. Figure 8.12 summarizes these relationships and indicates the final destinations of the nuclear axons. The vestibular nuclei are included because, although they are not intracerebellar, they do receive direct projections from Purkinje cells.

Cerebellar afferents

Mossy fibres send information from the spinal cord to the cerebellum by direct and indirect pathways. Of at least 10 tracts, four are sufficiently large and well-known to be described here (Fig. 8.13). The **dorsal spinocerebellar tracts** in the dorsal quadrants of the spinal cord and the **ventral spinocerebellar tracts** in the lateral quandrants arise in the thoracolumbar segments of the cord and terminate in the cerebellar cortex as mossy fibres. The **cuneocerebellar** and **rostrospinocerebellar tracts** are the rostral equivalents of these tracts and also terminate in the cerebellar cortex as mossy fibres. The information carried by the dorsal spinocerebellar and cuneocerebellar tracts is about muscle length and tension, joint position and skin deformation (**proprioception**). Each half of the cerebellum thus receives information during limb movement about the phase and strength of contraction of individual muscles, the joint angles and the time at which the limb extremity touches the ground. In contrast, the ventral spinocere-bellar and rostrospinocerebellar tracts are thought to convey information mainly about the activity of the anterior horn cells in the execution of motor commands, such as the rhythm generation which underlies walking. Other spinocerebellar tracts give rise to mossy fibres indirectly by terminating on the lateral reticular nucleus in the medulla. The axons of this nucleus form mossy fibres.

Information is sent from many areas of the cerebral cortex, including the visual and auditory areas and the motor cortex. The pathways project along with the corticospinal and corticobulbar fibres and terminate on the ipsilateral pontine

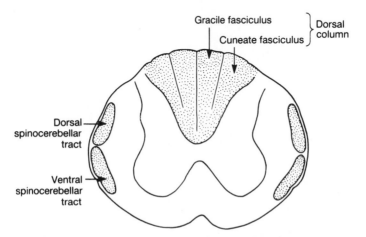

Fig. 8.13 Diagram of the ascending pathways of the spinal cord in transverse section.

nuclei. The axons of the **pontine nuclei** form mossy fibres and are the largest source of such fibres. Collaterals from corticospinal and corticobulbar fibres also terminate here. These inputs are excitatory.

Vestibular input is both direct from the vestibular ganglion and indirect, arising from cells in the vestibular nuclei which give rise to mossy fibres.

Climbing fibres are axons of neurons of the **inferior olivary nucleus**, a prominent nucleus in the medulla at the level of the pyramids. This nucleus has many subdivisions, each with a particular pattern of input, and each subdivision projects to Purkinje cells in a particular part of the cerebellar cortex. The largest inputs to the inferior olivary nucleus come from the premotor cortex, the vestibular nuclei and the spinal cord, but there are other inputs, such as from the superior colliculus which conveys information about the direction of visual movement.

Functional zones of the cerebellum

Although the histology of the cerebellar cortex is uniform, it is divided functionally into longitudinal zones, shown in Fig. 8.14. First, the division is made on the projection of the Purkinje cells to the cerebellar nuclei. In the most medial zone, the Purkinje cells all project to the fastigial nuclei; in the intermediate zone, they project to the globose and emboliform nuclei; and in the lateral zone they project to the dentate nucleus. Second, the longitudinally running climbing fibres and the Purkinje cell axons form zones which can be distinguished in the white matter by systematic differences in axon diameter with large-diameter axons towards the centre of each zone and smaller-diameter axons at the boundaries where zones adjoin. Recently, it has been shown that in these longitudinal zones there are narrower longitudinal bands which are the ultimate functional microzones.

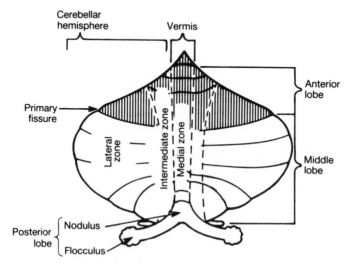

Fig. 8.14 Transverse and longitudinal subdivisions of the cerebellum. The surface has been unfolded and laid out flat. (From Noback, C.R. & Demarest, R.J. (1975) *The Human Nervous System*, 2nd edn, p. 290. McGraw-Hill, New York.)

Functions of the cerebellum

The cerebellum, like the basal ganglia, has long been thought to be concerned with movement because disorders of movement are associated with structural changes in the organ. There are three disturbances that follow damage to particular parts of the cerebellum:

1 disorders of balance and gait;
2 hypotonia; and
3 incoordination (ataxia).

With incoordination the rate, range and direction of movement are disturbed. An intention tremor, due to a timing disorder of the relations between agonist and antagonist muscles in a movement, is often present. Signs of incoordination are also seen in eye movements (nystagmus) and in speech muscles, producing peculiar speech patterns (staccato or scanning speech).

The region of the cerebellum affected determines the nature of the defect. Impaired balance without incoordination or hypotonia is a sign of flocculonodular lobe damage, i.e. in medulloblastoma, a tumour of children. Lesions of the vermis and intermediate area, particularly in the anterior lobe, are a feature of alcoholic cerebellar degeneration and are associated with incoordination of proximal muscles resulting in disorders of gait and of limb movement. Disorders of the cerebellar hemispheres more laterally cause hypotonia, incoordination and intention tremor of the limbs, particularly of the distal rather than the proximal muscles. Cerebellar disorders are always on the same side as the lesion.

Chapter 9
Higher Nervous Functions

9.1 INVESTIGATING THE BRAIN

The structure and function of the brain can be investigated in living people. Computer tomography (CT) and magnetic resonance imaging (MRI) are used to obtain well-defined images of brain structure; positron emission tomography (PET) is used to visualize areas of metabolic activity and electroencephalography to measure electrical activity.

Imaging the brain

Methods of looking at the living brain have improved immeasurably over the past 20 years. One of these techniques, **CT**, uses X-rays to examine brain anatomy. CT scanning enables the brain to be imaged without discomfort to the subject. A number of fixed detectors are used, each forming an image at a different angle as the X-ray source moves round the patient. Instead of films, the scanner uses electronic detectors to form the picture. The computer reconstructs the image, which can be displayed on a television screen or printed.

Another technique, **MRI**, reveals brain structure by detecting radio waves emitted by atomic nuclei (e.g. hydrogen) that are aligned in a strong magnetic field and pulsed with radio waves. MRI can provide cross-sectional images of the brain with much better soft-tissue contrast than CT scanning. A further advantage is that there is no associated ionizing radiation.

PET scanning detects areas of increased or decreased metabolic activity in the brain. Patients are given positron-emitting radionuclides of important metabolic substances (e.g. glucose, O_2) which become concentrated in active brain regions. The emitted positrons react with electrons to produce γ-rays which are detected by the PET scanner, and the signals are processed by a computer in the same manner as in CT scanning. Increased electrical activity at synapses results in increased consumption of glucose and O_2, which is closely coupled to increased blood flow to the region. (The brain uses glucose almost exclusively as an energy source.) Thus the PET scan effectively detects localized changes in metabolic activity and blood flow. It can also locate radioactively labelled neurotransmitters or receptor agonists and thus determine the number of functional synaptic terminals or postsynaptic receptors in a region of the brain.

The electroencephalogram

The earliest technique for examining brain activity was the **electroencephalogram** (EEG). The EEG may be recorded from the scalp or directly from the cortical surface (**electrocorticogram**) and is a record of the minute electrical currents spontaneously generated by neuronal activity. The EEG was first recorded in the 1930s by Berger, who noted that changes in the EEG occurred with sensory stimulation and with age and were associated with the seizures of patients with epilepsy.

Origin

These electrical records are the algebraic summation, at the point on the surface at which the measurements are made, of the changes in potential resulting from current flow between dendrites and cell bodies in the underlying brain. This current is a consequence of the excitatory and inhibitory postsynaptic potentials. Note that the magnitude of the potentials recorded (μV) is much attenuated, reflecting the fact that the electrical changes in membrane potential have been transmitted through a volume conductor.

Characteristics

In conventional EEG recording, 8–16 leads are placed in a standard manner on the scalp over specific cortical areas. The normal EEG in the awake and alert state shows irregular, low-voltage waves (30–80 μV) of high frequency. The trace, a typical example of which is shown in Fig. 9.1 (eyes open), is said to be *desynchronized*. If the subject is bored or the eyes are closed, waves with a frequency of 8–10 Hz are recorded, particularly in occipital leads (Fig. 9.1, occipital trace), indicating *synchronized* activity of neurons. This is termed the α-**rhythm**, which is thought to be induced by activity of the thalamic nuclei. In the frontal areas the dominant activity is faster and of lower amplitude and is called the β-**rhythm**.

Use

The EEG remains the simplest method to examine brain activity and provides the best temporal resolution. It finds clinical use mainly in the diagnosis of epilepsy. It may also be used to detect an infarction or tumour, as indicated by a flat trace in one or more leads, or to assess brain death, which is indicated by a continuous flat trace from all leads, repeated on several occasions. The latter occurs if the brain is

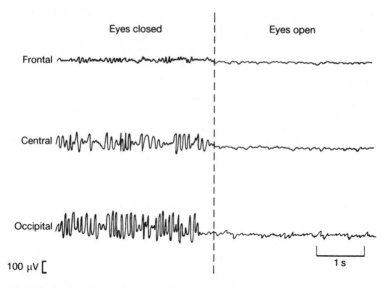

Fig. 9.1 EEG recorded from a human scalp. The traces show typical waveforms recorded over the frontal and occipital poles and in the central region between them. Note the appearance of the 8–10 Hz α-rhythm in the central and occipital traces when the subject has his eyes shut and its disappearance (α-blocking) when the eyes are opened.

deprived of blood for more than 3–8 minutes, although electrical activity in the brainstem may survive longer (7–10 minutes). Another use for EEG recording is in sleep research (p. 250).

By averaging the EEG response to many repetitions of a sensory stimulus (evoked potentials; p. 151) or voluntary movement (motor and readiness potentials; p. 226 and Fig. 8.8), it is possible to detect very small but consistent voltage patterns. These are generated by synchronized activity within sensory or motor areas of the cortex respectively.

9.2 CONSCIOUSNESS

The subjective experience of consciousness is difficult to define physiologically but regions of the brainstem are known to be vital for consciousness and for regulating its level. Consciousness can be divided into certain aspects of experience which are physiologically related to different parts of the brain. The corpus callosum plays an important role in connecting the hemispheres and producing the unity of consciousness that is normally subjectively apparent.

Each morning we wake up—that is, we become conscious of the external world and of our own thoughts and emotions. This state of being conscious has so far defied psychological and physiological explanation. Indeed, all we can do is first define what is meant by being conscious, and then describe what is known of the relationship between brain state and consciousness.

Being conscious means being aware of our identity, of our past and present, and possessing the notion of the future. It evidently includes outward-looking aspects—awareness of others, aesthetic and ethical judgements, the ability to express one's thoughts and ideas—as well as inward aspects—our emotions and fantasies.

There is a variety of levels of consciousness. We can pay close attention (a heightening of consciousness), or daydream (a turning inward which to the observer is a diminution of consciousness). Finally, consciousness is regularly lost every night as we go to sleep and is miraculously regained as we wake up.

Modulation of the level of consciousness

Consciousness may be modulated up (**arousal**) or down (drowsiness, sleep). At a certain level of arousal we may pay **attention** to certain signals to the exclusion of others.

Arousal

Role of reticular formation

There is considerable evidence that the level of consciousness is determined by neural activity in the midbrain grey matter. This region, together with similar regions in the pons and medulla, comprises the **reticular formation**. Reticular means net-like, and refers to the nets of cells (small brainstem nuclei) found enmeshed in fibre tracts in this region.

Modern research in this area started with the observation of patients who had a particular form of sleeping sickness in the aftermath of the influenza epidemic which followed the First World War. Victims of this disease, encephalitis lethargica, initially suffered from insomnia but as the disease progressed they fell into a deep sleep from which they could only briefly and with difficulty be aroused. Many died, and at postmortem there were lesions in the midbrain reticular formation.

Repetition of this experiment of nature in which lesions were made in the same region in animals had the same effect as in humans—the animals became somnolent and developed a synchronized EEG pattern. Furthermore, it was found that electrical stimulation of this region would arouse normal sleeping animals, at the same time converting their EEG to the waking pattern.

Mechanism of arousal

The afferent connections of the reticular formation are very widespread. It is possible to record from reticular nuclear cells and find that they are excited by skin, visual, auditory and olfactory inputs—a truly polysensory input. Such a system cannot convey precise information about a stimulus. It does however convey the message that a stimulus has occurred.

One modern interpretation of reticular function (the **reticular activation theory**) is that the signals from the midbrain reticular formation are used to vary the activity of the **thalamic** pacemakers, and thus cortical neuron excitability. The non-specific nuclei of the thalamus (including the intralaminar and anterior thalamic nuclei), some of which receive input from the reticular formation, project to wide areas of the cerebral cortex. Microelectrode studies show that the thalamic neurons of the non-specific system have two effects. First, they excite cortical cells. Second, they appear to suppress interneurons which in the drowsy state inhibit the specific thalamic nuclei. Upon arousal or alerting, this inhibition is removed so that the specific thalamic nuclei conduct their sensory information more easily to the cortex. One sign of this is that a bigger evoked potential is generated upon arousal.

Attention

Attention is a selective process in which we choose to deal more effectively with one type of sensory stimulus at the expense of other stimuli. For example, we can focus on auditory events and neglect somatosensory events and within one modality we can attend to one attribute and not to another, such as look for red clothing only.

Attention is thought to be a cortical process involving the **association cortex** adjacent to the primary sensory areas in which the incoming signals receive further processing. The cortical sites of attention are revealed by PET scanning. When subjects are presented with stimuli to the three senses—visual, auditory and somatosensory—but asked to pay attention only to one, a number of changes are noticed in the brain blood flow. Attention to the visual modality increases activity in the visual association cortex, the frontal eye fields and part of the parietal cortex. When attention is switched to other modalities, the primary and immediately adjacent association cortex of each modality are active.

Projections to the cerebral cortex from several nuclei in the brainstem and basal forebrain are also important in attention. They utilize specific transmitters, including noradrenaline (locus coeruleus), serotonin (raphe nuclei), dopamine (substantia nigra, ventral tegmental area) or acetylcholine (basal forebrain), which modulate the activity of their target neurons.

Disorders of attention

If particular association areas are damaged, patients are unable to pay attention to and consequently **neglect** stimuli in the affected sensory modality. Neglect has also been described following injury to other parts of the brain presumed to be concerned with attention, particularly the frontal eye fields. Neglect is most commonly seen following right-sided cortical lesions, with ensuing neglect of the left side. Neglect following left cortical lesions has also been described, but is generally less severe and is not as persistent. An interesting feature of neglect is that patients are unaware of and deny the problem.

One of the most dramatic demonstrations of neglect results from damage to the inferior parietal cortex. Patients with such damage ignore stimuli in their contralateral field and often fail to acknowledge all or part of the contralateral sides of their bodies. They fail to dress half of themselves, leave their contralateral foot out of a car and will often claim that their affected limb is someone else's. Parietal lesions that extend laterally to the temporal lobe produce a neglect of auditory stimuli orginating from the contralateral side.

Unification of the hemispheres

We have two cerebral hemispheres, each of which is capable of being aroused and paying attention. However, it is known that language functions are largely confined to one side of the brain, referred to as the **dominant hemisphere**. In most people this is the left hemisphere. Because of its verbal ability, there is no doubt that the left hemisphere is conscious. In patients where the two hemispheres are disconnected by section of the corpus callosum (an occasional therapy for intractable epilepsy), it is the left hemisphere which is able to respond to verbal commands, and to give spoken or written statements about consciousness as experienced by the patient. However, careful testing of such patients has revealed that the non-dominant (usually right) hemisphere also has memory, can pay attention and has the ability to perform complex tasks. While it cannot control speech or writing, a few simple words or emotions can be expressed in response to stimuli that are private to it. Moreover, this hemisphere can be shown by non-verbal means to understand, to some extent, spoken or written words; for instance, matching tests can be performed in which a patient is shown a word through the right visual field and uses the left hand to pick out the matching object from a tray. It should be noted that in such tests the dominant hemisphere is ignorant of the whole transaction.

More recent investigations indicate that there is a fundamental difference in the way that the two hemispheres process information, such that they have complementary roles in normal consciousness. The dominant hemisphere codes verbally and performs poorly if this coding is difficult. The non-dominant hemisphere, in contrast, appears to deal with information that is best expressed in a non-verbal, pictorial way, i.e. in the form of images, and may be important in emotional colouring of language. Figure 9.2 shows the drawings made with the right and with the left hand by a man with a complete transection of the corpus callosum and anterior commissure. The patient had always been right-handed and had never written or drawn with the left hand. As Fig. 9.2 shows, when he copied five sample figures his performance with the left hand was consistently better than with the right hand. Faces, which are difficult to describe in words, are recognized by the non-dominant hemisphere. **Prosopagnosia**, which is usually the result of a non-dominant hemisphere lesion, is a disorder in which familiar faces cannot be recognized.

Role of corpus callosum

Despite the specializations of the two hemispheres we are not aware of any discontinuity in consciousness. There is considerable evidence that this is because the two cerebral hemispheres are interconnected by the **corpus callosum**. The corpus callosum (see Fig. 3.13b) is a large bundle of, for the most part, myelinated axons which have their origin in one cerebral hemisphere and pass to equivalent points in the opposite hemisphere. All regions, except the primary visual, primary acoustic, and foot and hand areas of the pre- and postcentral gyrus, are linked in this way and this linking is part of the mechanism by which consciousness is unified.

Sperry and his colleagues in the 1960s clarified the function of the corpus callosum when they devised techniques which tested each hemisphere separately. In animal experiments this was achieved by sectioning the optic chiasma. The animal then had one eye blindfolded and was taught that, if it pushed a lever when it identified one pattern on a screen rather than another, it would be rewarded. If the blindfold was then put on the other eye, the animal performed as well as it did,

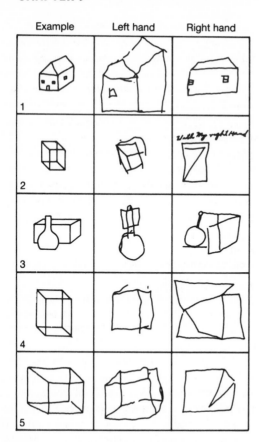

Example	Left hand	Right hand

Fig. 9.2 Drawings with the right and left hand by a man with complete transection of the corpus callosum and anterior commissure. The patient had been right-handed but now his performance with his left hand appears to be better. (From Gazzaniga, M.S., Bogen, J.E. & Sperry, R.W. (1965) *Brain* **88**, 221–236.)

after training, with the first eye. The information as to the correct pattern had clearly been transferred between the hemispheres (Fig. 9.3). That this transfer was due to the corpus callosum was supported by repeating the experiment in another animal in which the corpus callosum had been sectioned together with the optic chiasma. Now when the blindfold was placed on the trained eye no transfer could be demonstrated.

These results have been confirmed in patients with a sectioned corpus callosum. These patients were examined by a procedure which enabled information to be confined to the visual field of one hemisphere, even though the optic chiasma was intact. The patients fixated on a marked point on a screen and a picture of an object was projected on to the screen to the left or right of the fixation point so that it fell within one hemifield and was thus represented only in the contralateral hemisphere. The patients' task was to reach under the screen and select the matching object. Patients consistently succeeded with the hand ipsilateral to the field and failed with the contralateral hand. A normal subject performed the task with either hand. The corpus callosum thus transfers visual information between the hemispheres; other experiments have shown that this transfer can be of acoustic and somatosensory input as well. However, not all interhemispheric

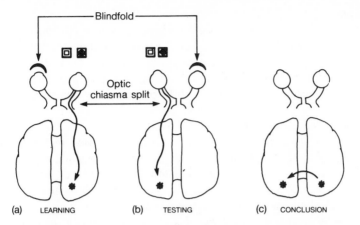

Fig. 9.3 Experiments demonstrating the role of the corpus callosum in transfer of visual information between the hemispheres. A visual discrimination task is learned through one eye and the animal is retested through the naïve eye. (a) Animal with transection of the optic chiasm learns the task. (b) Testing with the naïve eye shows that the left hemisphere knows the task. (c) Conclusion is that information is transferred through the corpus callosum to the naïve eye. (From Cohen, D.H. & Sherman, S.M. (1983) In: (Berne, R.M. & Levy, M.N. (eds) Physiology, pp. 288–313. C.V. Mosby, St Louis.)

transfer is through the corpus callosum. Discrimination of brightness is transferred through midbrain commissures and the anterior commissure can also serve as a link for functions mediated by the prefrontal and olfactory cortex.

Disturbances of consciousness

Physical lesions

Damage to the midbrain, pons, medulla or cortex may result in unconsciousness due perhaps to disruption of the reticular activation system. Such unconsciousness may be transitory, in which case it may be a form of epilepsy or of concussion. **Concussion** implies a brief loss of consciousness due to head and brain injury. Often brain function is impaired for several hours after the patient apparently returns to consciousness so that he or she may later not remember anything that happened during that time (**post-traumatic amnesia**). Alternatively, unconsciousness may be prolonged, in which case it is known as **coma** and the patient is said to be *comatose*. The term coma implies a loss of consciousness, which can be distinguished from sleep in that the subject cannot be totally aroused by strong external stimulation. Coma implies failure of either the association areas of both cerebral hemispheres, if the damage is very extensive, or of the ascending reticular formation of the brainstem and of the thalamus, the structures which keep the cortex 'awake'. Therefore, damage to a single cerebral hemisphere does not produce coma unless it secondarily affects the brainstem reticular formation.

Metabolic lesions

Disturbances of brain-cell metabolism, as in hyper- and hypoglycaemia, anoxia or following a drug overdose, may produce coma by affecting both the brainstem and cerebral cortex. Most forms of coma are accompanied by a fall in oxygen consumption of the brain. For instance, the blood supply may fall to 60% of normal during anaesthesia and to about 50% of normal in a diabetic coma.

Testing the level of consciousness

Consciousness may be graded, as the recovery from concussion implies. If patients can speak, it is usual to find out whether they know who they are, where they are and what time it is. A patient who fails such tests is *disorientated*. Patients may be unconscious but still be aroused by a shout. The patient's eyes may then focus on the source of the stimulation. A very deeply unconscious patient may be unresponsive to anything save a painful stimulus to which he or she may react with grimaces, grunts, movements of arms and legs, and an increase in pulse rate. Such tests, e.g. the Glasgow Coma Scale, are the basis of clinical assessments which provide a rough measure of the level of failure of brain function.

Epilepsy

Epilepsy may be defined as a recurrent, paroxysmal, transitory disturbance of the central nervous system (CNS) that is characterized by uncontrolled neural discharge. Epileptic seizures generally involve total or partial loss of consciousness and may be accompanied by uncontrolled motor activity.

The site of origin of epileptic seizures in the brain is of great interest in that those of focal origin may be treatable. Both cortical and subcortical (limbic and reticular) foci have been found but in many patients there are no signs or symptoms indicative of a focus. **Generalized** or **centrencephalic** epilepsy is the term used to cover this very large group of patients. The term centrencephalic refers to the theory that generalized epilepsy has its origin in disorders of the activating system in the brainstem or thalamus. However, the finding that in such patients the epilepsy may be successfully treated by cutting the corpus callosum suggests they may instead have a focus in one hemisphere from which action potentials spread across the corpus callosum to the other hemisphere. **Focal** epilepsy arises when the seizures begin in a localized area of the cerebral cortex.

Generalized epilepsy

Two varieties of generalized epilepsy are relative common—grand mal and petit mal. **Grand-mal** seizures are characterized by an abrupt loss of consciousness and violent involuntary contractions of skeletal muscles. Together these phenomena are termed a **convulsion**. Many patients have mild symptoms which precede the attack—the **aura**. Many forms of aura have been described—tingling or numbness in the limbs, visual or auditory hallucinations or sudden emotional changes, such as fear. The aura is followed by the convulsion and is usually the last thing the patient remembers. The convulsion has several prominent phases. First, the patient stiffens and is apparently thrown to the floor, respiration stops and

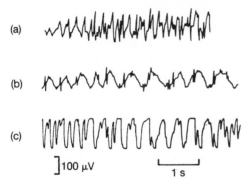

(a)

(b)

(c)

$]100\ \mu V$ $\underbrace{\quad\quad}_{1\ s}$

Fig. 9.4 EEG during epileptic seizures. (a) Tonic and (b) clonic phases of grand-mal seizure; (c) petit-mal seizure.

the pupils dilate. This is the **tonic phase**. It lasts 10–30 seconds, during which the patient often lies with the legs fully extended and the arms flexed and abducted as if decorticate.

It is followed by a **clonic phase** with severe jerking movements and, commonly, emptying of the bladder. After the convulsions the patient appears relaxed and drowsy, and may complain of headache. This stage lasts 30–60 minutes and is often followed by sleep. During a seizure, the EEG usually shows high-voltage spiking during the tonic phase which, during the clonic phase, becomes mixed with a high-voltage slow component (Fig. 9.4a). Electroconvulsive shock therapy (ECT) produces convulsions of a grand-mal type but the motor components are in practice prevented by anaesthetizing the patient and administering a muscle relaxant.

Petit mal usually affects children rather than adults; there is no warning, no aura, but brief periods of unconsciousness or altered consciousness during which the patient may stare blankly; the eyes may roll upwards until the pupils are hidden under the lid. Episodes last 5–30 seconds and a patient may stop what he or she is doing and restart after the seizure without apparently being aware of what has happened. The EEG of petit mal shows a characteristic wave-form (Fig. 9.4c). There is an alternation between high-voltage waves of short and of long duration (spike-and-dome complex).

Focal epilepsy

The commonest form of epilepsy developing in adult life is **temporal lobe** or **psychomotor** epilepsy. Here the epileptic focus is located in the temporal lobe. The seizure is characterized by automatic purposeful reactions such as chewing and smacking the lips. There are no convulsions and the patient may or may not lose consciousness. Focal motor seizures beginning in the motor areas of the cerebral cortex are known as **Jacksonian** epilepsy, after Hughlings Jackson who first clearly described them in 1870. Such seizures usually begin with a twitching, most often of the thumb or of a finger, a toe or the angle of the mouth. These regions are the most widely represented in the motor cortex. Thereafter, the 'march' of the seizure is that expected if the lesion had radiated out from these areas over the rest of the motor cortex. Focal sensory seizures are also known.

International classification

An international system for the classification of seizures was introduced in 1982 to take account of the finding that many patients with recurrent paroxysmal transitory disturbances of the CNS do not fall into the traditional categories. This system concentrates on the patient's signs and symptoms, grouping them as follows:

1 Generalized seizures. Patients in this class have bilateral signs and symptoms without local onset. Classical grand mal and petit mal both fall into this class.

2 (a) Partial seizures. Patients in this class have their signs and symptoms beginning locally. Jacksonian epilepsy falls into this class.

(b) Complex partial seizures. Patients in this class have impaired consciousness, complex hallucinations and may demonstrate automatism (a series of apparently purposive movements of which the patient is unaware) during the seizure. Temporal lobe epilepsy falls into this class.

3 Unclassified. Patients fall into this class if there are incomplete data for classification into classes 1 and 2.

9.3 SLEEP

Sleep is a state of being, distinct from the waking state. It is actively controlled by the reticular formation of the brainstem and does not resemble the unconsciousness resulting from brain damage. There are two types, slow-wave (SW) and rapid eye movement (REM) sleep. We all spend our sleep every night in these states or cycling between them. The EEG is used to define the sleep states.

SW sleep

As one goes to sleep, the α-rhythm, characteristic of the drowsy state, appears and is later replaced by low-amplitude θ waves of a slower frequency (4–6 Hz) which characterize stage 1 sleep (Fig. 9.5). As sleep becomes deeper (as judged by the difficulty of awakening the sleeper), occasional bursts of fast waves (12–15 Hz), called sleep spindles, are seen, signalling stage 2 sleep. Stage 3 is characterized by the presence of high-amplitude δ-waves with a frequency of 1–2 Hz and the presence of K complexes (bursts of more rapid waves on top of δ-waves). Stage 4 is characterized by a trace consisting almost entirely of large δ-waves. Similar phenomena are seen in recordings from all mammals and are referred to as **SW** sleep. When people are awake their eyes are constantly moving (Fig. 9.6a) but during SW sleep, eye movements are much reduced in amplitude and so too is muscle tone (Fig. 9.6b). Also during SW sleep the pupils of the eye are constricted (miosis).

REM sleep

At times during sleep the EEG shows a desynchronized pattern resembling that recorded in the waking state (Fig. 9.5). When the EEG from the cortex shows this desynchronized pattern, large-amplitude rapid (60–70 Hz) saccades in eye movement are recorded, which have given the name **REM** sleep to this stage of sleep (Fig. 9.6c). A total postural relaxation is also found due to an almost complete loss of muscle tone (Fig. 9.6c), though sudden jerks of limbs are very common.

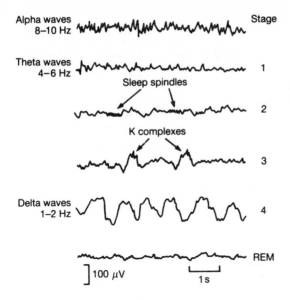

Fig. 9.5 Typical EEG traces recorded during the successive stages (1–4) of slow wave (SW) sleep, and during REM sleep.

Sleepers are more difficult to arouse from REM sleep than from SW sleep and sudden dilatation of the pupils (mydriasis) may accompany REM.

Closer examination has shown that the REM state is accompanied by a number of signs of excitement and stress. For instance, there is a marked variation in the pulse and blood pressure and irregularity in respiration, accompanied by secretion of corticosteroids. Furthermore, in males a cycle of penile erections occurs, beginning and ending with each REM period, and testosterone secretion is increased at night in association with REM sleep.

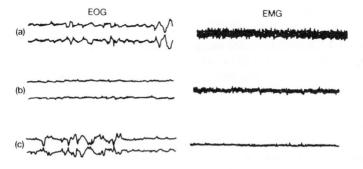

Fig. 9.6 Records of electrical activity from extraocular muscles (electro-oculogram, EOG) and from limb muscles (electromyogram, EMG): (a) in the awake state, (b) during stage 2 of SW sleep and (c) during deep REM sleep. Note in REM sleep the characteristic activity in the EOG and the flat trace in the EMG.

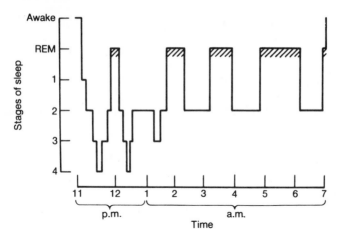

Fig. 9.7 The sleep cycle in a young man. The stages of sleep are judged from EEG criteria. Periods of REM sleep are indicated by cross-hatching. (From Hubbard, J.I. (1975) *The Biological Basis of Mental Activity*, p. 135. Addison-Wesley, Reading, USA.)

Duration of SW and REM sleep

The majority of people sleep 7–8 hours a night but there are also short (5–6 hours) and long (8–9 hours) sleepers. When compared with long sleepers, short sleepers go to sleep more quickly, spend less time in stages 1 and 2 and in REM sleep but spend the same time in the deepest stages (3 and 4) of SW sleep. The first REM period normally lasts about 20 minutes; later periods become progressively longer (Fig. 9.7). SW sleep develops with maturation of the nervous system. Those newborn animals whose CNS is not completely developed at birth show only alternations between waking and the REM state. Human babies at birth spend some 18 hours asleep, of which 45–65% is in the REM state. A 2-year-old is still spending 40% sleeptime in REM sleep, but a 5-year-old approaches the young adult figure of 20%. For those over 50 the period spent in REM sleep forms about 15% of total sleep.

Dreaming and mental activity during sleep

Subjects awakened from REM sleep almost invariably report that they have been dreaming. Apparently, we all dream, although we do not always recall our dreams.

It is attractive to suppose that the eye movements, limb twitches and penile erections, together with the alternations of pulse, arterial blood pressure and respiration which occur during the REM state of sleep, are associated with the acting-out of dreams. There is some evidence for this view largely drawn from correlation of the content of dreams and the observed signs. Dreams, of course, may be distressing and it is of interest that angina pectoris may also occur at this time, while sufferers from duodenal ulcers, migraine and partial seizures, especially those of frontal or temporal lobe origin, tend to have exacerbations of their symptoms during periods of REM sleep. Presumably because of the lack of muscle tone (atonia), sufferers from obstructive lung disease are also adversely affected by REM sleep.

Mental activity goes on during SW sleep. About 20% of subjects awakened from this state say they have been thinking or even deny being asleep. Evidently, the mental activity is akin to everyday mental activity and lacks the hallucinatory aspect of dreaming. Though it has been suggested that learning is possible during sleep, research shows that this may occur only when the EEG shows α-waves.

The need for sleep

The working hypothesis of many investigators is that SW sleep is needed for bodily repair. There is evidence, for instance, that marathon runners have more SW sleep after their race than before. It is well-established that growth hormone is only secreted during SW sleep. The idea has also been investigated that REM sleep, on the other hand, is concerned with recovery from mental tiredness.

Neural basis of sleep

Since the reticular formation is involved in maintaining consciousness (p. 243), we might assume that if it was inactive then an animal would go to sleep. Indeed, radical deafferentation does make an animal somnolent, with a synchronized EEG for a week or so following the operation. Thus, sleep has been attributed to the absence of arousing stimuli to the reticular formation and therefore to thalamus and cortex. However, sleep-waking rhythms reappear after a few weeks. Moreover, in humans removal of all external stimuli leads to hallucinations rather than sleep. More probably, sleep is accompanied by active inhibition of that part of the reticular formation responsible for arousal. In 1953 the Swiss physiologist Hess showed in a conscious animal that stimulation of a number of sites in the brain through implanted electrodes will induce the animal to go to sleep in a natural manner, and will also evoke the appropriate synchronized EEG. The best characterized of these sites lies in the forebrain in the region rostral to the hypothalamus between the anterior commissure and the optic chiasma—the **preoptic region**. Lesions here make animals permanently sleepless and the region contains a population of neurons which discharge maximally in both SW and REM sleep. The timing of SW sleep appears to be linked to the pacemaker activity of the nearby **suprachiasmatic nucleus**.

The preoptic region has substantial efferent projections to the midbrain, particularly to its reticular formation. These are known to be predominantly inhibitory, suggesting that the preoptic region, as well as inducing the behavioural and EEG signs of sleep, also puts the machinery for wakefulness and attention out of action.

Sleep disorders

The most common sleep disorder is **insomnia**, the complaint of not enough sleep. This may affect 20% of the population at one time or another, but in most cases no physiological problem with the sleep-generating mechanism can be identified. Insomnia is usually a symptom of an underlying pyschological disturbance (e.g.

depression, anxiety or excitement) or due to a physical cause such as the discomfort associated with disease or injury. The opposite to insomnia is **narcolepsy**, a disease of unknown cause in which there is an irresistible urge to sleep during daytime activities.

Some people have severely disturbed sleep due to **sleep apnoea**, periods where respiration stops completely during sleep. The sufferer awakes, takes a few breaths and falls asleep again, only to awaken soon after with another apnoea attack. Sleep apnoea can arise from intermittent airway obstruction, as in snorers, or from lesions of the CNS. Another disorder leading to disturbed sleep is **nocturnal myoclonus**, in which there are sudden repeated contractions of muscles, commonly in the legs, during sleep. This disease is thought to be akin to epilepsy.

Sleep-walking (**somnambulism**) and bed-wetting (**nocturnal enuresis**) have been shown to occur during periods of SW sleep. Somnambulists walk with their eyes open and avoid obstacles, but when awakened they cannot recall the episode. These problems are more common in children than in adults.

9.4 ORGANIZATION OF BEHAVIOUR

Certain behavioural patterns such as those related to eating, drinking and reproduction are so essential to the survival of the species that they are genetically programmed. The limbic system is responsible for the expression of such behaviours and includes a reward system which reinforces those behaviours favourable to the animal. Thus the limbic system has an important role in learning; it is also involved in the generation of emotional state. The prefrontal cortex has an 'executive' role in regulating the behavioural drives generated by these systems, so that actual behaviour is appropriately suited to the prevailing circumstances.

Ingestive and sexual behaviours

Ingestive and sexual behaviours are central to the survival of the organism and the species. They are generated in the **hypothalamus** but can be influenced by other brain regions, such as the **limbic system** (p. 69; Fig. 3.15) and **neocortex**. Electrical stimulation of the hypothalamus can evoke ingestive and sexual behaviours, and decortication does not prevent such behaviour. Interestingly, acting-out of the behaviour outlasts the stimulation, which appears to be only a trigger. Furthermore, the hypothalamic neurons control not only autonomic neurons and the pituitary but must also have access to motoneurons, probably through reticulospinal pathways originating in the midbrain. Certainly, it is possible to evoke behavioural acts by midbrain stimulation, but these displays differ from those evoked by hypothalamic stimulation in that they do not outlast the stimulation.

Control of hypothalamic function

Limbic nuclei (e.g. amygdala, septum) control and organize hypothalamic functions so that behaviours are expressed in appropriate situations and ways. For instance, the **amygdaloid nuclei** appear to have, as one function, control of what is eaten and when. Electrical stimulation of the medial amygdala in cats, dogs and monkeys facilitates eating, while stimulation of the lateral amygdala prevents eating even in starved animals. After lesions of the amygdala, animals and humans will put anything in their mouths. Rats, normally very conservative eaters, will for instance eat unfamiliar foods and even food which had made them ill in the past and which they previously avoided. The **septal nuclei** too, appear to have functions connected with coupling of food intake to metabolism and the regulation of taste preferences. Normally, when rats exercise, they eat appropriately to maintain body weight and drink enough to wash their food down. After septal lesions, increased activity leads to persistent weight loss and water intake becomes abnormally high, remaining so even if the animal is deprived of food. Further, the rats show an exaggeration of taste preferences as if some inhibitory control had been removed.

Reward system

In addition to spontaneously generating behaviours resulting from inbuilt drives, animals are able to learn from experience to select those behaviours which produce favourable outcomes (**reward**) and to avoid those that are unfavourable. This learning is promoted by activation of a brain pathway, the 'reward system', which signals to the animal that the outcome of behaviour was favourable. The reward system has been mapped in animal studies in which electrical stimulation through implanted electrodes was used to reward appropriate behaviour. In 1954 Olds and Milner found that rats thus prepared could be trained to press a lever which turned on stimulating current. Regions of the brain could be classified according to the percentage of time the animals spent pressing the lever. Brain areas with high scores were found in the septal nuclei and to a smaller extent in the cingulate cortex and the hypothalamus. Self-stimulation of the hippocampus occurred less often but was still at rates above control levels. The most remarkable area was the **median forebrain bundle** in the lateral hypothalamic area which, among its other functions, connects the hypothalamic nuclei with the septal nuclei rostrally and with the midbrain caudally. If a rat had a stimulating electrode implanted there, it continually pressed the bar, suggesting that the ensuing brain stimulation was experienced as a reward. Indeed, given a choice between feeding and pressing, the rat would rather stimulate its median forebrain bundle than eat.

A number of questions are raised by these studies, notably the meaning of 'reward' in this context. Studies on people are few but have particular value, for only in people can the meaning of 'reward' be investigated. Such studies indicate that 'rewarding' brain stimulation is pleasurable. In a patient provided with a small stimulator connected to several electrodes implanted in his brain, the highest score was recorded from one particular point in the septal region, which the patient declared made him feel 'good'.

Emotion

Emotions seem to have a similar role to the reward system in directing behaviour. The emotional consequences of an action or situation signal some aspect of the value and seriousness of that event, which is used to govern subsequent behaviour. Furthermore the outward *expression* of emotion is very important in the organization of social interactions between individuals of the same species. It allows the value and seriousness of one individual's situation to be understood by others and so guides their behaviour.

The **limbic system** (p. 69; Fig. 3.15) is thought to be the site of generation of emotional tone but emotional expression resulting from limbic system activity is mediated by the hypothalamus through the autonomic and somatic systems. In people, stimulation of the limbic structures may enhance and lesions reduce emotional tone. For instance, aggressive behaviour has been reported in a patient during stimulation of one of the amygdaloid nuclei through implanted electrodes, while bilateral amygdalectomy has resulted in an apathetic individual. Nowadays surgeons who attempt to treat psychiatric disorders use stereotactic techniques to make very localized brain lesions. Undercutting of the cingulate gyrus (see Fig. 3.15a; cingulotractotomy) has been used in severe chronic depression which does not respond to any other form of treatment. Enthusiasm for such operations should be tempered by the possibility that the improvement may not be permanent. In some studies, half of the patients initially improved by cingulotractotomy relapsed within 3 years of the operation and were left with permanent brain damage.

Role of prefrontal cortex

Animals are capable of adapting their internal drives to the realities of the immediate physical and social environment, and also to the historical context of remembered and predicted outcomes. These aspects of behavioural regulation seem to rely on the integrity of the **prefrontal cortex**, that part of the frontal lobe anterior to the premotor area (see Fig. 8.7). In this region information about the external world (from visual, auditory and somatosensory pathways) is integrated with information about the internal state of the organism relayed from the limbic system and hypothalamus.

This role of the prefrontal cortex was first suspected in the 19th century as a result of a study of Phineas Gage, a railroad foreman injured in an explosion when an iron bar penetrated his skull below the left orbit and emerged in the midline vertex. He survived despite (as a postmortem later showed) the severing of all connections between the prefrontal cortex and the rest of the brain, but was described as being changed from being a careful, conscientious, mild-mannered man to a facile, careless and irritable drunkard. He was said to be impatient of restraint or advice when it conflicted with his activities which were no sooner arranged than they were abandoned for others that appeared more feasible.

Prefrontal lobotomy

In 1930 Fulton described the effects in the chimpanzee of cutting the connections of the prefrontal cortex with the rest of the brain. The animals showed a significant reduction in emotional responses to failure and stress and, where previously aggressive, became tame. A similar operation (**prefrontal leukotomy** or **lobotomy**) thereupon became popular for treatment of severe depression and, initially, of schizophrenia. Following such lesions of the prefrontal cortex in people speech, movement, vision, hearing and intellectural skills appear normal. Behaviour is, however, abnormal in subtle ways.

Mood changes

The more obvious changes are in mood. For instance, in one series of 32 patients, 20 became euphoric. Those patients in whom this trait was most prominent were spontaneously happy and contented no matter what their circumstances, laughing and joking upon the slightest provocation. In other cases, particularly when patients were very shy before operation, they became calmer, found it easier to meet other people and were less embarrassed. A few patients who were very depressed for years before operation did not become euphoric but became even more despondent.

Intellectual changes

Such changes after prefrontal lobotomy are difficult to detect by ordinary psychological tests. Patients have difficulty in following instructions involving constant changes. For instance, when told to draw a circle, the patient will do so, but when then asked to draw a cross he or she is likely to repeat the circle (**perseveration**). This seems to be due to enhanced proactive inhibition (p. 259). Reading and writing are not affected but there is a weakening of more complicated intellectual function, as is demonstrated by the less satisfactory outcome of the operation in patients with more complex jobs. In patients with prefrontal lesions there is increased distractability, leading to an inability to concentrate for long periods on a task, while animals with similar lesions become unable to perform delayed-response tests. This may be categorized as a disturbance of goal-directed behaviour.

Personality changes

There is loss of social restraints, together with loss of the normal ability to evaluate one's own actions and profit from mistakes. Superficially, the patients appear satisfactory after operation but, upon questioning in detail, reveal that they cannot feel deep happiness or deep sorrow, that they forget things, and that they have lost most of their interests because they cannot concentrate on any one topic. Moreover, close relatives are convinced that there has been a personality change. Interestingly, patients with seizures confined to the prefrontal cortex also show associated transient personality changes.

9.5 LEARNING AND MEMORY

Memory can be divided into sensory, short-term and long-term memory on the basis of psychological testing and what is known of the underlying physiology. Sensory memory refers to persistent activity in sensory structures lasting about 250 ms, short-term memory involves changes in brain activity lasting a few seconds, and long-term memory involves more or less permanent changes in the brain. Memories are probably distributed in many sites in the brain but the hippocampus has a special role in forming and retrieving memories.

When we talk about memories or remembering, we use the word in a variety of senses. We may say 'I remember how to ride a bicycle', 'I remember that picture', 'I recall that song'. Research has shown that each of these memories is of a different sort and relies on different brain mechanisms. Motor memories, such as the ability to ride a bicycle, have not been much studied but it is clear that they may survive and be added to when other types of memory are badly impaired. The recognition and recall types of memory have been studied in human subjects mostly by means of verbal material and in animals by use of recognition tasks such as learning of mazes. It has become clear that these types of memory are not akin to a photographic record but what is remembered is a reconstruction based as much on what was expected as on what actually happened. Accordingly, witnesses are notoriously fallible.

People with so-called photographic memory can recall at will an image of what they have seen, apparently localized in space in front of their eyes. The image is known as an **eidetic image**. Detailed investigation suggests that eidetic imagery is an alternative means of storing information, distinct from, but not more effective than, ordinary memory. Nothing is presently known of the physiological basis of this ability.

Studies of the normal type of memory for verbal material indicate that there are short-term (primary) memories for snippets of information looked up immediately before use (e.g. telephone numbers), longer-term (secondary) memories of much-rehearsed material, and memories which last a lifetime (tertiary memory), such as one's name and important life events. Recently, after-images have been brought into this scheme, it being realized that they are epiphenomena of the initial step in memory (sensory memory).

Sensory memory refers to that brief period when information has reached receptors and is about to be sent on to the CNS. Information presented to the eyes can be evaluated and gives rise to reflex responses before it is perceived. The after-images seen when one looks at an object and then away from it may last about 250 ms, decaying over this period. Apparently, the memory trace is destroyed by fresh incoming information. The capacity of the visual sensory memory for after-images appears to extend to about 6 or 7 items. A sensory memory has also been demonstrated in the auditory system.

It seems obvious, although it is still a matter of much investigation and dispute, that for material to be remembered it must have been sequentially in sensory, primary, secondary and perhaps tertiary memory.

Short-term memory

The characteristics of short-term or primary memory are small capacity, short duration and storage as words.

1 Relatively few items can be recalled immediately after a short exposure (visual or aural). For instance, when photographs, printed letters or drawings are flashed on a screen for very short, precisely controlled intervals of time (tachistoscopic experiments) subjects can recall only five to nine items.

2 In this type of experiment, primary memory decays in a few seconds.

3 A study of errors in primary memory suggests that the material is coded in words. Volunteers in tachistoscopic experiments were found to be rehearsing the

material they had learned, by mumbling either overtly or covertly. In remembering, mistakes were made between letters that sounded alike. The sounds were confused, although the subjects were shown only the letters. The same mistakes were made, in fact, if the subjects were shown the material or if it was spoken. Thus, mistaking 'cat' for 'hat' was possible, but not 'cat' for 'kitten'.

Primary memory is thought to be represented in the nervous system by circulating nerve impulses, because an insult which disrupts the functioning of neurons prevents any memory of events which take place a short time before. Thus, after head injury causing loss of consciousness or after ECT there is a short but persistent loss of memory (**amnesia**) for the period immediately preceding. A permanent reduction in the capacity of short-term memory may occur after lesions in the cerebral cortex. For instance, a reduction in verbal and short-term memory has been described in a patient with a parietal lobe lesion.

Long-term memory

The characteristics of long-term, i.e. secondary and tertiary memory, are large capacity, long duration and organization.

1 The storage capacity of secondary and tertiary memory is presumably very large, though no quantitative estimates are available. There are accounts of people with phenomenal memories, perhaps the most sensational being that of a man observed by Luria over a 30-year period. This man had no obvious limits to the capacity of his secondary memory for he could never be given a list of words so long that he could not recall it perfectly.

2 People vary greatly in their ability to retain secondary memories. For instance, Luria's subject could reproduce any lengthy series of words even if, as was several times the case, the test was held about 15 years after the man originally learned the words. His memory was such, indeed, that he could recall the place where the test was first presented to him, how Luria was dressed, where he sat, and so on. It appears that tertiary memory may be permanent, in that some information is retained for the life. Often this information has a very strong emotional association.

3 With secondary memory, confusion has a semantic rather than an acoustic basis; that is, words which have a similar meaning are confused rather than words with a similar sound, implying that the organization is semantic and relational. This was utilized by classical orators, training themselves in the art of memory. They were advised to imprint on the memory first a visual picture containing a series of identifiable places, commonly a building with its rooms and ornaments. In each place was then put, in imagination, a mark or pointer to a particular part of the speech. When it came to making the speech, the orator, using visual memory, simply went through the building in an orderly fashion picking up each pointer on the way.

Secondary memory is notoriously difficult to access. There is much evidence that interference from material learnt before (**proactive inhibition**) or afterwards (**retroactive inhibition**) produces forgetting. Proactive inhibition appears to be the more troublesome. Tertiary memory appears to be easy of access and may be permanent, in that some information is retained for the life of the subject. Often this information has a very strong emotional association.

Secondary and tertiary memories are thought to be represented in the actual structure or pattern of the CNS for they remain after neural insults which have destroyed primary memories. Nevertheless, damage to certain parts of the brain does affect long-term memory, perhaps by destroying the mechanisms by which material to be remembered is recovered or laid down.

Neurological basis of long-term memory

The integrity of the limbic system is essential for secondary memory. Bilateral damage to certain parts of the hippocampus, fornix, mammillary body, medial hypothalamus or those parts of the thalamus connected with the mammillary bodies has the same effect, namely a remarkably severe and persistent amnesia.

Role of the hippocampus

Of the limbic structures, damage to the hippocampus causes the most severe amnesia. There are patients suffering from epileptic seizures apparently originating in the temporal lobes who have had the medial part of one temporal lobe cut away, together with the underlying hippocampus, to try to cure the epilepsy. Unfortunately, in one patient the operation was performed bilaterally and he developed global amnesia. He could not form any new memories and had only patchy memories of the decade preceeding the operation. When asked about his house, his family or the city he lived in, the patient did not even seem to understand these questions, as if their objective was entirely unknown to him. He appeared to feel completely isolated, with neither a past nor a future. Long after the operation, a grave memory defect persisted, although there was some improvement in the ability to pay attention.

Analysis of many case histories confirms that patients with damage to the hippocampal system show no loss of motor skills acquired preoperatively and that intelligence, as measured by formal tests, is unimpaired. But, with the possible exception of the acquiring of new motor skills, they seem largely incapable of adding new information to their long-term store. The immediate registration of new input (short-term memory) appears to take place normally and material that can be rehearsed verbally is held for many minutes. Interruption of the rehearsal, however, produces immediate forgetting of what has gone before. Material already in long-term storage is unaffected by the lesion.

Further evidence for the importance of the hippocampus comes from the effects of stimulation of the brains of conscious patients during operations. For instance, in 1954 the Canadian neurosurgeon Penfield described that, when he stimulated particular points in the temporal lobes overlying the hippocampus, his patients reported that fragments of past experiences seen, heard or felt long ago were being re-experienced and, moreover, at the same rate as originally. Things not in the focus of the patient's attention at the time of the original experience were not present in the hallucinations. Many of these patients suffered from epilepsy. Their attacks were preceded by flashbacks consisting of the recall of experiences from their past similar to those evoked by brain stimulation. It is important to note that patients reported that the material either elicited by stimulation or coming into consciousness before an epileptic attack was much more distinct than anything they could normally recall. Penfield was convinced that his patients had records of their experiences in their brain which were normally not completely available to them.

These striking observations might suggest that each hippocampus is a site at which memories are stored. However, this cannot be completely true, because patients with bilateral hippocampal damage can acquire motor skills and also verbal long-term memories by special techniques in which they are given part of the information but have to supply the rest. In both types of learning, they are as effective as normal subjects. Memory formation is now thought to require concurrent activity in the hippocampus and other brain sites.

Hippocampal lesions may prevent that suppression of old information which is apparently needed for the storage of new information. After bilateral removal of the hippocampus, rats and monkeys react to new stimuli more sluggishly. Furthermore, they are very slow to reverse their responses in the type of study in which the subject has to respond to one of two situations, e.g. the right or the left. They also continue to respond for much longer in the absence of a reward than do normal animals. Experiments with rats also show that the hippocampus plays a major role in the encoding of spatial memory. For instance, in freely moving rats hippocampal pyramidal cells display both selectivity and memory for specific locations.

Recently, a phenomenon known as **long-term potentiation** has been described in the hippocampus and cerebral cortex which may be the cellular mechanism of memory. Long-term potentiation was first discovered following brief, high-frequency stimulation of one of the inputs to the granule cells of the hippocampus (the **perforant pathway**). This pathway normally excites granule cells which have a particular type of glutamate receptor—the N-methyl-D-aspartic acid (NMDA) receptor. Following two to five trains of conditioning stimuli at frequencies of 20–100 Hz, the number of granule cells responding to single testing stimuli of this same pathway was increased for periods varying from hours up to days, depending on the duration of the conditioning trains. Blockade of the NMDA receptors with drugs prevents this development.

Classification of amnesias

Damage to the hippocampus and related structures results in memory impairment in which there is an inability to add new memories following the lesion, although older memories remain intact. This is called **anterograde amnesia**. It is also seen in some chronic alcoholics (**Korsakoff's psychosis**). In contrast, temporary disruption of brain function, for example after severe head injury, may result in **retrograde amnesia**, in which there is loss of memory of events that occurred over a considerable period prior to the injury. The events before the accident are often very dramatic and must have been registered by the nervous system, yet quite often the subject has no memory of them. During recovery the length of the period of retrograde amnesia shrinks markedly.

Clinical experience also suggests that **suppression** of memory occurs, for example, following ECT for depression. After treatment many patients forget their anxieties and are able to function normally again, although when interviewed later under the influence of a barbiturate they can recall their troubles, showing that these were not permanently forgotten but simply not recalled in the normal waking state.

9.6 SPEECH

Speech, together with its associated activities (reading and writing), is a most complex phenomenon. Basically, it is the conveying of meaning by the spoken word. It has a motor output which has specialized control regions in the brain, usually in the dominant hemisphere (the left side in most people).

It is necessary to distinguish the production of voice (**phonation**), the shaping of voice into words (**articulation**) and speech proper, which implies meaning and understanding. In one sense, speech is a motor activity involving control of the labial, lingual, pharyngeal, palatal and respiratory muscles. Motor activities in general result from plans made in the frontal lobe, which are further elaborated by the basal ganglia and cerebellum and relayed to the ventrolateral thalamus. A final relay to the motor cortex brings appropriate motor neurons and thus muscles into action. The motor aspects of speech do not differ from other motor activities except in the planning stage, which is complex.

The sounds made by animals as the result of strong emotion are akin to the expressions made by humans in similar excited states. The production of such sounds is controlled from the limbic system in animals.

Dominant hemisphere

It has long been known that speech is unique amongst motor activities in that its higher control depends on the integrity of only one of the two cerebral hemispheres, the **dominant hemisphere**, which, in most people, is the left. This hemisphere also controls the right hand but cerebral dominance is not necessarily paralleled by appropriate handedness. Tests have shown that 95% of right-handed and 50% of left-handed people have their speech centre on the left.

Dysarthria

The production of speech thus involves subcortical and motor cortical components and also a component from association areas. Disturbance of any component will cause a difficulty in speaking. Such difficulties are termed **dysarthrias** if they involve the motor apparatus in the strict sense and **dysphasias** if they involve the unique one-sided cerebral control. A characteristic dysarthria occurs in many diseases, such as the scanning speech of cerebellar disease and the slow, slurred speech of Parkinson's disease. While there may be some temporary disturbance of articulation, a permanent dysarthria does not commonly occur with the hemiplegia following a cerebral vascular accident because the muscles of the palate, pharynx and vocal cords are bilaterally innervated through the vagus and glossopharyngeal nerves. Bilateral loss of cerebral control of the medullary cranial nerves, on the other hand, causes a severe dysarthria as well as numerous other signs and symptoms (**pseudobulbar palsy**).

Dysphasia

Following brain lesions in the cortical areas shown in Fig. 9.8, patients have a difficulty in speaking which differs from dysarthria. There is no impairment of the muscular apparatus or its control. The difficulty is in the higher processes connected with the selection and ordering of words. This type of disturbance is commonly called an aphasia, which literally means 'without speech'. However, strictly speaking, such patients should be said to have a **dysphasia**, since they can speak, although with difficulty.

Non-fluent aphasia

A French physician Broca noticed in 1861 that patients with a right-sided hemiplegia had a poorly articulated speech produced slowly and with great effort and abnormal in rhythm and intonation. For these reasons, it is termed a **non-fluent** asphasia. This is the commonest form of aphasia. The speech content has been compared to a telegram in that patients can name objects and produce single words but have difficulty with sentences, coming to grief on the connecting words. These patients have trouble in reading and find writing difficult or impossible. Their comprehension appears unimpaired but formal testing reveals deficits. At postmortem, such patients may have lesions involving the frontal lobe (Fig. 9.8a). An older name for this sort of aphasia is **motor** aphasia.

Fluent aphasia

In 1874 a German neurologist Wernicke described patients with normal articulation, rhythm and expression, and correct grammar who had great difficulty in finding the correct words. The patients were unable to understand spoken or written language, that is, unlike the patients with non-fluent aphasia, their comprehension was obviously severely impaired. In severe cases, they might produce a grammatically correct but meaningless flow of language. This form is termed **fluent** or sensory aphasia. At postmortem such patients may have lesions involving the posterior part of the temporal lobe (Fig. 9.8b).

Other forms of aphasia are known in which patients are thought to have lesions in the white matter underlying and connecting the areas described by Broca and Wernicke (Fig. 9.8). Patients with such **conduction** aphasia can, for instance,

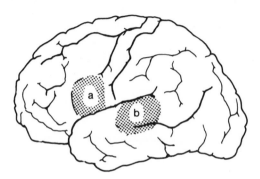

Fig. 9.8 A lateral view of a human brain showing areas important for the generation of speech. (a) Broca's area; damage here may cause motor aphasia. (a) Wernicke's area; damage here may cause sensory aphasia.

comprehend the speech of others but cannot repeat it. Their own speech is fluent but full of circumlocutions and of errors in word use.

More restrictive language defects arise when the speech area is disconnected from the cortical representation of one particular modality. In these cases, a patient may be able to name an object presented in one sensory modality but not when presented in another (**anomic** aphasia). Speech itself is fluent and comprehension intact.

Chapter 10
Endocrine System

The endocrine system is one of the two coordinating and integrating systems of the body, the other being the nervous system. It acts through chemical messengers (**hormones**) carried in the circulation and is involved in:
1 maintenance of the internal environment (homeostasis);
2 the control of the storage and utilization of energy substrates;
3 the regulation of growth, development and reproduction; and
4 the body's responses to environmental stimuli.
In contrast to the nervous system, the actions of the endocrine system are slower in onset, more prolonged and generally more diffuse. The two systems are linked, however, through the **hypothalamus** which controls the secretion of many of the endocrine glands. Consequently, influences acting on or through higher centres of the brain, e.g. emotion, can affect endocrine secretion.

Evidence that an organ functions as an endocrine gland can be obtained by studying the specific effects of its removal from the body, by transplantation of the gland back into the body and by injections of extracts of the gland. The first experiment in endocrinology is attributed to Berthold, who in 1849 showed that transplantation of a testis into the abdomen of a castrated cockerel restored its secondary sex characteristics. The principal endocrine glands in mammals are the **hypothalamus, pituitary, thyroid, parathyroids, adrenals, pancreatic islets,**

gonads and **placenta**. Hormones and hormone-like substances are also produced in the heart (p. 443), kidney (pp. 288 and 327), liver (p. 280), lungs (p. 651), pineal gland (p. 318), thymus (p. 344) and certain cells of the gastrointestinal tract (p. 588). Many substances which were first identified as hormones produced in endocrine glands are now also known to be synthesized in the nervous system, where they act as neurotransmitters or neuromodulators (p. 100).

10.1 HORMONES

Hormones are carried in the circulation at low concentrations to their target tissues where they bind at low concentrations to specific receptors to elicit specific cellular responses. Peptide hormones and adrenaline bind to receptors in the plasma membrane and their actions are usually mediated by intracellular messengers (e.g. cyclic adenosine monophosphate (cAMP), cyclic guanosine monophosphate (cGMP), inositol trisphosphate, diacylglycerol). These stimulate protein kinases, which regulate specific cellular responses. Steroid and thyroid hormones bind to nuclear receptors to induce gene activation and the synthesis of specific proteins, which in many cases are enzymes.

Basic concepts

A hormone can be defined as a chemical substance which is synthesized and secreted by a specific cell type, is transported in the circulation and at very low concentrations elicits a specific response in distant target tissues. Not all hormone-like substances meet these criteria. For example, some cells release substances which diffuse into surrounding regions and act locally. These are referred to as **paracrine** secretions (p. 31). Sometimes cells secrete enzymes that act on plasma proteins to produce hormones, e.g. the renin–angiotensin system.

Hormones may be classified according to their chemical structure into:

1 **peptide hormones**, e.g. growth hormone, insulin and antidiuretic hormone;
2 **steroid hormones**, e.g. aldosterone, oestrogen and testosterone; and
3 **tyrosine derivatives**, e.g. thyroxine and adrenaline.

The majority of hormones are peptides—either proteins, glycoproteins or polypeptides. With the advent of **recombinant DNA technology** (Box 10.1), it is now possible to synthesize a number of human protein hormones in sufficient quantities for use in the treatment of hormone deficiencies.

The concentrations of hormones in the blood are extremely low, e.g. peptide hormones may range from 10^{-10} to 10^{-12} mol L^{-1} and steroid hormones from 10^{-6} to 10^{-9} mol L^{-1}. Because of the low concentration of hormones in the blood, their presence is difficult, if not impossible, to detect by chemical analytic techniques and so **bioassays** and **radioimmunoassays** have been devised for measuring hormone levels (Box 10.2). The development of the radioimmunoassay technique, which is highly sensitive and usually less cumbersome to perform than the bioassay, has led to rapid advances in endocrinology.

Hormone synthesis and secretion

The amino acid sequence of a peptide hormone, like other polypeptides, is translated from its mRNA species, which in turn is transcribed from the DNA of its

Box 10.1 Recombinant DNA technology

This technique is important not only for the synthesis of proteins but also for determining the sequence of amino acids in proteins. Prior to its development, the sequencing of a protein was carried out by stepwise analysis of the individual amino acids. However, it is much easier to sequence the nucleotides in DNA or RNA, because the process can be readily automated. Now, with the aid of recombinant technology, it is possible to elucidate the structure of the messenger RNA (mRNA) that codes for a particular protein by synthesizing and sequencing its complementary DNA (cDNA). Specific mRNA is isolated and copied 'backwards' to generate cDNA. The cDNA is inserted into a vector DNA and then into a bacterium which multiplies the number of copies of the recombinant DNA (Fig. B10.1). Clones of the bacteria transformed in this way can be used to prepare the cDNA and its products (i.e. RNA or protein). The cDNA can also be labelled and used as a probe to identify cells expressing the messenger for a particular protein.

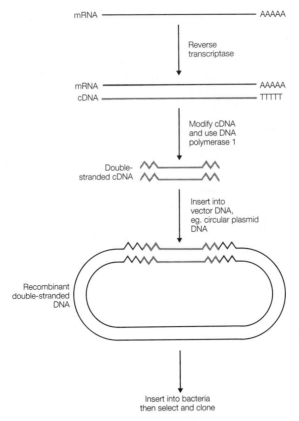

Fig. B10.1 Outline of the scheme used to prepare multiple copies of cDNA and its products. (From O'Riordan, J.L., Malan, P.G. & Gould, R.P. (1988) *Essentials of Endocrinology*, 2nd edn, p. 13. Blackwell Scientific Publications, Oxford.)

Box 10.2 Hormone assay

There are several methods for assaying hormone concentrations in serum or urine—bioassay, radioimmunoassay and enzyme-linked immunosorbent assay (ELISA)—although some hormones, e.g. catecholamines and steroid hormones, may be estimated by chemical analysis.

Bioassay

This involves quantifying the responses of tissues to various concentrations of a hormone (standard or unknown) either *in vitro* or *in vivo*. For example, chorionic gonadotrophin was formerly assayed by its ability to induce ovulation in rabbits. A standard curve is plotted and the concentration of the unknown is extrapolated from it. Such assays are often cumbersome to perform and variable in their response. However, they do measure the biological activity of the hormone.

Radioimmunoassay

This assay, now widely used, depends on the availability of radioactively labelled hormone and of antibodies that react with it. Peptide, thyroid and steroid hormones can all be labelled with ^{125}I to a high degree of specific activity. Antibodies to the larger peptide hormones are prepared by immunizing experimental animals against these molecules. Antibodies to smaller polypeptides, thyroid and steroid hormones (and even to cAMP) can also be produced if those molecules are first made antigenic by attaching them to large proteins.

A competition assay is set up in which the unlabelled standard or unknown hormone competes with the labelled hormone for a site on the antibody (Fig. B10.2); the more unlabelled hormone present, the less labelled hormone is bound to the antibody. The hormone-bound complex is then separated from the free hormone by various physicochemical means (e.g. filtration, precipitation, charcoal exclusion) and the radioactivity of the hormone-bound complex determined. A standard curve is prepared and the concentration of the unknown is extrapolated from this.

One of the advantages of the radioimmunoassay is that it is extremely sensitive (in the femtomole range). However, it does not necessarily measure the biological activity of a hormone since it can detect precursors and degradation products of a hormone, thus leading to overestimation of hormone concentration.

Enzyme-linked immunoadsorbent assay (ELISA)

The ELISA, as it is commonly known, was developed to eliminate the need for radioactive labelling and to automate procedures of hormone assay. Instead of the hormone (antigen) being labelled with a radioisotope it is linked to an enzyme marker, e.g. horseradish peroxidase or alkaline phosphatase, which generates a coloured product when the appropriate substrates are added. Alternatively, an antibody against the hormone may be linked to an enzyme marker. A further refinement of this type of assay is that it is usually carried out with the reactants bound to a 'solid phase' (e.g. polyvinylchloride plastic) to facilitate the washing steps involved. Small quantities of proteins, such as hormones and antibodies, are readily adsorbed to the surfaces of plastic tubes or wells when solutions are incubated in them. Excess adsorption sites can then be blocked and after each reaction step the tubes are washed with detergent solutions.

Box 10.2 *Continued*

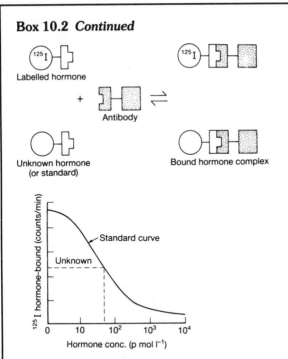

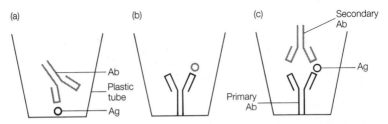

Fig. B10.2 Radioimmunoassay.

There are a number of variations of such assays (Fig. B10.3). In an antibody-capture assay the hormone is immobilized on a solid support and an enzyme-linked antibody is allowed to bind; in an antigen-capture assay an antibody is attached to a solid support and the labelled hormone (antigen) is allowed to bind. A third type of assay is the two-antibody assay in which an antibody is immobilized on a solid support and the hormone allowed to bind; a second antibody labelled with an enzyme marker is then allowed to bind to the hormone and quantified.

Fig. B10.3 Various types of ELISA: (a) antibody-capture assay; (b) antigen-capture assay; (c) two-antibody assay. The molecule labelled with an enzyme marker is indicated in grey.

gene. In eukaryotes the gene itself contains not only sequences called **exons** that code for the protein but also other sequences called **introns** interspersed among the exons. The entire sequence is first transcribed into immature mRNA and then the introns are removed by enzymes that cut and splice the message. After splicing, the mature mRNA migrates from the nucleus to the ribosomes in the cytoplasm where it is translated into a polypeptide. As discussed in Chapter 2, peptide hormones, and other peptides destined for secretion, are synthesized on the rough endoplasmic reticulum as part of larger precursor proteins called **prepropolypeptides**, which are subsequently modified by deletion of peptide sequences and, in the case of glycoproteins, by the addition of carbohydrate moieties (p. 33; Fig. 2.1). The peptide hormones are stored in vesicles and released by **exocytosis** in response to appropriate stimuli.

Steroid hormones are lipophilic; they are not encapsulated in vesicles for secretion and their rate of release is determined by their rate of synthesis. They are synthesized from cholesterol in steps that take place in the mitochondria, smooth endoplasmic reticulum and cytoplasm (p. 294; Fig. 10.14). As they are synthesized, they diffuse across the plasma membrane of the cell into the interstitial fluid and then into the blood stream.

Synthesis of the hormones which are derivatives of tyrosine occurs by specific pathways in thyroid (p. 282) and adrenal medullary cells (p. 292), as discussed later.

Hormone transport and inactivation

Once released into the blood stream, lipid-soluble hormones, e.g. thyroid and steroid hormones, are carried bound to various plasma proteins. The **free** form is usually only a small fraction of the total hormone in the blood and exists in dynamic equilibrium with the **bound** form. Generally speaking, only the free hormone can bind to or enter the target cell. The concentration of the active hormone in the blood is therefore determined by the dynamic relationship between its rate of secretion, its rate of inactivation and the degree to which it is bound to plasma proteins.

Hormones have a half-life in the body of minutes to days. Inactivation may occur in the blood, in the liver or kidney, or in some cases in the target tissues. Hormones may be inactivated by degradation, oxidation, reduction, methylation or conjugation to glucuronic acid, and excreted in the urine or bile. Peptide hormones generally have short half-lives (minutes) because they are rapidly degraded by peptide cleavage.

Hormone actions

Hormones affect the growth, development, metabolic activity and function of tissues. The responses are often the result of the actions of several hormones. Actions may be **stimulatory** or **inhibitory** and **additive** or **synergistic**. A hormone, which has no effect *per se* but is necessary for the full expression of the effects of other hormones, is said to have a **permissive** action. Hormones may alter:

1 membrane permeability;
2 activity of rate-limiting enzymes in reaction pathways;

3   protein synthesis (blocked by puromycin or cycloheximide); or

4   gene activation leading to the transcription of new mRNA species (blocked by actinomycin D).

These actions are not mutually exclusive and hormones may act in one or more of these ways.

Some hormones may act on several tissues, others only on a particular tissue or organ. In each case, the responsive cells must have the appropriate hormone **receptors**. Many hormone receptors have been identified by binding studies with radioactively labelled hormones and shown to be discrete chemical entities, either protein or glycoprotein in nature. Their characteristics have been discussed in Chapter 2, p. 35. The first step in the action of a hormone is its binding to a specific cell receptor. Peptide hormones, which do not penetrate cells readily, act by binding to specific receptors in the **plasma membrane**; so too does adrenaline. Recently it has been shown that some of the peptide hormone–receptor complexes may be internalized, i.e. taken up into the cell by endocytosis. The reason for this is not clear; it could be concerned with the regulation of receptor numbers or, in some cases, a necessary step in hormone action. In most cases, once the hormone is bound to its receptor it activates a membrane-bound enzyme, via a G protein (p. 37), that stimulates the production of an intracellular messenger (e.g. cAMP, cGMP, inositol trisphosphate, diacylglycerol, Ca^{2+}). The intracellular messengers then act, either directly or indirectly by stimulating protein kinases, to regulate specific cellular processes (p. 42).

In contrast, steroid hormones readily cross the plasma membrane and bind either to **cytoplasmic** receptors, which are then translocated to the nucleus, or to **nuclear** receptors in their target tissues (p. 45; Fig. 2.11). The hormone–receptor complex induces gene activation leading to the transcription of new mRNA species and, consequently, to specific protein synthesis. Thyroid hormones, which also cross the plasma membrane readily, regulate gene transcription by binding to **nuclear** receptors which are already bound to the DNA in their target tissues.

Control of hormone secretion

The immediate stimulus for the secretion of a hormone may be neural, hormonal, or the level of some metabolite or electrolyte in the blood. In the long term, secretion rates are usually maintained at fairly constant levels by **negative feedback** mechanisms, whereby increased levels of hormone in the blood lead to the inhibition of further hormone secretion. Positive feedback control is much less common, but an example is to be found in the hormonal control of ovulation during the female reproductive cycle.

10.2 THE PITUITARY GLAND

The anterior pituitary secretes at least six hormones, some of which (thyroid-stimulating hormone (TSH), adrenocorticotrophic hormone (ACTH), follicle-stimulating hormone (FSH) and luteinizing hormone (LH)), are trophic hormones which regulate the secretion of hormones from other endocrine glands, and others, growth hormone (GH) and prolactin, which act directly on target tissues. Secretion of the anterior pituitary hormones is controlled by

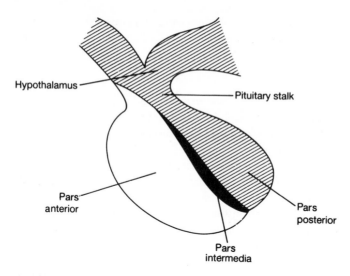

Fig. 10.1 The pituitary gland.

hypothalamic neurohormones, which reach the anterior pituitary via the hypophyseal portal system. Release of the trophic hormones is also controlled by negative feedback from their target gland hormones. The posterior pituitary secretes two neurohormones (antidiuretic hormone and oxytocin), which are synthesized in the cell bodies of neurons in the hypothalamus and transported down their axons to the posterior pituitary for storage. Their release is triggered by action potentials initiated in the hypothalamus in response to appropriate stimuli.

The **pituitary** (hypophysis) is a small gland, weighing approximately 0.5 g in humans, situated at the base of the skull and connected to the brain by the pituitary stalk. It consists of an anterior part (adenohypophysis), and a posterior part (neurohypophysis; Fig. 10.1). During embryonic development the **neurohypophysis**, which includes the pars posterior (pars nervosa), is derived from a downward evagination of the brain. The **adenohypophysis**, which includes the pars anterior (pars distalis) and the pars intermedia, comes from an outgrowth of the roof of the mouth known as Rathke's pouch. In the adult human, the pars intermedia is only a remnant and the pars anterior and pars posterior may be equated with the terms anterior and posterior pituitary respectively.

The pituitary secretes at least eight hormones, six from the anterior pituitary and two from the posterior pituitary. Four of the anterior pituitary hormones regulate other endocrine glands and are referred to as **trophic** (or tropic) hormones. The pituitary is subject to control by the hypothalamus and by the secretions of its own target glands.

Anterior pituitary

The anterior pituitary synthesizes and secretes at least six hormones (some of which unfortunately are referred to by different names)–**GH** (growth hormone, somatotrophin), **prolactin**, and the trophic hormones, **TSH** (thyroid-stimulating hormone, thyrotrophin), **ACTH**, (adrenocorticotrophic hormone, corticotrophin), **FSH** (follicle-stimmulating hormone) and **LH** (luteinizing hormone). FSH and LH are known collectively as the gonadotrophins. Each of these peptide hormones is produced by a specific cell type, except for FSH and LH which are both secreted by a single cell line. These different cell types are interspersed throughout the pars anterior and their hormones are secreted in response to stimulation by the corresponding hypothalamic neurohormones (releasing hormones).

Many of the hormones are synthesized as larger molecules or **prohormones**. The prohormone may contain the sequences of a number of hormones, e.g. the prohormone for ACTH also contains the sequences of melanocyte-stimulating hormones, of lipotrophins and of endorphins (p. 278). The hormone secreted by a particular cell type will then depend on where the prohormone is enzymatically clipped.

The actions of each of the anterior pituitary hormones are discussed in more detail later in the appropriate sections. Briefly, growth hormone acts on many tissues to regulate growth and metabolism, while prolactin stimulates lactation in the mammary glands. The trophic hormones stimulate the growth and secretion of other endocrine glands in the body—TSH acts on the thyroid gland, ACTH acts on the adrenal cortex, and the gonadotrophins, FSH and LH, act on the testes and ovaries (Table 10.1). Not only do the trophic hormones influence the secretion of their target glands, they also influenze their size and development. Removal of the pituitary (hypophysectomy) therefore causes:

1 in children, inability to grow due to lack of growth hormone;

2 atrophy of the thyroid gland and hypothyroidism due to lack of TSH;

3 atrophy of the adrenal cortex and corticosteroid deficiency due to lack of ACTH; and

4 infertility and deficiency of sex hormones due to lack of FSH and LH.

However, removal of the pituitary is not incompatible with life, although hypophy-

Table 10.1 Anterior pituitary hormones

Hormone	Structure	Target tissue	Main action
Growth hormone	Protein (M_r 22 000)	Several tissues	Growth and metabolism
Prolactin	Protein (M_r 23 000)	Mammary gland	Milk production
Thyroid-stimulating hormone	Glycoprotein (M_r 28 000)	Thyroid	Secretion of thyroid hormones
Adrenocorticotrophic hormone	Polypeptide (M_r 4500)	Adrenal cortex	Secretion of corticosteroids
Follicle-stimulating hormone	Glycoprotein (M_r 32 000)	Testes and ovaries	Secretion of sex hormones and production of gametes
Luteinizing hormone	Glycoprotein (M_r 30 000)		

sectiomized animals have a low tolerance to cold and stress. (In hypophysecto-mized animals, the posterior pituitary hormones may continue to be secreted from the cut end as the neurons that synthesize these hormones have their cell bodies in the hypothalamus).

Control of anterior pituitary hormone secretion

Hypothalamic neurohormones

Release of the anterior pituitary hormones is regulated by **neurohormones** which are elaborated in small neurons (parvicellular cells) in the hypothalamus. These neurohormones are released at the level of the median eminence. They diffuse into a primary plexus of capillaries and are transported down large portal vessels in the pituitary stalk to a secondary set of capillaries or sinusoids in the anterior pituitary—usually called the **hypophyseal portal system** (Fig. 10.2). They comprise both releasing and release-inhibiting hormones and some of the anterior pituitary hormones may be subject to dual control. It has been suggested that the name of the releasing hormone should end in '-liberin' and the inhibiting hormone in '-statin' but this nomenclature has not yet been universally adopted and the most commonly used nomenclature is given below. The following releasing hormones have been identified—**GH-releasing hormone** (GHRH), **thyrotrophin-releasing hormone** (TRH), **corticotrophin-releasing hormone** (CRH) and **gonadotrophin-releasing hormone** (GnRH). GnRH stimulates the release of LH and the synthesis of FSH, which appears to be passively secreted. The release-inhibiting hormones secreted by the hypothalamus are **GH release-inhibiting hormone** (GHRIH, somatostatin) and **prolactin release-inhibiting hormone**, which now appears to be dopamine. Thus, secretion of GH is subject to dual control by the hypothalamus. The structure of TRH was the first to be elucidated and it was found to be a tripeptide (pyroglutamylhistidyl-proline amide). The other hypothalamic neurohormones which have been identified are

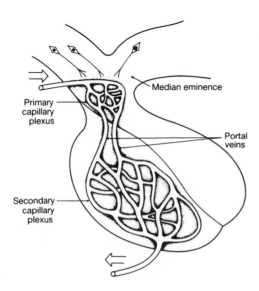

Median eminence

Primary capillary plexus

Portal veins

Secondary capillary plexus

Fig. 10.2 Hypophyseal portal system.

also small peptides, except dopamine, which is a tyrosine derivative related to noradrenaline.

Negative-feedback pathway

The secretion of releasing and inhibiting hormones by the hypothalamus may be influenced by emotional and environmental factors acting through the central nervous system. In the long term, however, secretion of the hypothalamic neuro-hormones and anterior pituitary hormones is principally controlled through **negative-feedback** mechanisms triggered by the blood level of the anterior pituitary hormones (short feedback loop) and the target gland hormones (long feedback loop). Figure 10.3 shows the possible negative-feedback pathways that are postulated to exist, and in many cases do exist, for the regulation of the anterior pituitary hormones.

Melanocyte-stimulating hormone

The pars intermedia secretes **melanocyte-stimulating hormone** (MSH), which in amphibia and fish causes the skin to darken by dispersing the melanin granules within the melanophores and thus enables these animals to blend their skin colour with the environment. In the human adult, the pars intermedia occupies only 1% of the pituitary and its role is unknown. There are at least two types of MSH in animals, α-MSH and β-MSH, both of which are polypeptides ($M_r \sim 2000$). The structural sequence of α-MSH corresponds to residues 1–13 of ACTH and excess production of MSH, or ACTH, in humans can cause an increase in melanin synthesis and hyperpigmentation. MSH secretion is under dual control by both

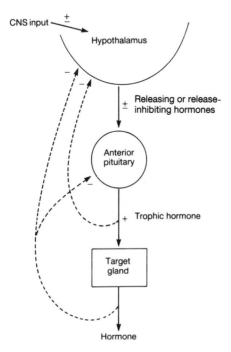

Fig. 10.3 Control of secretion of the anterior pituitary hormones. The broken lines indicate negative-feedback pathways.

releasing and release-inhibiting hypothalamic neurohormones, the inhibiting hormone playing the dominant role.

Posterior pituitary

The posterior lobe of the pituitary secretes two peptide hormones, **antidiuretic hormone** (ADH) and **oxytocin**, each with a molecular mass of approximately 1000. These hormones are synthesized in the hypothalamus in the cell bodies of large neurons (magnocellular cells), which are organized into discrete groups called the supraoptic and paraventricular nuclei (Fig. 10.4). The hormones are found in both nuclei, but in separate cells. Each hormone is synthesized as part of a prohormone containing a corresponding **neurophysin** and packaged into granules in the Golgi apparatus. Neurophysin is cleaved from the hormone as the granules are transported down the axons to their terminals in the posterior pituitary, where they are stored prior to their release into the circulation. Release of each hormone, together with its corresponding neurophysin, is triggered by nerve impulses originating in the hypothalamus. This process is known as **neurosecretion**.

ADH

ADH is a nonapeptide with a six-amino ring formed by a disulphide bridge between two cysteine residues. It controls the reabsorption of water by the kidneys and also constricts arterioles, hence the alternative name **vasopressin**. In mammals, there are two types of ADH which differ by a single amino acid—arginine (humans) or lysine. ADH has a half-life of about 5 minutes in the plasma and is metabolized by the liver and kidneys.

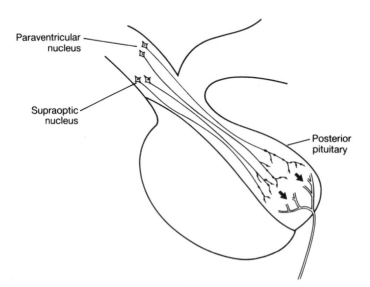

Fig. 10.4 Neurosecretion from the posterior pituitary.

Actions of ADH

Two types of receptor for ADH have been identified: V_1-receptors found in vascular smooth muscle and V_2-receptors found in the distal tubules and collecting ducts of the kidney.

ADH increases the **reabsorption of water** by the kidneys and so reduces the excretion of water from the body. It acts on the distal portions of mammalian nephrons, increasing their permeability to water (p. 635). Water moves passively out of the nephrons along an osmotic gradient and so urine volume is decreased. This action of ADH (via V_2-receptors) appears to be mediated by cAMP

In addition to its role in renal function, ADH is a potent **vasoconstrictor**. It acts on vascular smooth muscle (via V_1-receptors) to increase the concentration of inositol trisphosphate (p. 43), which increases intracellular Ca^{2+} and triggers muscle contraction. Under normal conditions, this action of ADH does not significantly alter arterial pressure because it is counteracted by the baroreceptor reflexes. However, the vasoconstrictor effect of ADH is important in the maintenance of arterial pressure in potentially hypotensive conditions, such as haemorrhage (p. 442), and may play a role in some types of hypertension.

Control of ADH secretion

Consistent with the dual role of ADH, several factors influence its release.

1 **Osmotic pressure**. Osmoreceptors in the hypothalamus, or areas close to it, respond to an increase in osmolality of the extracellular fluid leading to the release of ADH. Subsequent renal conservation of water thereby restores the normal osmolality of the body fluids. Conversely, the ingestion of a large volume of water reduces the osmolality of the extracellular fluid, leading to a reduction in ADH release and an increase in the renal excretion of water.

2 **Blood volume and pressure**. Haemorrhage promotes ADH release in response to:

(a) decreased stimulation of atrial stretch receptors;

(b) decreased stimulation of arterial baroreceptors;

(c) activation of arterial chemoreceptors; and

(d) increased release of angiotensin II. The decreased stimulation of atrial stretch receptors and arterial baroreceptors is also thought to underlie the transient effect on ADH release of changes in body position such as from a supine to a sitting position.

3 **Other factors**. Pain, exercise, stress, sleep and drugs such as morphine induce ADH secretion, while alcohol is a powerful inhibitor of secretion. The effect of morphine reflects the fact that endogenous opioids (see below) control the release of ADH (and oxytocin) both centrally and locally. The well-known diuretic effect of alcoholic beverages results not simply from increased fluid intake but also from a direct suppression of ADH release.

Diabetes insipidus

Damage to the ADH-producing neurons in the hypothalamus may result in the condition known as **diabetes insipidus**, which is characterized by the voiding of large volumes of dilute urine (polyuria) and excessive thirst (polydipsia). This condition can be treated satisfactorily with synthetic ADH. Diabetes insipidus may also arise from failure of the kidneys to respond to ADH, and inappropriate levels of ADH can occur because of ectopic production of the peptide by a tumour.

Oxytocin

Oxytocin is a ring-like nonapeptide similar in structure to ADH but differing in two amino acid residues. Its synthesis and metabolism are similar in many respects to those of ADH.

Actions of oxytocin

Oxytocin stimulates the myoepithelial cells of the mammary gland, causing **milk let-down**. It also causes **uterine contraction** in the oestrogen-stimulated uterus during parturition and is used clinically for the induction of labour. It has no known function in males.

Control of oxytocin secretion

Milk let-down is a reflex action in response to suckling at the breast. A neuroendocrine reflex pathway is involved, in which impulses initiated by suckling are relayed to the hypothalamopituitary axis, causing the release of oxytocin. This is carried by the circulation to the mammary glands. Pain, embarrassment and anxiety can cause inhibition of oxytocin release. There is a marked elevation of plasma oxytocin levels during parturition in response to stretching of the uterus and cervix. There are no known disorders due to abnormal oversecretion of oxytocin.

Opioid peptides

Research on opiates (e.g. morphine) and their actions on the brain and gut led to the isolation and identification of endogenous peptides called **opioids**. These were first discovered in brain (and later in the gut, adrenal medulla, testes and placenta) and found to be pentapeptides which were named **met-enkephalin** (Tyr–Gly–Gly–Phe–Met) and **leu-enkephalin** (Tyr–Gly–Gly–Phe–Leu). Subsequently, larger opioid peptides were discovered in the pituitary and named **endorphins** and **dynorphins**. The former are found in the pars anterior and pars intermedia and the latter in the pars posterior. These opioids are also present in brain and gut.

The family of opioid peptides has grown to at least nine members, all of which contain either the met-enkephalin or leu-enkephalin sequence at their N-terminals. Complementary DNA-gene cloning techniques have shown that the family has three branches which stem from:

1 pro-opiomelanocortin, the precursor of β-endorphin and related peptides:
2 proenkephalin, the precursor for met- and leu-enkaphalin; and
3 prodynorphin, the precursor for dynorphins and neoendorphins.

The pro-opiomelanocortin precursor contains the sequences of several active peptides including ACTH, β-lipotrophin, β-endorphin, and α- and β-MSH. This precursor molecule is processed differently depending on its location. In the pars anterior of the pituitary, ACTH and β-lipotrophin are produced (some of the β-lipotrophin is split to produce β-endorphin). In the pars intermedia, ACTH is further cleaved to form α-MSH and β-lipotrophin is split to produce β-endorphin.

Opioid peptides are powerful analgesics when injected into the ventricles of the brain (p. 157). There are at least three subclasses of opioid receptors: μ, δ and κ. The enkephalins bind more strongly to δ than to μ receptors; β-endorphin binds to μ and δ receptors, while dynorphins have a greater affinity for κ receptors.

The opioid peptides in brain appear to act as neurotransmitters or neuromodulators and play a role in the perception of pain and in behaviour, cardiovascular regulation, respiration and endocrine responses. The roles of the pituitary and adrenal opioids are not well-understood but they may be involved in local control of hormone release.

10.3 GROWTH HORMONE (GH)

GH is a protein secreted by specific cells in the anterior pituitary, called somatotrophs. Not only does it promote the growth of muscle and bone but it has metabolic effects on most of the tissues of the body. Its action are counterregulatory to insulin, as it increases blood glucose and mobilizes free fatty acids. Its growth-promoting actions (and some of its other actions) are mediated by the insulin-like growth factor, IGF-1, a polypeptide produced mainly in the liver. The secretion of GH occurs in bursts in response to dual control by GHRH and GHRIH; it is stimulated by low blood glucose and high blood amino acids, and during certain stages of sleep.

The conditions of dwarfism and gigantism are caused by disturbances in the function of the pituitary gland. Young animals cease to grow when their pituitary glands are removed but growth can be restored by injections of anterior pituitary extracts. GH was first isolated from bovine anterior pituitaries in the 1940s. It is secreted by certain cells (somatotrophs) of the anterior pituitary and is a protein of M_r 22 000, there being some differences in the amino acid sequence between species. It has close structural similarities with a placental hormone, chorionic somatomammotrophin, and, to a lesser extent, with prolactin, which suggests their evolution from a common progenitor molecule.

Growth and hormones

The major function of GH is to stimulate the growth of bones and other tissues. The growth process is not simple and, in addition to an adequate food supply and to genetic endowment, a number of hormones are involved, including **GH, sex hormones, thyroid hormones** and **insulin**. There are two periods of accelerated growth: the first occurs in the first 2 years of life and the second at the time of puberty (Fig. 10.5). The period of accelerated growth at puberty is associated with increased levels of sex hormones, including androgens of adrenal origin. Paradoxically, the cessation of growth around 18–20 years of age is also due to the sex hormones, which cause fusion of the growing ends of bones (**epiphyseal closure**). Thus, the effect of excess or deficiency of GH secretion depends on whether it occurs before or after closure of the epiphyses. Excess GH secretion in young people causes **gigantism**, but in adults it leads to **acromegaly** in which there is a general coarsening of the features due to thickening of bone and soft tissue. On the other hand, a deficiency of GH in children leads to **dwarfism**. Humans respond only to GH of human (or other primate) origin. This can now be synthesized using

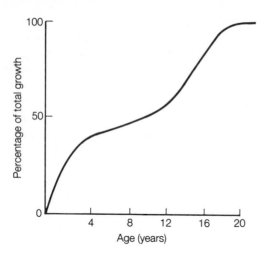

Fig. 10.5 Rate of growth in humans. Note the accelerated periods of growth in the first 2 years of life and at puberty.

recombinant DNA technology and is used to treat children of short stature due to GH deficiency.

Actions of GH

GH affects the metabolic activity of most of the tissues of the body. Its growth-promoting effect is in part due to its ability to stimulate the uptake of amino acids and their incorporation into proteins in muscle and bone. The effect on amino acid uptake is a direct one but its other growth-promoting actions are mediated by an **IGF-1**, formerly known as somatomedin C. IGF-1 is a polypeptide produced mainly in the liver. GH also has effects on carbohydrate and fat metabolism which in general are antagonistic to those of insulin, i.e. it is counter-regulatory to insulin. GH increases blood glucose (after an initial decrease) by stimulating hepatic gluconeogenesis and by inhibiting glucose uptake by muscle. it also raises free fatty acid levels by increasing their mobilization from adipose tissue. The plasma half-life of GH is about 20 minutes but its actions continue for much longer.

Control of GH secretion

The level of GH in the blood as measured by radioimmunoassay fluctuates continuously due to pulsatile release of the hormone, the burst frequency being higher in adolescents. The physiological control mechanism of GH secretion are not well-understood but secretion is increased during fasting and in stress and exercise. GH release from the anterior pituitary is under hypothalamic control by **GHRH** and **GHRIH**, which stimulate and inhibit secretion respectively, and by IGF-1 which has a negative-feedback effect on GH production. Factors that stimulate GH secretion are low blood glucose and high blood amino acid concentrations. Bursts of hormone secretion also occur during stages 3 and 4 of sleep, and associated with the nocturnal rise in GH is a characteristic early-morning increase in blood glucose levels.

10.4 THE THYROID GLAND

Hormones of the thyroid gland, thyroxine (T_4) and triiodothyronine (T_3), are essential for optimal metabolic activity in the body. They also have other important functions, particularly in the control of growth and the development of the nervous system. They are synthesized in the lumen of the thyroid follicle by iodination and coupling of tyrosine molecules attached to thyroglobulin, which is then taken up by the surrounding follicle cells and the thyroid hormones released to the circulation. The synthesis and release of thyroid hormones are controlled by TSH, which in turn is controlled by TRH and by feedback inhibition of the thyroid hormones.

Thyroid hormone deficiency in the fetus produces the condition of **cretinism**, which is characterized by a failure to grow and severe mental retardation. This condition is caused by failure of the thyroid gland to develop properly. A closely related condition occurs in certain parts of the world, e.g. New Guinea, where the content of iodine in the soil and in the water is low, and in these areas it is associated with enlargement of the thyroid gland (**goitre**). This rarely occurs in coastal regions because of the high iodine content of fish and other sea foods. Endemic cretinism and endemic goitre have become less common since the introduction of dietary iodine supplementation, for example, in the form of iodized salt.

The thyroid gland is composed of two lobes joined by a narrow connection called the thyroid isthmus (Fig. 10.6). The gland is bright red because it has a very rich blood supply. It lies in front of the trachea just below the larynx. Each lobe is composed of many spherical follicles which are about 50–500 µm in diameter and which are surrounded by a dense network of capillaries. Each follicle is composed of a single layer of epithelial cells surrounding a cavity in which the thyroid hormones are stored as integral components of the storage protein, **thyroglobulin**.

The main hormones secreted by the thyroid gland are iodinated derivatives of tyrosine, T_4 and T_3, collectively known as the thyroid hormones (Fig. 10.7). The

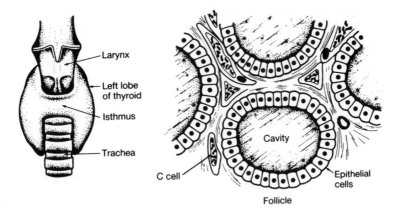

Fig. 10.6 The thyroid gland.

Thyroxine (T$_4$)

3,5,3'-triiodothyronine (T$_3$)

3,3',5'-triiodothyronine (reverse T$_3$)   **Fig. 10.7** The thyroid hormones.

thyroid gland also secretes the peptide hormone, **calcitonin**, which lowers plasma calcium concentrations and is discussed later (p. 290). Sufficent T$_3$ and T$_4$ is stored in the gland to last 2–3 months. The only structural difference between T$_3$ and T$_4$ is that T$_4$ contains four iodine atoms whereas T$_3$ has three. There is approximately 50 times more T$_4$ than T$_3$ in the plasma. In the tissues, T$_4$ may be converted to T$_3$ or to reverse T$_3$ (3,3',5'-triiodothyronine) by deiodination (Fig. 10.7). T$_3$ is the major hormone active in target cells; reverse T$_3$ is inactive.

Synthesis and secretion of thyroid hormones

The thyroid gland actively concentrates iodide to a level normally some 25 times that in the plasma. The iodide is oxidized by a peroxidase in the follicle cells to atomic iodine which immediately iodinates tyrosine residues contained in thyroglobulin (Fig. 10.8). Thyroglobulin is a large protein (M_r 670 000) which is synthesized in the follicle cells and secreted into the follicular cavity. The iodinated tyrosine residues in thyroglobulin undergo coupling to form T$_4$ and T$_3$. Iodination and coupling are thought to take place at the cell surface bordering the follicular cavity, and the thyroid hormones are stored in the cavity as integral amino acids in the thyroglobulin molecule.

The synthesis of the thyroid hormones can be summarized as follows.

1 Iodide uptake and concentration (blocked by thiocyanate and perchlorate).

2 Oxidation of I$^-$ → I + e (blocked by propylthiouracil and carbimazole).

3 Iodination of tyrosine molecules in thyroglobulin by atomic iodine (also blocked by propylthiouracil and carbimazole).

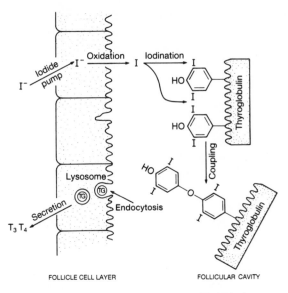

Fig. 10.8 Synthesis of thyroid hormones (T_3, T_4), their storage in association with thyroglobulin and their secretion.

4 Coupling of either two diiodotyrosine residues to form T_4, or of a diiodotyrosine (DIT) and a monoiodotyrosine (MIT) residue to form T_3.

As indicated above, the action of the antithyroid drugs—thiocyanate, perchlorate, propylthiouracil and carbimazole—is due to their interference with specific steps in the biosynthetic pathway. Paradoxically, excess iodide also inhibits the biosynthesis of the thyroid hormones.

All the steps in the synthesis of the thyroid hormones are stimulated by **TSH** acting through cAMP. TSH also stimulates secretion of the thyroid hormones. The first step in secretion is the uptake of small globules of colloid into the follicle cells by endocytosis. The globules then fuse with lysosomes and their contents are digested, thus liberating the thyroid hormones which diffuse out of the follicle cells into the blood.

The proteolytic degradation of thyroglobulin within the follicular cells causes the liberation of MIT and DIT, as well as T_3 and T_4. A specific iodotyrosine deiodinase enzyme in the follicle cells allows the iodine to be removed from MIT and DIT (but not from T_3 and T_4) so that the iodine and the tyrosine can be recycled. This reaction is important for conserving supplies of iodine in the thyroid, and a genetic abnormality of the deiodinase can cause iodine deficiency because of loss of iodotyrosine in the urine.

Transport and inactivation of thyroid hormones

Most of the circulating T_4 is bound to plasma proteins, mainly to **T_4-binding globulin** and, to a lesser extent, to prealbumin and albumin. Total plasma T_4 and T_3 (protein-bound and free) can be determined by radioimmunoassay as measures of thyroid function. However, as the total concentration of hormone in the blood is influenced markedly by the concentration or binding capacity of T_4-binding globulin, an adjustment should be made for this, or preferably specific measurement of free T_4 or free T_3 be made. T_3 is less firmly bound to plasma proteins than is T_4, which is reflected in the finding that, although the concentration of total T_4 in the plasma (about 100 nmol L^{-1}) is much higher than that of total T_3 (about 1.5 nmol L^{-1}), the concentrations of the free hormones are similar (about 0.04 nmol L^{-1} for T_4 and 0.02 nmol L^{-1} for T_3).

The thyroid hormones are broken down in several tissues, particularly the liver and skeletal muscle. T_4 has a half-life of 7 days, T_3 about 1 day. Much of the iodide that is released is reclaimed, but about 150 µg of iodide is lost in the urine and faeces daily and must be replaced in the diet.

Actions of thyroid hormones

1 **Thermogenesis**. One of the principal effects of the thyroid hormones is to stimulate oxidative metabolism and thereby increase the production of heat in warm-blooded animals. They increase oxidative metabolism in all tissues of the body except the brain, lungs, spleen and sex organs. The increase in basal metabolic rate produced by a single injection of T_4 begins after a latency of several hours and lasts 9 days or more. The basal metabolic rate may increase by as much as 100% while after thyroidectomy it may fall by 50%. T_3 and T_4 may cause a slight increase in body temperature but increased secretion of hormone does not seem to occur in humans as part of an acute response to cold. However, the hormones may act in a permissive manner to allow other mechanisms such as increased sympathetic activity to accelerate production of heat. It appears that the action of thyroid hormones on oxidative metabolism is due, at least in part, to an increase in the synthesis of Na^+-K^+ ATPase and hence in the activity of the Na^+-K^+ pump.

2 **Effects on growth and development**. As already mentioned, thyroid hormones are essential for normal growth in childhood. In amphibia, the thyroid hormones act as a trigger for metamorphosis and premature metamorphosis can be induced in tadpoles by feeding them pieces of thyroid gland. T_3 and T_4 appear to stimulate growth by a direct effect on tissues but they also have a permissive role on GH action.

3 **Effects on the nervous system**. The thyroid hormones are essential for normal myelination and development of the nervous system in childhood. In the adult, a deficiency of thyroid hormones may lead to listlessness and blunting of intellect; an excess to restlessness and hyperexcitability.

4 **Other effects**. Excess production of thyroid hormones causes an increased cardiac stroke volume and tachycardia. These effects are partly a direct response to the thyroid hormones, which increase sensitivity to catecholamines, and are partly secondary to increased demands for oxygen associated with their thermogenic action. The thyroid hormones have less well-defined effects on carbohydrate and liplid metabolism, including the lowering of plasma cholesterol concentration. The metabolism of protein is also affected by thyroid hormones, and hyperthyroidism may lead to wasting of skeletal muscle and a negative nitrogen balance. The thyroid hormones influence calcium metabolism, and demineralization of the skeleton is common in severe hyperthyroidism. Adequate

secretion of hormones is also necessary for gonadal function and lactation.

Thyroid hormones exert their varied effects upon their target tissues by binding of T_3 to a specific receptor for this hormone in the cell nucleus. Such binding modulates transcription of mRNA from a variety of genes coding for enzymes and structural proteins involved in the expression of thyroid hormone action.

Control of thyroid hormone secretion

For normal thyroid hormone secretion, there must be adequate intake of iodine in the diet. The immediate stimulus for the release of thyroid hormones is **TSH** secreted by the anterior pituitary. TSH acting through cAMP stimulates every step in the production and secretion of T_3 and T_4. In addition, it controls the size and vascularity of the gland; if the pituitary is removed, the thyroid atrophies.

Secretion of TSH is stimulated by **TRH**, which is secreted by neurons in the hypothalamus and transported to the anterior pituitary in the hypophyseal portal vessels. The actions of TRH appear to be mediated by a rise in cytosolic Ca^{2+}, probably triggered by the second messenger inositol trisphosphate, although some of them are mimicked by cAMP. TSH secretion is also regulated by feedback inhibition of T_3 and T_4 on the anterior pituitary (Fig. 10.9) where they may reduce the number of cell receptors for TRH. The thyroid hormones may also act at the level of the hypothalamus to affect the output of TRH, but the significance of this mechanism is unclear.

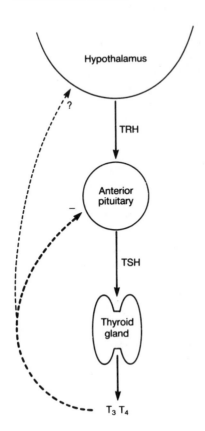

Fig. 10.9 Control of thyroid hormone secretion.

Disorders of thyroid function

Hypothyroidism may result from disease of the pituitary or of the thyroid gland (autoimmune thryoiditis or **Hashimoto's disease**), or from severe deficiency of iodine in the diet. Severe hypothyroidism in the adult is called **myxoedema** because of the puffiness of the hands and face due to an abnormal accumulation of mucoproteins in the subcutaneous layers. Other signs are low metabolic rate, bradycardia, cold intolerance, mental and physical lethargy and slow hoarse speech. Severe hypothyroidism in the fetus results in the condition of **cretinism**.

Hyperthyroidism (thyrotoxicosis) results from the overproduction of thyroid hormones and is characterized by a high metabolic rate, tachycardia, heat intolerance, hyperexcitability, restlessness and weight loss. A common form of hyperthyroidism is **Graves' disease** which is also characterized by protruding eyeballs (exophthalmos) and goitre. Graves' disease is caused by abnormal thryoid stimulators in the blood. Long-acting thyroid stimulator was the first of these to be discovered and it is now clear that this is one of a group of autoantibodies called **thyroid-stimulating antibodies** which are responsible for this condition.

Goitre is often associated with hyperthyroidism. However, it may also be a manifestation of hypothyroidism in which there is a compensatory increase in TSH secretion, as occurs within iodine deficiency (endemic goitre) or in diseases of the thyroid gland which result in defective T_3/T_4 synthesis or secretion. Natural goitrogens found in plants of the Brassica family (e.g. cabbage) block the incorporation of iodine into tyrosine residues of thyroglobulin.

In the diagnosis of disorders of thyroid function the physician is guided by estimates of plasma levels of free T_3 and T_4 and of TSH, by responses to test doses of TRH, or by the pattern of thyroid uptake of radioactive iodine measured externally. Radioactive technetium, given as the pertechnetate, is now commonly used instead of radioactive iodine for imaging of the thyroid gland.

10.5 THE PARATHYROID GLANDS AND CALCIUM METABOLISM

In mammals, endocrine tissue important for controlling calcium balance is located in the region of the thyroid gland. Parathyroid glands on the dorsal surface of the lobes of the thyroid gland secrete parathyroid hormone, which raises plasma calcium and lowers plasma phosphate concentration. In addition, the parafollicular or C cells of the thyroid gland secrete calcitonin, which lowers the plasma calcium concentration. Another important regulator of calcium metabolism is an active metabolite of vitamin D known as calcitriol, which is produced by the kidneys. This stimulates calcium and phosphate absorption in the gut and promotes bone mineralization.

Calcium metabolism

Calcium has a number of essential physiological functions in the body, including:
1 maintenance of normal permeability of cell membranes;
2 maintenance of normal excitability of nerve and muscle;
3 release of neurotransmitters, many hormones and exocrine secretions;
4 muscle contraction;
5 formation of bone and teeth;

6   coagulation of blood;
7   production of milk; and
8   activity of many enzymes.

About 99% of body calcium and 80% of body phosphorus is found in bone. The concentration of calcium in plasma is about 2.4 mmol L^{-1} (2.1–2.6 mmol L^{-1}), of which about 1.2 mmol L^{-1} is ionized, about 1.1 mmol L^{-1} is bound to protein and a small amount is present as complexes with anions, such as citrate, bicarbonate or phosphate. Phosphorus occurs in plasma in a number of organic constituents, including lipids and nucleotides, and is also present as inorganic phosphate at a concentration of about 1 mmol L^{-1}. The solubility product of calcium and phosphate is such that the product of the concentrations of the free ions is low and remains relatively constant.

On a typical diet, about 25 mmol (1 g) of calcium is ingested daily, but individuals vary greatly in their intake. Most of this calcium is in the form of calcium phosphate derived especially from milk and other dairy products. There is abundant phosphorus in the diet to meet daily needs. For calcium balance to be maintained, an intake of 12.5–20 mmol per day is recommended for infants and adults. A higher daily intake (25–37.5 mmol) is recommended for rapidly growing adolescents and for women during pregnancy and lactation, and after the menopause. Net intestinal absorption of dietary calcium depends heavily upon nutritional needs and vitamin D (calcitriol) effects, and is influenced by faecal losses of calcium in intestinal secretions (Fig. 10.10). Urinary excretion is also important in regulating calcium homeostasis. Intestinal absorption occurs mainly in the duodenum by active transport (p. 600), but some concentration-dependent passive absorption of calcium occurs throughout the small intestine. Both intestinal

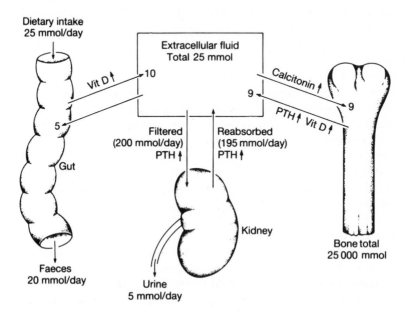

Fig. 10.10 Calcium balance in the body and its regulation by parathyroid hormone (PTH), calcitonin and vitamin D.

absorption and urinary excretion of calcium are under hormonal control, as is the exchange of calcium between the extracellular fluid and the skeleton.

Bone

Bone is composed of an organic matrix of collagen in a ground substance consisting largely of mucopolysaccharides and non-collagen proteins, within which crystals of a complex salt of calcium and phosphate very similar to hydroxyapatite $(Ca_{10}(PO_4)_6(OH)_2)$ are deposited. A more amorphous form of calcium phosphate is also present, as well as small amounts of Na, Mg, Cl and F.

Three types of cells appear to function in the formation and resorption of bone, namely osteoblasts, osteoclasts and osteocytes. The **osteoblasts** synthesize and secrete collagen fibres and promote the deposition of calcium phosphate crystals, while **osteoclasts** cause resorption of bone. Bone resorption depends on the destruction of collagen by lysosomal enzymes and phagocytosis, and on the dissolution of bone mineral by an increase in lactate and citrate production. **Osteocytes** are the most numerous cells in mature bone and are formed from osteoblasts. They appear to play an essential role in the exchange of calcium between extracellular fluid and bone. This role depends upon the activity of parathyroid hormone, calcitriol and calcitonin. Only about 1% of the calcium and phosphate of bone is in equilibrium with the extracellular fluid through the activity of osteocytes, the so-called exchangeable pool of bone mineral which acts to buffer short-term changes in blood calcium ion concentration. The remaining 99% comprises the non-exchangeable pool, from which the mineral can be released only by osteoclastic resorption, as occurs in normal remodelling or in certain disease states. The remodelling of bone is influenced by many hormones, not only parathyroid hormone, calcitonin and calcitriol, but also GH, thyroid hormones, gonadal hormones and cortisol. The role of vitamin D and its active metabolite, calcitriol, has achieved prominence in recent years.

Vitamin D

Vitamin D is essential for proper bone development, and a deficiency of this vitamin in children causes **rickets**, a disorder characterized by stunted growth and bowing of the limbs due to failure of mineralization of bone. In adults, vitamin D deficiency can cause a failure of minerialization of bone (**osteomalacia**). Vitamin D is a steroid found in a limited number of foodstuffs, e.g. cod-liver oil, and is also synthesized in the skin by the action of ultraviolet light on 7-dehydrocholesterol. The vitamin occurring naturally in animals is vitamin D_3 (cholecalciferol) and that in plants is vitamin D_2 (ergocalciferol). As the two forms are equally active, the generic term vitamin D will be used to designate either.

Conversion to calcitriol

Vitamin D is now regarded as a *prohormone* because it is converted to an active metabolite **calcitriol** which controls calcium and phosphorus metabolism. It is first converted to a 25-hydroxy vitamin D in the liver and then, in the kidney, to 1,25-dihydroxy vitamin D (calcitriol), if extracellular concentrations of calcium or phosphate are low. The renal enzyme 1α-hydroxylase, which catalyses the final conversion, is stimulated by parathyroid hormone, low plasma phosphate, gonadal steroid hormones and prolactin. Synthesis of calcitriol is also controlled by negative feedback of calcitriol itself. If extracellular concentrations of calcium and phosphate are normal, most of the vitamin is transformed in the kidney to the 24,25-dihydroxy and 1,24,25-trihydroxy derivatives which are inactive.

Actions of calcitriol

Calcitriol acts on the **small intestine** to promote the **absorption of calcium and phosphate**, which are necessary for bone formation, and it facilitates bone miner-alization by increasing the extracellular fluid concentration of calcium and phosphate. It has also been shown, together with parathyroid hormone, to have the reverse effect on bone mineralization *in vitro*, causing release of these ions from bone. This action may be important in bone remodelling. It appears to promote gene expression like other steroid hormones (p. 45), and in the intestine this leads to an increase in the synthesis of a calcium-binding protein, which may promote calcium absorption. It can also reduce parathyroid hormone synthesis by directly inhibiting expression of the parathyroid hormone gene.

Calcitriol acts directly on osteoblasts to stimulate the synthesis of *osteocalcin* and *osteopontin*, and osteoclast-mediated resorption of bone.

Parathyroid hormone

Parathyroid hormone is an 84-amino acid polypeptide (M_r 9500) secreted by the 'chief' cells of the parathyroid glands (Fig. 10.11). Four **parathyroid glands** are usually found embedded in the dorsal surface of the lobes of the thyroid gland and accessory parathyroid tissue sometimes occurs elsewhere in the neck or mediastinum. Removal of the parathyroid glands (and accessory tissue) causes plasma

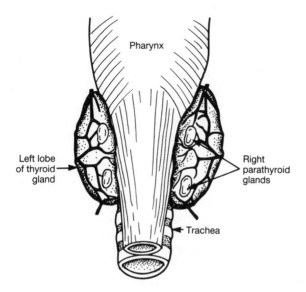

Fig. 10.11 The parathyroid glands viewed from behind.

calcium to fall and leads to **hypocalcaemic tetany**. This is characterized by extensive spasms of skeletal muscle and can lead to asphyxiation due to laryngeal spasm.

The hormone is synthesized as the preprohormone containing additional amino acids which are removed before the hormone is secreted. Further cleavage of the hormone into two or more fragments occurs in the liver and kidney. This complicates the interpretation of radioimmunoassays used to measure the concentration of circulating hormone. As only the first 34 amino acids in the N-terminal portion of the molecule are necessary for full biological activity, some of these fragments may retain biological activity.

Actions of parathyroid hormone

Parathyroid hormone **increases ionized plasma calcium** and **lowers plasma phosphate** concentration. It acts on the bone, the kidney and, indirectly, on the gastrointestinal tract. Its actions on bone and kidney appear to be mediated by cAMP. Parathyroid hormone increases the rate of bone resorption by stimulating the activity of osteocytes and osteoclasts. (The presence of vitamin D is needed for these actions.) These effects are important in long-term regulation of plasma calcium while its actions on the kidney appear to be more important in compensating for short-term changes. In the kidney, parathyroid hormone *increases* the tubular reabsorption of calcium and *decreases* that of phosphate. The resulting fall in plasma phosphate stimulates calcium release from the bone and prevents the deposition of calcium phosphate. Parathyroid hormone also stimulates the formation of calcitriol in the kidney. Thus, the absorption of calcium in the gastrointestinal tract is increased as a consequence of an increase in calcitriol. In the long term, increased secretion of parathyroid hormone may result in a net loss of calcium from the body through the kidney. Under these conditions, the increased ionized plasma calcium increases the filtered load by an amount that exceeds the additional calcium reabsorbed in the tubules.

Control of parathyroid hormone secretion

Parathyroid hormone secretion is regulated by plasma calcium acting on the parathyroid glands so that it varies inversely with the level of plasma ionized calcium level (Fig. 10.12). A decrease in plasma calcium concentration causes an increase in secretion of parathyroid hormone and vice versa. Since parathyroid hormone increases extracellular calcium, further release is inhibited—a typical negative-feedback mechanism.

Calcitonin

Calcitonin is a polypeptide (M_r 3500) secreted by the parafollicular or C cells of the **thyroid gland** (Fig. 10.6). It decreases plasma calcium and phosphate levels by inhibiting bone resorption. It does so by acting on osteoclasts to inhibit their function. It also increases the urinary loss of calcium and phosphate by decreasing their reabsorption in the kidney. Its actions are mediated by cAMP. The release of calcitonin is stimulated by an increase in plasma calcium (Fig. 10.12) and it has

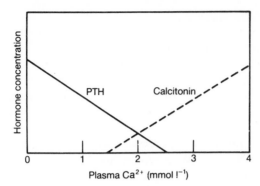

Fig. 10.12 Effect of plasma calcium on the concentrations of parathyroid hormone (PTH) and calcitonin in blood. (After Copp, D.H. (1968) In: Talmage, R.V. & Belanger, L.F., *Parathyroid Hormone and Thyrocalcitonin (Calcitonin)*, p. 25, Excerpta Medica, Amsterdam.)

been suggested that calcitonin protects animals against hypercalcaemia. Removal of the thyroid gland does not cause hypercalcaemia in humans, whereas hypocalcaemia regularly follows removal of the parathyroid glands. Therefore, the physiological significance of calcitonin is uncertain. It is used clinically in the treatment of Paget's disease to reduce the accelerated rate of bone turnover.

Disorders of calcium metabolism

Hypocalcaemia causes excessive neuromuscular irritability. A rapid decrease of ionized plasma calcium to below 1 mmol L^{-1} results in spontaneous firing of peripheral nerves. On the afferent side, this causes unusual sensations such as tingling (paraesthesia); on the efferent side, muscular twitching, spasm and cramps can develop, causing **manifest tetany**. If ionized calcium decreases more slowly or to a lesser extent, stimuli such as localized ischaemia, hyperventilation or pressure over a nerve may be required to elicit the motor events. This is termed **latent tetany**. A low ionized plasma calcium may be consequent upon decreased levels of parathyroid hormone or calcitriol in the body. It may also result from increased plasma pH, e.g. in respiratory alkalosis, due to the release of hydrogen ions from plasma proteins making additional negatively charged binding sites available for calcium. Tetany may also occur when ionized magnesium concentration in plasma is reduced, and in metabolic alkalosis associated with potassium depletion.

Osteomalacia, due to a deficiency of calcitriol, is characterized by a normal bone mass but the newly synthesized matrix fails to calcify. In **osteoporosis**, the bone is normally calcified but bone mass is decreased to a greater degree than in **osteopenia** which accompanies ageing. Oestrogen deficiency leads to osteoporosis, which most commonly occurs in women after the menopause. Osteoporosis also occurs with prolonged immobilization and with excess glucocorticoid secretion or administration.

Hypercalcaemia is seen most frequently in hyperparathyroidism. Increased calcium mobilization from bone leads to painful softening and fractures of bones, and the increased calcium and phosphate excretion in urine may result in nephrocalcinosis and renal stones. Increased plasma calcium may also cause headaches and decreased tone (hypotonia) in skeletal and intestinal muscle.

10.6 THE ADRENAL GLANDS

There are two adrenal glands, situated one on top of each kidney. Each adrenal gland comprises two endocrine organs–the adrenal medulla and the adrenal cortex (Fig. 10.13). The two parts of the adrenal gland have different embryonic origins and are anatomically quite distinct. The adrenal medulla secretes **catecholamines** while the adrenal cortex secretes a number of **steroid hormones**.

Adrenal medulla

The chromaffin cells of the adrenal medulla secrete the catecholamines, adrenaline and noradrenaline. These hormones are released by splanchnic nerve stimulation in response to emergencies. They circulate in the blood stream and complement the noradrenaline released locally in tissues. Their actions are numerous and include the stimulation of heart rate and contractility, the inhibition of gut motility, bronchodilatation and the stimulation of glycogenolysis.

The medulla is a modified nervous tissue derived from the neural crest and can be regarded as a collection of postganglionic sympathetic neurons in which the axons have not developed. The catecholamines are produced in chromaffin cells which are of two types, one secreting **adrenaline** (epinephrine) and the other **noradrenaline** (norepinephrine). In humans, adrenaline constitutes about 80% of the catecholamines produced by the medulla. Small collections of chromaffin cells are also located outside the adrenal medulla, usually adjacent to the chain of sympathetic ganglia. The organization of the blood supply of the adrenal gland is such that the medulla receives blood rich in corticosteroids, which regulate the synthesis of the enzyme that converts noradrenaline to adrenaline (p. 108).

In addition to catecholamines, the adrenal medulla also contains enkephalins, dynorphins, neurotensin, somatostatin and substance P. The role of these adrenal peptides has not yet been elucidated, although it is known that met- and leu-

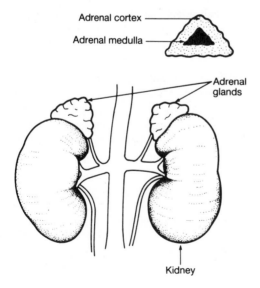

Fig. 10.13 The adrenal glands.

enkephalins are located within the adrenaline-containing chromaffin cells and are co-secreted with adrenaline.

The synthesis of catecholamines from tyrosine has been dealt with in Chapter 4 (p. 108). The amines are stored in membrane-bound granules and their secretion is initiated by acetylcholine released from **preganglionic sympathetic fibres** that travel in the splanchnic nerves. Acetylcholine depolarizes the chromaffin cells causing Ca^{2+} to enter and trigger the release of the granular contents by exocytosis. Once released into the blood stream, the catecholamines have only a short half-life (minutes). They are rapidly taken up into extraneuronal tissues to be degraded by catechol-O-methyltransferase, or into nerve terminals to be recycled or degraded by monoamine oxidase. The degradation products eventually appear in the urine.

Actions of catecholamines

The actions of adrenaline and noradrenaline are complex and depend on their binding to various adrenoceptors. Both bind to α- and β-adrenoceptors, but their affinities differ according to the type of adrenoreceptor and its subclass (p. 109). Noradrenaline causes widespread vasoconstriction and a marked increase in peripheral resistance, while adrenaline causes vasoconstriction in skin and viscera but vasodilatation in skeletal muscles so that total peripheral resistance may decrease. Both catecholamines increase heart rate and contractility directly, but in the intact animal the increase in peripheral resistance and mean arterial pressure caused by noradrenaline administration leads to reflex bradycardia. Adrenaline has a more pronounced effect on metabolic processes and increases the basal metabolic rate, stimulates glycogenolysis and mobilizes free fatty acids. Catecholamines also cause bronchodilatation and relaxation of the gastrointestinal tract.

Control of catecholamine secretion

The secretion of catecholamines is initiated by sympathetic activity controlled by the hypothalamus and occurs in response to such stimuli as pain, excitement, anxiety, hypoglycaemia, cold and haemorrhage. Increased secretion is part of the 'fight or flight' reaction described by Cannon in 1921. In an emergency, catecholamines released by the adrenal medulla are disseminated in the blood stream while noradrenaline is released from sympathetic nerve terminals locally. In frightened or stressed animals, there is a general increase in sympathetic activity in which the sympathetic nerves appear to play the dominant role, because removal of the adrenal medulla does not seriously impair an animal's ability to cope with stress.

Tumours of chromaffin cells called **phaeochromocytomas** can secrete catecholamines in excess. This may cause hypertension (noradrenaline) or hyperglycaemia (adrenaline).

Adrenal cortex

The adrenal cortex secretes glucocorticoids (cortisol and corticosterone), which affect the metabolism of carbohydrates, fats and proteins and are important in mediating the response of the body to fasting and stress; mineralocorticoids (mainly aldosterone), which are essential for the maintenance of sodium balance and consequently extracellular fluid volume; and androgens, which may play a minor role in reproductive function, particularly as a source of androgens in the female. The secretion of glucocorticoids and androgens is controlled by ACTH, while that of aldosterone is controlled mainly by the renin–angiotensin system.

The cortex is organized histologically into three zones—the outer **zona glomerulosa**, which secretes **aldosterone**, the middle **zona fasciculata**, which secretes mainly **glucocorticoids**, and the inner **zona reticularis**, which secretes mainly **androgens**. Removal of the pituitary causes the fasciculata and reticularis zones to atrophy but has little effect on the zona glomerulosa. During fetal life, the cortex has a large inner fetal zone which produces precursor steroids for the synthesis of oestrogens by the placenta (fetoplacental unit; p. 321).

The adrenocortical steroids have the same basic structure as other steroids—the glucocorticoids and aldosterone contain 21 carbon atoms and the androgens 19 (Fig. 10.14). They are derived from cholesterol, which may be synthesized directly by the adrenocortical cells or taken up from low-density lipoproteins in the circulation and stored as cholesterol esters in lipid droplets in the cytoplasm. Cholesterol is released from the lipid droplets by the action of cholesterol esterase which is stimulated by ACTH. It is then converted to pregnenolone in the mitochondria. This step, which is also regulated by ACTH, is the rate-limiting step in steroid biosynthesis. The actions of ACTH appear to be mediated by cAMP. Pregnenolone is then transferred to the smooth endoplasmic reticulum. It undergoes further modifications here and in the miotochondria to form the three main classes of steroids (Fig. 10.14).

The steroid hormone produced depends on which hydroxylases and other enzymes are active in the tissue concerned. For example, formation of cortisol requires 17-hydroxylation of pregnenolone to 17-hydroxypregnenolone or of progesterone to 17-hydroxyprogesterone (Fig. 10.14). These steps require NADPH and O_2. On the other hand, the synthesis of corticosterone (and aldosterone) does not require 17 α-hydroxylase. Thus, the activity of this enzyme determines the proportion of the two glucocorticoids secreted. There is no appreciable storage of these hormones in the adrenal cortex and the rate of release corresponds to the rate of synthesis. In humans, approximately 20 mg of cortisol, 3 mg of corticosterone and 0.2 mg of aldosterone are secreted per day. The adrenal cortex also secretes significant amounts of androgens, particularly dehydroepiandrosterone and androstenedione. These are only weakly androgenic but can be converted in other tissues to the more potent androgens, testosterone and dihydrotestosterone (p. 307).

The corticosteroids are transported in the circulation mostly bound to plasma proteins such as **corticosteroid-binding globulin** (transcortin) and albumin. About 90% of cortisol and 60% of aldosterone exists in the bound form, which acts as a reserve and protects the steroids from degradation. The half-life of cortisol is about 1 hour, whereas that of aldosterone is about 20 minutes. Inactivation of the steroid hormones occurs principally in the liver followed by conjugation with glucuronic acid or sulphate to form water-soluble forms which are rapidly excreted in the urine.

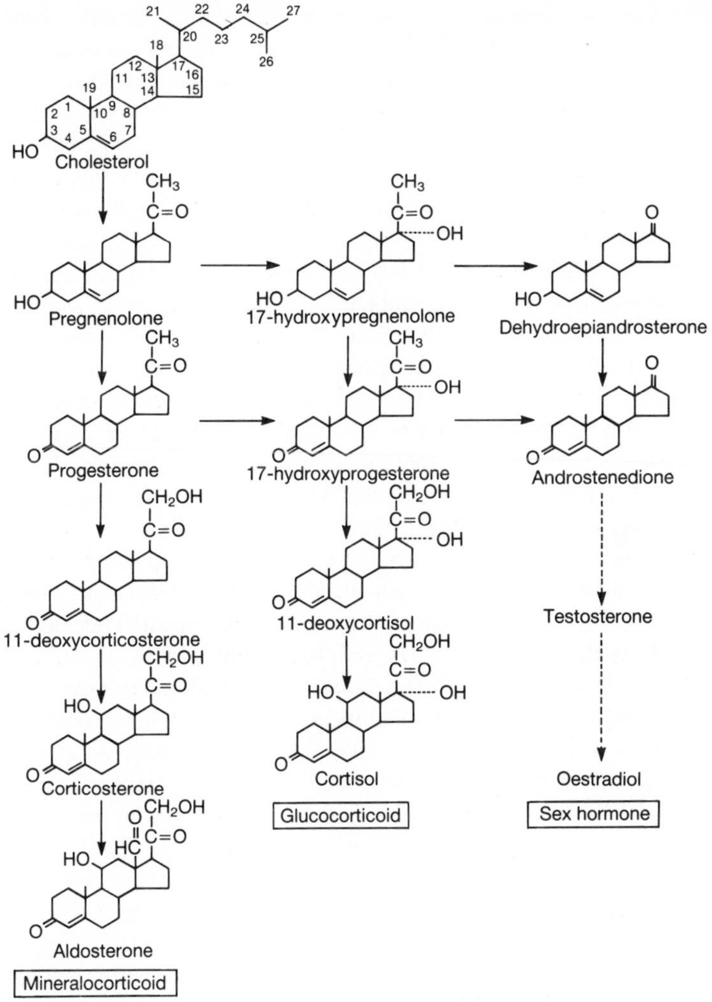

Fig. 10.14 Synthetic pathways for steroid hormones.

Actions of glucocorticoids

Glucocorticoids play an important role in the control of the intermediary metabolism of carbohydrate, fat, protein and purines throughout the body. Their modes of action are incompletely understood but regulation of the synthesis of specific enzymes is important. In many instances these enzymes are key regulators of metabolic functions or are required to mediate the actions of other hormones, such as adrenaline, glucagon, GH, vasopressin and angiotensin II. Hence they are said to have a 'permissive' action on the effects of other hormones. The following actions occur at physiological concentrations of these hormones.

1 Effects on intermediary metabolism. In the normal state they promote **glycogen storage** in the liver by stimulating glycogenesis. During fasting they stimulate **gluconeogenesis** in the liver to provide glucose for brain metabolism. The main substrates for gluconeogenesis are amino acids derived from **protein catabolism** in skeletal muscle. Glucocorticoids also stimulate this process and excess production of these hormones causes severe muscle wasting. The glucocorticoids are also counter-regulatory to insulin in that they **raise blood glucose**, effectively by inhibiting glucose uptake in muscle and adipose tissue. They also enhance **fatty acid mobilization** from adipose tissue by potentiating the lipolytic effects of catecholamines and GH.

2 Maintenance of normal circulatory function. Glucocorticoids are essential for the maintenance of normal **myocardial contractility** and **vascular resistance.** Their action on the vasculature is a permissive one in that they potentiate the vasoconstrictor effects of catecholamines. Glucocorticoids also decrease the permeability of vascular endothelium and therefore help maintain blood volume.

3 Adaptation to stress. The way in which the body adapts to stress is not well-understood but it is known that the release of cortisol increases during stress and that, in patients with adrenocortical deficiency, stress factors such as heat, cold, infection or trauma can cause hypotension and death.

In addition, the glucocorticoids do have some mineralocorticoid activity and this may be of significance because of their comparatively high secretion rate. In large doses, the glucocorticoids **suppress the immune response.** They decrease the number of circulating lymphocytes and eosinophils, cause involution of the thymus and lymph nodes, and may depress the antibody response. Therefore, synthetic corticosteroids are used therapeutically to suppress rejection of transplanted organs and to treat allergies. They also have **anti-inflammatory** properties and are used in the treatment of rheumatoid arthritis and related diseases. They may also be used to increase the synthesis of surfactant in fetal lung.

Control of glucocorticoid secretion

The secretion of cortisol and corticosterone (and androgens) is controlled by **ACTH** produced by the anterior pituitary. The secretion of ACTH is regulated by hypothalamic secretion of **CRH** and **ADH** into the hypophyseal portal system. In this instance, the ADH is secreted by the same cells as those that release CRH, which are distinct from the ADH-secreting cells that project to the posterior pituitary. Cortisol exerts a negative feedback effect on the hypothalamus and on the pituitary (Fig. 10.15). The level of plasma cortisol follows a diurnal pattern, peak levels occurring in the morning just before waking. This variation is related to the sleeping and waking pattern and interruption of this pattern may be partly responsible for the fatigue (jet-lag) experienced by air travellers. The effect of stress can increase cortisol secretion at any time.

Actions of mineralocorticoids

Aldosterone is the main mineralocorticoid produced by the adrenal cortex. It acts chiefly on the distal tubules of the **kidney** to promote the **reabsorption of Na$^+$** in exchange for K$^+$ and H$^+$ ions which are excreted (p. 652). Excess production of aldosterone with retention of Na$^+$ leads to expansion of extracellular fluid volume and hypertension. Adrenalectomy leads to a fall in extracellular Na$^+$, hypotension and eventually death.

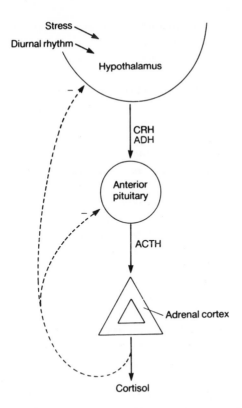

Fig. 10.15 Control of glucocorticoid secretion.

Control of mineralocorticoid secretion

ACTH is not the major regulator of aldosterone secretion, in contrast to the other corticosteroids, but it does play a supportive role. The primary regulator of aldosterone secretion appears to be **angiotensin II** produced by the renin–angiotensin system (p. 652). An increase in plasma K^+ concentration also stimulates the release of aldosterone by the adrenal cortex.

Disorders of adrenocortical function

In **Addison's disease** there is deficient secretion of all adrenocortical hormones, commonly due to autoimmune destruction of the gland. This disease usually develops slowly and is characterized by lethargy, weakness, weight loss and hypotension. Sudden stress can precipitate a crisis requiring emergency medical treatment. A common feature of this disease is hyperpigmentation of the skin due to excessive secretion of ACTH, leading to stimulation of melanocyte activity in the skin (p. 275). The elevation of ACTH levels is brought about by removal of the negative feedback provided by cortisol. Patients with Addison's disease require replacement therapy with both a glucocorticoid and a mineralocorticoid.

    Cushing's syndrome is a general term for conditions associated with prolonged elevation of plasma glucocorticoids resulting from excess ACTH secretion (a specific condition known as **Cushing's disease**), tumours of the adrenal cortex or overadministration of glucocorticoids in the course of therapy. Cushing's syndrome is characterized by redistribution of body fat, a 'moon face', severe muscle wasting, osteoporosis and a predisposition to diabetes and hypertension. A diagnosis of Cushing's disease is indicated if plasma cortisol is elevated throughout the day and if ACTH secretion is not suppressed by low-dose administration of the potent glucocorticoid drug dexamethasone.

Primary aldosteronism is due to excess mineralocorticoid secretion caused by a tumour of the adrenal cortex. This leads to K^+ depletion and Na^+ and water retention, resulting in hypertension, muscle weakness, tetany and hypokalaemic alkalosis.

Adrenogenital syndrome is associated with excessive androgen secretion which may cause masculinization in the female and precocious puberty in the male. This may be due to an androgen-secreting tumour or it may be congenital. The latter is known as congenital adrenal hyperplasia in which one of the enzymes involved in cortisol synthesis is deficient. This leads to increased ACTH secretion by the pituitary and hence excess production of adrenal androgens. Treatment with glucocorticoids corrects the deficiency and also suppresses excess ACTH secretion.

10.7 THE PANCREATIC ISLETS

The pancreatic islets produce a number of hormones, the most important being insulin and glucagon. Insulin is secreted by the islets in response to high blood glucose; it lowers blood glucose levels and stimulates the synthesis of glycogen, fat and protein. In contrast, glucagon, which is secreted in response to low blood glucose, increases blood glucose by stimulating glycogenolysis and gluconeogenesis. During feeding insulin promotes glucose utilization and storage as glycogen and fat, while during fasting the actions of glucagon help to maintain adequate supplies of glucose which are essential for the function of the brain.

The pancreas is both an endocrine and an exocrine gland. The endocrine cells of the pancreas are localized in the **islets of Langerhans** which constitute only 2% of the mass of the pancreas. **Insulin** is produced in the $B(\beta)$ cells and **glucagon** in the $A(\alpha)$ cells of the islets (Fig. 10.16). More recently it has been discovered that a third hormone, **somatostatin**, identical to GHRIH secreted by the hypothalamus, is also produced in the islets from $D(\delta)$ cells. Another polypeptide produced in the islets is **pancreatic polypeptide**, but its role is uncertain. Insulin was successfully extracted by Banting and Best in 1921 from the dog pancreas after they had first depleted it of proteolytic enzymes by ligating its exocrine ducts. The disease known as **diabetes mellitus** is due to a deficiency of insulin or to insulin resistance.

Insulin

Insulin is a small protein (M_r 5800) consisting of two peptide chains, called A and B, which are linked by two disulphide bonds. The A chain contains 21-amino acid residues and the B chain 30 residues. Beef and pig insulin differ from human insulin in only a few residues and both are used in the treatment of diabetes mellitus. Human insulin can now be made by chemical manipulation of porcine insulin and by DNA recombinant technology. The therapeutic potencies of these three types of insulin are remarkably similar. Insulin is synthesized as a larger single polypeptide preproinsulin, which is cleaved soon after synthesis to form

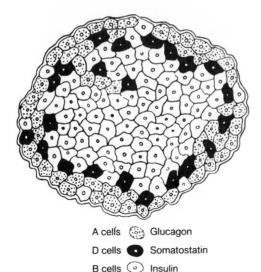

A cells ⊙ Glucagon
D cells ● Somatostatin
B cells ⊙ Insulin

Fig. 10.16 Islet of Langerhans.
(From Orci, L. & Unger, R.H.
(1975) *Lancet* **2**, 1243–1244.)

proinsulin (Fig. 10.17). Proinsulin is packaged into vesicles and converted to insulin by cleavage of a connecting peptide to form the two peptide chains. Insulin is released from the cell by exocytosis in response to an increase in blood glucose. It is secreted into the hepatic portal vein and thus reaches the liver directly. Its half-life in the blood is only a few minutes.

Actions of insulin

Insulin **lowers blood glucose** by facilitating glucose uptake in muscle and adipose tissue and by inhibiting hepatic glucose output. The most important effects of insulin are in the liver, where it stimulates glycogen and fat synthesis and inhibits glycogen breakdown and ketone body formation. In muscle, it stimulates glucose and amino acid uptake, and glycogen and protein synthesis, and in adipose tissue it stimulates glucose uptake and triglyceride synthesis. Insulin also increases K^+ uptake into cells and consequently can lower plasma K^+.

The mechanism by which the binding of insulin to its specific cell membrane

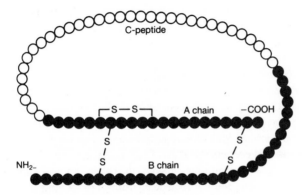

Fig. 10.17 Structure of proinsulin (insulin is shown in black).

receptor is translated into these specific events is not known. The receptor is composed of two α-subunits, which bind the hormone, and two β-subunits, which contain tyrosine-specific protein kinases (p. 45; Fig. 2.10). Binding of insulin results in autophosphorylation of the receptor, the role of which has yet to be elucidated. No specific intracellular second messenger has been identified, and it is probable that different mechanisms are involved for different kinds of cell response. Intracellular cAMP levels decrease following insulin binding to its receptor, but the significance of this is not known. Insulin stimulates glucose transport by causing the translocation of transporter proteins from an intracellular store to the plasma membrane.

Control of insulin secretion

An increase in plasma **glucose** concentration provides the major stimulus for insulin secretion. There is an initial rapid phase of secretion, due to release of preformed hormone, followed by a second slower phase of sustained secretion due particularly to release of newly synthesized hormone. Some **amino acids**, such as arginine, are also potent stimulators of insulin release. After feeding, the level of insulin may rise even before that of blood glucose because **gastric inhibitory peptide** can also stimulate insulin release. Other stimuli for insulin release are **glucagon** and the **sulphonylurea drugs**, such as tolbutamide, which are used in the treatment of mild cases of diabetes. **Adrenaline** inhibits insulin release, and so too does **somatostatin**. The latter, which is produced in the islets of Langerhans, probably plays an important role in the local control of insulin secretion. Neural control is also important; parasympathetic activity enhances insulin release, while sympathetic activity inhibits it.

Glucagon

Glucagon is a polypeptide (M_r 3500) which is secreted by the pancreas in response to low blood glucose. In contrast to insulin, it **increases blood glucose** levels. Glucagon acts on the liver (via cAMP) to stimulate **glycogenolysis** (breakdown of glycogen) and **gluconeogenesis** (synthesis of glucose from lactate, amino acids or glycerol). Glucagon secretion is also stimulated by amino acids, as is insulin release. This ensures that, when there is a high intake of amino acids, a precipitous fall in blood glucose due to elevated insulin is prevented by the action of glucagon. Glucagon is also a potent stimulator of fat mobilization but an infusion of glucagon does not normally increase free fatty acid levels, probably because its actions are offset by a concomitant rise in insulin.

Control of energy utilization and storage

The brain uses glucose almost exclusively as an energy source and it is necessary to maintain an adequate blood glucose concentration at all times or convulsions and coma will ensue. Blood glucose is normally maintained at fairly constant levels of 4–6 mmol L^{-1} by the interactions of several hormones, namely **insulin, glucagon, GH, adrenaline** and **cortisol**. Of these, insulin acts to lower blood glucose and is known as the regulatory hormone, whereas the other four act to raise blood glucose and are known as the counter-regulatory hormones. During the **absorp-**

tive state following a meal, there are adequate glucose supplies and this causes the secretion of insulin which enhances the utilization of glucose and the storage of energy as glycogen and fat. During the **postabsorptive** or **fasting state** blood glucose falls and insulin secretion decreases in relation to that of the counter-regulatory hormones. The ratio of insulin to glucagon is probably the most important factor in controlling the shift from the absorptive to the postabsorptive state. This ensures that during fasting adequate glucose levels are maintained for the brain, initially by glycogenolysis, and then by gluconeogenesis in the liver and by glucose-sparing reactions in other tissues. A further account of energy storage and utilization is given in Chapter 20 (p. 676).

Effects of insulin deficiency

Diabetes mellitus is a major health problem which affects about 2% of the population. A predisposition to diabetes is inherited but the genetic factors are complex. Two types of diabetes are recognized clinically—**insulin-dependent** or juvenile-onset diabetes (type I) and **non-insulin-dependent** or maturity-onset diabetes (type II). Type I patients (10–12% of diabetics) have low plasma insulin and require injections of insulin. It is thought that type I diabetes is an autoimmune disease possibly triggered in childhood by a viral infection. Type II patients may have normal or elevated levels of insulin but show decreased sensitivity to insulin, often correlating with a reduction in insulin receptor concentration. Type II patients are often obese and generally show improvement with weight reduction. Of particular interest is the recent discovery of high concentrations of the polypeptide amylin in the pancreatic B cells of diabetic patients. Amylin decreases insulin secretion but its precise role in diabetes is not known. For experimental purposes, diabetes can be simulated in animals by treatment with alloxan, which destroys the B cells of the islets of Langerhans.

By 'diabetes mellitus' is meant the passing of sweet urine. The capacity of the renal tubules to reabsorb glucose actively is exceeded and glucose spills over into the urine. The loss of so much solute causes an **osmotic diuresis**, resulting in polyuria and polydipsia. Large amounts of salts are consequently lost and this can lead to dehydration and hypotension. The blood concentration of free fatty acids is raised and their metabolism produces **ketones** which cause **metabolic acidosis**. If left untreated, the patient may become unconscious (**diabetic coma**). The administration of too much insulin can also lead to coma (**insulin coma**) because of the sudden lowering of blood glucose and the brain's dependence on glucose as an energy source.

There are also long-term effects of the disease: in particular, atherosclerosis in the coronary and leg arteries, and capillary abnormalities in the eyes, kidneys and nerves. The latter effects may be caused by the high glucose levels leading to abnormal protein glycosylation and to the production of sorbitol which causes osmotic damage. Diabetics are also prone to infection and injuries take longer to heal than normal.

In the diagnosis of diabetes, the physician is guided by the fasting blood glucose concentration and the presence of glucose and ketones in the urine.

Chapter 11
Reproduction

From a physiological point of view a fundamental difference between the female and the male is that the former undergoes obvious cyclic variations in reproductive activity. **Female** mammals have an **ovarian cycle** with a characteristic mean frequency for each species, e.g. 4 days in rats and 28 days in women. In women and in some other primates, the cycle is marked by a period of menstrual bleeding, which is due to the shedding of the endometrial lining of the uterus. Other female mammals do not menstruate, although they show a phase called **oestrus**, or 'heat' as it is commonly known, when the female becomes receptive to the male just before ovulation. In some species, e.g. the cat and rabbit, ovulation is triggered by copulation. In most mammals, however, ovulation is regulated by an intrinsic rhythm controlled by interactions between the hypothalamus, pituitary and gonads.

11.1 SEXUAL DEVELOPMENT

At an early stage of development the fetus has primordial genital ducts for both the male and the female forms. In the genetic male testosterone promotes the development of the male ducts, while Müllerian-inhibiting hormone causes the regression of the potential female ducts. In the genetic female the absence of these hormones allows the opposite to occur. Both sexes initially have the potential for cyclic reproductive activity, but in the male this is abolished by testosterone at a critical period of development just before birth. At puberty sexual maturity occurs, and is accompanied by increased secretion of the gonadal hormones.

The sexual differences between male and female depend ultimately on differences in their chromosomes. In most mammalian species, the female has two X sex chromosomes while the male has one X and one Y sex chromosome, in addition to the autosomal chromosomes (22 pairs in humans). It is possible to determine the sex-chromosomal pattern by laboratory examination of cell smears, since the female pattern can be conveniently recognized by the presence of an extra piece of chromatin (an inactive X chromosome), called the sex chromatin or Barr body. Most sex-chromosome abnormalities, such as XXY (Klinefelter's syndrome) and XO (Turner's syndrome) result in incomplete development of the gonads. The XXX (superfemale) and XYY (supermale) patterns, however, do not seem to be characterized by abnormal sexual development. Both departures from the normal sex-chromosomal pattern and hormonal disturbances can result in defective development of the reproduction system.

In humans, the 'indifferent' gonads of both sexes are identical until differentiation begins at about the sixth week of fetal life. In the genetic male, the presence of the H-Y gene, a specific sex-determining gene located on the Y chromosome, causes differentiation of the gonads into the testes. Genetic females lack the H-Y gene and as a result the indifferent gonads develop into the ovaries. At this stage, the fetus has primordial genital ducts for both the male (**Wolffian ducts**) and the female (**Müllerian ducts**). In the male fetus, testosterone secreted by the developing testes causes the Wolffian ducts to develop into the internal genitalia (epididymis, vas deferens, seminal vesicles and prostate) and a peptide hormone called **Müllerian-inhibiting hormone** (or anti-Müllerian hormone), also secreted by the testes, inhibits the development of the Müllerian ducts. In the female fetus, the absence of testosterone and **Müllerian-inhibiting hormone** allows the opposite to occur. The external genitalia are similarly bipotential at this stage. In males, androgenic hormones derived from the testes cause the external genitalia to develop into the male form, whereas external genitalia develop into the female form in the absence of these androgens.

The secretion of testosterone and Müllerian-inhibiting hormone is necessary but not sufficient for the development of the male phenotype. The tissues must also have the necessary androgen receptors to respond. Otherwise a defect known as **testicular feminization** (androgen-insensitivity syndrome) will result, in which genetic males appear as phenotypic females with abdominal testes. Moreover, some fetal tissues require testosterone for differentiation while others, e.g. the prostate and the penis, require dihydrotestosterone. The conversion of testosterone to dihydrotestosterone is catalysed by the enzyme, 5α-reductase. In **5α-reductase deficiency**, the affected males at birth have testes but lack a prostate gland, and their external genital organs resemble those of the female. Another defect caused by an inborn error of metabolism is **congenital adrenal hyperplasia**, which is associated with masculinization of the female fetus (p. 298).

Animal studies indicate that the potential for developing cyclic reproductive activity is present in both sexes at the fetal stage. The secretion of testosterone in the male appears to abolish this intrinsic rhythm, because if testosterone is injected into the genetic female at a critical period the sexual rhythm is not generated. This critical period occurs within a few days after birth in the rat, while in the monkey it occurs before birth.

After birth reproductive development is dormant until **puberty** when the reproductive organs in both sexes are reactivated by increased secretion of gonadotrophins due to maturation of the hypothalamopituitary axis. Puberty coincides with accelerated growth of the body and the development of the **secondary sexual characteristics**. The onset of puberty usually occurs at about 11 in girls and 12 in boys, although this may vary considerably depending on genetic and environmental influences. Sexual maturity is signified by the **menarche** or first menstrual bleeding in the female and by the first ejaculation in the male. At about 50 years of age in women the **menopause** occurs: the ovary ceases to respond to gonadotrophins, the sexual cycles gradually disappear and menstruation ceases. In contrast, the production of sperm and testosterone in males continues throughout life, although it may gradually diminish.

11.2 THE MALE REPRODUCTIVE SYSTEM

Once puberty has been reached, testosterone production and spermatogenesis in the testes occur continuously. These two functions are controlled by the hypothalamopituitary axis through the release of the gonadotrophins, and folliclestimulating hormone (FSH) and luteinizing hormone (LH). Testosterone has a negative-feedback effect on the secretion of the gonadotrophins; it also promotes the development of the male reproductive systems, the production of sperm, and the characteristics of maleness.

The primary reproductive organs or gonads of the males are the **testes**, which produce **spermatozoa** and also secrete the male sex hormone, **testosterone**. In addition, there are the **accessory** reproductive ducts (Fig. 11.1) and secretory glands (**seminal vesicles, prostate gland** and **bulbourethral glands**) which are involved in the transport and delivery of spermatozoa to the female. A few weeks before birth the two developing testes pass out of the abdominal cavity into the scrotal sac. Failure of the testes to descend into the scotum (**cryptorchidism**) may result in infertility, because the testes will not develop properly and production of spermatozoa (**spermatogenesis**) depends on a temperature about 4 °C below body temperature. The scrotum can contract or relax to move the testes closer to or further away from the body so that this temperature is maintained.

Spermatogenesis

Spermatozoa are produced in the **seminiferous tubules** of the testes (Fig. 11.2a). These tubules are lined with germ cells and **Sertoli cells**. The latter closely surround the developing germ cells (Fig. 11.2b) and provide them with nutrients and stimulating factors, hence they are sometimes referred to as nurse cells. They also secrete a number of substances, including androgen-binding protein and the hormone inhibin.

At puberty, in response to gonadotrophic hormones released by the anterior pituitary, spermatogenesis is initiated and occurs continuously thereafter. Primordial germ cells called **spermatogonia** divide to form **primary spermatocytes** which undergo the first meiotic division to form **secondary spermatocytes**. The

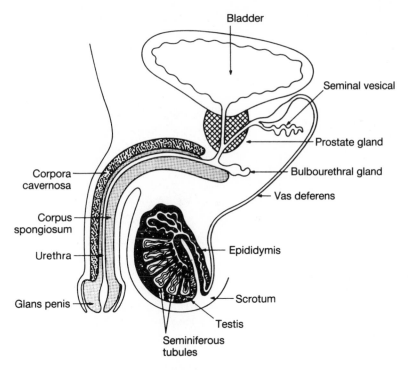

Fig. 11.1 The male reproductive system.

secondary spermatocytes, which contain a haploid number of duplicated chromosomes, then undergo a second meiotic division to form haploid **spermatids**. During meiosis, there is a reduction in the number of chromosomes from 46 to 23. Finally, spermatids differentiate to give rise to **spermatozoa**. The process of spermatogenesis from spermatogonium to the release of spermatozoa into the lumen of the seminiferous tubule takes about 70 days.

When spermatozoa are released into the semiferous tubules, they are non-motile and incapable of fertilizing an ovum. From the seminiferous tubules, spermatozoa pass into the **epididymis** where they mature and are stored until ejaculation takes place. The production of spermatozoa is a continuous process, and spermatozoa not ejaculated eventually deteriorate and are reabsorbed by phagocytosing epididymal cells.

The **mature sperm** consists of a head, middle piece and long tail (Fig. 11.3). The head is composed mainly of the nucleus and is covered by a cap known as the **acrosome**. The acrosome is a large vesicular structure containing lytic enzymes that help the sperm to penerate the outer coat of the ovum. The middle piece consists of a helical sheath of mitochondria surrounding a core of contractile filaments which extend into the tail. The mitochondria provide energy for motility of the spermatozoon, which depends on a wave-like movement of the tail.

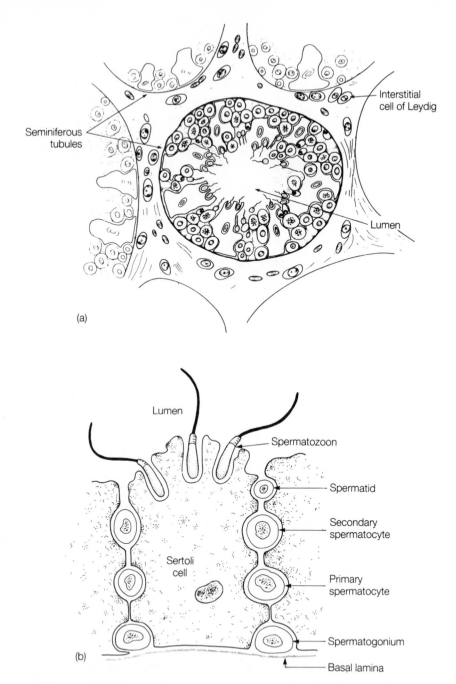

Fig. 11.2 (a) Diagram of the testis in cross-section illustrating the seminiferous tubules and interstitial cells of Leydig. (b) Structure of the wall of a seminiferous tubule showing the intimate relationship between the Sertoli cells and the developing germ cells.

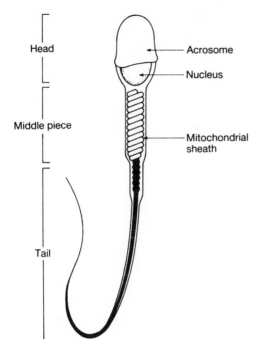

Head — Acrosome

— Nucleus

Middle piece

— Mitochondrial sheath

Tail

Fig. 11.3 The mature spermatozoon.

Testosterone

The principal androgenic hormone produced by the testis is **testosterone** (Fig. 11.4). Testosterone is a C19 steroid synthesized from cholesterol, as illustrated in Fig. 10.14 (p. 295). This steroid hormone is secreted by the **interstitial cells of Leydig** which lie scattered between the seminiferous tubules (Fig. 11.2a). Most of the androgens in the male are synthesized here, although a small amount is produced in the adrenal cortex (p. 294). Once released into the blood, testosterone is bound to a specific carrier globulin, called sex hormone-binding globulin. In most of its target tissues (except muscle), testosterone appears to be converted to a more potent androgen called **dihydrotestosterone** (Fig. 11.4). It acts via nuclear receptors to promote gene activation and the synthesis of specific proteins (p. 45). The hormone is metabolized in the liver by reduction and conjugation with glucuronic acid, and the metabolites are excreted by the kidneys.

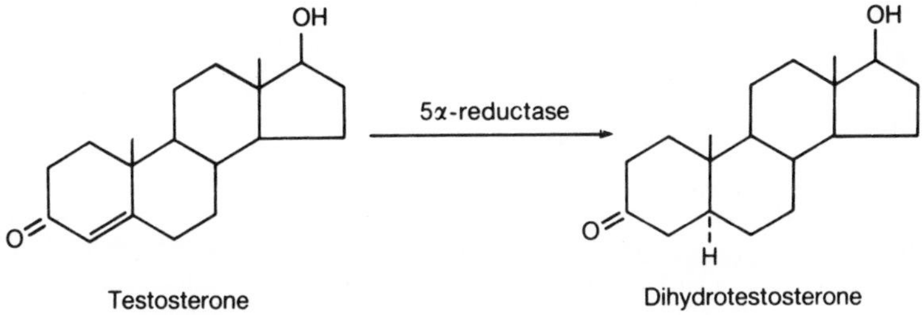

OH

5α-reductase

OH

Testosterone

Dihydrotestosterone

Fig. 11.4 Testosterone and its conversion to dihydrotestosterone.

Testosterone promotes the development of the reproductive system and of the **secondary sexual characteristics** of the male and has important anabolic effects in skeletal muscle and bone. The most obvious effects of testosterone are seen at puberty, namely, enlargement of the penis and testes, increased rate of growth of muscle and bone, appearance of facial, axillary and pubic hair, and change in the pitch of the voice. Castration or removal of the testes in childhood prevents most of these changes from occurring. Castrated males do not become bald and the presence of testosterone seems necessary for baldness to occur in males genetically predisposed to this condition. Testosterone also promotes libido (sexual drive).

Control of the male reproductive activity

The anterior pituitary gonadotrophic hormones—**FSH** and **LH**—control, respectively, spermatogenesis and synthesis of testosterone (Fig. 11.5). The release of the gonadotrophic hormones occurs in a pulsatile manner and is controlled by a common hypothalamic neurohormone—**gonadotrophin-releasing hormone** (GnRH), formerly known as **luteinizing hormone-releasing hormone**. Both testosterone and FSH are necessary for normal spermatogenesis. Testosterone inhibits secretion of LH in a typical negative-feedback manner, acting mainly on the hypothalamus, but has less effect on FSH secretion. **Inhibin**, a polypeptide

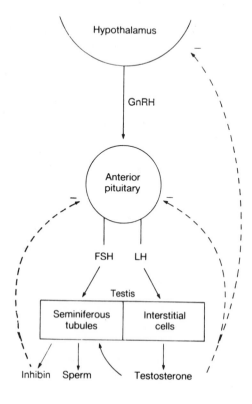

Fig. 11.5 Control of spermatogenesis and testosterone secretion.

hormone secreted by the Sertoli cells in the testes, has a negative-feedback effect on secretion of FSH.

FSH and LH are glycoproteins and each can be dissociated into two subunits called α and β. The α-subunits of FSH and LH are identical in structure, and specificity is determined by the β-subunit. Their actions are mediated by cyclic AMP. Inhibin is a protein composed of two peptide chains, α and β, which are joined by disulphide bonds. There are two forms of the β-chain, β_A and β_B, both of which in combination with the α-chain are equally active in inhibiting FSH secretion. However, dimers of the β-chains, $\beta_A\beta_B$ and $\beta_A\beta_A$, are both potent stimulators of FSH secretion and hence $\beta_A\beta_B$ is called activin. The physiological role of activin has yet to be established.

Semen

The ejaculated fluid or semen contains spermatozoa and secretions of the seminal vesicles, prostate gland and bulbourethral glands. The average volume of ejaculate in man is about 3 ml and this contains approximately 100 million spermatozoa per ml. The secretions of these accessory glands comprise the bulk of the semen and assist in the transport and nourishment of the sperm. The seminal fluid contains high concentrations of **fructose**, which serves as an energy substrate for the spermatozoa, and also high concentrations of **prostaglandins** which may increase motility of the uterus, thus promoting transport of spermatozoa in the female genital tract. Fertility depends on the quality of the semen, the two most important factors being the number and motility of the spermatozoa. Decreased sperm production (**oligospermia**) is a common cause of infertility in the male and a count of less than 20 million spermatozoa per millilitre is generally considered incompatible with fertility.

Erection

The erection of the penis, which is necessary for coitus and delivery of semen to the female, is due to the engorgement of the penis with blood. Erection may be initiated by psychic stimuli and by tactile stimulation of the glans penis. Failure of erection, i.e. **impotence**, may be due to psychological as well as organic disturbances. In erection, a spinal reflex arc is involved in which impulses pass along afferent nerves to integrating centres in the **sacral spinal cord**, there to initiate impulses which travel back along **parasympathetic fibres**. Excitation of the parasympathetic fibres causes **arteriolar dilatation** in the penis so that the venous sinusoids of the corpora carvernosa and corpus spongiosum (Fig. 11.1) become engorged with blood. This, together with compression of draining veins by the ischiocavernosus muscle, produces penile erection.

Ejaculation

This is a reflex action involving movement (**emission**) of spermatozoa and glandular secretions into the **urethra** followed by the sudden ejection of the semen from the urethra. Emission of the glandular secretions occurs in a definite sequence. During erection, the secretion of the bulbourethral glands is discharged to lubricate the urethra. During ejaculation, the alkaline secretion of the prostate is discharged first to neutralize the acidity of the male urethra and the vagina. This is followed by the discharge of spermatozoa and finally the secretion of the seminal vesicles is added.

Ejaculation is triggered by stimulation of tactile receptors in the glans penis causing impulses to pass along afferent nerves to centres in the **lumbar spinal cord** and initiate impulses which return along sympathetic fibres. This **sympathetic activity** leads to contraction of the smooth muscle of the epididymis, vas deferens and secretory glands propelling spermatozoa and glandular secretions into the urethra. At the same time, the internal sphincter of the urethra constricts, preventing semen from entering the bladder. Contraction of the bulbospongiosus and ischiocavernosus muscles due to reflex activity in **somatic motor nerves** then leads to pulsatile emission of the seminal fluid from the urethra.

Vasectomy

This procedure, in which the vasa deferentia are cut and tied, is an effective means of contraception in the male. Spermatogenesis continues after vasectomy but, without an outlet, the spermatozoa degenerate and are reabsorbed in the epididymis. Since less than 5% of semen consists of spermatozoa, the volume of the ejaculate is little affected by vasectomy.

11.3 THE FEMALE REPRODUCTIVE SYSTEM

Activity in the female reproductive system is regulated by periodic changes in the secretion of FSH and LH by the anterior pituitary, and subsequently oestrogen and progesterone by the ovaries. A central event in the female cycle is ovulation, which is triggered by a surge in LH secretion. An external manifestation of the cycle is menstruation, when a fall in oestrogen and progesterone causes the shedding of the endometrial lining of the uterus. Since the ovarian hormones suppress the release of FSH and LH, their removal coincides with a rise in FSH which allows a new cycle to begin again.

The primary reproductive organs of the female (Fig. 11.6) are the two **ovaries**, which produce ova and secrete the sex hormones, **oestrogen** and **progesterone**. The accessory reproductive structures comprise the two **oviducts** (fallopian tubes), and the **uterus, cervix** and **vagina**. At the opening of the vagina lie the external genitalia (the vulva).

Oogenesis and follicular development

The formation of primordial germ cells or **oogonia** is completed during fetal development of the ovary. Before birth many of the oogonia begin the first stage of meiosis but become arrested without completing division to form diploid **primary oocytes**; others degenerate. At birth, each human ovary contains approximately 1 million oocytes. The number surviving in both ovaries at puberty is less than about 400 000 and continues to decline so that at the time of menopause few viable cells remain. Although the first stage of meiosis begins before birth, it is not until just before ovulation that the first meiotic division is completed with the formation of the **secondary oocyte** containing a haploid number of duplicated chromosomes. In this process, most of the cytoplasm is retained by the oocyte and a smaller rudimentary cell, called the first polar body, is split off. The second meiotic division

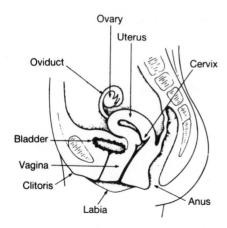

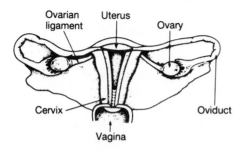

Fig. 11.6 The female reproductive system.

occurs after fertilization in the oviduct; a second polar body is eliminated and the mature **ovum** is formed.

In the ovary, the primary oocytes are arranged in **primary follicles**. Each primary follicle (Fig. 11.7) consists of an oocyte surrounded by a single layer of **granulosa cells**. From puberty onwards, in response to a rise in the gonadotrophic hormone FSH at the beginning of each cycle, several follicles begin to develop by proliferation of the granulosa cells. Normally, only one of these reaches the stage of ovulation; the rest degenerate. Only about 400 of the primary follicles develop into ova during the reproductive life of a woman. As a follicle matures, the granulosa cells secrete mucopolysaccharides which form a translucent halo called the **zona pellucida** around the oocyte. Soon after, the developing follicle becomes surrounded by a capsule of ovarian tissue made up of an inner cellular layer, the **theca interna**, and a more fibrous outer layer, the **theca externa**. With further maturation fluid accumulates amongst the granulosa cells to form a central cavity filled with fluid called the **antrum**. In the mature **Graafian follicle** the oocyte, embedded in a mass of granulosa cells, protrudes into the antrum (Fig. 11.7). The theca interna cells synthesize androstenedione (see Fig. 10.14, p. 295), which is converted by the granulosa cells into the oestrogen hormones, **oestradiol** and **oestrone**. About the middle of the ovarian cycle, ovulation occurs: the follicle ruptures and the secondary oocyte, together with its surrounding granulosa cells, is extruded into the peritoneal cavity. It is then swept by the movement of cilia into the open end of the oviduct, which is closely applied to the ovary.

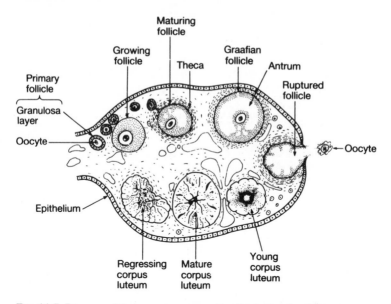

Fig. 11.7 Diagram of the ovary in cross-section illustrating the different stages in the development of a follicle and corpus luteum during one ovarian cycle. (The sequential arrangement is not an actual representation.)

After ovulation, the granulosa cells remaining in the ruptured follicle, together with cells of the theca interna, proliferate to form a new endocrine structure—the **corpus luteum**. This continues to secrete oestrogen but also produces the hormone **progesterone**. The corpus luteum is functional for about 12 days after ovulation, after which it regresses, due probably to a decline in LH secretion. Thereafter, unless fertilization of the ovum and implantation have occurred, the regression of the corpus luteum causes a decline in the secretion of oestrogen and progesterone, the onset of menstruation and the initiation of a new cycle of ovarian activity.

Gonadotrophic hormones

Maturation of follicles in the ovary requires the presence of the **gonadotrophic hormones, FSH** and **LH**, which are secreted by the anterior pituitary. The secretion of both gonadotrophic hormones is stimulated by **GnRH**, which is produced in the medial basal area of the hypothalamus and is released in a pulsatile manner with a periodicity of 1–2 hours. Secretion of the gonadotrophins is regulated by feedback actions of oestrogen and progesterone (see later) and of inhibin (see Fig. 11.10). Inhibin, which is secreted by the granulosa cells of the follicles, acts on the pituitary to inhibit the release of FSH.

FSH, together with oestrogen, promotes **development of the follicle** and LH is needed for the **secretion of oestrogen** by the ovary. The sudden peak in the LH secretion mid-cycle appears to **trigger ovulation**. Ovulation can be induced in infertile women by treatment with gonadotrophic extracts from the pituitary or urine. To be effective, such extracts must contain FSH for follicular development and LH for the production of oestrogen and ovulation. LH is also required for normal corpus luteum function.

Menstrual cycle

The average length of the menstrual cycle is 28 days but it may vary considerably among different women. The first signs of bleeding signal the start of a new menstrual cycle and bleeding continues for about 3–5 days. During this time and until ovulation, the **ovarian follicles** develop and secrete increasing quantities of **oestrogen**. Oestrogen acts on the uterus to stimulate regeneration and growth of the **endometrium** from the remnants left after the previous menstrual cycle, causing a two- to threefold increase in the thickness of the endometrium. The first 2 weeks of the menstrual cycle are therefore referred to as the **follicular phase** with respect to the ovary and as the **proliferative phase** with respect to the uterus (Fig. 11.8). The variation that occurs in the duration of the menstrual cycle is usually due to variation in this first half of the reproductive cycle. Ovulation occurs at about the mid-point of the cycle, i.e. around day 14.

During the second half of the reproductive cycle, the **corpus luteum** develops and secretes both **oestrogen** and **progesterone**. Oestrogen continues to promote proliferative activity in the endometrium while, under the action of progesterone, the endometrial glands become distended with secretory products, including **glycogen**, which is an important nutrient for the developing embryo should implantation take place. Endometrial blood flow increases and the spiral arteries become more tightly coiled and twisted. The second half of the cycle is therefore referred to as the **luteal phase** with respect to the ovaries and as the **secretory phase** with respect to the uterus. If implantation does not occur, the corpus luteum regresses, there is a rapid fall in secretion of oestrogen and progesterone

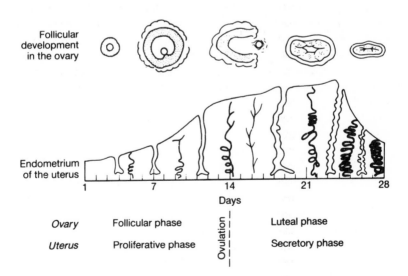

Fig. 11.8 Diagram showing the changes occurring in the endometrium and in an ovarian follicle during the menstrual cycle.

and, for reasons which are unclear, the endometrium undergoes shrinkage due to the loss of extracellular water and constriction of the spiral arteries. This causes a reduction in blood flow to the endometrium with cell death and weakening of the walls of blood vessels. As the phase of vasoconstriction wears off, blood leaks from the damaged vessels to initiate **menstrual bleeding** and eventually all but the basal layer of endometrium is detached from the uterus. A second phase of vasoconstriction of the spiral arteries minimizes loss of blood.

Ovarian hormones

The ovary secretes both steroid hormones, **oestrogen** and **progesterone** (Fig. 11.9), and the protein hormone **inhibin**. Oestradiol is the main oestrogen produced but oestrone, which has less biological activity than oestradiol, is also secreted in significant amounts.

Oestrogen and progesterone are C18 and C21 steroids, respectively, and are synthesized from cholesterol as outlined in Fig. 10.14 (p. 295). These steroid hormones are transported in the blood bound to plasma proteins. Like other steroid hormones, oestrogen and progesterone produce their effects in responsive tissues by combining with nuclear receptors to promote gene activation with a subsequent increase in synthesis of the effector proteins (p. 45). They are metabolized by reduction and hydroxylation in the liver and their metabolites (particularly oestriol and pregnanediol in conjugation with glucuronic acid) are excreted by the kidney.

Oestrogen is secreted by the granulosa cells of the follicles and is also produced after ovulation by the corpus luteum. The peak of LH that occurs at ovulation is preceded by a rise in oestrogen secretion which stimulates LH secretion, i.e. a high level of oestrogen has a **positive-feedback** effect on LH (and FSH) secretion at this point in the cycle (Fig. 11.10). During the secretory phase of the menstrual cycle, oestrogen and progesterone act synergistically to exert a **negative-feedback** effect on the secretion of the gonadotrophic hormones. It is now believed that the primary site of the negative-feedback action of oestrogen is the pituitary, although the positive-feedback of oestrogen is probably modulated by the hypothalamus (Fig. 11.10). The negative-feedback effect diminishes as the concentrations of oestrogen and progesterone fall with regression of the corpus luteum.

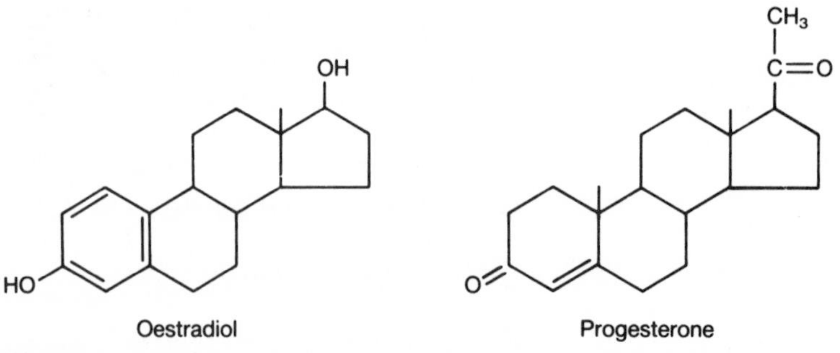

Fig. 11.9 Structure of oestradiol and progesterone.

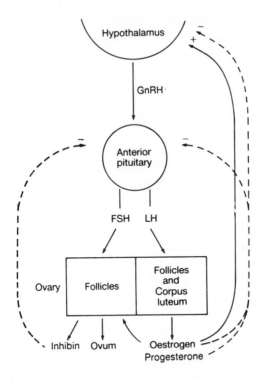

Fig. 11.10 Control of follicular development and oestrogen secretion.

The present model of positive oestrogen feedback is that oestrogen in the absence of progesterone acts by increasing the frequency and amplitude of GnRH secretory pulses, thus stimulating LH release and both FSH and LH synthesis. This effect is inhibited by progesterone; thus, there is no positive feedback in the luteal phase, despite comparable or greater levels of oestrogen.

In addition to its positive and negative effects on gonadotrophin secretion, oestrogen:
1 sensitizes the ovaries to the effects of gonadotrophins by stimulating the synthesis of FSH and LH receptors;
2 stimulates growth of the endometrium and contractility of the myometrium;
3 stimulates the output of mucus from the cervical glands and causes changes in the properties of the mucus which assist entry of spermatozoa;
4 causes the vaginal epithelium to proliferate and show increased cornification;
5 stimulates the growth and development of the breasts, particularly of the lactiferous ducts;
6 promotes the growth of bones and skeletal muscle and helps to bring about the characteristic female patterns of distribution of body hair and adipose tissue;
7 promotes closure of the epiphyses at the end of the period of linear skeletal growth; and
8 helps to conserve bone (p. 291); osteoporosis is a common condition in postmenopausal women.
Progesterone, which is present in significant amounts only during the luteal phase of each menstrual cycle, acts on tissues which have already been stimulated by

oestrogen. In addition to having inhibitory effects on gonadotrophin secretion and follicular development during the luteal phase, progesterone:

1 transforms the endometrium to its secretory phase and decreases the spontaneous electrical activity of the myometrium;

2 modifies the composition of cervical mucus, making it more viscous and resistant to penetration by spermatozoa;

3 causes further changes in the vaginal epithelium with regression of cornification;

4 promotes development of the breasts, particularly of the secretory units; and

5 causes an increase in basal body temperature after ovulation, which may be useful clinically to indicate that ovulation is occurring.

The principal role of **inhibin** appears to be to suppress FSH secretion by pituitary gonadotrophs.

Control of the female reproductive cycle

The blood levels of the gonadotrophic and ovarian hormones throughout the ovarian cycle are shown in Fig. 11.11. A marked peak in the level of gonadotro-

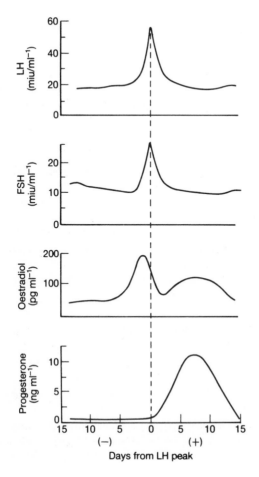

Fig. 11.11 Plasma levels of LH, FSH, oestradiol and progesterone during the ovulatory cycle centred on the mid-cycle LH peak (day 0). (After Thorneycroft, I. A. *et al.* (1971) *Am. J. Obstet. Gynecol.* **111**, 947–951.)

phins occurs at the mid-point of the menstrual cycle and coincides with the time of ovulation. (Because it is difficult to pinpoint the time of ovulation, the peak in blood LH has been arbitrarily designated as day 0 in Fig. 11.11, but in clinical practice the beginning of menstruation is taken as day 1 of the cycle.) Just before the beginning of each cycle there is a small rise in FSH. This rise in FSH stimulates follicular development and, together with LH, leads to an increase in oestrogen secretion. Oestrogen plays an important role in the maturation process because it sensitizes the granulosa cells to the effects of gonadotrophins and thus increases their capacity to produce oestrogen. The surge in oestrogen secretion that occurs just prior to ovulation has a positive-feedback effect on the anterior pituitary, probably by way of the hypothalamus, causing a marked rise in LH and FSH secretion. The peak in LH secretion triggers ovulation. After ovulation, the increase in oestrogen and progesterone that coincides with the development of the corpus luteum prepares the endometrial lining of the uterus for implantation. If implantation does not occur, the levels of oestrogen and progesterone fall in parallel with the demise of the corpus luteum. The fall in ovarian hormones causes menstruation to occur and also removes the negative-feedback influence on the secretion of the gonadotrophic hormones. The resultant rise in output of FSH triggers development of a new batch of follicles and the beginning of a new cycle of uterine and ovarian function.

Menopause

The termination of reproductive function in the female occurs at the menopause, which is the time of the last menstrual period. The period of time leading up to this is termed the **climacteric**, which is often characterized by irregular menstrual cycles, hot flushes, sweating, vaginal dryness and urethral irritation. The menopause is thought to result from the exhaustion of follicles leading to a fall in oestrogen and progesterone levels. The reduction in the negative-feedback effects of oestrogen and progesterone causes plasma FSH and LH to rise to high levels. In postmenopausal women bone loss is accelerated and may develop into osteoporosis, a disorder characterized by a reduction in bone density and an increased incidence of bone fractures. **Hormone replacement therapy** with oestrogen helps to conserve bone mass, the benefits being greatest soon after menopause. It also reduces or eliminates menopausal symptoms and reduces the risk of coronary heart disease. A progestagen should also be administered because of the increased risk of endometrial cancer associated with unopposed oestrogen therapy.

Hormonal contraception

The oral contraceptive pill is a very effective means of preventing pregnancy. Hormonal contraception is the outcome of research into the negative-feedback effects of the sex hormones. However, natural oestrogens and progestagens (progestins) are not effective when taken orally, because they are rapidly degraded in the gut and liver, and it was not until the development of synthetic oestrogens and progestagens that hormonal contraception was widely adopted. The most commonly used contraceptive pill is a combination of a synthetic oestrogen and a synthetic progestagen taken daily but withdrawn towards the end of the month. They act synergistically, primarily through feedback inhibition of FSH and LH secretion, to suppress maturation of the follicles and ovulation. Preparations containing a progestagen also produce changes in the cervical secretions, making it more difficult for sperm to penetrate the uterus and to undergo capacitation (p. 319). These appear to be the main actions of the progestagen-only contraceptive which is administered continuously, either orally or by the slow release of hormone from a vaginal suppository or intramuscular injection. The synthetic steroids may also induce changes in the endometrium, preventing successful embryonic implantation.

Hyperprolactinaemia

Prolactin is required in mammals for breast development and lactation (p. 322). In some rodents it prolongs the life of the corpus luteum and so has been called luteotrophic hormone. Although the production of prolactin is controlled principally by the inhibitory influence of dopamine, a stimulatory influence is exerted by thyrotrophin-releasing hormone and by oestrogen during pregnancy. Hyperprolactinaemia occurs when prolactin secretion is excessive, as for example in certain pituitary tumours. Elevated blood prolactin concentrations cause anovulation and amenorrhoea in women, and impotence and infertility in men. In these conditions, prolactin appears to inhibit the actions of LH and FSH on the ovary and testis, respectively.

Pineal gland

This gland, situated in the centre of the brain, secretes the hormone **melatonin**. In experimental animals, the administration of melatonin inhibits ovulation by preventing the release of GnRH and consequently LH. In some animals, the secretion of melatonin fluctuates in relation to light (and hence day length) and it is thought that the pineal gland plays a role in determining the seasonal breeding patterns of such animals.

Melatonin is synthesized from serotonin and its synthesis is controlled by the enzyme hydroxyindole-O-methyltransferase (HIOMT). Light suppresses the synthesis of this enzyme in those animals in which there is a 24-hour (circadian) rhythm of melatonin release. Light energy is tranduced to nerve impulses in the eye and relayed by a complex pathway to the spinal cord and thence to the pineal gland via sympathetic nerve fibres relaying in the superior cervical ganglion. The release of noradrenaline by these postganglionic sympathetic fibres during daylight suppresses the synthesis of HIOMT and hence of melatonin.

The pineal gland may also play a role in reproductive activity in humans. Hypersecretion of melatonin due to a pinealoma can suppress gonadal activity and delay the onset of puberty. Conversely, hyposecretion of melatonin may lead to early sexual maturity.

11.4 PREGNANCY

After fertilization in the oviduct, the fertilized ovum is transported down the oviduct and implanted in the uterine wall. Implantation occurs about 7 days after fertilization, during which time the fertilized ovum develops to form the blastocyst. Part of the blastocyst forms the chorion, which becomes incorporated into the placenta. The placenta forms not only a transporting organ between mother and fetus, but also is an endocrine organ, which secretes a number of hormones essential for the maintenance of pregnancy.

The duration of pregnancy (or gestation) in women is approximately 38 weeks, or 40 weeks when taken from the last menstruation. Sperm can survive for a few days in the female reproductive tract; the ovum remains fertile for less than a day. Therefore, fertilization can occur if sperm are deposited in the female reproductive tract a few days before ovulation. Sperm need to spend some period of time in the female reproductive tract before they are capable of fertilizing the ovum. This process is known as **capacitation** of the sperm. Only one of the many hundreds of thousands of sperm deposited in the vagina can fertilize the ovum.

Fertilization occurs in the oviduct and the **zygote** begins to divide as it makes its way to the uterus. By the time it reaches the uterus (about 3 days), it is a small mass of cells called a **morula**. Identical twins may arise if the morula separates into two parts during this stage. The morula develops into the blastocyst, which is composed of an outer layer of trophoblastic cells separated from an inner mass of embryonic cells (Fig. 11.12). Implantation of the blastocyst in the uterine wall occurs about 7 days after fertilization.

The trophoblastic layer forms part of the **chorion** which gives rise to the fetal part of the placenta. At about 9 days after fertilization, the trophoblastic layer begins to secrete the hormone **chorionic gonadotrophin**. This phormone prolongs the life of the corpus luteum so that it continues to secrete oestrogen and progesterone, which are necessary for the continuation of pregnancy during the

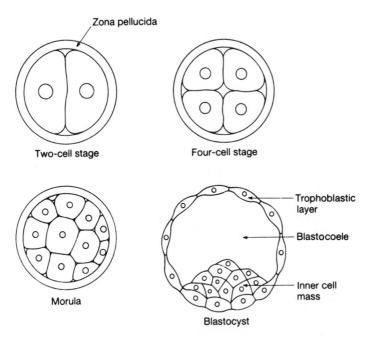

Zona pellucida

Two-cell stage

Four-cell stage

Morula

Blastocyst

Trophoblastic layer

Blastocoele

Inner cell mass

Fig. 11.12 Stages in the early development of the embryo.

first trimester. Thereafter, the role of the corpus luteum is supplanted by the placenta. The **placenta** is not only a means for exchanging respiratory gases, nutrients and waste products between fetal and maternal circulations, it also serves as an endocrine gland in its own right. The placenta secretes both protein and steroid hormones necessary for the continuation of pregnancy during the second and third trimesters.

Placental hormones

The placenta secretes at least four hormones. Two of these hormones are proteins, namely chorionic gonadotrophin and chorionic somatomammotrophin, while two are steroids, namely oestrogen and progesterone. The maternal plasma levels of these hormones during pregnancy are shown in Fig. 11.13.

Chorionic gonadotrophin (CG)

CG is a glycoprotein (M_r 38 000) secreted by the trophoblastic cells of the placenta and which is chemically and biologically similar to LH. It is secreted in large quantities during the first trimester (Fig. 11.13) and its main role is to maintain the corpus luteum during the early part of pregnancy. Since CG is secreted as early as 9 days after fertilization, its detection by radioimmunoassay provides a simple test for pregnancy.

Chorionic somatomammotrophin (CS)

This was originally named placental lactogen. It is a protein (M_r 18 500) secreted by the trophoblastic cells and is structurally similar to human growth hormone. The level of CS in the maternal circulation increases steadily throughout preg-

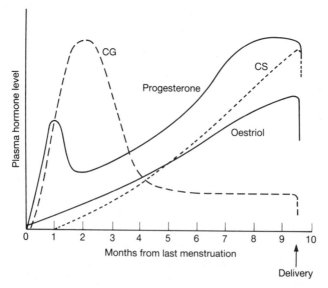

Fig. 11.13 Maternal plasma levels of placental hormones during pregnancy. CG, Chorionic gonadotrophin; CS, chorionic somatomammotrophin.

nancy (Fig. 11.13) but its functions are not well-defined. Like growth hormone, it has a counter-regulatory action to insulin, promoting free fatty acid mobilization and inhibiting glucose uptake in the mother. As its name implies, it also promotes mammary development in preparation for lactation.

Progesterone

During pregnancy the maternal plasma levels of progesterone reach a peak 3 weeks after fertilization and then decline before increasing up until the time of parturition (Fig. 11.13). The initial peak reflects the development of the corpus luteum and the later rise is due to secretion by the placenta. Progesterone is required throughout pregnancy to maintain the endometrium and to suppress spontaneous contractions of the myometrium of the uterus. Progesterone has relaxant effects at other sites in the body and also stimulates the development of the mammary glands.

Oestrogen

Oestriol is the main oestrogen secreted by the placenta; the maternal plasma levels of this hormone are shown in Fig. 11.13. The placenta cannot form the precursor steroid, 17-hydroxyprogesterone, and depends on the availability of steroid precursors synthesized by the fetus. Thus, the fetus and the placenta complement each other in the synthesis of oestrogen and are said to function as the **fetoplacental unit**. It is possible to monitor the well-being of the fetus by measuring the maternal plasma levels of oestrogen but better information is obtained by ultrasound scans. Although the total plasma levels of oestrogen may increase some 50-fold during pregnancy, the levels of free oestrogen are not greatly increased because of a corresponding rise in the sex hormone-binding globulin. Oestrogen is required during pregnancy for the uterus to develop to accommodate the growing fetus and also for the development of the mammary glands.

Parturition

The exact trigger for parturition in women has not been established. A fall in **progesterone** secretion would be expected to enhance uterine contractions but such a fall does not appear to take place immediately prior to delivery. An increase in **oxytocin** release would likewise be expected to stimulate uterine contractions. Oxytocin is indeed released at parturition in response to stimulation of stretch receptors in the uterus and cervix and is used clinically to induce labour. Nevertheless, parturition can still be initiated in women with hypothalamic damage who lack oxytocin. **Prostaglandins** will also induce labour but again there is insufficient evidence that they provide the trigger for parturition in women. In sheep, however, it has been established that the fetus determines the time of delivery by increasing its release of **adrenocorticotrophic hormone** and hence cortisol. The increase in fetal cortisol secretion causes a change in placental steroid synthesis that somehow initiates uterine contractions through the release of prostaglandins. The relevance of such a mechanism to parturition in women is not fully understood.

Towards the end of pregnancy **relaxin**, a polypeptide hormone, can be extracted from the ovary, uterus and placenta. Relaxin and other similar polypeptides appear to play a role in parturition by promoting relaxation of the birth canal.

Lactation

Mammary glands provide both milk for nourishment and antibodies for protection of the young. Milk is produced by epithelial cells lining the alveoli which drain into slender ducts. Development of the alveoli and ducts requires the presence of several hormones—oestrogen, progesterone, cortisol, chorionic somatomammotrophin and growth hormone. During the latter part of pregnancy, prolactin is required for full maturation of the mammary glands and milk production. It also stimulates the production of calcitriol (p. 289) and so may help in maintaining calcium balance during pregnancy and lactation.

The initiation of copious lactation occurs shortly after parturition. Although the reason for this is not fully understood, it requires a fall in the concentrations of oestrogen and progesterone and a rise in prolactin. While prolactin stimulates milk production, milk let-down depends on the release of oxytocin, which is induced by suckling (p. 278). Suckling is also necessary for the continued release of prolactin because it stimulates nerve endings in the nipples which in turn send impulses to the hypothalamus that inhibit the release of prolactin release-inhibiting hormone. In the absence of suckling, prolactin secretion is reduced and lactation ceases. The suckling stimulus also inhibits secretion of gonadotrophins and this presumably accounts for the suppression of ovulation that often occurs in nursing women.

Chapter 12
Blood

Blood is that part of extracellular fluid within the cardiovascular system. It consists of a yellowish fluid, **plasma**, in which cellular constituents are suspended. These are mostly red cells (**erythrocytes**), with much smaller numbers of white cells (**leukocytes**) and platelets (occasionally called **thrombocytes**). Antigenic components of erythrocyte membranes allow the blood of people to be classified into various **blood groups**.

Blood circulates through the body carrying O_2 and nutrients to the tissues and removing CO_2 and other waste products. The red cells are essential for this transport of the respiratory gases. The circulation of blood also dissipates heat generated by oxidative reaction in cells and it distributes hormones, thus helping to maintain homeostasis and to coordinate the activities of the organs of the body. By transporting white cells and antibodies (**immunoglobulins**) to the tissues, the blood promotes defence against pathogenic organisms and foreign substances. The white cells have important phagocytic and immunological functions. In descending order of frequency in blood of adults (see Table 12.6), there are five types of white blood cell: neutrophil, lymphocyte, monocyte, eosinophil and basophil. The platelets and coagulation proteins in the plasma provide a means of self-defence to stop bleeding (**haemostasis**) after injury.

If a sample of blood is collected in a tube containing anticoagulant and the tube is centrifuged, the blood will separate into its main components (Fig. 12.1a). It is possible to estimate, from the lengths of the columns of red cells and plasma, the fraction of a volume of blood occupied by the red cells ($x/(x + y)$ in Fig. 12.1a). This is the **haematocrit** or **packed cell volume** (PCV; see Table 12.4).

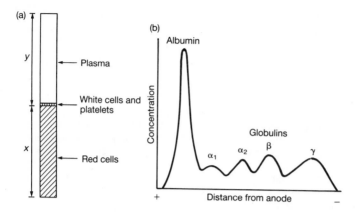

Fig. 12.1 (a) Fluid and cellular constituents of a sample of blood as seen after centrifugation. The fraction occupied by the red cells $(x/(x + y))$ is the haematocrit. (b) Separation of serum proteins by electrophoresis.

The actual volume of circulating red cells (Table 12.1) can be estimated from the dilution of an intravenous injection of the subject's red cells labelled with a radioactive isotope (e.g. ^{51}Cr). This **dilution principle** (p. 4) can also be used to estimate plasma volume by measuring the extent of dilution in a subject's plasma of albumin labelled with ^{125}I. Blood volume can then be calculated by adding red-cell volume and plasma volume or it can be estimated from one of these two values by means of the haematocrit, e.g. blood volume = red-cell volume/haematocrit.

Some 90% by weight of plasma is water, about 8% is plasma proteins (Fig. 12.1b)—albumin, globulins (α, β, γ) and fibrinogen—while 2% consists of organic compounds and electrolytes (see Table 1.1). Apart from the γ-globulins, the plasma proteins are synthesized in the liver. Albumin is the major contributor to the colloid osmotic (oncotic) pressure of the plasma and also transports several hormones, bilirubin, fatty acids and Ca^{2+}. Globulins also transport various substances and most of the antibodies in blood are γ-globulins, made by plasma cells in lymphoid tissue (p. 345). When plasma coagulates, the soluble plasma protein fibronogen is converted into insoluble fibrin (p. 360), with consumption of some other coagulation proteins. The liquid remaining is called **serum**.

12.1 BLOOD CELL PRODUCTION

The cells of the blood are produced in the red bone marrow through lines of developing precursors, which can be identified in samples of bone marrow. These

Table 12.1 Normal blood volumes (ml kg^{-1}) as the 95% range (mean $\pm$ 2 SD) (data from Dacie, J.V. & Lewis, S.M. (1991) *Practical Haematology*, 7th edn, p. 12. Churchill Livingstone, London)

	Men	Women
Red-cell volume	25–35	20–30
Plasma volume	40–50	40–50
Total blood volume	60–80	60–80

immature cells arise in turn from populations of self-renewing, progenitor stem cells. Production of blood cells is controlled by cytokines released by supporting cells in the marrow. Production of red cells also depends on the hormone erythropoietin, secreted by the kidneys.

Sites of production

The production of blood cells (**haematopoiesis**) begins in the first few weeks of gestation with the appearance of primitive cells in the mesodermal tissues of the yolk sac. From about 6 weeks to 7 months of fetal life, all types of blood cell are formed mainly in the liver but also in the spleen. During the fifth month, production of blood cells begins in the red marrow of the bones and, at birth, the marrow is the main site of haematopoiesis, with lymphocyte production continuing also in the spleen and other lymphoid tissues.

At first, red marrow occupies the cavities of all bones but from the age of 5–7 years, fat begins replacing the marrow of the limb bones. In adults, red marrow is normally confined to the bones of the trunk, the skull and the upper ends of the humerus and femur. Even at these sites about 50% of the marrow consists of fat but if there is increased demand for blood cells, the red marrow can expand into these fatty spaces and back into the limb bones.

In an adult, the red bone marrow consists of fat and developing blood cells in clusters held together by fine reticulin fibres and separated by a network of sinusoidal capillaries, which drain into a central venous sinus. New cells are released into the blood through gaps in the endothelium of the marrow sinusoids. Samples of red marrow can be obtained for cytological examination by **aspiration** with a syringe and marrow-puncture needle, or by means of a **trephine**, with which a small core of marrow is cut from the bone for histological examination.

Stem cells and haematopoiesis

Blood cells are produced in the bone marrow from precursors called **stem cells**. These cells can maintain their number by mitosis and can also develop along a number of different lines leading to mature blood cells. The earliest stem cell in the marrow has the potential to develop into all the types of blood cell and is hence called **totipotent**. Other later generations of stem cells have more limited proliferative capacity and also become **committed** to develop along some particular line(s). Thus the **myeloid stem cell** gives rise to other progenitor cells which eventually differentiate into red cells, granulocytes (neutrophils, eosinophils, basophils), monocytes or platelets. In films of bone marrow it is possible to identify and name fairly well-defined types of developing cell (Fig. 12.2) but the stem cells, which probably resemble lymphocytes, cannot be distinguished. The totipotent stem cells also give rise to **lymphoid** stem cells which are found in lymph nodes, spleen and thymus, as well as in the bone marrow. The lymphoid stem cells develop into **B lymphocytes** and **T lymphocytes** (p. 344).

The study of haemopoietic stem cells is important because it is becoming clear that a number of haematological disorders which involve more than one type of blood cell probably arise because of defects in multipotent stem cells able to develop along the affected pathways.

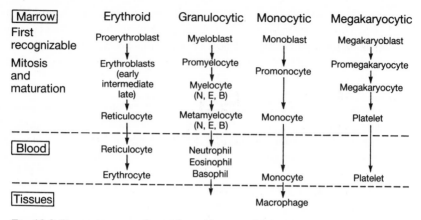

Fig. 12.2 Bone-marrow production lines. N, neutrophil, E, eosinophil; B, basophil.

Although stem cells cannnot be recognized in films of bone marrow by microscopy, they can be grown in culture where they give rise to colonies of descendant cells. They are therefore often referred to as **colony-forming units** (CFUs) and are specified by the name of the cell(s) each colony comprises. For example, the myeloid stem cell is also known as CFU-GEMM (i.e. colony-forming unit with the potential to develop along the granulocytic, erythroid, monocytic and megakaryocytic lines).

Stimulation of multipotent and committed stem cells, and proliferation and maturation down a particular cell line, requires various haematopoietic **growth factors**. Many of these are **cytokines** (p. 345), small proteins with paracrine activity (p. 266), which are produced by various cells in the bone marrow, e.g. endothelium, macrophages, fibroblasts and lymphocytes. Thus the production of platelets requires **megakaryocyte colony-stimulating factor**, which acts on the specific stem cell concerned, and also **thrombopoietin**, which stimulates development of megakaryocytes (Fig. 12.2). Similarly, production of red cells depends on various local growth factors, e.g. interleukin-3, insulin-like growth factor-1 (p. 280), in addition to the hormone, erythroprotein (p. 327).

12.2 THE RED BLOOD CELL

Production of red cells in the bone marrow involves overlapping stages of mitosis and maturation, with release of reticulocytes into the blood. Renal secretion of erythropoietin stimulates the rate of red-cell production to keep pace with the rate of destruction. Erythropoiesis requires essential dietary constituents, particularly iron, vitamin B_{12} and folic acid, and depletion of the body stores of these substances can reduce red-cell production.

Erythropoiesis

The process of red-cell production is called **erythropoiesis** (Fig. 12.3). In a Romanowsky-stained film of marrow, the earliest recognizable red-cell precursor is a large nucleated cell, the **proerythroblast** (pronormoblast), with intensely basophilic cytoplasm reflecting its content of RNA and ribosomal protein. This cell develops into an **early** (basophilic) **erythroblast** (normoblast) and the cells subsequently become smaller (**intermediate** and **late** erythroblasts), with increasingly acidophilic cytoplasm as haemoglobin is synthesized, and the capacity for mitosis is lost as nuclear pyknosis occurs. Extrusion of the nucleus then leaves cells, **reticulocytes**, which contain remnants of RNA and ribosomes and continue

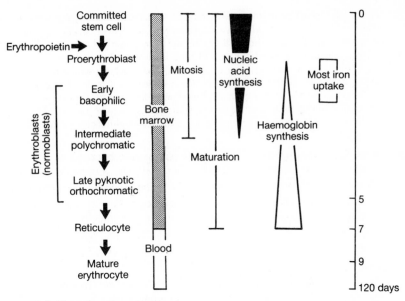

Fig. 12.3 Normal erythropoiesis.

making haemoglobin. Reticulocytes mature for 1–2 days in the marrow and are then released into the blood where, after a further 1–2 days, they lose their remaining ribosomes to become mature red blood cells.

Reticulocytes can be identified in blood films stained with a dye such as new methylene blue which reacts with the ribosomes to form precipitates visible as dark blue granules or filaments. The **reticulocyte count** of blood (see Table 12.4) is a useful index of the erythropoietic activity of the bone marrow.

The maturation of red-cell precursors in the marrow is accompanied by 3 or 4 mitotic divisions up to the stage of the intermediate erythroblasts, so that one proerythroblast can give rise to about 16 mature red cells. Erythropoiesis is partly *ineffective*, however, because normally 5–10% of developing red cells do not survive to reach the circulation. This premature loss of cells is increased in certain disorders of red-cell production. Each mitotic division, and its effects, lasts about 1 day so that one erythropoietic cycle, from proerythroblast to release of reticulocytes, takes about 6 or 7 days. This development cycle involves overlapping biochemical events, with early uptake of iron by erythroblasts, decreasing synthesis of nucleic acids and increasing synthesis of haemoglobin (Fig. 12.3). Up to 35% of the total content of haemoglobin in the mature red cell is made during the reticulocyte stage.

Control of erythropoiesis

In health, the rate of production of new red cells by the bone marrow balances the rate at which old cells are destroyed. This steady state is controlled by a glycoprotein hormone, **erythropoietin** (M_r 33 000), which is normally present in the plasma at low concentration (about 10 pmol L^{-1}) and with a half-life of about 5 hours. Erythropoietin is secreted by the kidneys in response to local hypoxia. The hormone acts on the red marrow to cause increased output of erythrocytes until the rise of haemoglobin concentration of the blood restores normal delivery of O_2 to the tissues. Growth factors secreted by cells in the bone marrow also stimulate erythropoiesis.

About 10–15% of the erythropoietin in blood seems to be made by the liver and this organ may be the source of the plasma erythropoietin detectable in patients with non-functional kidneys. The mode of action of the hormone is uncertain but it is needed at all stages of erythropoiesis. It increases the rate of mitosis of red-cell precursors and shortens their maturation time in the marrow. Erythropoietin has now been synthesized by recombinant DNA techniques and is available for therapeutic use, e.g. for treating the otherwise intractable anaemia which occurs in chronic renal failure. Other hormones also appear to stimulate erythropoiesis: corticosteroids, androgens, growth hormone and thyroxine. Some of these hormones may act by stimulating release of growth factors in the marrow.

Synthesis of haemoglobin and erythropoietic factors

Erythroblasts in the bone marrow undergo mitosis and begin making **haemoglobin**, first by synthesizing **haem** and the polypeptide chains of **globin** separately and then by combining them to form haemoglobin (Fig. 12.4). Certain dietary constituents (e.g. iron, vitamin B_{12}, folic acid, vitamin B_6 (pyridoxine), amino acids) are required for normal production of DNA and haemoglobin (Fig. 12.4). Deficiencies of any of these can decrease haemoglobin production and erythropoiesis but of the various possibilities, by far the most common is deficiency of iron. Synthesis of haemoglobin may also be *defective* because of a fault (often genetic) in the production of haem or of globin.

Synthesis of haem begins in mitochondria (Fig. 12.4) with the condensation of succinyl-coenzyme A and of glycine to form δ-aminolaevulinic acid (δ-ALA). This is the rate-limiting step in haem synthesis and is controlled by the enzyme ALA-synthase, which requires pyridoxine (vitamin B_6) as cofactor and is inhibited by haem. After several further reactions, protoporhyrin IX is formed and the enzyme ferrochelatase catalyses the insertion of ferrous iron into the protoporphyrin ring to form haem. Iron is brought to the erythroblasts attached

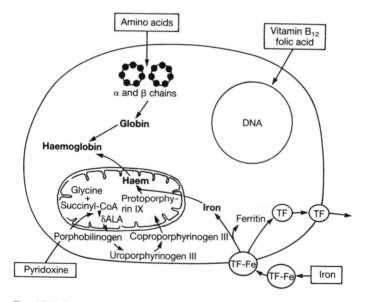

Fig. 12.4 Dietary requirements for synthesis of haemoglobin and DNA in erythroblasts δALA, δ-aminolaevulinic acid; TF, transferrin.

to a plasma protein, **transferrin**. It is probable that, after combination with a membrane receptor, the transferrin is taken up into the cytoplasm of the erythroblast where the iron is released and the transferrin is returned to the extracellular fluid (Fig. 12.4). While haem is being made in the mitochondria, amino acids are used for the synthesis on the ribosomes of matching quantities of the α and β polypeptide chains of globin. Synthesis of DNA by the dividing erythroblasts requires an adequate supply of vitamin B_{12} and folic acid, which are cofactors in the synthesis of nucleotides.

Iron

Adults contain between 35 and 90 mmol (2–5 g) of **iron**, women generally having rather less than men. A 60-kg woman would have about 48 mmol of iron in her body and a 70-kg man about 70 mmol. Some 60–70% of the body iron is in haemoglobin in the blood (Table 12.2) and 4–5% in myoglobin of muscle. Small amounts of iron are present in some enzymes and in the cytochrome proteins of mitochondria. Most of the rest of body irons is stored in hepatocytes and in macrophages of the liver, spleen and bone marrow, as **ferritin** and **haemosiderin**. Ferritin is also present in the blood (Table 12.2) at concentrations which reflect the level of iron stores.

Ferritin is a water-soluble compound consisting of a variable amount of ferric hydroxyphosphate enclosed by a protein shell of apoferritin. Synthesis of apoferritin is stimulated by the presence of iron in the cell. Haemosiderin, an insoluble complex of protein and iron, is probably formed by the partial lysosomal digestion of aggregates of ferritin.

About 70 μmol of iron circulates in plasma bound to the β-globulin, transferrin. This protein is synthesized in the liver at a rate which varies directly with the level of stored iron. One molecule of transferrin can bind two atoms of iron in the ferric state. Transferrin is normally only one-third saturated with iron so that **serum iron** concentrations are lower than the total **iron-binding capacity** of the serum (Table 12.2).

Serum iron concentrations are lower in women than in men and decrease in pregnancy, iron deficiency and chronic infections. Serum iron-binding capacity increases in iron deficiency and in pregnancy but decreases with iron overload and chronic infections. The concentration of serum ferritin is usually a more accurate index of the state of the body iron stores than the serum iron and iron-binding capacity.

Intake and absorption of iron, and the factors affecting these, are discussed in Chapter 17 (p. 600). Note that the handling of iron by the body is unique in that it

Table 12.2 Representative values of iron in adult

	Men	Women
Haemoglobin iron (mmol L^{-1} blood)	8–11	7–10
Stored iron (mmol)	5–27	0–18
Serum iron (μmol L^{-1})	14–32	11–30
Serum iron-binding capacity (μmol L^{-1})	50–70	50–70
Serum ferritin (μg L^{-1})	40–340	15–140

is absorption of iron, rather than excretion, which is regulated to maintain iron balance.

Absorption of iron from the small intestine varies directly with the rate of erythropoiesis and inversely with the amount of iron in the body stores so that 20–30% of dietary iron may be absorbed in iron deficiency but less than 5% when the iron stores are full. Iron not released to the blood after entry into the small intestinal mucosal cell is sequestered as ferritin and is ultimately lost from the body when the mucosal cell is shed from the tip of the villus (Fig. 17.2). This is the main route of excretion of iron.

Iron exchanges and requirements for balance

The iron in plasma, about 70 μmol in a 70-kg man, turns over between five and 10 times a day. Of this turnover, some 80–85% arises from unidirectional movement of iron from extracellular fluid to bone marrow for the synthesis of haemoglobin, and from phagocytic cells in the liver, spleen and bone marrow to the extracellular fluid, as iron is released from degraded haemoglobin (Fig. 12.5). The remaining movements of iron through the plasma are accounted for by exchanges of iron with non-erythroid tissues, including absorption from the small intestine and loss in exfoliated intestinal cells. Small amounts are also lost in sweat and urine. Iron from the breakdown of red cells is thus reused for erythropoiesis, stored iron is available to meet extra demands and absorption normally balances loss from the body.

The inevitably daily loss of iron in adults, mainly from the gut, is between 9 and 18 μmol. Women lose, in effect, an extra 9–18 μmol a day due to menstruation and require an additional 18–36 μmol per day during pregnancy. Adolescents and infants require extra iron for growth. **Negative iron balance** occurs if these daily needs are not met by absorption of iron from the diet.

Deficiency of iron

Continuing net loss of iron will lead to exhaustion of the stores and a fall in serum iron despite a compensatory rise in iron-binding capacity. When the saturation of

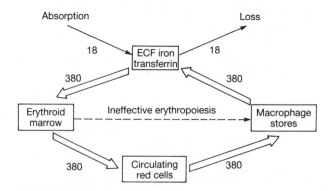

Fig. 12.5 Daily iron exchanges. Values (μmol) for a 70-kg man with a haemoglobin concentration of 150 g L^{-1} and red-cell life-span of 120 days. Iron exchanges from ineffective erythropoiesis are omitted. ECF, extracellular fluid.

transferrin is less than about 15%, delivery of iron to the marrow is impaired and an **iron-deficiency anaemia** develops.

There are three main causes of a negative iron balance and iron deficiency. **Increased physiological demands** for iron occur during periods of rapid growth, especially in infants and young children, while the need for extra iron in pregnancy and for menstruation makes women liable to iron deficiency. **Inadequate absorption** of iron can arise after surgery (e.g. gastrectomy) and in diseases causing malabsorption. Low dietary intake in adults may contribute to iron deficiency due to other causes (e.g. pregnancy) and in infancy may be due to improper feeding. The most common cause of iron deficiency in adults is pathological **loss of blood**, especially when the bleeding is chronic. Blood with a haemoglobin concentration of $150 \, \mathrm{g \, L^{-1}}$ contains about 9 µmol of iron per ml. Therefore, an average daily loss of more than about 10 ml of blood is likely to cause a negative iron balance because this volume of blood will contain more iron than can be maximally absorbed (about 90 µmol) from a normal diet containing about 270 µmol of iron per day. Chronic loss of blood occurs most commonly from the uterus and from the gastrointestinal tract.

Vitamin B_{12}

Vitamin B_{12} (cobalamin), a cobalt-containing molecule, is synthesized by bacteria and the chief dietary sources are protein-rich foods of animal origin, especially liver and kidney, but also muscle meats, with smaller concentrations in eggs, cheese and milk. The ingested vitamin combines with **intrinsic factor** made by parietal cells in the stomach and is absorbed in the terminal ileum (p. 603).

Aspects of vitamin B_{12} balance are summarized in Table 12.3. An adult loses about 1.5 nmol of vitamin B_{12} each day, mainly in the urine and faeces, and needs to absorb a similar amount daily to stay in balance. Body stores, mainly in the liver, are enough to meet requirements for 3–4 years if dietary intake were to cease completely.

Deficiency of vitamin B_{12} is almost always the result of impaired absorption; only rarely does it arise from inadequate intake (e.g. in strict vegetarians). Impaired absorption may be due to lack of intrinsic factor (e.g. in pernicious anaemia or after gastrectomy), reduced absorptive capacity of the ileum (e.g. with ileal disease or after resection) and competition for the vitamin by intestinal parasites or bacteria.

Table 12.3 Vitamin B_{12} and folic acid

	Vitamin B_{12}	Folic acid
Daily diet supplies	4–22 nmol	900–2300 nmol
Absorption limit	$2.5 \, \mathrm{nmol \, day^{-1}}$	50–100% of intake
Absorption site	Ileum	Duodenum, jejunum
Daily needs	1.5 nmol	220 nmol
Adult store	1500–2200 nmol	14–45 µmol
Time for depletion	3–4 years	2–7 months
Serum concentration	$150–700 \, \mathrm{pmol \, L^{-1}}$*	$4–20 \, \mathrm{nmol \, L^{-1}}$*
Red-cell concentration	—	$300–1500 \, \mathrm{nmol \, L^{-1}}$*

* Radioimmunoassay; values for normal range depend on method.

Folic acid

Folic acid (pteroylglutamic acid) is a vitamin present in foods mainly as a polyglutamate. Folates occur in most foods, the highest concentrations being found in liver, yeast, green vegetables, nuts and fruit. The diet may provide as much as 2.3 μmol daily but the amount available varies widely depending on the type of food eaten and the method of preparation, since folic acid is easily destroyed by cooking. Absorption of folate is mainly from the duodenum and jejunum (p. 603).

Handling of folate in the body is summarized in Table 12.3. An adult requires about 220 nmol of folic acid daily and up to 660 nmol or more per day during pregnancy. Folate and its breakdown products are lost in the urine and sweat. In contrast to the ample storage of vitamin B_{12}, the stores of folic acid, mainly in the liver, are small relative to the daily requirement. In the absence of intake, body stores can become exhausted in several months (Table 12.3), so that severe deficiency of folate can develop rapidly.

Inadequate dietary intake of folate is probably not uncommon and will contribute to deficiency due to defective absorption (e.g. in small intestinal mucosal disease), to drugs (e.g. anticonvulsants, alcohol) or to increased demands for the vitamin when cellular turnover is rapid, as in physiological states (e.g. pregnancy) or pathological conditions (e.g. malignant tumours).

Megaloblastic erythropoiesis

Deficiency of vitamin B_{12} or folic acid leads to decreased synthesis of DNA in proliferating tissues and produces characteristic morphological abnormalities in the bone marrow. The red-cell precursors are larger than normal (**megaloblasts**) and impaired synthesis of DNA leads to a delay in maturing of the nucleus relative to the cytoplasm. The circulating red cells are usually also normally large (**macrocytes**). Production of white cells and platelets is also affected and a macrocytic anaemia is often accompanied by low granulocyte and platelet concentrations in the blood.

It is thought that impaired synthesis of DNA due to deficiency of folate results from slowing of a rate-limiting step, the methylation of deoxyuridylic acid to thymidylic acid. Tetrahydrofolate, as the 5,10-methylene derivative, is involved as coenzyme in this reaction so that deficiency of folate would decrease the rate of synthesis of thymidylate and hence of DNA.

Vitamin B_{12} is probably necessary for the formation of tetrahydrofolate from methyltetrahydrofolate, which is the form in which folate is released from the small intestine (p. 603). In this view, deficiency of vitamin B_{12} leads to reduced synthesis of DNA, and hence megaloblastic anaemia, because of inadequate supply of tetrahydrofolate.

Characteristics of red cells

The mature, anucleate erythrocyte is a biconcave disc which has structural proteins to help preserve its functionally important shape and a simplified metabolism which acts to maintain cellular volume and to protect cellular constituents from oxidant damage by oxygen. The haemoglobin in erythrocytes undergoes conformational changes which account for important properties of oxygen transport. Red cells have a life-span of about 4 months before being degraded, with most constituents being recycled in the body.

The mature red cell is a biconcave disc, 7–8 μm in diameter, 2.5 μm thick near the rim and 1 μm thick at the centre. This curious shape enhances diffusion of the respiratory gases into and out of the cell since there is both a greater surface area and a shorter distance to central cellular regions than would be present in a spherical cell of the same volume. The biconcave shape also increases the flexibility of red cells which allows them to bend easily as they squeeze through narrow capillaries and it keeps tension on the membrane minimal when the red cells swell after they take up CO_2 from the tissues (p. 536).

Table 12.4 lists ranges of normal values of variables relating to the concentration of red cells in the blood: the **red-cell count, haemoglobin concentration** and **haematocrit**. These variables are maximal immediately after birth, fall to minimal values by about the age of 6 months, then rise slowly from the age of about 1 year to puberty, when the adult ranges are reached. In the elderly, there is a tendency for the values to decline a little. Changes in the haemolgobin concentration, red-cell count and haematocrit usually reflect changes in red-cell mass but this relationship can be distorted by variations in plasma volume, e.g. increase in the haematocrit with dehydation or decrease in the haematocrit as plasma volume rises in pregnancy.

The red-cell count, haemoglobin concentration and haematocrit can now be measured very accurately by electronic cell counters. These instruments also compute useful red-cell indices which indicate the average size and haemoglobin content of the cells. The **mean cell volume** (calculated as haematocrit/red-cell count) reflects red-cell size, which may be normal (**normocytic**), increased (**macrocytic**) or decreased (**microcytic**). The **mean cell haemoglobin** (calculated as haemoglobin concentration/red-cell count) and the **mean cell haemoglobin concentration** (haemoglobin concentration/haematocrit) indicate, respectively, the average amount and concentration of haemoglobin in the cells. Red cells are

Table 12.4 Normal red-cell values as the 95% range (mean ± 2 SD) (data from Dacie, J.V. & Lewis, S.M. (1991) *Practical Haematology*, 7th edn, p. 12. Churchill Livingstone, London)

	Men	Women	Children (1 year)
Red-cell count (× 10^{12} L^{-1} blood)	4.5–6.5	3.8–5.8	3.6–5.2
Haemoglobin (g L^{-1} blood)	130–180	115–165	105–135
Haematocrit (packed cell volume)	0.40–0.54	0.37–0.47	—
Mean cell volume (fl)	76–96	76–96	70–86
Mean cell haemoglobin (pg)	27–32	27–32	23–31
Mean cell haemoglobin concentration (g L^{-1} cells)	300–350	300–350	300–350
Reticulocyte count (× 10^9 L^{-1} blood)	25–85	25–85	25–85
Reticulocyte count (%)	0.2–2.0	0.2–2.0	0.2–2.0

called **normochromic** when they have a normal concentration of haemoglobin; if it is reduced, they are **hypochromic**.

Variations in size and haemoglobin content of red cells are also detected by microscopic examination of a blood film stained with a Romanowsky stain. Normal erythrocytes appear as round cells, staining an even pink with sometimes an area of central pallor (corresponding to the biconcavity) which occupies about one-third of the area of the cell.

In diseases affecting red cells, morphological abnormalities are often apparent in the blood film and arise in general from three main causes: abnormal erythropoiesis, damage to circulating red cells and changes associated with compensatory increases in red-cell production. Thus, **anisocytosis** (variation in size) and **poikilocytosis** (variation in shape) are non-specific features of abnormal erythropoiesis, while **hypochromic** erythrocytes are seen when there is inadequate synthesis of haemoglobin because of impaired production of haem or of globin. The presence of an abnormal haemoglobin may cause a diagnostic abnormality, e.g. **sickle cells** in people with haemoglobin S. Fragmented red cells (**schistocytes**) are caused by damage in the circulation, e,g. by passage of blood through strands of fibrin. Increased release of reticulocytes from the marrow results in **polychromasia** due to greyish-blue staining of the reticulocytes.

The red-cell membrane

The membrane of the red cell is a lipid bilayer closely associated with proteins situated on one side of the membrane or passing right through it (p. 8). Carbohydrates occur mainly as glycolipids and glycoproteins on the external surface of the membrane and often have blood-group specificity (e.g. A, B, H antigens; p. 351).

The membrane proteins form two main groups: **peripheral proteins** and **integral membrane proteins** (Fig. 12.6). Just inside the membrane, the peripheral proteins form an extensive filamentous network called the **red-cell membrane skeleton** (cytoskeleton). This is essential for the determination of its shape and deformability. The major constituent of this cytoskeleton is **spectrin**, a dimer consisting of α- and β-subunits. The membrane skeleton also contains filaments of **actin**. The integral membrane proteins, e.g. **protein 3** and **glycophorin A**, are tightly bound to the membrane and extend through it, to act as anchors for the underlying membrane skeleton (Fig. 12.6). Defects of this protein cytoskeleton are thought to explain some inherited abnormalities of erythrocyte shape, e.g. hereditary spherocytosis and hereditary elliptocytosis.

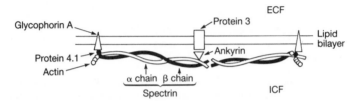

Fig. 12.6 Red-cell membrane skeleton. Peripheral spectrin dimers are connected to integral membrane protein 3 by ankyrin and to glycophorin A by protein 4.1. ECF, extracellular fluid; ICF, intracellular fluid.

The integral membrane proteins also subserve important cellular functions. One has binding sites for several glycolytic enzymes and provides transport facilities for HCO_3^- and Cl^- ions entering and leaving the cell. Other proteins include the Na^+-K^+ ATPase of the sodium pump and a $Ca^{2+}-Mg^{2+}$ ATPase, which mediates active efflux of calcium from the cell.

Haemoglobin

The red respiratory pigment of the erythroycte, **haemoglobin**, is composed of four polypeptide chains of the protein globin, each containing a molecule of haem. Haem consists of a protoporphyrin ring with a central atom of ferrous iron which can combine reversibly with a molecule of O_2.

There are four types of globin chains, α, β, γ and δ, which differ in their constituent amino acids. In an adult, 96–98% of the circulating haemoglobin is **haemoglobin A**, the globin of which has two α-chains and two β-chains. Adults also have small amounts of **haemoglobin A_2** ($\alpha_2\delta_2$) and of **haemoglobin F** ($\alpha_2\gamma_2$), which is the main haemoglobin of the fetus.

The tetrameric structure of haemoglobin allows conformational changes in the molecule which account for important characteristics of O_2 transport. These are discussed later (p. 524).

Red-cell metabolism

Without the benefit of mitochondria or of ribosomes for synthesizing protein, the erythrocyte survives for more than 4 months in the face of repeated oxidant stress from the high concentrations of O_2 and repeated mechanical stress from many daily passages through capillaries of diameter smaller than that of the cell. This cellular longevity seems to depend on a simplified metabolic organization with three main functions: to provide energy for maintaining cellular volume, to provide reducing power to protect the cell against oxidation and to help control the affinity of haemoglobin for O_2.

About 95% of the glucose consumed by red cells is metabolized by anaerobic glycolysis; 5% is used by the pentose–phosphate pathway. As well as providing energy for the Na^+-K^+ ATPase, the glycolytic pathway also maintains a supply of reduced nicotinamide adenine dinucleotide (NADH). This is a coenzyme for a reductase enzyme which helps to maintain the iron of haemoglobin in the ferrous (Fe^{2+}) state. If the iron is oxidized to the ferric (Fe^{3+}) state, the **methaemoglobin** so formed cannot combine reversibly with O_2. Methaemoglobin may also be reduced with the aid of the coenzyme nicotinamide adenine dinucleotide phosphate (NADPH), which is produced by the pentose-phosphate pathway. The reducing power of NADPH is also made available to the cell through its linkage with the tripeptide, **glutathione**. This reducing agent provides protection against the oxidation of sulphydryl groups of enzymes, globin and constituents of the membrane. It also counteracts auto-oxidation of membrane lipids and helps dispose of any hydrogen peroxide that forms.

A side-reaction of the glycolytic pathway in erythrocytes leads to the synthesis of 2,3-bisphosphoglycerate (2,3-BPG) from 1,3-bisphosphoglycerate. The 2,3-BPG combines with haemoglobin to reduce its affinity for O_2, thus promoting delivery of O_2 to the tissues. Conditions causing hypoxia (e.g. high altitude, anaemia) lead to increased synthesis of 2,3-BPG and therefore increased release of O_2 from haemoglobin.

Life-span and breakdown of red cells

Mature red cells normally remain in the blood for between 100 and 130 days. Thus, a little under 1% of the circulating red cells is destroyed each day and must be replaced by reticulocytes released from the bone marrow. For a man of 70 kg, this turnover of cells amounts to more than 200×10^9 per day or, on average, 2.6 million every second.

As red cells age in the circulation, they become smaller and more dense and there is a decline in the activity of their glycolytic and other enzymes. At the end of their lives, probably in response to degenerative changes in the cell membrane, effete erythrocytes are removed from the circulation by macrophages in the bone marrow, spleen and liver. Within the macrophage, the red cell is broken down with release and degradation of haemoglobin (Fig. 12.7). The amino acids of globin are returned to the general amino acid pool of the body. The haem groups are broken down by microsomal enzymes, with release of iron to the extracellular protein transferrin and its transport to erythroblasts for insertion into new haem groups (Fig. 12.4) or to iron stores.

The remainder of the haem group is converted to **bilirubin** which enters the blood where it is attached to albumin and is carried to the liver. Here the bilirubin is conjugated to glucuronic acid and secreted in the bile. In the intestine bacteria convert the bilirubin into

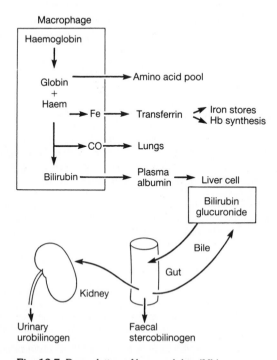

Fig. 12.7 Degradation of haemoglobin (Hb).

various products which are excreted in the faeces or are absorbed and excreted in the urine or release again into the bile (Fig. 12.7 and see also p. 586).

Anaemia

A reduction in the concentration of circulating haemoglobin, anaemia, can arise because of decreased production of red cells, increased destruction of red cells and bleeding. Such disturbances may cause characteristic changes in appearance of circulating erythrocytes, while physiological adjustments occur to maintain oxygenation of the tissues despite the reduced oxygen-carrying capacity of the blood.

In functional terms, anaemia is a disorder in which a decrease in the amount of circulating haemoglobin reduces the oxygen-carrying capacity of the blood so that this is insufficient to meet the needs of the tissues for oxygen.

For clinical purposes it is usual to define anaemia on the arbitrary basis of the concentration of circulating haemoglobin. Thus anaemia is a reduction in the concentration of haemoglobin in the blood below the lower limit of the normal range for the age and sex of the patient. A woman is therefore anaemic if her haemoglobin concentration is less than 115 g L^{-1}. The corresponding value for men is 130 g L^{-1} and for children aged 1 year, 105 g L^{-1} (Table 12.4).

This way of defining anaemia has obvious limitations. Thus, people may have apparently normal haemoglobin concentrations which are, nevertheless, suboptimal or pathologically low for them, while others, defined as anaemic, may in fact belong to that part of the population whose haemoglobin concentration is below the lower limit of a 95% range. Interpretation of haemoglobin concentration may thus require caution. However, anaemias severe enough to cause symptoms are usually associated with unequivocally low concentrations of haemoglobin in the blood.

Causes of anaemia

Anaemia, with a decrease in red-cell mass, may arise from one (or more) of three fundamental disturbances: decreased production of red cells, increased destruction of red cells and bleeding (Table 12.5).

Decreased production of erythrocytes may be due to a reduction in the rate of proliferation of precursors in the marrow, the **hypoproliferative anaemias**, or may arise from **defective maturation** of red cells, often associated with an abnormal degree of ineffective erythropoiesis (p. 327).

Reduced proliferation of red-cell precursors is commonly caused by deficiency of iron and, in many chronic inflammatory disorders, by an apparent failure of release of iron by macrophages. Decreased output of erythropoietin from diseased kidneys, infiltration of the bone marrow by abnormal cells (e.g. leukaemia) and failure of the marrow due to drugs, toxic chemicals and ionizing radiations also cause anaemia due to decreased erythropoiesis. These hypoproliferative anaemias are associated with a blood reticulocyte count which is disproportionately low for the degree of anaemia and the bone marrow is characteristically hypocellular with evidence of depressed erythropoiesis.

Abnormalities of maturation may affect nuclear function or synthesis of haemoglobin in erythroblasts. Deficiency of vitamin B_{12}, or of folate, or both, interferes with synthesis of DNA, which leads to megaloblastic erythropoiesis (p. 332). Genetic defects can cause

Table 12.5 Main causes of anaemia

Decreased production
Reduced proliferation
    Iron deficiency, chronic disorders, low erythropoietin, marrow
    infiltration, marrow failure
Defective maturation
    Nucleus: vitamin B_{12}, folate deficiency
    Haemoglobin: defective synthesis of:
        Globin—thalassaemias, haemoglobinopathies
        Haem—sideroblastic anaemias

Increased destruction
Cellular defect
    Membrane, haemoglobin, enzyme
Extracellular defect
    Immune: autoimmune disease
    Non-immune: trauma, drugs, bacteria, etc.

Bleeding
Acute
Chronic

---

reduced production of globin chains (**thalassaemias**), or variations in structure of globin chains (**haemoglobinopathies**), with resultant instability of the haemoglobin molecule or abnormal functioning in the transport of oxygen. Genetic and acquired anomalies may also lead to defective synthesis of haem, with a striking abnormal accumulation of iron in erythroblasts (**sideroblastic anaemias**). These disorders of maturation are accompanied by a low reticulocyte count and there is often marked erythroid hyperplasia in the bone marrow, with increased ineffective erythropoiesis.

An abnormally high rate of destruction of red cells may arise either because of some defect in the cells (often genetically determined) or because of some injurious extracellular factor (usually acquired; Table 12.5).

In such **haemolytic** anaemias, the life-span of the red cells is reduced and the increased destruction results in a raised output of bilirubin, which may be visible in the tissues as **jaundice** (p. 586). The tissue hypoxia causes increased secretion of erythropoietin, which leads to compensatory erythroid hyperplasia in the bone marrow and a raised blood reticulocyte count.

Acute loss of blood and O_2-carrying capacity may result in the medical emergency of hypovolaemic shock (p. 462) but chronic bleeding, especially if it is intermittent, may cause an anaemia to develop very slowly. The anaemia will usually occur only after the iron stores in the body are exhausted (p. 331).

Red-cell indices and anaemias

Anaemias may also be classified in terms of the red-cell indices—mean cell volume, mean cell haemoglobin and mean cell haemoglobin concentration (Table 12.4). This classification has the advantage that it immediately suggests a short list of likely disorders and hence any additional tests required. It may also allow identification of an important defect before overt anaemia has developed.

Macrocytic normochromic anaemias are characteristically those associated with megaloblastic erythropoiesis (vitamin B_{12} or folate deficiency) but also rise in alcoholism, liver disease and hypothyroidism. Increased numbers of circulating reticulocytes may also give a macrocytic blood picture.

Microcytic hypochromic anaemias are most often due to iron deficiency but also occur with thalassaemias and with sideroblastic defects. **Normocytic** and **normochromic** anaemias are commonly secondary to some chronic disorder (e.g. infection, renal failure, rheumatoid arthritis) but also occur in haemolysis, after acute bleeding, in bone-marrow failure and as a result of infiltration of the marrow by abnormal cells (e.g. metastatic carcinoma).

Compensatory adjustments in anaemia

Delivery of O_2 from the lungs to the tissues involves a number of steps, of which transport by haemoglobin is only one (Chapter 16). Ventilation, cardiac output, differential control of peripheral resistance to blood flow and diffusion from the capillaries to cells are other links in the chain. In general, a decrease in effectiveness of any of these components of the O_2-transport system may be counteracted by increased activity of the others. Thus, in anaemia, various adjustments occur in order to compensate for the reduced O_2-carrying capacity of the blood so as to maintain oxygenation of the tissues.

As the haemoglobin concentration falls, there is greater deoxygenation of haemoglobin in the capillaries. There is also increased synthesis of 2,3-BPG in the red cells, which promotes deoxygenation of haemoglobin, shifting to the right the oxyhaemoglobin dissociation curve (see Fig. 16.30). While the rise in 2,3-BPG allows greater unloading of O_2 from arterial blood, this leaves a smaller venous O_2 content and thus a diminished reserve available for extra demands (e.g. in exercise). The total blood volume is maintained, despite the decreased red-cell mass, by an increase in plasma volume. There is also a redistribution of the flow of blood away from organs like the kidneys and the skin to more vital areas such as the myocardium, brain and skeletal muscle.

When the blood haemoglobin concentration falls below about 70–80 g L^{-1}, cardiac output rises, at rest and during exercise. This is due to an increase in both stroke volume and heart rate. A net vasodilatation, with more rapid flow of blood of reduced viscosity (because of the decreased red-cell mass) gives rise to the characteristic **hyperkinetic** circulation of chronic anaemia.

The extent to which these adjustments allow adaptation to anaemia will depend upon a number of factors, including the severity of the anaemia, its speed of onset and the adequacy of myocardial oxygenation and function.

12.3 THE WHITE BLOOD CELLS

White blood cells defend the body by phagocytosis of foreign organisms and by specific immune responses to non-self invaders. The five types of white cell, which can be identified in stained blood films, develop in the bone marrow under the stimulus of several growth factors. Abnormal white blood cell counts occur in infections, allergies, disorders of the bone marrow and other conditions.

The white cells of the blood form two main groups both with defensive functions: **phagocytes**, which can engulf and destroy bacteria and other foreign matter, and

Table 12.6 Normal white cell counts ($\times 10^9$ L^{-1} blood) as the 95% range (mean ± 2SD) (data from Dacie, J.V. & Lewis, S.M. (1991) *Practical Haematology*, 7th edn, pp. 12–13. Churchill Livingstone, London)

Cell	Adults	Children (6 years)
Neutrophils	2.0–7.5	2.0–6.0
Lymphocytes	1.5–4.0	5.5–8.5
Monocytes	0.2–0.8	0.7–1.5
Eosinophils	0.04–0.4	0.3–0.8
Basophils	0.02–0.1	0.02–0.1
Total	4.0–11.0	5.0–15.0

lymphocytes, the effector cells of the immune system (p. 344). The phagocytic cells comprise **polymorphonuclear leukocytes**, in which the nucleus is often divided into several lobes, and **monocytes**, the precursors of **macrophages**, which are phagocytic cells, **fixed** or **free** in the tissues. All these phagocytic cells have cytoplasmic granules, which are often lysosomes. The granules of the polymorphonuclear leukocytes react with Romanowsky stains to give characteristic appearances in blood films and allow identification of three cell types: neutrophil, eosinophil and basophil, often known collectively as **granulocytes**. Normal ranges of total and differential white-cell counts in the blood are listed in Table 12.6. The appearance of each type of white cell in a Romanowsky-stained blood film is summarized in Table 12.7.

The **neutrophil** is the most frequently occurring white blood cell in adults. Its cytoplasmic granules are lysosomal in nature and liberate enzymes able to kill bacteria when the granules fuse with **phagosomes** (vacuoles containing bacteria or foreign material taken up by phagocytosis). Neutrophils act as phagocytes in acute inflammation (p. 342).

The granules of the **eosinophil** contain lysosomal enzymes and a protein called **major basic protein**, which can neutralize heparin. Eosinophils appear to have special roles in combating parasitic infestations, in phagocytosing antigen–

Table 12.7 Types of white cell in Romanowsky-stained blood film

Cell	Diameter (μm)	Nucleus	Cytoplasm	% of total (adults)
Neutrophil	12–15	2–5 lobes	Pink, granular; fine purple granules	40–75
Lymphocyte	6–8 (small) 12–16 (large)	Round, heavy chromatin	Thin rim, pale blue; occasional granule	20–45
Monocyte	12–20	Large, irregular; fine chromatin	Bulky, pale blue-grey	2–10
Eosinophil	12–15	Two lobes	Many large, oval, orange-red granules	1–6
Basophil	12–15	Large; irregular lobes	Few dark-blue granules; often overlie nucleus	<1

antibody complexes and in modulating the effects of histamine and leukotrienes in allergic reactions.

The most infrequent white cell in the blood is the **basophil** (Table 12.6). Its granules contain heparin and histamine. There are some similarities between basophils and the mast cells in tissue spaces but the precise relationship is not clear. Both types of cell have membrane receptors for immunoglobulin E (IgE) and the combination of cell and antibody in immediate hypersensitivity reactions (p. 347) leads to release of the contents of the granules.

The **monocyte** and the macrophage it becomes contain lysosomal granules. Macrophages have important phagocytic functions and promote stimulation of the immune system by antigens (p. 344).

Most **lymphocytes** in the blood are only a little larger than red cells. They comprise two mains groups, **T cells** and **B cells**, which mediate immune reactions triggered by antigens (p. 344).

Granulocyte production and life-span

Granulocytes and monocytes develop in the red bone marrow from a common progenitor, the myeloid stem cell (p. 325). The earliest recognizable granulocyte precursor is the **myeloblast**, which subsequently matures through the stages of **promyelocyte** and **myelocyte** (Fig. 12.2). The specific cytoplasmic granules first appear in the myelocyte and mitosis continues up to the end of this stage. Thereafter, with loss of mitotic capacity, the cells get smaller and there is progressive condensation of the nucleus in the **metamyelocyte** and **band** stages. Nuclear segmentation occurs in the mature granulocyte.

After mitosis, granulocytes spend up to a week maturing further before release to the blood. There are large numbers of metamyelocytes, band forms and mature cells kept in reserve in the marrow.

This relative abundance of developing granulocytes is indicated by the ratio of white-cell to red-cell precursors, the **myeloid to erythroid ratio**. In normal marrow, this varies between 2.5 : 1 and 15 : 1, and a normal marrow stores between 15 and 20 times as many granulocytes as are present in the blood.

Control of granulocyte production depends on a number of growth factors (p. 326). Some of these growth factors may be specific for a particular type of developing cell but others appear to act at various points in a number of different cell lines. A number of growth factors have now been made by recombinant DNA technology and one at least is used to accelerate production of neutrophils in patients who have had a bone-marrow transplant.

Following release into the blood, neutrophils are present in about equal numbers either freely circulating or rolling in a marginating pool along the walls of capillaries and venules. These latter cells are not included in a white-cell count of blood but there is a rapid and free exchange of cells between the circulating and marginal pools. Neutrophils spend about 10 hours in the blood before being lost at random into the tissues. Here they survive for probably 4–5 days before removal by macrophages after they have performed their phagocytic functions or have become senescent.

Eosinophils remain in the circulation longer than neutrophils and show diurnal fluctuations in concentration which are related inversely to the secretion of the hormone cortisol (p. 294). Monocytes are stored to some extent in the marrow and spend 20–40 hours in the blood before leaving to become macrophages in the tissues, where they may survive for months or years.

Abnormal white-cell counts

An increase in circulating neutrophils, a **neutrophil leukocytosis** (neutrophilia), is usually due to increased output from the marrow in response to bacterial infections, to inflammation and necrosis of tissues, or to acute haemorrhage. Very high counts, with precursor marrow cells in the blood, occur in **chronic granulocytic leukaemia**. A neutrophil count below the lower limit of normal, **neutropenia**, can occur due to the action of drugs on the marrow, in severe infections when output of neutrophils may fail to keep pace with demand, and as part of a **pancytopenia**, e.g. in bone-marrow failure.

An increased lymphocyte count, **lymphocytosis**, occurs in viral illness, in chronic bacterial infections (e.g. tuberculosis) and commonly in young children as a reaction to infections which produce a neutrophilia in adults. A high lymphocyte count in the middle-aged and elderly is commonly due to **chronic lymphocytic leukaemia**. A reduced lymphocyte count, **lymphopenia**, is uncommon but occurs in severe bone-marrow failure, in **acquired immunodeficiency sydnrome** (AIDS) and in patients taking immunosuppressive drugs. Raised concentrations of eosinophils, **eosinophilia**, are seen in allergic disorders (e.g. hay fever), parasitic infestations, reactions to drugs and in certain skin diseases (e.g. psoriasis). A **monocytosis** occurs in chronic bacterial infections while a raised basophil count (**basophilia**) is uncommon and is usually an accompaniment of a myeloproliferative disorder such as chronic granulocytic leukaemia.

Acute inflammation

Acute inflammation is the local response of living tissues to injury which may be due, for example, to infection, trauma, extremes of heat and cold, chemical agents, ultraviolet light, ionizing radiation and sometimes to antigen–antibody complexes. Tissue injury is also associated with other local reactions (chronic inflammation, repair and regeneration) and with general reactions (fever and leukocytosis). The **acute inflammatory reaction** is characterized by **local vasodilatation** causing redness and heat, and by **increased vascular permeability**, with the consequent increased accumulation of a protein-rich exudate and swelling. These reactions are mediated by a number of substances, such as plasma **kinins** activated in tissues and **histamine** released from mast cells and basophils. Kinins and histamine also stimulate nerve endings in the infected area, producing the sensation of pain. Concomitantly with the exudation of fluid, leukocytes migrate from the circulation into the infected area in response to **chemotaxins** released by micro-organisms, by damaged tissues, or as products of complement activation (see below). The cells involved in acute inflammation are **neutrophils** (polymorphs) and **monocytes**, the latter being transformed into **macrophages** on entering the tissues. These cells phagocytose micro-organisms and help to remove the tissue debris. The inflammatory excudate facilitates these processes and also

serves to dilute noxious agents. The immune system also contributes to inflammation by virtue of:
1 antibacterial and antitoxic antibodies which may be present in the exudate;
2 derivatives of the complement cascade; and
3 the antiviral substance **interferon.**

The immune system

The immune system, like the inflammatory reaction, deals with foreign invaders but discriminates very precisely between one invader and another. Once learned, its lessons are long remembered.

A further line of defence is provided by the immune system, the relative importance of which is well-illustrated by the high mortality accompanying the major immunodeficiency diseases. Immune responses can be distinguished from the inflammatory reaction by virtue of the following characteristics:
1 Specificity—an immune system is tailored to deal with the initiating agent **(antigen)** through specific **antibody** and **cell-mediated** reactions. In contrast, inflammation is non-specific.
2 Memory—a first encounter with an antigen is subsequently remembered, so that a second and subsequent encounters with the same antigen provoke a more effective response.
3 Discrimination between self and non-self—the host does not normally react with itself. Indeed, specific immunological non-reactivity (self-tolerance) normally protects the tissues of the host against immunological attack. Thus immune responses are characteristically directed against chemical groupings recognized as foreign.

Antigens

Most antigens are proteins or polysaccharides of M_r above 5000–10 000. Smaller molecules called **haptens** sometimes provoke an immune reaction if they become attached to body proteins or cells (e.g. nickel allergy). Only part of an antigenic molecule, the **antigenic determinant**, reacts specifically with the corresponding antibody. A determinant may be repeated several times on a single antigenic molecule or on a cell. Antigens important in clinical medicine include:
1 parts of the surface membrane of micro-organisms;
2 toxic products of micro-organisms, e.g. tetanus toxin;
3 parts of the surface membrane of human cells (these antigens are important in blood transfusion and transplantation);
4 substances used therapeutically which sometimes act as haptens giving rise to drug allergies; and
5 plant and animal antigens (e.g. pollen, animal dander, bee venom) which may provoke aberrant (allergic) immune responses.
Proteins from species other than humans will in general be recognized as foreign. The ability to provoke an immune response is not an absolute attribute of a molecule; host factors ultimately determine the extent to which an antigen will be seen as such.

The effector cells of the immune system

Lymphocytes are the main effector cells of the immune system and originate from bone-marrow precursor cells. They comprise two main populations: **T lymphocytes** (~80%) and **B lymphocytes** (15–20%). A small population of circulating cells apparently belongs to neither group (**null cells**).

T lymphocytes depend on the thymus for their maturation (hence T cells), while B lymphocytes probably mature in the bone marrow. B lymphocytes were originally identified in birds, in which the bursa of Fabricius (hence B cells) is necessary for their maturation. This organ has no exact mammalian counterpart. While T and B cells are morphologically indistinguishable in routine histological preparations, they differ with respect to:

1 certain cell-surface proteins which can be used as markers in clinical pathology;

2 the nature of their receptors for antigen;

3 their migration patterns and the sites which they occupy in the lymphoid organs; and

4 their functions.

All lymphocytes bear receptors for antigen. In B cells, these are membrane-bound immunoglobulin (antibody) molecules. The T-cell receptor is made up of two polypeptide chains and has structural features reminiscent of antibody molecules. Both the T- and B-cell receptors probably stem from a common ancestral molecule. One T or B lymphocyte and its mitotic progeny (i.e. one lymphocyte **clone**) can respond to only one antigen. The immune system as a whole, however, can cope with a vast array of different antigens. It will thus be apparent that the cells making up the immune system comprise an enormously varied population, which between them carry a huge repertoire of different receptors. An educated guess puts the number of different B-cell receptors at 5×10^8 different clones of B cells, each of which reacts with a different antigen. The number of T-cell clones is probably of comparable size. Thus the immune system contains lymphocytes bearing complementary (or near-complementary) receptors for virtually all of the foreign antigenic molecules likely to be encountered. 'Holes' in the repertoire are infrequent.

Lymphocyte circulation

Lymphocytes are found in large numbers in the lymph nodes and spleen, and in mucosal lymphoid aggregations in the gastrointestinal and respiratory tracts (e.g. tonsils, Peyer's patches). A substantial proportion of all lymphocytes constantly recirculates between the tissues, the lymph and the blood. If the immune system is to function efficiently, antigens entering the body must make contact reasonably quickly with the lymphocytes that express the complementary receptors. Before the immune system will recognize and respond to it, a new antigen must be displayed on either **macrophages** or **dendritic (interdigitating)** cells. Both cell types are nearly ubiquitous in their distribution. The antigen presented on the surface of these cells is, in effect, displayed before a stream of circulating lymphocytes. Those lymphocytes bearing receptors specific for the antigen will react with it and will proliferate; those bearing inappropriate receptors will not respond.

T-lymphocyte subsets

T cells comprise several subsets, each with its own distinct function, that collectively result in **cell-mediated immunity**. About 80% of the circulating T cells are regulatory, i.e. they facilitate or amplify other T-cell or B-cell responses (**T-helper cells**), or they dampen down or eliminate the same responses (**T-suppressor cells**). These regulatory mechanisms are still poorly understood. The ratio of helper to suppressor cells in the circulation is about 2 : 1 and is characteristically reduced in **AIDS**, in which the virus preferentially infects T-helper cells. Because T-helper cells are the first lymphocytes to interact with antigen displayed on macrophages and dendritic cells, they are sometimes referred to as T-helper/ inducer cells. Another subset of T cells (**T-effector cells**) subserves two main functions. One group (**cytotoxic T cells**) is able to destroy cells infected with viruses. A second group mediates **delayed-type hypersensitivity** (DTH), an immune response which is important in the defence against those infections in which the causative organisms proliferate intracellularly (e.g. tuberculosis, leprosy), and also in the rejection of foreign tissue grafts. In DTH, the T cells produce several different chemical substances, collectively known as **cytokines** (see below). The term hypersensitivity in DTH implies a pathological reaction. It is, however, with the exception of its extreme variants, a normal response.

Cytokines

Cytokines (previously known as lymphokines) are proteins of low molecular mass that regulate many cellular activities non-specifically. These include:

1 enhancing the inflammatory response;
2 inducing the growth and differentiation of cells of the immune system; and
3 establishing cellular and antiviral activity. (One type of cytokine with antiviral activity is **interferon**.)

A wide variety of cells can transiently synthesize and secrete cytokines. Very small concentrations of cytokines are able to activate cells via receptors present on the cells that produce them (i.e. autocrinally) or on those in close proximity (i.e. paracrinally).

B lymphocytes and antibody production

B cells proliferate on contact with antigen and ultimately develop into **plasma cells** which synthesize and secrete antibody. While B lymphocytes make up 15–20% of blood lymphocytes, plasma cells are located in lymphoid and other tissues rather than in the circulation. The first encounter with an antigen (the **primary response**) is slow and ineffectual. Antibody levels are low and the reaction takes 10–14 days to reach its peak (Fig. 12.8). The host's memory cells are however primed, with the result that a second and subsequent exposures to the same antigen will quickly (within 1–2 days) stimulate the production of high levels of antibody (the **secondary response**).

The antibodies secreted by plasma cells are found predominantly in the γ-globulin fraction of plasma, with small quantities in the β-globulin fraction (p. 324). Collectively they are referred to as **immunoglobulins**. The basic monomeric form of all antibody classes consists of four polypeptide chains—two heavy chains

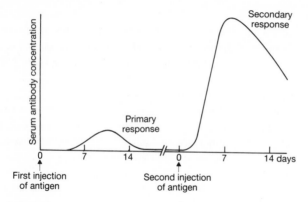

Fig. 12.8 Antibody concentration in serum in response to a first and a second injection of antigen.

and two light chains—held together by disulphide bonds (Fig. 12.9). The immunoglobulins comprise five structural classes of antibody which subserve different functions. These are named accordingly to their heavy chains, both of which are identical in any one molecule. Thus the immunoglobulin classes IgG, IgA, IgM, IgD and IgE (listed in descending order of their plasma concentrations) are characterized respectively by the heavy chains γ, α, μ, δ and ε. There are two types of light chain, κ and λ, both of which are found in all five antibody classes. Within one antibody molecule, however, the two light chains are always the same.

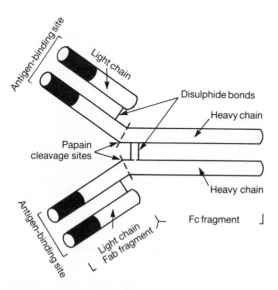

Fig. 12.9 Schematic representation of the IgG molecule showing the four polypeptide chains, the variable regions (shaded) of the chains and the fragments produced by enzymic digestion.

The chemical structure of the antigen-binding site varies from one antibody (and one B-cell clone) to another, and forms part of the variable region of the molecule. The structure of the rest of the molecule is relatively constant. It is possible with papain to cleave antibody molecules into two antigen-binding fragments (Fab fragments) and a crystallizable fragment (Fc fragment) which contains part of the heavy chains. The Fc fragment is responsible for all the biological attributes of antibodies other than antigen binding (e.g. the ability to cross the placenta, to fix to mast cells and to activate the complement cascade). Because of differences in the structure of the variable region, relative molecular masses for the five different antibody classes given below are approximate.

IgG (M_r 150 000) makes up about 80% of the immunoglobulin in plasma. It is also found in extravascular tissues and can cross the placenta. It is the main antibody synthesized during the secondary response and is of major importance in the defence against micro-organisms and their toxins.

IgM (M_r 900 000) is a pentamer. It is the first immunoglobulin to appear in both phylogeny and ontogeny. It is also the first antibody to be produced in the primary response and the only antibody made by the fetus.

IgA (M_r 160 000) usually exists as a monomer in the plasma and as a dimer in surface secretions (tears, saliva, colostrum, gastrointestinal and bronchial secretions). It provides an important defence mechanism at these sites, particularly against viral infections.

IgD (M_r 185 000) is present mainly on the surface of B lymphocytes where it acts as a receptor for antigen. (A given B cell may have IgD surface receptors while secreting IgM and IgG. Even though the cell produces antibodies of more than one class, the antigen-binding site will be identical in all of them.)

IgE (M_r 200 000) is present at very low concentrations in plasma. It attaches to mast cells and basophils via receptors on the Fc portion of the molecule, i.e. it sensitizes these cells so that they release their granules if the attached antibody combines with antigen. This type of reaction exemplifies type I hypersensitivity, a common example of which is hay fever. IgE levels increase significantly in response to infestation by certain parasites.

Functions of antibodies

The binding of antigen to antibody facilitates the removal of the antigen from the body. In the case of bacteria, antibodies either facilitate phagocytosis of whole organisms or neutralize their toxic products. Antibodies are capable of coating viruses, thereby reducing their pathogenicity. Certain types of antibody, such as IgM and some IgG subclasses, fix (or activate) **complement** on combining with antigen. Complement is a collective term encompassing a series of proteins which circulate in an inactive precursor form. They are activated sequentially giving rise to the production of several different factors, the functions of which include:

1 increasing vascular permeability;
2 enhancing phagocytosis; and
3 exerting a chemotactic effect towards polymorphs.

Complement is also capable of bringing about lysis of certain bacteria and cells. Complement is activated by IgM and IgG antibody (the **classical pathway**) and in a slightly different fashion by several other substances, e.g. bacterial endotoxin (the **alternative pathway**).

Active and passive immunity

Immunological memory is exploited in prophylactic immunization, which aims to provoke a primary response using vaccines (organisms or their toxic products treated to render them non-pathogenic). This is an instance of **active immunity** as the antibodies are made by the host's own cells. A subsequent encounter with the corresponding live organisms or their toxic products evokes a secondary response, which will prevent or significantly attenuate the associated disease. Memory is a function of modified T and B cells but the nature of the underlying changes is still uncertain. Its duration varies with the nature of the antigen and it can be lifelong for certain diseases. It is, for example, very uncommon to get measles twice.

It is possible to transfer antibodies from one person to another or even from animals to humans. Immunity gained in this way is referred to as **passive immunity**. Antibodies transferred in the maternal milk provide significant protection to breast-fed infants (the antibodies are not extensively degraded in the gut at this age). Antiserum may also be useful as a stop-gap measure; for example, human serum containing antihepatitis A antibodies is sometimes given to short-term travellers to countries where the hygiene is poor. Passive immunity has a short duration of a few weeks at best. Animal serum is prone to cause hypersensitivity reactions and is now seldom used in clinical practice.

Tissue transplantation and the major histocompatibility complex

An individual will reject tissue or organ grafts from a genetically non-identical member of the same species. The cell-surface proteins that account for these individual differences are called **histocompatibility** (or **transplantation**) **antigens** and they are both numerous and complex. The most obtrusive of these antigens in transplant surgery belong to a genetic system known as the **major histocompatibility complex** (MHC), which in humans is called the **human leukocyte** antigen (HLA) **system**. The MHC, which is present in all higher species, is a recognition system which controls a variety of cell–cell reactions. It plays a crucial part in the interaction between T and B cells, and also between antigen-presenting cells and T cells. Susceptibility to a variety of human diseases can be correlated with the HLA antigenic make-up.

Abnormalities of the immune response

Many disorders of the immune system have been described. Broadly these fall into three groups.

1   Immunodeficiencies affecting one or other component of the immune system. AIDS is the best publicized and now the commonest, but there are many others. The preventing and treatment of graft rejection involves the administration of immunosuppressive drugs (e.g. azathioprine and cyclosporin) and some measure of immunodeficiency is the price of a successful transplant. Immunodeficiency of varying severity also complicates other conditions, e.g. uraemia and severe burns.

2 Aberrant or excessive function of the immune system. In hypersensitivity (which occurs in several forms, one of which is allergy) the immune response to extraneous antigens (i.e. from outside the body) is deleterious to the host. In autoimmune diseases the immune system reacts against the host's own tissues, e.g. in **Graves' disease** against the thyroid-stimulating hormone receptor of the thyroid gland and in **myasthenia gravis** against acetylcholine receptors in skeletal muscle.

3 Neoplasms (cancers) arising in cells of the immune system. All lymphocytes can be triggered to proliferate in the uncontrolled fashion that characterizes cancer. The overproduction of one cell type at the expense of the others will sooner or later result in immunodeficiency. In **multiple myeloma** there is excessive synthesis of a single *aberrant* immunoglobulin arising from a malignant change in a single plasma cell and this **monoclonal antibody** appears in the serum and in some cases in the urine. Characteristically, the levels of the normal immunoglobulins are markedly reduced in myeloma, resulting in increased susceptibility to infection.

12.4 BLOOD GROUPS

Blood groups are systems of genetically determined antigenic substances on the membranes of red cells, e.g. the ABO and Rhesus blood-group systems. These antigens on red cells can be detected by reactions with the corresponding antibodies. Cross-matching of blood before transfusion aims to avoid dangerous reactions between donor blood-group antigens and recipient blood-group antibodies. In pregnancy, maternal antibodies to fetal blood-group antigens can lead to destruction of fetal red cells.

Human red cells have on their membranes **antigenic** substances which permit the classification of blood into **groups**. If red cells with a particular antigen are mixed with the corresponding antibody, the red cells can clump together or **agglutinate**, a phenomenon which forms the basis of **blood-grouping** tests. If the combination of red cells and antibodies occurs in the body, the reaction can lead to dangerous breakdown of red cells within the circulation (**intravascular haemolysis**) or outside it due to phagocytosis by macrophages (**extravascular haemolysis**). This is important in blood transfusion and in pregnancy.

If a patient receiving a blood transfusion has antibodies in the plasma which react with the transfused red cells, potentially lethal haemolysis may follow. The aim of the strict **cross-matching** procedures used in blood-banks is to ensure that such **incompatible** transfusions do not occur. During pregnancy, following leakage of fetal red cells across the placenta into maternal blood, a woman may be stimulated to produce antibodies to an antigen on the fetal red cells if her own erythrocytes lack this antigen. This may occur if her lymphocytes have been sensitized to the antigen in a previous pregnancy. The maternal antibodies may cross the placenta and react with fetal red cells, causing them to be destroyed. This haemolytic anaemia, and its consequences, may be severe enough to cause death of the fetus or to lead to complications after birth (**haemolytic disease of the newborn**).

Blood-group antigens and genes

The antigens on red cells which define blood groups are genetically determined. When it is shown that a set of red-cell antigens is inherited independently of others, these antigens are said to form a particular **blood-group system**. In people of European orgin, there are, at present, 15 well-defined blood-group systems. Antigens in some of these systems are shown in Table 12.8.

The genes determining blood-group antigens are carried on pairs of autosomal chromosomes, except in the case of the Xg system, in which the gene concerned is on the X chromosome. The genes of blood-group systems generally behave as co-dominants, so that if a person has a particular gene, the corresponding antigen can be detected on the red cells. In such cases the genetic constitution of a person for the blood-group system concerned, the **genotype**, can be determined directly from the recognizable characteristics the genes produce in the red cells, the **phenotype**. However, genes of some blood-group systems have no detectable effect on the red cells. These are **amorphic** genes (amorphs). When such genes are present, it may not be possible to determine a person's genotype directly from the phenotype but it may be possible to work it out by study of phenotypes in the family.

In those blood-group systems that have been investigated, the genes have been found to control synthesis of enzymes which modify the composition of the antigens which are glycoproteins or glycolipids on the red-cell membrane.

Blood-group antibodies

Blood-group antibodies are of two kinds, naturally occuring and immune. Naturally occurring antibodies are those found in the blood of people who have not been exposed to red-cell antigens concerned (Table 12.8). The most common are those of the ABO system, which occur regularly in people whose own red cells lack the corresponding antigens. Naturally occurring antibodies are usually IgM in type and, probably because of their large size, they cannot cross the placenta and thus cannot cause haemolytic disease of the newborn.

Immune blood-group antibodies arise as the result of an immune response to red-cell antigens not normally possessed by an individual but acquired by blood transfusion or by transplacental passage of fetal red cells during pregnancy. Thus most antibodies of the Rhesus, Kell and Duffy systems (Table 12.8) are formed in this way. Immune antibodies are most often IgG in type, although IgM antibodies

Table 12.8 Examples of antigens and antibodies in blood-group system

System	Antigens	Antibodies	Antibody type
ABO	A, B, H	Anti-A, anti-B	N
Rhesus	C, D, E, c, e	Anti-C, -D, -E, -c, -e,	I
MNS	M, N, S, s	Anti-M, -N, -S	N
P	P, P_1	Anti-P_1	N
Lutheran	Lu^a, Lu^b	Anti-Lu^a	N, I
Lewis	Le^a, Le^b	Anti-Le^a	N
Kell	K, k	Anti-K	I
Duffy	Fy^a, Fy^b	Anti-Fy^a	I
I	I, i	Anti-I	N

I, immune; N, naturally occurring.

also occur, often early, in an immune response. The IgG antibodies can cross the placenta and can thus cause haemolytic disease of the newborn.

IgG antibodies are often **incomplete**, i.e. they fail to agglutinate red cells having the corresponding antigens when the cells are suspended in physiological saline. This is because the IgG molecule, in contrast to IgM (**complete antibody**), is too short to bridge the gap between adjacent red cells kept apart by a net negative surface charge. Agglutination will occur if these electrostatic repulsive forces between the red cells are reduced, e.g. by treating the cells with enzymes to remove charged groups from the membrane. Incomplete antibodies or complement (p. 347) attached to red cells can also be detected by the **antiglobulin test** (Coombs test).

ABO blood-group system

The four main groups (phenotypes) of the ABO system, **A, B, AB** and **O**, are defined by the presence or absence of two red-cell antigens, A and B (Table 12.9). Either antigen may be present on the cells (group A or group B), both antigens may be present (group AB) or neither (group O). In addition, the naturally occurring antibodies, anti-A and anti-B, are present in the serum if the red cells lack the corresponding antigens (Table 12.9).

The A and B antigens are determined by the allelomorphic genes *A* and *B*. These genes, together with an amorphic third gene, *O*, are inherited as a pair, one from each parent. There are thus six genotypes but only four phenotypes (Table 12.9). The gene frequencies are not equal and, in people of European origin, groups O and A are much more common than B and AB (Table 12.9). Different phenotypic frequencies are found in other races. In some populations of Australian aborigines, for example, the groups B and AB do not occur.

The groups A and AB can each be divided into two subgroups because of two common variants of the A antigen, A_1 and A_2. About 80% of people of European origin with red cells of groups A and AB belong to the subgroups A_1 and A_1B, respectively. The remainder belong to the subgroups A_2 and A_2B. The practical importance of these subgroups is that A_2 and A_2B red cells react more weakly with anti-A antibody than A_1 and A_1B cells. A_2 and A_2B cells may react so weakly with some anti-A antibodies that they can be wrongly grouped as being O and B, respectively.

The A and B antigens of red cells are formed from substance H, which reacts with the antibody, anti-H. The H antigen occurs on all red cells irrespective of their ABO group but the amount present depends upon whether or not a person also has the *A* or *B* genes or both. These genes determine synthesis of the enzymes

Table 12.9 ABO blood-group system. Frequencies (UK) are approximate

Phenotype	Genotype	Antigen on cells	Antibody in serum	Frequency (%)
A	*AA, AO*	A	Anti-B	42
B	*BB, BO*	B	Anti-A	8
AB	*AB*	A, B	None	3
O	*OO*	None*	Anti-A, anti-B	47

* Almost always, red cells have H antigen, with maximal amount on group O cells.

which change the composition of the H antigen to A and/or B antigenic substances.

A, B and H substances are not confined to red cells; they occur also on white cells, platelets, epidermal cells and endothelium. In addition, in people called **secretors**, these substances are found in many body fluids and secretions, including saliva and sweat. About 80% of people of European origin are secretors. All have H-antigenic substance in their body fluids, while those of groups A, B or AB, also have the A, B or A and B antigens, as well as H. The ability to secrete A, B and H substances in body fluids is determined by a dominant gene, *Se*, inherited independently of the *ABO* and *H* genes, and active in people homozygous or heterozygous for *Se*. **Non-secretors** are homozygous for the allele, *se*.

Rhesus blood-group system

The five main antigens of the Rhesus system are known in Fisher's notation as **C, c, D, E** and **e**. According to Fisher's theory, these antigens are determined by three pairs of allelic genes: *C* and *c, D* and *d, E* and *e* (Table 12.10). These genes are inherited in sets of three, one set from each parent, and the *d* gene is amorphic.

The term Rhesus-positive is sometimes used for a person whose red cells have the D antigen and rhesus-negative for a person whose red cells lack it. However, in transfusion practice, it is usual to reserve the term Rhesus-negative for persons of genotype *cde/cde*, whose red cells lack the antigens C, D and E.

Naturally occurring Rhesus antibodies are very rare, e.g. some forms of anti-E. Immune antibodies are common following sensitization by transfusion or by pregnancy. Anti-D has been responsible for many problems and causes haemolytic transfusion reactions and haemolytic disease of the newborn, often of a severe degree. Fortunately, this latter disorder, due to anti-D, can now be avoided if, after transplacental leakage of fetal red cells into the maternal circulation (e.g. during parturition), anti-D immunoglobulin is injected into the woman. This will prevent her lymphocytes being sensitized by the fetal red-cell D antigen. Transfusion reactions and haemolytic disease of the newborn are sometimes caused by antibodies with other Rhesus specificities, e.g. anti-c and anti-E.

Other blood-group systems are occasionally implicated in transfusion reactions and haemolytic disease of the newborn, e.g. the immune antibodies anti-Fy^a

Table 12.10 Rhesus blood-group system. Frequencies (UK) are approximate

Phenotype	Most common genotype		Frequency of genotype (%)
	Fisher notation	Short notation	
CcDee	CDe/cde	R_1r	32
CCDee	CDe/CDe	R_1R_1	17
ccee	cde/cde	rr	15
CcDEe	CDe/cDE	R_1R_2	14
ccDEe	cDE/cde	R_2r	13
ccDEE	cDE/cDE	R_2R_2	3
Others			6

and anti-K (Table 12.8). However, such reactions are usually of much less clinical importance than those due to the ABO and rhesus systems.

Blood transfusion

Blood for transfusion can be kept for several weeks if it is collected from donors under aseptic conditions into sterile plastic packs containing a suitable preservative solution, and it is stored at 4°C. A commonly used preservative, citrate–phosphate–dextrose (CPD) solution, provides citrate as anticoagulant and glucose (dextrose) as metabolic substrate for the red cells. Adequate numbers (i.e. not less than 70%) of red cells remain viable after transfusion when previously stored in CPD solution for 3–4 weeks at 4°C. Adding adenine to the solution can increase this period to 5 weeks.

During storage at 4°C, the red cells show a progressive decrease in content of adenosine triphosphate and 2,3-BPG, while with decreased activity of the Na^+–K^+ pump, the cells gradually lose K^+ to the surrounding plasma and gain Na^+ from it. The concentration of K^+ in the plasma may reach values as high as 30 mmol L^{-1} after storage of blood for 4 weeks. The pH of the plasma also decreases with time of storage and its concentration of ammonia rises. These changes can make stored blood dangerous for transfusion in certain patients, e.g. those with renal or hepatic failure. The decline in Na^+–K^+ pump activity makes some red cells spherocytic, with loss of deformability. These effects may be irreversible and after transfusion the abnormal red cells are destroyed very rapidly by macrophages in the spleen and elsewhere. Other constituents of blood do not withstand prolonged storage. Granulocytes begin to lose their phagocytic capacity within 6 hours of collection and they are functionally inert after 24 hours. Platelets lose their haemostatic effect (p. 357) within 48 hours at 4°C, while the labile coagulation factors, V and VIII (p. 362), also rapidly deteriorate in chilled blood.

Before donated blood is made available for issue from a blood-bank, the ABO and rhesus groups of the cells are determined and commonly the serum is screened for atypical antibodies. Serological tests are also done for syphilis, hepatitis and human immunodeficiency virus (HIV). Before transfusion, the ABO and rhesus groups of the patient's red cells are determined, the serum is checked for unexpected antibodies and red cells from the donor are tested against the patient's serum by **cross-matching** tests (compatibility tests). These cross-matching tests are essential for checking that there has been no error in ABO grouping of donor and recipient, and for ensuring that the recipient's serum does not contain naturally occurring or immune antibodies active against the donor's cells.

Transfusion of whole blood is sometimes necessary but, over recent years, the use of cell-separator machines and large-scale production of plasma constituents have made it increasingly possible to transfuse specific components of blood which the patient lacks. Thus **red-cell concentrates**, often resuspended in a small volume of electrolyte solution, are used to restore the haemoglobin concentration in an anaemic patient in whom the plasma volume may already be expanded. **Platelet concentrates** are of use in patients with severe thrombocytopenia (p. 365). A variety of plasma fractions is also available to supply coagulation factors, e.g. **cryoprecipitate**, which is rich in factor VIII and fibrinogen, and for expanding plasma volume, e.g. **stable plasma protein solution**. A useful source of antibodies against common viruses is **pooled normal immunoglobulin** and various specific immunoglobulins are also available, e.g. anti-D and antibodies against tetanus, hepatitis B and diphtheria.

However much care is taken in cross-matching and administering blood, transfusion carries definite risks of unpleasant or even fatal complications. Major red-cell incompatibility can lead to lethal intravascular haemolysis or delayed extravascular breakdown of donor cells. Transfusion of blood contaminated with bacteria can cause profound shock with hyperpyrexia, while allergic reactions to transfused white cells, platelets and plasma proteins can also be severe. Circulatory overload, air embolism and changes in plasma electrolyte concentrations (e.g. hyperkalaemia) may occur and there may be direct transmission of disease, e.g. HIV and cytomegalovirus infections, hepatitis and malaria.

12.5 HAEMOSTASIS

Blood vessels, platelets and the blood coagulation system react together to arrest bleeding (haemostasis) after injury. Vascular damage promotes vasoconstriction, which is reinforced by substances released by platelets. Platelets adhere to exposed subendothelial elements, then aggregate to form a platelet plug. Endothelial damage activates the intrinsic pathway of coagulation, which is catalysed by phospholipid in platelet membranes, while tissue factor reaching the blood triggers the extrinsic pathway of coagulation. Thrombin is generated and this causes formation of fibrin and production of the haemostatic plug. Free thrombin is rapidly inactivated by various antithrombins. Eventually, during repair, plasmin dissolves fibrin and antiplasmins inactivate any plasmin reaching the blood. Abnormalities in these processes can cause bleeding disorders and thrombosis.

Haemostasis is the process by which bleeding from an injured blood vessel is arrested or reduced. Similar mechanisms appear to confine blood to the cardiovascular system by keeping vessels free of leaks, while other agents maintain the fluidity of the blood.

A haemostatic defect, if it is severe enough, can cause a **bleeding disorder** and activation of haemostasis within the circulatory system can lead to the formation of a solid mass of blood constituents, a **thrombus**, within an intact blood vessel. In arteries, thrombosis can cause **ischaemia**—reduction or cessation of the supply of blood to a tissue—and this can lead to death of cells (**necrosis**), with areas of ischaemic necrosis (**infarcts**) often seriously, or fatally, impairing the function of an organ (e.g. myocardial infarction). In veins, such as the deep veins of the leg or the veins of the pelvis, fragmentation and onward movement of thrombi (**embolism**) can also have fatal consequences (e.g. pulmonary embolism).

Haemostasis involves interlocking reactions of blood vessels, platelets and the coagulation system (Fig. 12.10). The **haemostatic response to injury** comprises two main phases. In the **primary** phase, reactions of blood vessels and of platelets promote slowing of flow and formation of an aggregate of platelets at the site of injury. In the **secondary** phase, activation of coagulation by tissue factor released from damaged cells and by contact of plasma with disrupted vascular surfaces leads to the formation of **fibrin**, which stabilizes the platelet mass to yield a **haemostatic plug**.

Reactions of blood vessels

Many blood vessels respond to trauma by **vasoconstriction** and this reduces the rate of bleeding. Capillaries are probably not able to contract but their endothelial cells seem able to adhere to one another so as to seal small gaps. Contraction of

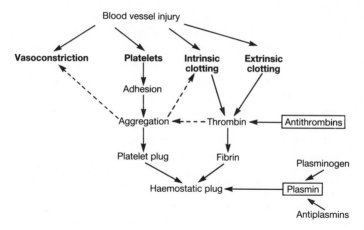

Fig. 12.10 Summary of normal haemostasis.

smooth muscle in larger vessels, such as arterioles and venules, occurs promptly after injury as a result of depolarization of smooth-muscle cells. Contraction is mediated also by local reflexes, by **thromboxane A₂** and possibly **serotonin** released by activated platelets, and also by vasoconstrictor substances liberated from vascular endothelium. Intense vasoconstriction of large arteries in avulsion injuries can sometimes produce surprisingly effective, although temporary, haemostasis.

Vascular damage also promotes haemostasis if the endothelial lining of vessels is lost or disrupted. When injury removes endothelium, exposure of blood to subendothelial constituents of a vascular wall, particularly collagen, allows adhesion of platelets to the damaged area and activation of the **intrinsic** pathway of coagulation. Damaged endothelial cells also release **tissue factor**, which triggers the **extrinsic** pathway of coagulation, and **von Willebrand factor**, which is necessary for platelet adhesion.

While endothelial damage activates haemostasis, a normal intact endothelium can *limit* haemostatic reactions. The luminal surface of intact endothelium has a coating, about 50 nm thick, of **glycosaminoglycans**. These are polysaccharides made by endothelial cells and the main one, **heparan sulphate**, may act like the anticoagulant, heparin, in accelerating inactivation of coagulation factors by a plasma protein, **antithrombin III** (p. 363). **Protein C**, another inactivator of coagulation factors, is also activated by an endothelial component (p. 363). The endothelial glycosaminoglycans are strongly negatively charged and are thought to repel platelets, the surfaces of which are also negatively charged. Platelets may also be prevented from aggregating on normal vascular walls by **prostacyclin**, a prostaglandin synthesized by endothelial cells. Endothelium also liberates **tissue plasminogen activator**. This stimulates generation of the protease **plasmin**, which can lyse fibrin. When these various antihaemostatic actions of endothelium are impaired by vascular injury, the balance will swing towards the haemostatic reactions generated by exposure of blood to subendothelial tissue.

Platelets

Platelets are cytoplasmic fragments, without a nucleus, which circulate as biconvex discs, 2–4 µm in diameter, 0.6–1.2 µm in thickness, and with a volume of 6–9 fl. The normal platelet count is $150–400 \times 10^9$ per litre of blood. Platelets are derived from megakaryocytes in the bone marrow (Fig. 12.2) and survive in the circulation for 8–10 days. At any one time, up to a third of the platelets released from the marrow may be sequestered in the spleen.

Electron microscopy has revealed structural components of platelets important for their haemostatic functions (Fig. 12.11). The plasma membrane is associated with glycoproteins, which are receptors for agents activating platelets. The membrane also contains phospholipids which provide a catalytic surface for coagulation (**platelet factor 3**) and yield **arachidonic acid** for synthesis of prostaglandins. Just under the plasma membrane is a circumferential band of **microtubules**, which probably helps maintain the discoid shape of the unactivated platelet. Extensive invaginations of the plasma membrane, the **open canalicular system**, increase the reactive surface area of the platelet and provide a route to the exterior for substances released from granules in the cytoplasm. A **dense tubular system**, probably analogous to the sarcoplasmic reticulum of skeletal muscle (p. 118) stores Ca^{2+} ions, which move into the cytoplasm when platelets are activated.

The cytoplasm of platelets contains contractile filaments of **actin** and **myosin** which enable activated platelets to change their shape. There are various organelles. **Dense bodies** store adenosine diphosphate (ADP), 5-hydroxytryptamine (serotonin) and Ca^{2+}. The α-**granules** contain a number of proteins including fibrinogen, von Willebrand factor, coagulation factor V, heparin-neutralizing factor (platelet factor 4), β-thromboglobulin, thrombospondin and platelet-derived growth factor. **Lysosomes, mitochondria** and **glycogen granules** are also present.

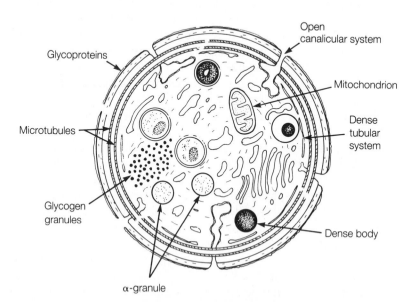

Fig. 12.11 Ultrastructure of the platelet. (After White, J. G. (1987). In: Bloom, A.L. & Thomas, D. P. (eds) *Haemostasis and Thrombosis*, 2nd edn. Churchill Livingstone, London.)

Haemostatic reactions

Platelets take part in a sequence of actions in haemostasis (Fig. 12.12). After damage to the endothelium of a blood vessel, platelets *adhere* to subendothelial collagen fibres. Adhesion depends upon **von Willebrand factor**, part of the coagulation–factor VIII complex. This protein is thought to bind to collagen and to a glycoprotein (Ib) receptor in the platelet membrane. The adhering platelets then *change shape* from smooth discs to spiny spheres, with many projecting filopodia (spines). The loss of discoid shape is thought to be due to depolymerization of the microtubules, while formation of the filopodia probably results from polymerization of actin.

The sequence may stop at this stage but usually activation by collagen goes on to trigger the **release reaction**, a Ca^{2+}-dependent step in which the contents of the platelet granules are secreted to the outside of the platelet. Thromboxane A_2 released in this way, and to some extent serotonin, reinforce local vasoconstriction, while ADP and thromboxane A_2 cause more platelets to become attached to those already adhering, so that a clump forms. This process of **aggregation** depends upon Ca^{2+} and also upon fibrinogen which is thought to link the platelets together by attaching to glycoprotein receptors (IIb and IIIa). At this point aggregation is reversible.

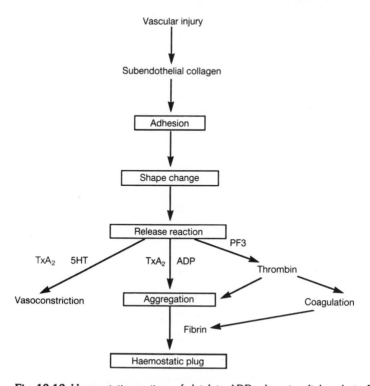

Fig. 12.12 Haemostatic reactions of platelets. ADP, adenosine diphosphate; 5HT, 5-hydroxytryptamine (serotonin); PF3, platelet factor 3; TxA_2, thromboxane A_2.

During these events, platelets also act as catalysts of coagulation, with local generation of thrombin and conversion of fibrinogen to fibrin. Negatively charged phospholipids become exposed on the plasma membrane of the platelets. These phospholipids act as catalytic sites to which complexes of the vitamin K-dependent coagulation factors are bound by Ca^{2+} bridges in optimal orientation for the generation of factor Xa and thrombin (p. 360).

Thrombin causes further aggregation of platelets, which is irreversible (platelet fusion) and this effect may be mediated by thrombospondin from the α-granules, which stabilizes the links of fibrinogen between platelets. Local production of fibrin reinforces the platelet aggregates. Eventually the mass of platelets and fibrin is compacted to form the definitive **haemostatic plug** (Fig. 12.12) by the action of actin and myosin filaments, causing contraction of the platelets. A similar phenomenon can be seen as **clot retraction** in freshly collected blood allowed to coagulate in a test-tube.

Prostaglandin synthesis and actions

An important mediator of the haemostatic reactions of platelets is thromboxane A_2, which is formed from the unsaturated fatty acid, **arachidonic acid** (Fig. 12.13). A phospholipase (A_2), activated by agents such as collagen and thrombin, releases arachidonic acid from phospholipids in the plasma membrane of the platelet. Arachidonic acid is then converted by cyclo-oxygenase to the unstable prostaglandins (PG), PGG_2 and PGH_2, and these are transformed to thromboxane A_2 by a synthase enzyme. Thromboxane A_2 is very unstable, with a half-life of 30 seconds, and it spontaneously breaks down to an inactive product, thromboxane B_2. Thromboxane A_2 is a powerful vasoconstrictor, it induces the release

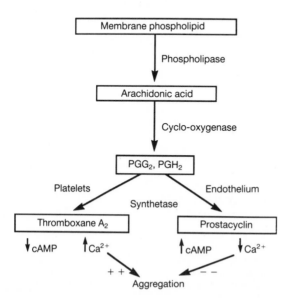

Fig. 12.13 Prostaglandin (PG) metabolism in platelets and endothelial cells. cAMP, cyclic adenosine monophosphate.

reaction and is one of the most potent natural platelet aggregators known. The mode of action of thromboxane A_2 appears to involve inhibition of adenylate cyclase, with decrease in the concentration of the second messenger cyclic adenosine monophosphate, cAMP) (p. 41) and a resultant rise in the concentration of free Ca^{2+} in the cytoplasm of the platelet.

A similar synthetic pathway produces prostaglandins in the endothelial cells (Fig. 12.13) but PGH_2 is converted to **prostacyclin**. Prostacyclin is a vasodilator and a potent **inhibitor** of platelet aggregation. This inhibitory effect seems to depend on stimulation of adenylate cyclase, with an increase in platelet cAMP and a fall in the concentration of Ca^{2+} in the cytoplasm. Prostacyclin may help to prevent deposition of platelets on normal endothelium but the total amount produced per day probably allows only partial control of platelets in this respect. However, release of prostacyclin by endothelium near an area of vascular damage may prevent unlimited accretion of a platelet plug.

Other actions of platelets

Platelets appear to take part in the process of **repair** after vascular injury. **Platelet-derived growth factor** is mitogenic for smooth muscle, fibroblasts and glial cells. It is probably also chemotactic for neutrophils and macrophages (p. 342) and may be involved in the development of atherosclerosis.

Blood coagulation

The coagulation system consists of cofactors and a series of zymogens (pro-enzymes) which sequentially activate one another, leading to the formation of fibrin clot at a site of vascular injury. The coagulation factors (Table 12.11) are

Table 12.11 Coagulation factors

Factor	Name	Approx. plasma concentration $(mg\,L^{-1})$	Function
I	Fibrinogen	3000	Fibrin polymer unit
II*	Prothrombin	100	Protease
III	Tissue factor	—	Cofactor
IV	Calcium	100 (2.5 mmol L^{-1})	Cofactor
V	Proaccelerin	10	Cofactor
VII*	Proconvertin	0.5	Protease
VIII	Antihaemophilic factor, VIII: C	0.1	Cofactor
IX*	Christmas factor	5	Protease
X*	Stuart–Power factor	10	Protease
XI	Plasma thromboplastin antecedent	5	Protease
XII	Hageman factor	40	Protease
XIII	Fibrin-stabilizing factor	10	Transamidase
—	High molecular weight (kininogen Fitzgerald factor)	80	Cofactor
—	Prekallikrein (Fletcher factor)	35	Protease

*Hepatic synthesis requires vitamin K.

mostly globulins which are made by the liver and circulate in plasma at low concentrations. Most are known by capital Roman numerals, although the first four are usually referred to by name. The active forms of the zymogens are serine proteases, which activate the next factor in the sequence by splitting a limited number of specific peptide bonds to uncover the active enzymic site containing serine. The chain of proteolytic reactions in the coagulation system produces a **cascade** effect with amplification and acceleration at each step, so that even though the initiating stimulus may have been trivial, there is eventually an explosive production of large amounts of fibrin at the site of injury. The sequences of amino acids around the active sites and the mode of action of the coagulation enzymes are very similar to those of the pancreatic serine proteases—trypsin, chymotrypsin and elastase (p. 582). These enzymes probably all have a common evolutionary origin.

The end-stage of blood coagulation (Fig. 12.14) is conversion of the soluble plasma protein, **fibrinogen** (factor I) into insoluble **fibrin** by the protease, thrombin (factor IIa).

Fibrinogen consists of three pairs of polypeptide chains, α, β and γ. Thrombin splits arginyl–glycine bonds near the N-terminus of each α- and β-chain to form **fibrin monomer** and two pairs of small peptides, fibrinopeptides A and B. Fibrin monomer molecules spontaneously polymerize to form a weak gel held together by electrostatic bonds. Thrombin activates factor XIII and, by transamidation in the presence of Ca^{2+}, this enzyme (XIIIa) causes strong peptide bonds to form between glutamine and lysine residues of the chains in the fibrin polymer, converting it to the insoluble, stable fibrin clot.

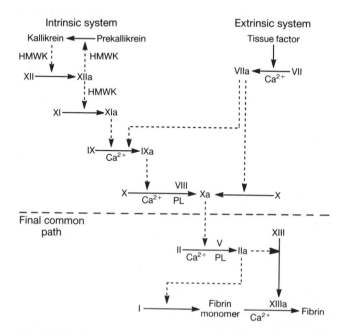

Fig. 12.14 The intrinsic, extrinsic and final common pathways of blood coagulation. HMWK, high-molecular-weight kininogen; PL, platelet phospholipid.

Prothrombin is converted to thrombin by sequential cleavage of two peptide bonds in the prothrombin molecule by activated factor X, in the presence of Ca^{2+} and platelet phospholipid, and with factor V as cofactor (Fig. 12.14).

By itself, factor Xa generates thrombin slowly but the reaction is about 100 000 times faster when factor Xa is bound by Ca^{2+} to the phospholipid surfaces of platelets, together with factor V and prothrombin.

The cofactor activity of factor V is initiated by splitting of peptide bonds in the molecule by thrombin, which thus has an autocatalytic action on its own production. This effect is limited, however, because thrombin also triggers inactivation of factor V by protein C (p. 363).

Intrinsic and extrinsic coagulation systems

Activation of factor X is the culmination of preceding enzyme reactions in two pathways, the **intrinsic** and **extrinsic** systems (Fig. 12.14). The components of the intrinsic pathway are already present in circulating blood; the extrinsic pathway requires, in addition, **tissue factor** (thromboplastin), which is released by damaged cells.

The intrinsic system is activated when plasma comes into contact with constituents of subendothelial tissues, e.g. collagen fibrils or negatively charged, 'foreign' surfaces such as glass or particles of kaolin. Three zymogens—factor XII, factor XI and prekallikrein (Fletcher factor)— and one cofactor—high molecular weight kininogen (HMWK)—are involved in this phase of **contact activation**.

The first stage (Fig. 12.14) is reciprocal proteolytic activation of factor XII and prekallikrein, in which factor XIIa, in the presence of HMWK, converts prekallikrein to kallikrein. Kallikrein, in the presence of HMWK, then activates factor XII. With HMWK as cofactor, factor XIIa then converts factor XI to its active form. It is uncertain what triggers this cyclic activation of factor XII and prekallikrein. Binding of factor XII to foreign surfaces may cause a conformational change in the molecule which could either expose the proteolytic site able to activate prekallikrein or make factor XII susceptible to a low level of protease activity in prekallikrein.

The next reaction in the intrinsic pathway is activation of factor IX in two proteolytic steps by factor XIa in the presence of Ca^{2+}. Factor IXa then activates factor X in a complex formed with factor VIII as cofactor, and with Ca^{2+} binding the reactants together on the phospholipid matrix of the platelet membrane. Thrombin acts on factor VIII, as it does on factor V, to increase the reactivity of the cofactor and eventually to cause its inactivation through the protein C system and by direct proteolysis.

Factor X may also be activated by the extrinsic pathway. Here, a complex formed of factor VII, tissue factor (III) and Ca^{2+} directly activates factor X (Fig. 12.14). Tissue factor is a lipoprotein found in microsomal preparations of various tissues and on the membrane of injured endothelial cells. The extrinsic pathway shows positive feedback with a reciprocating activating cycle between factor Xa and the activated form of factor VII.

The idea of separate intrinsic and extrinsic pathways leading to activation of factor X is useful as a basis for laboratory tests but from a functional point of view the distinction is probably artificial since there are cross-links between the pathways. Thus the factor VII–tissue factor complex directly activates factor IX (Fig. 12.14) and a product of the activation of factor XII can generate VIIa. Although the physiological importance of these reactions is uncertain, they may indicate why both intrinsic and extrinsic systems are necessary for normal haemostasis. Moreover, it is also known that platelets can directly activate factor XI without the intervention of factor XII, HMWK or prekallikrein. This may explain why people with a deficiency of these three coagulation factors do not have a bleeding tendency, whereas patients deficient in factor XI do.

The coagulation factors

The plasma protein coagulation factors fall into three main groups. The **contact factors** (XII, XI, prekallikrein and HMWK) are stable in blood or plasma stored at 4°C, do not need Ca^{2+} for activation to serine proteases and are also concerned in pathways other than coagulation (see note below). The **vitamin K-dependent** factors (II, VII, IX, X) require vitamin K for their synthesis and Ca^{2+} for their activation, are stable in plasma kept at 4°C and, except for prothrombin, are not consumed in coagulation so that they are present in serum. The **fibrinogen group** are those coagulation factors which react with thrombin (fibrinogen, factors V, VIII and XIII) and they are all consumed or inactivated during coagulation. Factors V and VIII rapidly lose their activity in blood or plasma stored at 4°C and special blood products must be used for transfusing these factors. The concentrations of the fibrinogen group of plasma proteins increase in pregnancy, in women taking the contraceptive pill and in inflammatory states.

The contact factors XIIa, and kallikrein with HMWK, can activate the fibrinolytic system (p. 364) and HMWK is both a cofactor for the activation of prekallikrein and a substrate for kallikrein. Kallikrein splits off from HMWK the nonapeptide, **bradykinin**, one of a group of **kinins**, which increase vascular permeability, cause smooth muscle to contract and generate chemotactic activity in leukocytes (see also p. 342).

In the case of the vitamin K-dependent coagulation factors, vitamin K is needed for a final post-ribosomal step in their synthesis, in which an extra carboxyl group is added to glutamate residues to form γ-carboxyglutamic acid side-chains. These carboxyl groups allow the coagulation factors to bind to platelet phospholipid by Ca^{2+} bridges.

Factor VIII in the blood is a complex of two proteins. Factor VIII:C is the coagulation cofactor and its activity is deficient in patients with haemophilia A. Factor VIII:C is a single-chain glycoprotein (M_r 330 000), which is made in the liver, and possibly in other tissues, under the control of a gene on the X chromosome. It has been synthesized by recombinant DNA techniques.

The other part of the factor VIII complex is von Willebrand factor (VIII:WF), which is required for the adhesion of platelets to collagen and also acts as a carrier for VIII:C. This factor derives from a monomer of M_r 250 000, and circulates as a population of multimers (M_r 500 000 to more than 10×10^6). The von Willebrand factor is synthesized by endothelial cells and by megakaryocytes under the control of a gene on autosomal chromosome 12. This factor is deficient or defective in patients with von Willebrand's disease. It is not known how or where the two components of factor VIII become associated.

Inhibitors of coagulation

Thrombin is a very potent enzyme: there is potentially enough thrombin in 10 ml of blood to coagulate all the circulating plasma in an adult in less than 20 seconds at 37°C. Various inhibiting agents limit the action of thrombin.

Phagocytic cells in the liver, and elsewhere, remove particulate thromboplastins from blood and hepatocytes can degrade activated coagulation factors (e.g. IXa, Xa, XIa). Flow of blood past an area of injury dilutes activated intermediates of coagulation and disperses loose aggregates of platelets. These actions limit the

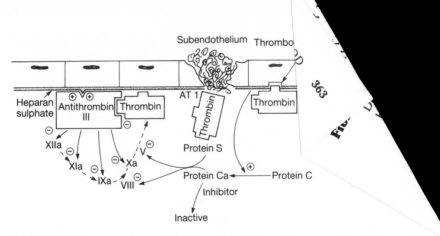

Fig. 12.15 Inactivation of coagulation by antithrombins and the protein C system. AT1, Antithrombin I.

generation of thrombin. Thrombin is also removed from the blood during coagulation by adsorption of fibrin. This process, which is called **antithrombin I**, inactivates only small amounts of thrombin. More important inhibitors are certain plasma proteins, particularly **antithrombin III** and **protein C** (Fig. 12.15).

Antithrombin III is a single-chain glycoprotein, which reacts on a mole-to-mole basis with thrombin to form an irreversible complex in which both molecules are inactivated. Antithrombin III also inactivates factor Xa and the other serine proteases in the intrinsic pathway (IXa, XIa and XIIa). Two other proteins, α_2-**macroglobulin** and α_1-**antitrypsin**, also contribute to the antithrombin effect of plasma.

Thrombin is also bound by a specific receptor on endothelial cells, **thrombomodulin**. The result of this interaction is conversion of circulating protein C to its active form, Ca (Fig. 12.15). Protein Ca is a serine protease which, in the presence of phospholipid, Ca^{2+} and a cofactor, **protein S**, inactivates factors V and VII and thus limits the generation of thrombin. Proteins C and S require vitamin K for their synthesis in the liver and protein Ca also enhances fibrinolysis. Protein Ca is itself inactivated by a specific inhibitor in plasma.

Anticoagulants

Ca^{2+} ions are needed at a number of steps in the coagulation pathways (Fig. 12.14) and agents which remove Ca^{2+} from solution keep blood fluid *in vitro*. Ethylenediamine tetraacetic acid (EDTA) as the sodium or potassium salt and sodium oxalate are used for this purpose while trisodium citrate is the anticoagulant constituent of various solutions used for preserving blood for transfusion.

Heparin is a naturally occurring glycosaminoglycan which is used therapeutically as an anticoagulant *in vivo* when rapid onset of action is desired. **Warfarin**, and similar drugs, are also used as anticoagulants *in vivo*.

Heparin acts by binding to antithrombin III, thereby inducing in the molecule a conformational change which greatly accelerates inactivation of factor Xa in particular, but also of thrombin. Thrombin is also inactivated to a small extent by the circulating glycoprotein, **heparin cofactor II**. Heparin accelerates this reaction.

Coumarin-indanedione drugs, like warfarin, interfere with the action of vitamin K and their anticoagulant effect takes time to develop. By depleting the amount of functional in the liver, they prevent formation of the γ-carboxyglutamate residues in factors and X. This effectively renders the proteins functionally inactive.

sis

During the repair of blood vessels and healing of wounds, fibrin deposited in haemostatic plugs and in extravascular sites is removed by the **fibrinolytic system**. Fibrin is broken down to soluble fragments (fibrin degradation products) by a serine protease, **plasmin**. Plasmin, a two-chain polypeptide, is derived by cleavage of a single peptide bond in plasminogen, a single-chain β-globulin in the plasma. This proteolytic reaction is brought about by **plasminogen activators**.

There are extrinsic plasminogen activators in many tissues, in endothelium and in various body fluids, including urine (**urokinase**), tears and saliva. Intrinsic plasminogen activator activity arises in the blood itself and is generated by the coagulation factors involved in contact activation (factor XII, kallikrein and HMWK). The importance of this form of activator is probably slight since physiological fibrinolysis seems to be due largely to release of **tissue plasminogen activator** from endothelial cells. This activator has now been synthesized by recombinant DNA techniques and is being used to treat coronary thrombosis.

Plasmin seems able to lyse fibrin selectively because plasminogen binds avidly to fibrin as the polymer forms. Tissue plasminogen activator, also incorporated into the fibrin as it forms, or diffusing in later from adjacent damaged tissue, lets plasmin be generated in close association with its natural substrate, fibrin. As the fibrin in a haemostatic plug is laid down, it is thus provided with an enzyme system for its subsequent dissolution.

While fibrin is the physiological substrate for plasmin, this protease will also attack other proteins, especially fibrinogen, factor V and factor VIII. This non-specific proteolysis is normally prevented, since any free plasmin in the blood is rapidly inactivated by irreversible binding to α_2-**antiplasmin**, a circulating glycoprotein. Control of the activation of plasminogen is provided by **plasminogen-activator inhibitor 1**. This is released by endothelial cells and rapidly inactivates tissue plasminogen activator.

Small amounts of plasminogen activator can be detected in venous blood and this seems to originate from endothelium of capillaries and venules in the microcirculation. Flow of blood through these vessels may at times be sluggish, and slowly flowing blood coagulates easily. A continuous slow release of plasminogen activator and the subsequent formation of plasmin may therefore be important for maintaining the fluidity of blood in small vessels. Increase in the level of circulating plasminogen activator occurs in exercise, acute stress and in response to adrenaline. Fibrinolysis may also be important for maintaining the patency of other tubular systems in the body, e.g. the urinary tract.

Bleeding disorders

People with haemostatic defects typically suffer from two kinds of bleeding: they bleed for an *abnormally long time* after injury and they bleed *spontaneously*, without preceding trauma. In defects of platelets and small blood vessels, certain features of this bleeding differ from those found in disorders of coagulation (Table 12.12). Thus **purpura**, i.e. multiple small bruises and haemorrhagic spots (pete-

Table 12.12 Characteristics of bleeding in haemostatic defects. (Adapted from World Health Organization (1972) *Inherited Blood Clotting Disorders*, p. 17. World Health Organization, Geneva)

Type of bleeding	Platelet/vascular defect	Coagulation defect
Spontaneous	Purpura Mucosal bleeding	Single deep bruises/bleeds Bleeds in joints
From superficial cuts	Profuse/prolonged	Not excessive
From deep injuries	Immediate onset Often responds to local treatment	Delayed onset then prolonged/episodic Poor response to local treatment

chiae) in the skin, and bleeding from mucosal surfaces, are characteristic of the spontaneous bleeding seen with platelet and vascular defects. Severe coagulation defects are typically associated with single, spreading bleeds in deep tissues or joints. The responses to mild and moderate trauma also tend to differ in these two groups of haemostatic disorders (Table 12.12).

Purpuras due to acquired vascular defects may be benign local phenomena, e.g. simple easy bruising and senile purpura, or may be more generalized haemorrhagic lesions associated with, for example, severe infections, hypersensitivity reactions (e.g. Henoch–Schönlein syndrome) and drug allergies. Inherited malformations of small vessels (hereditary haemorrhagic telangiectasia) and defective perivascular connective tissue (e.g. scurvy) can also cause troublesome bleeding. A low platelet count and **thrombocytopenic purpura** can arise from *decreased production* of platelets by the bone marrow, because of drugs, ionizing radiations, viral infections, failure of the marrow or infiltration of it by abnormal cells, e.g. in leukaemia. Thrombocytopenia may also be due to *increased destruction* of platelets by immune reactions, as in immune thrombocytopenic purpura, after viral infections (e.g. measles) and as an allergic response to drugs. Purpura can also occur in people with normal numbers of circulating platelets but with functional defects, e.g. failure of aggregation in thrombasthenia.

Inherited deficiencies of coagulation factors are rare but spectacular disorders when severe, and usually involve only one factor (e.g. deficiency of factor VIII:C in haemophilia A). In von Willebrand's disease, in addition to a deficiency of factor VIII:C, there is also an inherited lack of von Willebrand factor (VIII:WF) leading to defective adhesion of platelets and bleeding typical of that seen with platelet defects. Acquired disorders of coagulation factors are due usually to multiple deficiencies, commonly of the vitamin K-dependent factors because of lack of vitamin K. Since this vitamin is fat-soluble, deficiency of it and defective vitamin K-dependent coagulation factors, can arise in disorders impairing absorption of fats. Similar effects can also occur in liver disease, in premature babies and during treatment with anticoagulants like warfarin (an overdose of which often causes bleeding). Disease of the liver can also be associated with other haemostatic abnormalities, including thrombocytopenia, reduced synthesis of fibrinogen and increased fibrinolytic activity.

Diffuse intravascular thrombosis (disseminated intravascular coagulation) can arise because of release of procoagulant material (e.g. amniotic fluid) into the circulation or because of widespread endothelial damage (e.g. severe bacterial infections). There may be such gross depletion of coagulation factors and platelets that generalized bleeding occurs.

Tests of haemostatic function
Certain simple screening tests are useful for assessing haemostatic function

quickly. These are a platelet count, examination of a blood film, a bleeding time and tests of coagulation.

A **platelet count** will demonstate a thrombocytopenia and examination of a **blood film** will often confirm a **low platelet count** and may reveal its cause, e.g. leukaemia. The **bleeding time**, the time taken for cessation of bleeding from small punctures in the skin (made by a standard technique), is an index of the integrity of platelets. In people with a normal platelet count, a prolonged bleeding time would suggest defective platelet function or von Willebrand's disease.

Two simple coagulation tests are used to monitor the extrinsic and intrinsic pathways. The **prothrombin time** (PT) tests the extrinsic system and the final common path (Fig. 12.14), by measuring the time taken for a sample of citrated plasma to coagulate when tissue factor and Ca^{2+} are added. If a test plasma takes longer to coagulate than a known normal plasma, this suggests deficiency (or inhibition) of one or more of the factors VII, X, V, II and I.

In the **partial thromboplastin time with kaolin** (PTTK), the intrinsic system of a sample of citrated plasma is activated by incubating it with kaolin for several minutes, then a substitute for platelet phospholipid and Ca^{2+} are added. If the test sample takes longer to coagulate than a control sample, this indicates deficiency (or inhibition) of one or more of the factors XII, XI, IX, VIII, X, V, II and I. The combined results of the two tests, PT and PTTK, will usually indicate if a coagulation defect is present and will suggest the likely possible defect(s), depending on whether either or neither test, or both, give abnormal results.

If thrombin and Ca^{2+} are added to a sample of citrated plasma, the time taken for the plasma to coagulate, the **thrombin time**, is a useful test for rapidly assessing the concentration and reactivity of fibrinogen.

Chapter 13
Cardiovascular System: Introduction and the Heart

13.1 THE CARDIOVASCULAR SYSTEM

The heart and blood vessels comprise the cardiovascular system whose function is to circulate blood around the body. This provides a transport system subserving homeostasis. The heart consists of two pumps lying side by side; each pump has an atrium and a ventricle. The right atrium receives venous blood from the body and the right ventricle pumps it to the blood vessels of the lungs (the pulmonary circulation) for oxygenation. At the same time the left atrium receives blood from the lungs and the left ventricle pumps it into the arteries of the rest of the body (systemic circulation). Arteries branch into arterioles, the calibre of which determines blood flow to the tissues and arterial and capillary blood pressure. Networks of narrow capillaries supplied by the arterioles provide a large surface area in the tissues through which diffusional exchange of essential substances occurs. Capillaries drain into a large-capacity venous system returning blood to the heart.

Functions

The cardiovascular system in the adult contains a **blood volume** of 4–6 litres. It is composed of a cardiac pump and a series of distributing and collecting vascular tubes linked by very thin capillaries which permit rapid **diffusion** of substances (p. 5) between blood and the interstitial fluid bathing the cells. Thus the cardiovascular system is a **transport system** that links the external environment to the tissues and distributes substances essential for metabolism. The essential substances are O_2 from the lungs and nutrients from the gastrointestinal tract. At the same time, the cardiovascular system removes from the tissues CO_2 and other byproducts of metabolism, carrying them to the lungs, kidneys and liver. Thus the cardiovascular system is essential for **homeostasis** of the extracellular fluid (p. 2) by ensuring the appropriate distribution of available water and solutes to all parts of the body.

There are additional homeostatic functions of the cardiovascular system. Not only does it **circulate hormones** (p. 270) but the heart itself produces a hormone, atrial natriuretic peptide (pp. 443 and 653). The cardiovascular system is also concerned with heat distribution and **temperature regulation** (p. 686). It also transports the agents involved in **haemostasis** (p. 354) and the cells and antibodies concerned with the body's **defence mechanisms** (p. 339).

Basic characteristics

The **heart** is divided into two pumps lying side by side, pumping in phase but distributing blood in series: the right side receives blood from the body and then propels it at low pressure through the vascular system of the lungs (the **pulmonary circulation**); the left side receives blood from the lungs and then propels it at high pressure to all other tissues of the body (the **systemic circulation**; Fig. 13.1). In one cycle all the output of the right heart has to circulate through the lungs but, since the circulations of the different tissues are arranged in parallel, only some of the output of the left heart circulates through any one systemic tissue. Note that there is normally no direct transfer of blood between the two pumps.

Each side of the heart has two chambers; the **atrium** receives blood from the *veins* and aids its flow into the **ventricle** which propels it into the *arteries* (Fig. 13.2). At the end of a contraction the heart always contains some blood which is added to during its relaxation phase (*diastole*). Pressure is then generated through muscular contraction to expel some of that blood (*systole*). Thus the pumps generate pulsatile pressures, 0 to ~25 mmHg in the right ventricle and 0 to ~120 mmHg in the left ventricle. When the body is at rest, diastole occupies two-thirds of the total cycle. The product of the frequency of pumping (**heart rate**) and the volume ejected at each contraction by any one side (**stroke volume**) is the **cardiac output**. Typical values for a resting adult person are 60–70 beats per min, 70–80 ml per beat and 5–6 L min^{-1}, respectively.

The distribution of **blood flow**, at rest, to each of the systemic tissues is given in Table 13.1. This is related to tissue weight, to its level of metabolism and, in the case of kidney and skin, to the flow required for filtering excretory products and for temperature regulation, respectively. During moderate exercise the cardiac output increases and its distribution changes (Table 13.1). Heart muscle and skeletal muscle need this increased blood flow to satisfy their increased metabolism and the skin receives a higher blood flow to dissipate the extra heat production. During exercise some blood is shunted away from the gut and kidney but a constant blood flow to the brain is maintained.

The distributing vessel leaving the right ventricle is the **pulmonary artery** and

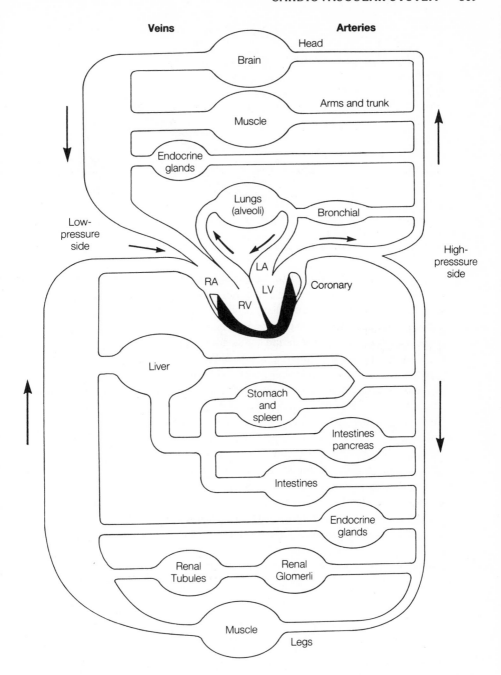

Fig. 13.1 Schematic representation of the cardiovascular system. LA, Left atrium; LV, left ventricle; RA, right atrium; RV, right ventricle.

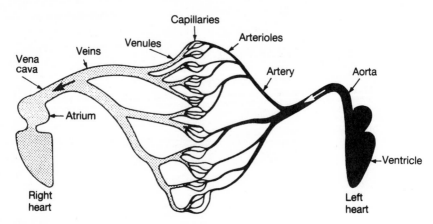

Fig. 13.2 Blood vessels of the systemic circulation.

that leaving the left is the **aorta**. To supply a tissue or organ, these branch into smaller *arteries* and finally *arterioles* before entering the vast *capillary* network of fine tubes (Fig. 13.2). The capillary network is drained by *venules* which collect into *veins* and finally into the **pulmonary veins** entering the left atrium, or into the large **superior** and **inferior venae cavae** entering the right atrium.

The high-pressure distributing vessels of the systemic circulation, the aorta and **arteries**, have elastic fibres in their walls which stretch, storing energy as the vessel distends to accommodate about half of the blood ejected during systole. During diastole, elastic recoil of the walls releases energy which sustains aortic and arterial pressure and thereby maintains blood flow towards the periphery. In this way, intermittent flow from the heart is converted into continuous pulsatile flow through the arteries. Backwards flow into the heart is prevented by a valve guarding the entrance to the aorta, the aortic valve.

Arterioles have a narrower lumen than arteries and are the **major site of resistance** to blood flow. This high resistance results in a considerable fall in blood pressure as blood flows through arterioles. There is also a large damping of the pulsatile flow as it is converted to a continuous steady flow. Contraction of circular

Table 13.1 Distribution of blood to systemic tissues at rest and in moderate exercise

	Rest			Exercise		
	(L min^{-1})	%	$(\text{L min}^{-1}$ $100\,\text{g}^{-1}$ tissue)	(L min^{-1})	%	$(\text{L min}^{-1}$ $100\,\text{g}^{-1}$ tissue)
Total flow	5.8	—	—	17.5	—	—
Brain	0.75	13	0.050	0.75	5	0.050
Cardiac muscle	0.25	4	0.083	0.75	5	0.250
Skeletal muscle	1.2	21	0.004	12.5	71	0.042
Skin	0.5	8	0.010	1.9	11	0.038
Kidney	1.1	19	0.367	0.6	3	0.200
Gut	1.4	24	0.050	0.6	3	0.021
Other	0.6	10	0.002	0.4	2	0.001

smooth muscle in the walls of arterioles increases their resistance and thus decreases blood flow through all blood vessels. Inevitably arteriolar constriction will also elevate the pressure in the arteries and decrease the pressure in the capillaries. The opposite changes occur when the smooth muscle relaxes. The adjustment of arteriolar calibre regulates tissue blood flow, aids in the control of arterial blood pressure and, by altering capillary pressure, influences the net flow of water across the capillary wall.

Capillary networks provide a very large cross-sectional area through which blood flows slowly, giving ideal conditions for diffusional exchange between blood and interstitial fluid. Some fluid leaks across the capillary wall but it is returned slowly as lymph plasma to the cardiovascular system by a set of collecting tubes called the **lymphatic system**.

The **venules** and **veins** are the major set of collecting conduits returning blood at low pressure from the capillaries to the heart. Systemic veins have a relatively large capacity. Indeed, in the resting supine position, veins hold four times as much blood as do the arteries. They are also very distensible. Smooth-muscle contraction in the walls of veins causes a reduction in venous distensibility and hence in the volume they can accommodate, allowing blood redistribution to other parts of the cardiovascular system, when necessary.

In the wall of the aorta and carotid arteries, there are nerve endings known as **baroreceptors** which respond to stretch and hence monitor the arterial blood pressure. This information, together with other sensory inputs, is relayed to the **cardiovascular centres** (the coordinating centres in the brain) which alter the **autonomic nervous activity** to the heart, arterioles and veins in order to maintain arterial blood pressure at a certain level whilst altering total blood flow and its distribution. Pressure, flow and distribution are also controlled by **local mechanisms** intrinsic to the heart and arterioles. In addition, the sensory inputs directly or indirectly release certain **hormones** which act on the heart, arterioles and veins to control blood flow and pressure but in particular blood volume.

13.2 ANATOMY OF THE HEART

There are four chambers in the heart: right and left atria communicate through their respective atrioventricular (AV) valves with right and left ventricles. Most of the filling of the ventricles is by passive flow-on of blood through the atria from the veins; the final part of filling is aided by atrial contraction. The AV valves close as soon as the ventricles contract, preventing reflux of blood back into the venae cavae and pulmonary veins. When the ventricular pressures generated exceed those in the pulmonary artery or aorta, the pulmonary and aortic valves open, allowing blood to be ejected from each ventricle. Cardiac muscle is a network of striated muscle cells specialized for synchronous activity by gap junctions which electrically connect the cells.

Functional gross anatomy

The heart consists of four chambers: the **right atrium** leading into the **right ventricle** and the **left atrium** leading into the **left ventricle** (Fig. 13.3). The aperture between each atrium and its respective ventricle is guarded by an

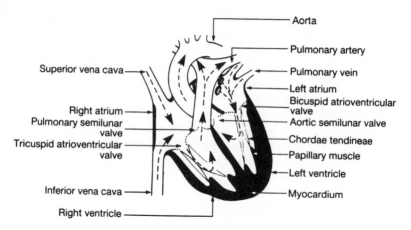

Fig. 13.3 Structure of the chambers and valves of the heart. The atrioventricular rings of the connective tissue lie in the plane of the valves. Arrows indicate direction of blood flow.

atrioventricular (AV). valve. The right valve has three cusps (the *tricuspid valve*) and the left valve has two cusps (the bicuspid or *mitral valve*). Attached to the free margins of these valves are tendinous cords—the **chordae tendineae**—which are attached to projections of ventricular muscles known as **papillary muscles**. The exits from the right ventricle into the pulmonary artery and from the left ventricle into the aorta are guarded by the pulmonary and aortic **semilunar valves**, respectively. Rings of connective tissue, the **atrioventricular rings**, separate the atria from the ventricles and act as a fibrous skeleton for the origin and insertion of atrial and ventricular muscle and for the attachment of the bases of the heart valves.

Venous blood continuously flows from the systemic circulation via the superior and inferior venae cavae into the right atrium and from the pulmonary circulation via four pulmonary veins into the left atrium. When pressure in the atrium is greater than in the ventricle, the AV valves are open and blood flows onwards into the respective ventricles. When the ventricles are about 80% full, the atria contract and propel blood into the ventricles to complete ventricular filling.

After a very short pause (~0.1 second) both ventricles contract (systole) and ventricular pressure very rapidly exceeds atrial pressure, causing the AV valves to close, thereby preventing backwards flow from ventricles to atria. The papillary muscles simultaneously contract during systole, tensing the chordae tendineae which prevent the AV valves from everting into the atria. Continued contraction of the ventricles raises the ventricular pressure above that in the pulmonary artery or aorta, thereby opening the pulmonary and aortic valves. Blood is then ejected from the right ventricle at low pressure into the pulmonary circuit and from the left ventricle at high pressure into the systemic circuit. When the ventricles stop contracting, the ventricular pressures drop below the pulmonary arterial and aortic pressures. This causes the pulmonary and aortic valves to close, terminating blood ejection and preventing reflux of blood back into the ventricles. When the ventricular pressures fall below atrial pressures, the AV valves open and the ventricles start to refill with blood (diastole).

The magnitude of pressure generated by each heart chamber during contraction is reflected in the thickness of its muscle wall—known as the **myocardium** (Fig. 13.3). The atrial myocardium is thin whilst ventricular myocardium is thick, especially that of the left ventricle which generates the highest pressure. The inner surface of the myocardium is lined with **endocardium** which is a continuation of the endothelium of blood vessels. The outer surface of the myocardium is covered with mesothelial tissue called **epicardium**. Enclosing the heart is a thin fibrous sac—the **pericardium**—which is lined by a parietal layer of mesothelium. The pericardium is relatively stiff and limits excessive acute enlargement of the heart. The **pericardial space** between the epicardium and pericardium contains just enough interstitial fluid to act as a lubricant.

Cellular structure of cardiac muscle

The myocardium is comprised of **cardiac muscle cells**. Compared to skeletal muscles these are very much shorter (150 µm cf. a few millimetres or even many centimetres) and somewhat narrower (20 µm cf. 40 µm). Also, in contrast to skeletal muscle, cardiac cells are branched, abut end-to-end to form a network (**syncytium**), are richer in mitochondria and each has at its centre a single nucleus (Fig. 13.4; cf. Fig. 5.2). Just as in skeletal muscle, actin and myosin filaments are arranged into sarcomeres with their characteristic A, I and H bands and M and Z lines. In cardiac muscle, however, the sarcoplasmic reticulum is less extensive and the transverse (T) tubular system is wider and located at the Z line, not at the A–I junction, where it forms a dyad, not a triad (p. 117) with the sarcoplasmic reticulum.

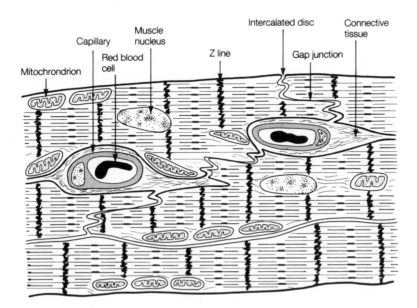

Fig. 13.4 Network arrangement of cardiac muscle cells.

The apposition of one cardiac cell with another coincides with one of the Z lines and is specialized into a dense **intercalated disc** (Fig. 13.4). Within this disc, adjacent areas of membrane are bridged by low-resistance pathways called **gap junctions** which allow electrical current to pass to adjacent cells. The **desmosomes** in the discs provide sites of adhesion between one cardiac cell and another and ensure that the tension developed by one cell is transmitted through to the next.

13.3 ELECTRICAL PROPERTIES OF THE HEART

Cardiac muscle has a myogenic rhythm, that is, it has the ability to contract rhythmically without nervous input. The action potential for each heart beat is generated by a pacemaker in the right atrium and transmitted through the heart along specialized conducting pathways. Pacemaker cells have a resting membrane potential which slowly depolarizes spontaneously and on reaching threshold triggers an action potential. Cardiac action potentials are typically of long duration (200–400 ms) and involve changes in Na^+, K^+ and Ca^{2+} conductance. There is considerable overlap in time between the cardiac action potential and the contraction it initiates so that, in contrast to skeletal muscle, two contractions cannot summate; nor can a fused tetanic contraction occur. Pacemaker firing rate, and therefore heart rate, is increased by sympathetic and decreased by parasympathetic nerve activity. The autonomic nervous system also influences conduction velocity through the heart and the duration of the cardiac action potential. The electrocardiogram (ECG)—comprising the P wave and QRS complex, which reflect atrial and ventricular depolarization, respectively, and the T-wave reflecting ventricular repolarization—can reveal abnormalities in electrical activity (cardiac arrhythmias).

Origin and conduction of electrical activity through the heart

A denervated (or transplanted) heart continues to beat in an orderly sequence of atrial then ventricular contraction followed by a passive filling phase. This ability to depolarize and contract rhythmically without innervation is called **myogenic rhythmicity**. All parts of the heart have this ability, although the intrinsic contraction rates of atrial and particularly ventricular muscle is quite slow (when separated from each other and the nodes).

Normally, the initiation of each heart beat and its sequential coordination involves a **specialized conduction system** of cardiac cells that have few myofibrils and contract only weakly. In the absence of nervous control, action potentials originate at a rate of about 100 per minute in the **sinoatrial (SA) node**, a small area located in the wall of the right atrium near the entrance of the superior vena cava (Fig. 13.5). The SA node is the **pacemaker**, dictating the rate of beating of the entire heart. From the SA node action potentials are conducted along the plasma membrane from one atrial cell to another through the low electrical resistance of the intercalated discs. Conduction velocity through the atrial muscle is enhanced by three conducting pathways (anterior, middle and posterior **internodal bands**). This ensures that both atria depolarize and therefore contract almost simultaneously.

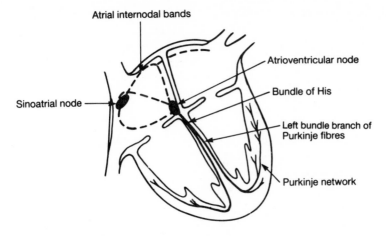

Atrial internodal bands

Atrioventricular node

Bundle of His

Sinoatrial node

Left bundle branch of Purkinje fibres

Purkinje network

Fig. 13.5 The SA and AV nodes and the specialized conducting pathways of the heart.

The internodal bands merge as they approach the next node, the **atrioventricular (AV) node**, which is located in the atrioventricular fibrous ring on the right side of the atrial septum. *The AV node is the only electrical pathway through the insulating fibrous ring.* Conduction through the AV node is slow (0.05 m s^{-1}) compared with atrial or ventricular muscle (0.5 m s^{-1}). This effectively delays transmission for about 0.1 second at resting heart rates (see Fig. 13.6) and ensures atrial contraction is finished before ventricular contraction begins.

From the AV node, action potentials travel at a speed of 1 m s^{-1} down the ventricular septum in the **bundle of His** and along the right and left bundle branches to enter the **Purkinje network** which ramifies throughout the ventricular muscle. The conduction velocity through the Purkinje network is very fast (5 m s^{-1}). Because of extensive branching of the Purkinje network, excitation reaches all parts of both ventricles rapidly, causing them to depolarize and therefore contract almost simultaneously.

Characteristics of cardiac potentials

Cardiac action potentials have different characteristics in different regions of the heart. Three patterns are found—one for Purkinje fibres and ventricular muscle, one for atrial muscle, and one for the SA and AV nodes (Fig. 13.6). Note that the **action potential duration** also varies in different regions of the heart. It is shortest (200–250 ms) in the SA and AV nodes and atrial muscle, longest (300–400 ms) in the Purkinje fibres and intermediate in duration (250–300 ms) in the bundle of His and ventricular muscle.

In **ventricular muscle** (Fig. 13.7), the resting membrane potential (phase 4) is steady at about -90 mV due to stability of the membrane conductance to K$^+$ (G_K) and to Na$^+$ (G_{Na}). The action potential has an initial rapid depolarization (phase 0) resulting from a sudden increase in G_{Na} as *fast* Na$^+$ channels open, together with a sharp decrease in G_K. The action potential reaches a peak of about $+20$ mV and then has a rapid but short decline (phase 1) which is mainly due to inactivation of fast G_{Na} channels. Phase 2 of the action potential has a prolonged

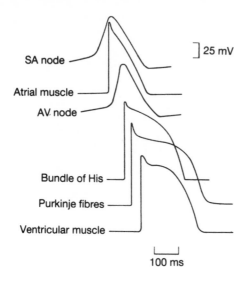

SA node

Atrial muscle

AV. node

Bundle of His

Purkinje fibres

Ventricular muscle

$\rbrack$ 25 mV

100 ms

Fig. 13.6 Shape, duration and sequence of cardiac action potentials. Note also the delays caused by the anatomical sequence of depolarization and by the relative conduction velocities down the conducting system. (After Hoffman, B.F. & Cranefield, P.F. (1960) *Electrophysiology of the Heart.* McGraw-Hill, New York.)

shoulder or plateau which is mainly due to a delayed opening of Ca^{2+} channels (increased G_{Ca}) together with a delayed increase in the opening of *slow* Na^+ channels (increased G_{Na}). The membrane then repolarizes quite quickly (phase 3) as G_K, G_{Ca} and G_{Na} return to normal.

These events contrast with those in axons (see Fig. 4.4) or skeletal muscle in which G_{Ca} is not involved in the action potential and G_K does not decrease during the initial depolarization. The involvement of slow Ca^{2+} and slow Na^+ channels in cardiac muscle is responsible for the long duration of the cardiac action potential (about 250 ms compared with about 5 ms in skeletal muscle).

In **atrial muscle**, the action potential has the same ionic bases as in the ventricle but the time-course is a little different (Fig. 13.6); there is a less obvious plateau (phase 2) and therefore a longer repolarization phase (phase 3).

In the **SA** and **AV nodes** (Fig. 13.7), the cells have a high membrane G_{Na} and hence a less negative resting membrane potential than other heart cells. In between action potentials their membrane potential (phase 4) is also unstable and slowly depolarizes from about -70 mV to about -50 mV. This slow depolarization is called the **prepotential** or **pacemaker potential**. The instability is attributable to a gradual fall in G_K, but towards the end of the prepotential an increase in the conductance of slow Ca^{2+} and slow Na^+ channels also contributes. When G_K is sufficiently low, the threshold potential of -50 mV is reached and this triggers the action potential. However, nodal cells lack functionally fast Na^+ channels and phase 0 is therefore a slow depolarization, partially due to activation of slow Na^+ channels but even more importantly due to a further increase in G_{Ca}. In contrast to ventricular muscle, G_K also increases during the action potential. The action potential reaches a peak of about 0 mV, has no plateau and repolarizes (phase 3) at a speed similar to atrial muscle.

The slope of the pacemaker potential in the SA node is steeper than in the AV node. Therefore, the SA node triggers its action potential first and is the pacemaker from which each heart beat originates. The action potential propagates

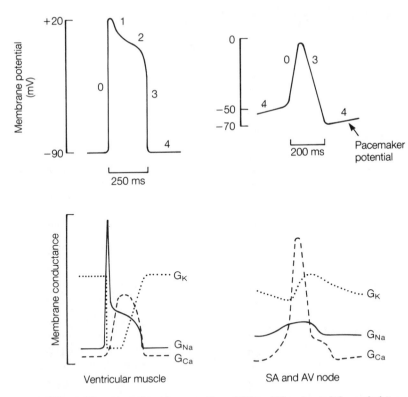

Fig. 13.7 Cardiac potentials in the ventricle and SA or AV nodes and the underlying changes in membrane conductance to K^+, Na^+ and Ca^{2+}. Note that membrane conductance is plotted on a logarithmic scale.

from the SA node through the atria, causing the atrial action potential, and arrives at the AV node well before the pacemaker in the AV node has reached its threshold. Thus the action potential that occurs in the AV node is initiated by the action potential conducted from the SA node. The action potential then propagates through the bundle of His and ventricular muscle.

Cardiac refractory period and temporal relationship between electrical and mechanical events

During the **absolute refractory period** (ARP; about 200 ms) of the action potential the cardiac cell is inexcitable and during the subsequent **relative refractory period** (RRP; about 50 ms) there is a gradual recovery of excitability (Fig. 13.8). A second action potential cannot be generated during the ARP, whereas a strong stimulus can elicit an action potential in the RRP. The strength required decreases progressively during the RRP. An action potential generated during the RRP has a slower rate of depolarization, a lower amplitude and shorter duration than usual (Fig. 13.8). Immediately after the RRP, the next action potential is still of a shorter duration and results in less Ca^{2+} entry.

When measured from the beginning of the action potential there is a latency of

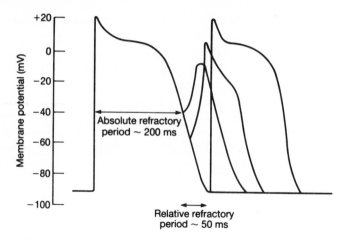

Fig. 13.8 The absolute and relative refractory periods of the ventricular action potential.

about 10 ms before the muscle starts to contract (Fig. 13.9) compared with about 2 ms in skeletal muscle (see Fig. 5.7). The peak of developed tension occurs just before the end of the ARP and the muscle is half-way through its relaxation phase by the end of the RRP. The total duration of the mechanical event is about 300 ms. Thus electrical and mechanical events in cardiac muscle overlap considerably in time (in contrast to skeletal muscle where the short-duration action potential is virtually over before the contraction begins; see Fig. 5.7) and it is therefore impossible in cardiac muscle during high-frequency stimulation to produce the summation and tetanus typical of skeletal muscles. This electrical–mechanical overlap in cardiac muscle means that when a second action potential is triggered at

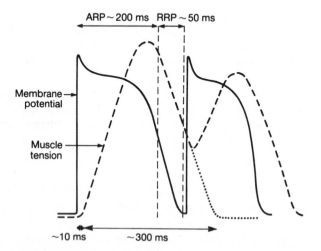

Fig. 13.9 The temporal relationship between the first action potential and the resulting muscle tension and between a second action potential (triggered immediately after the RRP) and its resulting muscle tension. Note that summation of muscle contractions is impossible.

the very start of the RRP, the second contraction is superimposed on the semi-relaxed phase of the first contraction (Fig. 13.9). It is a relatively weak contraction because the shorter duration of the second action potential has resulted in less Ca^{2+} than usual entering the cell (p. 391); thus the force developed by the cross-bridges between actin and myosin filaments will be reduced.

The physiological importance of the prolonged cardiac refractory period is that it protects the ventricles from too rapid a re-excitation which would impair their ability to relax long enough to refill adequately with blood. Furthermore, since the total refractory period is longer than the time taken for conduction through the atria or ventricles, recycling of excitation in the muscular network is not seen in the normal heart.

Alteration of heart rate and action potential duration

The **cardiovascular centres** in the brain normally alter the intrinsic discharge rate of the **SA node** via the autonomic nervous system and the hormone adrenaline. The **sympathetic** nerves *increase* and the **parasympathetic** nerves *decrease* the **heart rate**. At rest, there is activity in both sympathetic and parasympathetic nerves but the latter are more active in humans, reducing the rate from its intrinsic value of about 100 to about 70 beats per min. Since the parasympathetic fibres project to the heart in the vagus nerve (cranial nerve X), their normal activity is often referred to as **vagal tone**. Intense parasympathetic activity can actually stop the pacemaker of the heart for a brief period. Normally, a low heart rate (**bradycardia**), for example during sleep, is due to an increase in parasympathetic discharge and a decrease in, or even absence of, sympathetic discharge. The opposite changes in autonomic activity lead to an accelerated heart rate (**tachycardia**) and in acute anxiety or severe exercise the heart rate can reach up to 200–220 beats per minute. Alterations in heart rate are referred to as **chronotropy**. Pacemaker rate is also sensitive to temperature and in humans increases approximately 10 beats per minute per degree C increase in body temperature.

Noradrenaline, the transmitter released by postganglionic **sympathetic** fibres, and **adrenaline**, the hormone released from the adrenal medulla in response to increased sympathetic activity, bind to plasma membrane β_1-**adrenoceptors** in the SA node. This increases the rate of pacemaker potential depolarization such that threshold potential is reached earlier; the earlier triggering of action potentials allows more to occur per unit time (Fig. 13.10). The effects of β_1-adrenoceptor activation appear to be mediated through the intramembrane G_s-protein and increased intracellular cyclic adenosine monophosphate (cAMP; p. 41) resulting in a faster increase in both G_{Na} and G_{Ca} during the pacemaker phase of the potential; it is unclear whether there are changes in G_K.

Acetylcholine, the transmitter released by **parasympathetic** fibres, acts on **muscarinic cholinoceptors** in the SA node to hyperpolarize and slow the rate of pacemaker potential depolarization (Fig. 13.10). The hyperpolarization is achieved by a decrease in cAMP due to activation of muscarinic receptors, which inhibit adenylate cyclase via the G_i-protein. The decrease in cAMP causes additional K^+ channels to open. The cause of the reduced slope of the pacemaker potential is still uncertain; it appears to be due to reductions in G_{Ca} and G_{Na}. The cardiac response to parasympathetic stimulation is quicker than to sympathetic

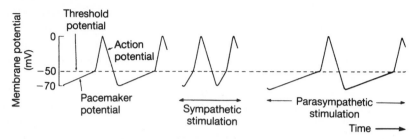

Fig. 13.10 Effects of sympathetic and parasympathetic stimulation on the rate of depolarization of the pacemaker potential and the frequency of action potentials.

stimulation and so vagal stimulation can alter the heart rate within the time required for a normal heart beat.

Sympathetic and parasympathetic fibres also innervate the atria and AV node; in addition the sympathetic also affect the Purkinje fibres and ventricular muscle. These autonomic fibres alter the **conduction velocity**; for instance, sympathetic activity decreases and parasympathetic activity increases the AV node delay. They also alter the **action potential duration** which is shortened by sympathetic activity and lengthened by parasympathetic activity. Such alterations in duration and conduction velocity are unique to cardiac muscle and at the higher heart rates ensure that the heart functions effectively by making all phases of the heart's cycle shorter. More rapid changes in sarcolemmal G_K and G_{Ca}, as well as a more rapid uptake of Ca^{2+} into the cisternae of the sarcoplasmic reticulum (p. 393), underlie the shorter cycle.

The minimum duration of the atrial action potential is about 120 ms. Thus the theoretical maximum rate of atrial contraction is about 400 per minute. However, the AV node cannnot conduct more than about 220 action potentials per minute, so that a coordinated heart cannot beat above this rate.

Electrocardiogram

The synchronized depolarization spreading through the heart causes currents in the extracellular fluid that establish field potentials over the whole body. These potential differences can be detected by electrodes placed on the body's surface. The signal (about $\Delta1$ mV) has to be amplified and the record produced is called the **electrocardiogram (ECG)**.

Characteristics of the ECG

The pattern of the ECG varies depending on the electrodes' position but certain features are always present (Fig. 13.11). The **P wave** is produced by the spread of electrical activity during atrial depolarization. The **QRS complex** is produced by ventricular depolarization and the **T wave** by ventricular repolarization. When no depolarization or repolarization is occurring, there is no potential difference in the ECG—the **isoelectric line**. Atrial repolarization does not produce any detectable

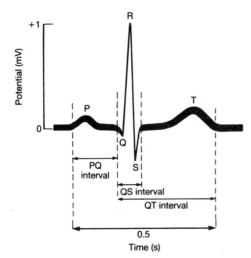

Fig. 13.11 Standard ECG record taken from limb lead II for a resting heart rate.

wave because it occurs during the much larger QRS complex. Since ventricular repolarization is less well-synchronized than ventricular depolarization, the T wave is longer in duration but smaller in amplitude than the QRS complex.

Depending on the electrode position, the QRS complex may have one, two or sometimes three components (see Figs 13.14 and 13.15). If after the P wave the first deflection from the isoelectric line is negative (by convention downwards), it is called a Q wave; if positive, it is called an R wave and if the next deflection falls below the isoelectric line, it is called an S wave.

The **PQ** or **PR interval** (Fig. 13.11) is the time required for excitation to spread through the atria, AV node and bundle of His. The **QS interval** is the time required for excitation to spread through the ventricles. The **QT interval** and **PS interval** are measures of the duration, respectively, of the ventricular and atrial action potential (Fig. 13.12).

ECG vectors

At a given electrode, the polarity of the voltage change depends on the direction in which the wave of potential change is moving in the heart. If a wave of depolarization is travelling towards a positive electrode, a positive potential is recorded and if the wave is travelling away from the electrode a negative potential is recorded. A wave of repolarization travelling towards or away from the positive electrode causes a negative or positive potential, respectively. At any instant, the total electrical activity can be thought of as a **vector** having both direction and magnitude (Fig. 13.13).

Depolarization of the atria commences at the SA node and spreads through atrial muscle to the AV node. The net result is a vector directed downwards and to the left (P wave). Ventricular depolarization starts in the interventricular septum which depolarizes from left to right, resulting in a vector directed downwards and to the right (first wave of the QRS complex). Depolarization then spreads through the ventricular muscle from inner to outer surface. Since the left ventricle has more

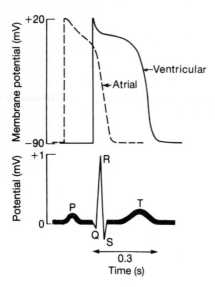

Fig. 13.12 Relationship between the atrial and ventricular action potential and the ECG.

bulk than the right, the mean direction of the vector is downwards and to the left (second wave of the QRS complex). Activation of the last segments of the myocardium results in a vector directed upwards and to the right (third wave of the QRS complex). Ventricular repolarization spreads from the outer to the inner surface of the ventricular myocardium. Since it is opposite in both direction and sign from the second wave of the QRS complex, it produces a vector in the same direction, i.e. downwards and to the left (T wave).

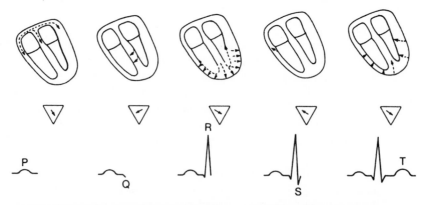

Fig. 13.13 Sequence of depolarization and repolarization of the heart, the resulting vector in the Einthoven triangle and the development of the ECG waves (as recorded by a limb lead, usually lead II, looking directly towards the vector arising from the middle stage of the sequence).

Standard and augmented limb leads

It is conventional to regard the heart, which is oriented downwards and to the left, as situated in the centre of an equilateral triangle, the **Einthoven triangle**, with the upper corners corresponding with the shoulders and the lower corner with the pubic symphysis. Recording electrodes (referred to as leads), when attached to right and left arms (RA and LA) and to one or other foot, conventionally the left (LF), are effectively placed at the corners of this triangle (Fig. 13.14).

From **standard limb leads**, bipolar recordings are obtained between any two corners of the triangle (Fig. 13.14) either between LA and RA (lead I), between LF and RA (lead II) or between LF and LA (lead III), with a fourth electrode on the right foot acting as an earth. From **augmented limb leads**, unipolar recordings

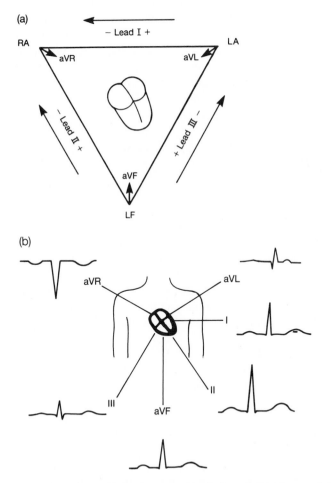

Fig. 13.14 (a) Einthoven triangle for recording the ECG from limb leads. LA, left arm; LF, left foot; RA, right arm. Arrows indicate the direction in which a particular lead looks at the heart. (b) ECG contours as recorded from standard (I, II and III) and augmented (aVR, aVL and aVF) limb leads which examine the heart, when standing, in the vertical plane.

are obtained, one of the three corners of the triangle acting as the active (positive) electrode, the remaining two leads connected together as the indifferent (negative) electrode and the right foot earthed (Fig. 13.14). When the active electrode is the RA, LA or LF, the lead is designated aVR, aVL or aVF, respectively. The term aV refers to the augmented voltage (about 50% more) obtained by these leads compared with standard limb leads.

Each lead can be regarded as looking at the heart in a particular direction. Standard limb leads look from the positive to the negative electrode and augmented limb leads look at the heart from the limb by which they are named (Fig. 13.14). Lead II normally looks most directly at the main vectors and therefore has the largest as well as positive P, R and T waves with small negative Q and S waves. In leads I and aVF, the deflections are all positive but of smaller amplitude than lead II because they do not look so directly at the main vectors. Lead aVL looks at the heart from the side and detects best the vector associated with interventricular septal depolarization. This vector moves away from the lead, causing a negative Q wave. Neither lead III nor aVL detects the P wave well and the T wave is very small and positive. Lead aVR is the only one looking at the heart such that the main vectors are moving away from the electrode. Thus, P and T waves are negative and the QRS complex is seen as a large negative Q wave.

Chest leads

Unipolar recordings with the active electrode in one of six positions are obtained from **chest leads** (Fig. 13.15). The indifferent electrode consists of the LA, RA and LF connected together and the right foot is earthed. Lead V_1 is placed over the fourth right intercostal space near the sternum; V_2 is in a similar position on the left; V_3 is on the left midway between V_2 and V_4; V_4 is over the left fifth intercostal space in the mid clavicular line; and V_5 and V_6 are in the same transverse plane as V_4 in the anterior axillary line and mid axillary line, respectively.

Chest leads give larger ECG deflections than limb leads (Fig. 13.15). The P and T amplitudes depend on how directly the electrodes look at the P and T vectors. Leads V_1 and V_2 show a small positive R wave and large negative S wave; lead V_4 shows a small Q and S wave and a large R wave (the full QRS complex); and leads V_5 and V_6 show mainly a large R wave in the QRS complex.

Mean electrical axis and transition point

The direction of the vector at any instant is the electrical axis of the heart at that instant. The **mean electrical axis** is defined as the direction of the largest vector in the limb leads and, since it correlates well with the long axis of the heart (i.e. the interventricular septum), it can be used to infer the orientation of the heart. In Fig. 13.14, limb lead II has the biggest QRS vector so that in this case the mean electrical axis for the ventricle is about 30° left of vertical. (The normal range is anywhere in the lower left quadrant.) Tall, thin people with a narrow thorax tend to have a more vertical heart and hence a more vertical electrical axis. Hypertrophy of the left ventricle shifts the electrical axis to the left (left axis deviation) while hypertrophy of the right ventricle produces right axis deviation.

The chest lead that gives equal-amplitude R and S waves, usually V_3 or V_4,

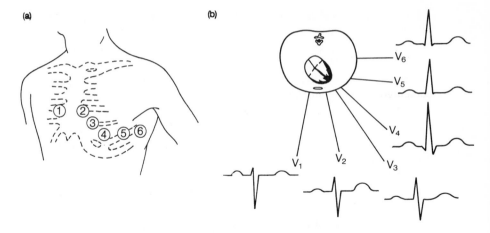

Fig. 13.15 (a) Positions for the ECG leads (V_1–V_6) on the chest which examine the heart, when standing, in the horizontal plane. (b) Note how the shape of the ECG varies according to the chest lead position.

overlies the anterior edge of the interventricular septum and is said to mark a **transition point**. Rotation of the heart about a vertical axis in the transverse plane depends on body build and also occurs with hypertrophy of one ventricle. It can be detected as a right or left shift of the transition point.

Cardiac arrhythmias

One clinical use of the ECG is to diagnose **cardiac arrhythmias** which reflect disturbances in the initiation or propagation of action potentials. *Ischaemic heart disease*, in which there is a reduced blood flow in the heart's coronary vessels, is a common cause of arrhythmias.

In the conducting system the AV node is the most susceptible to damage, developing a very slow initial depolarization and prolonged action potential. This decreases the speed of conduction through the AV node and causes a longer interval than usual between atrial and ventricular depolarizations (**first-degree heart block**). In partial (**second-degree**) heart block, only some of the atrial impulses reach the ventricle. The action potentials that do not reach the ventricle arrive at the AV node when it is refractory. In complete (**third-degree**) heart block, the AV node fails to transmit any impulses. The atria then beat at the rhythm dictated by the SA node which, because of compensating reflexes, may be faster than usual and *latent pacemakers* in the ventricle initiate their own rhythm of 20–40 beats per minute.

Latent pacemakers may develop in the absence of damage to the SA or AV nodes and become sites known as **ectopic foci**. Arrhythmias can also occur because a cardiac impulse may re-excite some region through which it has already just passed; this phenomenon is known as **re-entry**.

If ectopic foci or re-entry cycles are located in the atria, a **premature atrial systole** may result, causing an extra atrial and ventricular contraction (*extrasystole*) such that the pause between the extra beat and the next normal beat is slightly longer than the usual beat interval. Other atrial arrhythmias are **atrial tachycardia** (intermittent high heart rates up to 220 beats per minute), **atrial flutter** (regular atrial contractions greater than 220 beats per minute with some degree of partial heart block as the AV node cannot transmit every impulse) and **atrial fibrillation** (irregular rapid atrial contractions with the ventricle contracting at a totally irregular rate, often before it is adequately filled with blood).

If ectopic foci or re-entry cycles are located in the ventricles, a **premature ventricular systole** (*extrasystole*) may result and the long interval between it and the next natural beat (nearly two normal beats in length) is called the *compensatory pause*. It occurs because, when the SA node's action potential arrives at the ventricle, the ventricular muscle is still refractory from its extrasystole. Other ventricular arrhythmias are **ventricular tachycardia** (regular) and **ventricular fibrillation** (irregular). The latter is the most serious arrhythmia because the ventricular contractions are rapid, uncoordinated and ineffective. Ventricular fibrillation can sometimes be treated successfully by electrical *defibrillation* in which the heart is exposed to a brief pulse of external electric current. This depolarizes instantly the entire myocardium, which may then be able to repolarize as a coordinated unit without redeveloping fibrillation.

13.4 MECHANICAL ACTION OF THE HEART

A knowledge of the temporal relationships between the ECG, ventricular volume, and atrial, ventricular and arterial pressures is essential to understand fully how the heart works. For each heart beat, the cardiac cycle comprises a filling period when the ventricles are relaxed (diastole) and an ejection period when the ventricles are contracting (systole). Each diastolic and systolic period commences with a phase where the volume in the ventricles does not change (isovolumetric) because all the valves are closed; hence at these times the ventricular pressures are rapidly falling in diastole or rising in systole. The closure of the AV valves at the beginning of systole can be heard with a stethoscope on the chest wall as the first heart sound; closure of the aortic and pulmonary valves at the end of systole causes the second heart sound. Rapid filling of the ventricles can cause a third heart sound. Sounds at any other time are abnormal. Pulsations in the jugular vein of the neck indicate events occurring in the right atrium.

Cardiac cycle and heart sounds

The relationships between electrical, mechanical (hydrostatic pressure and volume) and valvular events during one complete heart beat, referred to as the **cardiac cycle**, are illustrated in Fig. 13.16. Events on both sides of the heart are similar, though slightly asynchronous. The organization of the electrical conduction system is such that contraction of the right atrium precedes that of the left atrium and the left ventricular contraction precedes that of the right ventricle. Right ventricular ejection, however, begins before left ventricular ejection because pulmonary arterial pressure is lower than aortic pressure. Furthermore, since the pulmonary circuit offers less resistance to blood flow than the systemic circuit, right ventricular ejection goes on for longer and hence the pulmonary valve closes after the aortic valve.

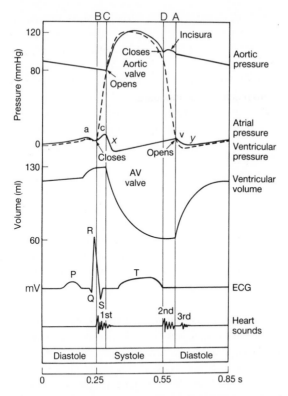

Fig. 13.16 The events occurring in the left side of the heart during one cardiac cycle at rest. A and B indicate when the AV valve opens and closes respectively. C and D indicate when the aortic valve opens and closes respectively.

At a resting heart rate of 70 beats per minute, each cardiac cycle lasts 0.85 second and is composed of two periods: **systole**, when the ventricles contract and eject blood into the aorta or pulmonary artery, and **diastole**, when the ventricles are relaxed and fill with blood from the veins. The duration of ventricular muscle contraction and therefore, by definition, systole is 0.3 second (Fig. 13.9). Diastolic filling of the ventricles occupies almost two-thirds of the cycle (0.55 second) at rest. However, when heart rate increases to its maximum of ~200 beats per minute, the cycle is reduced to 0.3 second. A reduced action potential duration can shorten systole to about 0.15 second, leaving only 0.15 second for diastolic filling.

It is convenient to begin a description of the cardiac cycle prior to atrial contraction, i.e. in *mid-diastole* (Fig. 13.16). At this point both atrial and ventricular pressures are low, the AV valves are open and the ventricles contain nearly 80% of their final filled volume. Atrial pressure is slightly higher than ventricular so that the direction of blood flow is from atria to ventricles but only a little further filling of the ventricle occurs. As the blood that enters the heart cannot leave it (because aortic and pulmonary valves are closed), both atrial and ventricular pressures slowly increase. At this stage, the aortic pressure remains high, for reasons explained later.

In *late diastole* the P wave of the ECG occurs, reflecting atrial depolarization. Towards the end of the P wave, the atria begin to contract, causing an increase in atrial pressure (the **a wave**; Fig. 13.16) which propels much of the blood within the atria into the ventricles. The addition of this final 20% to ventricular filling is accompanied by a small increase in ventricular pressure. As the atria begin to relax, atrial and ventricular pressures drop slightly. The volume contained in each ventricle at the end of diastole (about 130 ml when standing and about 160 ml when lying) is the **end-diastolic volume**.

Towards the end of diastole the QRS complex of the ECG begins, indicating that ventricular depolarization has started. By the end of the QRS complex, the ventricle starts to contract. In this phase—*early systole*—ventricular contraction immediately generates a rapid increase in ventricular pressure which rapidly exceeeds atrial pressure. This causes the AV valves to shut, setting up vibrations which are transmitted to the chest wall and with the aid of a stethoscope can be heard as a low-pitched sound (**the first heart sound**). With microphones and recording apparatus (phonocardiography), the waves composing this sound can be displayed (Fig. 13.16).

As the ventricles continue to contract with their valves closed, both tension in the ventricular wall and pressure in the ventricular lumen increase markedly. This is associated with vibrations in ventricular muscle fibres which contribute to and prolong the first heart sound. Since blood can neither enter nor leave the ventricles, this period is referred to as the **isovolumetric** or **isometric phase of ventricular contraction**. Because of the high ventricular pressure, the AV valves bulge back into the atria causing a sharp rise in atrial pressure to about 10 mmHg (left atrium) or 5 mmHg (right atrium)—the **c wave**.

When ventricular pressures first exceed aortic and pulmonary arterial pressure, the aortic and pulmonary valves open and blood is accelerated and ejected very rapidly from the ventricles into the arteries. Aortic and pulmonary arterial pressures follow very closely their respective ventricular pressures and in young adults rise from the diastolic minima of about 80 mmHg (systemic) and 8 mmHg (pulmonary) to their systolic peaks of about 120 mmHg (systemic) and 25 mmHg (pulmonary). During this rapid ejection phase, the shortening of the ventricles pulls the AV fibrous rings downwards and, since the venous openings into the atria are in a fixed position, the effect is to lengthen the atria and increase their capacities. Thus there is a sudden fall in atrial pressure, often to negative values, called the **x descent**.

In *mid-systole* the T wave of the ECG reflects ventricular repolarization. Towards the end of the T wave, in *late systole*, ventricular muscle starts to relax and ventricular pressures drop (Fig. 13.16). Although ventricular pressure actually drops below aortic pressure, blood continues to be ejected slowly because of the momentum imparted to it during the initial acceleration earlier in systole. Note that the total energy causing blood to flow is the sum of the potential energy from hydrostatic pressure gradients and the kinetic energy possessed by the moving mass of blood (*see* p. 408). In the aorta, pressure does not drop as quickly as ventricular pressure because, owing to the resistance to flow through peripheral arterioles, only about half of the ejected volume has been propelled through the aorta. The remainder has been accommodated by elastic distension of the aorta. As ejection into the aorta slows, the stretched aorta recoils, propelling more blood onwards to the arteries. In addition, at the end of ejection, there is a small transient retrograde aortic flow towards the ventricles, which causes the aortic valve to close. The volume contained in each ventricle at the end of the systole is called the **end-systolic volume** and is about 60 ml when standing. As the end-diastolic volume was 130 ml, about 70 ml (the **stroke volume**) has been ejected in systole. The proportion of the end-diastolic volume that is ejected (i.e. stroke volume/end-diastolic volume) is the **ejection fraction**.

In this phase of *early diastole*, since all valves are now closed, no blood can enter or leave the ventricles and this period of rapid ventricular relaxation and rapidly falling ventricular pressure is called the **isovolumetric** or **isometric phase of ventricular relaxation**. During this period the **second heart sound** (a shorter and higher-pitched sound than the first) is heard. It is generated by the vibration from closure of aortic and pulmonary valves. This sound is often split, especially during inspiration, because the aortic valve closes slightly before the pulmonary valve. The pressure wave associated with closure of the aortic valve is reflected along the aorta, giving rise to an increase in aortic pressure, and the pressure dip before it is called the **incisura**. Throughout systole the atrial pressure has gradually increased from below zero such that, by the end of isovolumetric ventricular relaxation, it has reached a peak of about 5 mmHg (left atrium) or 2 mmHg (right atrium)—the **v wave**. This gradual increase in atrial pressure results from blood from the veins accumulating in the atria against closed AV valves and from the return of the AV fibrous ring to its resting position in mid-systole.

Later in *early diastole* when ventricular pressures have dropped just below those of the atria, the AV valves open. As the atria empty passively and rapidly into the ventricles, the atrial pressure falls from its v-wave peak—the **y descent**. The *rapid filling* of the ventricles, with blood that had accumulated in the atria, sets up vibrations sometimes detectable as the **third heart sound**. After initially declining, the atrial and ventricular pressures then gradually increase in *mid-diastole* and ventricular filling now proceeds *slowly* (**diastasis**) due to continuing venous flow into chambers closed by aortic and pulmonary valves.

Throughout diastole, atrial pressure is slightly greater than ventricular pressure. The rapid initial filling of the ventricles early in diastole is very important because it means that when heart rates are high and the diastolic period short, ventricular filling will lose only the relatively small contribution from the later diastasis. (Similarly, the rapid large-volume ejection early in systole prevents a short-duration systole from markedly reducing stroke volume.) Throughout diastole the aortic pressure remains high due to both the continuing elastic recoil of aortic walls stretched during the previous systole and the resistance to flow downstream. There is a slow decline to a diastolic minimum as the blood flows from the aorta into the rest of the vascular system.

Abnormal heart sounds

A **fourth heart sound** can be heard just before the first heart sound if vibrations are set up during the rapid ventricular filling associated with atrial contraction. These vibrations may occur when the atrial pressure is high or the ventricle is stiff, as in ventricular hypertrophy.

Abnormal sounds called **murmurs** can occur at any stage of the cardiac cycle; some do not have a pathological basis; others do. When blood passes through a narrowed orifice or regurgitates back into a chamber, its flow becomes turbulent, usually generating a murmur. Pathological narrowing of a valve's orifice is referred to as **stenosis**. Stenosis of an AV valve results in turbulent flow into the ventricles (**diastolic murmur**) and stenosis of the aortic or pulmonary valve causes turbulent flow into the aorta or pulmonary artery (**systolic murmur**). If a valve is **incompetent** and fails to close properly, a backflow of blood occurs. Incompetent AV valves allow systolic regurgitation into the atria (systolic murmur) and incompetent aortic or pulmonary valves cause diastolic regurgitation into the ventricles (diastolic murmur). Defects in the atrial or ventricular septum also cause systolic murmurs, while a **patent ductus arteriosus** connecting the aorta with the pulmonary artery can be associated with a **continuous murmur** throughout the cardiac cycle. The frequency, character, duration and timing of the murmur, and the site on the chest wall where it is heard maximally are often diagnostic of the underlying abnormality.

Jugular venous pulse

Since there are no valves at the entry of the venae cavae into the right atrium, the three positive a, c and v waves in the right atrial pressure are transmitted into the large veins. They can be seen as pulsations in the jugular vein in the neck (p.459) and are called the **jugular venous pulse** (JVP). In fact it is the x and y descents that the observer sees most noticeably. Because the waves are transmitted backwards from the right atrium, each wave in the JVP occurs after the corresponding one in the atrium (Fig. 13.17). The **a wave** of the JVP occurs because of atrial contraction but it is not just a delayed reflection of the increase in atrial pressure. It is also contributed to by damming of blood in the large veins, which results from atrial contraction constricting the orifices of the venae cavae. The **c**

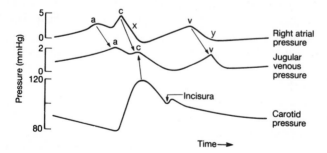

Fig. 13.17 Temporal relationships between right atrial, jugular venous and carotid arterial pressures.

wave is a delayed reflection of the increase in atrial pressure, occurring when the AV valve bulges into the atrium. The pressure pulse transmitted from the adjacent carotid artery during peak systole also makes a contribution—hence the name **c wave**. The **v wave** of the JVP has the same origins as the v wave in the right atrial pressure. The relative amplitudes of the JVP waves are variable because the jugular vein is subjected to rhythmical pressure fluctuations induced by breathing (p. 434) which occur at a rate much slower than the heart beat. The shape and magnitude of the JVP can indicate the presence of cardiac arrhythmias, valvular incompetence or stenosis in the right heart and, in particular, increases in right atrial pressure.

13.5 CONTRACTILE PROPERTIES OF CARDIAC MUSCLE

Heart muscle performance and therefore its stroke volume are affected by the extent to which it is stretched during diastole; its current state of contractile energy; and the arterial pressure (load) against which it must eject blood. Length–tension or volume–pressure curves are a means of examining cardiac performance. The initial stretch is the passive tension or preload of the heart muscle, which determines the number of actin–myosin cross-bridges that can form and thereby the tension generated by the subsequent isometric or isotonic contraction. For any particular initial stretch, the contractile energy can be increased by entry of extracellular Ca^{2+} induced, for example, by adrenaline. The load which is encountered after the contraction starts is the afterload; the larger it is, the less effective the velocity of ejection, as shown by load–velocity curves. The relationship between wall tension and lumen pressure can be most precisely examined using the law of Laplace.

The **contractile process** of cardiac muscle is similar to that of skeletal muscle (p. 119), except that **extracellular Ca^{2+}** as well as intracellular Ca^{2+} is required. Extracellular Ca^{2+} enters the cardiac cells, via the sarcolemma and T tubules, during each depolarization phase of the action potential (p. 375). It helps activate the contractile mechanism (p. 121) and serves as a trigger to release Ca^{2+} from the intracellular store in the cisternae of the sarcoplasmic reticulum. During repolarization the sarcoplasmic reticulum actively accumulates Ca^{2+} (Ca^{2+} AT-Pase pump). At the same time, through both a plasma membrane Ca^{2+} ATPase pump and Ca^{2+} exchanged for the 'downhill' movement of Na^+ into the cell (countertransport), Ca^{2+} is also returned to the extracellular fluid. The Na^+ gradient is maintained by the Na^+–K^+ ATPase pump. Because the movement of Ca^{2+} from the uptake to the release sites within the sarcoplasmic reticulum is slow, the quantity of Ca^{2+} available for release at the next contraction depends on the interval between contractions.

Length–tension and load–velocity curves

Length–tension curve

Contractile properties of muscles can be examined by length–tension curves, as described previously for skeletal muscle (p. 125). Muscles have elastic and

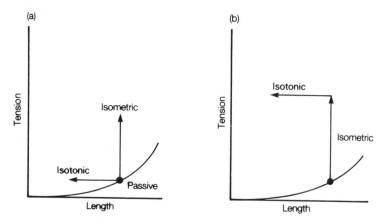

Fig. 13.18 Length–tension curves for a strip of cardiac muscle. (a) Tension developed during passive stretch and during an isotonic compared with an isometric contraction. (b) In the heart the isotonic contraction is preceded by an isometric contraction.

contractile elements that, when stretched, both recoil and generate a *passive tension* (or passive force) roughly proportional to their new length (*see* Fig. 5.8; Fig. 13.18a). When the muscle at a particular length contracts and is compelled to a lift a weight (load), it shortens at a constant force and the contraction is referred to as **isotonic**. If the muscle remains at the same length as it contracts, the contraction is referred to as **isometric** (Fig. 13.18a). In isometric contractions (p. 125) *active tension* plus *passive tension* comprise the *total tension* measured during the contraction (Fig. 13.19).

The maximal active tension in skeletal muscle occurs when the overlap of actin and myosin filaments is optimal (*see* Fig. 5.9) and coinicides with the muscle's natural resting length. In cardiac muscle the resting length is shorter than the optimal length (Fig. 13.19), so that cardiac muscle is normally operating on the ascending part of the length–tension curve. This curve is also steeper than for skeletal muscle. Thus, unless the cardiac muscle becomes overstretched, small increases in initial length (which increase passive tension only slightly) will result in large increases in active and total tension. This intrinsic ability of cardiac muscle to enhance its contractile energy whenever it is slightly stretched is the basis of the *intrinsic control of stroke volume*, considered later.

On stretching cardiac muscle, the increase in active tension is attributable to the more effective formation of cross-bridges between actin and myosin filaments, as in skeletal muscle, and to the extra steepness of the cardiac length–tension curve. The latter arises to some extent because myofibril lengthening increases the sensitivity of troponin to Ca^{2+} and induces an increase in free intracellular Ca^{2+}.

Since the myocardium is a syncytium, active tension from a large number of cardiac fibres cannot be graded by recruitment of motor units, as occurs in skeletal muscle (p. 130). In cardiac muscle, active tension can be slightly increased by a higher frequency of action potentials because this increases the influx of Ca^{2+} available for the contractile process. The main way by which contractile energy is enhanced is through sympathetic or adrenaline activation of β_1-adrenoceptors

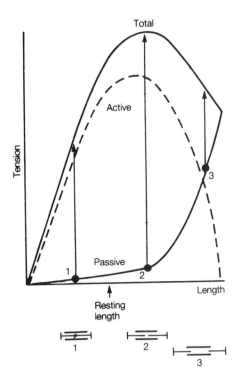

Fig. 13.19 Total and active tension developed during isometric contractions starting from different cardiac muscle lengths (i.e. different passive tensions). Active tension depends on the extent of overlap of actin and myosin filaments. Note that the resting muscle length is shorter than the length giving optimum active tension.

(Fig. 13.20). This leads to release of cAMP, which increases the membrane conductance to Ca^{2+} (G_{Ca}), again increasing the Ca^{2+} available intracellularly. An increase in extracellular Ca^{2+} will similarly enhance contractile energy. Such alterations in contractile energy which are *not* caused by a change in initial length are referred to as a change in **myocardial contractility**. The term **inotropic state** is synonymous with contractility. Changes in myocardial contractility are the basis of the *extrinsic control of stroke volume*, considered later.

Although parasympathetic stimulation via muscarinic cholinoceptors decreases atrial contractility, it does not affect appreciably ventricular contractility. Myocardial contractility is however depressed during cardiac failure, severe acidosis or severe hypoxia (low O_2).

Preload and afterload

The passive tension in a muscle before contraction is referred to as the **preload** and in the whole heart this is related to the *end-diastolic volume*. If a muscle has to lift a load after it has started contracting that load is called the **afterload**; the muscle will first develop sufficient isometric tension to match the total load and then shorten, lifting the load in an isotonic contraction (p. 127; Fig. 13.18b). For the ventricle the *aortic pressure* can be regarded as the equivalent of its afterload; the ventricle will first develop sufficient isometric tension, and hence ventricular pressure, just to exceed the pressure in the aorta, thus opening the AV valve. Then it will commence its ejection, shortening against the afterload by means of an isotonic contraction.

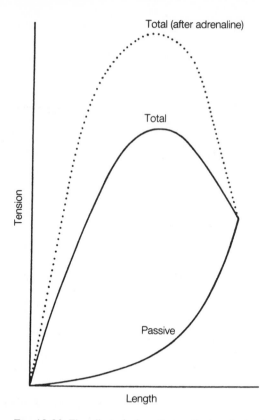

Fig. 13.20 The effect of adrenaline on the length–tension curve of a strip of cardiac muscle.

Load–velocity curve

As in skeletal muscle, if the afterload is increased, both the velocity of contraction and the amount of shortening are reduced (Fig. 13.21a). The effect of afterload on velocity for a range of afterloads is depicted by a **load–** or **force–velocity curve** (Fig. 13.21b; cf. Fig. 5.10). Maximum velocity of shortening. (V_{max}) occurs at zero load when the contraction is in fact isotonic. V_{max} is a measure of the state of myocardial contractility, as is the maximum (peak) rate of tension or force production. Note that contractile state is measured by the power of contraction, not the amount of developed tension. In an intact heart, as opposed to strips of cardiac muscle, a useful index of contractility is the peak rate of pressure developed in the left ventricle (dP/dt_{max}) during the isovolumetric (pre-ejection) phase of systole.

When the load–velocity relationship is examined after an increase in the initial preload, the velocity of shortening increases, especially with the larger afterloads (Fig. 13.21c). Adrenaline also increases the velocity of shortening for any particular afterload; this time the effect is more marked on the smaller afterloads (Fig. 13.21c). Velocity of shortening in the heart obviously relates to the *velocity of stroke volume ejection*.

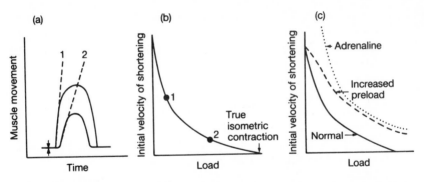

Fig. 13.21 (a) Measurement of initial velocity of shortening (distance/unit time, dashed line) for an isotonic contraction of a strip of cardiac muscle subjected to a small (1) and large (2) afterload. Arrow marks beginning of contraction. (b) Load–velocity curve. (c) Effect of increased preload or adrenaline on the load–velocity curve.

Volume–pressure relationships of the heart

By analogy with length–tension curves for strips of cardiac muscles, **volume–pressure curves** can be constructed for the heart (Fig. 13.22). Length is now analogous to ventricular volume, passive tension to diastolic ventricular pressure, and total tension to the maximum systolic ventricular pressure that could be developed isometrically at each volume. During diastole, the heart fills with blood, increasing in volume from an end-systolic volume of 60 ml to an end-diastolic volume of 130 ml, with an increase in pressure for the left ventricle from about 5 mmHg at A to about 10 mmHg at B. Note that the AV valve will be opening at A and closing at B.

During the isovolumetric (i.e. isometric) phase of ventricular contraction, the pressure increases to C. If the aorta was clamped so that blood could not escape, the pressure would continue to rise to C* on the curve for maximum systolic ventricular pressure. Note that for a normal end-diastolic volume, C* is well to the left of the maximum peak. Normally, the isovolumetric phase of ventricular contraction terminates at C when the aortic valve opens.

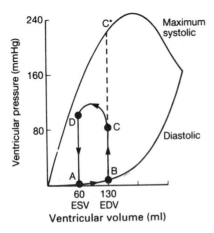

Fig. 13.22 Left ventricular volume–pressure curves depicting the normal cardiac cycle of ABCDA (see text). C* is the ventricular systolic pressure achieved by an isometric contraction if the aortic valve remains closed EDV, End-diastolic volume; ESV, end-systolic volume.

The pressure at C will depend on the diastolic pressures in the aorta. This is about 80 mmHg and during the ejection of blood the pressure increases to about 120 mmHg and then declines, reaching D at about 100 mmHg. From C to D, the contraction is referred to as **auxotonic**, not isotonic, because it is occurring against the afterload of a varying aortic pressure.

When the aortic valve closes at D, isovolumetric ventricular relaxation occurs and the pressure drops from D to A, at which point the AV valve opens because the ventricular pressure has dropped below that of the atria. The cardiac cycle is now completed. The *area* enclosed by the volume–pressure loop (ABCDA) is a measure of the **external work** done by the heart.

The volume–pressure loop of Fig. 13.22 has been obtained from the volume–time and pressure–time graphs of the ventricle given in Fig. 13.16. Note that in both figures positions A and B mark the opening and closing respectively of the AV valve, and positions C and D the opening and closing respectively of the aortic valve.

Law of Laplace

Length–tension curves of strips of cardiac muscle and volume–pressure curves of the whole heart are not identical because volume is proportional to length cubed (or radius cubed) and the tension within the muscle is not identical to the pressure in the chamber lumen. The relationship between wall tension (force per unit cross-sectional area of the wall) and lumen pressure (force per unit surface area) is described by the **law of Laplace** (Fig. 13.23).

For closed hollow organs with a circular cross-section, the total circumferential wall tension depends on the cross-sectional area of the wall (i.e. the circumference of the wall ($2\pi r$) multiplied by the thickness (u) of the wall), and on the wall tension (T). It equals $2\pi r u T$. The total pressure in the lumen depends on the cross-sectional area of the lumen (πr^2) and the transmural pressure (P_t), and equals $\pi r^2 P_t$. Transmural pressure is the inside pressure minus the outside pressure. At equilibrium, the total tension and total pressure must counterbalance each other, hence $2\pi r u T = \pi r^2 P_t$. Thus the law of Laplace states:

$$P_t = \frac{2Tu}{r} \quad \text{or} \quad T = \frac{P_t r}{2u}$$

With the law of Laplace, a number of properties of the heart can be explained:

1 The rise in ventricular pressure during the ejection phase (C to D in Fig. 13.22) is due not to increasing muscle tension but to the physical effect of a change in heart size. As the radius gets smaller and the wall thicker, additional ventricular pressure is generated for the same tension in the ventricular wall.

2 In a dilated heart, the large radius and thin walls contribute, together with the overstretching of the actin and myosin, to the decline in the maximum systolic ventricular pressure (Figs 13.19 and 13.22).

3 An excessively dilated heart caused, for example, by congestive cardiac failure, has to generate more total wall tension to develop a normal systolic ventricular pressure. This extra tension requires more O_2 consumption at a time when O_2 is being insufficiently transported

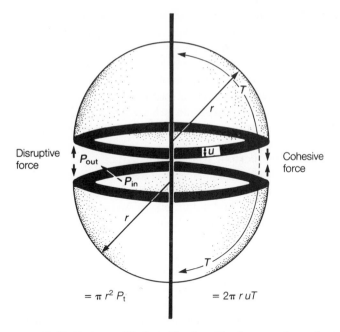

Fig. 13.23 The basis of the law of Laplace. The disruptive force, due to the lumen pressure (P_{in}), pushing the two hemispheres apart is opposed by the cohesive force, due to the wall tension (T), holding the sphere together. Note that $P_t = P_{in} - P_{out}$, where P_{in} is greater than P_{out}.

by the whole cardiovascular system and when the high cardiac wall tensions are compressing the coronary arteries more and for a longer period within the cardiac cycle (p. 450).

4 In trained athletes or in the chronically volume- or pressured-loaded heart, hypertrophy of the heart develops. This increase in myocardial wall thickness can compensate for the increased chamber dilatation, allowing a given pressure to be generated by the usual wall tension. However, since cardiac hypertrophy in the chronically loaded heart, in contrast to that of the trained athlete, is not accompanied by new coronary capillary formation, this may lead to insufficient O_2 delivery to the cardiac muscle.

13.6 CONTROL OF STROKE VOLUME

Stroke volume is controlled by mechanisms intrinsic to the heart and by the external influence of the sympathetic nervous system and adrenaline. Intrinsic control causes the heart to respond with a greater force of contraction, ejecting a larger stroke volume when the end-diastolic volume increases. This is Starling's law of the heart. End-diastolic volume can be altered by events in the chest and changes in blood volume or in venous capacity. Extrinsic control of stroke volume is dependent on increased Ca^{2+} entry, enhancing myocardial contractility.

Intrinsic control of stroke volume: Starling's law of the heart

Volume–pressure curves and the work loop concept were first described in 1895 by Frank using an isolated frog heart in which the aorta was clamped and the ventricles filled to different volumes between each beat. In 1914, Starling and his colleagues developed a mammalian heart–lung preparation in which the pulmonary circuit was intact and the lungs were mechanically ventilated (to keep the heart supplied with O_2) but the systemic circuit was replaced by a system of blood-filled tubes. The diameter of these tubes could be decreased to increase resistance and hence increase aortic pressure (afterload) and the height of the reservoir returning blood to the heart could be elevated to increase the filling pressure and hence increase end-diastolic volume (preload). Because the temperature of the blood was kept constant and the heart was denervated, heart rate remained constant.

An increase in end-diastolic volume (preload) causes the heart to begin its isovolumetric contraction at a higher pressure and volume (position B′ in Fig. 13.24). Thus the new volume–pressure loop is larger but with pressures at C′, D′ and A′ only slightly elevated. The end-systolic volume is slightly increased but the heart now operates at a larger end-diastolic volume and ejects a larger stroke volume. The ejection velocity will also be greater (Fig. 13.21c). The relationship between each end-diastolic volume and the resulting stroke volume (Fig. 13.25) is often referred to as the **Starling curve** or ventricular function curve and it illustrates the so-called **Frank–Starling law of the heart**. According to Starling, 'the mechanical energy set free on passage from the resting to the contracted state depends . . . on the length of the muscle fibres'. The heart thus has an intrinsic ability to control its stroke volume, responding with a greater force of contraction to the stimulus of increased diastolic stretch. The upper limit to this intrinsic ability is reached when the enlarged end-diastolic volume has resulted in the optimal myocardial length.

An increase in aortic pressure (afterload) requires that an equal increase in ventricular pressure must occur before the aortic valve can open (position C′ in (Fig. 13.26). Thus, with the remaining energy of this contraction the stroke volume is smaller in the first ejection against the increased afterload (dashed line). Hence, the end-systolic volume increases and with normal venous return *the end-diastolic volume subsequently increases*, stretching the ventricle so that at the next contraction more tension is developed and so on until a new steady state is reached. The final volume–pressure loop is shifted to the right, with A′, B′ and

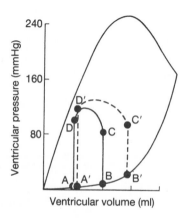

Fig. 13.24 Left ventricular volume–pressure curves depicting the normal cardiac cycle (ABCDA) and the effect of increasing end-diastolic volume from B to B′.

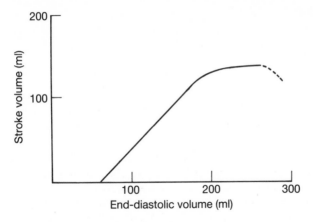

Fig. 13.25 The Starling curve—the relationship between end-diastolic volume and stroke volume (providing afterload is constant).

especially C' and D' pressures increased. The loop has a larger area, indicating more external work, and the heart now operates with a larger end-systolic and end-diastolic volume but with an unchanged stroke volume. However, this intrinsic mechanism of compensation is limited and with larger increases in afterload the eventual response is a decreased stroke volume and, as shown in Fig. 13.21c, depending on the increase in end-diastolic volume (preload) the ejection velocity may also be decreased.

The Starling curve is often plotted with other variables on its axes. On the x axis, end-diastolic muscle length and, providing ventricular compliance is unaltered, end-diastolic ventricular pressure or mean right or left atrial pressure are all used instead of end-diastolic volume. On the y axis, stroke work (i.e. stroke volume × mean aortic pressure), peak systolic ventricular pressure or cardiac output (if heart rate and afterload are constant) are all used instead of stroke volume. When cardiac output is plotted against mean right atrial pressure, the Starling graph is often called a **cardiac function curve** (p. 435).

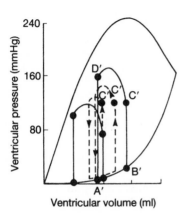

Fig. 13.26 Left ventricular volume–pressure curves showing final cardiac cycle A'B'C'D'A' and intermediate steps (dashed lines) in response to a sudden increase in afterload.

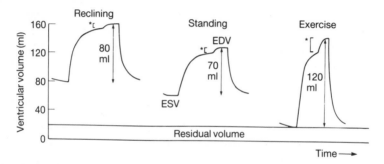

Fig. 13.27 End-systolic volume (ESV), end-diastolic volume (EDV) and the resultant stroke volume during reclining, standing and exercise in the upright posture. Note the residual volume that can never be ejected and the atrial contribution to ventricular filling (*).

The Starling mechanism operates whenever there is a change in diastolic filling of the ventricles to alter stroke volume and balance the outputs of the right and left ventricles. Its primary function is the latter—**to preserve the equality of right and left ventricular outputs**. Transiently these can be different but within a few beats this must be corrected.

For example, if right ventricular output into the pulmonary circuit were to exceed transiently that from the left into the systemic circuit, pulmonary filling of the left ventricle would increase and its distension, through the length–tension Starling mechanism, would raise left ventricular output. Indeed, at the beginning of inspiration the right ventricular output transiently exceeds the left because the right receives more blood due to the effect of the respiratory pump (p. 434). Another common situation that provokes a transient imbalance is the consequences of gravity (p. 459); on standing, the right output drops below that of the left for a few beats.

Stroke volume of both ventricles alters because of the Starling mechanism when passive mechanisms or venous contraction change the systemic **venous capacity**.

For example, the volume contained in the systemic venous system is decreased when reclining compared with standing (p. 415). This results in a greater central blood volume in the heart and lungs and the increased end-diastolic volume leads to an increased stroke volume (Fig. 13.27). Reflex contraction of smooth muscle in the veins also decreases venous capacity, for example in response to haemorrhage (p. 432) or in exercise where the effect is augmented by the skeletal muscle venous pump (p. 434). Decreased venous capacity contributes slightly to the increased end-diastolic volume of moderate exercise in the upright posture (Fig. 13.27).

Acute increases or decreases in **total blood volume** (transfusion or haemorrhage) also increase or decrease the end-diastolic volume and hence through the Starling mechanism alter stroke volume accordingly.

Factors influencing end-diastolic volume

The ventricular volume achieved by the end of diastole, the end-diastolic volume, can be altered by a number of factors, such as the time interval between contractions, the velocity of ventricular relaxation, the contribution of the atrial contraction to ventricular filling, the magnitude of atrial and ventricular pressures during diastole, the distensibility of the ventricle and the dilating effect of the negative intrathoracic pressure surrounding the heart.

The end-diastolic volume can be decreased by a **very high heart rate**, which reduces the time available for the slow phase of diastolic filling (*diastasis*). If at the same time there is an increase in the sympathetic drive to augment myocardial contractility, there will be a distinct increase in the rate of ventricular relaxation, which will increase the initial rapid phase of diastolic filling. Thus, some of the decrease in diastasis is compensated for by the rapid **relaxation recoil of the ventricles**. But, in this situation, as the high heart rate is coupled with an increase in cardiac output, the **resistance of the AV valves** to the large flow rate has a detrimental effect on all phases of filling. On balance, the end-diastolic volume in severe exercise remains close to resting values.

In moderate exercise there is an increase in end-diastolic volume, some of which is due to a sympathetic-induced increase in **atrial contractility** which, by increasing the a wave of atrial pressure, increases the atrial contribution to ventricular filling (Fig. 13.27).

A large increase in **atrial pressure** at any other time in the cardiac cycle reduces the pressure gradient, and hence blood flow, from capillaries to atria; this will tend to limit increases in end-diastolic volume. Similarly, an elevated **ventricular pressure** during diastole will restrict further filling of the ventricles (see diastolic pressure curve, Fig. 13.22). For any particular diastolic ventricular pressure, reduced distensibility or **compliance** of the ventricles—caused by hypertrophy, fibrosis or accumulation of fluid in the pericardial space (*cardiac tamponade*)—will result in a smaller end-diastolic volume.

The **intrathoracic pressure** (intrapleural pressure) surrounding the heart influences end-diastolic volume. This pressure is subatmospheric (p. 478); at rest it is about -3 mmHg at the end of expiration and about -5 mmHg at the end of inspiration but it becomes positive during large expirations and more negative during large inspirations. Positive intrathoracic pressure, particularly that produced by a forced expiration against a closed glottis (the **Valsalva manoeuvre**), for instance during defecation, reduces diastolic filling of the right ventricle. The negative intrathoracic pressure of inspiration enhances, in particular, right ventricular filling (p. 434).

Extrinsic control of stroke volume

In the 1950s, Sarnoff investigated the effect of **adrenaline** and **sympathetic** stimulation on the heart–lung preparation and demonstrated that the maximum systolic pressures which could be generated isometrically for each end-diastolic volume were increased (Fig. 13.28; note the similarity with the length–tension curve of Fig. 13.20). Activation of β_1-adrenoceptors increases Ca^{2+} availability intracellularly, thus enhancing the inotropic state, i.e. **myocardial contractility**, and **ejection velocity** (see also the load–velocity curve of Fig. 13.21c). An increase in sympathetic activity to the heart occurs during exercise, during circumstances when the mean arterial pressure has decreased, for example in haemorrhage, and when cardiac output has to be maintained against a high afterload.

The positive inotropic effect alters the volume–pressure loop in one of two ways (Fig. 13.28); one leads to an increase in the stroke volume and one maintains a normal stroke volume despite an increase in afterload.

In the first case (Fig. 13.28a), for a normal end-diastolic ventricular volume and pressure (B) and only a slightly larger than normal pressure (C') at the end of isovolumetric contraction, the left shift of the maximum systolic pressure curve

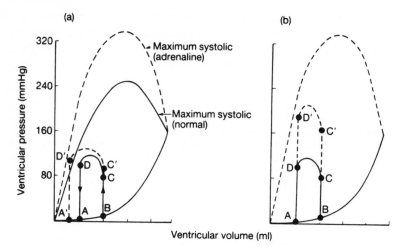

Fig. 13.28 Effect of adrenaline on the maximum systolic pressure of the ventricular volume–pressure curve permitting for the same end-diastolic volume at B either (a) an increased stroke volume with the cardiac cycle A'BC'D'A' or (b) a normal stroke volume maintained in the face of an increased afterload of the magnitude of C'.

allows position D' to occur at a slightly higher pressure and a very much lower volume. Thus, a *larger stroke volume has been ejected* without any prior change in end-diastolic volume. Its ejection velocity will also be high. The final end-systolic volume at D' (or A') is obviously smaller. Even at the highest level of sympathetic activity, the end-systolic volume will never become zero; there is a residual volume of about 20 ml that can never be ejected (Fig. 13.27). If the effects of different degrees of sympathetic activity on stroke volume are examined over a range of end-diastolic volumes, a family of Starling curves can be plotted (Fig. 13.29).

In the second case (Fig. 13.28b), if sympathetic stimulation has also caused an

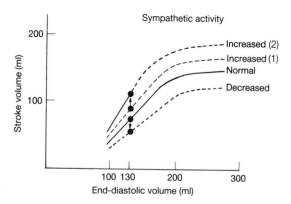

Fig. 13.29 Effect of alterations in sympathetic activity on the Starling curve and the resultant alteration in stroke volume for any particular end-diastolic volume (providing the afterload is constant).

increased afterload as a result, for example, of arteriolar constriction, the increased myocardial contractility will permit the high pressure of C' and D' to be reached without any change in end-diastolic volume or end-systolic volume. Thus the stroke volume is normal. Usually, the compensation for an increased afterload is a mixture of intrinsic (Fig. 13.26) and extrinsic control of stroke volume.

13.7 MEASUREMENT OF CARDIAC OUTPUT, STROKE VOLUME AND HEART RATE

Various invasive and non-invasive techniques are available for measuring cardiac output, stroke volume and heart rate so that, for instance, their values at rest and in exercise can be determined.

Cardiac output is the volume of blood pumped per minute by each ventricle and is thus the total blood flow through each of the pulmonary and systemic circuits. It is the product of **stroke volume** (volume per beat) and **heart rate** (beats per minute). Table 13.2 gives representative values for these variables at rest and near their maximum values. Note that increments in heart rate contribute more than stroke volume to the cardiac output of exercise, that the maximum heart rate and the minimum end-systolic volume are the same for non-athletes and trained athletes and that hypertrophy of the heart in the trained athletes allows their resting heart rate to be lower and their maximum cardiac output to be higher. Athletes can therefore supply more O_2 per minute to skeletal muscle and perform greater levels of exercise than non-athletes.

Cardiac output ($\dot{Q}$) can be measured experimentally by placing electromagnetic flowmeters or ultrasonic flow probes around the circumference of the aorta. Clinically, two alternative methods are used.

The **Stewart–Hamilton indicator dilution technique** involves injecting rapidly a dye, radioactive isotope or cold saline into the venous circulation in the right ventricle. The indicator quickly mixes with the cardiac contents and during the next few beats the entire blood–indicator mixture is pumped out of the heart into the circulation. The concentration of this mixture is measured by continual sampling from an artery in the arm for one complete circulation (about 30–40

Table 13.2 Representative cardiac variables at rest and during maximum exercise for non-athletes and highly trained athletes

	Cardiac output (L min^{-1})	Heart rate (beats min^{-1})	Stroke volume (ml)	EDV (ml)	ESV (ml)
Non-athlete					
Rest	5	70	70	130	60
Maximum exercise	21	190	110	130	20
Trained athlete					
Rest	5	40	120	200	80
Maximum exercise	34	190	180	200	20

EDV, end-diastolic volume; ESV, end-systolic volume.

seconds at rest). The mean concentration of the mixture for one complete circulation is determined and

$$\dot{Q} = \frac{\text{Amount of indicator injected}}{\text{Mean concentration of indicator} \times \text{duration of one circulation}}$$

The second method uses the **Fick principle**. This states that the amount of a substance taken up by an organ (or the body) per unit time is equal to the blood flow multiplied by the difference in concentration of that substance between arterial and mixed venous blood. In practice, the average steady-state oxygen consumption ($\dot{V}_{O_2}$) of the whole body is measured (p. 500) for about 15 minutes, during which time blood samples are taken from a systemic artery and pulmonary artery (the latter contains mixed venous blood). The samples are then analysed (p. 524) for the O_2 content of arterial (C_aO_2) and mixed venous ($C_{\bar{v}}O_2$) blood and $\dot{Q}$ is calculated from:

$$\dot{Q}\ (\text{L min}^{-1}) = \frac{\dot{V}_{O_2}\ (\text{L min}^{-1})}{C_aO_2 - C_{\bar{v}}O_2\ (\text{litres of } O_2 \text{ per litre of blood})}$$

Stroke volume can be determined by dividing the measured value of cardiac output by the average heart rate (from an ECG record or arterial pulse). With new techniques, stroke volume itself can now be measured by the non-invasive procedures of impedance cardiography, echocardiography, radionuclide imaging and by a method depending on the Doppler effect and pulses of ultrasound.

Chapter 14
Vascular System

14.1 PHYSICS OF BLOOD FLOW

The rate of blood flow (volume per unit time) through a vessel depends on the pressure gradient from one end to the other and the resistance encountered (Poiseuille equation). Resistance is determined by the vessel's dimensions (radius and length) and the blood's viscosity. Flowing fluid has three forms of energy—kinetic, potential and gravitational. Less resistance is offered to blood flow when vessels are arranged in parallel rather than in series. When the total cross-sectional area of parallel vessels is large, as in capillaries, the velocity of flow (distance per unit time) is slow. The cross-sectional area and length of each set of vessels determine the volume of blood that each contain. Because veins have very compliant walls their capacity is also affected by changes in wall compliance and by the pressure across the wall.

Before examining in detail the organization of the vascular system and the functions of each type of blood vessel, the fundamental physical concepts governing blood flow through tubes in series and in parallel will be considered.

Flow, pressure, resistance and velocity
Liquid flows along the lumen of a rigid tube from a higher to a lower hydrostatic

pressure (but see p. 408) and the **rate of flow** ($\dot{V}$, volume/unit time) is directly proportional to the **hydrostatic pressure gradient** (ΔP). In the vascular system, blood pressures (force/unit area) are usually expressed in mmHg, although other units are also used (1 mmHg ≈ 1.36 cmH$_2$O ≈ 133 Pa). The dimensions of the tube and the nature of the fluid determine the **resistance** (R) to flow; for a given ΔP the greater the resistance the less the flow. These determinants of flow can be summarized in an equation analogous to Ohm's law for electrical circuits:

$$\dot{V} = \frac{\Delta P}{R}$$

If $\dot{V}$ and ΔP are in units of L min^{-1} and mmHg, respectively, then the calculated values of R have the units mmHg L^{-1} min^{-1}.

Resistance

Resistance results from the *friction* (viscous forces) between the molecules or particles of the fluid as they move and from the friction between this fluid and the walls of the tube. During steady streamline flow, an infinitesimally thin layer of fluid in contact with the wall does not move whilst the next layer moves slowly with the most axial (central) stream moving at the fastest rate. Such layered or *laminar flow*, parallel to the axis of the vessel, thus has a *parabolic velocity profile* (see Fig. 14.2). The concept of **viscosity** (η) of a fluid expresses the fact that adjacent layers interact rather than slip with infinite ease over one another. The greater the viscosity of a fluid the greater the resistance to flow.

Since friction is greatest between the fluid and the tube wall, the dimensions of **length** (l) and **radius** (r) of the tube affect the resistance to flow. The longer the tube or the smaller its radius, the greater is the resistance offered. These determinants of resistance are summarized as:

$$R = \frac{8\eta l}{\pi r^4}$$

Note that resistance is affected far more by changes in radius than by changes in length or in viscosity; a two-fold decrease in radius will cause a 16-fold increase in resistance.

Hagen–Poiseuille equation

Introducing the resistance term into the Ohm's law equation gives the **Hagen–Poiseuille equation** which describes the relationships between flow, the pressure gradient and the determinants of resistance for an unbranched tube:

$$\dot{V} = \frac{\Delta P \pi r^4}{8\eta l}$$

Since for any blood vessel the vessel length and blood viscosity are relatively constant, flow through it can be increased only by increasing its radius, the pressure gradient or both. Only arterioles can achieve marked changes in radii and alter resistance. A change in flow through other types of blood vessel requires a change in the pressure gradient.

The Hagen–Poiseuille equation also indicates that at a constant flow an increase in an arteriole's radius will result in a decrease in the pressure gradient along it. This, as will be explained later (p. 421), may decrease the pressure upstream (arteries) and increase the pressure downstream (capillaries).

Velocity

Fluid also flows at a **mean linear velocity** (v, distance/unit time) which is proportional to the **rate of flow** ($\dot{V}$) through the tube and inversely proportional to the tube's **cross-sectional area** (πr^2). Thus:

$$v = \frac{\dot{V}}{\pi r^2}$$

For a given $\dot{V}$, the narrower the tube the smaller is its cross-sectional area and thus the faster will be the velocity through it.

Vessels in series and in parallel

In the vascular system, it is important to consider not only an individual vessel but also an entire network. Where the vessels are arranged **in series**, the total resistance to flow through all the vessels is the sum of all the individual resistances, whereas, where they are arranged **in parallel**, the reciprocal of the total resistance is the sum of all the reciprocals of the individual resistances (Fig. 14.1). For vessels in parallel therefore the total resistance is considerably smaller than the resistance of each individual vessel. Arteries, arterioles, capillaries, venules and veins are, in

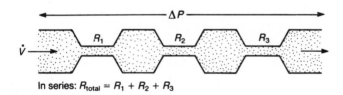

In series: $R_{total} = R_1 + R_2 + R_3$

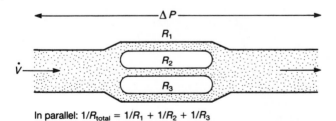

In parallel: $1/R_{total} = 1/R_1 + 1/R_2 + 1/R_3$

Fig. 14.1 Upper: resistances (R_1, R_2, R_3) arranged in series. Addition of a fourth resistance in series would increase the total resistance. Lower: resistances arranged in parallel. Addition of a fourth resistance in parallel would decrease the total resistance. (The resistance of the connecting tubes is assumed to be negligible.)

general, arranged in series with respect to one another (see Fig. 13.2). However the vascular supply to the various organs (see Fig. 13.1), and the vessels, e.g. capillaries, within any organ (see Fig. 13.2), are arranged in parallel.

For tubes in series, provided they are not branched, application of the Poiseuille equation reveals that the same volume/unit time ($\dot{V}$) will flow sequentially from one tube to the next with a pressure gradient along each tube that is inversely related to the fourth power of the radius. Thus the narrower the tube, the greater the pressure drop as fluid flows through it. As the velocity equation indicates, the velocity will be highest in the narrowest tube.

For tubes in parallel, through which fluid is flowing at the same $\dot{V}$ as in the series example above and whose total cross-sectional area is the same as the feeder tube, the combined resistance of the parallel tubes and the pressure gradient along them are smaller than in the series example. Furthermore, the pressure gradient across each of the parallel tubes will be the same and the flow through each tube will be directly proportional to the fourth power of the radius.

Since the flow through each parallel tube is different, the velocity in each is derived by substituting from the Hagen–Poiseuille equation for $\dot{V}$ in the velocity equation. Thus velocity = $\Delta P\, r^2/8\eta l$ and, in each individual parallel tube, velocity will be directly proportional to the square of its radius. Thus the narrower the parallel tube, the slower the fluid's velocity.

If the parallel tubes, as in the capillary network, constitute an increase in total cross-sectional area, the mean linear velocity through them will be slower.

Fluid energy

As described by **Bernoulli's equation**, the **total fluid energy** (E) per unit volume (in kg m^{-1} s^{-2}) in a fluid moving in a *horizontal vessel* is the sum of its **kinetic energy** and **potential energy**. Kinetic energy is dependent on the fluid's density (ρ, in kg m^{-3}) and the square of the mean linear velocity (v, in m s^{-1}) such that kinetic energy = $\frac{1}{2}\rho v^2$. Potential energy (P) is the hydrostatic pressure (mmHg converted to kg m^{-1} s^{-2}) at a particular point along the length of the vessel.

If a fluid suddenly flows from a narrower vessel into a much wider vessel, the large decrease in velocity (see velocity equation, above) causes a large decrease in kinetic energy. However, as there will be little change in total energy, there is an increase in the hydrostatic pressure (referred to here as **lateral pressure**) in the wider vessel. This becomes important in the cardiovascular system whenever there is a pathological dilatation (*aneurysm*) in a vessel.

If the fluid is moving in a *vertical vessel*, the total fluid energy in Bernoulli's equation is now the sum of the kinetic energy ($\frac{1}{2}\rho v^2$), the potential energy due to hydrostatic pressure (P) and the **gravitational potential energy**. Thus:

$$E = \frac{1}{2}\rho v^2 + P + \rho gh$$

in which gravitational potential energy equals ρgh where ρ is the density, g the acceleration due to gravity and h the height above (+) or below (–) the heart, the site of energy generation.

Thus in the cardiovascular system blood raised above the heart has a positive gravitational potential energy while its hydrostatic pressure potential energy becomes less by an amount related to its height above the heart (see p. 457). The converse occurs below the heart. Hence the change in gravitational potential energy is equal but opposite to the change in hydrostatic pressure potential energy. It follows that the total fluid energy of blood in a particular vessel is the same whether the vessel is aligned horizontally or vertically.

Deviations from predicted blood flows

The Hagen–Poiseuille equation was formulated for laminar flow of a homogeneous fluid with constant viscosity through a rigid unbranched tube. These characteristics are not typical of the vascular system.

1 *Flow is not always laminar.* Under certain conditions, flow can become **turbulent**. Turbulent flow is characterized by eddies of fluid moving not only parallel to the overall direction of flow but also across it and counter to it, and this flattens the velocity profile (Fig. 14.2). When there is turbulence, the volume flow becomes proportional to the square root of the pressure gradient. In other words, for turbulent flow the pressure gradient must increase fourfold to generate a doubling of flow (Fig. 14.2). Turbulence always occurs transiently in the aorta or pulmonary artery during early systole. It can occur in large arteries if the velocity is high and exceeds a critical value of about 40 cm s^{-1} (e.g. in severe exercise). Note also that in arteries, because of the rhythmic ejection of the heart, the laminar flow is also **pulsatile** and, because of the rhythmic pausing of the heart, the velocity profile is flat, rather than parabolic, which lowers mean velocity.

If blood viscosity is low (e.g. in severe anaemia) or there are pathological irregularities of the vascular wall (sclerosis), turbulence occurs at a lower critical velocity. The noise of turbulent flow can often be heard with a stethoscope.

2 *Blood is not a homogeneous liquid with a constant viscosity* inasmuch as it comprises cells and plasma; the former, in particular the red blood cells, are

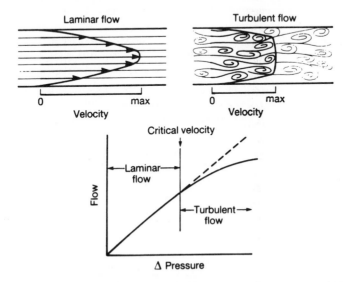

Fig. 14.2 Upper: velocity profiles of laminar and turbulent flow. Lower: relationship between pressure gradient and flow rate (or velocity) and the transition from laminar to turbulent flow once a critical velocity is reached.

responsible for most of the blood's viscosity. Since a greater proportion of blood cells travels in the axial (central) stream of flow, there is a greater proportion of the less viscous plasma near the vessel wall (which flattens the parabolic velocity profile). Axial streaming is more evident as velocity increases so the **effective viscosity** of blood is less at high velocities and greater at low velocities. At low velocities red blood cells aggregate into rouleaux, causing even larger increases in effective viscosity.

Under pathological conditions, low velocities can occur even in large vessels, either as a result of a failing heart or in a vessel distal to a pathological constriction or due to immobilization of limbs. The resulting rouleaux formation of red blood cells and consequential large increase in effective viscosity will reduce the velocity still further.

Since blood flows through arterioles and capillaries at a low velocity, the effective viscosity would be expected to be high. However, this does not occur because of an unexplained phenomenon in which the effective viscosity of any fluid suspension decreases considerably with decreasing tube radius when it flows through tubes smaller than 100 µm in radius (**Fahraeus–Lindqvist** effect). The effective viscosity of capillary blood is also decreased because erythrocytes travel through in single file, a phenomenon referred to as **plug flow**. Furthermore, in capillaries, the effective viscosity of blood depends on the **deformability** of erythrocytes since they are often larger than the capillaries they traverse. In sickle-cell anaemia, the desaturated haemoglobin can become crystalline and this reduces considerably erythrocyte deformability.

Note that blood viscosity increases in hyperproteinaemia and when there is an increased concentration of erythrocytes (increased haematocrit, e.g. at altitude, p. 551). In these situations the heart will have to do more work to maintain normal blood flow. In all types of anaemia (p. 339) when severe enough, blood viscosity decreases.

3 *Blood vessels are not rigid* but are elastic distensible tubes. When intraluminal pressure increases, vessels are **passively stretched** to varying degrees; this is most marked in systemic veins and in all pulmonary vessels. This passive increase in radius allows a disproportionate increase in blood flow as pressure increases (Fig. 14.3a). If such passive changes in radius are not taken into consideration, the pressure–flow relationship will appear to deviate from the predictions of the Hagen–Poiseuille equation.

In small blood vessels, especially in systemic arterioles, the pressure–flow curves do not pass through the origin but intersect at a positive pressure (normally about 20 mmHg for an arteriole) called the **critical closing pressure** (Fig. 14.3b). Below this pressure, flow ceases. The cause of this phenomenon is unknown. This critical closing pressure in arterioles alters according to the radius; it increases when arterioles constrict and decreases when they dilate.

Under pathological conditions, e.g. in shock, the pressure generated by the heart may not exceed the critical closing pressure of a vessel so that blood flow to an organ is compromised. Furthermore, concurrent constriction of the arterioles, as part of the reflex to maintain arterial blood pressure, may actually raise the critical closing pressure. This reduces the blood flow still further.

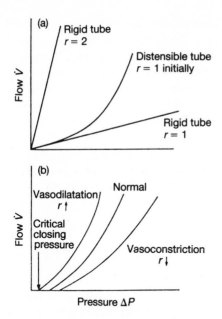

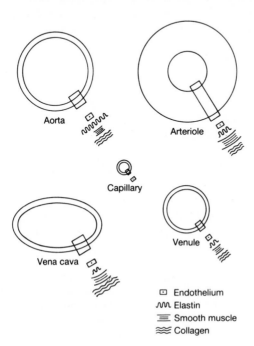

Fig. 14.3 Pressure–flow relationships (a) in rigid tubes of two different radii and in a distensible tube, and (b) in cutaneous arterioles as a function of arteriolar radius; note the critical closing pressure.

Properties of blood vessel walls

Elastance, compliance and capacitance

Blood vessels are lined by endothelium and in addition, except in capillaries, there are varying amounts of elastin, collagen and smooth muscle in the wall (Fig. 14.4).

Fig. 14.4 Relative dimensions of the internal lumen and wall thickness of blood vessels (arteriole, capillary and venule drawn to a scale 500 × that of the aorta and vena cava). The relative proportions of components of the vessel wall are also shown.

Elastin, and to some extent smooth muscle, allow the vessel to stretch while collagen limits the extent of stretching. These effects can be examined by taking an isolated segment of a blood vessel tied at both ends and determining the luminal pressure generated at various volumes (Fig. 14.5). As the vessel is stretched it exerts **elastic tension** which raises the luminal pressure. The elastic properties can be described by the term **elastance** ($\Delta P/\Delta V$), whereas the distensibility of the vessel is expressed as the **compliance** ($\Delta V/\Delta P$). When the wall is easy to stretch, its elastance is small and its compliance large.

Arteries and the aorta have a moderate elastance and compliance at normal physiological pressures but at pressures higher than 200 mmHg they become overfilled and their walls become rather rigid and incompliant. With ageing, the major arteries become infiltrated with fibrous tissue and therefore at all pressures they are less distensible, i.e. stiffer.

Veins have less elastin than arteries (Fig. 14.4) so that at high volumes where a vein wall is actually being stretched (Fig. 14.5) the vein is very stiff. However, below this range, the vein is very distensible because the cross-sectional profile changes from the flattened ellipse of low volumes to the circular shape of higher volumes. This ability of veins to increase markedly their volume at low pressures explains why veins are said to have a high **capacitance** and are described as the capacitance vessels. The term capacitance in the cardiovascular system has no strict definition and is sometimes used interchangeably with the term compliance. Venous compliance, and hence venous capacity, can be decreased by sympathetic activity.

In the range of pressures and volumes usually encountered in each vessel, arteries are about 10-fold (systemic) and twofold (pulmonary) less distensible than systemic and pulmonary veins. On the other hand, arterioles and capillaries, because they contain little or no elastin, have low distensibilities; they are therefore relatively rigid.

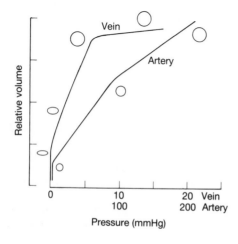

Fig. 14.5 Pressure–volume relationships in isolated segments of a vein and artery.

Blood vessels and the law of Laplace

Strictly speaking, in Fig. 14.5 one should consider the **transmural pressure**, P_t (i.e. the pressure on the inside minus the pressure on the outside of the wall), rather than the luminal pressure. An elastic tube will distend if the inside pressure is higher and collapse if the outside pressure is higher. In the body the outside pressure is the hydrostatic pressure of the interstitium and is usually close to zero (p. 428). Thus a change in the transmural pressure will alter the radius (r) of the tube and this affects the total circumferential wall tension. For hollow open-ended cylinders, as opposed to spherical organs like the heart, there is only one radius of curvature, u is the thickness of the wall and T is the **passive wall tension** per unit cross-sectional area of the wall. The total circumferential wall tension ($\pi r u T$) at equilibrium is counterbalanced by the total pressure in the lumen ($\pi r^2 P_t$) as described by the **law of Laplace** (p. 396). Thus:

$$T = \frac{P_t r}{u} \quad \text{or} \quad P_t = \frac{Tu}{r}$$

The tension in the wall of a blood vessel, therefore, depends on the radius, thickness of the wall and the transmural pressure. For a given transmural pressure, the tension in the wall is greater the larger the radius and the thinner the wall. The vessel's radius and wall thickness are appropriate for the transmural pressure to which the vessel is usually subjected (Table 14.1).

Note in Table 14.1 that a capillary with a small radius and low transmural pressure requires only a thin wall to sustain its lesser tension. On the other hand, an artery with its large radius and high pressure will have a higher tension and a need for a thicker wall. The thick wall of a vein is not related to a high transmural pressure as venous pressures are low but is related to the presence of elastin and smooth muscle that are required for changing venous compliance.

When a particular vessel is subjected to an increase in transmural pressure, its distensibility (Fig. 14.5) will result in an increase in the radius and a decrease in the thickness leading to increases in wall tension. On passive stretching, the elastin, collagen and smooth muscle withstand these increased tensions but only within the physiological range of transmural pressures; beyond this, the wall will tear.

Table 14.1 Passive wall tension and wall stress in blood vessels expressed in N m^{-1} ($T = P_t r$) and in kN m^{-2} ($T = P_t r/u$) respectively

Vessel	Internal radius	Wall thickness	Transmural pressure (mmHg)	Transmural pressure (kPa)	Wall tension (N m^{-1})	Wall stress (kN m^{-2})
Aorta	13 mm	2.0 mm	100	13.3	173	86.7
Artery	2 mm	1.0 mm	85	11.3	23	22.7
Arteriole	15 μm	20 μm	65	8.7	0.13	6.5
Capillary	4 μm	1 μm	25	3.3	0.01	13.3
Venule	15 μm	2 μm	15	2.0	0.03	15.0
Vein	2 mm	0.5 mm	10	1.3	2.7	5.3
Vena cava	15 mm	1.5 mm	5	0.7	10	6.7

If an area of an arterial wall becomes weakened (for example, in severe arteriosclerotic disease) it is more distensible, resulting in a bulge or aneurysm. The increasing radius and progressive thinning of this bulge will result in higher tensions (law of Laplace) in an already weakened wall. Furthermore, the slower velocity of flow at this point will result in a greater lateral pressure (Bernoulli). Thus rupture may eventually occur.

In arterioles and smaller veins, which possess sufficient smooth muscle for their contraction to decrease the radius of the vessel, the reduction in radius and increase in wall thickness will reduce the resulting passive wall stress (law of Laplace). Thus the active tension generated by contraction of the smooth muscle will not elevate the total tension as much as expected and the resultant effect of raising transmural pressure on total tension will be dampened.

Relationships between flow, pressure, resistance, cross-sectional area, velocity and volume throughout the vascular system

These variables have already been considered for vessels in series and in parallel (p. 407) and will now be examined in relation to the vascular system as a whole (Fig. 14.6). The *parallel branching* of the blood vessels is such that there is a rise

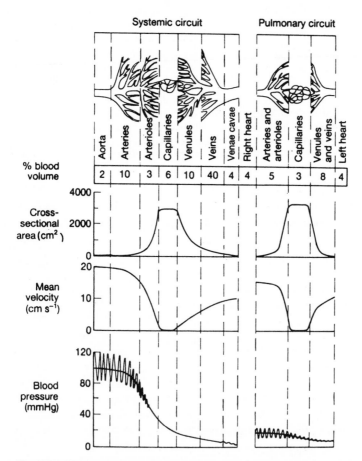

Fig. 14.6 Relationships between percentage required blood volume, total cross-sectional area, mean velocity and blood pressure in the systemic and pulmonary circuits at rest and in the supine position.

in the **cross-sectional area**, greatest in the capillaries and of moderate proportions in the venules and small veins.

The **percentage of total blood volume** that is accommodated in each set of vessels is determined by the cross-sectional area and the length of individual vessels. Total blood volume is about 5–6 and 4–5 litres, respectively, in average men and women (p. 324). In the supine position at rest, about 75% of this blood is in the systemic circuit, about 8% in the heart and about 16% in the pulmonary circuit. The volume in the heart and pulmonary circuit is referred to as the **central blood volume**. The aorta and systemic arteries contain about 12% of the blood but there is only about 3% in the systemic arterioles and about 6% in the systemic capillaries for, despite their large total cross-sectional area, these vessels are very short. Most of the blood (about 55%) is accommodated in the systemic venous system, indicating its importance as a **blood reservoir**. In the pulmonary circuit, about 5% of the total blood volume is in the arterial system, 3% in the capillaries and 8% in the veins.

During standing at rest (p. 457), the cross-sectional areas are slightly different and about 6% of the total blood volume is in the heart and about 9% in the pulmonary circuit, resulting in an increase to about 65% in the systemic venous system. During exercise (Chapter 21), there will be a greater proportion of blood in the capillaries, venules and, to a lesser extent, arterioles of skeletal musculature, as well as in the heart and in the pulmonary circuit. The proportion in the systemic venous system will be reduced correspondingly.

The **velocity** in a particular type of vessel will be the volume flow rate divided by the total cross-sectional area. Flow is *pulsatile* in the aorta and arteries and hence the velocity varies above and below a mean of about 20 cm s^{-1} (Fig. 14.6). Flow becomes *non-pulsatile* partway through the arterioles and is very slow (0.05 cm s^{-1}) in the capillaries. This low velocity is necessary to allow sufficient time for adequate diffusion between blood and cells across the capillary wall. In the venae cavae the mean velocity increases to about 10 cm s^{-1}, not 20 cm s^{-1} as in the aorta, because their total cross-sectional area is about twice that of the aorta.

During exercise, increases in cardiac output can cause up to five-fold increases in velocity in the arteries and veins but in the capillaries of skeletal musculature a concomitant increase in cross-sectional area results in very little increase in velocity (p. 455).

Blood pressure is pulsatile in the aorta and arteries reaching, in the systemic circuit at rest, a systolic peak of about 120 mmHg and a diastolic trough of about 80 mmHg in a healthy young adult (see Figs 13.16 and 14.6). Since pressure is related to volume flow and resistance (Poiseuille's equation), the drop in mean pressure along each set of vessels will indicate their relative **resistance** to flow.

Because of their large radii, the aorta and large arteries offer very little resistance to flow and the mean pressure along them drops only from 100 to 95 mmHg. As the arteries get narrower the pressure drop becomes bigger (from 95 to 75 mmHg). The greatest resistance and hence the largest fall in pressure (from 75 to 35 mmHg) occurs along the arterioles and there is a progressive transition from pulsatile to non-pulsatile pressure. Despite the even smaller radii of the capillaries, their total resistance is only about half that of the arterioles and thus the pressure drop is smaller (from 35 to 15 mmHg). The relatively smaller total resistance of the capillaries is a result of their vast parallel network. The pressure fall along the venous system is small (15 to 1–2 mmHg) because, although veins and arteries are of a similar size, the veins are more numerous and arranged in parallel. In the large veins, the blood pressure becomes slightly pulsatile from the action of the right atrium and nearby arterial pulsations (p. 390).

Thus, of the total resistance to flow in the systemic circuit at rest the aorta and arteries constitute about 25%, the arterioles about 40%, the capillaries about 20% and the venous system about 15%. The combined resistance to flow of all the parallel vascular beds of the systemic circuit is termed the **total peripheral resistance**. With a total pressure drop of ~100 mmHg, from aorta to right atrium, and a cardiac output of ~6 L min^{-1}, this resistance amounts to ~17 mmHg L min^{-1}. In the pulmonary circuit, in contrast, with a total pressure gradient of ~10 mmHg, there is **pulmonary resistance** of ~1.7 mmHg L min^{-1}.

All the pressures given in Fig. 14.6 are for the supine position when the vessels are at the level of the heart. In the vertical position the pressures are affected by gravity (p. 457). Pressures and resistances within the pulmonary circuit are considered in Chapter 15 (p. 447).

During exercise the profiles in Fig. 14.6 will change because the increased force of ventricular contraction and the consequent increase in cardiac output will elevate the mean arterial blood pressure (p. 691). There is also an increase in the radius of arterioles supplying skeletal musculature. This increase dominates the systemic response resulting in a decrease in total systemic arteriolar resistance. In the working muscle the smaller drop in pressure along its arterioles will elevate mean pressures in its capillaries and veins.

14.2 CHARACTERISTICS OF THE SYSTEMIC ARTERIAL CIRCULATION

The aorta and large arteries are highly elastic and their stretching in systole and recoil in diastole converts intermittent flow of blood from the heart into a continuous pulsatile flow through the vessels. The maximum and minimum pressure in the main arteries are referred to as the systolic and diastolic blood pressures, respectively. They can be measured with a sphygmomanometer. These pressures are determined to varying degrees by the pumping of the heart, arterial distensibility and the resistance of blood vessels, particularly the arterioles, further downstream.

Since blood enters the aorta at high velocity, *flow* tends to be *turbulent* and since it enters only during systole, *flow, velocity* are *pressure* are pulsatile (see Figs 13.16, 14.6 and 14.7). Although two-thirds of the stroke volume is ejected during the first third of systole, such that aortic flow and velocity reach a peak early in systole, the peak of the blood pressure pulse occurs later (Fig. 14.7). This peak in pressure is delayed because the aorta is not a rigid tube and its *elastic stretching* modifies the pressure pulse.

The aorta stretches in systole to accommodate about 50% of the stroke volume whilst the other 50% flows on into the peripheral vessels (Fig. 14.8). During diastole the stretched area *elastically recoils*, propelling the accommodated blood on to the next arterial segment which stretches as it fills, and so on. When the elastin stretches, the kinetic energy of liquid motion is converted into

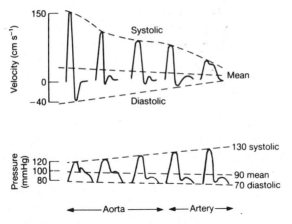

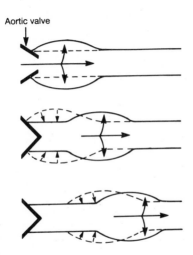

Fig. 14.7 Changes in velocity (flow pulse) and in pressure (pressure pulse) in the arterial system at increasing distances away from the heart. (After McDonald, D.A. (1974) *Blood Flow in Arteries*, 2nd edn. Edward Arnold, London.)

potential energy. When the elastin recoils it converts potential energy back to kinetic energy. This prevents the aortic pressure from dropping to zero and causes it to fall only gradually to a minimum of about 80 mmHg. The maintenance of a relatively high pressure during diastole thus ensures that the blood accommodated in systole is propelled out to the periphery during diastole. Elastic recoil thus converts the intermittent flow from the heart into a continuous, albeit pulsatile, flow through the arterial system.

The forward movement of blood itself (**flow pulse**) has a mean velocity of about 20 cm s^{-1} in the aorta, *decreasing* to about 15 cm s^{-1} in a small artery (Fig. 14.6). In comparison, the **pressure pulse** (pulse-wave) is transmitted through the column of blood and along the vessel walls at a very high mean velocity of about 4 m s^{-1} in the aorta, *increasing* to about 12 m s^{-1} in small arteries. The velocity of this pressure pulse is higher:

Fig. 14.8 Progression along aorta of alternating sequence of elastic stretching, accommodating about 50% of the stroke volume, followed by elastic recoil propelling blood to the next segment. Large arrows indicate direction of blood movement; small arrows indicate direction of elastic recoil.

1   the less viscous the blood;
2   the greater the mean blood pressure;
3   the more rigid or thicker the vessel wall; and
4   the smaller the lumen radius.

With increasing age, the greater stiffness of arterial walls results in an increased pulse-wave velocity, as also does the overstretching of arterial walls in hypertension.

The velocity, and therefore the amplitude, of the flow pulse decreases at increasing distances from the heart (Fig. 14.7) because branching increases arterial cross-sectional area. In contrast, the increase in velocity and amplitude of the pressure pulse occurs because first, the diastolic pressure decreases as a result of the pulse-wave losing energy from the alternating transfer between kinetic and potential energy and second, the systolic pressure increases as a result of complex fluid dynamics. The latter includes the reflection of energy back towards the heart due to the decreased elasticity of the smaller arteries and from reflections at branch points. Such factors also dampen the sharp incisura of the aortic pressure wave and convert it to the artery's smaller **dicrotic notch** and the distinct **dicrotic wave** (Figs. 14.7 and 14.9).

Clinically, these arterial pressure pulses can be felt, giving information not only about heart rate and its regularity but also about stroke volume and arterial distensibility. The tension or hardness of the pulse reflects the pulse pressure. Electromechanical transducers placed on the skin over an artery can yield details about the shape of the pulse and provide more precise information.

Measurement of arterial blood pressure

The maximum to which arterial pressure rises is called the **systolic blood pressure** (at rest the range is 100–140 mmHg at 20 years of age) and the minimum to which it falls is the **diastolic blood pressure** (range 50–90 mmHg).

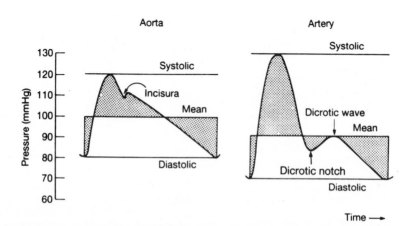

Fig. 14.9 Determination of mean arterial blood pressure by integrating the pressure fluctuation against time, i.e. the shaded areas above are equal to the shaded areas below the mean pressure.

With increasing age diastolic and, in particular, systolic pressures increase; the latter is due to loss of arterial elasticity. The difference between systolic and diastotic pressure is the **pulse pressure.**

The **mean arterial blood pressure** is calculated by integrating the pressure against time (Fig. 14.9). For the aorta the mean arterial blood pressure (about 100 mmHg) is approximately the arithmetic mean of the systolic and diastolic pressure or, expressed another way, the diastolic pressure plus half the pulse pressure. In a peripheral artery, which has a different pressure–time contour, it is approximately the diastolic pressure plus one-third of the pulse pressure (about 95 mmHg).

The systolic and diastolic arterial blood pressures can be measured directly by inserting a liquid-filled catheter into the appropriate artery and recording with a pressure transducer. Clinically, they are measured indirectly by a **sphygmomanometer** (Fig. 14.10). This comprises an inflatable rubber cuff, covered by a layer of non-distensible fabric, which is usually wrapped around the upper arm at the level of the heart and connected to a mercury manometer. The cuff pressure is altered by pumping air into the cuff and releasing it through a needle valve. A stethoscope is placed distal to the cuff over the branchial artery in the antecubital fossa. The cuff is then inflated to a pressure higher than the expected systolic pressure compressing the brachial artery so that no blood flows through it. When the cuff pressure is slowly reduced, the degree of arterial compression progressively decreases and varying sounds (**Korotkoff sounds**) audible with the stethoscope result from *intermittent* and *turbulent* blood flow through the artery. Four phases of sound followed by a fifth phase of silence can be distinguished.

When the cuff pressure has fallen to just below the systolic pressure, a clear, but often faint, tapping sound suddenly appears in phase with each cardiac contraction. The tapping sound

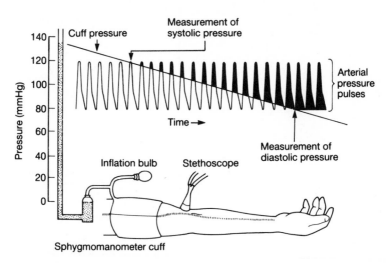

Fig. 14.10 Use of sphygmomanometer and stethoscope to measure indirectly systolic and diastolic pressure at the appearance and disappearance, respectively, of the Korotkoff sounds. (After Rushmer, R.F. (1970) *Cardiovascular Dynamics*, 3rd edn. W.B. Saunders, Philadelphia.)

is produced by the transient and turbulent blood flow through the artery during the peak of each systole. The systolic pressure is defined as that cuff pressure at which the tapping sound is first heard. During the next ~15 mmHg fall in the cuff pressure, the tapping sound becomes louder (phase I). In the following ~20 mmHg fall, the sound becomes quieter with a murmuring quality (phase II) and may suddenly disappear in the latter part of this phase (the auscultatory gap). In the next ~5 mmHg fall in cuff pressure, the sound of the murmuring becomes very loud and thumping (phase III). In the following ~5 mmHg fall, the sound becomes muffled and rapidly grows fainter (phase IV) and finally the sound disappears (phase V). When blood flow velocity is high, for example in exercise, the beginning of phases IV and V may be separated by 40 mmHg or more. The beginning of phase IV (muffling) and of phase V (disappearance) are used to measure diastolic pressure. The diastolic pressure is usually defined as the cuff pressure at which muffling, not disappearance, occurs, although this probably overestimates diastolic pressure. However, if there is an obvious difference at rest between these, both values are reported.

Determinants of arterial blood pressure

Mean arterial blood pressure depends on the product of cardiac output and total peripheral resistance (p. 439). **Systolic pressure** is affected mainly by stroke volume and in particular by ejection velocity. **Diastolic pressure** is affected mainly by total peripheral resistance and the time allowed for blood to flow out of the arteries (i.e. the duration of diastole which is determined by the heart rate).

Systolic pressure is seen to increase (Fig. 14.11) when there is:

1 an increase in diastolic pressure of the previous pulse;

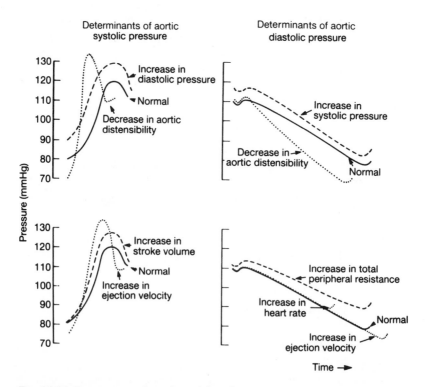

Fig. 14.11 Determinants of systolic and diastolic pressure in the aorta. Effect of each factor shown when all other factors are held constant.

2 an increase in stroke volume;

3 an increase in ejection velocity (without a change in stroke volume); or

4 a decrease in aortic or arterial distensibility.

Diastolic pressure is seen to increase (Fig. 14.11) when there is:

1 an increase in the systolic pressure of that particular pulse;

2 a decrease in ejection velocity;

3 an increase in aortic or arterial distensibility;

4 an increase in heart rate; or

5 an increase in total peripheral resistance.

14.3 CIRCULATION THROUGH SYSTEMIC ARTERIOLES

Arterioles are the site of greatest resistance to blood flow, which is controlled through contraction or relaxation of smooth muscle in their walls (vasoconstriction or vasodilatation). Changes in resistance alter both the amount of blood flowing through the arterioles and the upstream arterial and downstream capillary pressures. Arteriolar calibre, and hence blood flow, is altered by changes due to wall stretching, production of tissue metabolites and the local release of chemicals—intrinsic control. Extrinsic control is by hormones and the autonomic nervous system, mainly through sympathetic nerves.

Arterioles constitute at rest about 40% of the total peripheral resistance and are both the site of the largest fall in blood pressure (Fig. 14.6) and the site at which resistance is altered. As already mentioned (pp. 371 and 406), contraction of smooth muscle in arteriolar walls (**vasoconstriction**) decreases the arteriolar radius, thereby increasing their resistance. As a consequence there is a decrease in blood flow through the arterioles and (Fig. 14.12) a larger than usual fall in blood pressure along them due to an elevation in the upstream arterial pressure and a reduction in the downstream capillary pressure. The converse changes follows relaxation of the smooth muscle (**vasodilatation**).

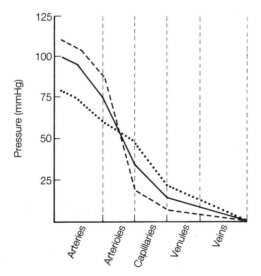

Fig. 14.12 Blood pressure profile in the vessels: control conditions (solid line), after vasoconstriction (dashed line) and after vasodilatation (dotted line).

Smooth muscle (pp. 135–141) is arranged around the circumference of arterioles. One type can contract spontaneously (**myogenic** contraction); the other type requires activity in **sympathetic nerves** in order to contract. There is normally a tonic discharge in these sympathetic nerves which, together with any myogenic contraction and contraction caused by circulating hormones or local humoral agents, gives a background degree of vasoconstriction referred to as **vasomotor tone**. Its degree varies from organ to organ and influences the proportion of the cardiac output that each organ receives (see Table 13.1). For example, at rest the vasomotor tone is high in arterioles of the skeletal musculature and low in those of the gut, kidney and skin. The higher the vasomotor tone in a person at rest, the greater the increase in blood flow during maximal vasodilatation.

Intrinsic control of arteriolar blood flow

Many organs, especially those with a small degree of neurogenic control, exhibit the phenomenon of **autoregulation**; that is, regardless of large changes in perfusion pressure their blood flow remains remarkably constant (Fig. 14.13). The mechanism underlying autoregulation remains uncertain—an intrinsic response of arteriolar smooth muscle to either stretch (myogenic) or local metabolites has been suggested. Both mechanisms may operate to varying degrees in different tissues.

Myogenic autoregulation

In the brain (p. 452) and possibly the kidney (p. 456), the smooth muscle of arterioles spontaneously contracts when, as described by the law of Laplace, the arteriolar wall tension is passively increased by an increase in blood pressure. Conversely the arterioles relax when the pressure decreases. It is thought that increased wall tension opens Ca^{2+} channels in the smooth muscle and the resulting Ca^{2+} influx increases the muscle tone. The reduction in radius caused by the contraction matches the increase in perfusion pressure such that (Poiseuille equation) there is no change in blood flow over a certain range of pressure. This is referred to as **myogenic autoregulation** of blood flow (Fig. 14.13). It allows blood flow to a tissue to be independent of variations in arterial blood pressure. Myogenic regulation is also present to varying degrees in the heart, gut and skeletal musculature (Section 15.2). It does not occur in the skin.

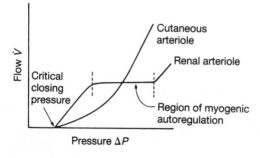

Fig. 14.13 Pressure–flow relationships in arterioles which do (renal) and do not (cutaneous) exhibit autoregulation.

Metabolic autoregulation

All organs need to receive a blood flow which can vary in proportion to their metabolic requirements. During increased organ metabolism there is a local decrease in the partial pressure of $O_2(Po_2)$, an increase in the partial pressure of CO_2 (Pco_2) and an increase in H^+ concentration in the interstitial fluid. These changes cause the smooth muscle to relax to a degree that is appropriate to the increased metabolism and ensure an increase in flow with little (or no) change in perfusion pressure. This is called **metabolic autoregulation** (or active hyperaemia). Increases in local temperature and in the local concentrations of other metabolites (such as adenosine triphosphate, adenosine diphosphate, adenosine monophosphate, adenosine, inorganic phosphate, lactate and pyruvate, as well as increases in K^+ concentration and interstitial osmolarity of exercising skeletal musculature) have also been found to cause vasodilatation. Metabolic autoregulation is well-developed in the skeletal musculature, heart and brain, but, in other organs and to some extent also in skeletal muscles, it can be overriden by nervous control of the arterioles.

Metabolic autoregulation may contribute to the overall phenomenon of autoregulation illustrated in Fig. 14.13, inasmuch as an increase in pressure by increasing blood flow, in the face of constant metabolism, will lower tissue metabolites and result in vasoconstriction.

When the blood supply to an organ is temporarily obstructed (for a period of seconds up to a few minutes), its restoration is accompanied by a large increase in blood flow (**reactive hyperaemia**) which depends on the duration of the obstruction and the metabolism of the organ over that time. The vasodilating effects of the metabolites that have accumulated during the obstruction contribute to this reactive hyperaemia.

When an increase in Pco_2 (hypercapnia) or decrease in Po_2 (hypoxia) occurs throughout the body (for example, due to impaired breathing) then all systemic arterioles will dilate. Both the fall in mean arterial blood pressure and the detection of arterial hypercapnia and hypoxia will elicit various cardiovascular reflexes (p. 443) which will oppose to varying degrees the local vasodilatation.

Usually small changes in the metabolism of an organ can be satisfied by metabolic autoregulation without affecting significantly the total peripheral resistance. However if the metabolism of the organ increases markedly there will be a fall in total peripheral resistance. The resultant initial fall in mean arterial blood pressure will then elicit cardiovascular reflexes (p. 440) to increase cardiac output and restrict the blood flow to other organs in order to maintain the eventual mean arterial blood pressure.

Humoral control

In addition to metabolites, other chemicals released locally in an organ also affect its blood flow. The sweat and salivary glands and gastrointestinal mucosa, when activated, produce not only their exocrine secretions but also an enzyme, kallikrein. This converts plasma kininogens into active **kinins**, such as kallidin and **bradykinin**, which have marked vasodilating effects in the glands as well as locally within the skin or gut. These kinins are inactivated by other tissue enzymes. During inflammatory or allergic reactions, kinins are also liberated from tissues and the vasodilator **histamine** is released from basophils and mast cells (p. 342). Inflammatory responses are also modulated by vasoconstricting **leukotrienes** released from leukocytes (p. 340). When a blood vessel is cut, **serotonin** and the prostaglandin derivative **thromboxane A$_2$** are released from activated platelets and cause vasoconstriction (p. 355). On the other hand **prostacyclin** released from endothelial cells, causes vasodilatation (p. 359). In general the F series of **prostaglandins** mainly causes vasoconstriction, while the E series are vasodilators.

Recently discovered is an **endothelium-derived relaxing factor** (EDRF), now identified as nitric oxide, which is released from the endothelium of *arteries* in response to both endothelial shear stress and the actions of, for example, bradykinin, adenosine diphosphate and substance P. EDRF causes the artery to dilate, facilitating blood flow to the arterioles. Whether similar relaxing factors are released from arteriolar endothelium is not yet known. The endothelium also appears to release vasoconstricting compounds.

Extrinsic control of arteriolar blood flow

Nervous control

In organs with a resting vasomotor tone caused by sympathetic activity, an increase in **sympathetic** discharge to arterioles causes further vasoconstriction whilst a decrease in discharge causes vasodilatation. The sympathetic neurotransmitter, **noradrenaline**, acts powerfully on α_1-**adrenoceptors** on the plasma membranes of vascular smooth muscles to cause contraction. Arterioles of the skin, gut, skeletal musculature and kidneys have a dense sympathetic innervation whereas those of the brain and heart are more sparsely innervated.

Noradrenaline also acts very weakly on the β_2-adrenoceptors of vascular smooth muscle, the activation of which causes relaxation. However, the strong α_1 response is dominant.

The vasoconstriction resulting from a certain level of sympathetic discharge is considerably greater in the skin, kidney and gut than in the skeletal musculature. Thus the blood flow to the skin, kidney and gut can be restricted in favour of other organs. This sympathetic pathway is controlled by cardiovascular centres in the brainstem and operates, as part of the baroreceptor reflex, to regulate mean arterial blood pressure (p. 440).

A special system of sympathetic nerves controlled from the motor cortex and hypothalamus (p. 446) innervates arterioles of the skeletal musculature. It releases acetylcholine at postganglionic nerve endings and causes vasodilatation. This **sympathetic cholinergic pathway** is usually silent but is activated during emotional reactions of alarm, rage or fear and during the initial phase of exercise.

Vasodilatation in most tissues is brought about by a reduction in sympathetic activity, but **parasympathetic** nerves cause vasodilatation in the erectile tissue of the external genitalia. Dilatation in the salivary glands, exocrine pancreas, and in the gastric and colonic mucosa is also influenced by parasympathetic nerves. The parasympathetic neurotransmitters mediating vasodilatation are acetylcholine and vasoactive intestinal polypeptide. Parasympathetic, as well as sympathetic, nerves supply arterioles of the brain, heart and lung, but their functions are not clear.

Hormonal control

The adrenal medulla continuously secretes **adrenaline** and a small amount of noradrenaline. The level of secretion by the adrenal medulla is proportional to its sympathetic input, which is mainly under the control of the hypothalamus (p. 293). Adrenaline activates, to an equal extent, the α_1-**adrenoceptors**, resulting in vasoconstriction and the β_2-**adrenoceptors**, resulting in vasodilatation. The net arteriolar response of a particular organ to adrenaline depends on the relative densities of α_1- and β_2-receptors. The density of β_2-receptors is higher than α_1-receptors in the heart and skeletal musculature. Thus during exercise the increased release of adrenaline contributes to the increased blood flow to the heart and skeletal musculature, whilst decreasing blood flow to the skin, gut and kidney where the density of α_1-receptors is greater.

The hormones **angiotensin II** and **antidiuretic hormone** (vasopressin) are powerful vasoconstrictors, while **atrial natriuretic peptide** decreases the sensitivity of vascular smooth muscle to these and other vasoconstrictor substances. These hormones are concerned primarily with the control of blood volume and thus the long-term maintenance of a normal arterial blood pressure (p. 442).

14.4 CIRCULATION THROUGH SYSTEMIC CAPILLARIES

Precapillary sphincters and arterioles control how much blood passes at a slow velocity through a capillary. Capillary walls provide a large surface area for the diffusion of substances between blood and interstitial fluid surrounding the cells. Diffusion is by far the most important process for exchange of gases, solutes and water across the capillary wall. The capillary wall is, however, somewhat leaky and allows a little water and solutes smaller than M_r 70 000 to move by bulk flow. Some capillaries are less permeable because of tight junctions between endothelial cells; others are more permeable because of fenestrations within the endothelial cells. The amount of fluid leaving a capillary by bulk flow depends on the permeability and surface area of its wall and the hydrostatic and osmotic pressure gradients between the blood and interstitial fluid. The hydrostatic pressure gradient results in ultrafiltration of fluid from the capillary while osmotic pressure gradients promote reabsorption of fluid; in most tissues there is on balance a slight excess of ultrafiltration—Starling equilibrium.

The entrance of an arteriole (or metarteriole) into the vast network of capillaries is guarded by a ring of smooth muscle, the **precapillary sphincter** (Fig. 14.14). These sphincters exhibit **myogenic rhythmicity** which results in intermittent and variable flow rates through any individual capillary. The direction of flow may change in some capillaries depending on both their location within the capillary bed and the degree of constriction of nearby sphincters. This can be seen, for instance, in the capillary bed of the rabbit's ear. The net degree of constriction or dilatation of these sphincters is controlled by **metabolic autoregulation**. For example, at any one moment in resting skeletal musculature about 10% of the

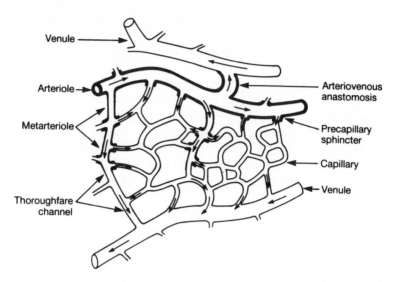

Fig. 14.14 A capillary network and its associated vessels. Thick walls denote considerable amounts of smooth muscle.

capillaries are open and the remainder contain blood that is stationary (these capillaries may even be completely empty). Blood may bypass the capillaries and flow through the **metarterioles**, which contain little smooth muscle, and through small vessels called **thoroughfare channels** (Fig. 14.14). In the skin and gut, short, relatively large-diameter vessels called **arteriovenous anastomoses** act as shunts between arterioles and venules (Fig. 14.14). They have thick walls of smooth muscle controlled by sympathetic nerves acting on α_1-adrenoceptors. Note that capillaries cannot contract nor distend.

The characteristics of capillary beds which have already been considered (Fig. 14.6) are the parallel arrangement of their narrow (4 µm radius) vessels, their contribution ($\sim$20%) to the resting total peripheral resistance, the blood pressure drop along them, the absence of pulsatile flow, their large cross-sectional area, the low velocity of flow in the capillary bed and the low flow rate through an individual capillary. It is important to remember that the blood flow and pressure in the capillaries are determined by the arterioles (Fig. 14.12). Under resting conditions, only about 25% of the capillaries are open, giving a total capillary cross-sectional area of about 3000 cm^2. Since the average length of a capillary is about 0.1 cm and the velocity of capillary blood flow at rest is 0.05 cm s^{-1}, blood will take about 2 seconds to traverse the length of a capillary. When a tissue increases its metabolic rate, more of its capillaries are open and their total cross-sectional area is increased. As arteriolar vasodilatation increases the pressure gradient (ΔP) between the beginning and end of each individual capillary (Fig. 14.12), the velocity of blood flow through each capillary, the radius of which does not change, will also increase (velocity is proportional to $\Delta P r^2$; p. 408). The **transit time** through a capillary is, however, rarely shorter than 1 second which is still sufficient to allow adequate **diffusion** (p. 7) of gases and nutrients across the capillary wall.

The vast network of capillaries also provides a **large surface area** for exchange with the tissue cells, with a **short diffusion distance** of no more than 50 μm between blood and cells. The area increases and the distance decreases whenever more capillaries are open.

Structure of capillary walls

Capillary walls are composed of a single layer of endothelial cells about 1 μm in thickness. In contrast to the plasma membrane of cells (p. 8), the capillary wall in most tissues is leaky for it lets through substances of M_r less than about 70 000. These include practically all the solutes in the plasma except the plasma proteins. But capillary permeability does differ in different tissues, and in liver, for example, proteins pass through to a greater extent than elsewhere while, in brain, movement of water-soluble solutes is markedly reduced. These differences reflect the organization of the capillary wall. In liver, obvious spaces between adjacent endothelial cells are seen in electron micrographs, whereas in most brain capillaries adjacent cells are held firmly together by **tight junctions** which restrict water and solute movement and result in a *blood–brain barrier* (p. 71). In some specialized capillaries in the gut and kidney (renal glomerulus), exchange of solutes is facilitated by **fenestrations**—areas within the endothelial cells where little or no cytoplasm separates the plasma membranes on the two surfaces of the cell. For most vascular beds, adjacent endothelial cells are attached to each other at their margins, though the attachment offers relatively little resistance to solute exchange. Furthermore, a continuous **basement membrane** encircles the periluminal surface of all capillaries, providing additional support to the endothelial cells.

It is now realized that the plasma membranes of endothelial cells, like those of epithelial cells (p. 24), contain a variety of specific pathways through which ions and other water-soluble solutes can be transported between blood and interstitial fluid. In brain capillaries, because of the tightness of the junctions holding adjacent cells together, the cellular pathway provides the dominant route for transendothelial movements of water-soluble solutes. In other capillary beds, however, much of the exchange of these solutes and of water occurs by diffusion between the cells (paracellular pathway). Proteins and other macromolecules may also be transported across endothelial cells by a process of transcytosis (p. 15). In contrast, lipid-soluble substances, including O_2 and CO_2, diffuse directly across the plasma membranes of endothelial cells.

Ultrafiltration and reabsorption of fluid across capillary walls

It must be emphasized that passive diffusional movements are responsible for virtually all of the exchanges of gases, solutes and water between capillaries and interstitial fluid. However these exchanges require a relatively permeable capillary wall. At the same time, to drive blood through the systemic circulation requires relatively high hydrostatic pressures. As a consequence of these two factors, it is inevitable that some fluid moves by **bulk flow** from capillary to interstitial fluid in a process called ultrafiltration. This fluid must be returned by bulk flow to the circulation to maintain the circulating blood volume—most is reabsorbed back into the capillary by osmosis, the rest is returned by the lymphatic system.

Colloid osmotic pressure

As the structure of the capillary wall allows free permeability to water and to solutes of M_r less than 70 000, small molecules diffuse through the capillary wall and achieve equal concentrations on both sides. Thus they make no contribution to the effective osmotic pressure (p. 21). As plasma proteins with M_r greater than 70 000 cannot cross the capillary wall (i.e. proteins with $\sigma = 1$ are reflected), they produce an **effective osmotic pressure**. Since plasma proteins are large enough to be classed as colloids, this effective osmotic pressure is known as the **plasma colloid osmotic pressure** (or plasma oncotic pressure).

Proteins are present in plasma at a concentration of about 1 mmol L^{-1} of plasma, whereas the protein concentration in the interstitial fluid is much lower (0.1–0.33 mmol L^{-1}) depending on the tissue. If one assumes, however, that the interstitial protein concentration is close to zero, the concentration difference for protein would cause an effective osmotic pressure for plasma of 16 mmHg (calculated using the van't Hoff equation, p. 21). However, proteins are negatively charged and therefore attract diffusible cations and repel diffusible anions. This leads to an uneven distribution of small ions with a small excess (about 0.5 mmol L^{-1}) inside the capillary. This uneven distribution is referred to as the **Gibbs–Donnan membrane distribution** and the small excess of diffusible ions in the plasma is known as the *Donnan excess* (p. 22). This excess contributes a further 9 mmHg to the effective osmotic pressure of plasma. Hence the total plasma colloid osmotic pressure is 25 mmHg, first measured by Starling. This pressure is very small compared with the total osmotic pressure of all the solutes in plasma (5800 mmHg) measured when plasma is separated from pure water by an ideal semipermeable membrane.

Starling equilibrium

Starling realized that the plasma colloid osmotic pressure of 25 mmHg was very important because this value lies between that of the blood pressure in arterioles and venules. The blood pressure (hydrostatic pressure) at the arteriolar end of a capillary is about 32 mmHg and at the venular end about 15 mmHg. Hydrostatic pressure in the interstitial fluid varies from place to place but on average it is a few mmHg below atmospheric, say about – 2 mmHg. The net hydrostatic pressure gradient between blood and interstitial fluid (ΔP) forces fluid out of the capillaries by **ultrafiltration** and this force is greater at arteriolar than venular ends of capillaries (Fig. 14.15). This ultrafiltration is opposed by **osmosis** returning fluid into the capillaries, the osmotic force being the difference in colloid osmotic pressure between plasma and interstitial fluid ($\Delta \pi$). (Interstitial colloid osmotic pressure has been assumed in the figure, as a simplification, to be zero; it is, in fact, about 2–8 mmHg depending on the tissue.) Note that the osmotic force remains constant along the length of the capillary. At the arteriolar end, the net difference between hydrostatic and osmotic forces causes ultrafiltration with fluid leaving the capillary. At the venular end, the net difference causes reabsorption with fluid returning to the capillary. This balancing of ultrafiltration and osmosis is referred to as the **Starling equilibrium**. The equilibrium is not complete as there is slightly more ultrafiltration than osmosis (Fig. 14.15). The small amount of fluid lost into the interstitium is removed by the *lymphatic system* (p. 431).

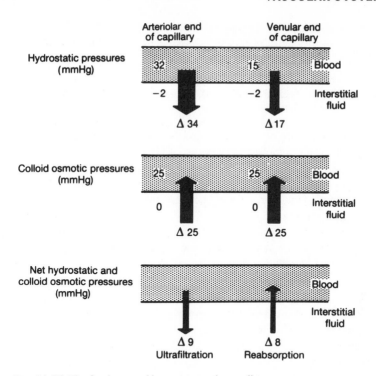

Fig. 14.15 The Starling equilibrium across the capillary.

In some capillary beds, blood perfusion is intermittent and ultrafiltration when capillary blood pressure is high is balanced by osmosis when capillary blood pressure is low. In particular organs, special conditions exist in capillary beds. For example, in the glomeruli of the kidneys capillary blood pressure is high and there is only ultrafiltration (p. 613). In the alveolar capillaries in the lungs, blood pressure is below the colloid osmotic pressure, hence there is only osmosis and liquid does not accumulate in the alveoli. In the liver, the interstitial fluid outside the capillaries is rich in protein and a very small capillary hydrostatic pressure is almost balanced by an equally small difference in colloid osmotic pressure.

Alterations in capillary blood pressure and in the colloid osmotic pressure of either plasma or interstitial fluid alters the degree of ultrafiltration or reabsorption. Net flow of fluid per unit time (J_v) across capillary walls is also affected by the number of capillaries that are open, which determines the surface area (A), and by the hydraulic conductivity (L_p), which is a measure of the ease with which water flows across. ($J_v = AL_p(\Delta P - \Delta \pi)$, see p. 21; a positive value for J_v indicates ultrafiltration.) Thus J_v increases in the skeletal musculature during exercise because ΔP increases and more capillaries are open. L_p and hence J_v are increased by substances such as histamine, bradykinin and substance P.

Excess ultrafiltration (Fig. 14.16) occurs when there is:
1   arteriolar vasodilatation;
2   an elevated capillary pressure (e.g. caused by venous obstruction or chronic right heart failure);
3   an elevated venous blood pressure (i.e. in the lower limbs in the erect posture);
4   an increase in interstitial protein concentration (due to an increase in protein permeabil-

ity of the capillaries in inflammatory or allergic reactions); and
5   a decrease in plasma colloid osmotic pressure (e.g. plasma protein deficiency).
An excess of reabsorption (Fig. 14.16) will result mainly from:
1   arteriolar vasoconstriction;
2   a decreased venous blood pressure; and
3   dehydration, which when severe, will increase the protein concentration in plasma more
than in interstitial fluid.

Moderate amounts of fluid produced by capillary ultrafiltration are normally
removed by the lymphatic system (p. 431). If ultrafiltration is excessive, the
volume of interstitial fluid increases. When clinically detectable this is called
oedema. Oedema (p. 654) can be local to an organ (*localized*) or widespread
(*generalized*). However, excessive oedema is usually limited by the distensibility of
the organ concerned.

After a haemorrhage (p. 460) fluid reabsorption into capillaries increases
because of a reduction in capillary blood pressure due to the haemorrhage itself
and because of reflex arteriolar vasoconstriction. Such responses will help to
restore the blood volume by drawing on the reservoir of the interstitial fluid
(p. 3). The Starling equilibrium thus provides an automatic control of the
constancy of circulating plasma volume at the expense of shrinkage (or expansion) of the interstitial fluid.

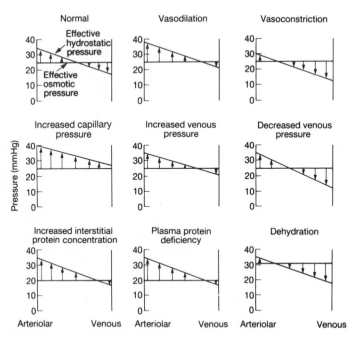

Fig. 14.16 The balance of ultrafiltration (↑) and reabsorption (↓) of fluid across the
capillary walls under various conditions. Effective hydrostatic pressure is the difference in
hydrostatic pressure between capillary blood and interstitial fluid; effective osmotic
pressure is the difference in colloid osmotic pressure between capillary blood and
interstitial fluid. The fall in blood pressure from arteriole to vein for each condition is kept
constant for simplicity (cf. Fig. 14.12).

14.5 CHARACTERISTICS OF THE LYMPHATIC SYSTEM

Plasma proteins that leak out of the capillaries and fluid that is not reabsorbed back into the capillaries are returned to the circulatory system via the lymphatic vessels.

The lymphatic system consists of a network of blind-ending **lymph capillaries** lying in the interstitial space near blood capillaries. The walls of lymph and blood capillaries are similar except that the spaces between endothelial cells in lymph capillaries are larger, making them readily permeable to protein and fluid. Lymph capillaries collect into thin-walled **lymph vessels** which eventually empty on the right side of the thorax via the **lymphatic duct** and on the left via the **thoracic duct** into the subclavian veins of the circulatory system. **Lymph** is composed of fluid (**plasma**) and cells called **lymphocytes** (p. 344) many of which are stored in **lymph nodes** found in the larger lymph vessels.

The **functions** of the lymphatic system are to return to the cardiovascular system the excess interstitial fluid that has resulted from more capillary ultrafiltration than reabsorption in the Starling equilibrium, and any **protein** that has leaked across the capillary wall. It also acts as a pathway for the **absorption of fats** from the gut (p. 603) and, by virtue of its lymph nodes and lymphocytes, is involved in **immune responses** (pp. 343–349).

Because of lymph flow, the **Starling equilibrium** should be restated as follows (with approximate values for a 70-kg adult):

Ultrafiltration   = Reabsorption   + Lymph flow
20 litres per day   16–18 litres per day   4–2 litres per day

Since the cardiac output amounts to at least 8000 litres per day for an adult man, or say 4000 litres per day of blood plasma, the 20 litres of fluid passing through blood capillary walls by ultrafiltration is less than 0.5% of blood plasma volume flowing per day. Hence these bulk movements of fluid through the blood capillary wall are of *little* importance for the exchange of nutrients, O_2 and CO_2. *These move predominantly by diffusion.* Furthermore, the daily exchanges of water by diffusion across the capillaries amounts to 80 000 litres, which is 4000 times the rate of ultrafiltration.

Lymph flow is sluggish and occurs at a rate of 2–4 litres per day. The pressure gradient driving this flow is dependent on the upstream pressure—the **interstitial hydrostatic pressure**, the magnitude of which depends on the amount of excess interstitial fluid and the distensibility of the interstitial space. The downstream hydrostatic pressure in the thoracic duct is negative because of its intrathoracic location. Lymph flow is aided by **myogenic** rhythmic contractions of **smooth muscle** in the walls of lymph vessels, retrograde flow being prevented by **lymphatic valves** similar to those found in veins (p. 434). During exercise, the rate of ultrafiltration from capillaries in skeletal muscle increases and the increased lymph flow required is aided by the rhythmic contractions of skeletal musculature around the lymph vessels. Similarly, increased contractions of gut musculature increase lymphatic return from the intestine. In addition, rhythmic alterations in lymphatic transmural pressure resulting from the cardiac and respiratory cycles will aid lymphatic return from the heart and lungs, respectively.

14.6 CHARACTERISTICS OF THE SYSTEMIC VENOUS CIRCULATION

The venous system has a large capacity, which, because of venous distensibility, can be altered passively by changes in transmural pressure. Venous capacity is also controlled by hormones and sympathetic activity which alter the degree of venoconstriction. Venous return is determined primarily by the gradient between mean systemic filling pressure (MSFP) and right atrial pressure. Venous return is also influenced by the suction effect of the heart, by intrathoracic pressure changes accompanying breathing and by skeletal muscle contractions, all of which rely on the presence of venous valves.

Several characteristics of the venous system have already been described (Fig. 14.6), namely its moderately large cross-sectional area, the fact that the veins accommodate 55% of the blood volume when supine, the moderate venous velocity, the venous contribution of ~15% to the total peripheral resistance, the resulting small pressure gradient (from 15 to 1–2 mmHg) and the small pressure pulses (a, c and v waves) in the largest veins. Two important characteristics of the venous system which require further discussion are the venous pressure gradient, which will determine **venous return** to the right atrium, and venous distensibility, which influences the **venous capacity** for accommodating blood, a factor which also affects venous return.

Alterations in venous capacity

The distensibility of the venous system allows increases or decreases in total blood volume to be accommodated passively with little detectable changes in venous pressure (Fig. 14.5). It also means that an increased or decreased transmural pressure will result in the passive distension or collapse of the veins.

   Smooth muscle in the walls of the venous system (Fig. 14.4) has α_1-**adrenoceptors** innervated by the **sympathetic system**. There is normally some degree of tonic discharge and hence some degree of constriction–**venomotor tone**. An increase in sympathetic activity will cause **venoconstriction**. This reduces the **capacity** of the venous system and makes its walls less distensible. Because of the relatively large radii in the venous system, venoconstriction has little effect on venous resistance or total peripheral resistance. Conversely, a decrease in sympathetic activity to the veins results in **venodilatation** and an increase in venous capacity. Venoconstriction occurs as part of the reflex response to haemorrhage and exercise coordinated by the cardiovascular centres. An increase in the circulating hormones **adrenaline** and **angiotensin II** will also cause venoconstriction.

Determinants of venous return

As the circulatory system is a closed circuit, *the venous return in all but moment-to-moment fluctuations must equal the cardiac output.* The magnitude of this flow depends on the pressure gradient from left ventricle to right atrium and on the resistance encountered in the blood vessels (Poiseuille equation). Venous return can be greater or smaller than cardiac output but only for a *short time*.

Primary determinant of venous return

The right atrial filling pressure is the pressure difference between *mean systolic filling pressure* (MSFP) and *mean right atrial pressure*. The greater the pressure difference, the greater the venous return.

MSFP was defined by Guyton in the 1950s as the weighted average of the pressures in all portions of the systemic circulation; the weighting is in proportion to the volume capacity of the vessel. It is also the *static blood pressure* which prevails throughout the systemic circulation if the heart is suddenly arrested experimentally and rapid equilibration of pressures in arterial and venous circuits has occurred. The MSFP is usually about 7 mmHg. An alternative term for MSFP is mean circulatory pressure.

For a given venous capacity, MSFP increases when the blood volume is expanded. For a given total blood volume, MSFP increases when the venous blood is compressed by venoconstriction (or, in other words, when the venous capacity is decreased). Arteriolar vasoconstriction will not alter MSFP because the arterioles only contain 3% of the total blood volume (p. 415). MSFP and venous return are considered further on p. 435. Note that MSFP can increase to a maximum of about 20 mmHg and the minimum MSFP can be close to zero.

Mean right atrial pressure is often referred to as **central venous pressure** and is about 1 mmHg. The filling of the external jugular vein, which can be observed visually (p. 459), is used clinically as a measure of central venous pressure. It is elevated slightly either when blood volume is increased or when a person adopts a supine compared with a standing posture. It can be elevated appreciably during right heart failure (p. 463).

Secondary factors influencing venous return

Suction effect of the heart

During each cardiac cycle the right atrial pressure fluctuates and the decreases in pressure, often to negative values, are called the x and y descents (see Fig. 13.16). They are caused respectively by the downward movement of the atrioventricular ring (early systole) and the rapid filling of the ventricle after the atrioventricular valve opens (early diastole). These decreases in atrial pressure constitute the so-called **suction effect of the heart** which aids venous return and is greater during increased cardiac activity. Note, however, that at negative atrial pressures there is a limit to the possible increase in venous return because the venae cavae collapse at the point of entry into the chest.

Respiratory venous pump

Venous return is assisted by the actions of breathing on intrathoracic and abdominal veins, the so-called **respiratory venous pump**, such that flow in the venae cavae increases during inspiration and falls during expiration. The effect of the respiratory pump on venous return is enhanced whenever the depth of breathing increases, for instance during exercise.

The intrathoracic (or intrapleural pressure) is subatmospheric and for the small tidal volumes of resting conditions is about -5 mmHg at the end of inspiration and about -3 mmHg at the end of expiration (p. 491). Since veins have distensible walls, these fluctuations in transmural pressure in the thorax cause dilatation of intrathoracic veins during inspiration and their compression during expiration. The descent of the diaphragm during inspiration will, at the same time, raise intra-abdominal pressure and compress the abdominal veins; the converse will occur in expiration. Thus the respiratory pump, within limits, aids venous return since the relative dilatation of intrathoracic veins in inspiration will decrease their resistance to flow and cause some movement of blood towards the heart from the abdominal veins or upper extremities. Compression of abdominal veins which can occur in both inspiration and expiration also propels blood towards the heart since retrograde flow from the abdomen into the limbs is prevented by valves in the limb veins. These **venous valves** are thin cup-like structures whose cusps obstruct the lumen at the threat of any retrograde flow and thus ensure unidirectional flow. (They occur in limb veins; a few valves which are not very effective are also present in abdominal and thoracic veins.)

Flow in the venae cavae can decrease markedly during forced expirations, coughing or Valsalva manoeuvres (p. 495) due to compression to both abdominal and intrathoracic veins. If venous return is excessively impeded fainting can result.

Since the heart and intrathoracic vessels have a lower pressure during inspiration, the venous return towards the right atrium from the extrathoracic veins will increase during inspiration. The resulting increase in right ventricular output will in turn increase the arterial flow into the pulmonary vessels and potentially the pulmonary venous flow into the left atrium. However, because the expansion of the lungs in inspiration increases the capacity of the pulmonary veins and arteries, the pulmonary pressure, i.e. the filling pressure of the left heart, may not increase until expiration reduces the lung volume. This explains why the preload of the right ventricle is higher in inspiration, and the preload of the left ventricle is higher in early expiration (p. 400).

Skeletal muscle venous pump

During exercise a further mechanism, the **skeletal muscle venous pump**, aids venous return. The exercise must not be a sustained contraction but one of alternating contraction and relaxation. The contraction of the skeletal musculature alters the transmural pressure of the veins within it and compresses them. This propels blood towards the heart since retrograde flow is prevented by the venous valves. When the skeletal musculature relaxes, the veins are dilated by inflow but only from below, as the venous valves again prevent retrograde flow. The further advantages of this pump when in the upright posture will be considered later (p. 459).

14.7 RELATIONSHIPS BETWEEN CARDIAC OUTPUT, VENOUS RETURN AND ATRIAL PRESSURE

On average, cardiac output and venous return are equal. Their graphical relationships to atrial pressure are called the cardiac function curve and vascular function curve respectively. The graphs allow one to visualize the effects of changes in myocardial contractility, vasomotor tone, blood volume or venous capacity on cardiac output and venous return.

The separate considerations of the control of cardiac output (in particular of stroke volume; p. 397) and of venous return can be combined in a graphical approach designed by Guyton in the 1950s. These graphs represent only steady-state responses and not momentary changes and the use of them requires some simplification of concepts and abstraction of ideas.

Cardiac function curve

Earlier, in Figs 13.25 and 13.29, the relationship between stroke volume and end-diastolic volume (the Starling curve) for varying degrees of myocardial contractility was described. Also on pp. 399–400 it was stated that cardiac output (at a constant heart rate) is proportional to stroke volume and that mean atrial pressure is directly related to end-diastolic volume. The relationship between cardiac output and mean right atrial pressure is referred to as the **cardiac function curve** (Fig. 14.17a). The operating point A moves up the 'normal' curve when venous return increases, since this results in a larger end-diastolic volume and hence a higher mean atrial pressure. Conversely, A moves down the normal curve when venous return decreases. A new steeper curve eventuates when there is increased sympathetic stimulation of the ventricle (i.e. an increase in myocardial contractility). A flatter curve eventuates when there is a decrease in myocardial contractility. A steeper curve also occurs (Fig. 14.17b) when there is a decrease in arteriolar resistance (i.e. vasodilatation) since this reduces the afterload and effectively allows a larger stroke volume for a particular end-diastolic volume. A flatter curve results from vasoconstriction.

Vascular function curve

The equivalent relationship between venous return and mean right atrial pressure is referred to as the systemic **vascular function curve** (Fig. 14.17c). The shape of this curve is due to the fact that, as right atrial pressure increases, venous return is opposed. As stated earlier, at negative atrial pressures, there is, however, a limit to the possible increase in venous return because the venae cavae collapse at the point of entry into the chest. The operating point A moves up or down the 'normal' curve as venous return increases or decreases, respectively. At a venous return of zero, the atrial pressure in this graph is + 7 mmHg. This is the static pressure which prevails throughout the systemic circulation when the circulation has stopped long enough for the pressures to have equalized. (It is also the MSFP, which was defined on p. 433.) The vascular function curve shifts upwards in a parallel manner (Fig. 14.17c) whenever there is an increase in blood volume or during venoconstriction. This indicates an increase in MFSP (p. 433). The converse occurs during the haemorrhage or venodilatation. A decrease in arteriolar resistance (vasodilatation) increases blood flow and causes a steeper curve (Fig. 14.17d) but does not alter the MSFP (p. 433); vasoconstriction flattens the curve.

Cardiac and vascular function curves

When these function curves are displayed together, the operating point A is the point of intersection (Fig. 14.17e). During exercise the new cardiac function curve

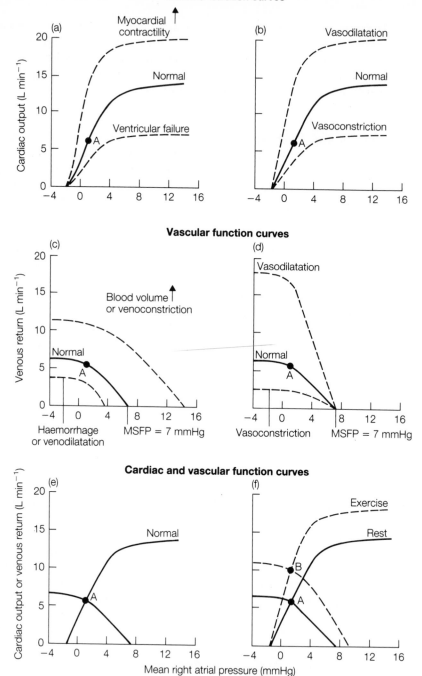

Fig. 14.17 (a, b) Cardiac function curves under different conditions. A, Normal operating point with a cardiac ouput of 6 L min⁻¹ and a mean right atrial pressure of 1 mmHg. (c, d) Vascular function curves under different conditions. A, Normal operating point with a venous return of 6 L min⁻¹ and a mean right atrial pressure of 1mmHg; MSFP, mean systemic filling pressure. (e) Cardiac and vascular function curves displayed together for normal (resting) conditions. (f) Cardiac and vascular function curves occurring at rest (operating point = A) and during a particular degree of exercise (operating point = B).

is steeper because of both increased myocardial contractility and the reduced afterload that results from vasodilatation (Fig. 14.17f). The vascular function curve in exercise is both steeper, due to vasodilatation, and shifted upwards with a higher MSFP, due to venoconstriction. Thus the operating point in exercise is at B. These curves will be used again in Chapter 15 to demonstrate the sequence of events occurring during haemorrhage and cardiac failure.

Chapter 15
Cardiovascular Regulation, Regional Circulation and Circulatory Adjustments

15.1 INTEGRATED REGULATION OF THE CARDIOVASCULAR SYSTEM

Short-term regulation of the cardiovascular system is concerned with rapid alterations in cardiac output ($\dot{Q}$) and total peripheral resistance (TPR) to meet the requirements of individual organs while at the same time maintaining arterial blood pressure. The arterial blood pressure is held constant at a level determined by a variety of receptor inputs. Long-term regulation maintains a constant blood volume and the correct balance of water between the blood and interstitial fluid. Baroreceptors continuously monitor arterial blood pressure and artial stretch receptors monitor the heart's end-diastolic blood volume. These sensory inputs and those from other receptors are integrated by the cardiovascular centres. These centres are found predominantly in the medulla oblongata and in other specific areas between the cerebral cortex and spinal cord. Central integration alters not only autonomic nervous activity to the heart, arterioles and veins but also production of hormones such as adrenaline and antidiuretic hormone.

The relationship $\dot{V} = \Delta P/R$, used on p. 406 to consider flow through a single vessel, also applies to the systemic circuit as a whole, so that:

Cardiac output ($\dot{Q}$) =

$$\frac{\text{mean arterial blood pressure (MABP)} - \text{mean right atrial blood pressure}}{\text{total peripheral resistance (TPR)}}.$$

Since the mean right atrial pressure is close to zero, this equation can be condensed to

$$\dot{Q} = \frac{MABP}{TPR}$$

$\dot{Q}$ is the product of heart rate and stroke volume. **Heart rate** is altered by parasympathetic and sympathetic nerves and by adrenaline (p. 379). **Stroke volume** is determined by end-diastolic volume (intrinsic control, Starling's law of the heart; p. 398) and is altered by sympathetic activity and adrenaline (extrinsic control; p. 401). In addition, end-diastolic volume and therefore stroke volume are altered by changes in venous capacity (p. 400) caused either passively or by adrenaline and sympathetic activity (p. 432).

TPR is controlled by intrinsic autoregulation and by sympathetic nerves and adrenaline (p. 421). But there is no direct control of **MABP**; it is determined entirely by $\dot{Q}$ and TPR. Thus:

MABP = $\dot{Q}$ × TPR

MABP is monitored by sensory receptors. Information from these receptors is integrated by cardiovascular centres in the central nervous system which reflexly control $\dot{Q}$ and TPR by efferent activity in autonomic nerves to the heart and blood vessels. This autonomic control interacts with the intrinsic mechanisms operating simultaneously in the heart and vessels.

Short-term regulation of the cardiovascular system is concerned with the rapid adjustments of these three variables, $\dot{Q}$, TPR and MABP. Compensatory changes occur or start to occur *within seconds* following perturbations induced, for example, by a change in posture, the onset of exercise or a haemorrhage.

Arterial baroreceptor reflexes

Characteristics of baroreceptors

The receptors which monitor MABP are called **arterial baroreceptors**. They are located in the wall of the aortic arch and in the **carotid sinus**, an enlarged part of the internal carotid artery just after it arises from the common carotid artery (Fig. 15.1). Baroreceptor nerve endings (both free and encapsulated) are embedded in the outer (adventitial) layer of the arterial wall which contains some elastic tissue. The sensory axons from the **aortic baroreceptors** travel in the **vagus nerve** and those from the **carotid baroreceptors** travel in the **glossopharyngeal nerve**. More than half of the baroreceptor axons are unmyelinated. Information from these receptors is relayed to the cardiovascular centres.

An increase in transmural pressure, usually brought about by an increase in arterial blood pressure, stretches the arterial wall and stimulates the baroreceptors. At normal MABP some baroreceptor afferents are tonically active, as shown by streams of action potentials recorded from their axons. The mean frequency of these action potentials changes in direct proportion to changes in MABP and, in addition, as the pressure gets higher, more afferents are recruited. Total baroreceptor activity is graded over the blood pressure range of 50–180 mmHg. Baroreceptors not only respond to MABP but also to the systolic to diastolic fluctuation in arterial pressure. The continuous monitoring of MABP by the

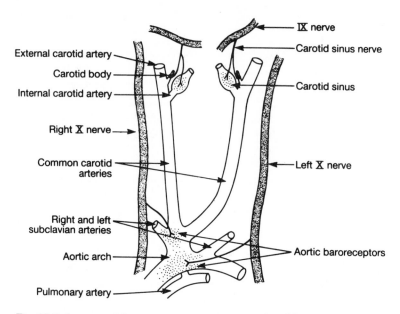

External carotid artery

Carotid body

Internal carotid artery

Right X nerve

Common carotid arteries

Right and left subclavian arteries

Aortic arch

Pulmonary artery

IX nerve

Carotid sinus nerve

Carotid sinus

Left X nerve

Aortic baroreceptors

Fig. 15.1 Location of the aortic and carotid baroreceptors and their sensory nerves.

baroreceptors permits adjustment of $\dot{Q}$ and TPR, via the autonomic efferent nerves, to maintain MABP constant at a particular set-point.

Baroreceptor reflex

An increase in MABP (caused, for instance, by a sudden change in posture from standing to lying down; p. 459) will increase the baroreceptors' tonic activity. The reflex changes initiated are as follows:

1 an increase in parasympathetic, and a decrease in sympathetic discharge, to the sinoatrial node causing a decrease in heart rate;

2 a decrease in sympathetic discharge to ventricular muscle causing a decrease in contractility and hence a reduction in stroke volume;

3 a decrease in sympathetic discharge to the veins, causing an increased venous compliance and capacity, and hence a reduced end-diastolic volume and subsequent stroke volume; and

4 a decrease in sympathetic discharge to the arterioles, causing a decrease in TPR. The arteriolar vasodilatation is greatest in the splanchnic area, modest in skeletal musculature and least in the kidneys.

Steps 1–3 will reduce $\dot{Q}$ and all of steps 1–4 take only a few seconds to occur. The vasodilatation in step 4 results in a greater capillary pressure which leads over the next 5–10 minutes to an increased ultrafiltration across the capillary wall and hence to some reduction in blood volume. These reflex reductions in $\dot{Q}$ and TPR will decrease MABP and oppose the initial elevation, thus **restoring MABP to or towards normal.**

A decrease in MABP initiates the converse changes. In addition, a large decrease in baroreceptor activity, for example after a haemorrhage, stimulates breathing (p. 542) and the secretion of adrenaline (p. 293) and antidiuretic hormone (p. 277). Adrenaline accelerates the heart and causes vasoconstriction; antidiuretic hormone augments the vasoconstriction.

The baroreceptor reflexes adjust the balance between $\dot{Q}$, TPR and MABP whenever there is a small change in the metabolism of an organ. If one organ increases its metabolism, dilatation of its arterioles by local metabolic autoregulation rapidly increases its blood flow. This local vasodilatation will slightly decrease the TPR and hence the MABP. The smallest decrease in MABP will be detected by the baroreceptors and will result in a reflex increase in $\dot{Q}$ and, to some extent, an increased vasoconstriction in all other organs, except the heart and brain. Thus, within a few seconds, the final decrease in TPR will be minimized and the extra blood flow required by that organ will be supplied by reducing blood flow to some of the other organs and by increasing the total blood flow. At the same time the decrease in MABP will be very small or even undetectable.

The ability of the baroreceptor reflex to regulate MABP tightly at the normal resting mean value depends on the magnitude of the initial deviation, the prevailing blood volume, the influence of other receptor reflexes and control of the cardiovascular centres by higher centres. The latter two factors determine the set-point for MABP and the role of the baroreceptor reflex is to ensure that any particular set-point is as precisely maintained as possible. Denervation of baroreceptors initially elevates the MABP but, with time, MABP returns to normal, although it is very labile around the set-point.

Reflexes intiated by cardiac stretch receptors

Atrial stretch receptors
Stretch receptors are found in the walls of the atria, mainly at their junctions with the venae cavae and pulmonary veins. Collectively they are referred to as **atrial stretch receptors**. Their afferent fibres are myelinated and travel in the **vagus** to the cardiovascular centres. Atrial stretch receptors are stimulated by the stretch resulting from an increase in **atrial (central venous) pressure.** Because the atria are very distensible, an increase in volume of the atria considerably stretches their walls with only a small increase in pressure. The atrial stretch receptors are thus low-pressure receptors which, in effect, monitor atrial blood volume (**atrial blood volume receptors**).

Atrial receptors are often categorized as atrial A and atrial B receptors because two distinct types of activity are seen in their afferent nerves. Atrial A receptors have a peak stimulation during the atrial pressure a wave (i.e. at the time of atrial contraction; p. 388); and the atrial B receptors show most activity during the v wave (i.e. at the very end of ventricular systole; p. 389). However, such nomenclature is too rigid as a change in heart volume or contractility may change the firing pattern of a particular receptor from one form to the other, and some receptors fire in phase with both the a and v wave.

Atrial receptor reflexes
A sustained increase in blood volume reflexly initiates an increase in parasympathetic and a decrease in sympathetic activity. This *complements* the reflex adjustments initiated by the baroreceptors which will also be stimulated by the increase in MABP that usually accompanies an increase in blood volume. These reflex

changes together result in a decrease in $\dot{Q}$ and TPR, an increase in venous capacity and an excess of ultrafiltration across the capillary wall. Thus MABP and blood volume decrease towards normal and the excess fluid is temporarily accommodated in the interstitial fluid.

When stimulated experimentally by the *rapid* infusion of blood, atrial receptors initiate an increase in heart rate mediated by the sympathetic system. This is the **Bainbridge reflex**; its physiological importance is uncertain.

Atrial receptors and hormone release

Whereas increased stimulation of atrial receptors reflexly alters autonomic activity, decreased stimulation affects also the release of certain hormones. When atrial receptor stimulation is reduced, there is a decrease in parasympathetic activity and an increase in sympathetic activity, in particular to the renal arterioles. This decreases renal blood flow and hence glomerular filtration rate, which reduces urinary **sodium excretion**, and increases the release of **renin** by the juxtaglomerular cells of the kidney (p. 651). The rise in plasma renin increases the concentration of circulating **angiotensin II** that:

1   causes further vasoconstriction and hence increases TPR;

2   stimulates the release of **aldosterone** from the adrenal cortex, which increases reabsorption of sodium;

3   stimulates **thirst; and**

4   increases the secretion of **antidiuretic hormone** from the posterior pituitary, which not only reduces urinary loss of water but also contributes to systemic vasoconstriction.

In addition, the atrial and baroreceptor afferents relay to the hypothalamus and, in response to the reduced sensory input, stimulate thirst and release of antidiuretic hormone (pp. 277 and 646). Thus the urinary loss of water and salt is restricted and decreases in blood volume are minimized.

These effects of renin, angiotensin II, aldosterone and antidiuretic hormone take from minutes to hours to develop. This **long-term regulation** of blood volume and hence of mean systemic filling pressure (p. 433) and MABP clearly involves the endocrine, renal and cardiovascular systems. It not only compensates for the decreased blood volume of a **haemorrhage** but also operates at all times to maintain **total body water** and the **correct distribution of extracellular fluid** between the interstitium and the blood.

Other cardiac receptors

1   There are stretch receptors located in the **ventricle** (predominantly the left) that have afferents (mainly myelinated) in the vagus. They normally increase their discharge at the same time as ventricular pressure increases. When their average discharge increases they cause reflex bradycardia and vasodilatation in parallel with the baroreceptor reflex.

2   Vagal receptors with unmyelinated axons are also present, predominantly in the left ventricle. These can be stimulated experimentally by various chemicals (e.g. veratrine and nicotine) causing reflex bradycardia, vasodilatation and depression of breathing (**Bezold–Jarisch reflex**). It is thought that prostaglandins, serotonin, bradykinin and the accumulation of metabolites, as in myocardial ischaemia, are the natural stimuli for this inhibitory reflex.

3 Afferent fibres from all regions of the heart travel with sympathetic nerves (so-called **sympathetic afferents**) and are also stimulated by the above chemicals and myocardial ischaemia. However, the reflex in this case is excitatory and thus tachycardia, vasoconstriction, increased breathing and angina (p. 463) result.

Atrial natriuretic peptide (ANP)

The hormones renin, angiotensin II, aldosterone and antidiuretic hormone act to increase extracellular fluid volume but a recently discovered family of peptides, ANP, can *decrease* effective circulating plasma volume within a matter of minutes (p. 653). The physiological role of ANP is, however, not established. It is known that ANP is released from secretory granules of the atria in response to atrial stretch—this could be caused physiologically by an increased blood volume. In pharmacological doses ANP causes ateriolar vasodilatation and also venodilatation by decreasing the sensitivity of vascular smooth muscle to vasoconstrictor substances such as noradrenaline and angiotensin II. In the kidney it increases glomerular filtration rate and inhibits tubular Na^+ reabsorption, leading to increased urinary loss of water (diuresis) and sodium (natriuresis). ANP also decreases the release of antidiuretic hormone and renin/aldosterone, which may contribute to the diuresis and natriuresis, respectively.

Reflexes initiated by other receptors

Input from the receptors described below alters the set-point for MABP.

1 Nasal receptors (p. 542) when stimulated by irritants result in a sneeze, bradycardia and vasoconstriction with variable changes in MABP. Stimulation by water causes a cessation of breathing, bradycardia and vasoconstriction in all but the heart and brain, leading to an increase in both TPR and MABP (this is the diving reflex which is well-developed in diving mammals).

2 Laryngeal and **tracheal receptors** that cause the cough reflex (p. 542) also produce marked bradycardia and increases in TPR and MABP. The sniff reflex caused by stimulation of the **epipharyngeal receptors** (p. 542) however is accompanied by tachycardia as well as increases in TPR and MABP. In contrast, lung irritant receptors of the lower airways have no known cardiovascular reflex.

3 Pulmonary C receptors (also known as **juxtacapillary (J) receptors**) in the alveoli of the lung (p. 543) are stimulated by pulmonary oedema. This causes not only rapid, shallow breathing but also a decrease in heart rate, TPR and MABP. The bronchial C receptors in the airways have no known cardiovascular reflex.

4 Lung stretch receptors (p. 542) in the lower airways are stimulated by an increase in tidal volume and not only help to determine the pattern of breathing but also cause an increase in heart rate and a decrease in TPR and MABP. The accelerated heart rate during inspiration (**sinus arrhythmia**) is partially due to this reflex.

5 Peripheral or **arterial chemoreceptors** (p. 545) in the carotid and aortic bodies are stimulated by low O_2 (hypoxia), high CO_2 (hypercapnia) and acidosis, which cause a decrease in heart rate and an increase in TPR and MABP. Activation of these chemoreceptors also increases the secretion of antidiuretic hormone (p. 277). However, stimulation of aortic bodies, in contrast to carotid bodies, may elicit slight tachycardia.

Stimulation of the peripheral chemoreceptors normally results in an increase in ventilation which at the same time stimulates the lung stretch receptors. Thus the bradycardia from the carotid body reflex can be demonstrated only during a breath-hold when the dominant tachycardia from the lung reflex is prevented. The increase in TPR and MABP from the chemoreceptor reflex is also tempered by local systemic vasodilatation (p. 423) and by severe hypoxia or acidosis decreasing myocardial contractility (p. 393).

6 Central chemoreceptors (p. 546) near the ventral surface of the medulla oblongata are stimulated by hypercapnia and reflexly increase heart rate, TPR and MABP. Most of the increase in RTP and MABP is opposed by local systemic vasodilatation and by the accompanying increase in ventilation, stimulating the lung stretch receptors.

7 Limb muscle chemoreceptors (metaboloreceptors) with small unmyelinated (group IV) afferents are stimulated by metabolites, notably H^+ and K^+ ions, released during exercise (p. 690) to cause increases in $\dot{Q}$, TPR, MABP and ventilation.

8 Limb muscle stretch receptors with small myelinated axons (group III) are stimulated by movement during exercise (p. 692) and also cause increases in $\dot{Q}$, TPR, MABP and ventilation.

9 Peripheral and **central temperature receptors** (pp. 154 and 684) during an elevated body temperature reflexly cause an increase in $\dot{Q}$, and possibly in MABP, and a decrease in TPR with preferential vasodilatation in the skin.

10 Pain receptors (p. 155) initiate increases in heart rate, TPR and MABP if the pain is mild, bradycardia if the pain is severe when, despite vasoconstriction, the MABP falls.

Cardiovascular centres

Early techniques of electrical stimulation indicated separate cardiac and vasomotor centres (areas) in the dorsal brainstem (pons and medulla oblongata) but these were not located within definable nuclei. This concept of cardiovascular centres still dominates many textbook accounts. However, in the 1970s microinjection of drugs, and discrete lesions, revealed little involvement of the pons in cardiovascular control but specific nuclei were now found to be actually in the ventral, as well as dorsal, medulla. By the 1980s it was also appreciated that the cardiovascular centres extend from the cerebral cortex to the spinal cord and include areas not only in the medulla oblongata but also in the cerebellum, hypothalamus and limbic system. The cardiovascular centres are responsible for integration of sensory information and subsequent modification of efferent autonomic nerve activity to the heart and blood vessels.

Spinal cord

The preganglionc sympathetic neurons supplying the arterioles, veins, myocardial muscle and the sinoatrial node are found in the **intermediolateral cell** column of the spinal cord (Fig. 15.2). It must be appreciated that sympathetic activation is not *en masse*; control of one organ can occur independently of another. Furthermore, sympathetic tone to the vessels is greater than to the heart. The intermediolateral cell column is controlled by three areas in the medulla but it is capable of some independent cardiovascular integration, predominantly some of the reflexes connected with pain and peripheral temperature.

Medulla oblongata

Information from most of the relevant receptors is transmitted in the vagus or glossopharyngeal nerve; their afferents synapse in the **nucleus of the tractus solitarius** of the medulla oblongata (Fig. 15.2) before reaching the main integrating areas. There is also an excitatory polysynaptic pathway within the medulla from the nucleus tractus solitarius to vagal preganglionic parasympathetic neurons which provides marked inhibitory tone to the Sinoatrial node. These parasympathetic neurons emanate mainly from the **nucleus ambiguus** and to a lesser extent from the dorsal motor nucleus of the vagus (Fig. 15.2); these vagal nuclei used to be called collectively the cardioinhibitory centre.

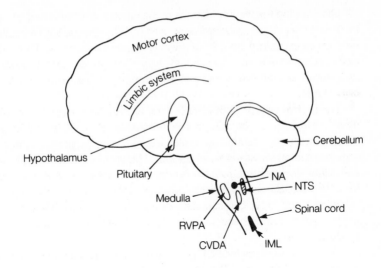

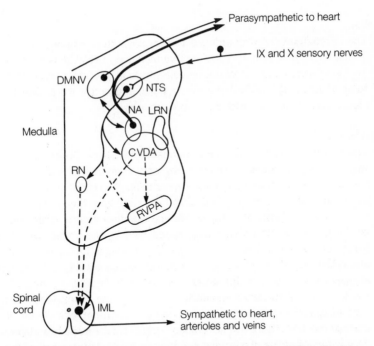

Fig. 15.2 Location in the brain of the cardiovascular centres. CVDA, Caudal vasodepressor area; DMNV, dorsal motor nucleus of the vagus; IML, intermediolateral cell column; LRN, lateral reticular nucleus; NA, nucleus ambiguus; NTS, nucleus of the tractus solitarius; RN, raphe nucleus; RVPA, rostral vasopressor area; solid lines, excitatory pathways; dashed lines, inhibitory pathways.

Overlapping the ventral margins of the nucleus ambiguus and lateral reticular nucleus is the **caudal vasodepressor area** (caudal ventrolateral medulla group).It receives an excitatory input from the nucleus tractus solitarius. Its γ-aminobutyric acid-containing neurons have an inhibitory effect on the rostral vasopressor area (see below) and on sympathetic output from the spinal cord's intermediolateral cell column.

The **rostral vasopressor area** (rostral ventrolateral medulla group) is located more rostrally and ventrally in the subretrofacial nucleus and its adrenaline-containing neurons send axons which excite the preganglionic sympathetic neurons in the intermediolateral cell column. This rostral area is tonically active but is under inhibitory control from the caudal vasodepressor area and polysynaptically from the nucleus tractus solitarius. Severe local hypoxia caused by brain ischaemia (reduced blood flow) can stimulate directly both the vasopressor area and the preganglionic sympathetic neurons (**central nervous system ischaemic** or **Cushing reflex**).

The **raphe nucleus** located ventrally near the midline is also involved, receiving excitatory input from the nucleus tractus solitarius and having an inhibitory influence on the spinal sympathetic neurons.

Cerebellum

The **cerebellum** helps coordinate the cardiovascular response to exercise. Its fastigial nucli and vermal cortex receive medullary projections from the nucleus tractus solitarius and the lateral reticular nucleus; the latter is an extra relay for some of the input from limb muscle afferents. In exercise, the cerebellum contributes to excitation of the rostral vasopressor area.

Hypothalamus

The hypothalamus contains many regions involved in cardiovascular control. The **hypothalamic depressor area** receives input from the nucleus tractus solitarius and, for instance, during the baroreceptor reflex, its output contributes to activation of the nucleus ambiguus and caudal vasodepressor area.

The hypothalamic **alerting** or **defence reaction centre**, often considered to be the hypothalamic pressor area, is activated by the limbic system (see also below) in alarm and defence responses. This results in increases in heart rate, $\dot{Q}$ and MABP, together with vasoconstriction in all organs except the heart, brain and skeletal musculature. In the latter there is vasodilatation because the defence reaction centre activates axons which descend from the cortex directly to the spinal cord to synapse with preganglionic cholinergic sympathetic dilator fibres. The defence reaction centre has an inhibitory effect on the nucleus ambiguus, caudal vasodepressor area and baroreceptor input to the nucleus tractus solitarius.

The **temperature-regulating centre** is also in the hypothalamus (p. 684). During heat stress, it receives peripheral and central thermoreceptor input and increases $\dot{Q}$, causes vasoconstriction in all organs except the heart and brain, and dilates the blood vessels of the skin (p. 454). The consequent large blood flow through the skin aids in the maintenance of a normal body temperature. The final decrease in TPR and MABP caused by vasodilatation in the skin depends on the degree of vasoconstriction elsewhere and the magnitude of the increase in $\dot{Q}$.

Finally, the hypothalamus contains the **paraventricular** and **supraoptic nuclei** (p. 276) which augment their release of antidiuretic hormone (vasopressin) in response to excitation of local osmoreceptors, stimulation of arterial chemoreceptors and depressed activity from atrial stretch receptors and arterial baroreceptors.

Limbic system

The **limbic system** integrates many emotional and behavioural reflexes (p. 254) and, via the hypothalamus, will elicit the pressor response of excitement or mild pain and the depressor response of emotional shock or severe pain.

Motor cortex

In exercise the **motor cortex** gives a central command (p. 692) which is relayed through the hypothalamus to provide much of the initial cardiovascular responses; the motor cortex may also contribute to steady-state cardiovascular responses. During the initial phase of exercise it also activates the cholinergic sympathetic dilator pathway to skeletal musculature.

Respiratory centres

The dorsal inspiratory cells of the **respiratory centres** (p. 539) in the medulla have an inhibitory effect on the cardiac vagal motor neurons of the nucleus ambiguus. This results in the tachycardia that usually accompanies inspiration (**sinus arrhythmia**).

15.2 CHARACTERISTICS OF BLOOD CIRCULATION THROUGH DIFFERENT ORGANS

The entire cardiac output flows through the pulmonary circuit, the alveolar capillaries of which subserve gas exchange with the external environment. Each of the organs supplied by the systemic circulation receives a different fraction of the resting cardiac output according to its mass, metabolic activity and how much extra blood flow it requires to satisfy its role in acquiring or disposing of nutrients, waste products, electrolytes or heat. The balance of the mechanisms controlling blood flow to each organ therefore varies—in some organs metabolic and myogenic autoregulation are dominant; in others nervous control takes precedence. In most circumstances flow to the brain and heart is protected at the expense of the other systemic organs.

Pulmonary circulation

Resistance and pressure

The pulmonary vessels have thinner, more compliant walls and are shorter in length and larger in diameter than the equivalent vessels in the systemic circuit. The entire cardiac output flows through the pulmonary circuit, the resistance of which is about a tenth that of the systemic circuit (p. 416). In the pulmonary artery, the systolic, diastolic and mean pressures at rest are about 25, 8 and 15 mmHg, respectively, and the pressure declines gradually to about 10 mmHg in the capillaries and to a mean of about 5 mmHg in the left atrium (see Fig. 14.6). Thus, in contrast to the systemic circulation, the arterial and venous parts of the pulmonary circulation contribute equally to the total pulmonary resistance. Since the pulmonary resistance is so low, flow remains pulsatile throughout, although the amplitude decreases as blood flows through the pulmonary vessels (see Fig. 14.6).

Capillaries

Pulmonary arteries deliver, from the rest of the body, systemic venous blood (which is low in O_2 and high in CO_2) to the pulmonary capillaries for exchange of gases between blood and alveolar air (p. 467); the oxygenated blood leaving in the pulmonary veins becomes the systemic arterial blood. The vast network of pulmonary capillaries in the alveolar walls provides the large surface area required for gas exchange between the alveolar air and pulmonary blood (pp. 471 and 500). The total cross-sectional area of the pulmonary capillaries, and hence the velocity of capillary flow, is similar to that of the systemic circuit (see Fig. 14.6). However, since the pulmonary capillaries are shorter, the transit time through each capillary at rest is about 1 second rather than 2 seconds. During severe exercise, there is about a fivefold increase in cardiac output and hence in pulmonary blood flow. This is associated with an increase in mean pulmonary arterial pressure, which opens up lung capillaries that at rest were unperfused, thus increasing their total cross-sectional area but by only 1.5 times. Hence, the velocity of flow through the pulmonary capillary increases considerably and the transit time can be as little as 0.3 second. Although this will impede diffusion of gases (p. 513), it is not normally the limiting factor in the supply of O_2 during exercise, at least at sea level (but see p. 694).

Capacity

In the erect posture, about 9% of the blood volume is in the pulmonary circuit. Since both pulmonary arterioles and veins are very compliant, the pulmonary circuit accommodates about 16% of the total blood volume in the supine position (see Fig. 14.6). Pulmonary blood volume also increases passively in exercise, generalized systemic vasoconstriction, left heart failure or mitral stenosis and decreases passively in generalized systemic vasodilatation and haemorrhage. Thus the pulmonary circulation functions as a blood reservoir (the **pulmonary reservoir**).

The compliant pulmonary vessels are also subjected to variations in transmural pressure induced by the fluctuating intrapleural pressure of ventilation. Thus their volume increases during inspiration and decreases during expiration. Powerful expiration or positive pressure breathing can markedly reduce pulmonary (and cardiac) blood volume and pulmonary flow, as well as impeding systemic venous return.

The effect of gravity in the erect posture (p. 475) and the low pressure and high distensibility of the pulmonary circuit results in less, or even intermittent, flow to the apex of the lung relative to the base. The consequence of this for gas exchange and the total ventilation/perfusion ratio of the lung is considered on p. 520.

Although pulmonary arterioles and veins are well-supplied with sympathetic vasoconstrictor and parasympathetic vasodilator fibres, the roles of these nerves are not clear. However, stimulation of the sympathetic system and administration of adrenaline can decrease the pulmonary blood volume by as much as 30%.

Hypoxic vasoconstriction

Neither myogenic nor metabolic autoregulation is demonstrable in the pulmonary circuit. In fact, the response of pulmonary arterioles to local hypoxia, hypercapnia and acidity is the opposite of that of systemic arterioles. **Hypoxia** is the more potent of the three stimuli. In poorly ventilated regions of the lung, where alveolar O_2 is low and CO_2 is high, local pulmonary **vasoconstriction** diverts blood flow to better ventilated areas. This improves gas exchange by helping to equalize the regional ventilation/perfusion ratios in the lung. However, this hypoxic vasoconstriction has the disadvantage that during the hypoxia of altitude or of chronic obstructive lung disease, all pulmonary arterioles are affected. The resultant increase in pulmonary resistance increases pressure in the right ventricle and pulmonary artery (pulmonary hypertension).

Ultrafiltration/reabsorption balance

Since it is essential for gas exchange that the alveoli do not accumulate liquid, the pulmonary circuit is accompanied by a lymphatic system more extensive than in any other organ. Furthermore, the normal colloid osmotic pressure of the plasma (25 mmHg) is greater than the difference in hydrostatic pressure between capillary (10 mmHg) and alveolar interstitium (– 3 mmHg; p. 485) and favours reabsorption of fluid across the entire length of the capillary. However, accumulation of fluid in the alveoli (pulmonary oedema) may occur and lead to dyspnoea (breathlessness; p. 542) when the left atrial pressure is increased during left ventricular failure or mitral stenosis or when the pulmonary vascular pressures increase during severe exercise.

Bronchial circulation

The metabolic needs of the airways (trachea to bronchioles) are met by blood delivered from the aorta through the **bronchial arteries** (see Fig. 13.1). This bronchial circulation constitutes only about 1% of the total cardiac output. Some venous drainage (about 25%) is via **bronchial veins** into the superior vena cava but the remaining 75% enters the **pulmonary veins**. There are also direct connections between the bronchial and pulmonary capillaries. This intermingling of the two circulations results in a very small degree of desaturation of the oxygenated pulmonary venous blood (p. 515).

Coronary circulation

Right and left **coronary arteries** branching from the aorta (see Fig. 13.1) supply the arterioles and capillaries of the myocardium. Venous drainage is via the **coronary sinus** (75%) and **anterior cardiac vein** (20%) which both empty into the right atrium. Some venous blood drains directly via **Thebesian veins** and small venules into all heart chambers. Venous blood entering the left side of the heart in this manner will cause a small reduction in the O_2 concentration of systemic arterial blood (p. 515).

Metabolism and control of blood flow

Since the heart at rest receives about 5% of the cardiac output and has a very high O_2 consumption relative to its mass (Table 15.1), the coronary venous O_2 content

Table 15.1 Blood flow, oxygen consumption, arteriovenous (a–v) O_2 difference and vascular resistance of tissues at rest

Organ	Weight (kg)	Blood flow (L min^{-1})	Oxygen consumption (ml min^{-1})	a–v O_2 difference (ml L^{-1})	Absolute resistance (mmHg L^{-1} min)	Specific resistance (mmHg L^{-1} min kg tissue wt)
Heart	0.3	0.25	30	120	400	120
Brain	1.5	0.75	45	60	133	200
Skin	5.0	0.50	10	25	200	1000
Skeletal muscle	30.0	1.20	55	60	83	2490
Gut	2.8	1.40	55	40	71	200
Kidney	0.3	1.10	20	15	91	27
Other	30.0	0.60	35	50	167	5010
Total	70.0	5.80	250	45	17	1190

is low and there is thus a very large arteriovenous O_2 difference. In contrast to other active organs, during increased cardiac activity there is little further decrease in its venous O_2 content and the increased demand for O_2 is satisfied mainly by a large increase (up to fourfold) in coronary blood flow. This is a prime example of **metabolic autoregulation**. Hypoxia and adenosine are more potent coronary vasodilators than an increase in $P\text{co}_2$ or blood acidity.

Coronary arterioles also exhibit another form of autoregulation such that coronary flow at rest is constant in the pressure range of 60–180 mmHg (see Fig. 14.13). (The important pressure here is aortic diastolic pressure; see below.) Such autoregulation is due to a combination of myogenic and metabolic mechanisms (p. 422).

Although the smooth muscle of the coronary arterioles receives both sympathetic and parasympathetic innervation and contains α_1-adrenoceptors (mediating vasoconstriction) and muscarinic cholinoceptors (mediating vasodilatation), the importance of these pathways is not clear. Any neurally induced vasoconstriction occurring during increased sympathetic stimulation is overridden by the metabolic vasodilatation resulting from the simultaneous increase in cardiac activity. Similarly, during increased parasympathetic stimulation, any neurally induced vasodilatation will tend to be overridden by the metabolic vasoconstriction. Note that adrenaline acting on β_2-adrenoceptors causes coronary vasodilatation.

Coronary artery compression and blood flow

Much of the resistance of the heart's vascular bed (Table 15.1) is due to the *compression* of the blood vessels during cardiac contraction. In systole, the peak pressure is about 120 mmHg in the aorta, coronary arteries and left ventricle and about 25 mmHg in the right ventricle (Fig. 15.3). Within the myocardial wall, contracting muscle will cause the intramyocardial pressures to be higher, particularly during the isovolumetric contraction phase. Thus, in systole, the coronary vessels of the left ventricle are compressed and flow in the left coronary artery is

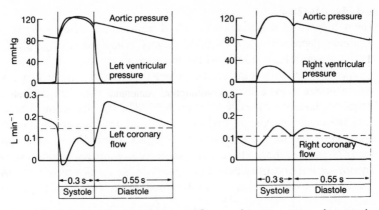

Fig. 15.3 Left and right coronary arterial flow in relation to aortic and ventricular pressures during a cardiac cycle of 70 beats per minute.

markedly reduced, whereas in the right ventricle the transmural pressure of the coronary arteries is reduced from 120 to only 95 mmHg and the right coronary flow is hardly affected. In diastole, the pressure in the coronary arteries falls to about 80 mmHg whilst that in both ventricles is close to 0 mmHg. Hence there is no compression of blood vessels in diastole. Left coronary arterial flow is therefore intermittent, ceasing and even reversing in early systole, and right coronary arterial flow is pulsatile, being slightly greater in systole than in diastole. Venous flow through the coronary sinus into the right atrium is greatest during the compression of systole and subsides during diastole.

Since coronary vessels are compressed during systole and since at rest diastole occupies about two-thirds of the cardiac cycle, about 85% of the left and 70% of the right coronary flow occurs during diastole (Fig. 15.3). (Note that as the left ventricle is the bigger, left coronary arterial flow is about 60% of the total flow.) As about 80% of the total coronary arterial flow occurs during diastole, a pressure somewhat nearer the *aortic diastolic pressure*, not the MABP, becomes the primary determinant of the pressure gradient for coronary flow. With tachycardia, which reduces the period of diastole considerably (p. 387), and with an increased myocardial contractility, a greater proportion of the coronary resistance arises from compression and coronary arterial flow is impeded for shorter but more frequent intervals.

Myocardial ischaemia

Reduction or cessation of blood flow (**ischaemia**) can be caused by atherosclerosis or thrombosis of the coronary vessels, by a low aortic diastolic pressure as in cardiac failure or haemorrhage, by a rise in atrial and ventricular diastolic pressures as in cardiac failure or by a high left ventricular systolic pressure, as in aortic valve stenosis. The resultant hypoxia of the myocardium affects mainly the left ventricle and, if severe, it leads to myocardial **infarction**, i.e. death (**necrosis**) of cardiac muscle. Treatment of myocardial ischaemia attempts to increase coronary flow (with vasodilating drugs), decrease myocardial O_2 consumption (by reducing preload and afterload with veno- and vasodilating drugs) and decrease heart rate and myocardial contractility (with β-adrenoceptor antagonists).

Cerebral circulation

The soft tissue of the brain, with its pial surface covered by a highly vascular arachnoid membrane and a tough dura mater, is encased in a rigid skull (p. 69). From the **vertebral** and **internal carotid arteries** are derived the arteries of the brain which, until they branch and penetrate the brain, lie close to the pia amongst the trabeculae in the subarachnoid space (see Fig. 3.16). Venules drain into the superficial veins of the pia which in turn drain either into the **venous sinuses** in the dura before entering the internal jugular veins or into the **vertebral veins** of the spinal cord and thence to the external jugular veins. Unlike other capillaries (p. 427), those of the brain (except in the circumventricular organs) are relatively impermeable to most substances except fat-soluble drugs, glucose, O_2 and CO_2. This impermeability is referred to as the **blood–brain barrier** and its significance is considered on pp. 71 and 547.

Cerebrospinal fluid (CSF) fills the four ventricles and subarachnoid space of the brain (see Figs 3.16 and 3.17). It is formed from plasma by brain capillaries, in particular those in the **choroid plexuses**, and is reabsorbed from the subarachnoid space through the **arachnoid villi** into the venous sinuses of the dura (p. 72). There is no lymphatic system in the brain; its drainage function is taken over by the CSF.

Cerebral blood flow (CBF) and its control

The brain receives about 13% of the cardiac output, has a moderately resistant vascular bed, a fairly high O_2 consumption and a large arteriovenous O_2 difference (Table 15.1). **CBF** remains remarkably *constant* at 0.75 L min^{-1} (see Table 13.1) over a wide range of bodily activities.

Although the cerebral arterioles are innervated by the sympathetic and para-sympathetic nervous systems, the autonomic nerves appear to have little physiological effect on CBF. In contrast, **myogenic** and **metabolic autoregulation** are important mechanisms. The former maintains CBF remarkably constant in the MABP range of about 50–150 mmHg (see Fig. 14.13) and the latter alters local blood flow to maintain a *constant O_2 supply* to individual regions of the brain according to their level of activity.

CBF is also affected when partial pressures of O_2 and CO_2 change throughout the body. Between the normal arterial P_{O_2} of 100 mmHg and a hypoxia of 50 mmHg P_{O_2}, there is little increase in CBF. Below 50 mmHg P_{O_2}, the cerebral arterioles are sensitive to the vasodilating effects of hypoxia and CBF increases about twofold as P_{O_2} decreases from 50 to 25 mmHg. However, the cerebral arterioles are much more sensitive to the slightest increase or decrease in arterial P_{CO_2} and acidity. Considerable vasodilatation occurs when P_{CO_2} is elevated (40–100 mmHg, hypercapnia) and considerable vasoconstriction occurs when P_{CO_2} is reduced (40–20 mmHg, hypocapnia). Within such limits, a doubling or halving of P_{CO_2} results in about a twofold change in CBF. The **dizziness** felt after excessive hyperventilation is the result of the reduced CBF which accompanies the resultant lowered P_{CO_2} (hypocapnia). Since a reduced O_2 supply is usually accompanied by an increase in arterial P_{CO_2} and acidity (except at altitude), the

CBF is regulated by hypercapnia rather than hypoxia to maintain a constant O_2 supply.

Insufficient O_2 delivery

If the O_2 content of arterial blood declines sufficiently or if the MABP drops, e.g. after a severe haemorrhage, below the range of myogenic autoregulation such that CBF decreases, the O_2 delivery to the brain is impaired. This precipitates sudden **unconsciousness** which, if it is of short duration, is a **faint (syncope)** but if prolonged is a **coma** (p. 247). This may be accompanied by necrosis of cerebral tissue (**cerebral infarction**). A thrombosis or a rupture in a major cerebral artery (i.e. a **stroke**) leads to local areas of brain **ischaemia** (reduced blood supply) and often to local areas of prolonged or permanent brain damage.

Intracranial pressure

Since liquid is essentially incompressible, the volume of blood, brain tissue and CSF within the rigid cranium must be constant (the **Monro–Kellie doctrine**). An increase in intracranial pressure (i.e. the pressure of the CSF) caused by insufficient reabsorption of CSF or a blockage in its normal route of flow will thus compress cerebral vessels and reduce CBF. The consequent hypoxia and hypercapnia will directly stimulate vasomotor sympathetic activity from the cardiovascular centres (p. 446) and the resultant rise in MABP, coupled with local cerebral vasodilatation, will tend to restore CBF. At the same time, the rise in MABP will reflexly decrease heart rate via the baroreceptors. These reflex responses to an increased intracranial pressure are called the **Cushing reflex**. There is obviously a limit to the increase possible in MABP and if intracranial pressure exceeds this (as in severe hydrocephalus) the CBF ceases.

A further consequence of the Monro–Kellie doctrine is that increases in cerebral blood pressure caused by **gravity** (p. 458) during a head-stand, together with any tendency to an increase in cerebral venous volume, are transmitted to the CSF such that intracranial pressure also increases. Thus little change occurs in the transmural pressure and cerebral venous return decreases only slightly. Conversely, in the erect posture, the decreases in cerebral blood pressure are compensated for by a corresponding fall in intracranial pressure. Furthermore, the major cerebral veins do not collapse because they are held open by their association with the dura.

Cutaneous circulation

At rest the skin receives about 8% of the cardiac output through its moderately resistant vascular bed. Since it has a very low O_2 consumption, the arteriovenous O_2 difference is also small (Table 15.1). The skin has an extensive superficial network of arterioles, capillaries and venules and an extensive **deep plexus of veins**. The latter may contain up to 1.5 litres of blood.

Control of cutaneous blood flow and volume

The arterioles and veins are richly innervated with **sympathetic** fibres acting predominantly via α_1-**adrenoceptors** and hence the vascular resistance and capacitance of the skin can be altered by reflexes integrated by the cardiovascular centres. For instance, during hypotension, haemorrhage, or exercise that does not generate a heat load, the blood flow to the skin can be restricted, making it feel cold, and the accompanying venoconstriction mobilizes, for the rest of the body, a sizeable volume of blood from the deep venous plexus of the skin. Cutaneous tissue, to a much greater extent than other tissues, can tolerate such a reduced blood flow by *reducing its O_2 consumption* proportionately. Metabolic autoregulation is therefore considered to be poorly developed in the skin (unless the blood flow has been reduced for some time). Note also that skin arterioles are incapable of myogenic autoregulation (see Fig. 14.13).

The α_1-adrenergic sympathetic pathway to cutaneous vessels, controlled via the hypothalamus, is also responsible for blushing or pallor during the **emotions** of embarrassment or fear, respectively.

Sympathetic control for temperature regulation

In the skin of the hands, feet and face (particularly the ears, nose and lips), there are also large numbers of **arteriovenous anastomoses** (see Fig. 14.14). These are vessels about 50 μm in diameter with thick walls of smooth muscle which contain α_1-**adrenoceptors** activated by sympathetic fibres. When **body temperature** increases, sympathetic activity to the skin decreases under the control of the hypothalamus (p. 446), resulting in vasodilatation particularly of these arteriovenous anastomoses. This vasodilatation is enhanced by **bradykinin** which is released by sweat glands due to hypothalamic activation of the cholinergic sympathetic fibres to the glands. The decrease in total peripheral resistance results in a fall in MABP and the resultant baroreceptor reflex causes a corresponding increase in cardiac output. The massive increase in skin blood flow is accommodated mainly within the arteriovenous anastomoses and the venous plexuses, which together provide a large area for heat exchange between the blood, skin and external environment. Thus the temperature of the body is regulated (p. 68) to a large extent by the amount of blood flowing through the skin which, from the thermoneutral state, can increase about 30-fold in heat stress and decrease about 10-fold in cold stress.

During prolonged exposure to the cold, the skin vasoconstriction changes to vasodilatation which produces a ruddy complexion on a cold day. This vasodilatation is not sympathetically mediated but is believed to be caused by **metabolic autoregulation**, which overrides sympathetic control, and by damaged skin cells eliciting an axon–axonal reflex (see below). **Frostbite** (skin necrosis) occurs when the skin becomes so cold that the tissue freezes.

The triple response

In the skin, the reactions of blood vessels can easily be observed. A pointed object drawn lightly over the skin causes the area of contact rapidly to become pale (the **white reaction**). The mechanical stimulation is thought to initiate local contraction of venules or precapillary sphincters. Greater pressure will cause the skin to become red (the **red reaction**), followed by a diffuse mottled reddening around the area of injury (the **flare**) and then a local swelling (the **wheal**). The latter three reactions comprise the so-called **triple response**. The red reaction probably results from the damaged cells of the skin releasing substances like histamine which rapidly cause a local vasodilatation or venodilatation. The flare results from the mechanical stimulation of nociceptive sensory endings of C fibres causing antidromic conduction in axon branches (axon–axonal reflex) that release substance P, which vaso-dilates nearby arterioles and causes local mast cells to release histamine. The wheal results from an increase in permeability of capillaries induced by histamine and substance P, with a consequent increase in interstitial fluid.

Skeletal muscle circulation

Skeletal muscle receives at rest about 20% of the cardiac output through a highly resistant vascular bed, the tone of which is controlled by **myogenic autoregulation** and by **sympathetic** activity predominantly affecting α_1-adrenoceptors. Because of the large mass of skeletal muscle (about 50% of body weight but about 90% of the total cellular mass), skeletal muscle consumes, even at rest, about 20% of the body's O_2 consumption (Table 15.1). Its large mass is also why skeletal musculature is an important contributor to overall vasoconstriction during reflexes initiated, for example, by hypotension or haemorrhage. In contrast to splanchnic and cutaneous veins, the veins in skeletal musculature are almost devoid of sympathetic nerve endings and so cannot alter neurogenically their venous capac-ity and be part of the blood reservoir.

During maximal exercise skeletal musculature receives, due to vasodilatation, about 90% of the cardiac output and accounts for about 90% of the total O_2 consumption (p. 690). The vasodilatation in skeletal muscles during established exercis results mainly from **metabolic autoregulation**. Although important at the beginning of exercise, the vasodilatation from adrenaline acting on β_2-**adrenoceptors** plays some role and that from the **cholinergic sympathetic pathway** plays no role once exercise is established.

In rhythmic exercise, blood flow in the muscles concerned can cease during the contraction phase but because of the accumulation of metabolites it is then augmented during the relaxation phase. In non-rhythmic exercise, especially that involving static contractions (isometric exercise), blood flow can cease entirely, thus limiting the duration of such exercise.

Splanchnic circulation

The splanchnic area receives through a low-resistance vascular bed about 25% of the cardiac output at rest (Table 15.1). A number of arterial branches from the **abdominal aorta** supply the stomach, intestine, pancreas and spleen which are drained by the **hepatic portal vein** into the liver. The liver receives about 75% of its blood from this source, the remainder coming from the **hepatic artery**. Blood from the liver drains via **hepatic veins** into the inferior vena cava.

Blood flow to the liver

In the splanchnic and hepatic arteries, the mean blood pressure is about 90 mmHg whilst in the portal and hepatic veins it is about 10 and 5 mmHg, respectively. The hepatic venous pressure is low because the hepatic arterioles are highly con-stricted due to an **intense vasomotor tone** from sympathetic fibres acting via α_1-adrenoceptors. Consequently blood enters the large sinusoidal liver capillaries at a pressure less than 10 mmHg. This vasoconstriction and hence the proportion of the blood flow to the liver from the hepatic artery are further controlled by metabolic and myogenic autoregulation. When the liver is more active, or there is a decrease in portal flow, **metabolic autoregulation** of hepatic arterioles causes the hepatic artery to supply up to 50% of the hepatic blood flow. Conversely, when the pressure and therefore flow in the portal system increase, the pressure is also raised (retrogradely via the capillaries) in hepatic arterioles which constrict through **myogenic autoregulation** to reduce hepatic arterial flow.

Note that the low hydrostatic pressure in the capillaries of the liver is matched by the high interstitial colloid osmotic pressure generated by the proteins which are being manufactured by the liver for the plasma. Hence the Starling forces for ultrafiltration and reabsorption across the capillary wall remain in balance (p. 429).

Blood flow to the gut

The arrangement of the blood vessels in the gut is described on (p. 557) and in Fig. 17.2. Splanchnic blood flow is adjusted to the degree of activity in the smooth muscle layers by **metabolic autoregulation**, and in the mucosal and submucosal layers by increased glandular activity releasing **bradykinin** which causes local vasodilatation.

Splanchnic contribution to control of MABP

The α_1-**adrenoceptors** in the splanchnic arterioles and veins mediate vaso- and venoconstriction in response to the generalized increase in sympathetic activity and adrenaline secretion that occurs, for instance, during exercise, hypotension and haemorrhage. This serves to counteract a fall in MABP and to redistribute blood to other more essential regions. Furthermore, the reduction in **splanchnic venous capacity** can make about 300 ml of blood available to the rest of the body. Venous reservoirs of blood are also found in the spleen of some animals (for example, the dog) but not in humans. However, since prolonged splanchnic vasoconstriction would lead to cell damage in the gut, accumulation of metabolites will eventually override some of the vasoconstriction.

Renal circulation

Because its role is to filter blood plasma and maintain electrolyte and water balance, the kidneys receive through a very-low-resistance vascular bed about 20% of the cardiac output (Table 15.1). Its blood supply and blood pressures are described on p. 609. The arterioles of the isolated kidney exhibit well-developed **myogenic autoregulation** such that renal perfusion is maintained constant at MABPs between 80 and 180 mmHg (see Fig. 14.13). Thus the renal functions of filtering plasma and regulating electrolyte and water balance are independent of fluctuations in blood pressure. However this myogenic autoregulation can be overridden by sympathetic activity and adrenaline acting on α_1-**adrenoceptors** of the kidney arterioles. Normally there is little sympathetic tone but, during exercise, hypotension, haemorrhage, systemic hypoxia and systemic hypercapnia, the renal arterioles participate in the vasoconstriction that ensures maintenance of an adequate MABP. Furthermore, during exercise or heat stress, renal vasoconstriction helps to compensate for the vasodilatation in skeletal musculature and skin.

(For renal blood flow changes in exercise, see Tables 13.1 and 21.1.) Metabolic autoregulation in the renal circulation is poor and hence when intense reflex sympathetic vasoconstriction occurs, the kidneys are vulnerable to damage (e.g. *tubular necrosis*).

15.3 CARDIOVASCULAR ADJUSTMENTS IN HEALTH AND DISEASE

The cardiovascular system has to adjust to the demands imposed by alterations in posture, thermal stress (p. 686) and exercise (p. 690). When standing, the effects of gravity cause an increase or decrease in hydrostatic pressures within the blood vessels below and above the heart, respectively. This has no effect on the energy gradients for blood flow but, at sites below the level of the heart, increases primarily the volume capacity of the veins and the amount of capillary ultrafiltration. Reflex mechanisms and the skeletal muscle venous pump counteract these problems. The cardiovascular system and its regulating mechanisms also compensate for the abnormalities induced by haemorrhage, hypotension, hypertension and cardiac failure.

Effects of posture

In Fig. 13.16 the pressures and volumes in the heart were described for the standing posture whilst in Fig. 14.6 these variables in the vascular system were for the supine position. A change in posture alters the hydrostatic pressures in the blood vessels, with resultant effects on transmural pressures and, in particular, on venous capacity. This in turn affects where the blood is distributed and, in particular, the size of the end-diastolic volume of the heart.

Hydrostatic pressure

In the supine or prone posture, all blood vessels are approximately at the level of the heart and the mean blood pressures in the aorta, the arteries and veins (of both the head and feet) and the right atrium are about 100, 95, 5 and 1 mmHg, respectively. When standing, the weight of the column of blood in the blood vessels results in an increase in arterial and venous pressure below the level of the heart and a decrease in these pressures above the level of the heart (Fig. 15.4). For each centimetre of blood (1 cmH$_2$O) above or below the level of the heart, pressure changes by 0.77 mmHg.

Thus, for a person 1.8 m tall whose head is about 50 cm above the heart in the upright posture, blood pressure in the head is about 40 mmHg less compared with the pressure in the supine position. Hence the arterial pressure at the entrance to the cranial cavity becomes 55 mmHg and, were it not for the fact that the veins in the neck collapse, the venous pressure at the cranial exit would be − 35 mmHg. Because of venous collapse, the venous pressure is in fact close to zero. In the feet, which are about 130 cm below the heart when standing, there is an increase in pressure of about 100 mmHg and hence the arterial pressure becomes 195 mmHg and the venous pressure becomes 105 mmHg provided that, in the case of the veins, the standing is motionless (see p. 459).

It must be noted that blood flows from a point of high to low total fluid energy.

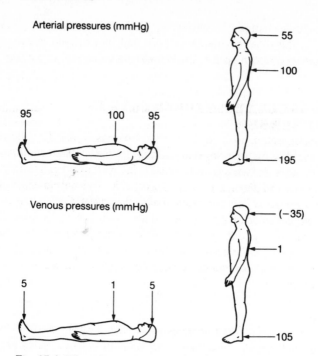

Arterial pressures (mmHg)

Venous pressures (mmHg)

Fig. 15.4 Effect of gravity on arterial and venous pressures.

The total fluid energy (p. 408) is the sum of the kinetic energy and the potential energy due to pressure plus that due to gravity. As blood flows towards the feet, it loses gravitational potential energy and gains potential energy due to pressure. Thus the total fluid energy in the arteries of the feet is the equivalent of 95 mmHg irrespective of a supine or standing posture. Therefore, although the changes in blood pressure on adoption of the standing posture mean that blood flows against a hydrostatic pressure gradient, for instance from the heart (100 mmHg) to the arteries in the feet (195 mmHg), the flow is not against a fluid energy gradient. Similarly, when total fluid energy is considered, it can be seen that it is not difficult for blood to return from the feet to the heart.

Transmural pressure, venous capacity and ultrafiltration

If blood vessels were rigid tubes, no cardiovascular adjustments for the standing posture would be required other than to compensate for decreased blood volume due to increased ultrafiltration across the capillary wall. However, the changes in transmural pressure affect, in particular, the veins which tend to empty and collapse above, and progressively fill below, the level of the heart.

The collapse of superficial veins in the neck breaks the column of blood and hence removes the effects of gravity, and the venous pressure approximates to zero. The veins, although collapsed, are not occluded so flow can continue unimpaired through spaces formed where the walls are not in total apposition. As blood accumulates behind the areas of collapse, every time the pressure rises sufficiently above zero some venous flow occurs. Deeper veins are better supported by surrounding tissue and do not collapse as much. Intracranial veins do not collapse at all because the CSF and blood are affected equally by gravity and because the venous sinuses are maintained patent by their structural association with the dura (p. 453). Thus when a person is standing the CBF depends only on the pressure in the carotid arteries, the venous pressures at the exit from the cranium, where the neck veins are collapsed, being zero at all times.

Irrespective of posture, if the mean right atrial pressure equals about 1 mmHg, the height of the uncollapsed column of blood in the external jugular vein is always about 1.5 cm *vertically* above the right atrium (which is approximately at the level of the sternal angle). A greater *length* of uncollapsed vein can obviously be seen the more tilted the subject is towards the supine position. The vertical height of this blood column above the sternal angle is used clinically as a measure of central venous pressure (p. 433) and the pulsations of the a, c and v waves and, particularly, the x and y descents (p. 390) can sometimes be distinguished at the top of the column.

Initially, on standing, venous valves prevent backflow from the heart. Veins below the level of the heart are then slowly (30–60 seconds) distended in proportion to the transmural force by the arterial inflow. This results in an increase of about 500 ml in venous blood volume, most of this being displaced from the thorax. Consequently, were it not for reflex compensations, there would be a reduction in end-diastolic volume which would reduce stroke volume (Starling mechanism) and produce systemic hypotension. Furthermore, the increased hydrostatic pressure in the capillaries below the level of the heart would cause considerably more ultrafiltration than usual and over a period of hours this would lead to loss of blood volume and, most noticeably, to **local oedema** in the ankles.

Reflex compensations

The decrease in end-diastolic volume caused by sudden standing is detected by the **atrial stretch receptors** (p. 441) and the decrease in MABP (from both the direct gravitational effect at the level of the carotids and the decrease in $\dot{Q}$) is detected by the **carotid baroreceptors** (p. 439). In the space of about 5 seconds, there are thus reflex increases in sympathetic activity and in adrenaline release, causing:

1 an increase in heart rate and in myocardial contractility tending to restore $\dot{Q}$ towards the supine value;

2 vasoconstriction in the skeletal musculature, skin, kidneys and gut, reducing blood flow to these organs, increasing TPR and decreasing capillary pressure; and

3 venoconstriction in all veins, but in particular in those of the skin and gut, which displaces some blood back to the thorax.

All of these responses contribute to **short-term regulation** and **restoration of MABP**.

Over a longer period of time, **long-term regulation**, the atrial stretch receptors will trigger an increase in antidiuretic hormone secretion. The reduced blood flow to the kidneys enhances the release of renin which increases circulating concentrations of angiotensin II and aldosterone. These responses, which are similar to those occurring after a mild haemorrhage, augment vasoconstriction and decrease urinary Na$^+$ excretion which helps **restore blood volume** towards normal.

One of the most important mechanisms counteracting the long-term effects of standing is the **skeletal muscle venous pump**. The rhythmic contraction of leg muscles reduces venous capacity and assists venous and lymphatic returns (p. 434). Intermittent closure of the venous valves also interrupts the vertical column of blood between the heart and limb and thus reduces the effective hydrostatic pressure. By these means the venous pressure in the feet can be lowered from about 105 mmHg to as little as 30 mmHg. The normal muscular contractions associated with moving around in the standing posture are sufficient to achieve this and hence there is less of an increase in venous capacity and no local oedema. Prolonged standing, venous obstruction or pregnancy can over-stretch the veins and lead to incompetent venous valves. The consequent ineffectiveness of the skeletal muscle venous pump results in the excessive distension of, in particular, superficial veins which fill with sluggishly moving blood—the condition known as **varicose veins**.

Restoration of the MABP may be inadequate during motionless standing or when suddenly standing (particularly in hypotensive patients or in patients with an impaired sympathetic system or in subjects in a hot environment where venodilatation in the skin predominates). If the MABP declines to a level (about 50 mmHg) below which myogenic autoregulation of cerebral blood flow cannot occur, the person becomes **dizzy**, has impaired vision and may even faint. **Fainting** has the advantage of returning the person to the supine position!

Haemorrhage

Loss of 10% of the blood volume (about that given by a blood donor) is easily compensated for and results after a few minutes in little change in MABP. Loss of up to 30% over a relatively brief period (e.g. 30 minutes) will cause a fall in MABP to about 70 mmHg and signs of **mild shock**. The cardiovascular reflexes initiated by such a fall will eventually compensate completely for the loss of blood volume. However, the rapid loss of a greater volume of blood will result in **severe shock** which may become irreversible if not treated by blood transfusion or by raising the blood volume by administration of a fluid with similar colloid osmotic pressure to plasma. The signs of shock are described in the next section. Prompt restoration of the lost blood prevents the adverse consequences of intense reflex vasoconstriction in the renal, splanchnic, muscle and skin vascular beds. Thus the drastic reduction in organ perfusion, which deprives the tissue of sufficient O_2 and can lead to permanent damage, in particular to the kidneys (**tubular necrosis**), is avoided.

Following a haemorrhage the decrease in blood volume reduces end-diastolic volume and hence stroke volume, $\dot{Q}$ and MABP. Detection of these reductions by atrial stretch receptors and arterial baroreceptors will result in an immediate increase in sympathetic activity and adrenaline secretion and a slightly slower increase in the secretion of angiotensin II, antidiuretic hormone and aldosterone. The consequent responses (p. 442) are the same as those which follow a change from the supine to the erect posture (p. 459) but are of greater magnitude. These responses may serve to restore $\dot{Q}$ and MABP towards normal. Further restoration can occur only by making good the deficit in blood volume. After about 15 minutes following a ~25% blood volume loss, at least 500 ml of plasma volume will have been provided by the transfer of interstitial fluid into the capillaries as a conse-quence of the reduced capillary pressure that has resulted from the hypotension itself and the vasoconstriction. In the next few hours, further restoration of circulating blood volume and replenishment of interstitial fluid occurs because of the increased water and Na^+ retention by the kidneys (urine production is minimal) and because of, when possible, the quenching of thirst. In contrast, replacement of the plasma proteins by hepatic synthesis requires 3–6 days and the formation of new erythrocytes takes a period of about 4 weeks.

The progress of these adjustments to haemorrhage can be examined using the cardiac and vascular function curves (Fig. 15.5), which were described on (p. 435). In Fig. 15.5, position A is the normal pre-haemorrhage operating point, $\dot{Q}$ being about 6 L min^{-1} and mean right atrial pressure about 1 mmHg. Immediately after the rapid loss of ~25% of the blood volume, the cardiac function curve is normal but, due to the reduction in blood volume, the vascular function curve is shifted substantially downwards and thus the operating point moves to position B. Position B depicts the resultant low right atrial pressure, reduced end-diastolic volume, small stroke volume and fall in $\dot{Q}$ which will produce hypotension. After about 5 seconds baroreceptor and atrial stretch receptor reflexes cause a sympathetic increase in myocardial contractility and vasoconstriction, which moves the cardiac function curve upwards and steepens it. At the same time reflex venoconstriction shifts the vascular function curve upwards a little. Thus position C is reached. After about 15 minutes, restoration of some of the lost volume (by capillary reabsorption of interstitial fluid) will move the systemic vascular function curve upwards a little further and the improved MABP will result in less sympathetic drive to myocardial contractility and less vasoconstriction and hence a flatter ventricular function curve. Thus position D will be reached. After a few more hours, the cardiovascular system will have returned to the original position of A.

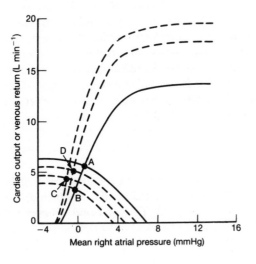

Fig. 15.5 Cardiac and vascular function curves in response to haemorrhage. Operating point: A, pre-haemorrhage; B, immediately after a rapid haemorrhage; C, after reflex increase in myocardial contractility, vasoconstriction and venoconstriction; D, after restoration of some of the blood loss.

Hypotension, fainting and shock

Hypotension

Arterial systolic pressures below 100 mmHg are taken to indicate **hypotension**. Such pressures are associated usually with decreases in $\dot{Q}$ but sometimes with decreases in TPR and occasionally in both. Hypotension may be the result of:

1 haemorrhage;
2 cardiovascular disorders such as aortic or mitral stenosis or cardiac failure;
3 loss of tone in resistance and capacitance vessels;
4 allergic or toxic reactions causing vasodilatation;
5 extreme pain;
6 standing;
7 coughing or straining (e.g. in defecation), if the intrathoracic pressure is raised sufficiently to impair venous return;
8 a $\dot{Q}$ during exercise that is insufficient to compensate for the vasodilatation in the skeletal musculature; and
9 endocrine disorders, for example Addison's disease (p. 297);

Fainting

The commonest manifestation of acute hypotension is **fainting (syncope)**, which is the sudden loss of consciousness due to cerebral ischaemia. It is usually of short duration since the resulting supine position helps elevate the MABP. Fainting that results from sudden bradycardia and diffuse vasodilatation is referred to as **vasovagal syncope**. If this is precipitated by strong emotion, the term **psychogenic syncope** is sometimes used.

Shock

Hypotension becomes critical only when inadequate blood flow disrupts organ function. This is referred to as **shock**. The signs of shock are pallor, coldness of the skin and sweating, collapse of superficial veins, hyperventilation, tachycardia, reduced urine formation, thirst, decreased body temperature, decreased metabolism and metabolic acidosis. If shock is untreated or sufficiently severe, it is likely to become irreversible with severe hypotension, unconsciousness, central nervous system damage, hypoventilation and further reductions in $\dot{Q}$, which will lead eventually to respiratory and cardiac failure and to death. Inadequate $\dot{Q}$ can arise either from sudden failure of the heart to pump, as in coronary occlusion, or from a severe reduction in blood volume due to dehydration, haemorrhage or plasma loss resulting from burns.

Hypertension

Repeated measurements of arterial diastolic pressure above 90 mmHg in the resting supine subject are taken clinically to indicate the condition of **hypertension**. It is associated with increases in $\dot{Q}$, TPR or blood volume. Clinically, hypertension is classified as primary or essential (95% of cases) and secondary or symptomatic.

The causes of **primary hypertension** are not clear but may include:
1 hereditary factors;
2 a large Na^+ intake in the diet; and
3 psychological factors.
   Secondary hypertension can be the result of:
1 renal disease causing renal vasoconstriction, reduced renal blood flow, the release of renin and a consequent increase in blood volume;
2 endocrine disorders leading to increased adrenaline secretion or to increased aldosterone or glucocorticoid secretion resulting in Na^+ retention; and
3 toxaemia in pregnancy.
   In hypertensive patients, the baroreceptor reflex modulates MABP around an elevated set-point. Hypertension can lead to further renal damage, more severe hypertension, ischaemic heart disease, congestive heart failure and strokes.

Cardiac failure

Substantial cardiac (ventricular) failure most commonly results from **impaired coronary perfusion** (ischaemic heart disease) usually caused by coronary atherosclerosis (thickening and hardening of the arterial wall). Cardiac failure may also result from heart valve disease or hypertension. However, a mild reduction in coronary perfusion may cause only **angina pectoris**, which is the term applied to the pain arising from myocardial hypoxia, which is commonly triggered by exercise.
   The sequence of events in an instantaneous cardiac failure can be examined using cardiac and vascular function curves (Fig. 15.6). Position A is the normal operating point where $\dot{Q}$ is about 6 L min^{-1} and mean right atrial pressure is about 1 mmHg. When the heart fails, myocardial contractility is insufficient to eject an adequate stroke volume and $\dot{Q}$ and MABP fall. As the heart becomes engorged with blood, end-diastolic volume and atrial pressures are raised. The

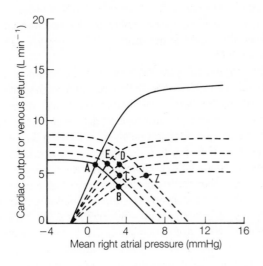

Fig. 15.6 Cardiac and vascular function curves in response to cardiac failure. Operating point: A, pre-failure; B, onset of sudden cardiac failure; C, after short-term reflex adjustments; D, after long-term adjustments; Z, position if short and long-term mechanisms fail; E, after clinical treatment.

cardiac function curve therefore becomes very flat but initially there is no change in the vascular function curve. The new operating point now occurs at position B with a low $\dot{Q}$ and high right atrial pressure.

Reflex mechanisms now come into play to try and minimize the decrease in $\dot{Q}$ and attempt to maintain CBF. Arterial baroreceptors detect the hypotension and their reflexes dominate the opposing reflexes triggered by the atrial stretch receptors detecting the increased end-diastolic volume. The reflex increase in sympathetic activity and adrenaline causes tachycardia, some improvement in myocardial contractility and vasoconstriction, which together steepen a little the cardiac function curve. Reflex venoconstriction causes a displacement upwards in the vascular function curve. Thus position C is reached within about 1 minute.

There now follows a sequence of longer-term adjustments over a week or so. The diminished blood flow and vasoconstriction have reduced renal perfusion and hence glomerular filtration. As a consequence there is Na^+ and water retention. Furthermore, the increased sympathetic activity and reduced renal blood flow promote the secretion of renin which increases circulating angiotension II and aldosterone and thus Na^+ retention is further enhanced. Meanwhile the baroreceptors have promoted antidiuretic hormone production, reinforcing the water retention. Fluid retention in the blood elevates capillary pressures, disturbing the Starling equilibrium and increasing interstitial fluid volume. When fluid retention reaches 3–4 litres, generalized oedema results and, if the left atrial pressure has also become raised, there may be pulmonary oedema too. Impaired lung gas exchange will result in cyanosis and hyperventilation, often with dyspnoea (breathlessness). The raised blood volume displaces the vascular function curve upwards even further. During this period of a week or so, some myocardial repair may have occurred, together with some hypertrophy of the undamaged myocardium in response to the prolonged increase in afterload caused by vasoconstriction. Thus an improved cardiac function curve may eventuate and the new operating point is now at D for this **stable state of chronic compensation** where $\dot{Q}$ is nearly back to normal. However, the heart remains congested, fluid retention is obvious and the ability to exercise is severely limited. Furthermore, the hypertrophied myocardium and dilated ventricles worsen the condition because they make greater demands on myocardial O_2 delivery (p. 397).

Note that when the initial cardiac failure is very severe, even position C cannot be reached because the failing ventricle cannot either respond to the increased sympathetic drive or cope with the increase in afterload from the vasoconstriction. The cardiac function curve either does not improve or gets even flatter. Renal function deteriorates further and fluid retention gets worse, moving the vascular function curve further upwards. Thus position Z rather than C eventuates. To prevent death from **congestive heart failure**, clinical intervention is required.

Administration of digoxin or other cardiac glycosides may be able to increase myocardial contractility and therefore steepen the cardiac function curve. They appear to act by inhibiting the sarcolemmal $Na^+–K^+$ ATPase so that the Na^+ gradient across the membrane falls, decreasing $Na^+–Ca^{2+}$ counter-transport (pp. 12 and 13). Thus Ca^{2+} expulsion from the cell slows down and intracellular Ca^{2+} accumulates, making more Ca^{2+} available for the contractile process.

Administration of vasodilating drugs will reduce the afterload and therefore also help steepen the ventricular function curve. Administration of diuretics reduces fluid retention and venodilating drugs increase venous capacity; these changes will move the vascular function curve downwards. After successful clinical treatment the operating point will shift to position E and eventually back to the normal position A.

Chapter 16
Respiration

16.1 INTRODUCTION

In biochemical terms, **cellular respiration** is the oxidation of organic substances by the passage of electrons through the coenzyme chain to molecular O_2. In the process O_2 acts as an electron acceptor and is converted to water. During oxidation, CO_2 is produced and energy is stored in the high energy bonds of adenosine triphosphate (ATP). Physiologists use the term respiration in a broader sense to include:

1 diffusion of O_2 and CO_2 between capillaries and cells;
2 carriage of these gases in the blood; and
3 gas exchange between blood and the external environment, which involves not only diffusion across the lung capillaries but also the bulk movement of gases in and out of the lung.

Diffusion of O_2 and CO_2 between the cells and systemic capillaries and between pulmonary capillaries and the gas of the alveoli occurs over short interstitial distances and through thin capillary walls. The quantity of gas transferred (**gas exchange**) is usually expressed as the rate of O_2 **consumption** and CO_2 **production** by the body as a whole.

For gases to be exchanged with the environment, they have to be moved into (inspiration) and out of (expiration) the alveoli of the lungs by **bulk flow**. This results from pressure gradients generated by the action of respiratory muscles and is aided in expiration by elastic recoil of the lungs. Bulk movement of gases, and the volumes and pressures involved, are the topic of **lung mechanics**. The volume of gas moved per minute is called the pulmonary **ventilation**. That part of each inspiration which reaches the **alveoli** is available for gas exchange by diffusion; the remainder occupies the conducting airways (**dead space**) across which no gas exchange occurs.

Gas transport around the body depends on **bulk flow** of blood in the cardiovascular system. For adequate and efficient gas exchange, not only must the alveoli be ventilated but also blood must flow through, i.e. perfuse, the alveolar capillaries. Thus the ratio of ventilation to **blood perfusion** must be reasonably uniform throughout the lung. Gas exchange at most tissues depends on a blood flow proportional to the metabolic need—much of this is provided by alterations in arteriolar tone in response to local metabolites (metabolic auto-regulation, p. 423).

The contents of O_2 and CO_2 in the blood are much higher than the amounts that can be carried in physical solution. **Haemoglobin** in the red blood cells plays a central role in the chemical combination of O_2 and CO_2 with blood. Carriage of CO_2 in the blood also leads to the formation of H^+ and HCO_3^- ions and thus affects **acid–base balance**.

In a normal healthy person at sea level, ventilation and the partial pressures of O_2 and CO_2 in arterial blood remain, at rest, remarkably constant. If metabolic demands change (e.g. in exercise), if the conditions under which O_2 is supplied alter (e.g. at altitude) or if the acid–base balance needs correction, then appropriate alterations in ventilation and blood flow occur by **reflex regulation**. The respiratory centres in the brain not only integrate chemoreceptor and mechanoreceptor information but also generate the rhythmicity of breathing.

Non-respiratory functions

The lungs also have functions other than gas exchange. For example:

1   the pulmonary capillary bed acts as a blood filter preventing particles like small clots, detached cells or air bubbles from reaching the systemic circulation where they may obstruct capillary blood flow within essential organs such as the brain;

2   the airways remove airborne particles by phagocytosis or mucociliary action aided by coughing;

3   ventilation of the airways contributes to heat loss (p. 679) and water loss (p. 646);

4   the pulmonary vessels are an important reservoir for blood (p. 445);

5   the larynx is used for the production of sound (phonation); and

6   the lung tissue has many metabolic functions, such as:

(a) conversion of angiotensin I to angiotensin II;

(b) synthesis and removal of bradykinin and certain prostaglandins;
(c) storage and release of serotonin and histamine;
(d) inactivation of noradrenaline and adrenaline;
(e) synthesis of peptides like substance P and opiates;
(f) secretion of heparin by mast cells; and
(g) secretion of immunoglobulins in the bronchial mucus.

These non-respiratory functions of the lungs will not be covered in any further detail.

Symbols and abbreviations

For use in respiratory physiology there is an agreed set of symbols and abbreviations (Table 16.1); those used for lung mechanics, however, vary.

The respiratory tract

On inspiration, air passes through the **upper respiratory tract** (nose, pharynx, larynx) into the trachea (Fig. 16.1); air can also gain entry to the pharynx via the mouth. The conchae of the paired **nasal cavities** create a vast surface area which is lined by ciliated columnar epithelium containing mucus-secreting cells; there are also coarse hairs or vibrissae at the nostrils and a dense vascular network in the submucosa. In the nasal cavities the air is **filtered** to remove foreign particles, **warmed** to 37°C and **humidified**. Sensory nerve endings of the trigeminal nerve detect irritants in the nasal mucosa and trigger sneezing. In the **pharynx**, sensory endings of the glossopharyngeal nerve detect irritants that cause the aspiration reflex. The pharynx serves as a common passageway for food and liquid entering from the mouth and for gas entering from the nasal cavities. During swallowing, food and fluid are deflected from the entrance to the larynx by a cartilaginous flap called the **epiglottis**.

The **larynx** is a cartilaginous 'box' that contains the **vocal cords** separated by an aperture called the **glottis**, which closes during swallowing. Vibration of these cords as air passes through the glottis produces sound, the amplitude and pitch of which can be altered by the speed of air movement and the size of the glottis respectively. Changes in glottal aperture also occur in normal breathing. The glottis dilates during inspiration and constricts during expiration; the constriction increases airway resistance and therefore prolongs (brakes) expiration. Laryngeal muscles controlling the glottis are skeletal and are innervated by the recurrent laryngeal nerve. Another vagal branch, the superior laryngeal nerve, contains afferent fibres from mucosal irritant receptors which initiate the cough reflex.

The **lower respiratory tract** commences with the **trachea** (diameter ~2.5 cm) which divides into the two main **bronchi**, one to each lung (Fig. 16.1). These repeatedly subdivide within the lung until, after some 23 'generations' or divisions, the alveoli are reached. The first 16 generations are the **conducting airways**. The walls of the trachea and bronchi down to the 11th generations (diameter ~1 mm) are supported by cartilage, comprising C-shaped rings in the trachea and incomplete plates in the bronchi, have bands of smooth muscle and are lined by a ciliated pseudostratified columnar epithelium. Mucoserous glands in the submucosa and

Table 16.1 Glossary of symbols

*Gas exchange**		
General variables	V	Gas volume in general. Pressure, temperature and percentage saturation with water vapour must be stated
	$\dot{V}$	Gas volume per unit time
	P	Gas pressure in general
	F	Fractional concentration in dry gas phase
	$\dot{Q}$	Volume flow of blood
	C	Content in blood phase
	f	Respiratory frequency—breaths per unit time
	R	Respiratory exchange ratio in general (volume CO_2/ volume O_2)
	D	Diffusing capacity in general (volume per unit time per unit pressure difference)
Symbols for the gas phase	I	Inspired gas
	E	Expired gas
	A	Alveolar gas
	T	Tidal gas
	D	Dead-space gas
	B	Barometric
Symbols for the blood phase	a	Arterial
	v	Venous
	c	Capillary
Special symbols and abbreviations	$\bar{x}$	Dash above any symbol indicates a mean value
	$\dot{x}$	Dot above any symbol indicates a time derivative
	x'	Apostrophe following a symbol indicates an 'end' value
	STPD	Standard temperature, pressure, dry (0 °C, 760 mmHg)
	BTPS	Body temperature, pressure, saturated with water
	ATPS	Ambient temperature, pressure, saturated with water
Examples	F_AO_2	Fraction of O_2 in the alveolar air
	$\dot{V}O_2$	Volume of O_2 consumed per minute
	P_cO_2	Partial pressure of O_2 in the capillary blood
Respiratory mechanics†	l	Total lung
	w	Chest wall
	rs	Total respiratory system, i.e. l + w
	mo	Mouth opening
	alv	Alveolar
	pl	Pleural surface
	bs	Body surface
	el	Elastic recoil
	res	Airflow resistance

* From Otis, A.B. (1964) In: Fenn, W.O. & Rahn, H.(eds) *Handbook of Physiology*, Section 3, *Respiration*, vol. 1, pp. 681–698. American Physiological Society, Washington.
† After Mead, J. & Milic-Emili, J. (1964) In: Fenn, W.O. & Rahn, H. (eds) *Handbook of Physiology*, Section 3, *Respiration*, vol. 1, pp. 363–376. American Physiological Society, Washington.

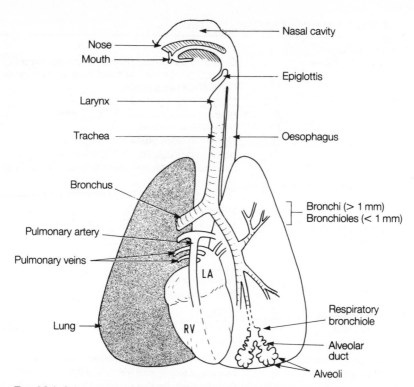

Fig. 16.1 Schematic representation of the upper and lower respiratory tract. LA, left atrium; RV, right ventricle of the heart.

mucus-secreting goblet cells in the epithelium produce a fluid which contributes to humidification and helps trap particles and soluble pollutants. These are then moved towards the pharynx by the action of the cilia (the so-called mucociliary escalator).

The 12th to 16th divisions of the airways are called the **bronchioles** (diameter <1 mm) which lack cartilage and therefore are susceptible to collapse when compressed by strong expirations. The lining epithelium is cuboidal and it contains many secretory neuroepithelial bodies and, in place of globlet cells, Clara cells—the functions of both are under investigation. The main components of the bronchiolar wall are elastic fibres and smooth muscle. The smooth muscle is innervated by parasympathetic fibres from the vagus nerves causing bronchoconstriction; the role of sympathetic innervation is unclear. Vagal nerve endings sensitive to stretch and irritants are found throughout the lower airways.

The 17th to 19th generations are the **respiratory bronchioles** which give rise to a few single alveoli and form the **transitional zone**. The 20th to 22nd generations are the **alveolar ducts** which are ~500 μm in diameter and are completely lined by alveoli (Fig. 16.2). The airways end blindly in clusters of numerous alveoli (the alveolar sacs) which can intercommunicate through the pores of Kohn. The alveolar ducts and alveoli comprise the **respiratory zone** where gas exchange occurs.

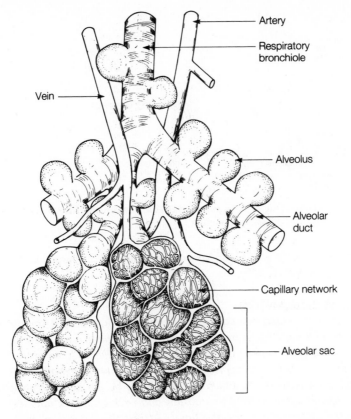

Fig. 16.2 Structure of the terminal airways and alveoli. The capillary networks are shown for only one alveolar sac.

There are 300–600 million alveoli in the lungs and, since each **alveolus** at the end of a normal expiration is ~100 µm in diameter, they provide 50–90 m² of surface area for gas exchange. The alveoli abut with one another so that the wall (interalveolar septum) of one is shared with another. The alveolar wall is, in effect, a huge dense capillary network and has a number of very thin layers through which gas molecules must diffuse. These layers comprise the fluid lining layer, the alveolar epithelium, the interstitium and the capillary endothelium (total average thickness ~1 µm) and are referred to collectively as the alveolar–capillary membrane or air–blood barrier (Fig. 16.3). The interstitium contains reticular and elastic fibres, which confer on the lungs some elasticity. The alveolar epithelial cells are of two kinds. Type I, by far the most numerous, are thin squamous cells. Type II are cuboidal in shape and have villi at their apical surface. They store surfactant in lamellar bodies and secrete it into the fluid lining layer. Alveolar macrophages rove over the surface of the alveoli and provide the last line of defence against foreign particles.

Alveolar capillaries receive blood low in O_2 and high in CO_2 through the pulmonary circulation (p. 447), while capillaries supplying the walls of the larger

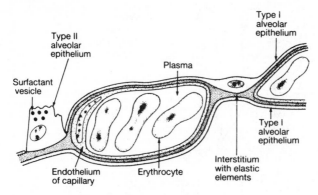

Fig. 16.3 Structure of the shared wall of two alveoli. The very thin fluid lining layer covering the alveolar epithelium is omitted.

airways receive oxygenated blood from the aorta through the bronchial part of the systemic circulation (p. 449).

16.2 LUNG MECHANICS

The thorax is a sealed compartment lined on the inside by a smooth moist membrane (the parietal pleura). There is a similar membrane (the pulmonary pleura) on the outer surface of the lungs and the cohesive force of the narrow liquid-filled space between the pleura ensures that the lungs do not collapse away from the chest wall. Air is drawn into the lungs by the action of inspiratory muscles (diaphragm, external intercostals and neck and back muscles) which enlarge the thorax and hence the lungs. At rest the diaphragm is the main inspiratory muscle. Expiration at rest is achieved entirely by elastic recoil of the lungs which pulls the diaphragm up and the chest inwards. More forceful expirations require contraction of the abdominal and internal intercostal muscles. The volume in the lungs at maximum inspiration is called the total lung capacity (TLC; ~6 litres) and its subcomponents are inspiratory reserve volume (IRV), tidal volume, expiratory reserve volume (ERV) and residual volume (RV). The first three volumes comprise the vital capacity (VC); the last two volumes comprise the functional residual capacity (FRC).

The thorax and respiratory muscles

The lower part of the trachea and the paired lungs are contained within the **thorax** which is a closed compartment. It is sealed at the top by connective tissue and muscles, which are attached to the sternum, upper ribs and vertebral column, and is totally separated from the abdomen by the upward dome of a thin sheet of skeletal muscle, the diaphragm (Fig. 16.4). In the midline, membranes associated with the pericardium of the heart, large blood vessels and the oesophagus create separate right and left compartments such that, should one lung collapse, the other can remain inflated. The inner surface of the ribcage and diaphragm is lined by a moist membrane, the **parietal pleura**, which is continuous with the midline membranes. Firmly attached to the outer surface of the lungs is another moist

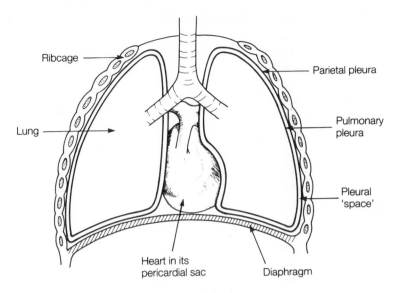

Fig. 16.4 Association between lungs and the thorax.

epithelial membrane, the visceral or **pulmonary pleura**. The **pleural space** (also called inter- or intrapleural space) between the parietal and pulmonary pleura is filled with a thin film of liquid which produces strong cohesive forces that prevent the pleura from separating but permit them to slide with ease over each other. If air is allowed to enter the pleural space (**pneumothorax**), the lungs no longer remain attached to the thoracic walls and collapse. Usually the cohesive attachment of the lung to the chest ensures that when the chest enlarges, so too do the lungs, and when the lungs elastically recoil, the chest correspondingly shrinks.

The dimensions of the chest are altered by the skeletal muscles of respiration (Fig. 16.5); those not active during quiet breathing are called the accessory muscles. During quiet breathing **inspiration** is achieved mainly by contraction of the **diaphragm,** which is innervated bilaterally by the **phrenic nerves** arising from cervical roots 3–5 of the spinal cord. Contraction of the diaphragm has the double action of depressing the floor of the thorax and raising the ribcage at its point of origin, thus respectively elongating and widening the thorax and hence increasing its volume. As the dome of the diaphragm becomes flattened by the contraction, the abdominal viscera are pushed against a relaxed abdominal wall. Between the ribs, extending diagonally downwards as they come forwards, are the **external intercostal muscles** which favour inspiration by elevating the anterior end of each rib, causing an increase in the width of the chest. Larger strenuous inspirations need the assistance of the **accessory muscles** of inspiration, such as the sterno-mastoid and scalene muscles of the neck and back, to raise the sternum.

During quiet breathing **expiration** is a passive event. No muscular contraction is involved and the thorax shrinks as a result of **passive elastic recoil** of elements in the lung stretched during the previous inspiration. Forceful expirations require the elastic recoil to be aided by muscle activity from the accessory muscles of

Inspiration

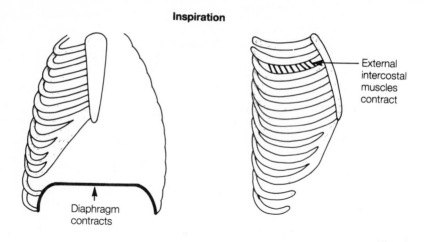

External
intercostal
muscles
contract

Diaphragm
contracts

Expiration

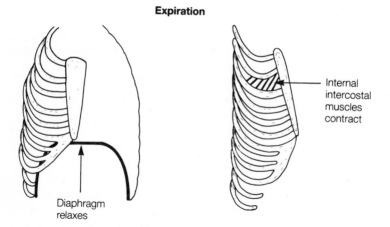

Internal
intercostal
muscles
contract

Diaphragm
relaxes

Fig. 16.5 Diaphragm and intercostal muscles.

expiration. The **internal intercostal muscles** extend diagonally upwards as they come forwards and their contraction depresses the ribcage. Contraction of **abdominal muscles** drives the diaphragm upwards into a highly domed shape expelling gas from the lungs—it is the more important expiratory muscle.

It is now known that contraction of the external and internal intercostal muscles is pronounced only during stimulated breathing or lateral bending of the trunk; their action is primarily to support the chest wall in the intercostal spaces. Furthermore there is now evidence that the muscles used for a quiet inspiration are the diaphragm aided by the **parasternal intercostal** and scalene muscles.

Expiration is more prolonged than inspiration, particularly in quiet breathing. It commences slowly because some inspiratory diaphragmatic activity continues into the first phase of expiration and it is prolonged by the braking effect that arises from expiratory constriction of the glottis. When the rate of breathing increases, the duration of expiration shortens markedly because the next inspiratory effort is timed to start earlier; furthermore there is now expiratory muscle activity and removal of some of the braking.

Lung volumes

The volume of air breathed in and out of the lungs is called the **tidal volume** (V_T). At rest this volume is about 500 ml in healthy young adults ($\sim$70 kg). At the end of a normal inspiration the extra volume that can be inhaled with maximal voluntary inspiratory effort is the **inspiratory reserve volume (IRV)**—about 3 litres (Fig. 16.6). At the end of a normal expiration the maximal volume that can be voluntary exhaled is the **expiratory reserve volume (ERV)**—about 1.3–1.5 litres. These volumes are readily measured by mouth-breathing in and out of a **spirometer** using breaths that are not rapidly forced. One cannot, however, expel all the gas from the lungs—there remains after maximal exhalation a volume of about 1.2 litres called the **residual volume (RV)**. This volume can only leave the lungs if they collapse away from the chest wall, for example in pneumothorax when air emboli break the cohesive film in the pleural space. (Even then about 200 ml remains trapped—the so-called minimal volume.)

The RV can be measured, after performing a maximal expiration, by breathing in a gas mixture containing some helium—a gas which is neither produced nor absorbed by the body. After exchanging a number of breaths with a closed system, the helium will become added to and mixed with the gases in the RV and hence the concentration of helium will be diluted (concentration = amount × volume). This is called the **helium dilution technique**.

Combinations of these volumes are referred to as capacities. Thus the volume remaining in the lungs at the end of a normal expiration, comprising the ERV and RV, is the **functional residual capacity (FRC)**—about 2.5 litres (Fig. 16.6). This is a substantial volume and, apart from the $\sim$150 ml dead space volume in the airways (p. 496), is the volume of gas normally filling the alveoli at the start of the inspiration. The volume of gas that can be shifted in and out of the lungs during maximal inspiration and expiration, about 5 litres, comprising the IRV, V_T and ERV, is the **vital capacity (VC)**. The total volume of gas (about 6 litres)

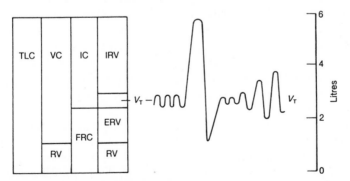

Fig. 16.6 The subdivisions of lung volume. ERV, Expiratory reserve volume; FRC, functional residual capacity; IC, inspiratory capacity; IRV, inspiratory reserve volume; RV, residual volume; TLC, total lung capacity; VC, vital capacity; V_T, tidal volume.

occupying the lungs at maximal inspiration includes in addition the RV and is the **total lung capacity (TLC)**. Terms such as inspiratory capacity (IRV + $\dot{V}_T$) and expiratory capacity (V_T + ERV) are also used.

These volumes and capacities increase with body size and are slightly smaller in females. With ageing there is a reduction in elastic recoil of the lungs and stiffening of the chest wall, which leads to a gradual increase in RV and FRC (particularly the former as ERV decreases) and a fall in VC with little change in TLC. Thus, in a younger person RV tends to be smaller than ERV; in an older person the RV is the larger. When lying down compared with standing, the RV does not change but FRC (and hence ERV) decrease because the outward elastic recoil of the chest is reduced (p. 480) as the weight of the abdominal contents no longer pulls the diaphragm downwards. When FRC is so reduced, more lung volume is available as the IRV. Because there is a greater blood volume in the thorax when lying down (p. 459), TLC and VC may decrease a little.

Lung volumes and capacities can also be affected by lung diseases. For instance, in **rstrictive lung diseases** like *pulmonary fibrosis*, the alveoli are so stiff that their expansion is restricted and all volumes and capacities are therefore less than normal. To compensate, the work of breathing is minimized (p. 494) by V_T becoming shallower and the frequency of breathing more rapid—such a pattern also maintains the product of the two, i.e. ventilation, close to normal. In **obstructive lung diseases**, like *asthma* (where the airways are narrowed by bronchoconstriction), *chronic bronchitis* (where plugs of mucus and inflammatory swelling of the bronchial mucosa obstruct the airways) and *emphysema* , vigorous expiration can raise pleural pressure so much that air is trapped in the alveoli, raising the RV and hence the FRC. There is little change in IRV but the ERV and VC are often reduced. In emphysema these volume changes are greater because destruction of alveolar tissue leads to loss of elastic recoil and the loss of radial traction this provides. The latter normally keeps small airways patent (p. 488). To compensate, V_T may be deeper and breathing frequency slower.

Lung-related pressures

Air enters and leaves the lungs by bulk flow down pressure gradients between the mouth and the alveoli. The required changes in alveolar pressure are created by outwards and inwards movements of the chest, which at the same time cause changes in pleural pressure. When the gas is stationary, the alveolar pressure is the same as at the mouth (atmospheric). Whether air is flowing (a dynamic event) or not (a static event), the pleural pressure is affected by the inward elastic recoil of the lungs and, at volumes below ~4 litres, by the outward elastic recoil of the chest. The recoil of lungs and chest in opposite directions creates a negative pressure in the pleural space (– 5 cmH₂O at FRC). The recoil forces are dependent on the degree of inflation. To examine these correctly, transmural pressures should be considered—the most important being the transpulmonary pressure (alveolar pressure minus pleural pressure).

There are four pressures related to breathing (Fig. 16.7a)—**mouth pressure** (P_{mo}), **body surface pressure** (P_{bs}, which is atmospheric unless in a pressure suit), intrapulmonary or **alveolar pressure** (P_{alv}) and intrapleural or **pleural pressure** (P_{pl}). (Cardiovascular physiologists also refer to pleural pressure as intrathoracic pressure.)

The gas that enters and leaves the lungs does so by bulk flow down the pressure gradient between the mouth (or nose) and the alveoli—thus the pressure gradient

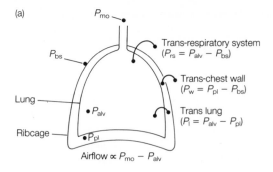

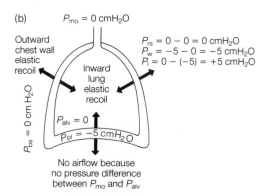

Fig. 16.7 (a) Lung-related pressures (on left: P_{alv}, alveolar pressure; P_{bs}, body surface pressure; P_{mo}, mouth pressure; P_{pl}, pleural pressure) and transmural pressures (on right: across respiratory system, P_{rs}; across chest wall, P_{w}; and across lung, P_{l}). Airflow is dependent on pressure gradient between mouth and alveoli. (b) Lung-related pressures at FRC due to outward elastic recoil of chest and inward elastic recoil of lungs.

for airflow is $P_{mo} - P_{alv}$. Airflow is a dynamic condition and the pressure gradient is affected by airflow rate, airway resistance and frictional resistance of lung tissue (pp. 486–490). Changes in P_{alv} required for airflow are created by outwards and inwards movement of the chest and are seen also as changes in P_{pl}.

When air is not flowing (a static condition), P_{pl} is a consequence only of the elastic recoil forces resulting from the degree of inflation or deflation of the respiratory system. Such elastic recoil creates pressure differences across the alveolar wall and across the chest wall. The relevant **transmural** pressures ($P_{in} - P_{out}$) are:

1 **trans-respiratory** system pressure (P_{rs}) between alveoli and the body surface, i.e. $P_{alv} - P_{bs}$,

2 **trans-chest-wall** pressure (P_{w}) between the pleural space and body surface, i.e. $P_{pl} - P_{bs}$; and

3 **trans-lung** or transpulmonary pressure (P_{l}) between alveoli and the pleural space, i.e. $P_{alv} - P_{pl}$.

Note that centrally located alveoli are subjected to the same pleural pressure as alveoli near the pleural surface because pressure is transmitted through the alveolar walls. These transmural pressures are rather small, and are measured with water manometers and expressed in cmH_2O rather than mmHg (1 cmH_2O = 0.76 mmHg).

Pressures at FRC

A **static** condition of no airflow occurs naturally at the end of a normal expiration at rest, i.e. at FRC. As no air is flowing, P_{mo} and P_{alv} are at the same pressure (atmospheric pressure, which on a standard day is 760 mmHg, but is used as a reference of zero pressure). At FRC P_{pl} is negative at about – 5 cmH_2O (Fig. 16.7b). This **subatmospheric** pressure arises from **elastic recoil** of the lungs and chest wall in opposite directions; the lungs recoil inwards and the chest recoils outwards. The two surfaces do not normally separate as they are held together by the cohesive action of the film of liquid in the pleural space. In this static condition at FRC;

P_l = + 5 cmH_2O (i.e. $P_{alv} - P_{pl} = 0 - (-5)$);
P_w = – 5 cmH_2O (i.e. $P_{pl} - P_{bs} = (-5) - 0$); and
P_{rs} = 0 cmH_2O (i.e. $P_{alv} - P_{bs} = 0 - 0$).

This zero pressure for P_{rs} indicates that at FRC the whole respiratory system is at its elastic equilibrium. Note also that at any lung volume, if there is no airflow, P_w = – P_l. Furthermore pressures, in contrast to volumes, are independent of body size.

P_{pl} of – 5 cmH_2O is an average value for a person in an upright posture. The weight of the lungs causes them to pull away from the thorax more at the apex (top) than at the base of the lung. Thus, when the trunk is upright, the P_{pl} is – 10 cmH_2O at the apex and – 2.5 cmH_2O at the base of the lung (1 cm of vertical height progressing down the lungs equals a P_{pl} of about + 0.2 cmH_2O).

Elastic properties of the respiratory system

The elastic recoil produced by the lungs alone, chest alone and lungs plus chest are estimated from volume and pressure measurements made when respiratory muscles are voluntarily relaxed and the volume prevented from escaping by occlusion of the mouth. The elastic recoil force of lungs plus chest is given by the pressure at the mouth, that for the chest alone by the pleural pressure and that for the lungs alone by the transpulmonary pressure (mouth pressure minus pleural pressure). The ease with which each structure stretches (compliance) is measured as the change in volume/change in pressure. Disease rarely affects chest wall compliance but diseased lungs can become either stiff or floppy. Elastic recoil is expressed as the elastance, which is the reciprocal of compliance (i.e. change in pressure/change in volume). Only a quarter to a third of the lungs' elastic recoil is caused by the elastic fibres of its alveolar walls; most is caused by surface tension at the alveolar tissue–gas interface.

Static pressure–volume relationships

Elastic properties of the lungs and chest wall, individually and when operating together (i.e. the respiratory system as a whole), are examined by determining pressures over the entire range of volumes that the lung can contain (Fig. 16.8).

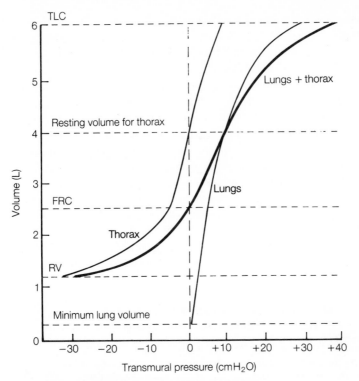

Fig. 16.8 Relaxation pressure–volume curves for thorax, lungs and the respiratory system as a whole (lungs + thorax).

These pressure and volume measurements must be made at different degrees of deflation and inflation at moments when breathing movements are voluntarily stopped, i.e. no air is flowing in or out (static conditions). At each step the spirometer tube attached to the mouth is occluded and, without closing the glottis, the thoracic muscles are then **relaxed**. Measurements are made of the volume (e.g. by a spirometer) and P_{pl} either by a catheter placed in the pleural space or, more simply, by one inserted, via the nose, into the lower third of the oesophagus. As this part of the oesophagus is in the intrathoracic compartment, its pressure reflects that of the pleural space provided that there are no oesophageal contractions. P_{alv} is technically impossible to measure because of the tiny size of the alveoli—under static conditions, however, it equals P_{mo} because no air is flowing. P_{bs} is atmospheric at all times.

P_{mo} is a measure of the **elastic recoil** of the **lungs and thorax in combination**. In this artificial situation where the airways are not open to the atmosphere, P_{mo} and therefore P_{alv} is negative, zero or positive depending on the lung volume as depicted in Fig. 16.8. The respiratory system is at its equilibrium point or resting volume—the FRC—when the P_{mo} it produces is zero. At volumes above this, P_{mo} is positive as the system attempts to collapse and recoil inwards, and at volumes below FRC it is negative as the elastic recoil forces are now directed outwards (Fig. 16.8).

More precisely, the transmural pressure should be measured but as P_{rs} equals $P_{alv} - P_{bs}$ and $P_{bs} = 0$, it follows that $P_{alv} = P_{mo}$.

The **elastic recoil force** of the **thorax** alone is estimated from P_{pl} since the P_w equals $P_{pl} - P_{bs}$ and $P_{bs} = 0$. The equilibrium position for the thorax occurs at a volume of about 4 litres; at volumes above and below 4 litres, P_{pl} is respectively positive (from inward recoil forces) and negative (from outward recoil forces; Fig. 16.8). Since the pleural film binds the chest to the lungs, the chest is at the same volume as the lungs and vice versa. At FRC (2.5 litres) this volume is much smaller than the equilibrium volume of the chest and thus the outwards recoil force of the chest creates a subatmospheric pressure of $- 5$ cmH$_2$O in the pleural space.

The **elastic recoil force** of the **lungs** alone is represented by P_l, that is $P_{alv} - P_{pl}$ or, in this case, $P_{mo} - P_{pl}$. At all volumes the elastic recoil force of the lungs is inwards so that P_l is always positive. By extrapolation (as in Fig. 16.18) the equilibrium position (zero pressure) for the lungs occurs close to zero volume (the minimal volume of ~200 ml) which can only be achieved if the RV is expelled (as can happen in a pneumothorax). At FRC the lungs' inward recoil force (P_l) is $+ 5$ cmH$_2$O. Since the chest recoil force (P_{pl}) at FRC is equal but opposite ($- 5$ cmH$_2$O) to the lung recoil force, the respiratory system as a whole is at equilibrium (P_{alv} or $P_{mo} = 0$).

When a person lies down, the chest's outward recoil force, and therefore that of the total respiratory system, is less because gravity is no longer pulling the abdominal contents away from the diaphragm. As a consequence the equilibrium point is at a lower FRC (at a P_{pl} and P_l of about $- 4$ and $+ 4$ cmH$_2$O, respectively).

A static pressure–volume relationship can also be produced by a second type of breathing manoeuvre that allows measurement of the elastic forces of the lungs alone. In this manoeuvre, at each volume the glottis and mouth are open and subjects no longer relax but hold their respiratory muscles in a sustained contraction so that the lung volume remains constant. Since the respiratory muscles are not relaxed, they are not producing any recoil forces; only the lungs are recoiling. As there is no air flowing and the airways are open to the atmosphere, P_{mo} and P_{alv} are both zero; only P_{pl} changes in proportion to the lung volume. As the lungs at all volumes recoil inwards and away from the pleural space, the P_{pl} is always negative during this manoeuvre (e.g. $- 30$ cmH$_2$O at the full inflation of TLC, $- 5$ cmH$_2$O at FRC and $- 2$ cmH$_2$O at RV). However, as it is transmural pressure that is used in Fig. 16.8, P_l are in fact all positive. For example, as $P_l = P_{alv} - P_{pl}$, the P_l at TLC is $0 - (- 30)$, which equals $+ 30$ cmH$_2$O.

Compliance and elastance

The ease with which an organ can be stretched is called its **compliance**. It is defined as the change in volume per unit change in pressure ($\Delta V/\Delta P$) and the more compliant an organ, the greater the volume achieved for a particular pressure change. Compliance can be measured for respiratory organs from static pressure–volume curves, usually from the steepest portions which occur around FRC. (At high volumes near TLC the inelastic elements limit the degree of stretch and thus the curves in Fig. 16.8 flatten near TLC and the compliances become very low.) The compliance of healthy adult lungs around FRC is about 0.2 L cmH$_2$O^{-1}. The slope of the thorax curve around FRC is similar to that of the lung curve (Fig. 16.8) so that chest compliance is also about 0.2 L cmH$_2$O^{-1}. For the respiratory system as a whole, its slope in Fig. 16.8, and therefore its compliance, is much less (about 0.1 L cmH$_2$O^{-1}). This is because the lungs and thorax are in series with each other and their compliances, like electrical conductances (p. 79), are summed as reciprocals.

Because lung volumes but not lung pressures vary with body size, compliance values, as measured above, will be large in an adult compared with a small child (or in a elephant compared with a mouse!). Measurements can be standardized by expressing them relative to the subject's FRC (i.e. ($\Delta V/\Delta P$/FRC)—this is called **specific compliance**. If the FRC is 2.5 litres and lung compliance is 0.2 L cmH_2O^{-1}, then the specific lung compliance is 0.08 cmH_2O^{-1} and in health this value would be similar for mammals of all sizes.

Changes in chest wall compliance are uncommon; however, lung compliance is altered by various **lung diseases**. In diseases associated with increased fibrous tissue in the alveoli (*pulmonary fibrosis*, an example of a restrictive lung disease), the lung becomes stiff—its compliance decreases. The stiffer lung requires a greater transpulmonary and therefore greater pleural pressure to achieve the same inspired volume as a normal lung; thus the inspiratory muscles have to work harder. In diseases where there is destruction of alveolar tissue creating alveoli with large air spaces (as in *emphysema*), only small pressure changes are required to inflate the very compliant lungs. An increased compliance is not, however, an advantage because much of expiration is usually produced by passive elastic recoil—in these floppy lungs there is little recoil. More expiratory muscle activity is self-defeating as airways collapse in the absence of radial traction (p. 488).

The elastic properties of the lung, as with blood vessels (p. 412), can be described by the **elastic resistance** or **elastance** ($\Delta P/\Delta V$), the reciprocal of the compliance. The elastance of the lungs and chest separately are both 5 cmH_2O L^{-1} and, as the lungs and chest are in series, their elastances, like electrical resistances (p. 79), are additive and the combined elastance is thus 10 cmH_2O L^{-1}.

The elastic resistance of the lung is due to two components. About a quarter to a third results from the **elastin fibres** in the alveolar interstitium; the remainder results from the **surface tension** at the interface between the gas in the alveoli and the liquid surface of the alveolar wall. (The interface is variously called the gas–liquid or air–tissue or alveolar interface.)

The magnitude of these two components can be demonstrated by pressure–volume curves determined in lungs removed from the body which are either gas-filled or liquid-filled (Fig. 16.9). The elastance of the gas-filled lung (5 cmH_2O L^{-1}) will be similar to that of the lung adhering to the thorax (provided that it has not been deflated below RV). When all the gas in the alveoli is experimentally replaced with liquid (saline), there is no surface tension because there is no longer a gas–liquid interface. Thus the elastance of the liquid-filled lungs is only 1.25 cmH_2O L^{-1}—a measure now only of the elastic resistance of the interstitial elastin fibres.

Alveolar surface tension and surfactant

According to the law of Laplace, alveolar surface tension should increase as the radius of the alveolus gets smaller and, for the alveolus to stay inflated, this surface tension must be opposed by an appropriate transmural pressure. This is the transpulmonary pressure. If the fluid lining the alveoli was purely interstitial fluid, the transmural pressure required for even moderate inflation would be enormous. However, the surface tension is lowered considerably by surfactant

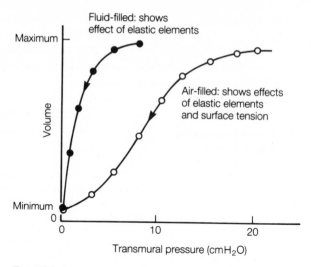

Fig. 16.9 Lung pressure–volume curves when gas-filled and liquid-filled, constructed during deflation of the lung.

secreted by the alveolar type II cells. Furthermore, surfactant lowers surface tension to a greater degree when it is aggregated into a smaller surface area and this permits small alveoli, at the same transmural pressure as nearby larger alveoli, to remain stable and not to empty into the larger ones. Alveolar stability is also achieved by mechanical support from the surrounding alveoli. Surfactant, by reducing the transmural pressure, also prevents the alveolar gas space from filling up with fluid ultrafiltered from the lung capillaries. Breathing is periodically interrupted by the large inflation of a sigh which serves to replenish some of the surfactant molecules and restore lung compliance to normal.

Surface tension at a gas–liquid interface is inversely proportional to the radius of curvature. The smaller the radius of a spherical alveolus, the greater the transmural pressure $(P_{in} - P_{out}$, i.e. $P_{alv} - P_{pl}$, which is the same as transpulmonary pressure) required to oppose the surface tension. This relationship between surface tension (T), transmural pressure (P_t) and radius (r) is described by the law of Laplace (p. 396) as $P_t = 2T/r$ in which the thickness (u) component has been omitted because the alveolar wall is so thin.

What is surface tension? Molecules in a liquid are powerfully attracted to each other such that any one molecule has equal attractive forces affecting it from all sides. However, when the molecule is at the surface of a liquid–gas interface, there are no attractive forces above the interface to balance those in the liquid below. This creates a tension at the interface parallel to its surface—the magnitude of this surface tension depends on the composition of the liquid. Furthermore, when the interface is curved as in a bubble or an alveolus, the surface forces are directed towards the centre of curvature, providing a collapsing force. To prevent a bubble or an alveolus from collapsing, positive pressure must exist in the gas phase (or negative pressure in the liquid phase) to balance the surface tension.

Reduction of surface tension by surfactant

If the liquid lining the alveoli was pure interstitial fluid, the surface tension would be very high, creating a P_{pl} as great as -30 cmH_2O, even at FRC. The alveolar fluid contains a substance called **surfactant** which has detergent-like properties and substantially lowers the surface tension to about one-sixth. Surfactant is a mixture of phospholipids (mainly disaturated phosphatidylcholine) secreted by the alveolar type II epithelial cells. Surfactant secretion is stimulated by alveolar expansion and β-adrenergic mechanisms; turnover is very rapid, with a half-life of about 2 hours.

The surface tension of interstitial fluid is ~70 dyn cm^{-1} (0.07 N m^{-1}) and with an alveolar radius of ~50 μm at FRC, the transmural pressure without surfactant would be $2 \times 70/0.005 = 28\,000$ dyn cm^{-1} (28 N m^{-1}), which is 28 cmH_2O. As the inside pressure P_{alv} is zero, the outside pressure P_{pl} is therefore -28 cmH_2O. However, because of surfactant, the alveolar surface tension (at FRC) is ~10 dyn cm^{-1} and thus the P_{pl} is actually -5 cmH_2O.

In the human fetus surfactant secretion by the type II alveolar cells is not established until the 30th week; thus premature babies are deficient in lung surfactant (**newborn respiratory distress syndrome**). Without surfactant the premature lungs have such a high surface tension that a lot of inspiratory effort is needed to generate enough negative P_{pl} even to inflate the lungs a little. As soon as inspiratory efforts stop, the high elastance of the lungs causes them to recoil very rapidly to very low volumes. Factors reducing production of surfactant or increasing its rate of destruction may contribute to **adult respiratory distress syndrome**.

Dependence of surface tension on area

The surface tension of a film of liquid relative to the surface area it occupies can be measured with a Wilhelmi balance. A film of interstitial fluid has a surface tension of 70 dyn cm^{-1}, irrespective of how small or large the area over which it is spread (Fig. 16.10). Similarly, although a household detergent has a low surface tension (e.g. 25 dyn cm^{-1}), its tension is also not area-dependent. A film of liquid

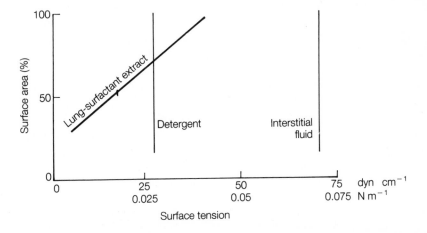

Fig. 16.10 Dependence of surfce tension on surface area for a film of liquid containing lung surfactant. Comparison made with detergent and intestitial fluid.

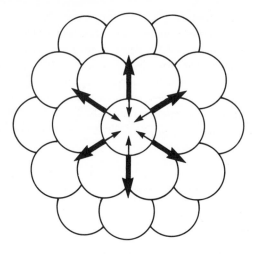

Fig. 16.11 Stability of alveoli produced by mechanical interaction with neighbouring alveoli.

extracted from the alveolar lining shows that its detergent-like surfactant has lowered the surface tension compared with the interstitial fluid film, and the presence of surfactant confers an area-dependence to the surface tension. As the film spreads over a larger area the surface tension increases, showing that surfactant molecules are more effective as a detergent when they are crowded together over a small surface area. This means that the alveolar surface tension will be lower in a deflated compared with an inflated lung and in a single alveolus as it gets smaller. Thus, the lung becomes stiffer, offering more elastic resistance as it becomes more inflated (e.g. Fig. 16.9) and alveoli of all sizes at the same transmural pressure have a similar stability.

Alveolar stability is an important consideration. In any one part of the lung the alveoli are unlikely to be exactly the same size and yet they will all be subjected at any one moment to the same transmural pressure. Imagine one larger and one smaller alveolus in the same alveolar sac or interconnected by an airway. If the surface tension of the two alveoli were the same, then the law of Lapace ($P_t = 2T/r$) would indicate that the small alveolus would require a larger P_t than the large alveolus to keep it open. The smaller alveolus would therefore be unstable and empty into the larger. Because the surface tension in the small alveolus reduces virtually in proportion to the reduction in its radius, its required P_t is the same as for the larger alveolus and thus the small alveoli do not collapse.

A second and more important factor contributes to alveolar stability. This is the mechanical interaction conferred by the geometry of adjacent alveoli (Fig. 16.11). Any alveolus tending to collapse will tend to pull away from its neighbours but at the same time will be held open by its attachment to them. This mechanical interaction is termed **alveolar interdependence**.

Reduction of ultrafiltration by surfactant

Not only does surfactant lower the overall alveolar surface tension and confer alveolar stability, but it also assists in preventing the air spaces of the alveoli from filling up with liquid. As the alveolar walls contain a massive network of capillaries, the forces causing ultrafiltration and reabsorption of liquid across the capillary wall (Starling equilibrium; p. 428) have to be considered. Surfactant, by lowering the pleural pressure, decreases the hydrostatic pressure gradient and therefore reduces the amount of fluid ultrafiltered.

In systemic capillaries the hydrostatic pressures in the plasma and interstitial fluid on average are 24 and – 2 mmHg, respectively, and their colloid osmotic pressures are 25 and 0 mmHg, so that ultrafiltration forces ($\Delta 26$ mmHg) are slightly in excess of reabsorption forces ($\Delta 25$ mmHg). The colloid osmotic pressures, and therefore reabsorption forces, are the same for the pulmonary capillaries; the hydrostatic ultrafiltration forces are, however, different. Pulmonary capillary pressure is low (10 mmHg; p. 447) and interstitial pressure is effectively that of the pleural space (at FRC this is – 5 cmH$_2$O or – 4 mmHg) so the ultrafiltration force ($\Delta 14$ mmHg) is considerably less than the reabsorption force ($\Delta 25$ mmHg) and the lungs remain dry. Only when pleural pressures reach – 16 mmHg (– 21 cmH$_2$O), during, for example, large inflations, will there be net ultrafiltration. The pulmonary circulation has a very extensive lymphatic network to prevent any permanent fluid accumulation. If surfactant was not present, the pleural pressure even at FRC would be – 30 cmH$_2$O (– 23 mmHg) and the ultrafiltration force ($\Delta 33$ mmHg) would be well in excess of reabsorption and the alveolar air spaces would be flooded with liquid.

Sighs and surfactant

A **sigh** is a reflexly generated (p. 543) single deep breath which occurs after a period of quiet breathing. Quiet breathing around FRC does not cause much breath-by-breath alteration in the alveolar surface area and, as a consequence, in the tightly compressed surface film the surfactant molecules are gradually squeezed away from the surface. The progressive lowering of surfactant concentration at the surface layer causes the surface tension to rise gradually so that the lung has more elastic resistance or, expressed another way, becomes less compliant. The purpose of the large inflation, which stretches and unfolds the alveolar surface area, is to spread out the surfactant molecules, allowing newly released molecules to come to the surface. After the sigh, when breathing is back at volumes near FRC, these extra molecules return the alveolar surface tension to its usual 10 dyn cm^{-1}.

Hysteresis

Most textbooks place undue emphasis on the phenomenon of a **hysteresis** loop present in the tension–area relationship of a surface film of surfactant (Fig. 16.12a) or in the pressure–volume relationship determined in an air-filled, but not a liquid-filled, isolated lung (Fig. 16.12b). Hysteresis is the term used to describe the fact that a particular area during expansion results in a higher surface tension than during compression; similarly, attaining any particular volume during inflation requires a larger pressure than during deflation. In the case of tension–area measurements, the hysteresis is greater the more it is compressed/expanded to its limits and is thought to be due to the time it takes even in these step-by-step (i.e. static) measurements for surfactant molecules to realign themselves at the changing surface area of the interface. In the case of the static pressure–volume relationship, hysteresis is only seen if the lungs are first deflated to their minimum volume. In other words, a collapsed lung does not significantly inflate until a fairly large transmural pressure, the critical opening pressure (~8 cmH$_2$O), is reached. The hysteresis is very small for volume changes between RV and TLC and virtually negligible for the normal tidal volume occurring at FRC; this hysteresis is caused by surfactant.

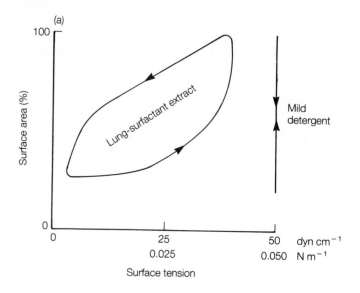

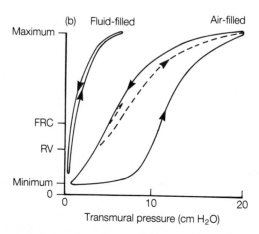

Fig. 16.12 (a) Surface tension–surface area curves for lung fluid extract compared with a mild detergent. (b) Inflation and deflation pressure–volume curves in air-filled and fluid-filled lungs. Note that there is virtually no hysteresis in the fluid-filled lung and little hysteresis if inflation starts at residual volume (RV) or functional residual capacity (FRC).

Resistance to airflow

The resistance to airflow in and out of the lungs is due mainly to friction in the airways with a very small contribution from friction within the lung tissue. Airway resistance is determined by the viscosity of air and the radii of the airways (Poiseuille equation). As the airways branch extensively in a parallel manner, the total resistance is very small and an alveolar-to-mouth pressure gradient of 1 cmH$_2$O is sufficient to cause airflow. The major sites of airway resistance are the upper respiratory tract and the medium-sized bronchi. Airways not supported *by cartilage can have their calibre enlarged by the radial traction which results from an increased lung volume, producing more alveolar elastic recoil. Conversely, they can be compressed to the point of collapse during forced expirations. Contraction of smooth muscle in the airway walls also produces changes in calibre. Bronchoconstriction is caused by parasympathetic activity and by local release of chemicals like histamine; bronchodilatation results from activation of smooth muscle β_2-receptors by substances like adrenaline.* Mucus secretion into the airway lumen can also increase airway resistance. Airway resistance is assessed by measuring the expiratory forced vital capacity (FVC), the volume so expired in the first second (FEV_1), the forced expiratory flow rate (FEF), the peak flow rate and the width of a dynamic pressure–volume loop.

Airflow into the lungs requires a pressure gradient between the atmosphere (i.e. the nose or mouth; P_{mo}) and the alveoli (P_{alv}). For a particular pressure gradient (ΔP), the magnitude of the airflow (volume/unit time, $\dot{V}$) depends mainly on the frictional resistance (R) encountered in the airways, as described by the Poiseuille equation where $\dot{V} = \Delta P/R$ and $R = 8\,\eta l/\pi r^4$ (p. 406). P_{mo} is normally zero and, for a 500 ml V_T typical of quiet breathing where airflow rises to about 0.5 L s^{-1}, P_{alv} reaches about -1 cmH$_2$O during inspiration, sucking gas into the alveoli, and about $+1$ cmH$_2$O during expiration, pushing gas out of the alveoli. These pressure gradients for airflow are tiny, particularly when compared with those for blood flow, because air has a much lower viscosity (η) than blood and flows through tubes that are much wider. Airflow resistance for quiet breathing is typically about 2 cmH$_2$O s L^{-1}, i.e. ($\Delta 1$ cmH$_2$O) / (0.5 L s^{-1}).

Airway resistance accounts for 80–90% of the resistance to airflow. The other factors that contribute are the inertia to movement of the respiratory system and the frictional (viscous) resistance encountered at the lung–chest wall surface and within the chest and lung tissues as the elements move against each other. All are negligible except for the frictional resistance of lung tissue, which in health causes only 10–20% of the airflow resistance. Lung tissue resistance can increase in such conditions as pulmonary fibrosis. The airway resistance plus the lung tissue resistance are often referred to as the **pulmonary resistance**— both are non-elastic resistances. Inertia becomes appreciable only during rapid respiratory actions like sneezing and coughing.

As with blood (p. 409), airflow can be turbulent, laminar or a mixture of the two (transitional). Airflow through the trachea is in general turbulent, while that through the major part of the bronchial tree is transitional. In the normal lung, laminar flow occurs only in the very small airways close to the alveoli where, as the total cross-sectional area is high, the velocity of flow is very slow. In the respiratory zone (alveolar ducts and alveoli), velocity of flow is so slow that diffusion of O_2 and CO_2 is actually more important than bulk flow.

Distribution of airway resistance

About 40–50% of airway resistance is found in the upper respiratory tract (nasal passages, pharynx and larynx) and this resistance is significantly reduced when one switches from nose to mouth breathing—this commonly occurs during the high airflow of exercise. Most techniques used for measuring breathing impose mouth breathing and therefore do not include the consequences of nasal resistances. Narrowing of the glottis during expiration increases laryngeal resistance and, as mentioned earlier, is part of the expiratory braking mechanism.

Below the larynx most of the resistance resides, not in the bronchioles, but in the trachea and bronchi down to about the seventh generation. Although the highest individual resistance is indeed to be found in the bronchiole with the smallest radius, the branching of the bronchial tree creates an enormous number of parallel tubes, the resistances of which add as reciprocals (p. 407) so that the total resistance is low. Thus, beyond the segmental bronchi, the airway resistance correspondingly falls as the total cross-sectional area of the airways progressively increases with each successive generation of branching towards the alveoli. The medium-sized bronchi (2–4 mm in diameter) around the fourth generation are normally the site of greatest resistance. The resistance of the airways beyond the 15th generation is virtually zero. Those pulmonary disorders which reduce the calibre of small airways are difficult to detect by measures of airway resistance as they contribute such a small proportion of the resistance.

Effect of lung volume on airway calibre and resistance

The pressure inside the airways is either zero or, when air is flowing, graded from mouth to alveoli; the pressure outside the airway wall (the pleural pressure) is caused by elastic recoil of the alveoli. Thus, larger lung volumes, by producing more elastic recoil, increase the transmural pressure ($P_{in} - P_{out}$) across the airway wall and, in those airways not stiffened by cartilage, the calibre of their lumen enlarges and so the resistance to airflow falls. (Contraction of airway smooth muscle (p. 489) can help stiffen the airway wall.) In effect, airways are supported and pulled outwards by the elastic recoil of the alveolar tissue surrounding them—an action referred to as **radial traction**. At small lung volumes, airway transmural pressure is low, the radial traction is less and the airways become narrower and hence airflow resistance increases. The greatest changes in resistance in fact occur at low volumes—those between FRC and RV.

In obstructive lung diseases breathing occurs at a higher FRC and, in asthma, where alveolar tissue is normal, this provides more radial traction in an attempt to open up the constricted airways a little. When the elastic recoil is reduced by destruction of alveolar tissue, as in emphysema, the airways are actually narrower not because the disease has affected the airway structure but because some of the radial traction has been lost.

Below FRC the increase in airway resistance can be dramatically exaggerated by collapse of airways not supported by cartilage. To expel air and reach volumes below FRC requires expiratory muscle contraction which, if done forcefully and rapidly, generates positive pleural pressures (as high as $+100$ cmH$_2$O). This positive pressure surrounding the airways is uniform along its length but in the lumen of the airway, pressure will fall from positive values at the alveolar end to zero at the mouth, so that at some point along the airway the positive pressure outside will exceed that inside. Depending on the pressure difference and the extent of cartilaginous support, the airway at this point will be compressed and possibly collapsed. This **dynamic airway compression** or collapse limits the maximum rates of expiratory flow. The lung volume at which airway collapse occurs during forced expirations after a maximum inspiration is called the **closing volume**.

In young healthy adults the closing volume occurs just before RV is reached but with ageing it occurs at volumes just above FRC. In emphysema where there is less radial traction, or in obstructive lung disease where airways are already narrow, dynamic airway compression occurs not only at lung volumes above FRC but also with only mild expiratory muscle effort. The wheezing noise during such expirations is due to vibration of the collapsed airways and the turbulent flow.

Effect of bronchial smooth muscle on airway resistance

Smooth muscle in all of the airways (trachea to alveolar duct) is under the control of the autonomic nervous system but since the medium-sized bronchi are the major site of airway resistance, it is control of their smooth muscle tone that physiologically alters airway resistance—hence the term **bronchomotor control.**

Bronchoconstriction is produced in a number of ways.

1 It is caused by increased activity in vagal **parasympathetic** fibres (via their postganglionic cholinergic fibres). The parasympathetic contribution to the resting bronchomotor tone can be abolished by atropine (an antagonist of muscarinic acetylcholine receptors) and as a result there is a 30% reduction in airways resistance. Parasympathetically mediated bronchoconstriction occurs as a reflex effect of stimulating irritant and cough receptors in the airways (p. 542)—the bronchoconstriction both limits the entry of the irritant and increases expiratory airflow velocity to dislodge and expel the irritant. Parasympathetic activity also **increases mucus secretion** into the airways, which in itself decreases the calibre of the airways.

2 Substantial bronchoconstriction and mucus secretion can also be caused *directly* by **local chemical mediators** such as histamine and leukotrienes released from mast cells in the airways, for example, during allergic **asthma** attacks.

3 The **decreased CO_2** levels in overventilated parts of the lungs also cause direct constriction of bronchial smooth muscle, *particularly in the very small airways,* thus increasing airways resistance in that locality with subsequent reduction in the local alveolar ventilation (p. 521).

Relaxation of bronchial smooth muscle (**bronchodilatation**) results from activation of **β_2-adrenoceptors** by circulating adrenaline or therapeutically administered sympathomimetics (e.g. isoprenaline and salbutamol). In humans few of these β_2-receptors are activated by transmitter release from sympathetic nerves. In addition bronchodilatation can be produced by a **non-adrenergic non-cholinergic** (NANC) innervation of the bronchial smooth muscle in which intestinal polypeptide (VIP) may be the neurotransmitter.

A second type of peptidergic innervation, releasing the neurotransmitter substance P and other neurokinins, is also present—it appears to be involved in the bronchoconstrictions accompanying bronchospasm. Mucus secretion into the airways is mainly under vagal cholinergic control and its volume and composition are modulated by sympathetic (α- and β-adrenergic) and peptidergic mechanisms and inflammatory mediators.

Forced expiratory assessment of airway resistance

As the direct measurement of resistance from $P_{mo} - P_{alv}$ and $\dot{V}$ is difficult, airway resistance is usually assessed indirectly, either from the width of the dynamic pressure–volume loop (p. 493) or from a forced expiration into a spirometer where expiration is preceded by a

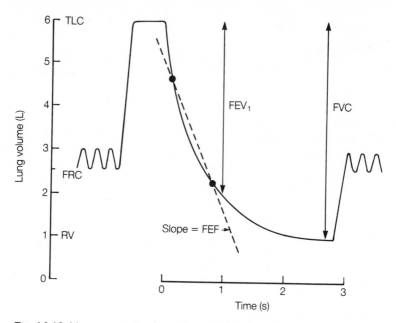

Fig. 16.13 Measurement from a spirometer of the volume expired in the first second (FEV$_1$), the forced vital capacity (FVC) and the forced expiratory flow rate (FEF).

maximal inspiration (Fig. 16.13). The volume exhaled is the **FVC** and even in health it is a little smaller than VC (measured without forceful breathing) because expiratory forces compress the small airways (p. 488). In obstructive diseases airflow is retarded by narrow airways, causing FVC to be quite a lot smaller than the VC. Loss of radial traction resulting in more small airway compression in emphysema exaggerates this difference further.

FEV$_1$ is a more sensitive measure of changes in airway resistance—in health it is about 80% of the FVC. In obstructive disease the **FEV$_1$/FVC** ratio is less than 80% as the decrease in FEV$_1$ is greater than the decrease in FVC. In restrictive diseases, where all lung volumes are smaller (p. 476), both FVC and FEV$_1$ will be smaller but their ratio does not decrease; if anything, it may increase.

As shown in Fig. 16.13, a similar indicator of airway resistance is **FEF**—also called the maximal mid expiratory flow rate (MMFR). This is simply the slope ($\Delta V/\Delta$time) of a straight line drawn between volumes at 25% and 75% of FVC on the expiratory curve. The fastest rate of airflow occurs soon after the beginning of a forced expiration and this is measured by expiring rapidly and forcefully into a peak flow meter. For a 70-kg healthy male, the typical peak expiratory flow rate is ~600 L min^{-1}.

Pressure and volume during the breathing cycle

Pleural pressure during the breathing cycle is the consequence of two forces, one due to elastic recoil which is proportional to the degree of inflation and the other due to the negative (inspiration) and positive (expiration) alveolar pressures generated to overcome the resistance to airflow. As air is flowing, the pressure–volume relationship is referred to as dynamic and is seen to be a loop. The width of the loop is an estimate of airflow resistance. The slope of the line drawn between the points of minimum and maximum volume of the breath, when airflow is zero, measures the elastic recoil, i.e. lung compliance. The area enclosed by the loop is a measure of the work of breathing (costing for a normal *breath at rest only about 5% of the body's total O_2 consumption). With slow deep breaths most of the extra work is against the elasticity of the alveoli, while with rapid shallow breaths most of the extra work is against the airway resistance. Maximal respiratory muscle forces can be estimated from the mouth pressures generated during maximal inspiratory (Müller's manoeuvres) and expiratory (Valsalva's manoeuvre) efforts against an obstructed mouth.*

During the **dynamic** event of air moving, pleural pressure (P_{pl}) is the consequence of two forces. The first is the elastic recoil of the lungs alone, as already described for static conditions in Fig. 16.8. The second is the additional dynamic pressure required to overcome the resistance to airflow; this is the negative P_{alv} of inspiration ($-1\,cmH_2O$ for a V_T of 500 ml) created by inspiratory muscle contraction, and the positive P_{alv} of expiration ($+1\,cmH_2O$) from recoil of lung tissue compressing the alveoli. (During deeper or more forceful breathing, expiratory muscle contraction will contribute to the compression of alveoli.) The apportioning of P_{pl} to its components of elastic recoil (P_{el}) and airway resistance (P_{res}) at different moments during inspiration and expiration is portrayed in Fig. 16.14.

When interpreting this diagram, note that:

1 as P_{el} is dependent on lung volume, the time-based shapes of these graphs are identical;

2 there is a brief moment of zero flow whenever airflow reverses direction, i.e. at the transition between inspiration and expiration and vice versa;

3 inward airflow reaches its peak near mid-inspiration and outward airflow peaks near mid-expiration at about $0.5\,L\,s^{-1}$;

4 given that mouth pressure is zero, P_{alv} is also the measure of P_{res} and its time-based shape is identical to that of the airflow it permits;

5 P_{pl} during dynamic breathing is due to the combined effect of P_{el} and P_{res}; and

6 P_{pl} at the beginning of inspiration (i.e. at FRC) is $-5\,cmH_2O$ and at the end of a 500 ml breath in is $-8\,cmH_2O$ (which in mmHg is -3 and -5 respectively).

In reality the duration of inspiration is shorter than expiration and airflow rates are somewhat greater than depicted, especially in inspiration, and occur earlier in each breathing phase. Airflow also increases whenever the same tidal volume is shifted in a shorter time or with larger tidal volumes (although a large volume taken over a longer time could leave the flow rate unchanged). As faster breathing usually accompanies deeper breaths, airflow rates can reach values in excess of $6\,L\,s^{-1}$ with a P_{alv} of $+12\,cmH_2O$ in expiration and $-12\,cmH_2O$ in inspiration. Even for 500 ml breaths, P_{alv} can be greater or smaller than $1\,cmH_2O$ if the airways are respectively bronchoconstricted or bronchodilated. A maximal forced inspiratory effort can produce a P_{pl} as low as $-80\,cmH_2O$ at peak inspiratory flow rates; for maximal forced expiratory efforts P_{pl} can be as high as $+100\,cmH_2O$.

Dynamic pressure–volume relationship

Figure 16.14 illustrated the time course of the volume and pleural pressure (P_{pl}) changes during one breath cycle. If for each moment the volume is now plotted against the P_{pl} (Fig. 16.15), the curve for expiration (CFA) does not follow the same path as for inspiration (ABC) and a loop results in which, at any particular volume, P_{pl} in expiration is less negative than in inspiration except at the end of

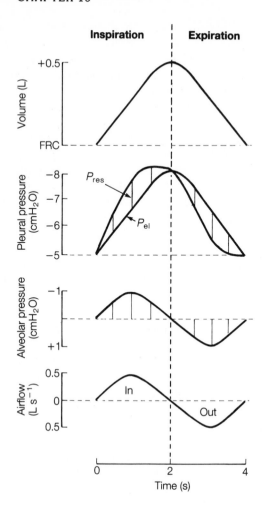

Fig. 16.14 Pressures and volume during the breathing cycle. P_{el}, Pressure due to elastic recoil; P_{res}, pressure required to overcome airway resistance and cause airflow.

inspiration (C) and end of expiration (A). If that part of the P_{pl} which is due only to P_{el} is plotted against volume, a single straight line results for both inspiration and expiration (AC). The slope of this line ($\Delta V/\Delta P$) is identical to that determined under static conditions for the lung alone (Fig. 16.8) and it is of course the measure of **lung compliance**. As the breathing manoeuvres required for static conditions are often difficult for the subject to achieve, lung compliance is usually measured quickly and simply by drawing a straight line between points A and C determined during dynamic conditions—A and C in fact represent static conditions because airflow at these points has ceased momentarily.

The **inspiratory–expiratory loop** (Fig. 16.15) is often called a hysteresis loop—very little, if any, hysteresis is due to tissue resistance (p. 487) or surfactant, the effect of which is only evident when preceded by deflation to near-zero volume (p. 485). The hysteresis in Fig. 16.15 is due almost entirely to airway resistance and the need to generate negative and positive P_{alv} in inspiration and expiration

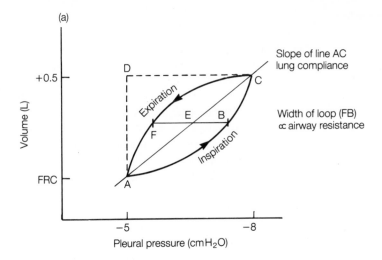

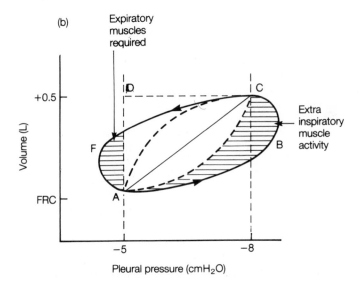

Fig. 16.15 Pressure–volume relationship during dynamic conditions. Slope of AC measures compliance; width of loop (FB) is an estimate of airway resistance; areas of loop indicate work of breathing. (a) A normal lung; (b) during asthma (depicted assuming volume has not increased).

(Fig. 16.14). Thus, P_{pl} is more negative in inspiration and less negative in expiration than the elastic recoil depicted by line AC. The width of the loop (measured in cmH_2O at its widest point parallel to pressure axis) indicates the magnitude of **airway resistance**—a true measure (cmH_2O L s^{-1}) requires flow rate to be measured too (p. 487). When airway resistance increases, e.g. during the bronchoconstriction of asthma, the loop is much wider (Fig. 16.15b).

Work of breathing

The area of the dynamic pressure–volume loop can be used to estimate the physical work that must be done by the respiratory muscles to overcome the elastic recoil of the lung and the non-elastic resistance of the airways. (Area is the product of pressure (g cm^{-2}) × volume (cm^3) and this has the same final dimensions (g cm^{-1}) as the strict measure of work, which is force × distance.) The area ABCDA in Fig. 16.15 represents the inspiratory work required to inhale 500 ml (a tidal volume typical of quiet breathing). This inspiratory work is divisible into two parts—that used to oppose the elastic recoil of the lung (AECDA) and that used to overcome airway resistance, allowing inspiratory airflow (ABCEA). Some of the energy used to stretch the elastic elements is stored as elastic energy whereas the rest, including the energy used to overcome the frictional resistance to airflow, is dissipated and lost as heat. During a normal 500-ml expiration no extra energy is required because the elastic energy is released from storage when the lungs are permitted to recoil. This passive elastic recoil finally vanishes as FRC is approached, but on the way was more than sufficient to provide the mechanical work (AECFA) required to overcome the airway resistance encountered by expiratory airflow. (Though not depicted here, expiratory resistance is usually greater than inspiratory due to narrowing of the tracheobronchial tree and glottis.)

Faster breathing rates require higher flow rates and therefore a larger P_{alv}—this will widen both sides of the hysteresis loop and is indicative of the increased work against resistive forces. When just the tidal volume increases, point C in Fig. 16.15a will be at a higher volume and pressure and thus the work (AECDA) done to overcome the elastic recoil of the lungs will have increased.

When airway resistance increases (e.g. in asthma) lung compliance does not change (Fig. 16.15b) but the wider inspiratory side of the loop indicates the greater inspiratory muscle work now required to overcome the airway resistance such that the pleural pressure during inspiration at B may even become more negative than at end-inspiration. Similarly, the width of the expiratory side of the loop will be greater and if the pleural pressure at F becomes less than at end-expiration, expiration can no longer be purely passive and the work component to the left of the vertical AD line has to be supplied by expiratory muscle contraction.

Effect of breathing pattern

At a particular level of ventilation the work of breathing depends on the pattern of ventilation—the same ventilation can be achieved by either slow deep breaths or rapid shallow breaths. With slow deep breaths most of the work is against the elasticity of the alveoli (static) while with rapid shallow breaths most of the work is against the resistance to airflow through the airways (dynamic; Fig. 16.16). Large tidal volumes increase the elastic work of breathing to a greater extent than high breathing frequencies increase the resistive work of breathing. Breathing is regulated to a pattern that is the most economical, i.e. produces minimum total work, explaining why at rest breathing frequency is usually about 12–15 breaths per minute.

Resistive work for a particular breath pattern increases in obstructive airway diseases so patients breathe more slowly to minimize the extra work of breathing. Elastic work increases

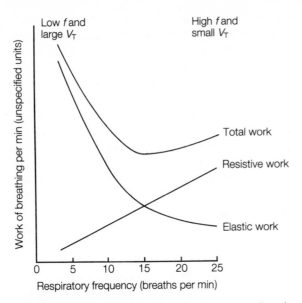

Fig. 16.16 Relationship of work of breathing to breathing frequency at a constant level of ventilation. Work is done against elastic resistance and airflow resistance.

in lung diseases which cause stiffer lungs and therefore to minimize the work patients take smaller breaths more frequently.

Metabolic cost of breathing

In a healthy person at rest respiratory muscles consume about 5% of the body's total O_2 consumption; this rises to about 30% in maximal voluntary hyperventilation. In lung or chest disease the metabolic cost can be this high even at rest, imposing a severe limitation on the amount of exercise such patients can perform.

Maximal respiratory muscle forces

The full strength of respiratory muscle contraction can be estimated by the maximum alveolar pressures generated during inspiratory and expiratory efforts against a closed glottis or an obstructed mouth. Respiratory muscle weakness can be detected by such measurements. Inspiratory effort (Müller's manoeuvres) at RV gives the greatest negative value (about $-120\,\mathrm{cmH_2O}$); expiratory effort (Valsalva's manoeuvre, as in straining during defecation) at TLC gives the greatest positive value (about $+200\,\mathrm{cmH_2O}$). These pressures are those in the alveoli but are measured at the mouth. As the lung's transmural pressure at RV and FRC would be its usual $2\,\mathrm{cmH_2O}$ and $30\,\mathrm{cmH_2O}$ respectively (Fig. 16.8), because this is dependent only on elastic recoil, the pleural pressure in maximum inspiratory and expiratory effort would be $-122\,\mathrm{cmH_2O}$ and $+170\,\mathrm{cmH_2O}$. The large positive pleural pressures generated in the Valsalva's manoeuvre, while aiding defecation, will compress intrathoracic veins, impeding venous return (p. 434), and may lead to fainting.

16.3 VENTILATION, GAS EXCHANGE, ALVEOLAR GASES AND DIFFUSION

The volume per minute that ventilates the lung is the product of the volume of each tidal breath and their frequency. Of the 500-ml tidal volume breathed in at *rest, about two-thirds reaches the alveoli (alveolar volume) and is available for gas exchange; the remaining third is not available because it fills the conducting airways (anatomical dead space) or does not come into contact with alveolar capillaries (alveolar dead space). The latter is very small. The two dead spaces comprise the physiological dead space which is estimated using the Bohr equation. The gases exchanged in the lungs and tissues are O_2 and CO_2 and the ratio of CO_2 production to O_2 consumption is the respiratory exchange ratio (or, at steady state, the respiratory quotient). The ventilatory requirement and O_2 extraction coefficient indicate how much of the ventilated air is actually utilized. Volumes are usually measured under ATPS conditions but should be stated in units of BTPS for ventilation and STPD for gas exchange (see Table 16.1).*

Ventilation and its components

In a healthy ~70-kg person at rest, the duration of each breath (total breath time, T_T) is about 4–5 seconds, of which about one-third is spent inspiring (inspiratory time, T_I) and two-thirds expiring (expiratory time, T_E). The **frequency** of breathing (f) ranges from 12–15 breaths per minute and the volume of each breath (V_T) from 500 to 600 ml. The product of the two, $f \times V_T$, is the **ventilation ($\dot{V}_E$)** of about 6–7 L min^{-1}. (Note that the dot above the V symbol indicates the volume is expressed per unit time and the subscripts identify the particular component of the gas phase; see Table 16.1 for full glossary of symbols.) $\dot{V}_E$ is also referred to as the pulmonary or minute ventilation. The subscript indicates that $\dot{V}_E$ is typically calculated from measurement of an expired (V_E) not an inspired (V_I) tidal volume. Strictly speaking, V_T is the average of V_I and V_E; the latter is slightly smaller because less CO_2 is produced by the body in exchange for the O_2 consumed.

At the end of a 500-ml inspiration, about a third of the breath (~150 ml) fills the conducting airways and is called the **dead space** (V_D)—'dead' because it cannot be used for gas exchange. Approximately 350 ml of the tidal volume has reached the alveoli (the **alveolar volume, V_A**) to mix with the rest of the gas in the alveoli and be available for gas exchange with the blood. As the volume in the alveoli at the end of expiration is about 2.5 litres (the FRC), the V_A adds only 12% more to the volume and thus serves as a source of alveolar gas replenishment rather than replacement. The inspired alveolar volume mixes by diffusion with the gas in the FRC (p. 487). At rest with a V_A of 350 ml and a breathing frequency of 12 min^{-1}, the amount of gas ventilating the alveoli per minute, the **alveolar ventilation ($\dot{V}_A$)**, is 4.2 L min^{-1} ($f \times V_A$). Similarly, the gas in the airways that is not available for gas exchange is 1.8 L min^{-1} ($f \times V_D$) — the **dead-space ventilation ($\dot{V}_D$)**.

All these volumes are dependent on gender, age, physical training and body mass—for instance, the larger the person the greater the volume, while in a smaller person the breathing frequency is higher. In severe exercise the values for a 70-kg person can become as great as $f \sim 50$ min^{-1}, $V_T \sim 3.0$ litres and $\dot{V}_E \sim 150$ L min^{-1}. As f becomes greater, the duration of each breath (T_T) becomes shorter and the times spent inspiring (T_I) and expiring (T_E) become about equal. Note that the maximum V_T of exercise is much smaller than the vital capacity of 5.0 litres, which can only be sustained for a few breaths. Although during exercise there is a small increase in V_D (to no more than ~350 ml) due to mainly passive and some active bronchodilatation, most of the increased V_T ventilates the alveoli (at least four-fifths in severe exercise compared with two-thirds at rest).

Different breathing patterns have a marked effect on the alveolar ventilation. For instance, deep slow breathing (e.g. $V_T = 1$ litre, $f = 6$ min^{-1}) still gives a resting $\dot{V}_E$ of 6 L min^{-1} but it is comprised of a $\dot{V}_D$ of 0.9 L min^{-1} (150 ml × 6 min^{-1}) and $\dot{V}_A$ of 5.1 L min^{-1} (850 ml × 6 min^{-1}). Compared with normal breathing, this provides proportionately more ventilation of the alveoli and less wastage per minute in the dead space; however, the work of breathing with such pattern is costly (p. 495). On the other hand shallow panting (e.g. $V_T = 250$ ml, $f = 24$ min^{-1}, $\dot{V}_E = 6$ L min^{-1}) not only has a high energetic cost but it is also very wasteful as it has a $\dot{V}_D$ of 3.6 L min^{-1} (150 ml × 24 min^{-1}) and a $\dot{V}_A$ of only 2.4 L min^{-1} (100 ml × 24 min^{-1}).

Anatomical, alveolar and physiological dead space

V_D comprises not only the volume of the conducting airways (the **anatomical dead space**) but also that part of the alveolar volume (**alveolar dead space**) occupying alveoli which are inadequately perfused with blood and therefore not fully participating in gas exchange. In health the alveolar dead space is less than 5 ml but in various lung diseases it can be larger due to inequalities between alveolar ventilation and blood flow (p. 520). **Physiological dead space** equals anatomical dead space plus alveolar dead space.

The anatomical dead space can be measured by plaster casts of the lungs of a cadaver or, in the living, by the **Fowler method** in which the gas (e.g. CO_2) composition of an expired breath is measured continuously with a rapidly responding analyser. The first part of the expirate is essentially atmospheric gas (virtually no CO_2 as it has come from the conducting airways where no gas exchange takes place. It is followed by gas from the alveoli, which usually contains about 5–6% CO_2 or 0.05–0.06 as a fraction (F_{CO_2}). Because there is gas mixing during expiration at the boundary between the dead-space gas and alveolar gas, there is no clear-cut change from one to the other (Fig. 16.17a). If, however, F_{CO_2} in the expirate is measured continuously and plotted against the volume expired, the resulting graph (Fig. 16.17b) permits measurement of the anatomical V_D. The dashed perpendicular line ACB, drawn so as to make the areas AXC and BOC equal, intersects the volume axis at a value (V_D) which represents where a clear-cut change would have occurred if there was no mixing.

The dead space, as well as alveolar volume and the alveolar and dead space ventilations, can be estimated using the **Bohr equation**. The principle behind this equation is that in an expirate the total amount of a gas exhaled equals the amount of that gas coming from both the dead space and the alveoli. The amount of a gas in each compartment = volume (V) of that compartment × fractional composition (F) of the gas concerned, so for CO_2:

$$V_E \times F_E{CO_2} = V_D \times F_I{CO_2} + V_A \times F_A{CO_2}.$$

This equation can be written either for the volumes (V) of these compartments or for their ventilations (volume/minute, $\dot{V}$). Since $F_I{CO_2}$ is virtually zero, the equation simplifies to:

$$V_E \times F_E{CO_2} = V_A \times F_A{CO_2} \tag{16.1}$$

V_E is measured as the volume of a single expirate (or a number of them collected over a known time in a bag), $F_E{CO_2}$ is a gas sample taken from the bag which is a

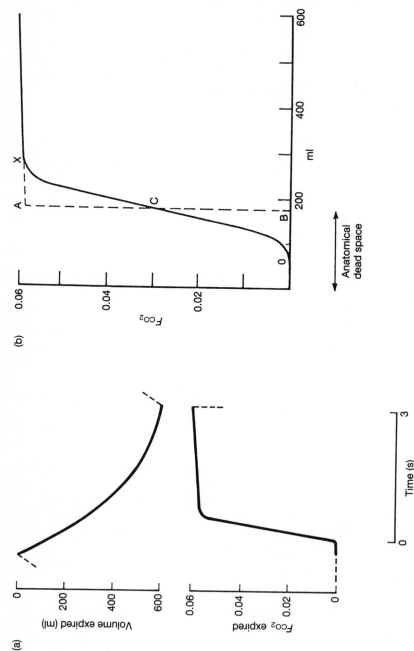

Fig. 16.17 (a) Expired volume and tidal fraction (F) of CO_2 measured at the mouth during expiration (dashed lines indicate end of preceding and beginning of next inspiration). Time axis is for a breathing frequency of 12 min^{-1}. (b) Fowler method for measuring anatomical dead space (V_D) from data collected in (a) during expiration. The line ACB is perpendicular to and intersects the volume axis at V_D. See text for further explanation.

mixture of alveolar and dead-space gases (hence the term mixed-expired) and $F_A CO_2$ is a gas sample representing the average alveolar gas value which is obtained about two-thirds of the way through expiration (see Fig. 16.18) or more easily but less acurately at the very end of an expiration (hence the term end-expired or end-tidal). The unknown V_A is calculated and V_D determined knowing:

$$V_E = V_D + V_A \tag{16.2}$$

When equation (16.2) is substituted into equation (16.1) and rearranged knowing $V_E = V_T$, the equation becomes:

$$V_D = \frac{F_A CO_2 - F_E CO_2}{F_A CO_2} \cdot V_T$$

This is the final format in which the Bohr equation is usually given but for the beginner it is harder to understand.

Since the partial pressure (P) of a gas is proportional to its fractional (F) concentration (p. 503), the Bohr equation is also written as:

$$V_D = \frac{P_A CO_2 - P_E CO_2}{P_A CO_2} \cdot V_T \quad \text{or as} \quad V_D = \frac{P_a CO_2 - P_E CO_2}{P_a CO_2} \cdot V_T$$

where the partial pressure of arterial (P_a) gas is substituted for alveolar (P_A) gas.

Strictly speaking, if in the Bohr equation the alveolar gas value is used, the V_D measured is closer to an estimate of anatomical V_D. A true measure of physiological V_D requires the arterial, not the alveolar, gas value to be used as this will permit alveolar V_D, which is due to unperfused alveoli, to contribute to the measurement. In healthy people the $P CO_2$ values in alveolar gas and arterial blood are virtually the same as there is little alveolar V_D, whereas in certain lung diseases arterial $P CO_2$ is higher than alveolar $P CO_2$ and the alveolar V_D is greater.

Gas exchange

At rest an adult consumes about 250–300 ml min^{-1} of O_2 (the **O_2 consumption,** $\dot{V} O_2$) and produces in exchange about 200–250 ml min^{-1} of CO_2 (the **CO_2 production, $\dot{V} CO_2$**). The ratio of CO_2 produced to O_2 consumed, $\dot{V} CO_2 / \dot{V} O_2$, is called the **respiratory exchange ratio** (R). In steady-state conditions, it is common to use the term **respiratory quotient (RQ)** which depends entirely on the chemical nature of the combusted food. If the substrates are solely carbo-hydrates the RQ is 1.0, while for a fuel of protein the RQ is 0.81 and for fat 0.7. Since our tissues oxidize a mixture of these three substrates, the RQ is on average 0.82. The R value is not the same as the RQ value during the transition from one steady state to another—for instance, at the beginning of hyperventilation the excreted CO_2 is greater than that produced metabolically because initially extra CO_2, derived from the large stores in tissue and blood (p. 533), is blown off. Thus, although the RQ may still be 0.82, the R value may have risen to as much as 1.4. An R value lower than 0.82 occurs under conditions where CO_2 stores are being replenished.

Gas consumption or production can be calculated from the difference in the amount of the gas that enters and leaves the lungs per minute. Therefore $\dot{V}_I$ and its gas fraction (e.g. $F_I O_2$) and $\dot{V}_E$ and its gas fraction (e.g. $F_E O_2$) have to be measured. Thus:

$$\dot{V}_{O_2} = \dot{V}_I \times F_{I_{O_2}} - \dot{V}_E \times F_{E_{O_2}}$$

and, as $\dot{V}_I$ is virtually the same as $\dot{V}_E$,

$$\dot{V}_{O_2} = \dot{V}_E \ (F_{I_{O_2}} - F_{E_{O_2}})$$

Similarly,

$$\dot{V}_{CO_2} = \dot{V}_E \times F_{E_{CO_2}} - \dot{V}_I \times F_{I_{CO_2}}$$

and, as $F_{I_{CO_2}}$ is close to zero,

$$\dot{V}_{CO_2} = \dot{V}_E \times F_{E_{CO_2}}$$

$\dot{V}_E$ is slightly smaller than $\dot{V}_I$ whenever less CO_2 is produced than O_2 consumed (i.e. the R value is <1.0). The exact difference can be calculated using inspired and expired N_2 fractions. Since N_2 is not consumed or produced,

$$\dot{V}_I \times F_{I_{N_2}} - \dot{V}_E \times F_{E_{N_2}} = 0$$

therefore:

$$\dot{V}_I = \dot{V}_E \times F_{E_{N_2}}/F_{I_{N_2}}.$$

This correction (called the **N_2 correction**) can be incorporated into the basic equation if required for precision.

The ratio of ventilation to O_2 consumption ($\dot{V}_E/\dot{V}_{O_2}$) is called the **ventilatory requirement** and at rest is ~25 (7.5/0.3 L min^{-1}). It shows what a large quantity of air enters the lungs relative to the amount of O_2 consumed. A better expression is the **O_2 extraction coefficient**, which takes into account how much of the incoming air is actually O_2. This coefficient is the amount of O_2 used/amount of O_2 inspired, expressed as a percentage:

$$\dot{V}_{O_2}/(\dot{V}_I \times F_{I_{O_2}}) \times 100$$

At rest the O_2 extraction coefficient is 15–20%, showing how little of the O_2 entering is actually used—there is in effect a large reserve available should the next few breaths be impeded. Neither of these ratios takes into account the O_2 reserves held in the 2.5 litres of the FRC.

In severe exercise $\dot{V}_{O_2}$ may be as great as 2–3 L min^{-1} depending on the person's ability to perform physical work—the more work the person is capable of, the greater the $\dot{V}_{O_2}$. Highly trained athletes may have a maximum $\dot{V}_{O_2}$ as high as 5–6 L min^{-1}. In severe exercise the stimuli to ventilation (p. 694) are such that breathing is far in excess of the $\dot{V}_{O_2}$ ensuring high enough alveolar P_{O_2} in a diffusion-limited situation (p. 513); thus the ventilatory requirement may be as high as 50 and the O_2 extraction coefficient as low as 10%. Furthermore, as the excessive breathing blows off CO_2 from its body stores, R values are greater than 1.0.

The gas laws and ATPS, BTPS and STPD conditions

As a gas expands when heated and shrinks when compressed, the numerical value ascribed to its volume (V) or the volume/unit time ($\dot{V}$) is dependent on the prevailing temperature (T) and pressure (P). Allowance must also be made for the volume occupied by water vapour. The partial pressure of water vapour (P_{H_2O}) depends entirely on the degree of saturation and the prevailing temperature.

In the lungs, gases are at body temperature $(273 + 37$ K), body pressure (which is at the same pressure as the ambient barometric pressure) and fully saturated with water vapour (P_{H_2O} 47 mmHg) at that temperature (these are **BTPS** conditions). However, expired volumes are usually collected at ambient temperature, which as a rule is lower than body temperature, so the volume shrinks and some of the water vapour condenses. In this situation the gas is at ambient temperature, ambient pressure and saturated with water vapour at that temperature (i.e. **ATPS**). Since it is the volume the gas occupied when it was in the lungs that is important, the volume at ATPS must be converted to its volume at BTPS. Similarly, the volumes of CO_2 produced and O_2 consumed are usually calculated from volumes collected at ATPS but in this case they are reported as they would be under standard conditions of physical chemistry — these are standard temperature (0°C), standard pressure (760 mmHg) and completedly dry ($P_{H_2O} = 0$ mmHg; i.e. **STPD**)—as it is the amount of gas that is relevant. The conversions are made using the **Boyle–Charles** formula where:

$$\frac{P_1 V_1}{T_1} = \frac{P_2 V_2}{T_2}$$

If water vapour is present, its partial pressure must be subtracted from the total pressure as it is the pressure of the 'dry' gas which is affected by the pressure and the temperature.

If in this equation subscript 1 represents ATPS and subscript 2 represents BTPS values and if the volume measured at ATPS is 5 litres, the barometric pressure is 765 mmHg and the temperature is 22°C (at which for fully saturated conditions $P_{H_2O} = 20$ mmHg), then:

$$\frac{(765 - 20) \times 5}{273 + 22} = \frac{(765 - 47) \times V_2}{273 + 37}$$

and thus the BTPS volume (V_2) = 5.45 litres. The difference in volume between the ATPS and BTPS condition is quite substantial and cannot be ignored.

Similarly, a $\dot{V}O_2$ of 300 ml min^{-1} under the ATPS conditions given above converts to:

$$\frac{(765 - 20) \times 300}{273 + 22} = \frac{(760 - 0) \times V_2}{273 + 0}$$

i.e. 272 ml min^{-1} at STPD.

One advantage of converting the volume consumed into STPD units is that as the molecular mass (in g) of any gas occupies 22.4 litres at STPD (**Avogadro's law**), a simple step ($\dot{V}O_2$ (ml min^{-1} STPD)/22.4) allows the O_2 consumption to be expressed as the mass consumed (mmol min^{-1}).

Note that the **ideal** (or **general**) **gas law** combines the principles of Avogadro's and Boyle–Charles' laws and states that $PV = nRT$ where n is the number of gram molecules and R is the universal gas constant.

Partial pressures of inspired, expired and alveolar gas

In the gas phase the partial pressure of a gas is proportional to its fractional share of the total volume of a dry gas mixture (Dalton's law). Expired gas is a mixture of that from both the dead space (which has the same composition as the inspired *air) and the alveoli. The partial pressurers of alveolar O_2 and CO_2 fluctuate very slightly during the breathing cycle as they are influenced by the tidal effect of inspired gas and the continuous effect of diffusional gas exchange between the blood and the alveoli. The mean alveolar P_{O_2} and P_{CO_2} are usually 100 and 40 mmHg, respectively, but can be raised or lowered by three fundamental determinants—the inspired gas composition, the magnitude of metabolism and the magnitude of alveolar ventilation—as described in the so-called alveolar gas equations.*

Each gas in a dry mixture exerts a **partial pressure** (P_{gas}) which is proportional to its fractional share (F_{gas}) of the total volume. This is **Dalton's law** and when applied to respiratory gases P_{H_2O} must be subtracted from the total pressure (the barometric pressure, P_B, in mmHg):

$$P_{gas} = F_{gas}\,(P_B - P_{H_2O})$$

Although the air breathed has a variable humidity, once it has been inspired beyond the nasal passages it virtually becomes fully saturated with water vapour at a value (P_{H_2O} 47 mmHg) dictated by the body's temperature (37°C). Thus

$$P_{gas} = F_{gas}\,(P_B - 47)\ \text{mmHg}.$$

The fractional composition and partial pressures for inspired (P_I), mixed-expired (P_E) and mean alveolar gases (P_A) are given in Table 16.2 for O_2, CO_2 and N_2, when P_B is 760 mmHg (standard barometric pressure at sea level). Note that the composition of the mixed-expired gas, at least for normal breathing (p. 496), is a mixture of one-third dead-space gas (and therefore equivalent in gas composition to inspired gas) and two-thirds alveolar gas. Inspired gases at sea level when fully saturated with water are $P_{I_{O_2}}$ 150 mmHg and $P_{I_{CO_2}}$ virtually zero (0.2 mmHg). Typical values for alveolar gases under normal healthy conditions are $P_{A_{O_2}}$ 100 mmHg and $P_{A_{CO_2}}$ 40 mmHg.

As shown in Fig. 16.18, the composition of the **alveolar gas** fluctuates very slightly during the breathing cycle. Alveolar gas is influenced by the tidal effect of inspired gas from the airways and the continuous effect of diffusional gas exchange with the blood perfusing the alveoli. The alveolar gas fluctuation at rest is small because these two effects are dampened by the size of the total alveolar gas volume (the 2.5 litres of FRC). At the beginning of inspiration the gas that first enters the

Table 16.2 Fractional composition (F) and partial pressures (P, in mmHg) of inspired (F_I or P_I), mixed-expired (F_E or P_E) and mean alveolar gas (F_A or P_A) for O_2, CO_2 and N_2 when P_B is 760 mmHg and all these compartments are fully saturated with water vapour at 37°C. Note that fractions for gases are given for a 'dry' gas mixture

	F_I	F_E	F_A	P_I	P_E	P_A
O_2	0.2093	0.163	0.140	149	116	100
CO_2	0.0003	0.037	0.056	0.2	26	40
N_2	0.7904	0.800	0.804	563	571	573
H_2O	—	—	—	47	47	47

alveoli is inspired from the dead space which contains alveolar gas left there from the preceding expiration—thus the P_{CO_2} in the alveoli continues to rise and P_{O_2} to fall. Only when the dead space is cleared does inspired fresh air enter the alveoli, diluting the alveolar gas to a small degree—thus, the P_{CO_2} now falls and P_{O_2} rises.

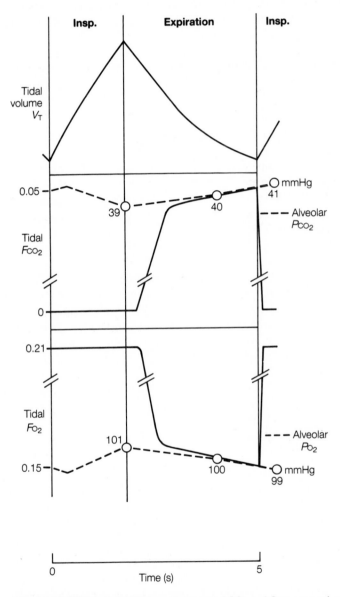

Fig. 16.18 Tidal volume (V_T) and fractions of CO_2 and O_2 measured at the mouth (F_{CO_2} and F_{O_2}) during inspiration and expiration. Dashed lines represent the changes in alveolar gas composition (expressed in mmHg) during the breathing cycle. Time axis is for a breathing frequency of 12 min^{-1}.

Throughout expiration, gas flows out of the alveoli and the alveolar gas is modified only by alveolar blood perfusion, which continually supplies CO_2 to, and removes O_2 from, the gas phase. The rise in P_{CO_2} and fall in P_{O_2} during expiration are often referred to as the **alveolar slope** (or less correctly, the alveolar plateau if the rise and fall are slight). The alveolar P_{CO_2} and P_{O_2} approach the mean value about two-thirds through expiration (Fig. 16.18) but if the alveolar slope is slight, the end-expired values can be used instead and this will only slightly overestimate $P_{A}CO_2$ and slightly underestimate $P_{A}O_2$.

The alveolar slopes for O_2 and CO_2 become steeper when tidal volumes are larger, particularly if the expiratory reserve volume is being utilized, or breathing frequency is slower than usual. Furthermore in exercise, increased gas exchange ($\dot{V}O_2$ and $\dot{V}CO_2$) also steepens the alveolar slope.

Determinants of alveolar O_2 and CO_2 and the alveolar gas equations

There are three fundamental determinants of alveolar gas composition—the inspired gas composition and the magnitudes of alveolar ventilation and metabolism. (Note also that the effect of metabolism on alveolar gases depends on both an adequate blood perfusion of the alveoli and the matching of this perfusion to the alveolar ventilation; p. 520.)

$P_{A}O_2$

$P_{A}O_2$ will decrease whenever there is a fall in the P_{O_2} inspired into the alveoli from the atmosphere. As $P_{I}O_2 = (P_B - P_{H_2O}) \times F_{I}O_2$, a fall in $P_{I}O_2$ can occur either at altitude where, although $F_{I}O_2$ is still 0.21, the P_B is less than 760 mmHg, or when the fractional composition of the atmosphere falls below the usual 0.21, as in poorly ventilated environments. $P_{A}O_2$ will also decrease if less than the usual volume of inspired gas ventilates the alveoli (i.e. if there is an impairment of $\dot{V}_A$), and if more O_2 than usual is extracted from the alveolar gas into the blood stream (i.e. if the body's $\dot{V}O_2$ increases). The relationship between these determinants is expressed in the **alveolar gas equation for O_2**:

$$P_{A}O_2 = P_{I}O_2 - \frac{\dot{V}O_2 \times 863}{\dot{V}_A}$$

where 863 is a conversion factor which, first, accounts for the $\dot{V}O_2$ being in STPD units and $\dot{V}_A$ in BTPS units (both are in L min^{-1}) and second, permits the fractional expression ($\dot{V}O_2/\dot{V}_A$) to be converted into partial pressures so that it can be subtracted from $P_{I}O_2$.

The above equation is derived in the following way. Step 1: $\dot{V}O_2 = \dot{V}_I F_I O_2 - \dot{V}_E F_E O_2$ (p. 500) which, because O_2 is only consumed from the alveolar, not dead-space, compartment, can also be written as $\dot{V}O_2 = \dot{V}_A F_I O_2 - \dot{V}_A F_A O_2$. Step 2: the last equation is rearranged and divided by $\dot{V}_A$ so that $F_A O_2 = F_I O_2 - \dot{V}O_2/\dot{V}_A$. Step 3: each part is multiplied by $(P_B - P_{H_2O})$ but, as $F_O_2 \times (P_B - P_{H_2O}) = P_{O_2}$ (p. 503), the equation is now written as $P_A O_2 = P_I O_2 - (P_B - P_{H_2O})\dot{V}O_2/\dot{V}_A$. Step 4: as $\dot{V}O_2$ is in STPD units and $\dot{V}_A$ in BTPS units, the Boyle–Charles law (p. 502) must be applied to convert $\dot{V}O_2$ into BTPS units. Thus $\dot{V}O_2$ (BTPS) = $\dot{V}O_2$ (STPD $\times$ 310 $\times$ 760/273 $\times$ $(P_B - P_{H_2O})$. Step 5: when the conversion in step 4 is inserted into the equation in step 3, the two $(P_B - P_{H_2O})$ cancel out and $310 \times 760/273 = 863$, so that the final form is derived: $P_AO_2 = P_IO_2 - (863 \times \dot{V}O_2/\dot{V}_A)$.

A graphical expression of the O_2 alveolar gas equation is shown in Fig. 16.19. Each curve portrays the relationship between P_AO_2 and one of its determinants, $\dot{V}_A$ for different values of P_IO_2 or of $\dot{V}O_2$. As P_AO_2 is the dependent variable, it is depicted on the vertical axis. The relationship between P_AO_2 and $\dot{V}_A$ is hyperbolic such that for resting conditions (middle curves), where P_IO_2 is 150 mmHg $(F_IO_2 = 0.21)$ and $\dot{V}O_2$ is 300 ml min^{-1} STPD, a normal alveolar ventilation ($\dot{V}_A$) of 5 L min^{-1} BTPS produces the normal alveolar Po_2 of 100 mmHg (point A). When hypoventilation occurs, P_AO_2 decreases; when hyperventilation occurs, P_AO_2 increases and approaches, but can never reach, the inspired Po_2 value of 150 mmHg as there is continuous O_2 consumption.

A raised hyperbolic curve results from either a decrease in $\dot{V}O_2$ (depressed metabolism) or an increase in P_IO_2—the magnitude of the change determines the degree of upwards shift (Fig. 16.19). Thus, for instance, if P_IO_2 were to increase, the P_AO_2 will be elevated (point B) without any need for an increase in $\dot{V}_A$. Similarly, a downward-shifted hyperbolic curve shows the lower P_AO_2 values for each particular level of $\dot{V}_A$ when either $\dot{V}O_2$ increases (e.g. during exercise) or P_IO_2 decreases (e.g. the hypoxia of altitude). Clearly, in these two situations, to prevent P_AO_2 from falling to point Y or C (Fig. 16.19), an increase in $\dot{V}_A$ is required. In the case of exercise the increase in $\dot{V}_A$ exactly matches the increase in $\dot{V}O_2$ and P_AO_2 remains constant (point Z), whereas at altitude the fall in P_AO_2 can only be minimized (point D). In intense exercise the increase in breathing may be such that P_AO_2 is greater than 100 mmHg.

P_ACO_2

P_ACO_2 is also determined by the inspired gas composition and the magnitudes of alveolar ventilation and metabolism. In this case P_ACO_2 will increase if there is CO_2 in the inspired gas (i.e. P_ICO_2 increases), if ventilation is depressed (i.e. $\dot{V}_A$ decreases) or if more CO_2 than usual is produced by metabolism and carried to the alveoli (i.e. $\dot{V}CO_2$ increases). Converse changes will decrease P_ACO_2 and the most obvious example is that caused by voluntary hyperventilation. The **alveolar gas equation for CO_2** is derived in a similar fashion to that for O_2 and is:

$$P_ACO_2 = P_ICO_2 + \frac{\dot{V}CO_2 \times 863}{\dot{V}_A} \quad \text{or} \quad P_ACO_2 = \frac{\dot{V}CO_2 \times 863}{\dot{V}_A}$$

as P_ICO_2 is frequently omitted because it is normally equal to zero (0.214 mmHg). (Note that the truncated form of this equation is independent of ambient pressure and is used clinically to measure $\dot{V}_A$, often with P_aCO_2 substituted for P_ACO_2.)

The graphical relationship between P_ACO_2 and $\dot{V}_A$ for normal conditions of a P_ICO_2 of 0 mmHg and a resting $\dot{V}CO_2$ of 250 ml min^{-1} STPD is shown in Fig. 16.20 such that a typical $\dot{V}_A$ of 5 L min^{-1} BTPS produces the normal P_ACO_2 of 40 mmHg (point A). Hypoventilation elevates P_ACO_2 while hyperventilation decreases P_ACO_2, though this never falls to the inspired Pco_2 value of 0 mmHg

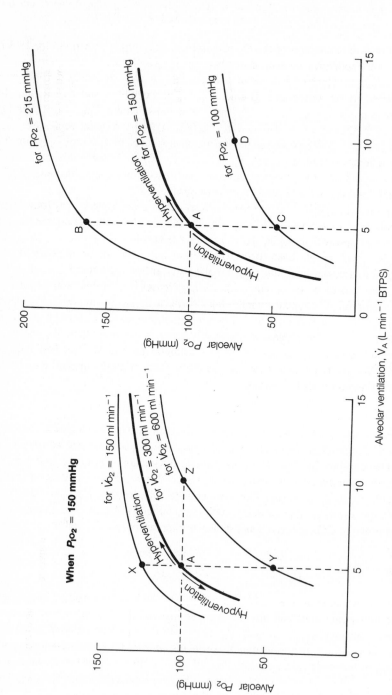

Fig. 16.19 $P_{A}O_2$ resulting from different alveolar ventilations (a) when $P_{I}O_2$ remains at the normal inspired level of 150 mmHg during three levels of $\dot{V}O_2$ and (b) when $\dot{V}O_2$ remains at the normal level of 300 ml min^{-1} STPD during three conditions of $P_{I}O_2$. In each case the middle curve has been calculated for normal metabolic and inspired O_2 conditions ($\dot{V}O_2 = 300$ ml min^{-1} STPD and $P_{I}O_2 = 150$ mmHg) with point A indicating that a normal $\dot{V}_A$ of 5 L min^{-1} BTPS produces a $P_{A}O_2$ of 100 mmHg; hypoventilation lowers and hyperventilation increases $P_{A}O_2$. The upper curve represents either decreased metabolism (a) or increased $P_{I}O_2$ (b) and the new $P_{A}O_2$ that results (point X and B) if there is no alteration in $\dot{V}_A$. The lower curve represents the situation in exercise (a) or at altitude (b), with points Y and C indicating the consequences if there is no reflex alteration in $\dot{V}_A$ compared with the actual reflex increase in $\dot{V}_A$ that occur in exercise (point Z) or at altitude (point D).

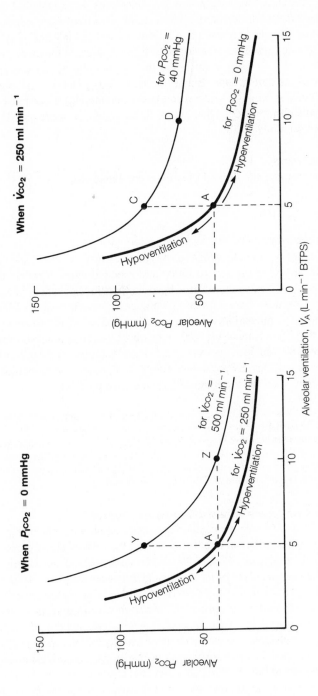

Fig. 16.20 $P_{A}CO_2$ resulting from different alveolar ventilations for normal metabolic and inspired CO_2 conditions ($\dot{V}CO_2 = 250$ ml min^{-1} STPD and $P_{I}CO_2 = 0$ mmHg) where point A indicates that a normal $\dot{V}_A$ of 5 L min^{-1} BTPS produces a $P_{A}CO_2$ of 40 mmHg and hypoventilation increases and hyperventilation lowers $P_{A}CO_2$. The upper curve represents either increased metabolism (a) or increased $P_{I}CO_2$ (b) and the new $P_{A}CO_2$ which would result (point Y and C) if there was no alteration in $\dot{V}_A$ compared with the actual $P_{A}CO_2$ achieved due to the reflex increase in $\dot{V}_A$ that occurs in exercise (point Z) or during hypercapnia (point D).

because of the metabolic production of CO_2. Upward-shifted hyperbolic curves (Fig. 16.20) result in situations where $\dot{V}_{CO_2}$ increases (as in exercise) or P_ICO_2 increases (hypercapnia); downwards-shifted curves (not illustrated) occur for decreases in $\dot{V}_{CO_2}$. Clearly, to prevent P_ACO_2 from rising to point Y or C (Fig. 16.20) when $\dot{V}_{CO_2}$ or P_ICO_2 increases, an increase in $\dot{V}_A$ is required. In the case of exercise the increase in $\dot{V}_A$ matches the increase in $\dot{V}_{CO_2}$ and P_ACO_2 remains constant (point Z), whereas during hypercapnia the rise in P_ACO_2 can only be minimized (point D). Clinically (for inspired PCO_2 conditions of 0 mmHg) the adequacy of $\dot{V}_A$ for tissue metabolism is judged by how much the P_ACO_2, or more usually P_aCO_2, deviates from 40 mmHg.

The alveolar gas equation

The equations for alveolar O_2 and CO_2 when combined give:

$$P_AO_2 = P_IO_2 - \frac{P_ACO_2}{R} + F$$

where R is the respiratory gas exchange ratio ($\dot{V}_{CO_2}/\dot{V}_{O_2}$) and F is a small correction factor (1–3 mmHg, which can be ignored except for exact calculations) derived from the N_2 correction (p. 500) and required when R is less than 1. Furthermore, P_aCO_2 frequently replaces P_ACO_2 in the above equation, particularly for clinical use, because there is rarely a difference between the two (p. 509). The advantage of this combined alveolar gas equation (often called **the** alveolar gas equation) is that P_aCO_2 is relatively easy to measure from an arterial blood sample, P_IO_2 is known (150 mmHg at sea level), the R value is assumed to be 0.8–0.85 and thus P_AO_2 can be calculated. P_aO_2 is always less than P_AO_2 and the magnitude of the $P_{A-a}O_2$ gradient can be an important clinical measure (p. 522).

The alveolar gas equation is derived in the following way. Step 1: since $R = \dot{V}_{CO_2}/\dot{V}_{O_2}$, one can substitute $\dot{V}_{CO_2}/R$ for $\dot{V}_{O_2}$ in the alveolar O_2 equation, $P_AO_2 = P_IO_2 - 863\ \dot{V}_{O_2}/\dot{V}_A$, which after rearranging becomes $P_IO_2 - P_AO_2 = 863\ \dot{V}_{CO_2}/R\ \dot{V}_A$. Step 2: further rearrangement produces $R\ (P_IO_2 - P_AO_2) = 863\ \dot{V}_{CO_2}/\dot{V}_A$ and since, according to the alveolar CO_2 equation, $P_ACO_2 = 863\ \dot{V}_{CO_2}/\dot{V}_A$, it follows that $R\ (P_IO_2 - P_AO_2) = P_ACO_2$, which when rearranged again, is the final format, $P_AO_2 = P_IO_2 - P_ACO_2/R$.

Partial pressures of gases in solution in the blood

In the lungs O_2 and CO_2 in the capillaries equilibrate with the alveolar gases so that the blood leaving the lungs, which becomes the arterial blood of the body, has a PO_2 of 100 mmHg and PCO_2 of 40 mmHg. Venous blood leaves the various tissues at different partial pressures and finally enters the lungs as mixed venous blood with a PO_2 of 40 mmHg and PCO_2 of 46 mmHg. Gas movement across the capillary walls of the lungs and tissues occurs by diffusion down a partial-pressure gradient and is dependent on the molecular weight and solubility of the gas, on the surface area available and on the distance travelled (Fick's law of diffusion). Usually most of the diffusion has occurred by the time the blood has traversed a third of the length of the capillary. Diseases that thicken the alveolar–capillary membrane or reduce the alveolar surface area cause a reduction in the pulmonary diffusing capacity for O_2. Successful gas transfer and maintenance of the *partial pressure gradients rely not only on alveolar ventilation but also on blood flow through the lungs and tissues. The relationship between the amount of O_2 transferred, the cardiac output and the arteriovenous difference in O_2 content is expressed in the Fick principle. Increased cardiac output and greater O_2 extraction from the blood, causing a decrease in O_2 content of venous blood, allow for increased O_2 consumption.*

When a liquid is exposed to a gas phase for long enough to reach equilibrium, the partial pressures of the gases in the liquid and gas phase are identical. (Note, however, that the content of the gas (millilitre of gas per litre of medium) may be very different between the two phases depending on the solubility of the gas in the liquid; p. 523). Given the large surface area and thinness of the air-to-blood barrier (p. 471), sufficient contact time occurs in the alveoli to permit the capillary blood leaving the alveoli to come to the same partial pressures as the alveolar gas. Thus, for resting normal conditions, $P_A O_2$ and end-pulmonary capillary blood Po_2 ($P_c \cdot O_2$) are both 100 mmHg and $P_A CO_2$ and $P_c \cdot CO_2$ are both 40 mmHg (Table 16.3 and Fig. 16.21). The end-pulmonary capillary blood leaves the lungs in the pulmonary veins to become the arterial blood for the rest of the body. Partial pressures in arterial blood are: $P_a CO_2$ 40 mmHg and $P_a O_2$ ~95 mmHg (Table 16.3 and Fig. 16.21).

The Po_2 in arterial blood is less than end-pulmonary capillary blood for the following reasons. The 1–2% of the cardiac output that comes through the bronchial and coronary circulations (p. 449) bypasses the alveoli and drains, respectively, into the pulmonary veins and directly into the left heart itself and this modifies slightly the arterial blood composition (such venous admixture is referred to as a shunt; p. 515). Because of differences in the respective shapes of CO_2 and O_2 blood dissocation curves (p. 533), this venous admixture has a negligible effect on $P_a CO_2$ but lowers by a few mmHg the $P_a O_2$. A further lowering of $P_a O_2$ occurs as a consequence of alveolar ventilation and alveolar perfusion not being in the same proportion throughout the lungs (p. 520).

The blood that enters the alveolar capillaries via the pulmonary arteries is a mixture of all the venous blood coming from all the organs, each metabolizing at different rates (p. 450). This mixed venous ($\bar{v}$) blood for a resting person typically has a $P_{\bar{v}} CO_2$ of 46 mmHg and a $P_{\bar{v}} O_2$ of 40 mmHg (Table 16.3 and Fig. 16.21). As sufficient time is allowed for equilibrium to occur as blood transits through the tissues capillaries, the interstitial fluid (ISF) bathing the 'typical' cell is said as an approximation to be at a Pco_2 of 46 mmHg and Po_2 of 40 mmHg.

Table 16.3 Partial pressure of gases in the alveoli and blood. Note that water cannot exist in a vapour form in a liquid (blood) phase

	Partial pressure (mmHg)			
	O_2	CO_2	N_2	H_2O
Alveolar gas (P_A)	100	40	573	47
End-pulmonary capillary blood ($P_c \cdot$)	100	40	573	—
Systemic arterial blood (P_a)	95	40	573	—
Mixed venous blood ($P_{\bar{v}}$)	40	46	573	—

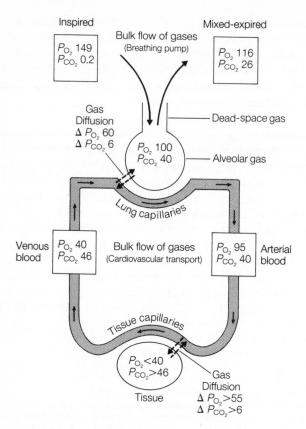

Fig. 16.21 Po_2 and Pco_2 mmHg in various parts of the respiratory and cardiovascular system. Bulk flow of gases occurs between atmosphere and lungs (breathing pump) and around the body in the blood stream (cardiovascular transport). Gases move across capillary walls by diffusion.

Diffusion

Gases move by diffusion down gradients of partial pressure (not concentration) between alveolar gas and pulmonary capillary blood and between systemic capillary blood and ISF. In Fig. 16.21 at both sites of diffusion the initial Po_2 and Pco_2 gradients are $\Delta 60$ and $\Delta 6$ mmHg respectively. As **Fick's law of diffusion** states (p. 6), the amount of gas diffusing in unit time through the resistance of a barrier is inversely proportional to the thickness of the barrier and directly proportional to the surface area of the barrier, the diffusion constant (D) for that gas through that barrier, and the partial pressure gradient of the gas across the barrier Thus:

$$\dot{V}_{gas} = \frac{area \times D_{gas}}{thickness} \times \Delta P_{gas}$$

As stated by Graham's law, the diffusion constant is directly proportional to the solubility of the gas, which is dependent on temperature, and inversely proportional to the square root of its molecular mass (M_r). For biological membranes at

37°C the diffusion constant for CO_2 is 23 times greater than for O_2 because, despite its larger M_r (44 vs. 32), CO_2 is very much more soluble than O_2 in plasma (0.65 vs. 0.031 ml L^{-1} mmHg^{-1}). Thus, given that the volume per unit time of CO_2 produced and O_2 consumed is similar, CO_2 diffusion occurs down much smaller partial-pressure gradients. Furthermore, because of its high diffusion constant, CO_2 diffusion is rarely impeded.

The diffusion of a substance in a liquid is described as being driven by its concentration gradient (p. 6), whereas the diffusion of a gas in both the liquid and gas phase is correctly described as being driven by a gradient of partial pressure. Concentration and partial pressure are simply related through the solubility of the gas. The use of partial pressure is convenient for it allows direct comparison of the forces driving diffusion in the gas and liquid phases. It also avoids the difficulty that more O_2 and CO_2 are carried in blood than would be possible in simple solution (p. 523). Thus the content of these gases in amount per litre of blood (often referred to as concentration) is much greater than the concentration in physical solution, which is the driving force for diffusion.

In the lungs, because of difficulties of their individual measurement, the surface area of the alveoli, the diffusion constant for O_2 and the thickness of the air–blood barrier (which is not only the alveolar–capillary membrane but also the distance within the capillary to the centre of the red blood cells) are all lumped together and called the **pulmonary diffusing capacity** for O_2 ($D_L o_2$). Thus:

$$D_L o_2 = \frac{area \times D o_2}{thickness} \quad \text{and so} \quad \dot{V} o_2 = D_L o_2 \times \Delta P o_2$$

This equation is *not* used to measure $\dot{V} o_2$ but is in fact rearranged to permit measurement of $D_L o_2$:

$$D_L o_2 = \frac{\dot{V} o_2}{\Delta P o_2} \quad \text{or} \quad D_L o_2 = \frac{\dot{V} o_2}{P_A o_2 - P_{\bar{c}} o_2}$$

where $P_{\bar{c}} o_2$ is the mean pulmonary capillary $P o_2$ averaged over the entire length of the capillary (Fig. 16.22). Again there is a problem of measurement—$P_{\bar{c}} o_2$ can only be deduced from a prior knowledge of all the other variables in the equation!

The problem is solved by measuring instead the pulmonary diffusing capacity for carbon monoxide (CO). This is because $P_{37\bar{c}} co$ can be taken as zero on account of the very high affinity between CO and haemoglobin, such that little CO is actually in solution in the plasma (p. 531)—thus $D_L co = \dot{V} co / P_A co$. Measurements are made after a subject inspires a gas mixture containing a small amount of CO (about 0.1%) for a few breaths. Then $P_A co$ is measured from the first end-expiratory sample and $\dot{V} co$ by the rate at which CO disappears from the alveoli into the blood over a 10-second breath-hold (estimated by measuring CO in the alveoli at the beginning and end of the breath-hold). $D_L co$ so measured at rest is about 25 ml min^{-1} mmHg^{-1}. Because of the M_r and solubility differences between CO and O_2, $D_L o_2$ is 1.23 × $D_L co$, i.e. ~30 ml min^{-1} mmHg^{-1}. Similarly, the pulmonary diffusing capacity for CO_2 ($D_L co_2$) is 23 times greater than $D_L o_2$.

Figure 16.22 depicts the calculated $P o_2$ profile of blood along the length of the pulmonary capillary from when it enters the capillary as mixed venous blood ($P_{\bar{v}} o_2$

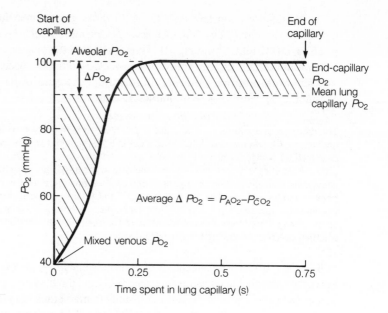

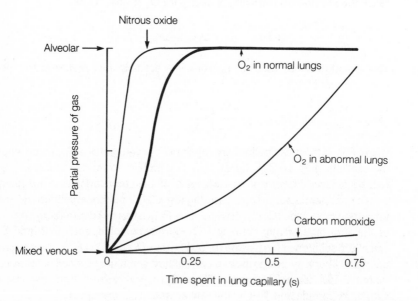

Fig. 16.22 Upper: Pulmonary capillary P_{O_2} as a function of time spent in the capillary of a normal lung. (Blood P_{O_2} changes slowly at the beginning of the capillary because at these low levels of P_{O_2} most of the O_2 diffusing into the blood combines with haemoglobin and causes very little increase in P_{O_2}.) Lower: Time-course of changes in capillary partial pressure for different gases in the normal lung (N_2O, O_2 and CO) and abnormal lung (O_2).

40 mmHg) to when it finally leaves 0.75 second later as end-capillary blood ($P_{c'}O_2$) at a composition identical to alveolar (P_AO_2 100 mmHg). The initial PO_2 gradient is Δ60 mmHg (100 – 40) and as O_2 rapidly diffuses into the blood elevating the capillary PO_2, the gradient gets smaller and smaller as the blood moves along the capillary. When breathing air, most of the diffusion in a healthy lung is complete about one-third of the way along the lung capillary. By equalizing the shaded areas below and above a particular PO_2 (Fig. 16.22), the mean pulmonary capillary PO_2 ($P_{\bar{c}}O_2$) can be estimated — normally ~90 mmHg. With reference to the equation $\dot{V}O_2 = D_LO_2 \times (P_AO_2 - P_{\bar{c}}O_2)$, one can see that at rest the appropriate values are: $\dot{V}O_2$ ~300 ml min^{-1}, D_LO_2 ~30 ml min^{-1} mmHg^{-1}, P_AO_2 100 mmHg and $P_{\bar{c}}O_2$ 90 mmHg, indicating that the average partial-pressure gradient for diffusion is 10 mmHg.

In exercise D_LO_2 increases by up to two- to three-fold because the larger tidal volumes, by inflating the alveoli more, increase the alveolar surface area and because more pulmonary capillaries are perfused. There is however no alteration in the thickness of the air-to-blood barrier. In heavy exercise P_AO_2 can be elevated to 110–120 mmHg because of high ventilation (p. 505, the $P_{\bar{v}}O_2$ of blood entering the alveoli can be as low as 20 mmHg (p. 530) and, because transit time through the pulmonary capillary can shorten to 0.25 second (p. 448), the mean $P_{\bar{c}}O_2$ can be some 30–40 mmHg lower than at rest. These will all increase the initial and average ΔPO_2 gradients for diffusion. In heavy exercise transit time may be too short to permit end-pulmonary $P_{c'}O_2$ to equilbrate with P_AO_2 and as a consequence P_aO_2 may fall to 80–90 mmHg (p. 694). This is detected as a widening of the alveolar–arterial PO_2 difference (p. 522).

Lung diseases, which either thicken the air–blood barrier (e.g. pulmonary oedema and pulmonary fibrosis, the latter caused for example by sarcoidosis or asbestosis) or reduce the alveolar surface area (e.g. alveolar collapse as in atelectasis or loss of alveolar septa as in emphysema), can reduce D_LO_2 considerably such that, even after a normal resting transit time of 0.75 second, end-pulmonary $P_{c'}O_2$ may not reach the P_AO_2 level (Fig. 16.22). Hence P_aO_2 may be well below the normal 100 mmHg so that the alveolar–arterial PO_2 difference widens. This is exacerbated even more during exercise or in conditions when the initial mixed venous to alveolar PO_2 gradient is already low (e.g. at altitude).

Transfer of gases across the alveolar–capillary membrane may be diffusion-limited or perfusion-limited. If the end-capillary is less than the alveolar partial pressure for the gas in question, its transfer is **diffusion-limited** (e.g. CO in all instances, O_2 in very strenuous exercise, O_2 at rest with a moderately severe abnormality of the alveolar–capillary membrane, and CO_2 with exceptionally severe alveolar–capillary abnormality). If the blood equilibrates with the alveolar gas early in its passage through the capillary (e.g. O_2, CO_2 and the anaesthetic gas, N_2O), the transfer of that gas is **perfusion-limited** and gas transfer would increase if pulmonary capillary flow increased.

D_LO_2 and D_LCO actually measure not only the diffusing capacity of the alveolar–capillary–red blood cell membrane (D_M) but also the effect of the rate at which O_2 or CO reacts with haemoglobin inside the red blood cell. The latter is given as θV_c where θ is the reaction rate in millilitres of gas per minute per mmHg partial pressure and V_c is the volume in ml of pulmonary capillary blood. The membrane and haemoglobin stages each offer, in series, a resistance to gas transfer, and the overall resistance from these components is derived as reciprocals of the capacities, i.e. $1/D_L = 1/D_M + 1/\theta V_c$. The two components of D_L are about equal in magnitude so that diseases which cause capillary blood volume to fall can reduce the diffusing capacity of the lung. As some of the CO_2 carried in the blood also reacts with haemoglobin, there is also a small θV_c component to D_LCO_2.

Fick's law of diffusion also applies to diffusion between cells and tissue capillaries. Although it is not easy to quantify, and varies from organ to organ, the concept of a tissue-diffusing capacity akin to that for the lung is important. As expected from Fick's law, such a tissue-diffusing capacity is proportional to the diffusion constant of the gas concerned and the surface area of the organ's capillaries and inversely proportional to the distance from the capillary to the cell's mitochondria. When O_2 and CO_2 metabolism by the tissue increases, the capillary surface area increases and, in contrast to the situation in the lung, the diffusion distance decreases. This is a consequence of metabolic autoregulation (p. 423) which ensures that more capillaries in the tissue are perfused and therefore 'open', which both increases the surface area for gas exchange and permits any one cell to be closer than it was to an open capillary. In contrast to the lung, capillary transit time through active tissues does not alter much (p. 426).

Fick principle

Another equation concerned with gas transfer states, in the case of O_2, that:

$$\dot{V}_{O_2} = \dot{Q}(C_aO_2 - C_{\bar{v}}O_2)$$

where $\dot{V}_{O_2}$ is the O_2 consumption (ml min^{-1}), $\dot{Q}$ is the cardiac output (L min^{-1}) and C_aO_2 and $C_{\bar{v}}O_2$ are, respectively, the arterial and mixed-venous content of O_2 (ml of O_2 per litre of blood). For CO_2 the equation is:

$$\dot{V}_{CO_2} = \dot{Q}(C_{\bar{v}}CO_2 - C_aCO_2)$$

These equations illustrate the **Fick principle**, which was described on p. 404. It is used to measure not gas transfer but $\dot{Q}$ (p. 404) and blood flow to individual organs (e.g. p. 616), and as a technique it is called the Fick method. (The equation is also referred to as the Fick equation and must not be confused with that for Fick's law of diffusion).

The Fick equation illustrates how dependent gas transfer is on the rate of blood flow through the lungs and tissues. At rest typical values are $\dot{V}_{O_2}$ 300 ml min^{-1}, $\dot{Q}$ 6 L min^{-1}, C_aO_2 195 ml L^{-1} and $C_{\bar{v}}O_2$ 145 ml L^{-1}. Were $\dot{Q}$ to fall due to various cardiovascular disorders (p. 530), $\dot{V}_{O_2}$ can only be maintained by extracting a greater amount of O_2 from the circulating blood—thus $C_{\bar{v}}O_2$ falls. C_aO_2 is dependent on alveolar P_{O_2} and the haemoglobin concentration (p. 523) and in these particular cardiovascular disorders neither of these will change. The arteriovenous O_2 difference ($C_aO_2 - C_{\bar{v}}O_2$) therefore widens, but only because $C_{\bar{v}}O_2$ falls. Some organs extract more O_2 out of the circulating blood than others, so the content of O_2 in their venous blood, and hence their arteriovenous O_2 difference, differ (see Table 15.1).

During exercise C_aO_2 remains virtually unchanged and the increase in O_2 delivery to the exercising muscles is achieved by an increase in $\dot{Q}$ and this, coupled with a greater extraction of O_2 from the blood (i.e. $C_{\bar{v}}O_2$ decreases), permits the increased consumption of O_2. Typical values for strenuous exercise are $\dot{V}_{O_2}$ 3000 ml min^{-1}, $\dot{Q}$ 20 L min^{-1}, C_aO_2 195 ml L^{-1} and $C_{\bar{v}}O_2$ 45 ml L^{-1}, with obvious widening of the arteriovenous O_2 difference.

The similarity of the Fick equation, $\dot{V}_{O_2} = \dot{Q}\,(C_aO_2 - C_{\bar{v}}O_2)$, to the ventilatory equation used on p. 500 to measure O_2 consumption, $\dot{V}_{O_2} = \dot{V}_E\,(F_IO_2 - F_EO_2)$, is obvious—the first is based on blood variables and the second on ventilatory variables. The version using alveolar ventilation, $\dot{V}O_2 = \dot{V}_A(F_IO_2 - F_AO_2)$, is more relevant here as gas transfer in the lungs is ultimately dependent on both $\dot{Q}$ and $\dot{V}_A$.

It was stated on p. 500 that R is the ratio of $\dot{V}CO_2/\dot{V}O_2$. From the $\dot{V}O_2$ equations just summarized and the equivalent ones for $\dot{V}CO_2$, it is apparent that R can also be calculated from:

1 $(F_ECO_2 - F_ICO_2)/(F_IO_2 - F_EO_2)$;
2 $F_ACO_2 - F_ICO_2)/(F_IO_2 - F_AO_2)$; and
3 $(C_{\bar{v}}CO_2 - C_aCO_2)/(C_aO_2 - C_{\bar{v}}O_2)$.

16.4 SHUNTS, REGIONAL VENTILATION AND PERFUSION, AND THE $\dot{V}_A/\dot{Q}$ RATIO

P_aO_2 (~95 mmHg) is always less than P_AO_2 or $P_{c'}O_2$ (100 mmHg). This alveolar–arterial PO_2 difference arises because of first, venous admixture from the coronary and bronchial circulations (normal anatomical shunts) which bypass the alveolar capillaries, and second, the consequences of an uneven base-to-apex distribution of the ratio of alveolar ventilation to blood perfusion ($\dot{V}_A/\dot{Q}$ ratio) which is due to gravity. The base of the lung receives 2.5 times more $\dot{V}_A$ and 6 times more $\dot{Q}$ than the lung apex when in the upright posture. Small spontaneous alterations in $\dot{V}_A/\dot{Q}$ are mitigated by local mechanisms, restoring either of the two components towards normal. For instance, too low a $\dot{V}_A$ would produce locally high CO_2 and low O_2, causing the nearby bronchioles to dilate and the pulmonary arterioles to constrict, the former improving local ventilation and the latter reducing local blood flow to match the ventilation. The alveolar–arterial PO_2 difference becomes wider in various lung diseases where there are intrapulmonary shunts, abnormal anatomical shunts, inequality of the $\dot{V}_A/\dot{Q}$ ratio or impairment of pulmonary diffusion.

Shunts

Any venous blood which is added to that in the systemic arteries (**venous admixture**) without going through ventilated areas of the lungs is referred to as a right-to-left (R–L) **shunt**. In the normal lung there are two main **anatomical** sources of such a shunt—some coronary venous blood drains via Thebesian veins into the left ventricle and some of the bronchial circulation returns via the pulmonary veins. Because of differences in the respective shapes of CO_2 and O_2 blood dissociation curves (p. 533), venous admixture from shunts has a negligible effect on P_aCO_2 but lowers P_aO_2 by a few mmHg relative to P_AO_2, thus creating an alveolar–arterial PO_2 difference (p. 522).

Normally only 1–2% of the cardiac output bypasses the alveoli. This can be viewed as wasted perfusion, analogous to the concept of wasted (dead-space) ventilation (p. 497). For wasted ventilation the basis of the Bohr equation (written usually for CO_2) states:

$$\dot{V}_E \times F_ECO_2 = \dot{V}_D \times F_ICO_2 + \dot{V}_A \times F_ACO_2$$

total ventilation = wasted ventilation + useful ventilation

For wasted or shunted perfusion the statement (written usually for O_2) is:

$$\dot{Q}_t \times C_aO_2 = \dot{Q}_s \times C_{\bar{v}}O_2 + \dot{Q}_c \times C_{c'}O_2$$

total perfusion = wasted perfusion + useful perfusion

Note that these equations can be condensed to $\dot{V}_E = \dot{V}_D + \dot{V}_A$ and $\dot{Q}_t = \dot{Q}_s + \dot{Q}_c$ but the same cannot be done for the fractions (F) or content (C) because they cannot be added without knowing the volume (or flow rate) of each. Furthermore, the O_2 must be expressed in units of content not partial pressure in the shunt statement because the shape of the O_2 haemoglobin dissociation curve (O_2 content vs. PO_2) is not linear but sigmoidal (p. 527).

Shunted blood flow ($\dot{Q}_s$), containing the unmodified mixed-venous content of O_2 ($C_{\bar{v}}O_2$), is added to the pulmonary capillary blood flow ($\dot{Q}_c$) which, as it has passed through ventilated alveoli, has a normal oxygenated pulmonary end-capillary content ($C_{c'}O_2$), to give the total pulmonary blood flow ($\dot{Q}_t$, i.e. cardiac output). The mixed blood comprises the final arterial content of O_2 (C_aO_2), which is less than $C_{c'}O_2$ by an amount dependent on the venous admixture. Flow is usually given in units of litres of blood per minute and content in millilitres of O_2 per litre of blood. A rearrangement of the equation given to illustrate the shunt statement produces the form known as the **shunt equation**:

$$\frac{\dot{Q}_s}{\dot{Q}_t} = \frac{C_{c'}O_2 - C_aO_2}{C_{c'}O_2 - C_{\bar{v}}O_2}$$

in which $\dot{Q}_s$ can be calculated after measurement of each of the other variables.

$\dot{Q}_t$ is first measured using the Fick principle (p. 404). $C_{c'}O_2$ is difficult to sample and is estimated from the O_2 content vs. PO_2 relationship of the O_2 haemoglobin dissociation curve and the assumption that its PO_2 is the same as alveolar PO_2. The C_aO_2 is measured from an arterial gas sample and $C_{\bar{v}}O_2$ from a venous blood sample which for accuracy should be from the pulmonary artery as this contains a true sample of the body's mixed–venous gases. As the shunted blood from the bronchial circulation may not have a composition identical to mixed-venous, the shunt equation has some limitations.

 The shunt equation is derived from the statements, $\dot{Q}_tC_aO_2 = \dot{Q}_sC_{\bar{v}}O_2 + \dot{Q}_cC_{c'}O_2$ and $\dot{Q}_t = \dot{Q}_s + \dot{Q}_c$ in the following way. The second equation is rewritten as $\dot{Q}_c = \dot{Q}_t - \dot{Q}_s$ and substituted in the first equation to produce $\dot{Q}_tC_aO_2 = \dot{Q}_sC_{\bar{v}}O_2 + (\dot{Q}_t - \dot{Q}_s) C_{c'}O_2$ which becomes $\dot{Q}_tC_aO_2 = \dot{Q}_sC_{\bar{v}}O_2 + \dot{Q}_tC_{c'}O_2 - \dot{Q}_sC_{c'}O_2$. This equation then rearranges as $\dot{Q}_s(C_{c'}O_2 - C_{\bar{v}}O_2) = \dot{Q}_t(C_{c'}O_2 - C_aO_2)$ and finally solves as $\dot{Q}_s/\dot{Q}_t = (C_{c'}O_2 - C_aO_2)/(C_{c'}O_2 - C_{\bar{v}}O_2)$.

 Pathological R–L shunts can occur on an anatomical basis in cardiovascular diseases (e.g. patent ductus arteriosus at birth, and atrial or ventricular septal defects—the so-called 'hole-in-the heart') and on a functional basis from intrapulmonary shunts in pulmonary diseases (e.g. blockage of bronchi or bronchioles with mucus, and bronchiolar or alveolar collapse in which the alveoli that are not ventilated but are still perfused constitute a shunt. A shunt-like effect can also occur if the alveolar ventilation to some of the alveoli is just reduced rather than abolished (see alveolar ventilation/perfusion ratio; p. 520)—when this occurs it contributes a hidden augmentation of the $\dot{Q}_s$ calculated by the shunt equation. A true shunt can be distinguished from the effect of a poor ventilation/perfusion ratio by breathing 100% O_2—the shunted blood will not come into contact with the high alveolar O_2 and so nothing will change (thus the final C_aO_2 will *not* improve), whereas the C_aO_2 in poorly ventilated alveoli will gain some benefit from exposure to high alveolar O_2 (thus the final C_aO_2 will increase).

The term **physiological shunt** was coined to refer to the presence of all types of shunts in health and disease and to be a term similar in concept to that of the physiological dead space (p. 497). However, some use it to refer only to those anatomical shunts normally present in health, as all other types of shunts are pathological.

Regional variations in ventilation

In the upright posture when breathing normally at FRC, the base (bottom) of the lungs receives ~2.5 times more ventilation per unit of lung volume than the apex (top) of the lungs (see Fig. 16.25). This can be measured by a bank of radioactive counters placed along the chest of a subject who is breathing tracer amounts of radioactive xenon gas. When in the supine position, the base and apex are in fact ventilated equally but it is the dorsal region of the lungs that now receives a higher ventilation relative to the ventral regions. The effects of gravity are the cause of these regional variations in ventilation—the lowermost regions are called the gravity-dependent regions and receive the better ventilation.

Gravity results in the weight of the lungs tending to pull them away from the chest very slightly at the base and very strongly at the apex. This causes P_{pl} at FRC to be -2.5 cmH$_2$O at the base and -10 cmH$_2$O at the apex of an upright lung (p. 478). The consequences of this are that the basal regions with the lower transmural pressure $(P_{alv} - P_{pl})$ will begin inspiration at FRC containing less gas than the apical regions (Fig. 16.23). However, because basal regions are on the steep part of the compliance curve, they will inflate more easily during inspiration and thus, for the same inspiration-to-expiration P_{pl} of $\Delta 3$ cmH$_2$O as in the apical

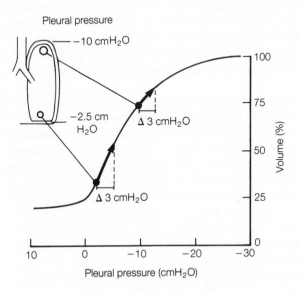

Fig.16.23 Effect of the upright posture on P_{pl} at the apex and base of the lung at FRC. Alveoli (depicted as an open circle in the lung) are more inflated at the apex than at the base and are therefore further up the compliance curve. Thick arrows indicate volume increment during each inspiration at the apex compared with the base.

regions, the Δ volume and resulting ventilation achieved will be greater (Fig. 16.23)

Effect of lung disease

When the lungs become stiffer because of age or lung diseases, the base of the lungs may be operating on the lower flat part of the compliance curve (Fig. 16.23), in which case ventilation of the basal may become *less* than that of the apical regions. Furthermore, as P_{pl} at the base may even become positive, small airways may be intermittently or totally closed due to compression (p. 488), thus producing intermittent basal ventilation or trapping gas in the alveoli. These effects can occur in healthy young people but only when deliberately breathing at low lung volumes below FRC.

Throughout the lungs uneven ventilation can occur if, due to disease, some areas have less compliant alveoli or the airways supplying some alveoli but not others have an increased resistance due to bronchoconstriction, inflammation, blockage or compression.

Regional variations in pulmonary blood flow

The regional distribution of pulmonary blood flow can also be measured by a bank of radioactive counters on the chest of a subject, this time after the subject has received an intravenous injection of radioactive xenon dissolved in saline. In the upright posture the base receives ~6 times more blood flow per unit of lung volume than the apex of the lungs (see Fig. 16.25). As with regional variations in ventilation, this base-to-apex gradation disappears on lying down and if supine, the blood flow is now greater in the gravity-dependent dorsal region.

Gravity does not affect the extravascular pressure in the region of the alveoli (i.e. the alveolar pressure, which is ~0 cmH$_2$O) but it does affect the hydrostatic pressure inside blood vessels (p. 457). In the lung at the level of the tricuspid valve, the pressure in the pulmonary artery is 25 mmHg systolic/8 mmHg diastolic (p. 447) falling, though still with a marked pulse (see Fig. 14.6), to ~12 mmHg at the arterial end and to ~8 mmHg at the venous end of the pulmonary capillaries. These pressures at the arterial and venous end of the capillary convert to ~16 and ~11 cmH$_2$O.

In the upright posture the base of the lung in an adult will be some 10 cm below and the apex some 15 cm above the tricuspid valve and so the blood pressures increase by 10 cmH$_2$O at the base and fall by 15 cmH$_2$O at the apex of the lung (Fig. 16.24). As pulmonary vessels have very thin walls, such changes in intravascular pressure relative to extravascular pressure (i.e. the transmural pressure, $P_{in} - P_{out}$) cause distension of vessels at the base and compression at the apex. Thus, for example, at the arterial end of the capillary the 16 cmH$_2$O becomes at the base 26 cmH$_2$O, the vessels distend and perfusion increases, while at the apex the pressure is only 1 cmH$_2$O and the vessels tend to collapse (Fig. 16.24). However, given the systolic/diastolic pulse, this pressure at the apex will become positive in systole and more negative in diastole and thus at the apex flow will actually be intermittent. Should pulmonary arterial pressure fall (e.g. after a haemorrhage), apical flow may cease entirely. Alveoli that are ventilated but not perfused constitute the alveolar dead space (p. 497). In exercise the elevation in pulmonary arterial blood pressure ensures that apical regions are perfused continuously.

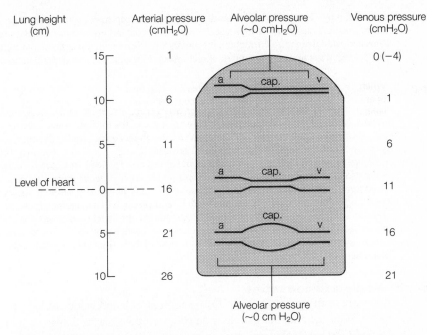

| Lung height (cm) | Arterial pressure (cmH₂O) | Alveolar pressure (~0 cmH₂O) | Venous pressure (cmH₂O) |

Fig. 16.24 Effect of gravity in the upright posture on pressure at the arterial and venous ends of pulmonary capillaries. Distension of vessels increases blood flow at the base of the lung; collapse of vessels reduces blood flow at the apex of the lung (the − 4 cmH₂O in parentheses indicates the pressure that would occur if the vessels were rigid).

Just below the apex arterial pressure will always exceed alveolar pressure but the gravitational effects on venous pressure are such (Fig. 16.24) that alveolar pressure is greater than venous pressure. Not only will there be some compression of the veins near the capillary but also the pressure gradient causing blood flow is actually the difference between arterial and alveolar pressure. Only in the middle and lower regions of the lungs will venous exceed alveolar pressure, thus permitting the arterial-to-venous pressure gradient to determine blood flow.

Effect of lung disease

Uneven perfusion of some alveoli with respect to others can occur because of local blockage by emboli and thrombi, local compression (e.g. by oedema or tumours), local damage to blood vessels by disease and local collapse or overexpansion of alveoli. Because pulmonary blood pressure is relatively low, uneven perfusion is easily precipitated by pulmonary hypotension.

Effect of breathing on lung perfusion

The extravascular pressure in the vicinity of the pulmonary capillaries is the alveolar pressure which fluctuates around 0 cmH₂O, becoming − 1 cmH₂O in inspiration and + 1 cmH₂O in expiration (p. 491)—greater airflow will cause greater fluctuations. Thus perfusion also depends on the phase and magnitude of the breathing cycle being augmented during inspiration by vessel dilatation. On the other hand, deep inspirations actually compress, not dilate, the alveolar vessels because the walls of the alveoli become so stretched that the vessels within them are compressed. During a normal expiration perfusion may cease at the apex because alveolar pressure may exceed arterial pressure. Furthermore, clinical treatment with positive-pressure ventilation can cause alveolar pressure to exceed arterial pressure throughout the breathing cycle, resulting in cessation of perfusion, particularly at the apex of the lung or, if recumbent, in the gravity non-dependent regions.

Pulmonary vascular resistance

In the pulmonary circuit a distinction between arteries and arterioles is not usually made and arteries, capillaries and veins each contribute one-third of the total pulmonary resistance (p. 447). For pulmonary arteries and veins away from the alveolar region, the extravascular pressure is not the alveolar but the pleural pressure. During inspiration towards TLC the increasingly negative pleural pressure and radial traction (p. 488) from surrounding lung tissue dilates arteries and veins, lowering their resistance and hence increasing pulmonary blood flow. This effect is, however, small and the overall pulmonary vascular resistance at large lung volumes is in fact raised markedly by compression of capillaries within stretched alveolar walls. During expiration from FRC towards RV the effect on capillaries is small but loss of radial traction and positive pleural pressures, particularly with a forced expiration, compress pulmonary arteries and veins, increasing their resistance considerably. Thus pulmonary vascular resistance is least near FRC.

Ventilation/perfusion ratios

Adequate exchange of O_2 and CO_2 between alveoli and blood, and the maintenance of partial-pressure gradients down which these gases diffuse, require that $\dot{V}_A$ and $\dot{Q}$ be reasonably well-matched in each of the alveolar units throughout the lung.

The ratio $\dot{V}_A/\dot{Q}$ is in fact a fourth determinant of alveolar gas partial pressures for a particular alveolar unit. It cannot be included in the format of the alveolar gas equations (p. 504) because these relate to the lung as a whole. The amount of O_2 (or CO_2) in an alveolus obviously depends on how much is added to the alveolus per minute, which is determined by local $\dot{V}_A$, and how much is removed (or added) per minute by the blood, which is determined by local $\dot{Q}$.

In adults at rest, typically $\dot{V}_A$ is 5 L min^{-1} and $\dot{Q}$ is 6 L min^{-1}, giving for the lung as a whole an overall $\dot{V}_A/\dot{Q}$ ratio of 0.83 (although in different individuals, as $\dot{V}_A$ and $\dot{Q}$ can range between 4 and 6 L min^{-1}, the $\dot{V}_A/\dot{Q}$ ranges from 0.8 to 1.2). These ratios of $\dot{V}_A/\dot{Q}$ produce on average on alveolar P_{O_2} of 100 mmHg and P_{CO_2} of 40 mmHg. If, however, a group of alveoli is unventilated (i.e. the $\dot{V}_A/\dot{Q}$ becomes zero), the gases in these alveoli will soon become the same as the blood entering them—that is, the mixed-venous values of P_{O_2} 40 mmHg and P_{CO_2} 46 mmHg. On the other hand, if these alveoli receive normal ventilation but are unperfused (i.e. the $\dot{V}_A/\dot{Q}$ becomes infinite) their gases soon become identical to inspired gases at a P_{O_2} of 150 mmHg and P_{CO_2} of 0 mmHg. These are of course extreme examples used to illustrate the fact that alveolar–capillary units with a low $\dot{V}_A/\dot{Q}$ ratio have a low alveolar P_{O_2} and high P_{CO_2} and those with a high $\dot{V}_A/\dot{Q}$ ratio have a high alveolar P_{O_2} and a low P_{CO_2}. Not surprisingly, even in the normal lung, all 300–600×10^6 alveolar units are not ventilated and perfused absolutely equally. Mismatching of $\dot{V}_A/\dot{Q}$ occurs in diseases where there is uneven ventilation (p. 518) or uneven perfusion (p. 519).

Local control of $\dot{V}_A/\dot{Q}$ matching

Two mechanisms work together to minimize mismatching of $\dot{V}_A/\dot{Q}$ in small groups of alveolar–capillary units. One affects the bronchioles to alter the local distribution of ventilation, the other affects the small pulmonary vessels to alter the local distribution of blood flow. If a group of alveoli are underventilated or overperfused (i.e. the $\dot{V}_A/\dot{Q}$ ratio has decreased) their alveolar Po_2 will be low and Pco_2 high. Bronchiolar smooth muscle is not particularly sensitive to Po_2 but in response to high airway Pco_2 it will relax. The resulting bronchodilatation reduces the local airway resistance and hence increases airflow to the alveoli those bronchioles supply. Thus local ventilation increases and the $\dot{V}_A/\dot{Q}$ ratio returns towards normal. Smooth muscle of small pulmonary arterioles is not particularly sensitive to Pco_2 but in response to low Po_2 it contracts. The resulting vasoconstriction increases vascular resistance, reducing the blood flow to the pulmonary capillaries and thus returning the local $\dot{V}_A/\dot{Q}$ ratio towards normal. Conversely, in a group of alveoli which are overventilated or underperfused (i.e. $\dot{V}_A/\dot{Q}$ has increased), the alveolar Po_2 will be high and the Pco_2 low and the $\dot{V}_A/\dot{Q}$ ratio will be adjusted back towards normal by local bronchoconstriction and vasodilatation.

Such a vasoconstricting response to low Po_2 is unique to pulmonary arterioles—in all systemic arterioles the response is one of vasodilatation. Pulmonary vessels respond more to the airway than blood Po_2 and the effect of low O_2 may be mediated by release of some vasoconstrictor substance rather than by a direct effect on the smooth muscle. The vasoconstricting response to low O_2 is, however, a disadvantage in the generalized hypoxia of high altitude or chronic obstructive lung disease because all pulmonary arterioles throughout the lung constrict.

Variation in $\dot{V}_A/\dot{Q}$ due to gravity

A normal cause of variation in $\dot{V}_A/\dot{Q}$ is based on regional variations in $\dot{V}_A$ and in $\dot{Q}$ (Fig. 16.25) that are induced by gravity and particularly evident in the upright posture. Both $\dot{V}_A$ and $\dot{Q}$ decrease from base to apex of the lungs but, as the reduction in $\dot{Q}$ (six-fold) is greater than the reduction in $\dot{V}_A$ (2.5-fold), their ratio *increases* from base to apex. Thus, the base of the lung, although well-ventilated and perfused, has a $\dot{V}_A/\dot{Q}$ ratio falling to about 0.6 and therefore contains alveoli

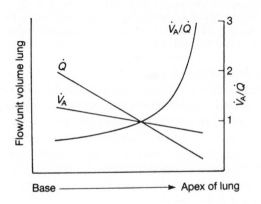

Fig. 16.25 Alveolar ventilation ($\dot{V}_A$) and perfusion ($\dot{Q}$) and the ventilation/perfusion ratio ($\dot{V}_A/\dot{Q}$) from the base to apex of the lung. (From West, J.B. (1977) *Ventilation/Blood Flow and Gas Exchange*, 3rd edn. p. 30. Blackwell Scientific Publications, Oxford.)

with a slightly lower $P_{A}O_2$ and higher $P_{A}CO_2$ than areas with a $\dot{V}_{A}/\dot{Q}$ ratio of 1.0. In contrast, at the apex of the lung there are very large increases in the $\dot{V}_{A}/\dot{Q}$ ratio rising to more than 3.0 over a short distance and the apex therefore contains alveoli with a very much higher $P_{A}O_2$ and lower $P_{A}CO_2$. The shape of the $\dot{V}_{A}/\dot{Q}$ line clearly shows that the variation in the ratio is in fact quite small in most of the lung, i.e. in the lower 75% of the lungs. In exercise the apex of the lung becomes better perfused so there is far less gravity-induced variation in the $\dot{V}_{A}/\dot{Q}$ ratios.

The poor perfusion of the apical regions causes them to contribute little to overall gas exchange. Furthermore the high $\dot{V}_{A}/\dot{Q}$ ratio of the apical regions means that the relative overventilation flushes out more CO_2 from the blood—in contrast, no extra O_2 is absorbed because, as the Po_2 is already high, the blood is already virtually fully saturated. Thus the $\dot{V}CO_2/\dot{V}O_2$, i.e. R, is high in the apical units and, conversely, low in the basal units.

As stated above, Po_2 is low (and Pco_2 high) in the alveolar gas and end-capillary blood of basal compared with apical regions. As total ventilation of basal is greater than apical regions, the alveolar gas from the basal region will make a greater contribution to the final expired alveolar gas mix. Similarly, the contribution from basal regions will dominate the final blood gas composition but to a much greater extent because the basal-to-apical gradation in blood flow is far greater. Thus the mixed blood leaving the lungs (the eventual arterial Po_2 for the rest of the body) will have a lower arterial Po_2 than the final alveolar Po_2. Moreover, when mixing within the blood phase is considered, one must sum the O_2 in units of content, not partial pressure, because of the non-linear shape of the oxyhaemoglobin dissociation curve (p. 527). Thus blood from the apical areas with a high $\dot{V}_{A}/\dot{Q}$ ratio, although having a high Po_2, will not have a higher O_2 content and therefore cannot compensate adequately for the low O_2 content of blood from the basal regions. As a consequence, the lowering of arterial Po_2 is exaggerated.

Assessment of $\dot{V}_{A}/\dot{Q}$ inequality

As the measurement of $\dot{V}_{A}/\dot{Q}$ inequality is difficult, its presence is indirectly assessed from the magnitude of the alveolar–arterial Po_2 difference. Often an 'ideal' alveolar Po_2 is calculated using the alveolar gas equation (p. 508) which assumes no $\dot{V}_{A}/\dot{Q}$ inequality. The presence of low $\dot{V}_{A}/\dot{Q}$ units is assessed with the shunt equation (p. 516) while pretending that there are no anatomical shunts and assuming that all the lowered $P_{a}O_2$ is due to blood passing through unventilated alveoli. The presence of high $\dot{V}_{A}/\dot{Q}$ units is assessed by calculating the physiological dead space using arterial and mixed-expired Pco_2 in the Bohr equation (p. 499) and assuming that the alveolar dead space component of physiological dead space (p. 497) is entirely caused by ventilation of unperfused alveoli.

Alveolar–arterial Po_2 and Pco_2 differences

In healthy upright lungs $P_{a}O_2$ is usually 5–10 mmHg lower than $P_{A}O_2$. Half of this alveolar–arterial Po_2 difference is caused by the gravity-induced variations in $\dot{V}_{A}/\dot{Q}$; the rest is a consequence of anatomical shunts and any alveolar units receiving no ventilation. The alveolar–arterial Po_2 difference increases in heavy exercise and in diseases which produce excessive $\dot{V}_{A}/\dot{Q}$ inequality, increased shunting or impaired diffusion across the alveolar–capillary membrane. The low $P_{a}O_2$ resulting from $\dot{V}_{A}/\dot{Q}$ abnormalities or impaired diffusion can be improved by high O_2 breathing, but not that due to shunts.

The effects of $\dot{V}_A/\dot{Q}$ and shunts also cause arterial P_{CO_2} to be greater than alveolar P_{CO_2} but the difference is less than 1 mmHg because the CO_2 blood dissociation curve (content vs. P_{CO_2}) in the physiological range is not only linear but also very steep (p. 533). Thus large changes in CO_2 content occur for only very small changes in P_{CO_2}. An arterial–alveolar P_{CO_2} difference may be just detectable if $\dot{V}_A/\dot{Q}$ mismatching is severe or the alveolar–capillary membrane is grossly thickened.

16.5 BLOOD GAS TRANSPORT

O_2 carriage

Nearly all of the O_2 in the blood is carried in the red blood cells bound to the haem part of haemoglobin; a very small amount (~1%) is carried in physical solution. The amount of O_2 per litre of blood thus depends on the haemoglobin concentration and its degree of saturation. This in turn depends on P_{O_2}. Typically in arterial blood with a P_{O_2} of 95 mmHg, the haemoglobin saturation is 97% and, if the haemoglobin concentration is 150 g L^{-1}, the O_2 content is 195 ml L^{-1}; in venous blood P_{O_2} is 40 mmHg, haemoglobin saturation 70% and O_2 content 140 ml L^{-1}. Each haemoglobin molecule binds four O_2 molecules to different degrees, giving rise to the sigmoidal shape of the oxyhaemoglobin dissociation curve in which O_2 content, or % haemoglobin saturation, is plotted against blood P_{O_2}. An increase in blood temperature and in the amount of CO_2, H^+ or 2,3-bisphosphoglycerate (2,3-BPG) bound to the globin part of haemoglobin decreases the affinity of haemoglobin for O_2 and thus shifts the curve to the right; opposing changes shift the curve to the left. Shifts are described as an increase (right shift) or decrease (left shift) in the P_{50} (i.e. the P_{O_2} at which 50% saturation of the haemoglobin molecule occurs). Shifts in either direction caused by changes in CO_2 and H^+ are called the Bohr effect.

The amount of O_2 physically dissolved in the blood depends on its solubility and is proportional to its partial pressure (Henry's law: gas content = solubility × partial pressure). At 37°C, 0.03 ml of O_2 is dissolved in each litre of blood for every mmHg increase in partial pressure (in other words, the **solubility** of O_2 is 0.03 ml L^{-1} mmHg^{-1}). Thus arterial blood at a P_{O_2} of 100 mmHg will contain in **physical solution** 3 ml of O_2 per litre of blood. If this was the only means of gas carriage, the body's resting O_2 consumption of 300 ml of O_2 per minute would require an impossibly high $\dot{Q}$ of at least 100 L min^{-1} with no O_2 left behind in the venous blood! Fortunately, the combination of O_2 with **haemoglobin** in the red blood cells results in about 200 m L^{-1} being carried in arterial blood; only about 1.5% of the total is therefore in physical solution.

The haemoglobin concentration in the blood is typically taken as 150 g L^{-1} (but see Table 12.4 for normal range) and each gram of haemoglobin when fully saturated with O_2 has combined with 1.34 ml of O_2. The term **O_2 capacity** of the blood is used to indicate how much O_2 per litre of blood is attached to the haemoglobin when fully saturated with O_2—it therefore depends on the individual's haemoglobin (Hb) concentration:

O_2 capacity = 1.34 × Hb concentration

$$(\text{ml } O_2 \text{ L}^{-1} \text{ of blood}) = (\text{ml } O_2 \text{ g}^{-1} \text{ of Hb}) \times (\text{g of Hb L}^{-1} \text{ of blood})$$

When the haemoglobin concentration is 150 g L^{-1} the O_2 capacity is 201 ml O_2 L^{-1}, whereas in an amaemic person with a haemoglobin of, say, 100 g L^{-1}, the O_2 capacity is much reduced, at 134 ml O_2 L^{-1}.

The blood, even arterial, is rarely fully saturated with O_2 as this requires a high P_{O_2} of nearly 300 mmHg. The expression **O_2 saturation of Hb** (%) describes whether the sites on the haemoglobin molecules which carry O_2 are fully or partially occupied:

$$O_2 \text{ saturation of Hb (\%)} = \frac{O_2 \text{ bound to Hb (ml } O_2 \text{ L}^{-1})}{O_2 \text{ capacity (ml } O_2 \text{ L}^{-1})} \times \mathbf{100}$$

Note that the amount of O_2 bound to the haemoglobin equals the total O_2 content of the blood sample minus the amount of O_2 in physical solution. Stated another way:

Total O_2 content = 1.34 × Hb concentration × % saturation + (P_{O_2} × 0.03)

Thus, for a healthy person with a typical arterial blood P_{O_2} of 95 mmHg, which causes 97% saturation, the O_2 bound to haemoglobin is 195 ml L^{-1} (assuming a haemoglobin concentration of 150 g L^{-1}), that in solution is 2.9 ml L^{-1} and the total is 197.9 ml L^{-1}. For venous blood with a P_{O_2} of 40 mmHg causing 70% saturation of haemoglobin, the O_2 bound to haemoglobin is 140.7, that in solution is 1.2 and the total is 141.9 ml L^{-1}.

Haemoglobin

The haemoglobin molecule is a globular protein (M_r 64 500) composed of four subunits. Each subunit comprises a haem group contained within a crevice in a polypeptide chain. **Haem** consists of a **protoporphyrin ring** surrounding an iron atom in the **ferrous** state (Fig. 16.26). The iron is bonded to four nitrogens of the protoporphyrin ring; iron can form two more additional bonds—one with O_2, and the other to the proximal histidine of a polypeptide chain. There are two α and two β polypeptide chains in each haemoglobin molecule in an adult (p. 355); the four chains together are referred to as the **globin**. Globin cannot combine with O_2 but it can bind to CO_2 and H^+ ions and, in the case of β-chains, to the molecule **2,3 biphosphoglycerate (2,3-BPG)**, formerly known as 2,3-diphosphoglycerate, which is a metabolite of red blood cells (p. 335).

Because there are four subunits, each molecule of haemoglobin can combine through its haem groups with up to four molecules of O_2. When no O_2 is bound the so-called deoxyhaemoglobin molecules give the blood a dark blue–purple colour. As O_2 binds to the haemoglobin converting it to oxyhaemoglobin, the blood colour changes to a red when finally well-oxygenated. The term **oxygenation** is used because the Hb–O_2 association is reversible; it is not an oxidation—this would consume O_2 permanently and therefore O_2 could not be released later to the tissues. The oxygenation process can be written in four stages as:

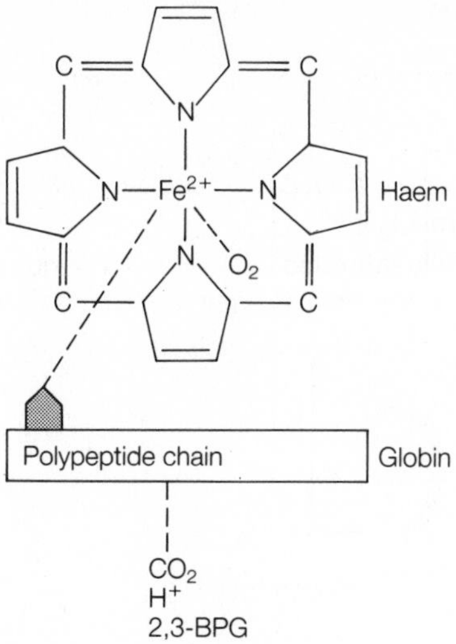

Fig. 16.26 Structure of one of the subunits of the haemoglobin molecule (note that polypeptide chain is actually coiled and much longer than shown). BPG, Bisphosphoglycerate; Fe^{2+}, ferrous ion.

$Hb + O_2 = HbO_2$

$HbO_2 + O_2 = Hb(O_2)_2$

$Hb(O_2)_2 + O_2 = Hb(O_2)_3$

$Hb(O_2)_3 + O_2 = Hb(O_2)_4$

Deoxygenated haemoglobin exists in a **'tense'** state compared with **oxygen-ated** haemoglobin which exists in a **'relaxed'** state (Fig. 16.27). In the tense deoxy state strong ionic bonds (salt-bridges) form between the four polypeptide chains, making them immobile and keeping them apart, particularly the β-chains which also have salt-bridge connections to the 2,3-BPG wedged between them. The consequence of this is that the ferrous ions do not lie in the plane of the protoporhyrin rings and these therefore adopt a bent shape. As O_2 attaches to the haem groups, the ferrous ions move into and straighten up the plane of the protoporphyrin rings. Furthermore, the ionic bonds between the polypeptide chains become broken, allowing the haemoglobin molecule to extrude the 2,3-BPG and change its configuration into its relaxed state. The attachment of one O_2 molecule facilitates the uptake of the subsequent ones. This **cooperative effect** on the **affinity of O_2 for haemoglobin** is referred to as the **haem–haem interaction** and gives rise to the sigmoidal shape of the oxyhaemoglobin dissociation curve.

Changes in the **O_2 affinity** of haemoglobin are also brought about by altered levels of $P\text{co}_2$, pH and 2,3-BPG. These directly affect the globin and through salt-bridges make the haemoglobin molecule more tense or more relaxed and thus change the affinity of the haem for O_2. As the initial site of action is at a spatially distinct site, the change in O_2 affinity is due to an **allosteric interaction**.

CO_2 decreases the affinity of haem for O_2 because it binds to terminal amino groups on all four polypeptide chains to produce carbamates $(R–NH_2 + CO_2 = R–NHCOOH)$ which form salt-bridges stabilizing the tense form. Similarly, H^+ ions bind to the polypeptides chains, specifically to the amino groups $(R–NH_2 + H^+ = RNH_3{}^+)$, carboxyl groups $(R–NHCOO^- + H^+ = R–NHCOOH)$ and, most importantly, to the nitrogen of the imidazole

Deoxyhaemoglobin (tense form)

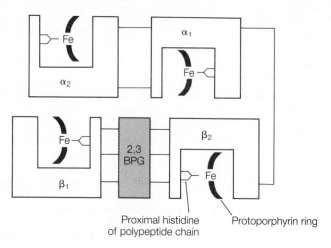

Proximal histidine
of polypeptide chain · Protoporphyrin ring

Oxyhaemoglobin (relaxed form)

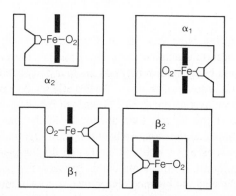

Fig. 16.27 The four subunits of haemoglobin in their deoxygenated (tense) and oxygenated (relaxed) forms. The ferrous ion and the protoporphyrin ring lie in a crevice created by the foldings of the α and β polypeptide chains. Thin lines indicate salt-bridges. BPG, Bisphosphoglycerate; Fe, ferrous ion.

groups ($N + H^+ = NH^+$) of the histidine residues, and thereby promote the tense form. An increased concentration of 2,3-BPG stabilizes the tense form by providing extra cross-links between the β-chains.

Thus hypercapnia (increased CO_2) and acidosis (increased H^+) decrease the haemoglobin affinity for O_2. An increased concentration of 2,3-BPG also decreases the haemoglobin affinity for O_2. As the production of 2,3-BPG by the red blood cells increases during chronic (long-term) hypoxia and chronic alkalosis, this counteracts the increased O_2 affinity associated with acute alkalosis.

Sickle-cell anaemia and a deficiency in the enzyme pyruvate kinase of red cells cause increased production of 2,3-BPG, while hexokinase enzyme deficiency results in decreased production. Blood becomes low in 2,3-BPG in chronic acidosis or when stored in blood banks for as little as 1 week (p. 353). This results in haemoglobin with a high O_2 affinity so that, when the blood is transfused, the haemoglobin will not offload its O_2 readily to the tissues.

Oxyhaemoglobin dissociation curve

The amount of O_2 bound to haemoglobin is determined by P_{O_2} and the relationship between the two is the **oxyhaemoglobin dissociation curve** (Fig. 16.28)— the term association curve is equally appropriate but rarely used. Such a curve is constructed by equilibrating (at the same temperature, pH and P_{CO_2}) a series of blood samples each with a gas mixture containing a different P_{O_2}. The vertical axis is expressed as either % saturation or O_2 content (ml L^{-1}). The values for O_2 content portrayed in Fig. 16.28 are for a person with a haemoglobin concentration of 150 g L^{-1} and do *not* include the O_2 content in physical solution.

The oxyhaemoglobin dissociation curve (Fig. 16.28) has a characteristic **sigmoidal shape** which is due to sequential binding of the four O_2 molecules, one to each of the four haem groups where each combination facilitates the next (the haem–haem interaction). The last binding proceeds very rapidly, which helps to counteract the tendency of the whole process to slow down as the haemoglobin molecule becomes saturated. The **flat upper portion** of the curve occurs around P_{O_2} values usually encountered in the alveoli (~100 mmHg) where O_2 is associating with the haemoglobin. The advantage of this flatness is that it permits the

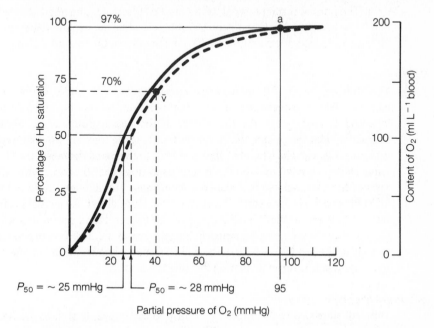

Fig. 16.28 The oxyhaemoglobin dissociation curve. Solid sigmoid curve for conditions of P_{CO_2} = 40 mmHg and pH = 7.4. Dashed sigmoid curve for conditions of P_{CO_2} = 46 mmHg and pH = 7.37. a, Arterial; v̄, mixed-venous value.

blood to attain a good saturation with O_2 even if the alveolar Po_2 were to fall to 60 mmHg (the haemoglobin would then be ~90% saturated). This offers some degree of safety for onloading O_2 when at altitude or in diseases producing low P_Ao_2 or P_ao_2. Conversely, increases in Po_2 above the normal 100 mmHg by voluntary hyperventilation or by breathing high O_2 will make very little difference to the O_2 content of the blood.

The **steep middle portion** occurs around Po_2 values encountered in the tissues (~40 mmHg) where O_2 is dissociating from the haemoglobin. The advantage of this steepness is that if the tissues were to increase their metabolism, larger quantities of O_2 can be offloaded to the tissues for only a very small decrease in blood Po_2. Thus, during activity decreases in venous Po_2 are relatively small and the blood-to-tissue Po_2 gradient required for diffusion can be maintained with little reduction.

Figure 16.28 also compares the O_2–Hb curve for samples of blood maintained at a pH of 7.4 and Pco_2 of 40 mmHg (normal arterial conditions) with a second curve for blood maintained at pH 7.37 and Pco_2 46 mmHg (resting venous conditions). Clearly, venous conditions have shifted the curve to the right and, for any given Po_2, less O_2 is bound to the haemoglobin or, stated another way, the affinity of haemoglobin for O_2 has decreased and thus more O_2 has been released. A change in O_2 affinity is usually described by the P_{50}, the **Po_2 at which the haemoglobin is 50% saturated**. The P_{50} for arterial conditions is ~25 mmHg; it increases to ~28 mmHg in venous blood. In other words, when blood has a decreased O_2 affinity, the P_{50} is increased because the curve itself is shifted to the right. The normal resting arterial values are Po_2 95 mmHg, haemoglobin saturation 97%, haemoglobin O_2 content 195 ml L^{-1}, while the venous values are Po_2 40 mmHg, haemoglobin saturation 70%, haemoglobin O_2 content 141 ml L^{-1}.

Bohr effect

The shift of the O_2–Hb dissociation curve to the right under conditions of increased Pco_2 and decreased pH (as in venous blood) shows how increased H^+ ions and CO_2 decrease the O_2 affinity of the haem group by an allosteric interaction with the globin, which promotes the tense form of the haemoglobin molecule. Conversely, when H^+ ions and Pco_2 in the blood decrease, the O_2–Hb curve shifts to the left and the P_{50} is decreased (Fig. 16.29) because the relaxed form of the haemoglobin molecule has an increased O_2 affinity. These right and left shifts due to H^+ ions and CO_2 are called the **Bohr effect**. An increase in blood temperature and in the 2,3-BPG produced by the red blood cells also shifts the O_2–Hb curve to the right; decreases in temperature and 2,3-BPG cause the curve to shift to the left (Fig. 16.29). The 2,3-BPG has a known allosteric effect and temperatures acts through altering the effective acidity.

Situations altering O_2 affinity

Different situations cause the curve, usually to advantage, to shift to the right or left. As blood enters the **lung** and CO_2 diffuses out of the blood into the alveoli, the reduction in blood Pco_2 and H^+ ions shifts the curve to the left (Fig. 16.28). Thus the affinity of haemoglobin for O_2 in the upper part of the curve increases at the

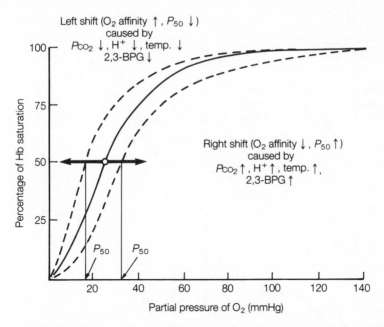

Fig. 16.29 Right and left shifts in the oxyhaemoglobin dissociation curve; solid line for P_{CO_2} 40 mmHg, pH 7.40 and blood temperature 37°C.

site where O_2 is available for onloading. As blood enters the **tissue** capillaries and picks up CO_2, the right shift specifically of the steep part of the curve decreases the affinity of haemoglobin for O_2 at the site where O_2 needs to be released to the tissues.

If the tissues become metabolically more active (e.g. muscles in **exercise**), CO_2 production will increase, more acid will form and more heat will be generated—all factors promoting further shifting of the O_2–Hb curve to the right, which automatically releases more O_2 from the haemoglobin to the tissues that need it in greater supply (Fig. 16.30a). Note that the difference in arterial–venous O_2 content (a–v O_2) is increased in exercise (p. 691). In intensely exercising skeletal muscles, lactic acid production will also raise the concentration of H^+ ions in the blood perfusing them—favouring even more offloading of O_2 for any given P_{O_2}. In prolonged severe exercise there is also an increased production of 2,3-BPG—again a factor promoting offloading of O_2 to the muscle. Acidosis would be expected to decrease production of 2,3-BPG (p. 526) but in this situation the excessively hypoxic state of venous conditions has a stronger effect and increases production.

In **circulatory failure**, if O_2 consumption is to be satisfied given the poor cardiac output, the a–v O_2 difference must increase (Fick principle)—the arterial O_2 is unchanged but the venous O_2 falls (Fig. 16.30a). Hypercapnia and acidosis shift the O_2–Hb curve to the right.

At high **altitude** or in diseases causing **chronic hypoxia** a shift of the curve to the right occurs due to increased production of 2,3-BPG (Fig. 16.30b). This ensures

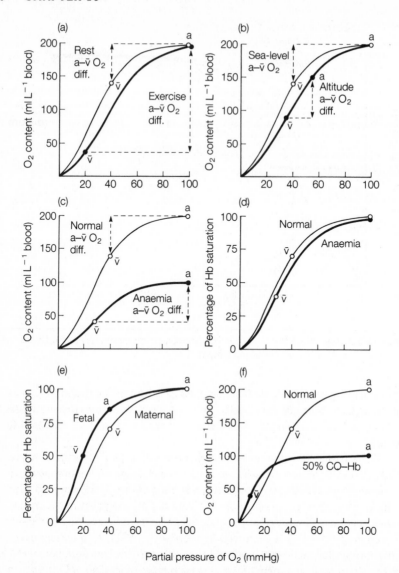

Fig. 16.30 The oxyhaemoglobin dissociation curve in the normal situation compared with (a) intense exercise or circulatory failure, (b) long-term altitude (chronic hypoxia), (c) O_2 content in anaemia, (d) percentage of haemoglobin saturation in anaemia, (e) fetal blood and (f) CO poisoning. The O_2 content is given for normal blood, assuming a haemoglobin concentration of $150\ g\ L^{-1}$. Arterial (a) and mixed venous ($\bar{v}$) values are shown.

that, despite the low $P_{a}O_2$ and lower haemoglobin saturation, the usual amount of O_2 will be released to the tissues from the blood for only a small fall in tissue (and hence) venous P_{O_2}. This effect of 2,3-BPG more than compensates for the effect of hypocapnia which would act to shift the curve to the left; the hypocapnia is the result of the stimulated breathing caused by the hypoxia of altitude. At altitude the a–v O_2 difference may, in accord with the Fick principle, actually decrease a little (not shown) because the cardiac output may increase.

In **anaemia** the haemoglobin concentration is low and therefore so is the O_2 content (Fig. 16.30c). The shape of the saturation curve, however, is very similar to normal, unless the anaemia is severe, in which case the curve is shifted to the right due to the increased synthesis of 2,3-BPG caused by chronic hypoxia (Fig. 16.30d). The increase in 2,3-BPG results in more O_2 being released from the haemoglobin to the tissues for a particular Po_2, thus compensating to some extent for the lower O_2 content. The a–v O_2 difference in mild anaemia will not change from normal but in severe anaemia there is a limit to how low the venous Po_2 can fall without impairing diffusion gradients between blood and tissues. Thus the a–v O_2 difference becomes smaller (not shown) and, if O_2 metabolism is not to fall, the cardiac output must increase.

Fetal haemoglobin has few β-chains; instead there are γ-chains (p. 335), which bind 2,3-BPG weakly thus creating fewer salt-bridges. The consequence is a relaxed state for the haemoglobin molecule and therefore, in effect, fetal haemoglobin has a curve which is shifted to the left, indicative of high O_2 affinity (Fig. 16.30e). This left shift more than compensates for the effect of fetal blood acidity which by itself would cause a right shift. The left shift facilitates the uptake of O_2 by the fetus from the placenta, where not only is the Po_2 rather low, but also the diffusion distances are greater than in the lung. O_2 transfer to the fetus is also enhanced by the higher haemoglobin concentration of fetal blood.

Carbon monoxide is poisonous because it combines with haem about 250 times more avidly than O_2. For instance, this gas at an environmental concentration of only 0.1%, if breathed for about 100 minutes, will come to occupy 50% of the haem sites. Given the now reduced O_2 capacity, the O_2 dissociation curve will be displaced below normal (Fig. 16.30f). It also shifts to the left because when CO combines with haem it forms **carboxyhaemoglobin** (CO–Hb), which pulls on the histidine of the polypeptide chains more strongly than oxyhaemoglobin, thus weakening the salt-bridges more and promoting a more relaxed form of haemoglobin. The left shift of the O_2–Hb curve makes the situation even worse, as what little O_2 was bound in the alveoli to the haemoglobin is not released very easily to the tissues. Carboxyhaemoglobin gives a distinctive cherry-red colour to the blood. The treatment for CO poisoning is a blood transfusion, or administration of 100% O_2 or, even better, hyperbaric O_2 (p. 553) which over a number of hours will very slowly displace the CO.

Methaemoglobin is formed when the ferrous ion in the haem is oxidized to the ferric state, a form which cannot combine with O_2. Formation of methaemoglobin is usually prevented by several enzymes in the red cells (p. 335), although it can occur congenitally, in nitrite poisoning or in toxic reactions to oxidant drugs. Furthermore, the presence of methaemoglobin causes the unaffected haem groups to produce a left shift to the O_2–Hb curve resembling that of carboxyhaemoglobin.

Cyanosis

Cyanosis is a blue–purple colour, most obvious in the skin, nail beds and mucosal membranes, caused by a low O_2 saturation of haemoglobin in the arteriolar blood and is indicative of blood with a low O_2 content. Its presence is detectable when there is at least 50 g L^{-1} of deoxyhaemoglobin but reliable recognition depends on skin pigmentation, suitable illumination and adequate capillary perfusion.

At a normal haemoglobin concentration of 150 g L^{-1}, cyanosis is evident at 66% saturation when the O_2 content is ~130 ml O_2 L^{-1}. However, in polycythaemia, in which the haemoglobin concentration is elevated to, say, 250 g L^{-1}, cyanosis can occur at 80%

saturation and would not be indicative of an abnormally low O_2 content. In contrast, in anaemia, where the O_2 content is always low, there may be insufficient deoxyhaemoglobin to allow cyanosis to be seen (e.g. if haemoglobin concentration was 100 g L^{-1}, the O_2 content is already low at 134 ml L^{-1} and would have to fall to 67 ml L^{-1} to be seen as cyanotic). **Central cyanosis** (blue mouth and tongue) is due to poor gas exchange in diseased lungs or to R–L shunts. In **peripheral cyanosis** (blue limbs but pink mucosa) the lungs are healthy but there is poor circulation.

Myoglobin

Myoglobin is not found in the blood, but in skeletal and cardiac muscle cells where it is a respiratory pigment which aids O_2 delivery. In contrast to haemoglobin, myoglobin only has one haem group and one globin chain and thus has a hyperbolic O_2 dissociation curve (Fig. 16.31) which is independent of pH or P_{CO_2}. Furthermore, its P_{50} is 5 mmHg and it is about 90% saturated at a P_{O_2} of 20 mmHg. O_2 is only extracted from the myoglobin store at the sudden start of exercise (p. 689) at a time when blood perfusion has not yet increased sufficiently to supply the required O_2. The myoglobin store of O_2 is replenished at the end of exercise, when the P_{O_2} of the muscle cells rises above 10 mmHg.

CO$_2$ carriage

The CO_2–blood dissociation curve (CO_2 content plotted against P_{CO_2}) is curvi-linear and does not show saturation. Typically, at rest, arterial blood has a P_{CO_2} of 40 mmHg with a CO_2 content of 490 ml L^{-1}; while venous blood has a P_{CO_2} of 46 mmHg and a CO_2 content of 535 ml L^{-1}. The CO_2 affinity of blood is decreased by oxygenation of the haem groups as this makes the globin less able to combine with CO_2 and H^+. Conversely, deoxygenated blood has a higher CO_2 affinity. These changes in CO_2 affinity caused by O_2 binding are called collectively the Haldane effect and result in downwards and upwards shifts in the CO_2–blood dissociation curve. CO_2 produced by the tissues is carried not only by the globin part of the haemoglobin molecule (as a carbamino compound, ~30%)

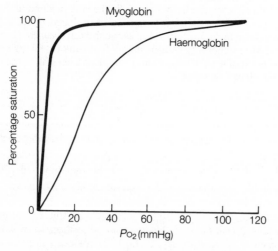

Fig. 16.31 Oxygen dissociation curves of myoglobin compared with haemoglobin.

but also in physical solution (~10%) and as bicarbonate (~60%). Due to the presence of carbonic anhydrase, most of the HCO_3^- (and H^+) is formed in the red cells but the HCO_3^- then diffuses out into the plasma in exchange for chloride. The H^+ ions are buffered by globin. An osmotic imbalance occurs so that water enters the red cells by osmosis, causing them to swell. The pH of the plasma is dependent on the ratio of HCO_3^- to P_{CO_2} (the Henderson–Hasselbalch equation), so that the pH of arterial and venous blood are 7.40 and 7.37 respectively. The P_{CO_2} is rapidly controlled by ventilation, whereas HCO_3^- is slowly altered by the kidneys.

CO_2–blood dissociation curve

The shape of the **CO_2–blood dissociation curve** (Fig. 16.32a) differs from that of the O_2–Hb curve—it is not sigmoidal but curvilinear. It rises very steeply in the initial part, is virtually linear over the P_{CO_2} values normally encountered in the blood and never reaches saturation. Thus CO_2 content cannot be expressed in terms of percentage saturation. The blood also contains about 2.5 times more CO_2 than O_2. Venous blood at a P_{CO_2} of 46 mmHg has a CO_2 content of 535 ml L^{-1} and arterial blood at a P_{CO_2} of 40 mmHg has a CO_2 content of 490 ml L^{-1}.

Blood contains only a small part (~2.5 litres calculated from ~500 ml of CO_2 per litre of blood × ~5 litres of blood) of the total amount of CO_2 in the body. CO_2 stores amount to over 100 litres, much of which is dissolved in fat or stored in bone. O_2 stores are minute, amounting to no more than 1.5 litres, and are held in the blood, alveoli and myoglobin.

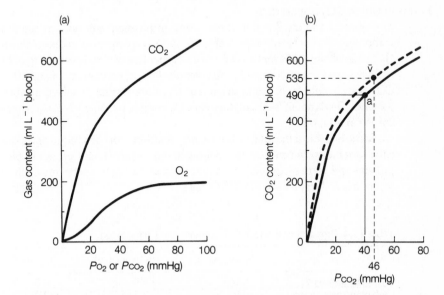

Fig. 16.32 (a) Comparison of O_2 and CO_2 blood dissociation curves. (b) The CO_2–blood dissociation curve, where a is the arterial and $\bar{v}$, the mixed-venous value. The solid curve is for a P_{O_2} condition of 100 mmHg when haemoglobin is nearly 100% oxygenated; the dashed curve is for a P_{O_2} condition of 40 mmHg when haemoglobin is ~70% oxygenated.

Haldane effect

For each O_2–Hb curve the P_{CO_2} and pH had to be defined (Fig. 16.28); similarly, for each CO_2–blood curve, the P_{O_2} must be held constant in the samples being equilibrated with CO_2 (Fig. 16.32b). Under arterial conditions (where P_{O_2} is 95 mmHg) the CO_2–blood dissociation curve is shifted down compared with venous conditions (where P_{O_2} is 40 mmHg). If high O_2 is breathed so that 100% O_2 saturation of haemoglobin is achieved, the CO_2–blood curve will shift just a little further down. Conversely, if P_{O_2} is lower than 40 mmHg (e.g. venous conditions in exercise or at altitude) the CO_2–blood curve will shift even further upwards. A downwards shift in the CO_2–blood curve means that for a given P_{CO_2} the blood can hold less CO_2 content—that is, the **CO_2 affinity** of the blood has decreased. Conversely, an upwards shift in the curve for a given P_{CO_2} indicates that the blood has a higher CO_2 affinity and can hold more CO_2 content. (As blood does not saturate with CO_2, an equivalent of the P_{50} of the O_2–Hb curve cannnot be used.)

The effect of changes in O_2 on CO_2 carriage is called the **Haldane effect** (whereas the effect of CO_2/H$^+$ on O_2 carriage is the Bohr effect). As with the Bohr effect, the Haldane effect is due to allosteric interactions (p. 525) ithin the haemoglobin molecule.

When the haem group is fully oxygenated, promoting the relaxed form of the haemoglobin molecule which has no salt-bridges, the imidazole groups in the globin chains are more dissociated (i.e. a stronger acid) and thus a weaker buffer. Globin is therefore less able to combine with H$^+$ ions and to form carbamino compounds (see below) with CO_2. The deoxygenated haemoglobin molecule, by contrast, is a weaker acid and thus a stronger buffer and can form more carbamino compound.

Forms in which CO_2 are carried

CO_2 is carried in the blood in three forms: in **physical solution**, as carbamino compounds and as bicarbonate (Table 16.4). CO_2 is very much more soluble than O_2—its solubility coefficient is 0.65 ml of CO_2 per litre of blood per mmHg of P_{CO_2}. At an arterial P_{CO_2} of 40 mmHg and a venous P_{CO_2} of 46 mmHg, 26 and 30 ml of CO_2 per litre of blood, respectively, are therefore carried dissolved in physical solution. This is about 5–6% of the total CO_2 content (cf. only 1% of O_2 content is carried in solution).

Some 5–10% of the total CO_2 is carried in the form of **carbamino compounds** which are formed when CO_2 combines with the terminal amine groups of plasma proteins and, in particular, of the globin chain of the haemoglobin molecule. The reaction is $R–NH_2 + CO_2 = R–NHCOOH$.

Table 16.4 Proportions in which CO_2 is transported in the blood

	% Arterial (a)	% Venous (v̄)	% (v̄ – a)*
Physically dissolved	5.5	5.8	8.0
Carbamino compound	4.9	7.2	30.0
HCO$_3$$^-$	89.6	87.0	62.0

* Percentage of newly produced CO_2 in each form.

TISSUE   PLASMA

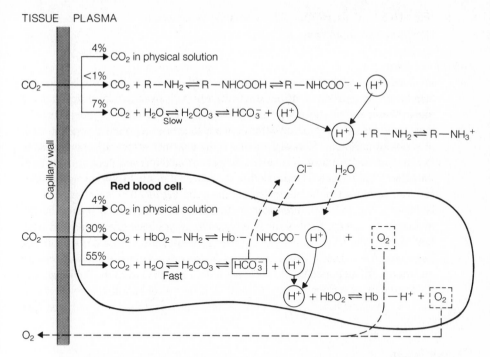

Fig. 16.33 The forms and proportions in which CO_2 are carried in the plasma and red blood cell. Also shown are the chloride shift, buffering, O_2 exchange and osmosis.

The remaining 85–90% of the total CO_2 is carried as **bicarbonate**. CO_2 is hydrated by water into carbonic acid which is a weak acid and thus it partially dissociates into H^+ ions and bicarbonate, i.e. $H_2O + CO_2 \rightleftharpoons H_2CO_3 \rightleftharpoons H^+ + HCO_3^-$. The hydration step is slow in the plasma but is accelerated (to as short as ~0.1 second) in the red blood cells by the enzyme **carbonic anhydrase**. Although most of the bicarbonate is formed in the red blood cells, only about a quarter of it remains there; the rest ends up in the plasma.

Table 16.4 shows not only the proportions in which the three forms of CO_2 are carried in arterial and venous blood but also, and more importantly, the proportions in which the CO_2 produced by the tissues is carried. As the newly produced CO_2 converts arterial into venous blood, 8% goes into physical solution, 30% forms carbamino compounds and 62% converts to bicarbonate. In other words, the carbamino form increases its contribution to the total. Figure 16.33 identifies how much of each is formed in the plasma compared with red blood cells.

Chloride shift

As bicarbonate is formed primarily in the red blood cell rather than in the plasma, the bicarbonate will diffuse out of the cell down its concentration gradient into the plasma (hence more bicarbonate is carried in the plasma, the proportional amount determined by the volume of plasma relative to red blood cells). To preserve electroneutrality, the anion chloride moves from the plasma into the red blood cell

(Fig. 16.33). This HCO_3^-/Cl^- exchange is known as the **chloride shift** or Hamburger phenomenon.

Fate of H^+ ions

H^+ ions are produced not only as carbonic acid dissociates into bicarbonate, but also by the dissociation of the carbamino compounds (weak acids) attached to the globin chains and plasma proteins (Fig. 16.33). The differential production of H^+ ions in the red cells compared with plasma also creates a potential concentration gradient for diffusion. However, not only is the red-cell membrane impermeable to H^+, but also a greater problem exists in the high acidity levels developing in both the plasma and red cell. The problem is solved by buffering. Plasma proteins buffer H^+ ions produced in the plasma where the buffering reaction is $R-NH_2 + H^+ = R-NH_3^+$. The H^+ ions produced in the red cell are buffered mainly by the imidazole groups (p. 525) of the haemoglobin polypeptide chains and this addition of H^+ (and CO_2) to the globin chain promotes the release of O_2 from the haem (Fig. 16.33). As the haem becomes deoxygenated it becomes a weaker acid and therefore a better buffer, so the concomitant loss of O_2 aids the buffering (and the formation of carbamino compounds). The degree of buffering is so good that the pH of venous blood (7.37) is only a little more acid than arterial blood (pH 7.40).

Osmotic balance

Each haemoglobin molecule can buffer many H^+ ions. This leaves many HCO_3^- (or Cl^-) ions as free ions constituting a pool of osmotically active particles and causing H_2O to move from the plasma into the cell (Fig. 16.33). As a consequence of **osmosis** the red cell swells; the increase in cellular volume can be detected as about a 3% greater haematocrit in venous compared with arterial blood.

Acid–base balance

Even though plasma proteins and haemoglobin act as blood buffers for H^+, the third buffer of the blood is the carbonic acid–bicarbonate system: $H_2O + CO_2 \rightleftharpoons H_2CO_3 \rightleftharpoons H^+ + HCO_3^-$. This may seem surprising as it is a producer of H^+ ions in the first place. However because it is a weak acid with a pK' of 6.1, it is a good physiological buffer. When there is an increase in H^+ the mass reaction is pushed to the left, effectively buffering the H^+ by producing CO_2. This CO_2 is then excreted very rapidly by the lungs through H^+-stimulated breathing (p. 543). If CO_2 in the body increases as a result of CO_2 in the atmosphere or impaired breathing, the mass reaction is pushed to the right and the H^+ so produced is buffered by increased plasma HCO_3^- under the regulation of the kidney, albeit it slowly (p. 670).

When the law of mass action is solved for H^+ and put into pH notation, the equation derived is called the **Henderson–Hasselbalch equation**. When applied to the carbonic acid–bicarbonate system this equation is written as:

$$pH = pK' + \log \frac{[HCO_3^-]}{0.03 \times P_{CO_2}}$$

where pK' is 6.1, [HCO_3^-] is in mmol L^{-1}, 0.03 is the solubility of CO_2 expressed in mmol L^{-1} mmHg^{-1} and P_{CO_2} is in mmHg. The denominator (0.03 × P_{CO_2}) is in this form as a substitute for [H_2CO_3] which is difficult to measure. At a P_{CO_2} of 40 mmHg, the pH is 7.40 as the [HCO_3^-] is usually 24 mmol L^{-1}; for venous conditions P_{CO_2} is 46 mmHg, pH 7.37 and [HCO_3^-] 25.7 mmol L^{-1}. The equation makes it clear that pH depends on the ratio of [HCO_3^-] to P_{CO_2} (the 'buffer pair'); the former is controlled slowly by the kidneys and the latter rapidly by the lungs. A primary change in one of the buffer pair is compensated for by a secondary change in the other in the appropriate direction (p. 670). If the primary change is an alteration in CO_2, it is called **respiratory acidosis** or **alkalosis**; if it is an alteration in [HCO_3^-] it is **metabolic acidosis** or **alkalosis**. Whole-body buffering also involves buffers in the interstitial and intracellular fluids (p. 666). Further details on acid–base balance are, at least in this textbook, the domain of the kidney (Chapter 19).

16.6 REGULATION OF RESPIRATION

Breathing occurs rhythmically. The rhythmicity is generated within the respiratory centres which are found in specific areas of the medulla oblongata (dorsal respiratory group (DRG), ventral respiratory group (VRG) and Botzinger complex) and pons (pontine respiratory group, PRG). The depth and frequency of the resting rhythm are determined by feedback to the respiratory centres from mechanoreceptors in the lungs and chest wall. Some stimuli, e.g. temperature and pain, affect breathing via higher brain centres (hypothalamus and limbic system) which then influence the respiratory centres. Breathing is also under appreciable voluntary control from the cerebral cortex (e.g. in breath-holding, taking large breaths or speaking) and this pathway bypasses the respiratory centres. Superimposed on the normal pattern of breathing are protective reflexes of short duration which are initiated, for example, by irritants as in coughing or sneezing. The principal control of ventilation is from chemoreceptors located peripherally (carotid and aortic bodies) and centrally (intracranial chemoreceptors). The peripheral chemoreceptors monitor arterial P_{O_2}, P_{CO_2} and pH, while the central chemoreceptors monitor brain P_{CO_2} and pH; together they regulate breathing to provide arterial P_{O_2}, P_{CO_2} and acid–base homeostasis. Breathing is more sensitive to stimulation by high CO_2 (hypercapnia) than low O_2 (hypoxia) so ventilation is matched to metabolism through the CO_2 produced rather than the O_2 consumed.

Respiratory centres

Historical background

In the early 1920s, Lumsden performed brainstem transections and concluded that respiratory control was provided by centres which he called the pneumotaxic centre (in the upper pons), the apneustic centre (in the lower pons), and the inspiratory and expiratory centres (in the medulla oblongata). The pontine centres were postulated because a transection between upper and lower pons, effectively removing the pneumotaxic centre, produces slow deep breathing. If done in conjunction with a bilateral vagotomy, to remove vagal sensory input from the lungs, inspirations become prolonged, separated by only brief expirations (a pattern referred to as apneustic breathing). This led to the conclusion that the pneumotaxic centre normally promoted expiration and the apneustic centre normally promoted inspiration; the two together finely tuned the normal breathing pattern.

Postulation of the medullary inspiratory and expiratory centres arose because, after a transection between the pons and medulla, breathing was dominated by prolonged expiratory spasms and irregular short inspiratory gasps (hence the terms gasping centre and inspiratory centre were used interchangeably). However, more refined techniques for pons ablation have since demonstrated that the medulla by itself will produce a relatively normal pattern of rhythmic inspiration/expiration. This pattern becomes slower and deeper when the incoming vagal sensory information is blocked. As breathing is abolished only by a transection between the medulla and spinal cord, the rhythmic drive to the respiratory muscles must be generated within the medulla.

Recent developments

No specific group of neurons has ever been found in the so-called apneustic centre so that it, and any theories of respiratory control involving it, have been abandoned. The pneumotaxic centre is now referred to as the **pontine respiratory group (PRG)** and it is now known to comprise expiratory neurons in the *nucleus parabrachialis medialis* and inspiratory neurons in the *nucleus parabrachialis lateralis* and the laterally sited *Kölliker–Fuse nucleus* (Fig. 16.34). Modern

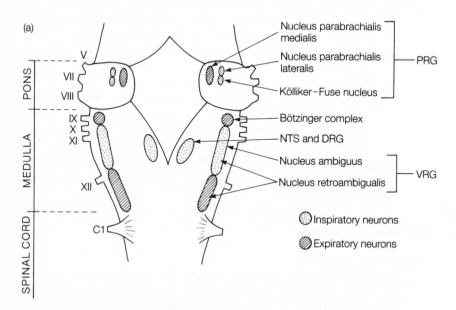

Fig. 16.34 (a) Location in the brain of neuronal groups constituting the respiratory centres. (b) Schematic representation of the well-established interconnections between neuronal groups, including sensory receptor inputs (solid lines, excitatory; dashed lines, inhibitory) Bötz C, Bötzinger complex; DRG, dorsal respiratory group; NTS, nucleus of the tractus solitarius; PRG, pontine respiratory group; VRG, ventral respiratory group.

evidence indicates that rather than inspiratory and expiratory centres, the medulla is composed of three bilateral aggregations of respiratory neurons known as the DRG, the VRG and the Bötzinger complex. They receive sensory information and after integration adjust appropriately the efferent activity to the respiratory muscles. The pontine nuclei somehow modulate DRG and possibly VRG activity.

The **dorsal respiratory group (DRG)** is part of the *nucleus of the tractus solitarius* from which it receives afferent information from respiratory-related mechanoreceptors and chemoreceptors. It is composed mainly of inspiratory neurons. These are upper motor (premotor) neurons projecting ipsilaterally down the spinal cord to the lower motor neurons of mainly the phrenic nerves which supply the diaphragm. The DRG inspiratory neurons also have inhibitory actions on the expiratory neurons in the VRG.

The **ventral respiratory group (VRG)** is located rostrally in the nucleus ambiguus and caudally in the nucleus retroambigualis. The *nucleus ambiguus*

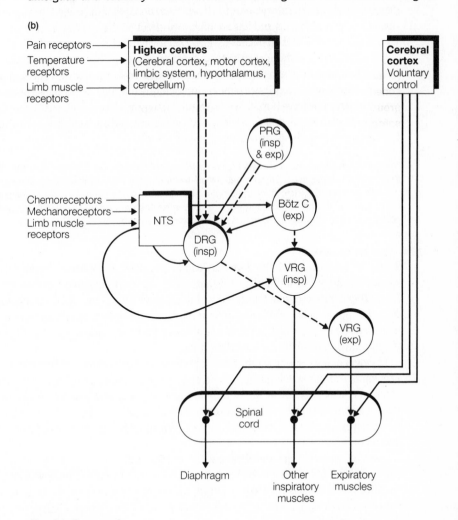

Fig. 34 *Continued.*

contains not only premotor inspiratory neurons but also motor neurons to the laryngeal muscles and parasympathetic neurons to the bronchioles and heart. The *rostral* part of the *nucleus retroambigualis* contains more inspiratory premotor neurons and, like those of the nucleus ambiguus, they mainly supply the external intercostals and accessory muscles rather than the diaphragm. These inspiratory VRG neurons receive sensory input and behave like those in the DRG. In the *caudal* part of the *nucleus retroambigualis* are mainly found the expiratory premotor neurons supplying the internal intercostal and abdominal muscles. During resting breathing, in which expiration is passive, the activity in these VRG expiratory neurons is insufficient to discharge their lower motor neurons. The VRG expiratory neurons appear to have inhibitory effects on inspiratory neurons late in the inspiratory phase, contributing to the termination of inspiration, the so-called inspiratory cut-off switch.

The **Bötzinger complex** is just rostral to the nucleus ambiguus and is composed entirely of expiratory neurons. It receives sensory input relayed through the nucleus tractus solitarius and has an inhibitory effect on the inspiratory neurons of the DRG and VRG and an excitatory effect on the expiratory neurons of the VRG.

It is not known where **respiratory rhythmicity** originates in the medulla. It may be that the DRG inspiratory neurons have pacemaker properties and spontaneously discharge in a phasic manner, inhibiting when they do so the expiratory neurons. An alternative hypothesis is a neural network of local re-excitation within each of the inspiratory and expiratory neuronal pools and an oscillating reciprocal inhibition between the phasic firing of the two pools.

Many other more complex models are currently under investigation. It seems clear that the final model will require a 'rhythm generator' to initiate the burst of inspiratory activity, a 'pattern generator' to determine the amplitude and timing of the burst (i.e. the final balance of V_T and f), and an 'input integrator' to synthesize the conflicting commands from the various sensory inputs prior to action on the pattern generator.

Higher brain centres

There is evidence that the **hypothalamus** provides a continual background excitatory drive to the DRG. Furthermore, the hypothalamic *temperature-regulating centres* receive inputs from cutaneous and deep-core thermoreceptors such that both hot and cold cutaneous stimuli increase breathing as do a fever and mild hypothermia (p. 684). Activation of the *defence reaction centre* (p. 446) in the hypothalamus also stimulates breathing.

Evidence also suggests that the **cerebral cortex** influences breathing by providing a continual inhibitory drive to both the hypothalamus and the DRG. The cerebral cortex and **limbic system** integrate *emotional* and *painful* stimuli and, via the hypothalamus, have both excitatory and inhibitory effects on the respiratory centres which produces a wide variety of respiratory expressions of emotion and pain.

During *exercise* the **cerebral motor cortex** activates with a feedforward command the DRG to increase breathing in proportion to the degree of exercise (p. 693). *Limb muscle receptors* (p. 444) appear to cause some of the ventilatory increase occurring in exercise (p. 693) and some of their input is integrated by **cerebellar nuclei** (p. 446).

All **voluntary** control of respiration, as in taking large breaths, performing forced expirations, hyperventilating deliberately, speaking or breath-holding, stems from the cerebral cortex but it bypasses the respiratory centres and travels in the pyramidal system directly to the lower motor neurons (Fig. 16.34). In these circumstances the cortex can for a short time override respiratory reflexes but eventually the reflex control takes precedence.

Mechanoreceptors and the pattern of breathing

Input in the vagus nerves from stretch and irritant receptors of the lung acts on the so-called inspiratory cut-off switch and helps determine the V_T and f that produce a particular level of ventilation. f is dependent on the time spent in T_I and T_E phases of breathing. If these lung receptor inputs are removed by a vagotomy (and the pontine nuclei are intact), breathing is slow and deep—this is the basic pattern generated by the respiratory centres. With vagal input the normal breathing pattern is achieved—one which is relatively fast and shallow.

Lung-stretch receptors are in the smooth muscle of the trachea and lower airways and are stimulated in proportion to the degree of lung inflation by the accompanying increase in transmural pressure. Their nervous discharge shows little adaptation to a sustained inflation but, during normal breathing, increases and decreases in phase with inspiration and expiration. This phasic afferent information inhibits directly, and via the PRG, the inspiratory neurons of the DRG while the inspiration is in progress. The outcome of this vagal input is a shorter T_I and smaller V_T; the vagal input also lengthens the duration of the next T_E by inhibiting the occurrence of the next inspiration.

Lung-irritant receptors are also stimulated during inspiration but they affect only the duration of the next T_E by shortening it via an excitatory input to the DRG, which promotes an earlier occurrence of the next inspiration. As the lung-irritant effect on T_E dominates that from the lung stretch receptors, the two receptor types together produce the normal breathing pattern of a shallow V_T and short T_I and T_E.

The vagal effect on breathing is more pronounced in the newborn and anaesthetized state; furthermore, the potency of the lung-stretch-receptor reflex is also species-dependent. In conscious adult humans this reflex appears to be weak in the resting V_T range but at the higher V_T of exercise or of breathing driven by hypoxia or hypercapnia, the effects of the lung-stretch-receptor reflex become more apparent. Thus, although the breathing pattern at rest is altered little by heart–lung transplantation, f in these patients does not increase as much during stimulated breathing.

The breathing pattern, particularly in humans, is also determined by **proprioceptors** (joint receptors, Golgi tendon organs and muscle spindles) that are present in the chest wall and diaphragm (the latter has mainly tendon organs) and provide information about thoracic inflation. Proprioceptors also provide feedback, which enables the strength of the contraction to be varied so that a certain V_T can be achieved if the airway resistance changes. The proprioceptor reflex is integrated within the spinal cord. Since one becomes aware of changes in chest compliance and airways resistance, it is assumed that the stretch afferents relay to the sensory cortex as well as to the respiratory centres. The distressful sensation of breathlessness is called **dyspnoea** but which receptors provoke this sensation is not known; chest-wall afferents have been implicated, but all respiratory mechanoreceptors may contribute.

Mechanoreceptors and protective reflexes

Mechanoreceptors having reflex effects on respiration are found in the upper airways (nose, epipharynx, larynx and trachea). In the lower airways (bronchi to alveolar ducts) there are lung-stretch, lung-irritant and bronchial C receptors and, in the alveoli, pulmonary C (or juxtacapillary) receptors. All of these mechanoreceptors provide protective reflexes affecting both ventilation and bronchomotor tone. As the cardiovascular system has a large role to play in gas transport, most of these respiratory mechanoreceptors also cause cardiovascular reflexes (p. 443). Similarly, some of the mechanoreceptors located in the cardiovascular system also elicit respiratory reflexes—baroreceptors or ventricular vagal afferents depress, while ventricular sympathetic afferents augment breathing.

Nerve endings of the trigeminal (Vth) nerves in the **nasal mucosa** are excited by chemical and mechanical irritants and reflexly cause a **sneeze** and broncholaryngeal constriction. A sneeze is a number of superimposed inspirations followed by a strong and rapid expiration and then a short pause in the expiratory position. Stimulation by liquids produces only a long pause in breathing, i.e. an apnoea, which is well-developed in diving mammals.

Glossopharyngeal nerve endings in the epithelium of the **epipharynx** are excited by mechanical irritants such as a blockage and this results in the **aspiration reflex** and bronchodilatation. The aspiration reflex is a powerful inspiration which attempts to pull the blockage into the pharynx.

Vagal nerve endings in the epithelium of the **larynx** and **trachea** are excited by mechanical and chemical irritants and reflexly cause a **cough** and broncholaryngeal constriction. A cough is a large slow inspiration followed, initially against a closed glottis, by rapid and powerful expirations. As in the sneeze, the broncholarygeal constriction increases expiratory airflow velocity, helping to expel the irritants. Mild stimulation of these receptors causes slow, deep breathing.

As described earlier, **lung-stretch receptors** are involved in determining the pattern of breathing and are slowly adapting vagal nerve endings in the smooth muscle of the trachea and lower airways. Experimentally, if these receptors are stimulated by a sustained inflation (followed by occlusion of the airways), there is a reflex cessation of diaphragmatic activity and the duration of the pause is proportional to the degree of inflation. Conversely, a sustained deflation with an occluded airway promotes strong and frequent inspiratory efforts. These responses are called the **Hering–Breuer reflex**, and in real life can be viewed as protective, preventing too much inflation or too much deflation. Stimulation of these receptors also causes bronchodilatation.

Lung-irritant receptors are vagal nerve endings in the epithelia of the trachea and lower airways that are stimulated by mechanical or chemical irritants such as mucus, noxious gases or tobacco smoke. The protective reflex initiated is an increase in f and V_T, together with broncholaryngeal constriction. Histamine released in **asthma** probably stimulates the lung-irritant receptors. These receptors can also be stimulated by large deflations and large inflations but the response rapidly adapts (hence their alternative name of rapidly adapting lung-stretch receptors). In this reflex the response is a short-lived, intense inspiration. Thus the lung-irritant receptors permit, and may even trigger, the augmented breaths or **sighs** that punctuate normal breathing. These sighs serve to reinflate and reset the compliance of alveoli which have gradually collapsed with time during quiet breathing (p. 485). The role of the lung-irritant receptors in determining the pattern of breathing has already been considered.

Bronchial C and **pulmonary C** receptors are vagal nerve endings of unmyelinated (C) fibres lying respectively in the bronchial interstitium or next to capillaries in the alveolar interstitium. Pulmonary C receptors, because of their position are also called **juxtacapillary** or **J receptors**. Bronchial and pulmonary C receptors are usually silent and therefore do not play a role in the regulation of normal breathing. Both are stimulated by mechanical distortion or increases in ISF pressure. Large inflations stimulate the bronchial C receptors, though with rapid adaptation, while lung congestion or oedema stimulates the pulmonary C receptors. Both receptor types are also stimulated by substances like histamine, bradykinin and prostaglandins, which are released in lung disease. The reflex initiated is broncholaryngeal constriction, apnoea followed by rapid shallow breathing and, if the stimulus is severe, the so-called J reflex which is an inhibition of spinal motor neurons, thus relaxing skeletal muscles. A general depression of somatic (and cardiovascular; p. 443) activity is a most appropriate response to serious lung damage.

Respiratory chemoreceptors

As some of the mechanoreceptors can also respond to certain chemical stimuli, the term chemoreceptors is usually reserved for those receptors monitoring, directly or indirectly, blood gases. Chemoreceptors maintain homeostasis of arterial Po_2, Pco_2 and pH and assist in ensuring that $\dot{V}_E$ is appropriate to the level of metabolism. They are stimulated by a rise in P_aco_2 (**hypercapnia**), a rise in arterial $[H^+]$ (**acidosis**) and a fall in P_ao_2 (**hypoxia**) and reflexly cause an increase in $\dot{V}_E$ (and of course in $\dot{V}_A$). As described previously, $\dot{V}_A$ is one of the determinants of the alveolar and hence blood gas composition (p. 504), so that the reflex increase in $\dot{V}_A$ lowers P_aco_2 and elevates P_ao_2, thus restoring alveolar and arterial gases back towards normal (Pco_2 40 mmHg and Po_2. Since $\dot{V}_E$ is more sensitive to increases in P_aco_2 than to decreases in P_ao_2 (Fig. 16.35), the controlling link between $\dot{V}_E$ and metabolism is the CO_2 produced rather than the O_2 consumed. The precision of control is such that in healthy individuals P_aco_2 and P_Aco_2 remain remarkably constant (± 2 mmHg) over a wide range of metabolic rates.

Ventilatory chemosensitivity

The sensitivity of breathing to hypercapnia and hypoxia can be experimentally determined by altering the composition of the inspired gas (the F_Ico_2 or F_Io_2). When, for example, the F_Ico_2 is raised, P_Aco_2 and hence P_aco_2 increase (p. 507). Some minutes later the final (i.e. steady-state) increase in P_aco_2 is less than the initial increase because the reflex stimulation of $\dot{V}_E$ lowers the P_aco_2. However, in

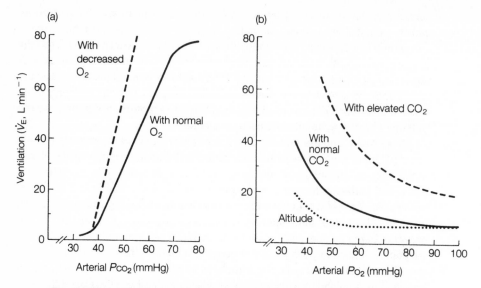

Fig. 16.35 (a) Ventilatory sensitivity to hypercapnia while O_2 is experimentally maintained constant (solid line for normoxia, $P_aO_2 = 100$ mmHg; dashed line for hypoxia, $P_aO_2 = 50$ mmHg). (b) Ventilatory sensitivity to hypoxia at simulated altitudes where P_aCO_2 falls progressively because of the $\dot{V}_E$ response (dotted line) and at sea-level while CO_2 is experimentally maintained constant (solid line for normocapnia, $P_aCO_2 = 40$ mmHg; dashed line for hypercapnia, $P_aCO_2 = 45$ mmHg).

the face of altered inspired gases, the $\dot{V}_E$ increase can never restore P_aCO_2 (or P_aO_2) to normal (see Figs 16.19 and 16.20).

In *hypercapnia* (respiratory acidosis) the steady-state responses show that the $\dot{V}_E$ is linearly related to P_aCO_2 (Fig. 16.35a, solid line). Note also that at lower P_aCO_2 values $\dot{V}_E$ becomes less sensitive to P_aCO_2 and at higher P_aCO_2 values $\dot{V}_E$ sensitivity is reduced due to CO_2 narcosis of the respiratory and other neural centres in the brain. During such hypercapnia tests, the P_aO_2 will progressively increase from the normal ~100 to up to ~120 mmHg because of hyperventilation. If the inspired O_2 is deliberately decreased experimentally to maintain the P_aO_2 constant at, say, 50 mmHg during the hypercapnia tests, the $\dot{V}_E$–P_aCO_2 curve not only shifts upwards but also steepens its slope (Fig. 16.35a, dashed line). This steepening indicates increased $\dot{V}_E$ sensitivity and is referred to as a multiplicative or potentiating interaction between hypoxia and hypercapnia; apparently it is well-developed only in humans. Although not illustrated here, $\dot{V}_E$ is also linearly related to arterial pH—as pH decreases (acidosis), $\dot{V}_E$ increases.

Voluntary hyperventilation removes CO_2 from the body and results in arterial hypocapnia (respiratory alkalosis). When such hyperventilation stops, automatic breathing proceeds at a reduced $\dot{V}_E$ (or there may be apnoea) as a consequence of the low P_aCO_2 reducing the activity of the chemoreceptors. This permits the P_aCO_2 to build up to normal levels so that normal breathing soon resumes.

In *hypoxia* $\dot{V}_E$ is inversely related to P_aO_2 with a curve that steepens progressively as the hypoxia becomes more severe (Fig. 16.35b). At P_aO_2 values below

30 mmHg, the low O_2 severely depresses brain activity, which decreases $\dot{V}_E$. Hypoxia occurs naturally and progressively as one ascends to high altitude; lowering the F_IO_2 can simulate this. As the reflex increase in $\dot{V}_E$ will also progressively lower P_aCO_2 and arterial $[H^+]$ which, in their turn, will depress $\dot{V}_E$, the $\dot{V}_E$ at a particular P_aO_2 at simulated altitude is less (Fig. 16.35b, dotted line) than when the P_aCO_2 is held constant at its normal ~40 mmHg by adding a little CO_2 to the inspired gas during the hypoxic tests (solid line). The potentiating interaction between hypoxia and hypercapnia can also be seen as a steeper curve (Fig. 16.35b, dashed line) when the P_aCO_2 is held elevated and constant at, say, 45 mmHg whilst the sensitivity to P_aO_2 is tested.

Peripheral chemoreceptors

The arterial or **peripheral chemoreceptors** are located in the **carotid** and **aortic bodies** and are stimulated by arterial hypoxia, hypercapnia and acidity. After removal of the peripheral chemoreceptors, all the $\dot{V}_E$ response to acute hypoxia, approximately 20% of the response to hypercapnia and nearly all the response to acute metabolic acidosis is lost. Peripheral chemoreceptors also cause bronchoconstriction, elicit cardiovascular reflexes (p. 433) and increase secretion of antidiuretic hormone (p. 277).

The **carotid bodies** are small nodules found bilaterally at the bifurcation of the internal and external carotid arteries (Fig. 16.36). The dense capillary network of each carotid body is supplied by a branch from the external carotid artery and the venous blood drains into the internal jugular vein. The blood flow of 20 ml g^{-1} min^{-1} is so great that the O_2 consumption of the carotid body causes only a tiny arterial–venous O_2 difference. Thus, their capillary gas tensions are very close to arterial values and any change reaches equilibrium in the carotid body within

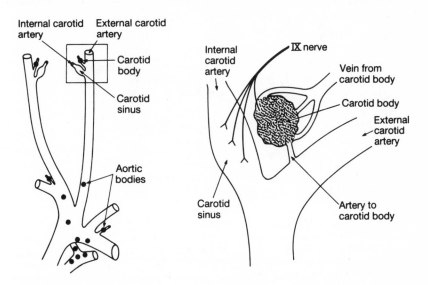

Fig. 16.36 Location of aortic and carotid bodies on the aortic arch and near the carotid arteries respectively. Enlargement shows the vasculature and innervation of the carotid body.

seconds. The carotid bodies are said therefore to monitor arterial P_{O_2}, P_{CO_2} and pH. Increased plasma $[K^+]$ is now known to also be an excitatory stimulus (p. 694).

The capillaries of the carotid body are surrounded by **glomus cells**, supportive **sustentacular cells** and sensory nerve endings of a branch of the IXth cranial nerve (**glossopharyngeal nerve**). Despite much research, it is still not known precisely how the carotid body senses its stimuli. Although some evidence suggests that the nerve endings themselves are chemosensitive, the glomus cells, which are metabolically very active and rich in a variety of neurotransmitters (e.g. dopamine, acetylcholine, substance P), are considered most likely to be the chemosensitive element.

The **aortic bodies** are scattered over the aortic arch and the main arteries of that area (Fig. 16.36) and are innervated by the Xth cranial nerve (**vagus nerve**). They are anatomically similar to the carotid bodies but less vascular and hence respond more sluggishly. Furthermore they are not as sensitive as the carotid bodies to hypoxia, hypercapnia and acidity. Their reflex effect on respiration is weak; their role is predominantly for cardiovascular reflexes (p. 443). As the aortic bodies have a lower blood flow, their capillary gas tensions and nerve activity are, in contrast to the carotid bodies, affected by anaemia, CO poisoning and hypotension. Thus in these conditions the aortic bodies may have important reflex effects on $\dot{V}_E$.

Central chemoreceptors

The intracranial or **central chemoreceptors** are located at a depth of about 500 μm in discrete areas near the **ventrolateral surface** of the **medulla oblongata** (Fig 16.37). Some evidence also exists for deeper, more widespread locations. Within any of these areas, no precise structure has yet been identified anatomically

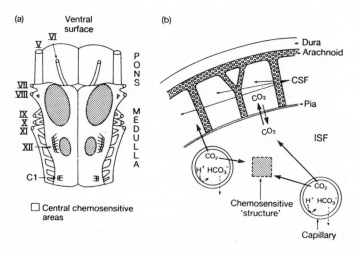

Fig. 16.37 (a) Location of the central chemosensitive areas on the ventral surface of the medulla oblongata. (b) The position of the chemosensitive structure in relation to the capillaries and the cerebrospinal fluid (CSF). ISF, interstitial fluid.

but it is presumed that nerve endings here monitor the brain ISF. The central chemoreceptors are stimulated by arterial hypercapnia, cause cardiovascular reflexes (p. 443) and are responsible for about 80% of the $\dot{V}_E$ response to hypercapnia. They are, however, insensitive to hypoxia.

Central chemoreceptors are only slightly stimulated by acute increases in arterial acidity. This is because of the presence of a blood–brain barrier (p. 71) which greatly restricts movements of ions (but not of O_2 and CO_2) and the extent to which arterial H^+ can influence the brain ISF. Since H^+ and HCO_3^- ions do not readily cross this barrier, most of the arterial $[H^+]$ changes in acute metabolic acidosis and alkalosis are not reflected in the brain ISF. During chronic metabolic acidosis or alkalosis, there is a slow leak of HCO_3^- down its concentration gradient between blood and ISF (see below), but equilibrium is never reached.

During acute hypercapnia (**acute respiratory acidosis**) the increase in blood P_{CO_2} immediately increases P_{CO_2} in the brain ISF. However, since the relative blood flow to the brain is about 40 times less than that to the carotid bodies, it takes 5–10 minutes to reach the final steady state (or even longer, considering the vast stores of CO_2 (p. 533) in fatty tissue and bone which are poorly perfused and contribute to the final CO_2 equilibrium). This ISF CO_2 is then hydrated to H_2CO_3 which dissociates into HCO_3^- and H^+. It is now believed that not just ISF H^+ but also ISF P_{CO_2} are the stimuli to the central chemoreceptors. The formation of H_2CO_3 may be accelerated by carbonic anhydrase, which has recently been found in capillary endothelia and large neurons within the ventrolateral location of the central chemoreceptors.

The pH and P_{CO_2} of brain ISF are influenced mainly by the composition of arterial blood but also to some degree by the rate of cerebral blood flow, the metabolism of the brain cells and by the slowly flowing cerebrospinal fluid (CSF) on the surface of the medulla (Fig. 16.37). Since the location of the central chemoreceptors was first identified by applying acidic CSF to the brain surface, it has been implied that the receptors monitor CSF pH. However, CSF is physiologically some distance from the chemosensitive structures and its pH is dependent on ISF pH, on blood P_{CO_2} and on nearby vascularity. Thus, the CSF pH changes slowly and may be somewhat different to that of the ISF pH at the site of the receptor.

During **chronic respiratory acidosis**, brain ISF pH is restored towards normal. This reduces the stimulus to the central chemoreceptors so that the chronic $\dot{V}_E$ response is less than the acute one. Such *homeostasis of brain pH* is essential for the viability of the neurons of the brain. Brain ISF pH is restored by a gradual increase in ISF $[HCO_3^-]$, resulting from three mechanisms.

The first mechanism is a consequence of hypercapnia elevating not only ISF but also glial cell P_{CO_2}. This leads to a very rapid hydration of glial CO_2 (catalysed by glial cell carbonic anhydrase) and formation of H^+ and HCO_3^-. The H^+ is buffered by cellular proteins but over some hours HCO_3^- leaks out of the cells into the ISF. This process is enhanced by the second mechanism, in which hypercapnia causes glial cells to reduce their basal anaerobic metabolism and, by producing less lactic acid, more HCO_3^- is free to diffuse into the brain ISF. (Lactic acid formation by the glial cells is also increased when the brain is hypoxic.) The third mechanism results from the slow diffusion of HCO_3^- from the blood into the brain ISF—the plasma HCO_3^- has been elevated by renal compensation (p. 672).

Rapid and slow changes in central chemoreceptor activity also occur during metabolic acidosis and hypoxia. Although acidosis and hypoxia stimulate the carotid bodies, the increased $\dot{V}_E$ produces arterial hypocapnia which rapidly decreases brain ISF H^+ and P_{CO_2} and thus decreases stimulation of the central chemoreceptors. Thus, the initial $\dot{V}_E$ response to the acidosis and hypoxia is quickly dampened to give the acute steady-state response. However, very gradually HCO_3^- will move from the ISF to the brain cells and, to a small extent, to the plasma. In hypoxia there is also an increase in lactic acid formation in the glial cells. All this will restore brain ISF pH towards normal and reduce the depression of the central chemoreceptors. Thus, the response to chronic metabolic acidosis and chronic hypoxia will become augmented compared with the acute response. This is the basis of the gradual increase in $\dot{V}_E$ that typifies acclimatization to altitude (p. 550).

16.7 RESPIRATION IN DISEASE AND UNUSUAL CONDITIONS

Hypoxia can be the result of respiratory diseases. Hypoxia also occurs in healthy individuals at high altitude where the decreased barometric pressue (P_B) lowers the inspired P_{O_2}. Physiological adjustments occur in response to short-term (acute) hypoxia and develop further during acclimatization to long-term (chronic) hypoxia. At high altitude the initial adjustment may be accompanied by acute mountain sickness and even after acclimatization chronic mountain sickness may suddenly appear. Diving is a condition where breathing must occur while submerged at increased P_B (hyperbarism). Breath-hold, snorkel and scuba diving create different problems—the first is limited by O_2 stores, the second requires breathing against an increased pressure on the chest and the third can lead, at depth, to nitrogen narcosis and, on ascent, to decompression sickness. Breathing pure O_2, either at normal atmospheric pressure or under hyperbaric conditions for the medical treatment of hypoxia, has dangers of O_2 toxicity, collapse of alveoli (atelectasis) and impaired blood CO_2 carriage.

Types of hypoxia

Hypoxia is a term used to denote a $P_{A_{O_2}}$ that is below the normal range of 90–100 mmHg. Blood and tissue, when low in P_{O_2}, can also be referred to as hypoxic, although there is a specific term, **hypoxaemia**, for a $P_{a_{O_2}}$ below the normal range of 80–90 mmHg. The ultimate consequence of any respiratory disease is a **deficiency of O_2 in the tissues**—this is also referred to as hypoxia, which can lead to confusion, as it may not necessarily be accompanied by a low $P_{a_{O_2}}$. The classification of hypoxia into four main types helps clarify the situation:

1   **hypoxic hypoxia**, in which $P_{a_{O_2}}$ is abnormally low (i.e. hypoxaemia);

2   **anaemic hypoxia**, in which $P_{a_{O_2}}$ is normal but the haemoglobin concentration and thereefore the O_2 content are reduced, and thus to compensate venous P_{O_2} is low (Fig. 16.30c)—CO poisoning (p. 531) and methaemoglobin formation (p. 531) are also categorized as anaemic hypoxia;

3   **stagnant** or **ischaemic hypoxia**, in which blood flow to a tissue is so low (e.g. in circulatory failure) that delivery of O_2, carried in otherwise normal blood, is

inadequate and thus venous blood is less oxygenated than usual (Fig. 16.30a); and

4 histotoxic hypoxia, in which $P_{a}O_2$ and perfusion are normal but due to a toxic agent (e.g. cyanide poisoning) the cells cannot utilize the O_2 properly and venous P_{O_2} is therefore more oxygenated than usual.

Hypoxic hypoxia (hypoxaemia)

This will occur if there is a decrease in $P_{A}O_2$ caused, as shown by the alveolar gas equation for O_2 (p. 504), either by a **low inspired P_{O_2}** (e.g. at high altitude) or by **hypoventilation**. Hypoventilation results from:

1 drugs like barbiturates or morphine which depress the respiratory centres;

2 damage to the respiratory centres or to respiratory muscles or their nerves;

3 interference with neuromuscular transmission as in myasthenia gravis;

4 injury to, or a stiff, chest wall;

5 stiff lungs;

6 obstructed airways; and

7 an imposed dead space from, for example, snorkel-breathing.

Hypoventilation leads to both hypoxaemia and hypercapnia. Although administration of high inspired O_2 will improve the $P_{a}O_2$, by removing the hypoxic stimulus the hypoventilation may get worse if the patient has a depressed central chemoreceptor response to CO_2. There are also other dangers if treatment is with pure O_2 (p. 553)

Hypoxaemia also occurs, with a normal $P_{A}O_2$, if there is impaired diffusion of O_2 across the alveolar–capillary membrane (p. 513), increased R–L shunts (p. 515) or increased inequality of the $\dot{V}_A/\dot{Q}$ ratio (p. 520). The hypoxaemia is not accompanied by arterial hypercapnia unless the disorders are very severe. Administration of high inspired O_2 will improve the $P_{a}O_2$ except if the disorder is a shunt (p. 516).

The low inspired P_{O_2} of high altitude

As given by Dalton's law $(P_{O_2} = F_{O_2} \times P_{B})$, a reduction in the inspired P_{O_2} can result from either a fall in $F_{I}O_2$ or a decrease in P_{B}. Hypoxia has long been studied in healthy individuals acutely and chronically exposed to the low P_{O_2} of high altitude. As one ascends a mountain, P_{B} and thus the inspired and therefore $P_{A}O_2$ fall (Table 16.5). Because hypoxia stimulates breathing, P_{O_2} in the table has been calculated using incremental increases in alveolar ventilation typical of that shown earlier in Fig. 16.35.

Acute hypoxia

A number of responses to hypoxia occur acutely.

1 Respiratory function: there is hyperventilation (Fig. 16.35) and consequential lowering of $P_{a}CO_2$ (hypocapnia or respiratory alkalosis). The exact magnitude is hard to predict because of the interaction of a number of factors. For example, hypoxia directly depresses the respiratory centres but stimulates ventilation via the peripheral chemoreceptors. However, the hyperventilation blows off CO_2 at a rate greater than it is being produced metabolically. Thus, the person becomes

Table 16.5 Effect of barometric pressure (P_B) on the partial pressure of inspired (P_IO_2) and alveolar O_2 (P_AO_2)

Height above sea level		P_B	P_IO_2	P_AO_2
Feet	Metres	(mmHg)	(mmHg)	(mmHg)
0	0	760	149	100
8500	2591	550	103	68
14 000	4267	450	84	48
21 000	6401	350	63	36
28 000	8534	250	42	30

* In moist tracheal air.

hypocapnia and arterial pH rises. These in turn, through peripheral and central chemoreceptors, inhibit ventilation. The net effect at moderate altitude is commonly an approximate doubling of ventilation at rest, but individual responses vary considerably.

2   Cardiovascular function: there is tachycardia and an increase in cardiac output, while arterial blood pressure usually remains about normal or is even decreased. Once more, in any individual, the effect is hard to predict quantitatively because of interactive processes (Section 15.1).

3   Cerebral function: this is very sensitive to the severity and speed of onset of hypoxia. Severe and sudden hypoxia will result in rapid loss of consciousness (p. 247). Moderate hypoxia of relatively gradual onset may cause only such changes as lack of judgement and psychomotor disturbance.

4   Blood: the degree of deoxygenation is detectable as cyanosis (p. 531). Venous Po_2 is low and the a–v O_2 difference (Fig. 16.30b) depends on the extent of any increase in $\dot{Q}$. Furthermore acutely, but not chronically, the hypocapnia causes a left shift in the oxyhaemoglobin curve, making it difficult to offload what little O_2 there is in the blood.

After 8–24 hours newcomers to high altitude frequently develop symptoms of **acute mountain sickness** which last 4–8 days. The symptoms are dizziness, headache, vomiting, nausea, breathlessness (dyspnoea; p. 542), irritability, insomnia, fatigue, loss of appetite and heart palpitations. The cause is not known but it appears to be associated with cerebral oedema.

Chronic hypoxia

Over subsequent weeks and months most people acclimatize to living at high altitude, although the ability to perform strenuous exercise may be limited. However, no human communities live permanently above 5500 metres (18 000 feet) because further acclimatization becomes impossible; instead there is a slow deterioration in all aspects of performance. **Acclimatization** is the process whereby the body tries to compensate and improve the O_2 delivery to the tissues. These same processes may occur in patients who have been hypoxic for a long time due to respiratory disease. The various compensating mechanisms of acclimatization are listed below (permanent residents born and bred at high altitude have some extra attributes which are not included here).

1 Respiratory function: the hyperventilation continues and in fact increases because of restoration of brain ISF pH (p. 548). Pulmonary diffusing capacity can increase up to three-fold as a result of more open pulmonary capillaries, increased pulmonary capillary blood volume and greater alveolar surface area. Cyanosis therefore may be absent at rest but will become apparent during exercise.

2 Kidney function: the kidneys excrete bicarbonate with eventual correction of the respiratory alkalosis; this takes 2–3 weeks.

3 Cardiovascular function: $\dot{Q}$ and heart rate tend to come back to sea-level values. However, hypoxic vasoconstriction of small pulmonary arteries (p. 521) leads to pulmonary hypertension and right ventricular hypertrophy—these responses have no apparent physiological advantage. The vasoconstriction also precipitates muscularization of pulmonary arterioles, which are normally devoid of smooth muscle.

4 Tissue function: there is an increase in the capillarity of many tissues and in the concentration of cytochrome oxidase; the former improves O_2 transfer and the latter permits the mitochondria to utilize the O_2 more effectively. The myoglobin content of skeletal muscle also increases, which raises the reserve store of O_2; myoglobin may also facilitate intracellular diffusion.

5 Blood: there is an increase in 2,3-BPG production, which shifts the oxyhaemoglobin curve to the right (Fig. 16.30b) to enable O_2 to be offloaded to the tissues more easily. Furthermore, hypoxia increases erythropoietin secretion (p. 327), leading to an increased red-cell count (polycythaemia). Depending on the altitude elevation, the haemoglobin concentration may increase to as much as 200 g L^{-1} and thus, despite the low P_aO_2 and percentage of haemoglobin saturation, the arterial O_2 content at altitude will not fall as much. There is a disadvantage to this—the increased blood viscosity increases quite considerably the work of the heart.

Some individuals, apparently successfully acclimatized, gradually or even suddenly fail in their compensating responses. Their ventialtion falls, producing dyspnoea and cyanosis. Haematocrit and pulmonary arterial pressure increase further, leading to right heart failure. In sudden failure, pulmonary oedema is the most dominant feature. These are all the symptoms of **chronic mountain sickness**. The only course of action is to remove the person to a lower altitude.

Increased ambient pressure and diving

During submersion in water, the pressure increases by 1 atm (1 atm = 760 mmHg) for every 10 metres (33 feet) of descent. The increased pressure (**hyperbarism**) itself is unlikely to have any effect on the solid tissues or liquids of the body. It is the gas-filled cavities of the body that are affected—the air-filled nasal sinuses, the middle ear, the gastrointestinal tract and the lungs. Gas will be compressed during the descent and will expand during the ascent. It is important for the cavities to communicate with the outside pressure (by opening the nose and mouth/glottis and, in the case of the middle ear, the eustachian tube), particularly during a rapid ascent. Submersion also poses the problems of hypothermia, distortion of information reaching the special sense organs, i.e. disturbance of vision and hearing, and the need to breathe under water.

A dive may be:

1 short enough to be accomplished on such air as can be inhaled at the surface—single-breath diving;

2 carried out at a shallow depth such that connection with the atmosphere can be maintained through a tube—snorkel diving; or

3 long enough to demand a continual supply of air, in which case the air must be supplied at whatever the increased ambient pressure—conventional and scuba diving.

Single-breath diving

This is, basically, **breath-holding** with the additional complicating factor of submersion. Suppose one takes a deep breath and goes down to 10 metres (+ 1 atm). The P_{O_2} and P_{CO_2} will approximately double—not quite, because both gases will pass along their gradients into the blood. After a brief period (20–30 seconds), the combined effect, through stimulation of respiratory receptors, of rising P_{CO_2} (the most potent stimulus), falling P_{O_2} and, as lung O_2 stores are used up, decreasing lung volume is sufficient to force the diver to the surface. The danger lies in the temptation to hyperventilate before submerging in an attempt to prolong the dive. The initial hypocapnia resulting from the prior hyperventilation may permit a longer stay at depth but by then the P_{O_2} has reached a rather low level. On ascent, as the ambient pressure drops, the P_{O_2} is very rapidly lowered further, to levels at which unconsciousness is likely to occur. Standard advice before such a dive is that four maximum breaths are allowable.

Snorkel diving

The problem in **snorkel diving** is essentially one of lung mechanics because the person is breathing to and from the surface pressure, which is 1 atm. Alveolar pressure is therefore, on average, ~1 mmHg (p. 491) either side of atmospheric pressure. The pressure on the chest wall is greater than atmospheric by an amount that depends on the depth of submersion. This must be counteracted by the action of the inspiratory muscles if the chest is not to collapse. The maximum subatmospheric P_{pl} that can be generated is of the order of 100 mmHg (p. 495), equivalent to a submerged pressure of ~1.2 metre. If one considers that maximum respiratory muscular effort can only occur for short periods and if one also takes into account the increased dead space of the snorkel tube, it is clear that a snorkel diver is confined to a snorkel length of about 0.4 metre.

Conventional and scuba diving

Conventional diving is being submerged within a closed chamber while breathing its compressed gases. **Scuba** (self-contained underwater breathing apparatus) **diving** involves breathing gases through a regulated valve system from a pressurized tank carried by the diver. In either case the pressure of the gas breathed and the pressure on the outside of the chest are equal but the problem is that of breathing N_2 (and O_2) at pressures greater than 1 atm. N_2 breathed at a pressure greater than ~4 atm (~ 30 metres below the surface) is at such a dissolved concentration in the tissues that it results in **nitrogen narcosis**—a condition also known as rapture of the deep. The mechanism causing this narcosis is not known but the symptoms are euphoria, loss of memory, clumsiness and impaired intellectual function resembling alcohol intoxication.

When diving at depths of more than 30 metres there are dangers, not at depth but as the diver ascends. The N_2, though inert, dissolves in the body fluids (especially the fatty tissues) in proportion to its partial pressure. If the ascent is too quick, the ambient pressure decreases too rapidly and the N_2 comes out of solution and forms tiny bubbles in the tissues and blood. Such bubbles can form emboli in the cerebral, myocardial or pulmonary circulations or in the tissue of the brain itself and in the joints. The symptoms of **decompression sickness** are paralysis, myocardial damage, coughing and dyspnoea which divers call 'the chokes', and severe pain in the joints ('the bends'). Treatment is by recompression in a pressure chamber, followed some time later by a gradual decompression.

The problem of N_2 narcosis can be avoided by breathing a mixutre of **helium** and O_2, because helium is only half as soluble as N_2 so that less is dissolved in the tissues. Helium has two other advantages—it has a low density, which decreases gas-flow resistance and therefore the work of breathing, and it is a much smaller molecule and can therefore diffuse more rapidly through the tissues, which makes it less likely to come out of solution and form bubbles. Even helium is a narcotic at very great depths, producing tremors, drowsiness and decreased manual dexterity.

Breathing pure O_2

Pure O_2 should not be breathed during diving or for prolonged medical treatment because it produces **oxygen toxicity**. At only 1 atm, 100% O_2 for even 12 hours produces tracheobronchial irritation and after a few days leads to thickening of the alveolar–capillary membrane. Neonates are even more susceptible because the lung damage is more extreme (bronchopulmonary dysplasia) and the developing vascular tisue of the retina becomes opaque (retrolental fibroplasia). Pure O_2 at pressures greater than 3 atm even for short periods affects the central nervous system and causes convulsions. Thus, when diving to great depths the fraction of O_2 in the mix delivered from the tanks is decreased to prevent the occurrence of high Po_2 in the blood and tissues.

Breathing pure O_2 at 1 atm can also cause alveolar collapse (**atelectasis**) of those alveoli beyond an obstructed airway. This is because O_2 is much more soluble than N_2, so it is absorbed out of the alveolus into the blood very much more quickly than N_2. The usual presence of 80% N_2 in the alveolus, by delaying alveolar gas absorption, acts as a 'splint' supporting the alveolus.

When breathing pure O_2 at 3 atm, the resting O_2 consumption can be satisfied entirely by the O_2 in physical solution. However, the carriage of CO_2 now has to take place without the assistance of desaturation of haemoglobin (the Haldane effect), so the usual amount of CO_2 loaded into the blood from the tissues (45 ml of CO_2 per litre of blood) raises the venous Pco_2 to ~56 instead of 46 mmHg (Fig. 16.32b).

Hyperbaric O_2, provided it is at pressures less than 3 atm for less than 5 hours, is of value for correcting CO poisoning as the high Po_2 helps displace CO from the haemoglobin molecule and raises the amount of O_2 dissolved in the blood.

Chapter 17
Digestive System

17.1 STRUCTURE AND ORGANIZATION

The role of this system is to digest food and absorb nutrients and water. This requires that ingested material passes in an orderly and controlled fashion from mouth to anus. The activity of the smooth muscle responsible for gastric and intestinal motility is regulated and coordinated by the enteric nervous system. During this passage, fluids containing enzymes that break down complex molecules are secreted into the lumen of the gut. This secretion is also regulated and coordinated by the enteric nervous system. Products of this digestion are absorbed across the epithelial cells, principally in the small intestine.

All the water, nutrient, vitamins, minerals and electrolytes required for growth and maintenance are absorbed from the digestive tract and transported to other tissues by the blood stream. However, only a small portion of the ingested solid material—small lipid- and water-soluble molecules—may be absorbed unchanged. The complex macromolecules that form the bulk of the diet must first be reduced to simpler, metabolizable forms by **digestion**. This is accomplished in the lumen of the digestive tract.

The **digestive tract** is a tube extending from the mouth to the anus and includes the pharynx, oesophagus, stomach, small intestine and large intestine (Fig. 17.1). Food entering the tract is mixed and propelled along the tract as a result of skeletal and smooth muscle activity. As ingested material progresses through the tract, the complex molecular structures may be broken down by hydrolytic enzymes which act in either acidic or neutral environments. The digestive fluids are secreted from glandular tissue that forms part of the intestinal wall or from glands that lie outside the intestinal tract and whose ducts enter the digestive tract (the salivary glands, pancreas and liver).

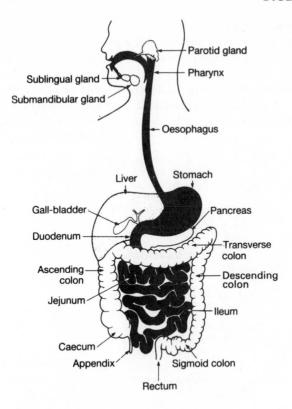

Fig. 17.1 Anatomy of the human digestive tract.

The **digestive process** starts in the mouth where the food may be ground and mixed with saliva. Here the addition of a salivary amylase to the food begins the breakdown of starches. After passage through the pharynx and oesophagus, this material enters the stomach where it is acted on by acid and gastric proteases. However, digestion in the stomach is limited and one of the main functions of the stomach is to store partially digested material (chyme) and slowly release this into the small intestine. The final phases of digestion occur in the small intestine and colon. In the small intestine the chyme is broken down by the actions of bile, pancreatic enzymes (e.g. protease, lipase, amylase) and enzymes released from desquamated epithelial cells. Most of the products of digestion are absorbed by the small intestine and the residue passes into the large intestine (colon). In the colon there is little digestion but the actions of the normal bacterial flora on dietary fibre (e.g. cellulose) result in formation of gases (e.g. methane and H_2) and in the synthesis of short-chain fatty acids and vitamins (e.g. vitamin K) which may be absorbed. In addition, the colon has some capacity to absorb water and to absorb and secrete ions. The solid waste products and undigested material are expelled as faeces.

Structure of the digestive tract

From the mouth to the anus, the digestive canal is lined by a mucous membrane (**mucosa**). In the mouth, oesophagus and anus, where there can be considerable abrasion, the epithelial layer is of a stratified squamous variety some 10–15 cells thick. In regions devoted to absorption and secretion, a single-layered columnar epithelium is found (Fig. 17.2). Absorption is enhanced in these regions by folding and by the formation of villi, which are finger-like projections of the mucosa.

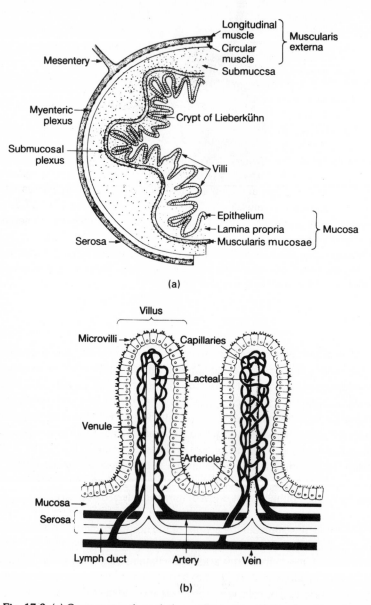

(a)

(b)

Fig. 17.2 (a) Cross-section through the small intestine. (b) Blood and lymphatic vessels supplying the villi.

Striated or smooth muscle envelopes the mucosa of the digestive tract. The striated muscle is found in the upper regions of the tract (the mouth, pharynx and upper oesophagus) and at the anus. In between, the tract is surrounded by an external layer of smooth muscle, the **muscularis externa**, composed of an outer layer with its fibres oriented in a longitudinal direction and an inner layer containing fibres with a circular orientation (Fig. 17.2). This part of the tract also contains a thinner, smooth-muscle layer, the **muscularis mucosae**, at the base of the mucous membrane and extending into the villi in the small intestine (Fig. 17.2a). The thicker outer coat is involved in the propulsion and mixing of food, while the muscularis mucosae may control the folding and shape of the mucosa.

The **blood vessels, lymphatics** and **nerves** to the intestinal tract travel in the mesenteries. The blood vessels entering the tract branch to supply the external muscle layer, the submucosa and the mucosa. The final vascular supply within these regions may vary from a simple rectilinear network in the muscle layers to a basket-like arrangement in the villi of the small intestine (Fig. 17.2b).

Intrinsic nerves in the tract between the lower oesophagus and the rectum comprise the **enteric nervous system**, in which the number (10^8) of neurons approaches that found in the spinal cord. These neurons are grouped in the **myenteric (Auerbach's) plexus** which lies between the outer longitudinal and circular muscle layer, and in the **submucosal (Meissner's) plexus** which lies between the circular muscles and the muscularis mucosae (Fig. 17.2a). These intrinsic networks are composed of excitatory and inhibitory motor nerves to the smooth muscle fibres, excitatory innervation to secretory cells and hormone-releasing cells, and regulatory nerves modifying secretion and absorption by the epithelial cells. The input of these plexuses comes from the autonomic nervous system and from intrinsic chemo- and mechanoreceptors which may have their sensory terminals in the epithelial sheet (Fig. 17.3).

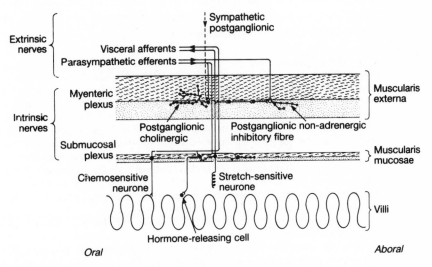

Fig. 17.3 Intrinsic and extrinsic nerves of the gastrointestinal tract.

Regulation of the digestive system

Both motor and secretory activity are regulated and coordinated by nerves and hormones. Nervous control has intrinsic and extrinsic components. Intrinsic nervous control involves local reflexes confined to the wall of the gut and effective over short distances (centimetres), whereas extrinsic control involves the central nervous system and allows regulation and coordination of gastrointestinal activity over greater distances. A variety of substances act as neurotransmitters in the enteric nervous system. Peptides secreted by cells in the epithelial layer act as hormones. Whereas the effects of nerves are sharply localized and can be rapidly initiated or terminated, the effects of hormones are slower in onset and in termination and tend to be more diffuse. Nervous control in the digestive system is found exclusively in the upper part of the system (mouth, oesophagus) and at its termination (anus) where rapid but short-lasting responses are required. Nervous and hormonal control play equally important roles in the control and regulation of gastric activity. In the intestine, motility is more dependent on neural mechanisms but, in the regulation of secretions of liver and pancreas, hormonal control is the more important.

In general, the motor and secretory activity of the gastrointestinal tract is regulated and coordinated by nerves and hormones. Nervous control can be subdivided into intrinsic and extrinsic components. **Intrinsic nervous control** involves local reflexes confined to the wall of the gut itself and effective over short distances (centimetres). Such intrinsic reflexes include local excitatory and inhibitory reflexes which act in combination to propel a bolus of food in a coordinated way along the gut—the peristaltic reflex. Receptors responsive to a variety of stimuli (e.g. stretch, pH, osmolarity, products of digestion) activate afferent fibres whose cell bodies are located in the intramural plexuses. These neurons synapse within the plexuses with efferent neurons supplying nearby smooth muscle, secretory cells or hormone-producing cells. A variety of substances have been suggested as neurotransmitters or neuromodulators in the enteric nervous system. As well as acetylcholine and noradrenaline, these include a variety of chemicals, many of them peptides (Table 17.1), and nitric oxide.

Some generalizations may be made in relation to this multiplicity of peptide neurotransmitters: peptide-containing neurons of the myenteric plexus project predominantly to the circular muscle layer; neurons of the submucosal plexus project into the mucosa and submucosa; there is often co-localization of peptides in the same neuron. Neurons containing vasoactive intestinal polypeptide/peptide histidine isoleucine (VIP/PHI) are inhibitory; neurons containing substance P or K are excitatory. Some neural peptides have no effects alone but modulate responses to other peptides. Table 17.2 summarizes some of the recognized roles of peptides as neurotransmitters. There is increasing evidence that most of the neurally induced relaxation of smooth muscle in the digestive tract that is attributed to non-adrenergic, non-cholinergic neurons (e.g. sphincter relaxation, gastric receptive relaxation and peristaltic relaxation) is caused by the neural release of nitric oxide (possibly in association with VIP).

Note that some gut tumours can produce these peptides that may then be released into the blood and have hormone-like actions. For example, some pancreatic islet cell tumours produce VIP, which stimulates pancreatic and small intestinal water and electrolyte secretion and can produce massive watery diarrhoea (p. 605).

Table 17.1 Gut peptides (after Makhlouf, G.M., Grider, J.R. & Schubert, M.L. (1989) In: Schultz, S.G. (ed.) *Handbook of Physiology*, Section 6, *the Gastrointestinal System*, pp. 123–132. The American Physiological Society, Bethesda.)

	Neurotransmitter	Endocrine	Paracrine
VIP/PHI	+		
Tachykinins (e.g. substance P, K)	+		
Bombesin	+		
Opioids	+		
Somatostatin	+	+	+
CGRP	+		
Neuropeptide	+		
Gastrin		+	
Cholecystokinin	+	+	
Secretin, GIP, enteroglucagon		+	
Motilin		+	
Neurotensin	?	+	
Pancreatic polypeptide		+	

CGRP, calcitonin gene-related peptide; GIP, gastric inhibitory peptide; PHI, peptide histidine isoleucine; VIP, vasoactive intestinal peptide.

Extrinsic nervous control involves the central nervous system and allows regulation and coordination of gastrointestinal activity over greater distances. Afferent fibres run in association with the autonomic nerves supplying the gut. They carry information from chemoreceptors and mechanoreceptors located in the gut wall. Those associated with sympathetic nerves may synapse in preverte-bral ganglia. Some of the information conveyed may reach areas within the brain. The efferent outflow conveyed by **sympathetic** or **parasympathetic** nerves from the central nervous system synapses within the intramural plexuses or secretory

Table 17.2 Functions of gut peptide neurotransmitters

Neurotransmitter	Functions
VIP/PHI	Relaxation of smooth muscle cells throughout the gut; involved in regulation of peristaltic reflex; regulation of intestinal water and ion transport
Tachykinins (e.g. substance P, K)	Contraction of smooth muscle, involved in regulation of peristaltic reflex
Bombesin	Regulation of gastrin secretion (stimulates release); contraction of smooth muscle
Opioids (e.g. met- and leu-enkephalin)	Modulation of activity of VIP and cholinergic neurons; involved in regulation of peristaltic reflex; inhibit intestinal secretion
Somatostatin	Modulation of release of VIP in myenteric plexus; involved in regulation of peristaltic reflex; regulation of intestinal water and ion transport
CGRP, neuropeptide y	No known functions

CGRP, calcitonin gene-related peptide; PHI, peptide histidine isoleucine; VIP, vasoactive intestinal polypeptide.

organs (salivary glands, liver, pancreas) directly. Examples of such extrinsic reflexes include **intestinointestinal inhibitory reflex, anointestinal inhibitory reflex** and the **gastrointestinal reflex** (see p. 568), as well as more complex coordinated behaviour (chewing, swallowing, vomiting).

In addition to nervous control, there is **hormonal regulation** of motility and secretion. A variety of peptides secreted by cells of neuroectodermal origin located within the epithelial cell layer function as hormones (Table 17.1 and p. 588). Some of these cells are exposed directly to the luminal contents and are presumed to be influenced by them; others lie deeper in the epithelial cell layer and would seem to be controlled by nerves or by other hormones. Released hormones (e.g. gastrin, secretin, cholecystokinin) diffuse into the wall of the gut and may act locally to alter gut activity (**paracrine** effect). In addition, they may enter capillaries and be transported through the circulation to exert their effects on more distant parts of the alimentary system or on other organs.

As well as these physiological peptides, bacteria in the gut lumen, cells involved in the immune system (e.g. phagocytes, mast cells, lymphocytes) and mesenchymal cells in the gut wall (e.g. endothelium, fibroblasts, smooth muscle) release chemicals that can modulate motility and secretion (p. 599).

Whereas the effects of nerves can be rapidly initiated or terminated and sharply localized, the effects of hormones are slower in onset and offset and tend to be more diffuse. Nervous control in the digestive system is found exclusively in the upper part of the system (mouth, oesophagus) and at its termination (anus) where rapid but short-lasting responses are required. Nervous and hormonal control play equally important roles in the control and regulation of gastric activity. In the intestine, motility is more dependent on neural mechanisms but, in the regulation of secretions of liver and pancreas, hormonal control is the more important.

17.2 MOTILITY OF THE DIGESTIVE TRACT

Both longitudinal movement and mixing of the luminal contents are essential if efficient digestion and absorption are to be maintained. In the upper regions of the digestive tract, where the movement of food is rapid, motility is controlled directly by the central nervous system. Where the movement is concerned mainly with mixing and slow propulsion (during digestion and absorption), the motility is regulated by a combination of mechanisms inherent in the muscle (**myogenic mechanisms**), by **neural mechanisms** and by **hormones**.

The general properties of the skeletal and smooth muscle that are responsible for the motility have been discussed in Chapter 5. The motility of the lower tract is largely under the influence of the myenteric plexus which in turn is influenced by the extrinsic nerves (parasympathetic and sympathetic) and the submucosal plexus (Fig. 17.3). Most **parasympathetic fibres** are **excitatory** and the **sympathetic fibres** are **inhibitory**.

Sympathetic inhibition appears to be mediated by presynaptic suppression of transmission at enteric cholinergic synapses. These sympathetic nerves were once believed to be the only fibres involved in the inhibitory control of motility but more recently stimulation of preganglionic parasympathetic fibres (and stretch receptors from the epithelium) has been found to inhibit smooth muscle. This effect is mediated by postganglionic non-adrenergic, non-cholinergic inhibitory fibres which release an unidentified transmitter, possibly VIP or, more likely, nitric oxide (p. 42).

Mouth and oesophagus

Chewing reduces the size of the ingested food and mixes it with saliva. The small, lubricated particles are then swallowed. Once initiated, swallowing is a reflex process coordinated in the medulla. It involves closure of the nasopharynx and glottis, to seal off the nasal passages and respiratory tract, and relaxation of the upper oesphageal sphincter. The pressure generated by swallowing is the major force propelling food through the oesophagus to the stomach. In the oesophagus a primary peristaltic wave assists in propelling solid food and, if particles remain, secondary peristaltic waves are initiated. The movement of the food bolus from the oesophagus to the stomach requires relaxation of the lower oesophageal sphincter, which is normally closed. It relaxes some 1–2 seconds after swallowing is initiated, remains relaxed for 8–9 seconds and then contracts. The resting pressure of the closed sphincter (10–15 mmHg) prevents reflux of the stomach contents.

Chewing

When the food enters the mouth, chewing movements (mastication) reduce the size of the particles and mix them with saliva; both of these actions contribute to the taste, odour and swallowing of food. The mechanical effects are accomplished by the cutting action of the incisor teeth and the crushing movements of the molar teeth.

Although the chewing movements are under voluntary control, much chewing occurs involuntarily. Chewing involves the coordinated action of the lower mandible, tongue, cheeks and palate. The lower jaw is usually held closed as a result of reflex (muscle spindle) activity but a conscious decision or the presence of food in the mouth may momentarily inhibit this reflex and initiate jaw opening. The alternate activation and inhibition of the nerves to the opening and closing muscles results in chewing. The appropriate rhythm, force and degree of occlusion are provided reflexly and as the reflex is unilateral the chewing force is exerted largely on the side containing the bolus of food.

Swallowing

Swallowing results in the movement of food from the mouth to the oesophagus (Fig. 17.4), movement down the oesophagus and movement from the oesopha-

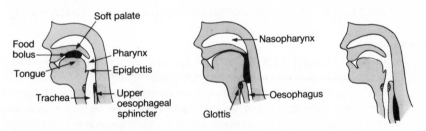

Fig. 17.4 Movement of a bolus of food from the mouth to the oesophagus.

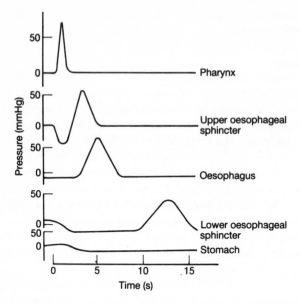

Fig. 17.5 Regional pressure changes during a normal swallowing movement. (Adapted from Davenport, H.W. (1975) *A Digest of Digestion*, p. 8. Year Book Medical Publishers, Chicago.)

gus to the stomach. Initially, the tongue forces a bolus of food to the rear of the mouth. The stimulation of receptors in the posterior wall and soft palate then results in activation of the **swallowing reflex**.

This activity is coordinated within a **swallowing centre** in the medulla oblongata and the reflex, once initiated, cannot be stopped. It involves the elevation of the **soft palate** to close the nasopharynx, the raising of the **larynx** and approximation of the vocal cords to close the **glottis**, the inhibition of breathing, and the

relaxation of the **upper oesphageal sphincter**. As the bolus enters the pharynx, the epiglottis is tilted back and this assists in the closure of the respiratory tract. Collectively these processes seal off the nasal passages and the respiratory tract. Lesions in the region of the swallowing centre which interfere with this coordinated activity may be fatal. Once the tongue has passed the food into the pharynx, the pharyngeal skeletal muscles contract and propel the bolus of food into the oesophagus. The pressure generated, which can be up to 100 mmHg above atmospheric, normally provides the major force propelling the food into the oesophagus. The pressure changes during swallowing are illustrated in Fig. 17.5.

Oesophagus

In the oesophagus, a **peristaltic wave** (a ring of contraction preceded by a region of relaxation) propels viscous material towards the stomach. Fluid material travelling under the influence of the pressure generated by swallowing and by gravity may arrive at the stomach before the contractile wave. The peristaltic wave travels at approximately 5 cm s^{-1} and takes approximately 8 seconds to travel the length of the oesophagus. The pressures generated by peristalsis in the oesophagus range from 30 to 120 mmHg. The **primary peristaltic wave**, which is initiated by the medullary swallowing centre, normally clears the oesophagus of food but if particles remain in the oesophagus **secondary peristaltic waves** may begin. These waves arise as a result of distension or irritation of the oesophagus and involve the swallowing centre. The secondary waves, which arise without any awareness, commence at the upper oesophageal sphincter and progress to the stomach.

The movement of the food bolus from the oesophagus to the stomach is accomplished after relaxation of the **lower oesophageal sphincter**. This sphincter is the terminal 4 cm of the oesophagus, is anatomically not much different from adjacent areas, and is normally closed. Pressure recordings from this region show that the sphincter relaxes some 1–2 seconds after swallowing is initiated, remains relaxed for 8–9 seconds and then contracts.

Between swallowing movements, the tone of this sphincter is maintained by slight contraction of the smooth muscle layer. Some tone is maintained in the denervated tissue, a reflection of intrinsic activity of the muscle and hormonal effects (gastrin, cholecystokinin (CCK) and motilin). The resting pressure of the closed sphincter (10–15 mmHg) prevents reflux of the stomach contents which would otherwise occur because the pressure in the stomach (5–10 mmHg) is higher than that in the thoracic oesophagus (– 5 mmHg, p. 478). If the acid contents of the stomach enter the oesophagus, the sensation of heartburn may be experienced. In the rare condition of achalasia, lower oesophageal motility is disordered as a consequence of destruction of neurons, the lower oesophageal sphincter fails to relax and passage of food from the oesophagus to the stomach is disrupted.

Stomach

The proximal region of the stomach acts as a reservoir for food, whereas the distal region is involved in mixing and breakdown of the food, a process aided by retropulsion, and the controlled release through the pyloric sphincter into the duodenum. The force of gastric contractions is regulated by neural and hormonal mechanisms. Vagal stimulation of cholinergic motor nerves, or gastrin released from the antral mucosa, stimulates contraction; non-adrenergic inhibitory nerve fibres inhibit gastric motility. By regulating gastric motility, gastric emptying is controlled very precisely so that the load presented to the duodenum at any one

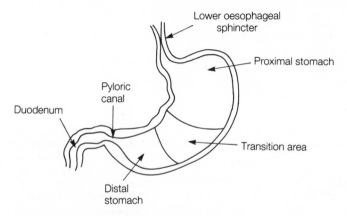

Fig. 17.6 Functional divisions of the stomach.

time is optimally matched to the digestive and absorptive capacities of the small intestine.

The stomach receives, stores, breaks down and partially digests food prior to its gradual release into the small intestine. The stomach has three outer smooth-muscle layers (the longitudinal, circular and oblique layers). Functionally, for both motility and secretion, the stomach can be divided into two distinct regions that overlap to some extent in a transition zone (Fig. 17.6). The proximal stomach is highly distensible, with a relatively thin layer of smooth muscle, has no basal smooth-muscle electrical activity, has a basal smooth-muscle tone, is lined by a secreting mucosa (see p. 575), and acts as a reservoir for food. The distal stomach is much less distensible, has basal smooth-muscle electrical activity, and displays phasic peristaltic contractions that serve to grind up food into smaller particles. The gastrin-secreting G cells are found in this portion of the stomach.

Gastric motility

As food enters the stomach, the intragastric pressure may rise but the increase is only slight due to the elasticity of the stomach wall and to the relaxation of the smooth-muscle cells. This latter is a consequence both of stimulation of stretch receptors in the oesophagus, which involves a vagovagal reflex (**receptive relaxation**), and also of a direct effect within the proximal stomach (**accommodation**). In both cases relaxation involves activation of postganglionic non-adrenergic, non-cholinergic myenteric inhibitory neurons.

The distal stomach generates continuous peristaltic waves at the rate of about 3 per minute. These originate in the transition zone and sweep towards the pyloric sphincter (Fig. 17.7). As the contractions pass over the antrum and pylorus, they progress more rapidly and become more forceful. During these contractions, the pressure in the antrum may rise to 30 mmHg or more. However, the **pyloric canal** or **sphincter**, which is normally relaxed, closes rapidly during each contraction

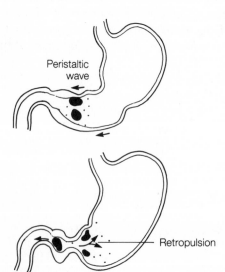

Peristaltic wave

Retropulsion

Fig. 17.7 Movement of a peristaltic wave (p.w.) into the pyloric region of the stomach, forcing a small quantity of chyme into the duodenum and the remainder back into the antrum (retropulsion).

and, at the most, no more than 1–2 ml of fluid containing small particles of food (<0.25 mm, referred to as **chyme**) enters the small intestine through the pyloric sphincter. Most of the antral contents are squirted back into the body of the stomach (Fig. 17.7). This produces a shearing effect that fragments larger food particles and is termed **retropulsion**.

During fasting quite different patterns of motility are observed. Under these conditions there is a coordinated pattern of alternating quiescence and activity that involves the stomach and small intestine, referred to as the **interdigestive motor cycle** or **migrating motor complex**. This develops 4–5 hours after a meal and recurs at about 2-hourly intervals until food is again consumed. A period of smooth-muscle inactivity is followed by a period of irregular contractions leading to a short period (5–10 minutes) of intense activity with regular and forceful contractions. In the stomach these periods involve an increase in muscle tone in the proximal stomach with superimposed phasic contractions, and phasic contractions at the rate of about 3 per minute in the distal stomach that can generate pressures exceeding 100 mmHg. In the human, these periods of activity precede, but are coordinated with, the activity patterns found in the small intestine that are initiated in the duodenum and spread to the ileocaecal junction. These coordinated bursts of smooth-muscle activity serve to empty the residual contents of the stomach, including basal gastric secretions, and refluxed duodenal contents, including biliary and pancreatic secretions, into the small intestine and to sweep the contents of the small intestine towards the colon.

In humans, with meals eaten at regular intervals during the day, the motility pattern associated with food intake occurs throughout the day and the interdigestive activity just described develops only at night.

Control of gastric motility

The contractile activity of the stomach is regulated by **myogenic, neural** and **hormonal** mechanisms.

The frequency of the contractions in the distal stomach (about 3 per minute) and their coordination depend on the spread of **slow waves** of depolarization (20–40 mV) throughout the smooth muscle layers (Fig. 17.8).

This rhythmic depolarization (**basic electrical rhythm**) is myogenic in origin.

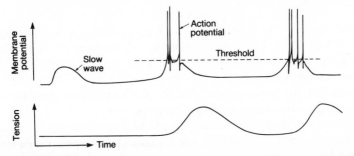

Fig. 17.8 Rhythmic depolarizations (slow waves) of the membrane potential of smooth muscle. In the small intestine, these initiate action potentials on reaching threshold, and cause the muscle to contract. (In the stomach, the slow waves are usually larger in amplitude and the spikes smaller than those illustrated here from the small intestine.)

Slow waves can be initiated in neighbouring regions by passive current spread through gap junctions. Thus the slow waves are conducted around and along the stomach wall. The slow waves of the stomach are larger than those in the small intestine and may have superimposed on them a number of small spike-like potentials (Fig. 17.8) which initiate contractions. Also, the slow wave itself may reach a potential at which contraction is initiated; the amplitude of the contraction is then dependent on the degree by, and the time for, which this threshold is exceeded. Thus, during normal digestion, it is the slow waves that determine the frequency and conduction velocity of the peristaltic contractions.

The ionic basis of the slow waves is not understood but it has been attributed to an electrogenic pump and to permeability changes in the plasma membranes of the smooth muscle.

The force of gastric contractions is regulated by neural and hormonal mechanisms. Vagal stimulation of cholinergic motor nerves, or gastrin released from the antral mucosa, lowers the membrane potential of the smooth-muscle cells. Subsequent slow waves thus lower the membrane potential further and for a longer time. The resulting potentials in the smooth-muscle cells then induce a more powerful contraction. In contrast to this, non-adrenergic inhibitory nerve fibres cause hyperpolarization, and thus inhibit gastric motility.

Gastric emptying

The rate at which the stomach empties is determined by the contents of the organ and by its motility. Liquids are emptied faster than solids, with the rate of emptying being approximately exponential. Solids empty more slowly since they must first be reduced to a small particle size.

By regulating gastric motility, gastric emptying is controlled very precisely so that the load presented to the duodenum at any one time is optimally matched to the digestive and absorptive capacities of the small intestine. The factors that regulate gastric emptying are the same as those that regulate gastric secretion—they are increased or decreased together. Essentially both are influenced by the physical and chemical nature of the gastric and duodenal contents and will be considered together later. (p. 578).

Vomiting

Vomiting results in the rapid expulsion of the stomach contents through the mouth. It is preceded by a large retrograde contraction, initiated and coordinated by central, peripheral and enteric nervous systems, that forces intestinal contents into the stomach. It involves:

1   forceful inspiratory movements when the glottis and nasopharynx are closed;
2   relaxation of the lower oesophageal sphincter; and
3   contraction of the abdominal and thoracic muscles. When the descent of the diaphragm coincides with contraction of the abdominal muscles, the elevation of abdominal pressure forces the gastric contents out through the mouth.

Integration of the vomiting sequence occurs in a **vomiting centre** in the medulla oblongata that is anatomically and functionally associated with the centres governing respiration. It is influenced by the adjacent **chemoreceptor trigger**

zone in the area postrema which may be stimulated by a variety of drugs (e.g. morphine and its derivatives) that induce vomiting, as well as being involved in the vomiting associated with radiation and with motion sickness.

Small intestine

The motility of the small intestine mixes the chyme from the stomach with the digestive juices secreted by the pancreas and liver, exposes the luminal contents to the intestinal wall across which absorption occurs, and propels the intestinal contents toward the large intestine, a process that normally takes some 4–6 hours. These functions are accomplished by three motility patterns— segmentation, pendular contractions and peristalsis. Both the intrinsic and extrinsic nerves play a role in regulating intestinal motility. Stimulation of the postganglionic cholinergic enteric nerves increases the amplitude of contractions, while stimulation of the sympathetic or non-adrenergic inhibitory nerves inhibits contractions. The extrinsic nerves are involved in reflexes such as the intestinointestinal inhibitory reflex, the anointestinal inhibitory reflex and the gastrointestinal reflex. The role of hormones in the control of intestinal motility is less clear but CCK stimulates motility.

Following a meal, most of the digestion and absorption of food takes place in the duodenum and jejunum. Thus the motility of this section serves to mix the chyme from the stomach with the digestive juices secreted by the pancreas and liver and also to expose the luminal contents to the intestinal wall across which absorption occurs. In addition, the intestinal contents are propelled toward the large intestine, a process that normally takes some 4–6 hours.

Motility of the small intestine

Following a meal, the above functions are accomplished by three motility patterns—**segmentation, pendular contractions** and **peristalsis**.

As in the stomach, the rhythmic contractile activity of the small intestine results from regular slow waves of depolarization of the smooth muscle cells (Fig. 17.8). Contraction in a segment of the small intestine is coordinated by the spread of slow-wave activity from cell to cell with consequent initiation of action potentials. The small intestine exhibits a decrease in the frequency of contractions along its length. This decrease is the result of successive portions of the intestine from the duodenum to the ileum having a lower intrinsic frequency of slow-wave genera- tion. In each portion, a pacemaker region drives the electrical activity. This is propagated to adjacent smooth-muscle cells. However, as the activity spreads, more distal portions are unable to follow this high frequency. A new pacemaker region operating at a slower rate then takes over.

Segmentation is the alternate contraction and relaxation of complete seg- ments of the small intestine and reflects the activity of the circular muscle (Fig. 17.9). As the chyme from one contracting segment is forced into adjacent relaxed areas, this motility pattern thoroughly mixes the luminal contents. The contracting segments may be short or long; the latter are more effective in displacing the contents into adjoining regions. The frequency of the segmental contractions decreases down the length of the human small intestine from about 12 to 9

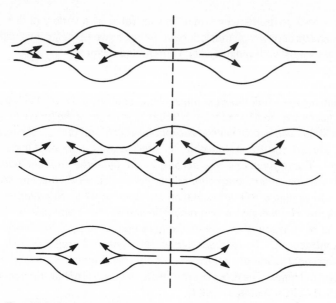

Fig. 17.9 Alternate relaxation and contraction of adjacent segments of the small intestine (segmentation) which cause thorough mixing of the contents.

contractions per minute. Although the contractile band does not progress, this pattern has some propulsive action because of the decreased frequency of segmentation from duodenum to ileum.

Pendular movements involve contraction of longitudinal muscle. This moves the wall of the intestine over the contents of the lumen and so serves to mix and propel the chyme through the small intestine.

Short **peristaltic contractions** travelling 10–15 cm are probably the main propulsive force in the small intestine. These moving, ring-like contractions are preceded by a more distal region of relaxation as they move towards the colon. They can be elicited by luminal distension.

During fasting a quite different pattern of motility, the **interdigestive motor cycle**, is observed (p. 564). In the small intestine, the short period (5–10 minutes) of intense activity involves up to 10–20 regular, forceful contractions per minute. Each of the cycles originates in the duodenum, reaches the terminal ileum in about 100 minutes and causes the intestinal contents to be propelled towards the ileocaecal sphincter. This interdigestive activity appears to play an important physiological role in that these patterns of motility effectively 'clean out' the small intestine.

Control of small intestinal motility

The motility of the small intestine depends on **myogenic, neural** and **hormonal** mechanisms.

Both the intrinsic and extrinsic nerves play a role in regulating intestinal motility. Stimulation of the postganglionic cholinergic enteric nerves increases the amplitude of contractions, while stimulation of the sympathetic or non-adrenergic inhibitory nerves inhibits contractions. Since the motility of the small intestine continues in the absence of extrinsic nerve supply, it has been argued that the neural regulation of motility is mainly intrinsic. This argument is supported by the observation that isolated segments of the small intestine increase their contractile activity in response to distension, acids, the products of digestion and placing of hypertonic solutions in the lumen.

The extrinsic nerves, however, also play a role and their importance is indicated by the presence of the following reflexes.

1   The **intestinointestinal inhibitory reflex**, in which distension of one intestinal segment causes complete intestinal inhibition.

2   The **anointestinal inhibitory reflex**, in which distension of the anus causes intestinal inhibition.

3   The **gastrointestinal reflex** in which food entering the stomach causes excitation of intestinal motility.

The role of hormones in the control of intestinal motility is less clear but **CCK** stimulates motility.

For chyme to enter the large intestine, it must pass through the **ileocaecal sphincter** which in humans is approximately 4 cm long. This sphincter at rest is closed by a pressure of some 20 mmHg. During normal movements, this pressure is reduced in association with peristaltic contractions in the terminal ileum and the chyme moves into the large intestine.

Large intestine and rectum

The large intestine stores undigested remnants prior to defecation, plays a role in salt and water balance and contains bacteria that break down cellulose and other complex carbohydrates, forming short-chain fatty acids, ammonia and gases. The progression of chyme through the large intestine is relatively slow (12–18 hours). As well as segmentation and local peristalsis, mass movements occur 2–3 times a day and drive the contents of the colon towards the rectum. The rectum is normally empty but its distension by the mass movement of faecal material from the colon induces relaxation of the smooth muscle of the internal anal sphincter and the urge to defecate. The motility of the colon is largely dependent on myogenic and neural mechanisms. Defecation is normally a voluntary act involving higher centres of the brain and also the medulla and spinal cord.

The large intestine stores undigested remnants prior to defecation, plays a role in salt and water balance and contains bacteria (largely anaerobic). These break down cellulose and other complex carbohydrates, forming short-chain fatty acids, ammonia and gases (methane and hydrogen). They also synthesize, and are an important source of, vitamin K. The progression of chyme through the large intestine is relatively slow (12–18 hours).

Colonic motility

The motility of the colon is in some ways similar to that found in the small intestine. The colon is inactive for a large proportion of the time but when contractions occur they may be either of a **segmental** or **peristaltic** nature. The segmental contractions have a slower frequency (3–4 per minute) than in the small intestine and the deep but restricted infoldings of the wall form distinct pouches called **haustra**. This movement slowly mixes the luminal contents and improves absorption of salt and water.

Propulsive activity, **mass movements**, occurs infrequently (2–3 times each day) and results in the development of relatively high pressures (80–100 mmHg) that drive part of the colonic contents towards the rectum. These contractions may travel 30 cm or more.

Defecation

The rectum is normally empty but its distension by the mass movement of faecal material from the colon induces relaxation of the smooth muscle of the **internal anal sphincter** and the urge to defecate. In paraplegics, this and the contraction of the rectum will lead to automatic defecation but normally contraction of the **external anal sphincter**, which contains skeletal muscle, allows retention of the rectal contents. When conditions permit, voluntary relaxation of the external sphincter allows defecation to proceed. The expulsion of the faecal mass may be assisted by raising the intra-abdominal pressure as a result of forced expiratory movements against a closed glottis.

Control of colonic and rectal motility

The motility of the colon is largely dependent on **myogenic** and **neural** mechanisms. Slow waves are propagated through the muscle layers but their activity has less directional coordination than that in the stomach and small intestine. Indeed, reverse slow-wave activity and contractions can be recorded in the distal colon and serve to keep the rectum empty between periods of mass peristalsis. The intrinsic nervous plexus plays an important role in the coordination and conduction of contractions as distension can induce coordinated local peristalsis. Loss of the plexus (as in Hirschsprung's and in Chagas' disease) results in marked constriction of the affected segment with pronounced colonic distension proximal to the segment. The motility of the colon is also influenced by the extrinsic nerve supply as illustrated by the **gastrocolic** and **duodenocolic** reflexes which stimulate motility after material has entered the stomach or duodenum, respectively. Mass movements may follow such reflex activity and result in the urge to defecate.

Defecation is normally a voluntary act involving higher centres of the brain and also the medulla and spinal cord. Rectal distension causes the rectal muscle to contract and the internal anal sphincter to relax. These actions involve intrinsic and extrinsic cholinergic pathways and, in the case of relaxation, also non-adrenergic inhibitory fibres. Afferent impulses from the rectum pass to the sacral cord and higher centres. Depending on the situation, the higher centres will either augment or inhibit the sacral centre. If defecation proceeds, the parasympathetic discharge along the pelvic nerves to the colon and rectum is augmented and the somatic motor activity in the pudendal nerve to the external anal sphincter ceases. This defecation reflex is facilitated by the tactile stimulation which accompanies passage of the faeces through the anus.

17.3 SECRETIONS OF THE DIGESTIVE SYSTEM

There are five major secretory tissues in the digestive system—the **salivary** glands, the stomach, the pancreas, the hepatic–biliary system and the intestine (Fig. 17.10). The first four of these have been studied extensively but rather less is known about the secretions of the intestine. Representative values for the compositions of the various secretions are given in Table 17.3.

Salivary secretions

About 1.5 litres of saliva are secreted per day, of which one-quarter comes from the submandibular glands and two-thirds from the parotids. Saliva contains mucus, water and ions, the enzymes, α-amylase (ptyalin), involved in initiating carbohydrate digestion, and a lipase (lingual lipase) which contributes to fat digestion. Water and mucin form a lubricant that moistens foods, helps swallowing and aids in speech. Water facilitates taste by partially dissolving ingestedmaterial. Saliva helps maintain the oral epithelium and teeth by preventing epithelial dehydration and inhibiting the proliferation of bacteria. Salivary gland acini generate a primary isosmotic secretion which is modified by duct cells so that saliva entering the mouth is hypo-osmotic, with less Na^+ and Cl^- but more K^+ than plasma. Increases in salivary flow reflexly result from stimulation of receptors in the mouth, pharynx and oesophagus via salivatory centres in the pons and medulla. The salivatory centres also receive impulses from higher centres of the brain. Parasympathetic stimulation causes a prolonged copious secretion of saliva accompanied by vasodilatation.

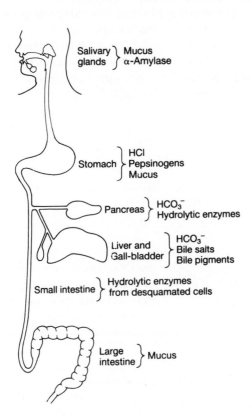

Salivary ⎱ Mucus
glands ⎰ α-Amylase

⎱ HCl
Stomach ⎰ Pepsinogens
⎰ Mucus

Pancreas ⎱ HCO_3^-
⎰ Hydrolytic enzymes

Liver and ⎱ HCO_3^-
Gall-bladder ⎰ Bile salts
⎰ Bile pigments

Small intestine ⎱ Hydrolytic enzymes
⎰ from desquamated cells

Large ⎱ Mucus
intestine ⎰

Fig. 17.10 Secretions of the digestive system. Note that secretions also contain electrolytes and water.

Table 17.3 Volumes and representative values for osmolarity, pH and ionic concentrations of plasma and gastrointestinal secretions

	Volume (litres per day)	Osmolarity (mosmol L^{-1})	pH	Na^+ (mmol L^{-1})	K^+ (mmol L^{-1})	Cl^- (mmol L^{-1})	HCO_3^- (mmol L^{-1})
Plasma	(3.0 total)	300	7.4	150	5	110	24
Saliva	1.5	100	7.5	40	15	25	30
Gastric juice	3.0	200	1.0	50	10	100	0
Pancreatic juice	1.5	300	7.8	140	10	70	80
Bile	0.5	300	7.5	140	5	80	20

Note:
1   Volumes given are average values in humans for 24-hour samples of secretion.
2   Values depend upon flow rates and therefore can vary quite widely.
3   In gastric juice HCO_3^- secreted by the surface cells reacts with H^+ and is converted to CO_2 and H_2O. Since H^+ and HCO_3^- are thereby removed from the solution, the measured osmolarity is less than the isosmolarity of the original secretions.
4   In the jejunum, pH rises from proximal to distal regions.
5   There is also an isosmotic secretion of NaCl from crypt cells in the small intestine. Normally, this may be up to 2 litres per 24 hours.

In the human, three pairs of salivary glands are found—the **parotid, submandibular** (also called the submaxillary) and **sublingual**. There are also numerous small mucus-secreting glands throughout the mouth and pharynx.

In the salivary glands, acini containing secretory cells drain into ducts that coalesce and finally open into the oral cavity (Fig. 17.11).

About 1.5 litres of saliva are secreted per day, of which one-quarter comes from the submandibular glands and two-thirds from the parotids. Secretion occurs at a continuous basal rate of 0.3 ml min^{-1} but, when stimulated by acidic foods, such as lemon juice, it can reach a maximum flow of 4–5 ml min^{-1}.

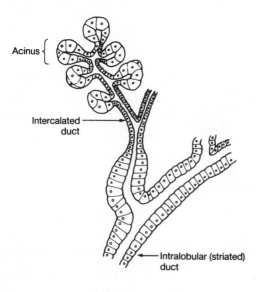

Acinus {

Intercalated duct

Intralobular (striated) duct

Fig. 17.11 Structure of a salivary gland. (Adapted from Leeson, C.R. (1967) In: Code, C.F. (ed.) *Handbook of Physiology*, Section 6, *Alimentary Canal*, vol. 2, pp. 463–495. Williams & Wilkins, Baltimore.)

Saliva contains a **mucous** secretion, mainly from the sublingual and submandibular glands, and a **serous** secretion of water and ions, mainly from the parotid and submandibular glands. It also contains the enzymes, α-**amylase** (ptyalin), secreted by the parotid gland and involved in initiating carbohydrate digestion, and a lipase (**lingual lipase**) secreted by glands within the mucosa of the tongue which contributes to fat digestion. Other substances found in the saliva are lysozyme, immunoglobulin (IgA) and blood group antigens (in 'secretors', (p. 352).

Functions of saliva

The water, salts and protein secreted by the salivary glands serve several functions. Water and mucin form a **lubricant** that moistens foods, helps swallowing and aids in speech. In addition, the water facilitates **taste** by partially dissolving ingested material. Saliva helps in the maintenance of oral epithelium and teeth by preventing epithelial dehydration and inhibiting the proliferation of bacteria. The bactericidal effect of the enzyme lysozyme may contribute to the latter effect. Dental caries is also inhibited by salivary HCO_3^- which neutralizes residual acid (ingested or produced by oral bacteria). The digestion of carbohydrates is initiated in the mouth by the secretion of salivary α-amylase. But the rapid passage of food through the mouth and oesophagus means that little digestion of carbohydrate occurs before it reaches the stomach. The effectiveness of the amylase depends on its dispersion through the food bolus and the rapidity of its inactivation by gastric acid. Estimates of its effectiveness range from 5 to 50%.

Formation of saliva

The serous secretion entering the mouth is hypo-osmotic when compared with plasma and has less Na^+ and Cl^- but more K^+ than plasma (Table 17.1). In the acini, Cl^- secretion (Fig. 1.13) results in the formation of an isosmotic, relatively protein-free secretion with an ionic composition similar to that of plasma, although HCO_3^- (40 mmol L^{-1}) is a little higher. As this primary secretion passes through the ducts, Na^+ is actively reabsorbed, accompanied by Cl^-, and small quantities of K^+ are secreted. Since the duct epithelium has a low hydraulic conductivity, the reabsorbed ions are not accompanied by water and so a hypo-osmotic secretion is produced. The actual composition of the serous secretion varies with the flow rate; the higher the rate, the closer the ionic composition is to that of the primary secretion (Fig. 17.12). Note that $Na^+–K^+$ exchange in the ducts is stimulated by aldosterone, a hormone involved in electrolyte balance.

Control of salivary secretion

Increases in salivary flow may result from stimulation of receptors in the mouth, pharynx and oesophagus. Food in the mouth stimulates taste receptors and irritates the oral mucosa, while chewing activates a variety of mechanoreceptors, such as pressure receptors adjacent to teeth and the muscle spindles of masticatory muscles.

These increases in salivation are reflex actions controlled *solely* by nervous activity. Afferent sensory fibres carry information to **salivatory centres** in the pons and medulla of the brain while efferent autonomic fibres supply the glands.

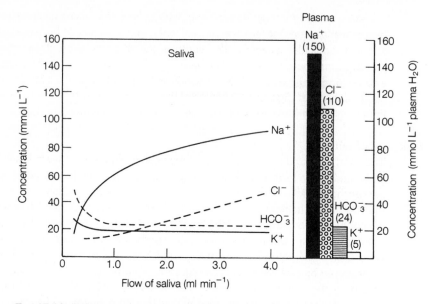

Fig. 17.12 Composition of saliva at different rates of flow. For comparison, the ionic composition of plasma is shown on the right. (Modified from Thayson, J.H., Thorn, N.A. & Schwartz, I.L. (1954) *Am. J. Physiol.* **178**, 155–159.)

The salivatory centres also receive impulses from higher centres of the brain and so the sight, smell and thought of food may also cause salivation, whereas sleep, fatigue, dehydration and fear can inhibit salivation.

In contrast to their actions in other viscera, the parasympathetic and sympathetic fibres have a similar, but not identical, effect in salivation. **Parasympathetic** stimulation, which is of major importance in the human, causes a prolonged copious secretion accompanied by vasodilatation.

The vasodilatation may be an indirect effect due to the release of an enzyme, kallikrein, which converts a plasma protein to the potent vasodilator, bradykinin. Alternatively, vasodilatation could involve VIP released from cholinergic nerves.

Acetylcholine released by parasympathetic fibres appears to initiate the secretion of electrolytes and water through a Ca^{2+}-dependent process initiated through the polyphosphoinositide system (p. 43). **Sympathetic** stimulation produces a small quantity of thick mucous saliva which is accompanied by vasoconstriction. It also increases enzyme secretion by stimulating β-adrenoceptors and elevating cyclic adenosine monophosphate (cAMP) in cells.

Gastric secretions

In total, 2–3 litres of an isosmotic gastric juice are secreted each day. This contains water and ions, mucus, HCl, the enzymes pepsin and gastric lipase, and intrinsic factor, which is essential for vitamin B_{12} absorption. HCl is necessary for the activation and optimum activity of pepsins; it denatures ingested proteins, breaking up connective tissue and cells, and plays a protective role, destroying bacteria and other potential pathogens. HCl is formed by active secretion from

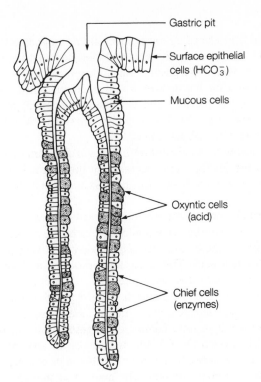

Gastric pit

Surface epithelial cells (HCO_3^-)

Mucous cells

Oxyntic cells (acid)

Chief cells (enzymes)

Fig. 17.13 Location and structure of chief and oxyntic cells in a gastric gland. (Adapted from Ito, S. & Winchester, R.J. (1963). *J. Cell Biol.* **16**, 541–577.)

stimulated oxyntic cells driven by an H^+–K^+ ATPase. Control of gastric secretion involves nerves and hormones. Acetylcholine from postganglionic parasympathetic nerves, the hormone gastrin and also histamine from the mucosa stimulate gastric secretion and motility. The control of gastric secretion is described classically in three phases—cephalic, gastric and intestinal.

Tubular glands in the mucosa of the fundus and body secrete the gastric juice containing pepsinogens, acid and mucus. **Chief cells** secreting **pepsinogens** tend to lie deep within the gland and **oxyntic (parietal) cells** secreting **acid** lie closer to the gland opening (Fig. 17.13). The glands secrete into numerous gastric pits (100 per mm^2) which drain into the stomach. The pits, the glands and the remaining mucosa contain large numbers of **mucus-secreting cells**. Surface epithelial cells lying between the pits produce an HCO_3^--rich secretion. The antral glands are essentially free of oxyntic and chief cells, but have numerous mucus-secreting cells and contain the G cells that synthesize and release gastrin into the interstitial fluid.

In total, 2–3 litres of an isosmotic gastric juice are secreted each day. At rest, the stomach secretes an HCO_3^--rich isosmotic fluid at a rate of 15–20 ml h^{-1}. But with the appropriate stimuli the stomach secretes gastric juice at rates which may exceed 150 ml h^{-1}. These stimulated secretions contain acid, pepsinogens, mucus, ions and intrinsic factor.

Functions of gastric secretions

Gastric **hydrochloric acid** is produced by the oxyntic cells as an isosmotic fluid (150 mmol L^{-1} HCl, pH 0.8). This secretion mixes with fluid within the stomach so that the final acid concentration of the gastric contents is somewhat lower. This acid environment is necessary for the activation and optimum activity of the gastric proteolytic enzymes, the pepsins, and it denatures ingested proteins, breaking up connective tissue and cells. The acid also plays a protective role, destroying bacteria and other potential pathogens.

The secreted **mucus** forms a viscid unstirred layer on the surface of the epithelial cells that traps the HCO_3^- ions secreted by these cells, and thus protects them from exposure to the acid and pepsin in the lumen of the stomach (see p. 595).

The **pepsins** have maximum proteolytic activities between pH 2 and 3 and are formed from precursor pepsinogens by the action of acid or by the autocatalytic action of pepsins. They initiate the breakdown of proteins by cleaving peptide bonds adjacent to aromatic amino acids. This results in the production of polypeptides of widely differing sizes. Note that since peptic digestion is incomplete, significant quantities of protein leave the stomach to be digested in the small intestine. A **gastric lipase** is also secreted and contributes to fat digestion.

There is one non-digestive but essential component of gastric juice—**intrinsic factor**. This heat-labile glycoprotein (M_r 60 000) is essential for absorption of **vitamin B$_{12}$** in the ileum. It is secreted by the oxyntic cells and its concentration rises concomitantly to the acid concentration. The amount of intrinsic factor secreted in 24 hours is sufficient to bind 50–200 µg of vitamin B$_{12}$, 10–50 times the amount of B$_{12}$ needed for the daily supply of the vitamin. The secretion of intrinsic factor is markedly depressed in gastric atrophy. This is associated with decreased body stores of vitamin B$_{12}$ and results, if untreated, in **pernicious anaemia** (p. 33).

Formation of gastric secretions

Mucus, ions and water make up the bulk of the gastric juice secreted by the fundus and antrum. The basal secretion is plasma-like but its composition varies with flow rate (Fig. 17.14); at higher rates, it is modified extensively by the secretion of acid from the fundus. Table 17.1 gives representative values for the average composition of gastric juice over a 24-hour period.

Acid (HCl) is formed by active secretion from stimulated oxyntic cells. The non-stimulated cells contain numerous intracellular tubulovesicles (Fig. 17.15a) whose membranes contain the H^+–K^+ ATPase but have only limited K^+ and Cl^- permeabilities. For this reason, the H^+–K^+ ATPase is relatively inactive within the vesicles. These vesicles lie adjacent to intracellular canaliculi that drain into the lumen of the gastric glands. Stimulation of the cells results in fusion of the membranes of the vesicles with the intracellular canaliculi (Fig. 17.15b). Thus the enzyme is inserted into the canalicular membrane. K^+ and Cl^- channels are inserted or activated at the same time but the precise mechanism responsible is not understood.

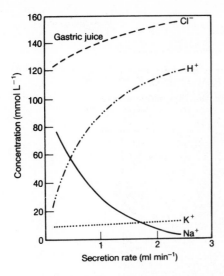

Fig. 17.14 Relationship between ionic composition of gastric juice and its rate of secretion. (After Nordgren, B. (1963) *Acta Physiol. Scand.* Suppl. 202, 1–83.)

An apical K^+ channel is required to allow some K^+ to leak from the cells into the canaliculi to provide the K^+ required for the activity of the H^+–K^+ ATPase. Any K^+ not recycled in this way contributes to the K^+ in the gastric secretions.

The H^+–K^+ ATPase moves H^+ from cell to the canalicular lumen. The OH^- remaining in the cell reacts with CO_2 to form HCO_3^-. This process is catalysed by intracellular carbonic anhydrase. The HCO_3^- passes across the basolateral membrane to the blood.

The Cl^- that accompanies H^+ into the lumen is transported as follows: Cl^- enters the cells across the basolateral membrane in exchange for HCO_3^-. Accu-

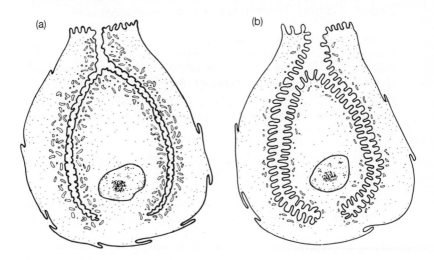

Fig. 17.15 Appearances of oxyntic cells in (a) the resting state and (b) after a secretory stimulus.

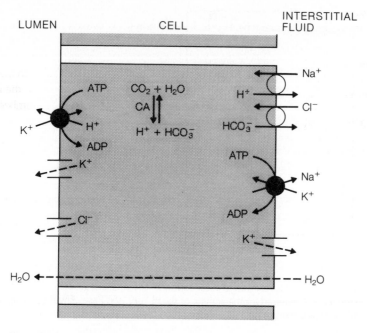

Fig. 17.16 A model of the isosmotic secretion of HCl by the oxyntic cells. CA, Carbonic anhydrase.

mulation of Cl^- within the cell provides a favourable electrochemical gradient for its diffusion across the apical membrane from cell to lumen. Thus Cl^- accompanies H^+ secretion to maintain electroneutrality and water flows down the osmotic gradient so created. The result is in an isosmotic secretion of HCl.

The basolateral membrane also contains an Na^+–H^+ exchanger. Normally, little of the H^+ produced in the cells leaves through this pathway.

Figure 17.16 illustrates one possible model for the formation and secretion of gastric acid. The high concentration gradient for H^+ between the lumen of the stomach and the plasma ($10^6 : 1$) appears to be maintained not only by H^+ transport but also by the general impermeability of the apical membranes and tight junctions of the mucosa.

Note that there is a specific inhibitor of the H^+–K^+ ATPase (omeprazole) which can be used to reduce acid secretion in patients with peptic ulcers.

Pepsinogens are synthesized in the chief cells, stored in granules and secreted by exocytosis as inactive precursors. This prevents digestion of the chief cells.

Control of gastric secretion

In general, stimulation causes an increase in secretion of both acid and enzymes and is accompanied by an increase in motility and hence gastric emptying (p. 566). Acetylcholine from postganglionic parasympathetic nerves, the hormone gastrin and also histamine from the mucosa stimulate gastric secretion and motility. These mechanisms, neural and hormonal, appear to interact. The control of gastric secretion is classically described in three phases.

1 Cephalic phase. Chewing and the presence of food in the mouth or pharynx stimulate gastric secretion. Receptors in these areas send information to the pons. Efferent parasympathetic fibres in the vagus activate secretory cells in the fundus and antrum directly via cholinergic mediation and also activate, via bombesin, the G cells in the pylorus that liberate gastrin. This hormone in turn further stimulates the secretion of gastric juice. The pontine centres of the brain are also innervated by higher centres so that the sight, smell or thought of food may increase secretion. The cephalic phase accounts for about 30% of the response to a meal.

2 Gastric phase. Stretch, products of protein digestion, and possibly Ca^{2+}, stimulate gastric secretion. They do so by stimulating receptors in the stomach which generate impulses through either the vagal or intrinsic nerves that synapse in the myenteric or submucous plexuses. These impulses stimulate gastric secretory cells directly. In addition, gastrin, released as a result of the same stimuli, increases secretion of acid and enzymes. The gastric phase accounts for about 60% of the response to a meal.

Gastrin release is inhibited by low intraluminal pH; thus gastric secretion is decreased as free acid accumulates. This is important between meals for, although the volume is small, the secretions are acidic. This ensures that gastrin release is inhibited and that the low level of secretion is maintained.

This inhibition of gastrin release is mediated via somatostatin-producing cells which are exposed directly to the luminal contents. These cells have extensive branching processes up to 40 μm long containing somatostatin granules which are released in the immediate proximity of gastrin-releasing cells when the luminal contents become acid—an example of a paracrine action. Note that vagal parasympathetic innervation of the somatostatin-producing cells inhibits release of somatostatin. Thus, the vagus enhances gastrin release by direct innervation of the G cells and by inhibiting somatostatin release.

Histamine released locally from mast cells directly beneath the epithelial layer in the stomach wall also stimulates acid secretion in this phase. Thus blockers of histamine H_2-receptors (e.g. ranitidine) in the mucosa can be used to decrease the production of gastric acid in people with peptic ulcers.

The precise interrelationship between the factors controlling gastric acid secretion and their mechanism of action at the cellular level are complex.

It seems that the basolateral membrane of the oxyntic cell has distinct receptors for acetylcholine, gastrin and histamine. Whereas histamine acts via a G_s protein and cAMP, acetylcholine acts through the polyphosphoinositide system with elevation of cellular Ca^{2+}. Gastrin also acts through Ca^{2+} elevation. In addition, prostaglandin E_2 acts via a G_i protein to inhibit cAMP production and thereby reduces acid secretion. Histamine seems to potentiate the effects of the other agents. At least some of the effects of gastrin are a consequence of its promotion of histamine release from cells within the epithelium. In addition, gastrin and acetylcholine, as well as noradrenaline, may release somatostatin from local endocrine cells which, in turn, inhibits acid secretion.

3 Intestinal phase. Chyme entering the intestine is associated normally with a decrease in gastric secretion and motility. This negative feedback thus matches the

delivery of chyme to the handling capacity of the upper small intestine and is of great physiological importance. A decreased pH, fat and hyperosmolarity in the duodenum cause the suppression of gastric activity and also inhibit gastric emptying. This is largely hormonally mediated— a variety of hormones released from the intestine, including secretin, CCK and gastric inhibitory polypeptide (GIP), have been implicated (Section 17.4).

Of less overall importance, hormones may be involved in the continuation of acid secretion, albeit at a reduced rate, during the intestinal phase. This may contribute up to 10% of the total gastric secretion following a meal.

Pancreatic exocrine secretions

The pancreas secretes about 1.5 litres of fluid each day. At rest, the secretion is plasma-like; with higher flow rates, an alkaline fluid rich in HCO_3 is secreted. This contributes to the neutralization of the acid chyme, which enters the duodenum from the stomach. With the appropriate stimulus the pancreas also secretes hydrolytic enzymes that act on all the major groups of nutrients. The pancreatic fluid is a mixture of an isosmotic Cl^- secretion from acinar cells and an isosmotic HCO_3^--rich secretion from duct cells. Although a cephalic and gastric phase can be recognized, the major control is exerted during the intestinal phase when nerves and, more importantly, intestinal hormones determine pancreatic control. Secretin stimulates secretion of HCO_3^--rich fluid from duct cells; CCK stimulates the release of an enzyme-rich secretion from acinar cells.

The pancreas consists of two portions, exocrine and endocrine, of which only the exocrine will be considered here. The digestive secretions of the pancreas are initially formed in acini that are similar to those in salivary glands. A system of ducts from the acini unites with the bile duct to form a common duct which enters the upper duodenum. The entrance of this common duct into the duodenum is controlled by the **sphincter of Oddi**.

The pancreas secretes about 1.5 litres of fluid each day. At rest (low flow rates) the secretion is plasma-like; with higher flow rates, an **alkaline fluid** rich in HCO_3^- is secreted. Representative values for its volume and composition are shown in Table 17.1. With the appropriate stimulus the pancreas will also secrete **hydrolytic enzymes** that act on all the major groups of nutrients.

Functions of pancreatic secretions

The HCO_3^--rich secretion of the pancreas and also, to an extent, of the liver contributes to the neutralization of the acid chyme, which enters the duodenum from the stomach. Thus a favourable pH is established for the pancreatic enzymes, which have optimal activities in the 6.7–9.0 pH range. A pH profile of the gastrointestinal tract is shown in Fig. 17.17.

The pancreatic enzymes or their precursors are secreted along with the aqueous secretion and include:

1 protease precursors—trypsinogen, chymotrypsinogen and procarboxypeptidases—which are secreted by the pancreas and activated in the small intestine. The activation is initiated by the conversion of trypsinogen to trypsin by the

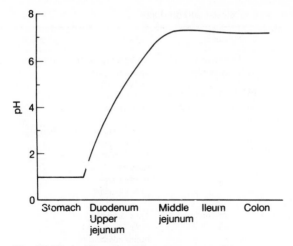

Fig. 17.17 A pH profile of the gastrointestinal tract.

intestinal enzyme, **enterokinase**, found on the apical plasma membrane of the intestinal epithelial cells. The trypsin in turn activates trypsinogen autocatalytically and also activates the other precursors;

2 an active **α-amylase** which is similar in action to the salivary α-amylase and splits α-1,4-glycosidic bonds and hydrolyses starch to mainly maltose;

3 lipases which act on triglycerides and phospholipids;

4 a number of other enzymes, e.g. **ribonuclease, elastase** and **collagenase**, that act on specific macromolecules.

The major pancreatic enzymes, their substrates, actions and products are listed in Table 17.4.

Formation of pancreatic secretions

Two groups of cells contribute to the secretion—the acinar cells and the duct cells. The basolateral membranes of the acinar cells contain receptors for both acetylcholine and for CCK. Their occupancy leads to an increase in cell Ca^{2+} via the polyphosphoinositide system (p. 43) and this in turn activates Ca^{2+}-sensitive K^+ channels in the basolateral membrane which therefore becomes hyperpolarized. Associated with these changes are exocytotic secretion of enzyme precursors and the production of an isosmotic NaCl secretion. In the human, this involves the Cl^- secretory process illustrated in Fig. 1.13.

The enzymes of the pancreatic secretions are all synthesized in the same cells and released in parallel. However, the ratio is not necessarily constant and examination of the pancreatic secretions of individuals on widely differing diets (e.g. high-protein or carbohydrates) shows that these secretions can change to suit the diet.

The cells lining the intra- and interlobular ducts produce an isosmotic HCO_3^--rich secretion which is stimulated by secretin via a cAMP-mediated mechanism. One simplified model of possible pathways is illustrated in the lower portion of Fig. 17.18.

Table 17.4 Pancreatic enzymes: their substrates and products

Precursor	Enzyme	Substrate	Action	Products
Trypsinogen	Trypsin	Protein and polypeptides	Endopeptidase that cleaves peptide bonds on the C-terminal side of arginine or lysine	Peptides
Chymotrypsinogen	Chymotrypsin	Protein and polypeptides	Endopeptidase that cleaves peptide bonds on the C-terminal side of tyrosine and phenylalanine	Peptides
Procarboxypeptidase	Carboxypeptidase	Protein and polypeptides	Exopeptidases that hydrolyse peptide bonds adjacent to C-terminal amino acids	
A	A	Protein and polypeptides	Hydrolyses nearly all C-terminal peptide bonds	Amino acids
B	B	Protein and polypeptides	Hydrolyses mainly bonds adjacent to C-terminal arginine or lysine	Amino acids
Proelastase	Elastase	Elastin and protein	Hydrolyses peptide bonds adjacent to neutral amino acids	Peptides
	Lipase	Triglycerides	Cleaves fatty acids from positions 1 and 3 of triglycerides	Free fatty acids and 2-monoglycerides
	Phospholipase A	Lecithin or cephalin	Cleaves off a fatty acid	Free fatty acid and lysolecithin or lysocephalin
	Cholesterol ester hydrolase	Cholesterol-fatty acids	Cleves off cholesterol	Cholesterol and free fatty acids
	Deoxyribonuclease	DNA	Hydrolyses DNA	Nucleotides
	Ribonucleases	RNA	Hydrolyses RNA	Nucleotides
	α-Amylase	Polysaccharides	Hydrolyses α-1, 4-glycosidic bonds	Maltose and glucose

Since the pancreatic fluid is a mixture of an isosmotic Cl^- secretion from acinar cells and an isosmotic HCO_3^--rich secretion from duct cells, the Cl^- and HCO_3^- composition of the secretion which emerges from the pancreas is not fixed but varies from about 40 mmol L^{-1} HCO_3^- and 110 mmol L^{-1} Cl^- at basal flow rates (~0.2 ml min^{-1}) to 120 mmol L^{-1} HCO_3^- and 30 mmol L^{-1} Cl^- at high flow rates (~2 ml min^{-1}).

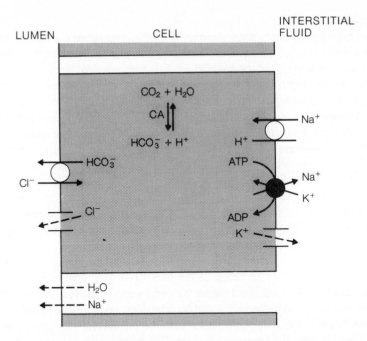

Fig. 17.18 A model of serous secretion by the duct cells of the exocrine pancreas. A relatively small volume of Cl⁻-rich secretion accompanies the enzyme-rich secretion of the acinar cells. A larger volume of HCO_3^--rich fluid is secreted by the duct cells. CA, carbonic anhydrase.

Control of pancreatic secretions

As with gastric secretion, a cephalic phase may be recognized during which pancreatic secretion of an enzyme-rich fluid is stimulated by vagal activity. This early phase of secretion is supplemented during the gastric phase of digestion by a stretch-activated vagally mediated reflex and possibly by the release of gastrin from the antrum. The major control is, however, exerted during the intestinal phase when nerves and, more importantly, intestinal hormones determine pancreatic control. The entry of chyme into the duodenum is generally followed by the secretion of an HCO_3^- and enzyme-rich fluid from the pancreas. The hormones mostly responsible for this secretion are secretin and CCK. **Secretin** (and a small quantity of CCK) is released from the mucosa of the duodenum and upper jejunum in response to acid chyme. It acts on the duct cells and stimulates secretions of the HCO_3^--rich fluid. The release of **CCK** is stimulated by the **products of digestion** in the upper small intestine. It stimulates the release of an enzyme-rich secretion from acinar cells. In addition, several peptides may inhibit secretion, including somatostatin, pancreatic polypeptide and glucagon (which inhibits duct secretion).

Finally, the mix of enzymes secreted is affected by diet. For example, pancreatic amylase and chymotrypsin levels increase when the diet is rich in carbohydrates and proteins.

Biliary secretions

Bile is a primary hepatocyte secretion to which duct secretions are added. About 0.5 litre of bile enters the duodenum each day. Bile is produced continuously but between meals is stored in the gallbladder. Bile contains two components which are important in digestion—bile salts and HCO_3^-. It also serves as an excretory route for bile pigments, cholesterol, steroids, heavy metals and a variety of drugs. Bile salts are essential for lipid digestion. The HCO_3^- assists in neutralizing the gastric acid that enters the duodenum. More than 90% of the bile salts secreted into the small intestine is reabsorbed, largely in the terminal ileum, and returned to the liver in the portal circulation. This recirculation of bile salts is economic in that the total pool (~2.5 g) is recycled as much as 6–8 times in a 24-hour period. Secretin increases the production of the HCO_3^--rich secretion from the duct cells. Jaundice, a yellow coloration of skin, sclera and mucous membranes, is detectable when plasma levels of bilirubin, a breakdown product of haem, are elevated above some 35 mmol L^{-1}.

Bile begins as a primary secretion of the **hepatic cells** (hepatocytes) into the **biliary canaliculi**. From there, it passes through a series of ducts whose secretions are added to this primary hepatic secretion. The ducts finally coalesce to form the common hepatic duct. The cystic duct from the gallbladder joins the common hepatic duct to form the common bile duct. This merges with the pancreatic duct as the two enter the wall of the duodenum. Their opening into the duodenum is controlled by the **sphincter of Oddi**.

About 0.5 litre of bile enters the duodenum each day. It is produced continuously but between meals the contraction of the sphincter of Oddi causes it to accumulate in the **gallbladder** (volume 20–50 ml). Following a meal, bile is ejected by contraction of the gallbladder and enters the duodenum after relaxation of the sphincter of Oddi. The bile contains two components which are important in digestion, namely, **bile salts** and HCO_3^-.

Functions of biliary secretions

As discussed on p. 602, the **bile salts** are important in the **digestion** and **absorption** of **fats** and in the absorption of **fat-soluble vitamins** because of their emulsifying action, particularly in the presence of lecithin and monoglycerides. The HCO_3^- secreted by both the hepatocytes and the duct cells aids in the **neutralization** of acid chyme which enters the duodenum from the stomach.

The biliary system provides an **excretory route** for bile pigments, cholesterol, steroids, heavy metals and a variety of drugs.

Formation of biliary secretions

At rest, the liver secretes a plasma-like fluid rich in HCO_3^- that contains the bile acids and bile pigments (mainly bilirubin). Bile acids (cholic and chenodeoxycholic acids in humans) are synthesized in the liver from cholesterol at a rate of 0.5 g per day and are conjugated with taurine or glycine to form bile salts. Conjugation, by lowering the pKs of the bile acids from about 7 to much more acid values, results in much of the bile acids being in the ionized (bile salt) form which is both more water-soluble and less lipid-soluble. This has two advantages: the solubility of the bile acids at the low pHs of the upper small intestine is enhanced and passive absorption from the upper small intestine in limited so that the bile salts remain within the intestinal lumen.

Of particular importance in the regulation of bile production is the recycling of bile salts. These are reabsorbed to a limited extent throughout the small intestine by passive diffusion. More importantly, in the terminal ileum they are reabsorbed by Na^+-dependent secondary active transport. Delaying a significant fraction of absorption until the terminal ileum ensures that a substantial amount of bile salts remains throughout the small intestine to promote fat absorption.

Reabsorbed bile salts are carried to the liver in the portal circulation. More than 90% of the bile salts secreted into the small intestine are reabsorbed and returned to the liver in the portal circulation, mostly bound to plasma proteins. This recirculation of bile salts is economic in that the total pool (~2.5 g) is recycled as much as 6–8 times in a 24-hour period and is referred to as the **enterohepatic circulation.**

The recycled bile salts consist not only of cholate and chenodeoxycholate but also deoxy-cholate and lithocholate that are formed in the lumen from cholate and chenodeoxycholate, respectively, by bacterial action.

The hepatocytes reaccumulate the bile salts from the portal blood by an Na^+-dependent co-transport mechanism in their basolateral membrane. The reabsorbed and also newly synthesized bile salts are then secreted across the canalicular membrane of the hepatocytes. HCO_3^- is also secreted by the hepatocytes. The tight junctions which hold adjacent hepatocytes together and form the only barrier between the plasma and the canaliculi are relatively leaky. Therefore, the secretion of the bile salts and HCO_3^- is accompanied by the movement of water and low-M_r solutes from the blood to the canaliculi and this primary secretion is isosmotic. The duct cells also secrete an HCO_3^--rich solution by a mechanism which may be similar to that in the pancreatic duct cells.

The gallbladder

The liver produces bile continuously, but bile is only required for the digestion of fats following a meal. Betwen meals the contraction of the sphincter of Oddi causes bile to accumulate in the **gallbladder** (volume 20–50 ml). Following a meal, bile is ejected by contraction of the gallbladder and enters the duodenum after relaxation of the sphincter of Oddi.

The bile salts are concentrated in the gallbladder by the active reabsorption of Na^+. In the human, Cl^- and HCO_3^- may follow passively. Water also follows passively driven by the osmotic gradient. The bile salts may be concentrated about threefold depending on the time between meals. This results in the formation of micelles which are cylindrical aggregates of bile salts that contain phospholipids and cholesterol as well as holding an appreciable number of inorganic ions. Because each micelle contains many molecules, the osmolarity of gallbladder bile is not substantially greater than that of plasma, despite the removal of the water and the concentration of the bile salts.

Bile pigments, phospholipids, principally lecithins, and cholesterol are also concentrated in the gallbladder bile. Too much cholesterol or insufficient bile salts or lecithin in bile can result in the precipitation of cholesterol and the formation of **gallstones**.

Control of biliary secretion

Biliary secretion is controlled by bile salts delivered to the liver by the portal circulation, circulating hormones and, to a small extent, by autonomic nerves. The most potent stimulators of bile salt secretion are the *bile salts themselves* which are reabsorbed in the terminal ileum and returned to the liver in the portal circulation. The greater the quantity of bile salts recycled, the smaller the production of new bile salts by the hepatocytes.

In health, about 500 mg of bile salts is synthesized each day—about 20% of the total pool. This replaces bile salts lost in the faeces. If bile salts are lost from the body, as much as 5 grams of bile salts can be synthesized in the liver each day.

Hormonal control of biliary secretion is exerted largely by **secretin**, released as a result of acid chyme in the duodenum, which increases the production of the $HCO_3{}^-$-rich secretion from the duct cells (as in the pancreas). CCK and gastrin may also stimulate the secretion of bile salts.

Nervous control plays only a minor role in regulating biliary secretion.

Activity in postganglionic parasympathetic fibres produces a small increase in secretion, whereas activity in sympathetic splanchnic nerves causes vasoconstriction and a decrease in secretion.

Bile pigments and jaundice

The breakdown of red cells in macrophages results in the degradation of the haem groups of haemoglobin with the formation of **biliverdin**. During this reaction, a molecule of CO is released. Biliverdin is reduced to **bilirubin**, which enters the blood, becomes attached to albumin and is carried to the liver. This is termed **unconjugated bilirubin**. Here, it is removed from the plasma by an anion transport system that extracts almost 100% of the load. Microsomal enzymes in the hepatocyte make bilirubin water-soluble by conjugating it with glucuronic acid, and bilirubin is secreted into the bile mainly as the diglucuronide. In the intestine, little of this **conjugated bilirubin** is absorbed. However, bacteria in the large intestine convert some of the bilirubin into a number of products. One of these, **stercobilinogen**, is excreted in the faeces as **stercobilin**. Another, **urobilinogen**, is readily absorbed from the gut, passes back to the liver and is taken up and released back into the bile. A small amount of the absorbed urobilinogen enters the systemic circulation and is excreted in the urine as **urobilin**.

Jaundice, a yellow coloration of skin, sclera and mucous membranes, is detectable when plasma levels of bilirubin are elevated above some 35 mmol L^{-1}. This can result from increased bilirubin formation (e.g. in haemolytic anaemia), decreased uptake by hepatocytes, disordered cellular conjugation, impaired secretion into the canaliculi or obstruction to biliary drainage. Increased production or failure of hepatocyte uptake results in increased unconjugated bilirubin which, being bound to albumin, is not excreted in the urine. In contrast, in duct obstruction, the level of conjugated bilirubin is increased and urinary bilirubin is increased.

Intestinal secretions

Brunner's glands in the duodenum produce a highly viscid secretion that lies on the surface of the duodenal epithelial cells trapping alkaline (HCO_3^--rich) fluid and protecting the cells from gastric acid. In the jejunum and ileum, there is a variable secretion from the intestinal crypt cells of an isosmotic NaCl solution (up to ~2 litres per 24 hours). A variety of enzymes important in the digestive process are released in the lumen from desquamated intestinal cells.

The intestinal epithelium consists of villi and crypts (Fig. 17.19). As well as mucus-secreting cells which are found throughout the intestinal tract, **Brunner's glands** in the **duodenum** produce a highly viscid secretion that lies on the surface of the duodenal epithelial cells and traps alkaline (HCO_3^--rich) fluid secreted by them. This may protect the duodenum from the acid chyme entering from the stomach. The rate of secretion is increased by parasympathetic activity and by secretin.

In the jejunum and ileum, there is also the variable secretion from the

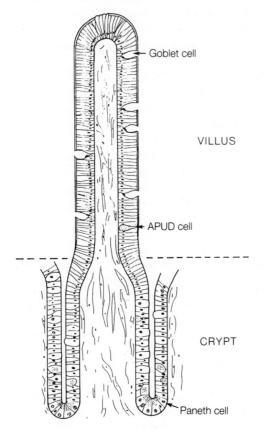

Goblet cell

VILLUS

APUD cell

CRYPT

Paneth cell

Fig. 17.19 A typical jejunal crypt and villus. At the base, the Paneth cells are thought to have some immunological role in protecting against infection from ingested organisms. The villous cells are constantly being renewed by cells derived from rapidly dividing cells in the middle of the crypt region. These migrate to the tip of the villus, a process that in the human takes about 2–3 days. A few cells along the villus become specialized to secrete hormones (APUD cells) and there are also some mucus-secreting goblet cells.

intestinal crypt cells of an isosmotic NaCl solution (up to ~2 litres per 24 hours). The cells accumulate Cl^- from the interstitial fluid across the basolateral membrane via the Na^+-K^+ $2Cl^-$ co-transporter. Secretion occurs when apical membrane Cl^- channels are activated and Cl^- then passes down its electrochemical gradient from the cells to the lumen. The leaky paracellular pathway allows Na^+ and H_2O to follow readily (Fig. 1.13).

This secretion is stimulated by a variety of peptides, one of which, VIP, acting through increased cell cAMP, may be the most relevant physiologically, and by cholinergic nerves which increase cytosolic Ca^{2+}. A variety of bacterial toxins (e.g. cholera toxin) can activate this secretory process by stimulating the G protein and increasing cAMP (p. 41), causing diarrhoea (p. 605).

Fluid collected from the small intestine, in the absence of gastric, pancreatic and biliary secretions, contains a variety of enzymes. Many of these, e.g. aminopeptidases, amylases and phosphatases, are important in the digestive process and are found in the brush-border region. They are not, however, intestinal secretions but are released from desquamated intestinal cells.

In the colon, mucus and a small volume of alkaline solution rich in HCO_3^- are produced. This may involve an $HCO_3^- - Cl^-$ exchange so that the colonic contents are higher in HCO_3^- than plasma but lower in Cl^-. Some of the HCO_3^- may be neutralized by organic acids (generated by bacterial action) in the lumen. In addition, there is an Na^+-K^+ exchange which results in the colonic contents being rich in K^+. This exchange is enhanced by aldosterone.

The control of intestinal secretion is discussed with absorption on p. 599.

17.4 GASTROINTESTINAL HORMONES

The major hormones are gastrin and CCK (which are related structurally), secretin, GIP and enteroglucagon (also related structurally) and somatostatin. Table 17.5 summarizes the properties and functions of each of the hormones.

The gastrointestinal hormones are **peptides** produced by enterochromaffin cells in the gastrointestinal mucosa and are involved in the control of gastrointestinal motility and secretion. The enterochromaffin cells are related to those endocrine cells (e.g. adrenal medulla) which are of neuroectodermal origin and are sometimes referred to as **APUD cells** (amine content, precursor uptake, ability to decarboxylate).

A number of gastrointestinal peptides are also found in central and peripheral neurons, where they may function as neurotransmitters or neuromodulators. For example, there is as much CCK in the brain as in the gut and, although its roles in the central nervous system are ill-defined, there is considerable evidence that it may play a part in the control of food intake.

Although a number of hormones have been postulated and many peptides identified in the gastrointestinal mucosa, to date only nine (Table 17.1) fulfil the criteria necessary to establish their importance as regulators of gastrointestinal function under physiological conditions.

1 The peptide should be released from a region of the gut into the blood in response to a physiological stimulus (e.g. one arising as a result of the intake of food) and produce an effect at a different region of the gut.

Table 17.5 Properties and functions of the major gastrointestinal hormones

	Released by	Effects
Gastrin	G cells in stomach and upper small intestine in response to products of digestion, extrinsic nerve stimulation, antral distension Release *inhibited by* low antral pH and somatostatin	↑Secretion of HCl, pepsinogen and intrinsic factor ↑Gastric motility Trophic effect on gastric mucosa
Cholecystokinin	I cells in upper small intestine in response to products of digestion	↑Secretion of enzyme-rich fluid from pancreatic acinar cells Contraction of gallbladder and relaxation of sphincter of Oddi ↓Gastric emptying Trophic effect on pancreatic acinar cells
Secretin	S cells in upper small intestine in response to ↓upper intestine pH, fatty acids in upper intestine	↑Secretion of $HCO_3{}^-$-rich fluid from pancreatic and hepatic duct cells ↓Gastric acid secretion Trophic effect on pancreatic acinar cells
GIP	K cells in upper small intestine in response to fat, amino acids and glucose	Releases insulin from pancreatic β cells ↓Gastric acid secretion and motility
Enteroglucagon	L cells in distal ileum and colon in response to glucose and fat in the lumen in these regions	↓Gastric and intestinal motility Trophic effect on intestinal crypt cells
Somatostatin	D cells in intestine and pancreatic islets in response to glucose, fats and bile salts in the intestinal lumen	↓Gastric acid and pepsin secretion ↓Gastrin release ↓Pancreatic enzyme secretion ↓Insulin and glucagon release Inhibits trophic effects of gastrin

GIP, gastric inhibitory peptide.

2   This effect must persist after all nervous connections between the two regions have been interrupted.

3   A substance isolated from the part of the gut to which the stimulus was applied must, when purified and injected into the blood in physiological concentrations, mimic the effects of the physiological stimulus.

4   This substance must be identified chemically.

The gastrointestinal peptides can be classified on the basis of their chemical structure into the following groups:

1   gastrin and CCK, the five amino acids at the carboxyl end of which (-Gly-Try-Met-Asp-Phe-NH$_2$) are identical. Whether peptides belonging to this group have gastrin or CCK effects at physiological concentrations is determined by a tyrosyl residue.

Such a residue at position 6 from the carboxyl end gives gastrin-like effects; at position 7 CCK-like effects are seen, provided that the tyrosyl is sulphated. At high enough concentrations, both gastrin- and CCK-like effects are produced, irrespective of the position and nature of the tyrosyl residue.

2 Secretin, GIP and enteroglucagon, which have sequences of amino acids in common. (The neurotransmitters VIP and PHI are also closely related structurally to this group.)
3 Others.

Because of their structural similarity, gastrin and CCK may interact at the same receptor site. The result will depend on the concentration of the hormones, on their affinity for the receptor and their efficacy in activating the target cell. Thus the combined effects of gastrin and CCK may be additive or, if one hormone is less effective than the other at a particular site, then its effect will be inhibitory.

The endocrine cells found in the gastrointestinal mucosa are exposed on their luminal surfaces to the contents of the gastrointestinal tract and constituents of the chyme can cause the release of their hormones into the blood stream. In addition, both intrinsic and extrinsic nervous reflexes can cause release of some of these hormones, which not only act through the blood stream but also may diffuse locally and have so-called **paracrine** actions.

These peptide hormones are removed from the circulation principally by the kidneys. They are filtered at the glomerulus and then endocytosed by the proximal tubular cells and degraded intracellularly; only their constituent amino acids are restored to the plasma. Hence the kidney effectively eliminates these hormones without excreting them in the urine. The continuous removal of freely filtered peptide hormones at rates proportional to their concentrations in the plasma ensures that their concentrations are largely determined by the rates at which they are released into the circulation.

The main actions of the individual gastrointestinal hormones on the motility and secretion of the digestive system are outlined below and summarized in Table 17.5. In addition, many of the hormones, e.g. GIP, gastrin, CCK and secretin, appear to stimulate the release of insulin. It should also be noted that the gastrointestinal hormones may have an important trophic effect on their target glands, e.g. gastrin stimulates the growth of the gastric mucosa and CCK stimulates the growth of acinar cells in the pancreas.

Gastrin

Two related gastrins, I and II, have been identified. Both are 17-amino-acid chains (G17) and differ only at position 12 depending upon whether the tyrosine residue there is sulphated (II) or not (I). As well as G17 gastrin, a larger form (G34; big gastrin) and a smaller form (G14) have been isolated from the human.

Between meals most circulating gastrin is in the G34 form. It is the G17 form that is released predominantly from the antrum following a meal.

The properties of gastrin are determined by the terminal tetrapeptide sequence (-Try-Met-Asp-Phe-NH$_2$). Pentagastrin is a synthetic product containing this peptide sequence that has all the properties of gastrin but is less potent. It is used clinically to stimulate gastric secretion, as for example in assessing inability to secrete HCl (**achlorhydria**).

The gastrins are produced by cells (G cells) which are in the mucosa lining the antral region of the stomach and the upper small intestine. Gastrin is released as a result of:

1 stimulation by products of digestion (especially peptides), caffeine and wine in the stomach, either as a direct effect on the cells or by local intrinsic nerve reflexes;

2 nervous activity in extrinsic pathways during the cephalic phase of gastric secretion. This is mediated via the vagus nerve and bombesin is the neurotransmitter involved. Local cholinergic reflexes also stimulate gastrin release; and

3 local distension of the antral region mediated by local intrinsic nerve reflexes. Release of gastrin is *inhibited* by increasing gastric acidity and by the intestinal hormones, secretin and GIP. Also, release of somatostatin from cells in the wall of the stomach, stimulated by both cholinergic and adrenergic extrinsic fibres, has a marked inhibitory effect on gastrin secretion.

Gastrin may be produced by certain tumours in the gastrointestinal tract (gastrinomas) and this can result in marked peptic ulceration (Zollinger–Ellison syndrome).

Gastrin is carried in the blood stream and **stimulates gastric secretion** of HCl, pepsinogen and intrinsic factor. It **enhances gastric motility** and increases the tone of the lower oesophageal sphincter (though the latter may not be observed at physiological concentrations). It produces a small increase in pancreatic secretion and causes contraction of the gallbladder. It also has a trophic effect, stimulating the growth of the gastric mucosa, and also duodenum and colon.

CCK

CCK has the same active terminal tetrapeptide sequence as gastrin but a total chain length of 33 amino acids, and a sulphated tyrosyl residue at position 7 from the carboxyl end. Other molecules with CCK-like action have also been isolated (e.g. CCK-8). The effects of CCK and of gastrin are qualitatively similar but differ quantitatively and may be competitive. It is produced by cells (I cells) found in the mucosa lining the duodenum and jejunum. It is released into the blood as a result of stimulation by **products of digestion** (especially fatty acids containing eight or more carbons, peptides and amino acids) in the duodenum and jejunum either as a direct effect on the cells or through local intrinsic nerve reflexes.

CCK is carried in the blood stream and stimulates enzyme-rich secretion from the acinar cells of the pancreas via the polyphosphoinositide system and Ca^{2+}. It also potentiates the effects of secretin on HCO_3^- secretion by duct cells. It causes contraction of the gallbladder and relaxation of the sphincter of Oddi. It inhibits gastric emptying and therefore delays the passage of fat, in particular, from the stomach to the duodenum. It also stimulates the growth of acinar cells in the pancreas.

Secretin

Secretin was the first hormone to be discovered (Bayliss and Starling, 1902). Secretin contains 27 amino acids; of these, 14 occupy the same relative position as in glucagon. The full molecule is required for activity. Its structure is unrelated to that of gastrin or CCK and it inhibits gastric acid secretion in a non-competitive way. Secretin is produced by cells (S cells) found in the mucosa lining the duodenum and jejunum. It is released into the blood as a result of increased acidity and by fatty acids in the duodenum and jejunum, either via a direct effect on the cells or indirectly through local nerve reflexes.

It is carried in the blood and stimulates the duct cells in the pancreas and liver, via cAMP, to produce increased volumes of HCO_3^--rich secretions. This effect is potentiated by CCK and by the neurotransmitter acetylcholine, both of which act through the polyphosphoinositide system and Ca^{2+}. In the stomach, secretin inhibits acid secretion by direct action on oxyntic cells and by inhibition of gastrin release. It also stimulates pepsinogen secretion. Like CCK, it stimulates the growth of acinar cells in the pancreas.

GIP

GIP, also called glucose-dependent, insulin-releasing peptide, is a 43-amino-acid polypeptide with some structural similarity to secretin. GIP is produced by cells (K cells) found in the mucosa lining the duodenum and upper intestine. Its release into the blood is stimulated by the presence of fat, amino acids and glucose in the upper intestine.

The major physiological effect of this hormone is to release **insulin** from β cells of the pancreas. It also inhibits gastric acid secretion (by a direct action on the oxyntic cells and by inhibition of gastrin release) and gastric motility. GIP may also stimulate secretion from the small bowel.

Enteroglucagon

Enteroglucagon, although structurally similar to glucagon, has quite distinct effects. It is produced by cells (L cells) in distal ileum and colon. Its release is stimulated by the presence of glucose and fats in the ileal lumen. Since most of these substances have normally been absorbed before the ileum is reached, blood concentrations of enteroglucagon are not normally high. It inhibits gastric and intestinal motility and has a trophic effect on intestinal crypt cells.

Somatostatin

Somatostatin, first identified as a hypothalamic inhibitor of the release of growth hormone from the anterior pituitary (p. 274), is produced by D cells in the intestinal epithelium and in the pancreatic islets as a 116-amino-acid precursor and is released either as a 14- or 28-residue peptide. Its release is stimulated by the presence of fats, glucose and bile salts in the intestinal lumen. It inhibits:

1   gastric acid and pepsin secretion and gastrin release;
2   pancreatic enzyme secretion;
3   acetylcholine release from enteric neurons and thus the intestinal responses to cholinergic stimulation;
4   insulin and glucagon release; and
5   the trophic effect of gastrin on the gastric mucosa. Somatostatin is also an important paracrine agent and neurotransmitter.

Other gastrointestinal hormones

Motilin is a 22-amino-acid polypeptide produced by M cells found in the mucosa lining the duodenum. Its release into the blood is stimulated by the presence of fat and acid in the duodenum and is inhibited by somatostatin. It appears to increase the responsiveness of gut smooth muscle to intrinsic neural activity and thus stimulates gastric emptying and also intestinal activity in the interdigestive motor cycle.

Neurotensin is a 13-amino-acid polypeptide produced by N cells found in the mucosa lining the ileum. Its release into the blood is stimulated by the presence of fat and it may be involved in the regulation of gastric motility.

Pancreatic polypeptide is a 36-amino acid peptide produced by F cells found in the pancreatic islets. Its release into the blood is stimulated by protein meals and it inhibits pancreatic enzyme and bicarbonate secretion.

As well as the hormones, other chemicals play a role in regulating and coordinating gastrointestinal function. For example, the **prostaglandins** are synthesized, released and degraded in the gastrointestinal tract and when ingested or injected can influence motility and secretion, and **histamine** has important effects on gastric acid secretion and in linking the immune system and the enteric nervous system (p. 606).

17.5 ABSORPTION

Absorption is the net passage of a substance from the lumen of the gut across the epithelium to the interstitial fluid. It is determined by the available surface area and the flux of molecules across the epithelium. The latter is affected by concentration and therefore by rate of digestion and also by rate of removal of absorbed substances in blood or lymph.

Absorption may be defined as the net passage of a substance from the lumen of the gut across the epithelium to the interstitial fluid. Two important factors influencing absorption are the available surface area and the flux of molecules across the epithelium (amount per minute per unit area), i.e.

Amount absorbed/minute = area × flux

In the gut, the surface area available for absorption is greatly expanded by loops or coils of the gut, by mucosal infoldings (Fig. 17.2a) which increase the surface area of the loops, by villi (Fig. 17.2b) and by microvilli which form the brush borders of the luminal surface of the cells. In an adult, the total surface area available for absorption in the gut is approximately 400 m^2, which is equivalent to 200 times the external surface area of the body!

The flux of molecules across the epithelium depends on their ability to penetrate the epithelial membranes and also on the driving force for their transport, i.e. their electrochemical potential gradient. The ability of a substance to cross the epithelium depends on its lipid solubility, its size, and on the presence of specific carrier molecules in the membrane (p. 10). Carrier-mediated transport maybe downhill, i.e. facilitated diffusion, or uphill, i.e. primary active transport, or secondary active transport (co-transport or counter-transport), as discussed on p. 10. Furthermore, the leakier the paracellular pathway. between the epithelial cells, the more readily organic solutes of small relative molecular mass, ions and water can cross the epithelium.

Both the number and types of carrier molecules in the luminal membrane can alter. For example, changing from a low- to a high-carbohydrate diet will result in an increase in the luminal membrane Na^+–glucose co-transporters in the jejunum, and an increase in energy needs (e.g. during pregnancy and lactation) results in an increase in all of the nutrient jejunal transporters.

The concentration of many substances in the lumen will be determined by their rate of digestion. Absorption will also depend on the time in contact with the absorptive surfaces, so control of motility is essential to allow normal absorption to take place. The concentration of the products of digestion in the interstitial fluid will depend on their rate of removal by the blood, which will depend on the blood flow. The blood vessels in the villi are arranged in loops (Fig. 17.2b) so that the distance for diffusion in the interstitial fluid from the basolateral epithelial cell membrane to the capillaries is minimized. The volume of blood flowing to the gastrointestinal tract represents 20–30% of the cardiac output at rest. The rate of capillary blood flow is a thousand times greater than that of the lymphatic flow. Thus substances such as glucose, amino acids, ions and water, which readily enter the capillaries, are removed from them and pass through the portal circulation to the liver. Only those substances, chylomicrons (small lipid aggregations; p. 602), which do not readily cross the capillary basement membrane are transported predominantly by the lymphatic circulation. Such substances enter the venous blood via the thoracic duct and are thus not exclusively transported directly to the liver after absorption. There is some recent evidence which suggests that the hydrostatic pressure in the capillaries in the villi is less than the plasma colloid osmotic pressure. If this is so, it indicates that the driving force for fluid movements in the villi always favours net absorption from the interstitial fluid to the plasma and that ultrafiltration does not normally occur across the capillaries. Therefore, absorption of water and solutes from the gut lumen will be favoured.

Absorption in the mouth

Normally, little of the constituents of a meal are absorbed in the mouth. However, some lipid-soluble drugs can be administered by placing them either under the tongue or next to the cheek.

Absorption in the stomach

The stomach has a tight epithelial lining, and relatively little absorption normally occurs from this organ. If the tight junctions are disrupted, leakage of acid and enzymes into the interstitial fluid can result in damage to the epithelium and underlying tissues.

In the stomach, the H^+ concentration is up to one million times greater than that in the plasma, while the Na^+ concentration can be as low as one-tenth of that in the plasma. This implies the existence of a gastric mucosal barrier in which both the apical plasma membranes of the epithelial cells and the tight junctions between these cells are very impermeable to ions. It is not surprising, therefore, that little absorption normally occurs in the stomach. However, lipid-soluble substances, e.g. alcohol (ethanol) and organic acids such as acetylsalicylic acid (aspirin) in their non-ionized form, are absorbed to some extent across the stomach wall, although never as fast as from the small intestine.

A number of agents can disrupt the gastric mucosal barrier. These include bile salts, short-chain fatty acids, acetylsalicylic acid, ethanol and perhaps corticosteroids. This has important clinical implications, for the entry of H^+ into the interstitial fluid acidifies the fluid and leads to damage to the epithelial cells and to the underlying capillaries, resulting in local haemorrhage.

Absorption in the intestine

The major site for absorption is the small intestine—a leaky epithelium. A variety of secondary active transporters in the luminal membrane enable sugars, amino acids, inorganic phosphate and some vitamins to enter the cell coupled to Na^+. About 90% of Na^+ absorption occurs in the small intestine, with the remaining 10% absorbed in the large intestine. Water absorption by osmosis follows the removal of solutes. Thus some 90% of water is absorbed in the small intestine. Normally salt and water absorption and secretion are coordinated so that little salt or water loss occurs in the faeces. There is both nervous and hormonal control of these processes. Calcium and inorganic phosphate (Pi) absorption are both stimulated by 1,25-dihydroxycholecalciferol (1,25-DHCC). Iron is unique in that absorption rather than excretion is regulated to maintain body iron balance. Iron not released to the blood after entry into the intestinal mucosal cell is sequestered as ferritin and is ultimately lost from the body when the mucosal cell is shed from the villus tip. Lipid absorption requires emulsification, digestion and micelle formation; diffusion from lumen to cell; esterification and synthesis of chylomicrons within the epithelial cells; passage from cell to interstitium by exocytosis; and removal from the gut interstitium by the lymphatic system.

The products of digestion are absorbed predominantly in the small intestine (Fig. 17.20). Absorption of salts and water occurs in the small intestine and also, to a more limited extent, in the large intestine. We shall deal first with the absorption of electrolytes and water and then with the absorption of minerals, sugars, proteins, lipids, vitamins and gases.

Table 17.6 gives typical volumes per day and ion concentrations found at the end of the different segments of the intestine.

Na^+

Under normal conditions, up to 1 mole of Na^+ (equivalent to 58 grams of NaCl) is absorbed per day. Only about one-sixth of this is derived from the diet, the remaining five-sixths having been secreted into the gastrointestinal tract. Since total exchangeable body Na^+ is approximately 3 moles, the importance of Na^+ absorption is obvious, as is the rapid depletion of extracellular fluid volume if Na^+ is lost from the gut in severe vomiting or diarrhoea. Approximately 5–10 mmol of Na^+ is excreted in the faeces daily.

About 90% of Na^+ absorption occurs in the small intestine; the remaining

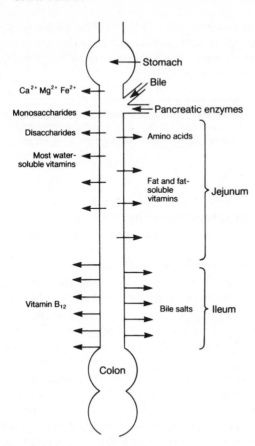

Fig. 17.20 Absorption in the small intestine. (After Booth, C.C. (1968) In: Code, C.F. (ed.) *Handbook of Physiology*, Section 6, *Alimentary Canal*, vol. 3, pp. 1513–1527. Williams & Wilkins, Baltimore.)

10% occurs in the large intestine. Absorption is thought to occur across mature cells in the villi.

In the small intestine, Na^+ absorption involves predominantly secondary active transport of Na^+ from the lumen to the cell and primary active transport from the cell to the interstitial fluid (Fig. 17.21). Thus Na^+ absorption in this segment is increased by glucose and amino acids in the lumen.

This coupling of Na^+ and organic solute movements has important therapeutic implications in the management of cholera and other secretory diarrhoeas. In such cases, the increased net Na^+, Cl^- and H_2O secretion can be offset by stimulating Na^+ absorption with glucose taken orally.

Note that in the small intestine, in particular, the paracellular pathway is leaky and therefore considerable passive movements also occur between the gut lumen and the interstitial fluid. These movements contribute to the rapid equilibration of luminal contents with interstitial fluid. Thus *luminal fluid in the small intestine, whatever its initial osmolarity, quickly becomes isosmotic with plasma.*

Na^+ absorption from the large intestine, particularly the distal colon and rectum, is dominated by Na^+ entry from the lumen through selective channels,

Table 17.6 Volumes per day and representative values for ionic concentrations of intestinal fluids at the end of each segment

At the end of the:	Volume (litres per day)	Na$^+$ (mmol L^{-1})	K$^+$ (mmol L^{-1})	Cl$^-$ (mmol L^{-1})	HCO$_3^-$ (mmol L^{-1})
Duodenum	9.0	60	15	60	15
Jejunum	3.0	140	6	100	30
Ileum	1.0	140	8	60	70
Colon	0.1	40	90	15	30

Note:
1 Volume leaving the duodenum includes ingested water. Therefore, fluid is somewhat hypo-osmotic at this stage.
2 Fluid has become isosmotic by the end of the jejunum.
3 In the ileum Cl$^-$ is reabsorbed in exchange for HCO$_3^-$.
4 In the colon, K$^+$ is secreted in exchange for Na$^+$, and HCO$_3^-$ is secreted in exchange for Cl$^-$. However, the buffering of H$^+$ from short-chain organic acids produced by bacterial action consumes some of this HCO$_3^-$, so that the HCO$_3^-$ concentration is lower than in the ileum. These organic anions now balance some of the charges on Na$^+$ and K$^+$ previously balanced by Cl$^-$ and HCO$_3^-$. By the end of the colon, the fluid may actually be hyperosmotic as a consequence of the production of organic solutes by bacterial action.

whose number is controlled by aldosterone. This hormone therefore stimulates the uptake of Na$^+$. This depolarizes the apical membrane and since this also contains K$^+$ channels, aldosterone also stimulates K$^+$ secretion in this segment of the gut.

K$^+$

K$^+$ is absorbed throughout the intestinal tract. In the jejunum, the mechanisms involved are not well-characterized but diffusion through the paracellular pathway may be the prime route. In the ileum and colon, there is both active K$^+$ movement through the cells and passive movement through a paracellular pathway. In the large intestine, K$^+$ is also secreted. This secretion is stimulated by Na$^+$ in the lumen and is increased by aldosterone. The consequence of large intestinal K$^+$ secretion is that the faeces normally contain more K$^+$ than Na$^+$ (Table 17.6).

Cl$^-$ and HCO$_3^-$

Cl$^-$ is absorbed from the intestine; HCO$_3^-$ is secreted. Absorption of Cl$^-$ may be passive, with Cl$^-$ driven through the paracellular pathway by the transepithelial electrical gradient generated by Na$^+$ absorption. Alternatively, it may involve secondary active transport across the luminal membrane (co-transport with Na or countertransport with cell HCO$_3^-$ which is thereby secreted) followed by exit across the basolateral membrane. The pathways involved in this Cl$^-$ loss from the cells remain to be defined but include specific Cl$^-$ channels (Fig. 17.21).

In the jejunum, all of these absorptive processes occur. The secondary active transport mechanisms seem to be confined to the villus cells. In contrast, the crypt cells secrete Cl$^-$ by the process illustrated in Fig. 1.13.

In the ileum and in the large intestine, the dominant pathway for Cl$^-$ absorption appears to be countertransport across the luminal membrane in exchange for cell HCO$_3^-$, which is thereby secreted. The secreted HCO$_3^-$ plays

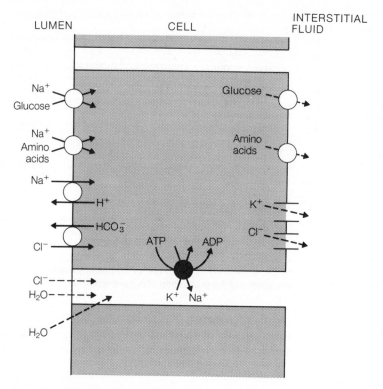

Fig. 17.21 A model of transport mechanisms and of isosmotic fluid absorption in small intestinal villous epithelial cells.

a role in the large intestine in buffering H^+ ions liberated from the short-chain organic acids generated by bacterial action on complex carbohydrates. Thus, although HCO_3^- secretion continues in the large intestine, the HCO_3^- concentration in the luminal fluid at the end of the colon is actually lower than at the end of the ileum (Table 17.6).

Water

Of the approximately 10 litres of water absorbed per day, about 8 litres (20% of total body water) have been secreted into the gastrointestinal tract (Fig. 17.22).

Approximately 90% of the water absorption occurs in the small intestine, with only about 1 litre absorbed in the colon, leaving 0.1 litre to be excreted in the faeces. As much as 2 litres of water can be absorbed by the colon per day.

There are rapid movements of water across the small intestine in response to osmotic gradients. Ions also move freely through the paracellular pathway. Thus the luminal contents rapidly attain the osmolarity of the interstitial fluid. Throughout the intestine, water absorption from an isosmotic luminal solution is secondary to solute absorption and is driven by the resulting osmotic gradient (Fig. 17.21). The coupling between solute and water movements probably occurs within the lateral intercellular spaces.

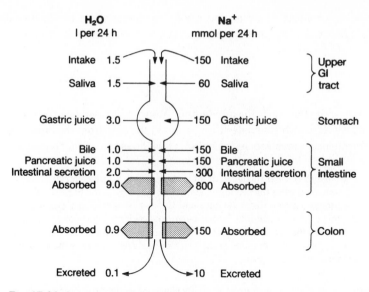

Fig. 17.22 Approximate H_2O and Na^+ movements across the gut over a 24-hour period.

Since the water permeability of the intestine is so high, a solute concentration in the lateral intercellular spaces only a little higher than in the gut lumen is sufficient to account for the water absorption.

Movement of this absorbed water from the interstitial fluid to the capillaries is driven by the plasma colloid osmotic pressure as expected. Changes in the capillary hydrostatic or plasma colloid osmotic pressure may influence net water movement between blood and lumen.

In the large intestine, water permeability is lower and osmotic gradients may be generated between the lumen and interstitial fluid. For example, because bacterial action in the large intestine generates osmotically active solutes (e.g. short-chain organic acids such as acetic acid, ammonia), faeces are normally hyperosmotic to plasma.

Regulation of intestinal absorption and secretion

It is important to stress that *normally salt and water absorption and secretion are coordinated so that little salt or water loss occurs in the faeces.* There is both nervous (Table 17.2) and hormonal control of these processes. However, as well as the physiological mediators, bacteria in the gut lumen, cells involved in the immune system (e.g. phagocytes, mast cells, lymphocytes) and mesenchymal cells in the gut wall (e.g. endothelium, fibroblasts, smooth muscle) release chemicals that can modulate intestinal secretion and absorption. Amongst these are inflammatory mediators such as prostaglandins (e.g. E_2 and I_2), adenosine, histamine and serotonin. These may act directly on the epithelial cells themselves. In addition, enteric nerve and endocrine cells have receptors for the compounds and the immune cells have receptors for neurotransmitters and hormones. Therefore, there is very close coordination and interaction between these systems in the regulation of intestinal water and electrolyte transport in health and disease. Failure to match absorption to secretion results in **secretory diarrhoea** (p. 605).

Minerals

Calcium

This is absorbed largely in the **duodenum**. Absorption is regulated by the hormone **1,25-DHCC**, which stimulates Ca^{2+} entry to the cells from the lumen (p. 288). The mechanisms involved in Ca^{2+} absorption, however, are not fully understood.

Ca^{2+} entry may be carrier-mediated or involve binding to a surface receptor and endocytosis. Ca^{2+} extrusion from the cell to the interstitial fluid across the basolateral membrane is an active process, with a Ca^{2+} ATPase and $Ca^{2+}-Na^+$ exchanger possibly involved. A Ca^{2+}-binding protein, **calbindin D**, has been isolated from intestinal epithelia. Its role in Ca^{2+} absorption may be to bind cytosolic Ca^{2+} so as to keep the free Ca^{2+} activity in the cell at the very low concentrations required for normal cell function while allowing the total quantity in the cell to be high.

Absorption of Ca^{2+} is impaired at an alkaline pH and by the presence in the gut of fat and of other substances such as oxalates and phytates which form insoluble complexes with Ca^{2+}.

Phosphorus

About 1 gram of inorganic phosphate (Pi) is absorbed from the small intestine each day. This absorption involves coupled Na^+-Pi co-transport at the luminal membrane, a process that appears to be stimulated by 1,25-DHCC. How Pi leaves the cells across the basolateral membrane is not understood, but a carrier-mediated process appears to be involved.

Iron

An average daily diet provides 180–270 µmol of iron. Red meats and sea foods are rich in iron, which is mainly in the ferric state and in organic form, e.g. as haem and as ferric–protein complexes. Iron is released from the food by acid and proteolytic enzymes in the stomach and small intestine. It is reduced to the ferrous state in the presence of a low pH and reducing agents, such as ascorbic acid and sulphydryl groups of proteins.

Absorption rather than excretion of iron is regulated to maintain iron balance.

Maximal absorption of iron occurs in the duodenum and upper jejunum and only very small amounts are taken up by the stomach and ileum. There are three pathways by which iron enters the epithelial cells from the gut lumen.

1 In its ionized form it may enter cells directly through some membrane transporter. Ferrous iron is absorbed more readily than ferric iron because it is more soluble.

2 Transferrin secreted into the lumen from epithelial cells and in bile chelates iron. The complex then binds to a transferrin receptor on the apical membrane of the epithelial cells and is taken into the cell by pinocytosis.

3 Haem binds to receptors on the luminal membrane of the cells and is absorbed as such. it is released from haem within the cells.

Iron absorbed by these three mechanisms enters a common pool within the cells. Absorption varies directly with the rate of erythropoiesis and inversely with the amount of iron in the body stores, so that 20–30% of dietary iron may be absorbed in iron deficiency but less than 5% when the stores are full.

The proportion of iron absorbed from different foods varies considerably, e.g. 20–30% of iron from meat, 5% of iron from eggs. In the case of inorganic iron (e.g. tablets of ferrous sulphate), the percentage absorbed decreases with increase in the total amount ingested. Absorption of iron is limited by dietary substances which form insoluble complexes with iron, e.g. phosphate, phytate and the tannin of tea.

Some of the absorbed iron appears in the blood within seconds. The mechanisms involved in movement of iron from the cells to the interstitial fluid are not understood. As it leaves the cells, much of it is complexed to transferrin in the interstitial space and it is carried in this form in the blood. Iron not complexed in this way is removed from the portal circulation by the hepatocytes.

Iron not released to the blood after entry into the intestinal mucosal cell is sequestered as ferritin and is ultimately lost from the body when the mucosal cell is shed from the tip of the villus. *This is the main route of iron excretion.* How intestinal mucosal cells regulate the amount of iron absorbed is not understood.

Carbohydrates

These are digested and then absorbed almost entirely as monosaccharides, the final conversion from the di- to the monosaccharide involving disaccharidases in the brush border of the epithelial cells themselves. Monosaccharides (principally glucose, galactose and fructose) are largely absorbed in the duodenum and upper jejunum. Fructose absorption is not dependent on Na^+ and occurs by facilitated diffusion. The rate of absorption of glucose (and galactose which competes with glucose for the co-transport carrier) is greatly influenced by Na^+ in the lumen. As already discussed, Na^+ and glucose appear to share a common mode of entry from lumen to cell, the diffusion of Na^+ down its electrochemical gradient providing the driving force for glucose entry into the cell which may be against its concentration gradient. The absorbed monosaccharides seem not to be metabolized by the cells but leave across the basolateral membrane by a carrier-mediated process, which is not dependent upon Na^+, and then enter the capillaries.

Proteins

Before absorption takes place, the proteins are broken down to peptides or amino acids. About 50% of the digested protein comes from ingested food, 25% from proteins in the gut secretions and 25% from desquamated epithelial cells. Gastric and pancreatic enzymes hydrolyse protein to short-chain peptides (up to 6 residues long) which are further hydrolysed at the brush border to free amino acids or to di- or tripeptides, all of which can enter the intestinal epithelial cells. Absorption normally occurs in the duodenum and upper jejunum. Amino-acid absorption, like glucose absorption, is increased by luminal Na^+. There seem to be at least four specific mechanisms by which amino acids are absorbed: one for neutral amino acids, one for basic amino acids, one for acidic amino acids and one for the imino acids (proline, sarcosine). Di- and tripeptides are absorbed rapidly from the lumen as such and are hydrolysed within the epithelial cells. Amino acids leave the cell either by simple diffusion or by carrier-mediated processes. In neonates, antibodies and other proteins contained in colostrum may be absorbed in their intact form by pinocytosis.

Lipids

The dietary lipids, up to 150 g per day, are primarily triglycerides, phospholipids, cholesterol and plant sterols. The absorption of lipid is normally an efficient process, and faeces contain less than 5 grams of fat per day. Since lipids are not water-soluble, digestion and absorption of these compounds are more complicated than for carbohydrates and proteins. There are in a number of stages.

1   Emulsification, digestion and micelle formation. Large lipid droplets must be converted to small lipid droplets (0.5–1 μm in diameter) and these stabilized to prevent coalescence. This process of **emulsification** requires a shearing force provided by gastric and intestinal motility and the presence of stabilizers, of which the physiologically most important are **bile salts**. In the stomach fat digestion is initiated by **lingual lipase** and **gastric lipase**. These liberate short-, medium-chain and unsaturated fatty acids. At the normal acidic gastric pH they have little effect on long-chain fatty acids.

In the duodenum, emulsification is enhanced by the combination of bile salts and lecithin so that droplets of about 0.5–1.0 μm are formed. This gives an enormous surface area. A protein, procolipase, secreted by the pancreas, is converted to its active form, **colipase**, by trypsin in the lumen. Colipase binds to **pancreatic lipase** and also to a bile acid and allows the pancreatic lipase to act at the oil–water interface of the emulsion to convert triglycerides to free fatty acids and 2-monoglycerides. These, together with bile salts, lecithin, cholesterol and fat-soluble vitamins, form **micelles**. Micelles are small aggregations (4–6 nm in diameter) of about 20 fat molecules whose hydrocarbon chains interdigitate within a non-aqueous fluid interior and whose polar groups form a negatively charged spherical shell surrounded by cations in aqueous solution. Other major enzymes involved in lipid digestion are listed in Table 17.4.

2   Passage from lumen to cell. The micelles diffuse to the luminal cellular membrane and then free fatty acids and monoglycerides leave the micelles and diffuse across the luminal membrane. The micelles themselves do not cross the membrane. The bile salts are absorbed to some extent in the duodenum and jejunum by passive diffusion. Absorption of at least half of the secreted bile salts occurs in the terminal ileum and involves an active, carrier-mediated process.

Note that glycerol and short- and medium-chain-length fatty acids (<12 carbons) are sufficiently water-soluble that they can diffuse directly to the luminal cellular membrane and do not require micelles to transport them.

3   Cellular metabolism and synthesis of chylomicrons. The long-chain fatty acids crossing the brush borders of the epithelial cells bind to a soluble protein and are transported to the smooth endoplasmic reticulum. Here, fatty acids are re-esterified to triglycerides. Then chylomicrons are assembled in the region of the Golgi apparatus of the cells. **Chylomicrons** are complexes, about 100 nm in diameter, of triglycerides (87%), cholesterol esters (3%) and fat-soluble vitamins,

all of which are enveloped in a hydrophobic coat composed of specific apoproteins (1%), phospholipid (9%) and free cholesterol.

4 Passage from cell to interstitium. Release of the chylomicrons from the cell occurs by exocytosis across the basolateral membrane.

5 Removal by the lymphatic system. The chylomicrons enter the lymphatic lacteals in the villi, not the blood capillaries. This reflects the fact that the endothelial cells lining the lacteals, unlike those in the capillary walls, are not held together by tight junctions and thus allow large molecules to penetrate. The chylomicrons pass through the lymphatic system to the systemic circulation via the thoracic duct.

The blood transports the chylomicrons throughout the body. Capillaries in adipose tissue and muscle contain binding sites for the chylomicrons. Apoprotein CII within the chylomicron activates the enzyme lipoprotein lipase and free fatty acids and monoglycerides are released from the chylomicrons and enter the adipose and muscle cells. The chylomicron remnants are then released from the capillary wall and pass on through the circulation to enter liver cells. Thus dietary triglycerides are delivered to peripheral tissues, whereas cholesterol is delivered to the liver.

In contrast, short- to medium-chain fatty acids (<12 carbons) are absorbed from the lumen without requiring prior hydrolysis or micelle formation and tend not to be re-esterified within the epithelial cells. They pass into the capillaries where they bind to albumin, and thence into the portal circulation rather than into the lymphatics. They are thus delivered preferentially to the liver.

Vitamins

The pathway followed by vitamins during absorption from the gastrointestinal tract is to a large extent dependent on their lipid and water solubilities. The **fat-soluble** vitamins (A, D, E and K) and their precursors are incorporated into micelles from which they diffuse into the epithelial cells. Like the dietary lipids, the absorption of these vitamins is dependent on the presence of adequate bile salts.

The absorption of many **water-soluble** vitamins is facilitated by the presence of specific metabolically dependent carriers. In some cases (e.g. thiamine (B_1), ascorbic acid (C)), these have an Na^+ dependence similar to that for the uptake of sugars and amino acids. The epithelial cells may modify the substance before releasing it to the plasma. For example, **folic acid** (pteroylglutamic acid) is a vitamin present in foods mainly as a polygultamate. Folates are absorbed mainly from the duodenum and jejunum. In the cells reduction and methylation to methyltetrahydrofolate occurs and this leaves the cells and enters the portal blood. The small intestine has a large absorptive capacity for folate.

The absorption of **vitamin B_{12}** is further specialized. This vitamin is released in the stomach from complexes with food proteins by acid and proteolytic enzymes. Within the gastric lumen, vitamin B_{12} binds to glycoproteins secreted into saliva and gastric juice (known as **R proteins**). In the duodenum pancreatic enzymes digest the R proteins and vitamin B_{12} then combines on a mole-to-mole basis with a glycoprotein, **intrinsic factor**, made by the oxyntic cells. This complex resists enzymic digestion and reaches the terminal ileum where it attaches to specific receptors for intrinsic factor on the brush border of the epithelial cells. (Note that these receptors do not recognize R proteins.) Mucosal uptake of the complex by pinocytosis is facilitated by luminal Ca^{2+} or Mg^{2+} at an alkaline pH. Uptake is followed by degradation of intrinsic factor within the cells and release of the vitamin to the portal blood. Absorption of vitamin B_{12} is limited by the number of receptors available in the ileum and a normal diet supplies rather more of the vitamin than can be absorbed. In the portal blood, vitamin B_{12} is bound to a transport protein, **transcobalamin II**, which readily releases the vitamin to the bone marrow and other tissues. However, most vitamin B_{12} in the blood is tightly bound to another protein, **transcobalamin I**, which gives up little of the vitamin to the tissues.

Gases

Gases in the gut come from either air that has been swallowed or are generated in the gut itself. At any time about 100 ml of gas is present in the intestines, ~60% of which is N_2, mainly from swallowed air. Of the remainder, hydrogen and methane are generated in the colon by bacterial action. CO_2 is formed when secreted HCO_3^- neutralizes gastric acid or organic acids formed by bacterial action in the colon. Odour-generating gases (e.g. H_2S) account for less than 1% of the total gas.

Gases are absorbed from the intestine, diffusing down their gradient of partial pressure to the blood and are then exhaled. Some of the gases generated in the large intestine are expelled through the anus, the total volume per day ranging between about 400 and 2000 ml.

Dietary factors influence this amount (e.g. consumption of large quantities of beans which provide fermentable carbohydrates for colonic bacteria results in increased gas formation).

Malabsorption and diarrhoea

A variety of diseases are associated with an increased need to defecate. Although increased frequency of defecation is often referred to as diarrhoea, it is necessary to distinguish between increased frequency associated with increased bulk of faeces (**malabsorption syndrome**) and increased frequency associated with increased fluid excretion (**watery diarrhoea**).

Malabsorption syndrome

This is usually a chronic disorder that arises as a consequence of impaired digestion or absorption of food. Thus it can arise from lack of digestive enzymes (e.g. chronic pancreatic inflammation—pancreatitis); lack of bile salts (e.g. obstruction of the common bile duct) which will affect fat digestion and absorption particularly; a loss of absorptive area (e.g. massive bowel resection; diseases involving the epithelium of the small intestine such as gluten-induced enteropathy); or failure of adequate removal from the interstitial fluid of absorbed substances (e.g. obstruction or disruption of the lymphatic drainage, chronic congestive heart failure).

As well as the increased mass of faeces, weight loss and symptoms and signs related to nutritional deficiencies of minerals and vitamins will develop as the underlying disease progresses.

Watery diarrhoea

Increased frequency of defecation associated with fluid faeces is usually acute and can have a number of causes.

1 Stimulation of gut secretion results in secretory diarrhoea. This may arise from stimulation of gastric secretion (e.g. by gastrin-producing tumours), pancreatic and biliary secretion (e.g. by secretin-producing tumours) or intestinal secretion (e.g. by VIP-producing tumours and bacterial enterotoxins such as cholera toxin). Mucosal injury with increased epithelial permeability (e.g. inflammatory bowel disease) can also result in increased net fluid movement from interstitial fluid to gut lumen.

2 The presence of **non-reabsorbable solutes** in the gut lumen results in **osmotic diarrhoea**. Causes include the ingestion of laxatives such as magnesium sulphate and impaired absorption of digested carbohydrates (e.g. lactose intolerance).

3 Disorders of motility (e.g. the increased motility that can be associated with diabetic damage to the autonomic nerves supplying the intestine) are uncommon causes of watery diarrhoea. However, any disease that increases the volume of fluid in the intestinal lumen will distend the lumen and stimulate stretch receptors, thereby resulting reflexly in increased smooth-muscle activity. In addition, a specialized motor pattern called power propulsion is triggered as part of an immune response, and results in forceful and rapid propulsion of the luminal contents over long lengths of bowel. Thus watery diarrhoea is often associated with increased frequency of defecation, and cramping abdominal pain.

Whatever the cause, the composition of the diarrhoea fluid is determined by the source of the primary secretion and the time spent in more distal segments of the intestine. For example, cholera enterotoxin stimulates small intestinal crypt cells to secrete. The secreted fluid is essentially isosmotic NaCl. As this flows through the ileum and colon, some Cl^- is exchanged for cell HCO_3^- and in the colon, some Na^+ is exchanged for cell K^+. This results in a faecal fluid that, while still approximately isosmotic, has lower concentrations of Na^+ and Cl^- and higher concentrations of K^+ and HCO_3^- than did the primary secretion. These exchanges mean that a person with chronic watery diarrhoea, as well as being depleted of extracellular fluid volume (with consequent cardiovascular manifestations, if severe), can also develop K^+ depletion, hypokalaemia (p. 660) and a metabolic acidosis (p. 671).

17.6 DEFENCE MECHANISMS IN THE DIGESTIVE TRACT

Large quantities of a variety of micro-organisms, many potentially pathogenic, are ingested with food each day. Several mechanisms provide protection from these.

1 Mucus throughout the tract traps organisms.

2 Gastric acid may destroy them. For example, the bacteria causing cholera (*Vibrio cholerae*) is extremely sensitive to acid.

3 Enzymes may digest organisms.

4 Products released by organisms (e.g. enterotoxins) may **stimulate intestinal secretion** and **inhibit fluid absorption**. As a consequence there will be increased luminal fluid, and an increased flow rate through the intestine that will wash organisms from the gut. This may result in frank secretory diarrhoea.

5 Antibodies are secreted into the lumen. Of these, **IgA** provides an important defence mechanism, particularly against viruses. The IgA dimer is secreted by plasma cells found throughout the tract in the lamina propria. This dimer binds to a receptor (the secretory component) on the basolateral membrane of the epithelial cells and the complex, **secretory IgA**, is endocytosed, diffuses through the cells and is released by exocytosis across the luminal membrane. **Aggregates of lymphoid tissue** (e.g. tonsils, Peyer's patches, appendix) are also found (p. 344). It is important to stress the intimate relationships in the gut between the immune system and neural pathways. Immunoglobulin bound to mast cell receptors recognizes sensitizing antigens and results in the release of a variety of chemicals from the mast cells including histamine, 5-hydroxytryptamine, prostaglandins, leukotrienes and cytokines. Of these, histamine appears to play a pivotal role in stimulating enteric neurons and thereby initiating the increased secretion and motility that may result in watery diarrhoea (p. 605). As more is learnt about the interplay between the immune system and the enteric and systemic nervous systems, it should become possible to understand on a rational basis the complex interrelationships between psychological and physical factors in a variety of gut diseases.

Chapter 18
Kidney

The kidneys are essential for life. Normally more water and ions are ingested than the body requires. This excess intake is excreted in the urine. The kidneys therefore regulate both the volume and the composition of the body fluids. As well as the surplus water and electrolytes, the urine contains metabolic waste products (including inactivated hormones), foreign substances and their metabolic derivatives.

The kidneys also produce humoral agents—erythropoietin, active metabolites of vitamin D, renin and prostaglandins.

Each human kidney has about a million functional units—the **nephrons**—arranged in parallel (Fig. 18.1).

The renal regulation of the volume and composition of the body fluids involves each of these nephrons in three processes.

1 filtration at the glomerulus;
2 tubular reabsorption; and
3 tubular secretion (Fig. 18.2).

18.1 RENAL BLOOD FLOW (RBF)

Organization
Each functional unit of the kidney—the nephron—is supplied with blood by an

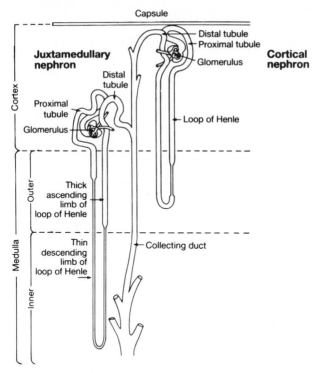

Fig. 18.1 An outline of the anatomical arrangement and organization of the nephrons within the kidney.

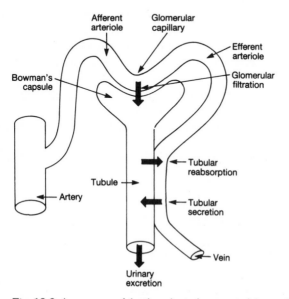

Fig. 18.2 A summary of the three basic functions of the nephron.

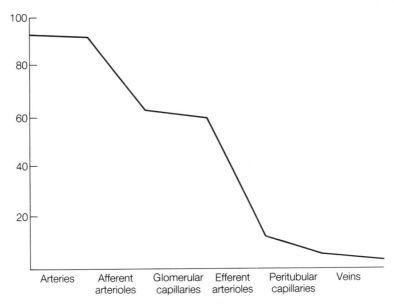

Fig. 18.3 The pressure profile of the renal vessels.

afferent arteriole that opens into a glomerular capillary bed. This drains into an efferent arteriole which supplies the peritubular capillaries and the medullary vasa recta.

Despite constituting only a small percentage of the mass of the body (< 0.5%), the kidneys receive 20–25% of the cardiac output at rest. The renal arteries arise directly from the abdominal aorta and within the kidney quickly branch into short interlobar, arcuate and then radial arteries from which arise the **afferent arterioles**. Each nephron is supplied with blood by an afferent arteriole which opens into a **glomerular capillary bed**. In turn this drains into an **efferent arteriole** which then breaks up into a second capillary bed that supplies blood to the rest of the nephron—the **peritubular capillaries**. In the juxtamedullary nephrons (p. 612) the efferent arterioles also supply the long venous loops—the **vasa recta**—that pass deep into the medulla. Blood drains into veins that finally merge to form the renal vein that enters the inferior vena cava.

Glomerular capillaries are thus unlike others in that they form a capillary bed in the course of an arteriole. This means that the hydrostatic pressure in these capillaries is determined by both afferent and efferent arteriolar resistances. This allows very precise regulation of capillary pressure and therefore of the driving force for glomerular filtration. The pressure profile of the renal vessels is illustrated in Fig. 18.3. Note that there is little pressure drop along the glomerular capillaries.

Regional variations in RBF

Total renal plasma flow (RPF) and RBF can be estimated in humans by measuring the clearance of para-aminohippurate (PAH; p. 622). But not all of the kidney is

Table 18.1 Distribution of blood flow within the kidney

	Weight (% total)	Vascular volume (ml 100 g^{-1} kidney)	Blood flow (% total)	Transit time (min)	Po_2 (mmHg)
Cortex	70	13.5	92.5	0.021	50–70
Outer medulla	20	3.8	6.5	0.086	20–30
Inner medulla	10	2.2	1.0	0.37	5–15

Note:
1 Although small in comparison with the flow in the cortex, outer medullary flow in absolute terms—approximately 30 ml 100 g^{-1} min^{-1}—is 10 times or more greater than the flow in resting muscle and about half that in brain!
2 The relatively low rate of flow in the inner medulla is important to the role of medullary circulation in preserving the osmotic gradient in the medulla and therefore in H_2O conservation by the kidneys.
3 Reflecting the low Po_2, some 70% of the inner medullary metabolism may be anaerobic.

perfused to the same extent. Experiments measuring transit times of dyes, wash-out of gases like ^{85}Kr and lodgement of microspheres in the renal vessels in experimental animals suggest that, of the total RBF, the cortex receives 90%, the outer medulla 7% and the inner medulla 1% (Table 18.1).

RPF and glomerular filtration rate (GFR) are both higher in the juxtamedullary nephrons, which appear to have a greater capacity than the cortical nephrons for salt and water reabsorption. There is evidence that when salt and water need to be conserved, a greater fraction of the RPF is delivered to the juxtamedullary nephrons; conversely, in salt- and water-loaded animals, there is a shift to the cortical nephrons. How such changes in regional RPF are mediated remains unclear.

Control of RBF

Intrinsic control—autoregulation—holds the glomerular filtration pressure relatively constant despite variations in mean arterial blood pressure over the range 80–180 mmHg. Superimposed on this is extrinsic control through the sympathetic nervous system and noradrenaline and angiotensin II. When a person is recumbent at rest, extrinsic control is minimal.

Intrinsic control

Arterial blood pressures between about 80 and 180 mmHg have little effect on RBF, which remains relatively constant over this range (Fig. 18.4).

This control of RBF, termed **autoregulation**, is independent of extrinsic nerve supply or extrinsic hormones. It is largely a consequence of afferent glomerular arteriolar resistance increasing as blood pressure increases. Autoregulation may be **myogenic**, the arterioles contracting when transmural pressure increases, as occurs in arterioles in other organs. However, **tubuloglomerular feedback** through the juxtaglomerular apparatus (p. 612) may play the more important role in this.

For example, any tendency for increased delivery of NaCl to the distal tubule will be sensed by the macula densa cells and result in increased afferent arteriolar resistance. Thus

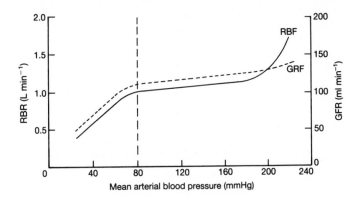

Fig. 18.4 Relationships between mean arterial blood pressure, renal blood flow (RBF) and glomerular filtration rate (GFR). Note the relative stability of both RBF and GFR within the physiological range of arterial blood pressure (autoregulation).

glomerular hydrostatic pressure will fall, GFR will decrease and the load delivered to the distal nephron will return towards normal. The precise nature of the link between the macula densa and the alterations in arteriolar resistance is unclear. Possible mediators include the renin–angiotensin system acting locally and prostaglandins.

Extrinsic control

Sympathetic vasoconstrictor activity is minimal at rest (warm, recumbent, relaxed) but increases with changes of posture, cold, pain, emotion and exercise, and so reduces RBF. The reductions associated with the erect posture and exercise are often exaggerated in patients with cardiac failure.

In addition to its direct vasoconstrictor effects, stimulation of renal sympathetic nerves causes renin release followed by increased circulating angiotensin II, which intensifies vasoconstriction, and release of prostaglandin (PGE_2), which attenuates vasoconstriction. In addition, angiotensin II stimulates release of noradrenaline from renal sympathetic nerve terminals and thereby magnifies the vasoconstriction response.

Severe reductions of arterial blood pressure, as in shock, will depress RBF and may precipitate acute renal failure. Chronic pathological processes that destroy nephrons reduce RBF.

18.2 STRUCTURE AND ORGANIZATION OF THE NEPHRON

Nephrons are tubes lined with epithelial cells supported on a basement membrane. At the start of a nephron in the renal cortex, the glomerular capillary bed is invaginated into the nephron. In this region, specialized epithelial cells—the podocytes—form Bowman's capsule, lining Bowman's space (Fig. 18.5).

Bowman's space drains into the **proximal convoluted tubule** within the renal cortex, then to the proximal straight tubule that runs from the cortex to the outer medulla. This leads within the medulla to the **thin descending** and **thin ascending limbs**, then the **thick ascending limb** of the **loop of Henle**. The thick

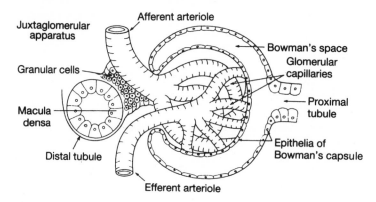

Fig. 18.5 The organization of the glomerulus. The juxtaglomerular apparatus includes modified smooth-muscle cells—granular cells—in the wall of the afferent arteriole and specialized epithelial cells—the macula densa—in the distal straight tubule.

ascending limb returns to the cortex as the **distal straight tubule** and makes contact with the afferent arteriole of its glomerulus, forming the **juxtaglomerular apparatus**. The distal tubule has several cell types and is heterogeneous in both structure and function. The distal straight tubule is followed by the **distal convoluted tubule** and then the **connecting segment** that joins with 6–8 connecting segments from other nephrons to form a **cortical collecting duct**. This runs into the medulla and becomes the **medullary collecting duct**. In the inner medulla these ducts merge, become progressively larger and finally terminate in the **ducts of Bellini** which drain into the ureter.

Nephrons are similar but not identical; the chief variants (Fig. 18.1) are:

1   superficial, **cortical nephrons**. These have short loops of Henle reaching only into the outer medullary zone and their efferent glomerular arterioles supply their peritubular capillaries. About 80% of the nephrons in the human kidney belong to this group;

2   deeper, **juxtamedullary nephrons**. These have long loops of Henle, which plunge deep into the inner medulla. Their efferent glomerular arterioles supply peritubular capillaries and also venous capillary loops (vasa recta) which course along the long loops of Henle and the collecting tubules deep in the medulla.

Note that all renal tubules receive only postglomerular blood, and that the vasa recta carry the sole blood supply to the inner medulla.

18.3 GLOMERULAR STRUCTURE AND FUNCTION

Structure

The glomerulus consists of a fenestrated capillary endothelium supported by a basement membrane on which sit the foot processes of the podocytes lining Bowman's space. The major diffusional barrier is the basement membrane; solutes of M_r below 10 000 are freely filtered. With larger molecules, diffusion is increasingly restricted a nd ceases at around 70 000 for negatively charged *albumin to 100 000 for neutral solutes. For the same hydrostatic pressure, gradient, an ultrafiltrate passes through this barrier about 100 times faster than through capillaries elsewhere.*

The energy required for filtration is supplied by the heart, *not by the kidney*. In the glomerular capillary tufts, blood at a pressure about 60% of mean aortic pressure is exposed to a filtering membrane of over 1 m² (more than half the external surface area of the body) that separates the plasma from Bowman's space (Fig. 18.5). The capillary endothelium is **fenestrated**; it lies on a basement membrane (0.2–0.3 μm thick) composed of loose fibrillar glycoproteins with fixed negative charges. The foot processes (**pedicles**) of the specialized epithelial cells (**podocytes**) contact the other surface of this basement membrane, which is synthesized by both the capillary endothelial cells and the podocytes. Interstitial contractile **mesangial cells** and the extracellular **mesangial matrix** support the glomerular capillaries and basement membrane. The mesangial cells may have a role in adjusting capillary blood flow and thereby affect glomerular filtration. Slit pores between the foot processes of investing podocytes provide a diffusion path from plasma to Bowman's space that passes between cells. Thus the major barrier to diffusion is the basement membrane. Solutes up to 10 000 M_r pass through the filter freely. For these solutes, which include the ions and metabolites of the extracellular fluid, the concentrations in the filtrate in Bowman's space equal those in the plasma (slightly modified for ions by the Gibbs–Donnan distribution). With larger molecules, diffusion is increasingly restricted and ceases around 70 000 M_r (for albumin, an elongated molecule with negative charge) to 100 000 M_r (for uncharged molecules).

The **GFR** is equivalent to the net flow of water across a membrane per unit time (J_v), given on p. 21 as

$$J_v = ALp \, (\Delta P - \Delta \pi)$$

where A is the area of the membrane available for flow, Lp is the hydraulic conductivity, which is a measure of the ease with which water flows through the membrane, and $(\Delta P - \Delta \pi)$ is the driving force, which is dependent on the differences in hydrostatic pressure (ΔP) and effective osmotic pressure $(\Delta \pi)$ across the membrane.

When applied to GFR, ΔP is the difference between the hydrostatic pressures in glomerular capillaries (P_{gc}) and in Bowman's space (P_t). Since normally only negligible amounts of protein are filtered, $\Delta \pi$ is the colloid osmotic pressure in glomerular capillaries (π_{gc}). Thus:

$$GFR = ALp \, [(P_{gc} - P_t) - \pi_{gc}]$$

Because it is difficult to estimate the area of the glomerular capillary bed with sufficient accuracy, ALp is usually represented as K_f, the ultrafiltration coefficient and

$$GFR = K_f [(P_{gc} - P_t) - \pi_{gc}]$$

Representative values for the human are P_{gc} 60 mmHg (with little drop from the afferent to the efferent end), P_t 20 mmHg and, at the afferent end of the capillary, π_{gc} 25 mmHg. However, as fluid is filtered, the concentration of plasma proteins rises and π_{gc}

approximates 35 mmHg by the end of the capillary. The average for π_{gc} is therefore ~30 mmHg. Substituting these values in the equation, the average driving force for fluid movement from capillary to Bowman's space is 9 mmHg.

Experimentally, a variety of hormones and other chemicals can alter GFR. Some do so by modifying arteriolar resistances but K_f may also be altered. One possible mechanism for this is that mesangial cell contraction is affected, leading to local, selective changes in capillary perfusion within the glomerulus.

Measurement of GFR

GFR can be measured using the relationship GFR = UV/P for any substance that is freely filtered and not reabsorbed or secreted. Inulin fulfils these criteria. Clinically the GFR is estimated using creatinine.

Assume that a solute is freely filtered at the glomerulus but is subsequently neither reabsorbed nor secreted. Let P and U be the concentrations in plasma and urine *in the same units*, and V be the volume of urine produced per minute.

Then, amount per minute in urine (UV) = amount entering Bowman's space per minute from plasma (i.e. the amount filtered per minute across the glomerulus). Since, for a freely filtered solute, concentration in Bowman's space = concentration in plasma (P), amount filtered per minute at the glomerulus = GFR $\times P$. Therefore, GFR $\times P$ = UV or

$$\text{GFR} = \frac{UV}{P}$$

What properties are required in a substance to be used for measurement of GFR? It must be:
1   freely filtered at the glomerulus (therefore $M_r \ll 50\,000{-}60\,000$);
2   not reabsorbed or secreted;
3   not metabolized by the kidney;
4   not toxic; and
5   easily measured in blood and urine.

Classically inulin (polyfructosan; M_r 5000) is used to measure GFR. But accurate measurement needs continuous infusion to establish a steady state with constant plasma concentration and rate of excretion.

Note that, since the amount of inulin excreted is directly proportional to plasma concentration (Fig. 18.6a), the plot of UV/P against plasma concentration gives a line parallel to the x axis (Fig. 18.6b). That is, UV/P for a substance which is filtered and neither reabsorbed nor secreted is independent of its concentration in the plasma.

Clinically, creatinine, which is produced by metabolism from creatine, is often used to measure GFR.

In the human there is a little proximal tubular secretion of creatinine, but the usual method for measuring plasma concentration overestimates creatinine because plasma contains a non-filtered chromogen which reacts as creatinine, and these errors tend to cancel. The best results are obtained by measuring the amount of creatinine excreted over 24 hours. Since plasma creatinine concentration stays relatively constant throughout the day, one plasma sample is usually all that is taken.

(a)

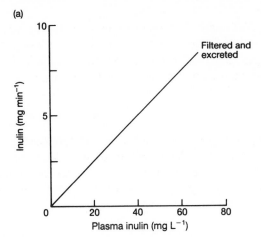

(b)

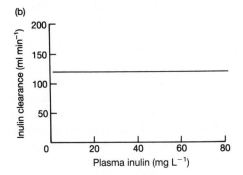

Fig. 18.6 (a) The relationship between plasma inulin concentration and the amounts of inulin filtered and excreted per unit time. (b) The relationship between plasma inulin concentration and the clearance (UV/P) of inulin.

A representative value for GFR in adults is about 125 ml min^{-1} or 180 litres per day, of the order of 50 times the volume of plasma in the body. This value represents the sum of the contributions of the individual nephrons from both kidneys.

GFR is better related to body surface area than to weight. Even so, it is quite variable from person to person. From about 2 years of age until after middle age, when the GFR declines slowly, the average value is 120 ml min^{-1} for each 1.73 m^2 (average body surface area). With 1 million nephrons in each kidney, the average filtration rate per nephron (single nephron glomerular filtration rate) comes to 60 nl min^{-1} or 90 µl per day.

Filtered load

If the GFR is known, it is a simple matter to calculate the amount of any solute filtered per minute—the **filtered load.**

Filtered load = GFR × plasma concentration

e.g. filtered load of glucose = 0.125 L min^{-1} × 5 mmol L^{-1}
$$= 0.625 \text{ mmol min}^{-1}$$

Filtration fraction

About 650 ml of plasma flows through the kidneys each minute (RPF, p. 622). Of this, about one-fifth is filtered; the remaining four-fifths pass into the peritubular capillaries. That is, the ratio GFR/RPF, called the **filtration fraction**, is about 0.20. Contrast this with the ratio of ultrafiltration to flow through typical systemic capillaries, which is about 0.005.

Variations in GFR

Results from early measurements with inulin probably overemphasized the constancy of GFR. Nevertheless, autoregulation of RBF (p. 610) results in a relatively stable glomerular hydrostatic pressure and therefore GFR. However, if arterial blood pressure falls below about 60 mmHg, as in shock, GFR ceases, leading to anuria.

The erect posture, emotion, pain, cold, exercise and loss of blood are all associated with reductions in GFR, especially the erect posture and exercise in some patients with cardiac failure. Pathological processes which destroy nephrons reduce GFR, and GFR may increase when blood and extracellular fluid volumes are expanded, especially with saline, which dilutes plasma protein and lowers colloid osmotic pressure.

18.4 CLEARANCE

The relationship UV/P can be estimated for any substance. It is referred to as the renal plasma clearance, or clearance for short. With inulin or creatinine, this relationship provides a good estimate of GFR. Any susbtance that is freely filtered and has a clearance value less than that of inulin must be reabsorbed. Any substance with a clearance greater than that of inulin must be secreted.

The expression *UV/P* can be calculated for any substance and is called **renal plasma clearance** or **clearance** for short. This has the dimensions of volume/unit time. In general, clearance allows comparison of the renal handling of different substances. However, in two situations where the clearance of substances with special properties is determined, it allows the calculation of a real rate of volume flow—GFR, as discussed above, and RPF (p. 622).

The clearance of inulin, as we have seen, estimates GFR. If a substance that is filtered has a clearance less than that of inulin, then there must be a net reabsorption of that substance within the renal tubules.

For example, a substance such as glucose that is freely filtered at the glomerulus but is normally completely reabsorbed from the tubules will have a clearance of zero.

If the clearance of a substance is greater than that of inulin, then there must be a net secretion by the tubular cells into the tubular fluid.

This comparison can be made formally by calculating the **clearance ratio**, i.e. clearance of $X(C_X)$/clearance of inulin (C_{In}), which compares the amounts of a substance in the urine and in the glomerular filtrate from which that urine was formed.

$$\frac{C_X}{C_{In}} = \frac{U \times V}{P \times GFR} = \frac{\text{amount of } X \text{ excreted}}{\text{amount of } X \text{ filtered}}$$

If $(C_X)/(C_{In})$ is less than 1.0, there is less X in the urine than was filtered, i.e. *X is reabsorbed*; while if $(C_X)/(C_{In})$ is greater than 1.0, there is more X in the urine than was filtered, i.e. *X is secreted* by the tubules as well as being filtered.

18.5 TUBULAR FUNCTION

Reabsorption and secretion occur in the tubules. Both processes may be either active or passive. The rate of epithelial transport is determined by surface area multiplied by flux. In the kidney, the available surface area is enormous and there is a very favourable ratio of surface area to volume. Normally, GFR is adjusted to allow adequate contact time with the epithelial cells in the different parts of the nephron. The flux of solutes depends upon the properties of the membranes of the epithelial cells and of the tight junctions between them. The proximal tubule is lined by a typical leaky epithelium and is specialized for the bulk reabsorption of filtered solutes and water. The epithelia lining more distal portions of the nephron show various degrees of tightness.

The whole volume of plasma is filtered many times daily. This effectively removes waste products from the blood, but in so doing removes water and all solutes of low M_r at the same time. *Hence a major task of the tubules must be recovery of water and solutes needed by the body.*

Reabsorption and secretion occur in the tubules. Both processes may be either active or passive. Before discussing some examples of each, the main factors affecting tubular function need to be summarized.

As discussed in on p. 26, the rate of epithelial transport is determined by (surface area × flux), where flux is defined as amount moved/unit time × unit area. In the kidney, the available surface area is enormous and offers a very favourable ratio of surface area to volume (each tubule handles only about $1/(2 \times 10^6)$ of the total volume filtered per minute—about 60 nl). In addition, especially in the proximal tubules, there is a very extensive brush border on the luminal surface. In contrast to the situation in the gut, there is no motility to control contact time. Instead, the rate of filtration is such that normally the load presented to the tubules does not exceed their capacity to deal with it. This may be accomplished, in part, through the juxtaglomerular apparatus (p. 610), which senses the load delivered to the distal tubule and alters afferent arteriolar resistance appropriately. However, if more solute than normal is filtered, or if tubular reabsorption is depressed, the increased flow rate downstream may overload the reabsorptive mechanisms for Na^+ in the distal nephron and result in an increased excretion of NaCl and water (an example of **osmotic diuresis**).

The flux of solutes depends upon the properties of the membranes of the epithelial cells and of the tight junctions. The proximal tubule is lined by a typical leaky epithelium (on p. 24). Glucose, amino acids, organic acids, phosphate, sulphate and some other solutes are co-transported across the luminal membrane coupled to Na^+ transport and driven by the energy inherent in the electrochemical gradient for Na^+ (Fig. 18.7). They pass across the basolateral membrane by either simple or facilitated diffusion. Isosmotic absorption occurs in this segment, as expected for leaky epithelia. The epithelia lining more distal portions of the

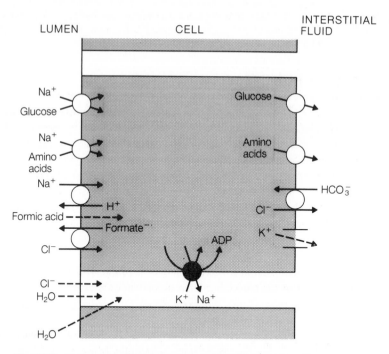

Fig. 18.7 A summary of the major transport mechanisms for solute movements across renal proximal tubules. Note that the cell formate exchanged for luminal Cl⁻ reacts with H⁺ in the lumen to form the weak, lipid-soluble formic acid which then diffuses back into the cell. The pathways for Cl⁻ loss across the basolateral membrane are not well-defined, but one possibility is the Cl⁻–HCO₃⁻ exchanger.

nephron show various degrees of tightness, as will be discussed later.

The gradient for reabsorption reflects, in part, the rate of removal of reabsorbed solutes by the peritubular capillaries. This is rapid and effective in the cortex but the arrangement of the vasa recta in long loops results in accumulation of reabsorbed solutes within the medulla (p. 637).

The same general principles apply to tubular secretion. Secreted solutes may be removed from the peritubular plasma or be synthesized within the tubular epithelial cells and transferred to the luminal fluid.

Some solutes are only reabsorbed, some are only secreted; some undergo both processes.

Organic solutes that are reabsorbed

Glucose, amino acids, organic acids, phosphate, sulphate and some other solutes are co-transported across the luminal membrane coupled to Na⁺ transport and driven by the energy inherent in the electrochemical gradient for Na⁺. They pass across the basolateral membrane by either simple or facilitated diffusion. A variety of smaller-molecular-weight proteins and peptides cross the glomerular filter. Most enter the proximal tubular cells by endocytosis and are then degraded to amino acids which are returned to the plasma. Urea is absorbed passively. Antidiuretic hormone (ADH) increases the permeability of the medullary collecting ducts to urea; hence, less urea is excreted at low urinary flow rates.

Glucose

Normally, all of the filtered glucose is reabsorbed and there is no glucose in the urine. Reabsorption involves co-transport with Na^+ at the luminal membrane (Fig. 18.7). The characteristics of glucose handling are illustrated in Fig. 18.8.

If the filtered load is increased by raising the plasma glucose concentration, the transport mechanism in some tubules becomes saturated and glucose begins to appear in the urine. The plasma glucose concentration at which glucose *first* appears is called the **plasma threshold** and in the human is some 10–12 mmol L^{-1} when GFR is normal at about 125 ml min^{-1}. Hence the transport mechanism begins to be saturated in some tubules when the filtered load reaches $10 \times 0.125 = 1.25$ mmol min^{-1}. At plasma glucose concentrations of about 15 mmol L^{-1}, the transport mechanism *in all the tubules* is saturated, and the maximal reabsorptive capacity of the tubules (**tubular maximum,** T_m) has been reached. This is thus about 1.9 mmol min^{-1} (15 mmol $L^{-1} \times 0.125$ L min^{-1}).

Note that the plasma threshold reflects the limit of the rate of tubular reabsorption; it is not a fixed plasma concentration. Because filtered load equals $P_{glucose} \times$ GFR, threshold concentration is inversely proportional to GFR. With GFR reduced from 125 to 60 ml min^{-1} it would take 20, not 10, mmol L^{-1} of plasma glucose to reach the threshold load of 1.25 mmol min^{-1}.

The commonest cause of glycosuria is **diabetes mellitus,** in which plasma glucose concentration is abnormally high. **Renal glycosuria** occurs as an uncommon anomaly when tubular reabsorptive capacity is subnormal. Here plasma threshold will be low, so that glucose appears in the urine, though its concentration in the blood is not abnormally high.

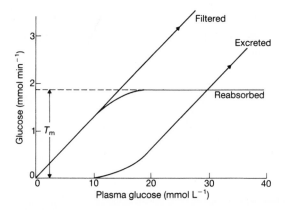

Fig. 18.8 Relationships between plasma glucose concentration and the amount of glucose filtered, reabsorbed and excreted by the kidneys (glycosuria) per unit time. T_m, Tubular maximum.

Amino acids

About 98% of the filtered load of amino acids is reabsorbed from the proximal convoluted tubules by co-transport with Na^+ across the luminal membrane (Fig. 18.7). There is evidence of competition within, but not between, five groups:

1 neutral amino acids;
2 imino acids;
3 basic amino acids and cystine;
4 glutamic and aspartic acids; and
5 glycine.

Normally about 0.5–2% of the filtered load is lost in the urine but **aminoacidurias**, in which large amounts of particular amino acids are excreted, result from deficiencies in specific enzymes, often genetically determined and frequently associated with abnormal metabolism of the excreted amino acids.

Organic acids

Metabolic substrates such as acetate, lactate, citrate, oxalate and keto acids (β-hydroxybutyrate and acetoacetate) are normally filtered in small quantities and reabsorbed in the proximal tubule. Again, they enter the cells from the lumen co-transported with Na^+.

In ketoacidosis and lactic acidosis, when plasma concentrations of the relevant organic acids are high, the T_m can be exceeded and these solutes are then lost in the urine. Under such conditions, they provide additional urinary buffer (p. 668).

Peptides and proteins

Although little plasma protein is filtered, there are a variety of smaller-molecular-weight proteins and peptides that do cross the glomerular filter. Most (including the very small quantity of albumin that leaks across) are endocytosed by the proximal tubular cells and then degraded; only their constituent amino acids are restored to the plasma. Hence the kidney effectively eliminates these substances without excreting them in the urine. The continuous removal of freely filtered peptide hormones (e.g. insulin, glucagon, vasopressin, parathyroid hormone and gut hormones) at rates proportional to their concentrations in the plasma ensures that their concentrations are largely determined by the rates at which they are released into the circulation.

In some situations plasma concentrations of specific low-molecular-weight proteins are elevated (e.g. release of myoglobin from damaged muscles or haemoglobin during excessive breakdown of red blood cells). The filtered load then exceeds reabsorptive capacity and, as fluid is removed, tubular concentration rises and protein precipitates, causing blockage of the lumen. If extensive, this can result in acute renal failure.

Urea

Normally about 0.6 mmol (5 mmol $L^{-1} \times 0.125$ L min^{-1}) or 900 mmol per day of urea is filtered at the glomerulus. By the end of the proximal tubule, some two-thirds of the filtered water has been reabsorbed. This raises the urea concentration in the tubular fluid and, since the proximal tubule is permeable to urea,

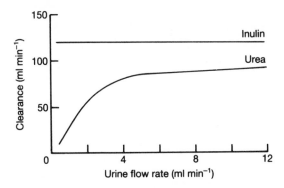

Fig. 18.9 Relationships between urine flow rate and clearance of inulin and urea.

about 50% of the filtered urea diffuses passively from the higher concentration in the tubular fluid to the lower concentration in the plasma.

In the absence of ADH, the urea permeability of the remainder of the nephron is low (although some urea enters the thin ascending limb of the loop of Henle, driven by the high medullary interstitial fluid concentration; p. 634). Thus at high urine flow rates (low levels of circulating ADH) up to 60–70% of the urea that enters the tubular fluid may be excreted in the urine. ADH increases the permeability of the medullary collecting ducts to urea, so at low urine flow rates (high levels of circulating ADH) net reabsorption of urea in this nephron segment can result in excretion of only about 30% of the filtered load of urea. (Recycling of urea in the medulla is discussed on p. 636.

At urine flow rates greater than about 2 ml min^{-1}, the clearance of urea is relatively independent of flow rate and is about two-thirds of the inulin clearance (Fig. 18.9).

Plasma urea concentration does not necessarily provide a good indication of renal function, for it depends upon the rate of production of urea (reflecting protein intake and catabolism) as well as its renal excretion (Fig. 18.10).

Organic solutes that are secreted

The kidney excretes a variety of metabolic end-products and potential toxins. Two distinct pathways have been identified: one for organic anions (acids), the other for organic cations. Up to 90% of some such solutes (e.g. PAH) can be removed in one passage through the kidneys. The clearance of PAH can therefore be used to estimate RPF. As well as secreting organic solutes removed from the blood, renal proximal tubular cells secrete ammonium—an essential process in the maintenance of acid–base balance.

The kidney, together with the liver, plays an essential role in removing from the circulation a variety of metabolic end-products as well as chemicals that would be toxic if allowed to accumulate in the body. In the kidney, these are taken up from the peritubular blood by the proximal tubular cells and then secreted into the tubular fluid.

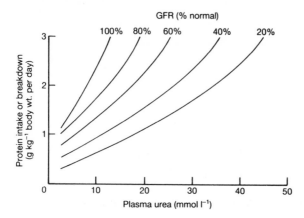

Fig. 18.10 Effects of alterations in renal glomerular function and protein metabolism on plasma urea concentration.

Two distinct pathways have been identified; one handles the organic anions (acids), the other organic cations (bases; p. 625). These systems can be extraordinarily effective in removing organic solutes from the plasma, with up to 90% of some solutes being removed in one passage through the kidneys.

Organic acids

A variety of chemicals (e.g. PAH) and drugs, including diuretics (e.g. frusemide and ethacrynic acid), penicillin, sulphonamides and salicylates, are secreted by the proximal tubule almost as rapidly as the plasma presents them to the epithelial cells. Secretion is active, steeply uphill, and shows saturation, competition between substrates, and inhibition—all typical of carrier-mediated transport (p. 10). A model of the pathways involved in organic anion secretion is illustrated in Fig. 18.11.

Use is made of this secretory pathway in estimating RPF with the organic anion PAH, which is removed avidly from the plasma.

The use of PAH to measure renal plasma flow

For any substance which is transferred from plasma to urine but is not produced, destroyed or stored within the kidney, the amount entering the kidney in the renal artery each minute must equal the amounts leaving the kidney each minute in the renal vein and in the urine. (This ignores the small amount that may leave through the lymphatic system.) These amounts are $P_a \times V_a$ for the renal artery, $P_v \times V_v$ for the renal vein and UV for the urine (Fig. 18.12), where P_a is the plasma concentration of such a substance entering the kidneys in the renal arteries, P_v is the concentration of the same substance leaving the kidneys in the renal vein, U is the concentration in the urine, V_a is the volume of plasma entering the kidneys in the renal arteries per minute (the **RPF**), V_v is the volume of plasma leaving the kidneys in the renal vein per minute and V is the volume of urine leaving the kidneys per minute.

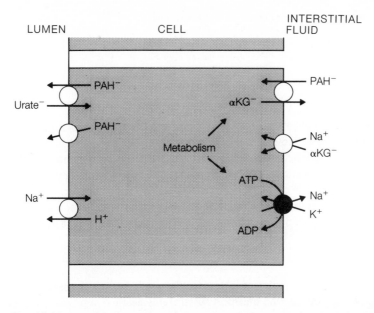

Fig. 18.11 A model of organic anion secretion. PAH $^-$ is shown as a representative compound. In this model, α-ketoglutarate (αKG $^-$) produced by metabolism exchanges for PAH $^-$ across the basolateral membrane and is recycled via a Na $^+$-coupled basolateral co-transporter. Other metabolic intermediates may substitute for αKG $^-$. At the luminal membrane, PAH $^-$ leaves the cell either by facilitated diffusion or in exchange for urate or other anions. Probenecid, which inhibits organic anion secretion, and also urate reabsorption, blocks this exchanger. (Adapted from Pritchard, J.B. & Miller, D.S. (1992) In: Seldin, D.W. & Giebisch, G. (eds) *The Kidney: Physiology and Pathophysiology*, 2nd edn. pp 2921–2945. Raven Press, New York.)

Thus:

$$(P_a \times V_a) = (P_v \times V_v) + UV$$

or:

$$(P_a \times V_a) - (P_v \times V_v) = UV$$

The volume of the urine per minute (about 0.5–1 ml min $^{-1}$) is so much smaller than that of the plasma entering the kidneys each minute (about 650 ml min $^{-1}$),

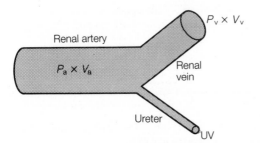

Fig. 18.12 Amount of a substance entering and leaving the kidney each minute. See text for meaning of symbols.

that V_v can be taken as equal to V_a. Therefore:

$$V_a (P_a - P_v) = UV$$

and

$$V_a \text{ or } \mathbf{RPF} = \frac{UV}{(P_a - P_v)}$$

This is an example of the use of the Fick principle (p. 404) to measure plasma flow rate in an organ. However, it has limited practical application, for although the amount of substance excreted in the urine is readily determined and plasma for measurement of arterial concentration can be obtained from any systemic artery (or vein), a sample of renal venous plasma is not easily obtained! However, if the kidneys could completely remove all of a substance entering each minute, then P_v would be zero and the equation would reduce to the familiar clearance equation:

$$\mathbf{RPF} = \frac{UV}{P_a}$$

Proximal tubular secretion of PAH is so efficient that at low plasma concentrations (that do not saturate the transport system) it is nearly completely removed from the plasma passing through the kidneys. Therefore, P_v for PAH approximates to zero, and the clearance of PAH approximates to the RPF.

The renal handling of PAH is summarized in Figs 18.13 and 18.14. Note that PAH is filtered at the glomerulus as well as being secreted. Thus, with a filtration fraction of 0.2, 20% of the PAH that enters the kidney each minute in the plasma is removed by filtration and much of the remaining 80% that flows on into the peritubular capillaries is then removed by secretion.

In reality, the removal of PAH is never complete and values for RPF and RBF obtained from PAH clearance in a person with normal renal function are approximately 90% of the true

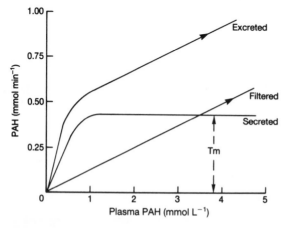

Fig. 18.13 Relationships between plasma PAH concentration and the amounts of PAH filtered, secreted and excreted by the kidneys per unit time.

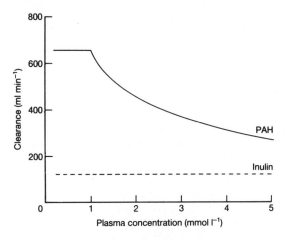

Fig. 18.14 The clearance of PAH as a function of plasma concentration (the clearance of inulin is also shown for comparison). Note that at low plasma concentrations the PAH clearance parallels the inulin clearance. Over this range the secretory system is not saturated and the PAH clearance can be used to estimate RPF, as discussed below.

value. If renal tubular function is impaired, less PAH may be secreted and an erroneous estimate of RPF will be obtained. In practice, PAH clearances are rarely measured clinically. An estimate of RPF can be obtained by following the time course of the removal of an injected bolus of Diodrast, which is also secreted by the organic anion system and (because of its high content of iodine) is opaque to X-rays.

RBF is readily estimated from the RPF if the haematocrit (haem) is known.

RBF – (haem × RBF) = RPF

Therefore:

RBF (1 – haem) = RPF

or:

$$\text{RBF} = \frac{\text{RPF}}{(1 - \text{haem})}$$

Values obtained in normal people are: RPF, 650 ml min^{-1}; RBF, 1200 ml min^{-1}. Thus, the kidneys, which constitute 0.5% body weight but account for nearly 10% of the total oxygen consumption of the body, receive about 20% of the resting cardiac output. Renal arteriovenous difference for oxygen is low. *The high rate of flow is required not to supply O_2 for renal metabolism but to provide the volume to be filtered.*

Organic bases

A model of the pathways involved in organic cation secretion is illustrated in Fig. 18.15. For the most part these are amines, and both endogenous substances (e.g. acetylcholine, adrenaline, histamine and some vitamins—riboflavin,

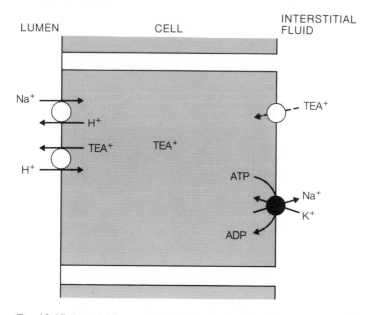

LUMEN   CELL   INTERSTITIAL FLUID

Fig. 18.15 A model for organic cation secretion. Tetraethylammonium (TEA$^+$) is shown as a representative compound. In this model, TEA$^+$ enters the cells from the plasma across the basolateral membrane by facilitated diffusion and leaves the cell across the luminal membrane in exchange for H$^+$. The H$^+$ is then recycled through the Na$^+$–H$^+$ exchanger. (Adapted from Pritchard, J.B. & Miller, D.S. (1992) In: Seldin, D.W. and Giebish, G. (eds) *The Kidney: Physiology and Pathophysiology*, 2nd edn. pp. 2921–2945. Raven Press, New York.)

thiamine) and a variety of drugs (e.g. amiloride, morphine, quinine) are secreted by this mechanism.

Ammonium ions

As well as secreting organic solutes removed from the blood, renal proximal tubular cells secrete ammonium (Fig. 18.16). This is an integral part of the role of the kidney in maintaining extracellular pH, discussed on p. 663. Ammonium is generated in the proximal tubular cells largely from glutamine, which enters the cells by Na$^+$-dependent co-transport across both luminal and basolateral membranes. Metabolism of glutamine produces α-ketoglutarate and NH$_4^+$. In turn, metabolism of α-ketoglutarate consumes H$^+$ ions, leaving HCO$_3^-$. Effectively, 2NH$_4^+$ and 2HCO$_3^-$ are produced for each glutamine metabolized.

Since the pKa of the reaction NH$_3$ + H$^+$ = NH$_4^+$ is 9.3, at pH 7.4 most of the ammonium is in the form NH$_4^+$. However, NH$_3$ is much more lipid-soluble than NH$_4^+$ and therefore, although the concentration of NH$_3$ is much lower, non-ionic diffusion of NH$_3$ also contributes to renal ammonium secretion.

Ammonium leaves the proximal tubular cell across the luminal membrane, either as NH$_4^+$ substituting for H$^+$ on the Na$^+$–H$^+$ exchanger or by non-ionic diffusion

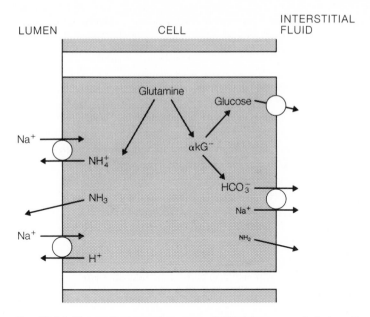

Fig. 18.16 The production and secretion of NH_4^+ by proximal tubular cells.

of NH_3, with the associated H^+ leaving on the $Na^+–H^+$ exchanger. Some NH_3 diffuses from the cell across the basolateral membrane and is removed in the peritubular blood. The HCO_3^- passes across the basolateral membrane. Several pathways may be involved, including a $Na^+–HCO_3^-$ co-transporter (Fig. 18.16).

Under normal conditions, most of the NH_4^+ synthesized by the proximal tubular cells passes into the luminal fluid. In the thick ascending limb of the loop of Henle, NH_4^+ is reabsorbed, probably by substituting for K^+ on the luminal $Na^+–K^+–2Cl^-$ co-transporter (see Fig. 18.21). As a consequence of the renal countercurrent system, this reabsorbed NH_4^+ is accumulated in the medulla and concentrated towards the papilla. As fluid flows through the collecting ducts, the epithelial cells secrete H^+ into the lumen (p. 630). This lowers luminal pH so that NH_3 diffusing from the medullary interstitium is converted to NH_4^+ and trapped in the lumen.

Organic solutes that are reabsorbed and secreted

Urate

Urate, the final product of purine degradation, is filtered at the glomerulus and both reabsorbed and secreted by proximal tubular cells. In humans, the balance between these processes is such that there is normally net reabsorption. A possible mechanism for reabsorption is illustrated in Fig. 18.11. In some species, urate is secreted by the same mechanisms as PAH (Fig. 18.11), but there may be a separate secretory pathway in the human.

Failure to excrete sufficient urate to balance production can result in precipitation of uric acid crystals in joints—gout. Among agents used to treat this disease is probenecid, which blocks urate reabsorption.

Inorganic ions that are reabsorbed

In addition to sodium and chloride, the reabsorption by the kidneys of magnesium, calcium, phosphate, sulphate and bicarbonate is of physiological importance. Magnesium, calcium and phosphate handling are all affected by parathyroid hormone. For phosphate and sulphate, the renal threshold is a little below the normal plasma concentration. Thus any increased production will result in increased secretion. The actual HCO_3^- ions that are filtered are not reabsorbed. Instead HCO_3^- reacts with secreted H^+ in the tubular lumen to form H_2O and CO_2. However, for every H^+ secreted, a HCO_3^- is formed in the cells and returned to the plasma, so the effect is to return to the plasma the same amount of HCO_3^- as was filtered.

Sodium and chloride represent the bulk of the filtered inorganic ions. Their renal handling is discussed on p. 637.

Magnesium

About one-third (or 0.3 mmol L^{-1}) of plasma Mg^{2+} is bound to proteins and cannot be filtered. The remainder, mostly ionized, is filtered. Of this, 25% is reabsorbed in the proximal tubule and 65% in the thick ascending limb. A further 5% is reabsorbed more distally. It remains to be established whether Mg^{2+} reabsorption occurs cellularly, paracellularly, or by both routes. In the thick ascending limb, Mg^{2+} reabsorption (and also Ca^{2+} reabsorption) is stimulated by **parathyroid hormone, glucagon, calcitonin** and **ADH**.

Calcium

About 50% (or 1.25 mmol L^{-1}) of Ca^{2+} is bound to plasma proteins and cannot be filtered. The remainder, mostly ionized, is filtered. Of the filtered load of about 200 mmol per day, less than 5 mmol is usually excreted. Some 60% is reabsorbed proximally, most of the remainder in the ascending limbs of the loop of Henle, distal tubules and collecting duct. Both cellular and paracellular routes are involved. In the proximal tubule and loops of Henle, alterations in Na^+ transport result in parallel changes in Ca^{2+} reabsorption. **Parathyroid hormone** stimulates reabsorption in the distal tubule, although it inhibits reabsorption proximally; but, since it raises plasma ionized Ca^{2+}, its net effect is often to promote urinary Ca^{2+} excretion.

Phosphate

Some 95% of the filtered inorganic phosphate (Pi) is reabsorbed, predominantly in the proximal convoluted tubule, with the threshold a little below the normal plasma concentration. Thus there is normally some phosphate, which is an important buffer, in the urine. Reabsorption involves luminal co-transport with Na^+; the mechanisms of basolateral exit are ill-defined. Reabsorption is inhibited by **parathyroid hormone**.

Sulphate

As with phosphate, most of the filtered sulphate is reabsorbed predominantly in the proximal convoluted tubule by a mechanism that includes luminal co-transport with Na^+. Again, the threshold is a little below the normal plasma concentration. Thus there is normally some sulphate in the urine and changes in plasma sulphate concentration will result in appropriate alterations in renal excretion. There is no known hormonal control over sulphate reabsorption.

Bicarbonate

Normally, about 4500 mmol of HCO_3^- is filtered each day (24 mmol L^{-1} × 180 L per day). Throughout the nephron, secretion of H^+ from the tubular cells into the lumen results in the conversion of filtered HCO_3^- to H_2O and CO_2 (Fig. 18.17). However, for every H^+ secreted into the lumen, a HCO_3^- is returned to the plasma across the basolateral membrane. Several transporters, including a $Na^+-HCO_3^-$ co-transporter (which may shift $3HCO_3^-$ for each Na^+) appear to be involved in this basolateral movement.

Although the HCO_3^- filtered from the plasma is actually destroyed in the tubular lumen, the effect is to replace it with an equivalent amount of HCO_3^- synthesized by the renal tubular cells, and the overall process is therefore referred to as the reabsorption of HCO_3^-. Of this reabsorption, 70–85% takes place in the proximal tubule (where the presence of carbonic anhydrase in the luminal membrane catalyses the conversion of H^+ and HCO_3^- to H_2O and CO_2 in the lumen), 10–20% in the loop of Henle (mostly in the thick ascending segment), 3–5% in

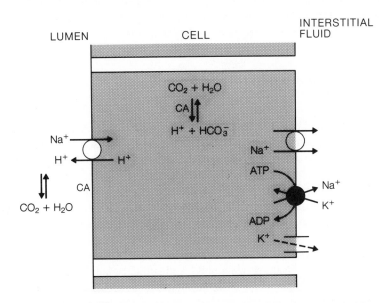

Fig. 18.17 Mechanisms involved in proximal tubule H^+ secretion and HCO_3^- reabsorption. Note the presence of carbonic anhydrase (CA) on the luminal membrane as well as intracellularly.

the distal tubule and 1–2% in the collecting ducts. Under alkalotic conditions, when plasma HCO_3^- concentration is elevated, the reabsorptive capacity of the nephrons is exceeded and HCO_3^- appears in the urine.

In the distal part of the nephron (late distal tubule and collecting duct), specialized type B intercalated cells may contribute to this HCO_3^- excretion by secreting HCO_3^- into the lumen via a coupled Cl^-–HCO_3^- exchanger.

The regulation of HCO_3^--reabsorption is discussed as part of acid–base balance on p. 663.

Inorganic ions that are secreted

Hydrogen ions

Renal epithelal cells secrete H^+ as an integral part of their role in maintaining extracellular pH, discussed in Chapter 19.

In the proximal tubule, luminal Na^+–H^+ exchange and an H^+ ATPase both contribute to this secretion (Fig. 18.17), which results in HCO_3^- reabsorption, as discussed above. Although the capacity for H^+ secretion is relatively large, the leaky nature of the epithelium limits the H^+ gradient that can be established in this nephron segment.

The thick ascending limb cells appear to secrete some H^+ by the same mechanisms as above. In the relatively tight distal part of the nephron (late distal tubule and collecting ducts), there is primary active transport of H^+ across the luminal membrane of specialized type A intercalated cells (Fig. 18.18). This combination of a tight epithelium and primary active H^+ transport enables the urine to be acidified to as low as pH 4.4 which, when the plasma remains at pH 7.4, is a ratio $[H^+]_{urine}/[H^+]_{plasma}$ of 1000 : 1.

Inorganic ions that are reabsorbed and secreted

Potassium

About 800 mmol of K^+ is filtered at the glomerulus each day. In the proximal convoluted tubule much of this filtered K^+ is reabsorbed. Some of this reabsorption may be a passive consequence of the absorption of NaCl and other solutes together with water, which concentrates luminal K^+ and creates a favourable gradient for diffusion through the paracellular pathway. Some may involve uptake into the cells across the luminal membrane.

In the medullary part of the proximal straight tubule and in the thin descending limb, there is K^+ secretion. This, combined with reabsorption of K^+ by the thick ascending limb (which involves co-transport across the luminal membrane with Na^+ and Cl^-), results in recycling of K^+ within the medulla.

The most distal parts of the nephron (late distal tubule and collecting duct) contain the **principal cells** that secrete K^+ (Fig. 18.19). In addition, K^+ reabsorption occurs in the medullary collecting duct, possibly via an H^+–K^+ ATPase in type A intercalated cells (Fig. 18.18).

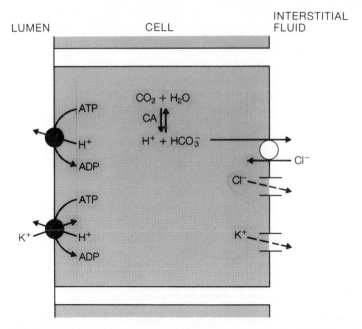

LUMEN   CELL   INTERSTITIAL
                                                FLUID

Fig. 18.18 H^+ secretion by type A intercalated cells in the distal nephron. As well as an H^+ ATPase, an H^+–K^+ ATPase, similar to that found in the gastric mucosa, is also shown and would contribute to K^+ reabsorption in this portion of the nephron. (For clarity, the basolateral membrane Na^+–K^+ ATPase is not shown.)

Figure 18.19 illustrates a model for K^+ secretion by the principal cells. Activity of the basolateral Na^+–K^+ ATPase keeps cell K^+ concentration high. Na^+ entry to the cells through luminal Na^+ channels depolarizes the luminal membrane and thereby increases the electrical potential driving force for diffusion of K^+ from cell to lumen through luminal K^+ channels. Thus, if the luminal Na^+ concentration is increased, there is increased K^+ secretion. There is also a neutral K^+–Cl^- co-transporter in the luminal membrane. Thus, reduction in luminal Cl^- concentration will result in increased K^+ (and Cl^-) secretion. Note that the luminal Na^+ channels are blocked by the diuretic amiloride. This, therefore, also inhibits K^+ secretion in this segment.

18.6 RENAL HANDLING OF SALT AND WATER

The human kidney can produce a urine ranging in volume from 0.5 or 20 litres to more per day, and containing from 5–10 to 600 or more millimoles of Na^+ per day, with an osmolarity of from 1200 mosmol L^{-1} (at the lowest volumes) to 50 mosmol L^{-1} with large volumes. This unique ability requires that about one-quarter of the filtered NaCl is reabsorbed in the medulla while the water associated with it returns to the cortex and is reabsorbed there. Urea is also accumulated within the medulla. The combination of medullary NaCl reabsorption and the arrangement of the vasa recta in loops results in the creation and maintenance of an osmotic gradient in the medullary interstitium. The collecting ducts are relatively impermeable to water in the absence of ADH. Under these conditions, a large volume of a dilute urine is produced (water diuresis). With ADH, water is free to move from lumen to interstitial fluid driven by the osmotic

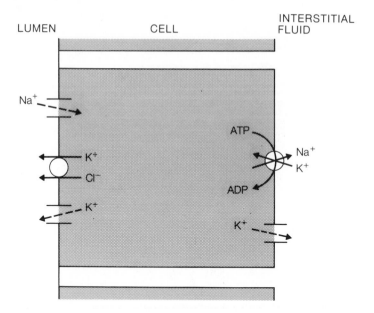

Fig. 18.19 Membrane pathways involved in K^+ secretion by principal cells.

gradient and a small volume of concentrated urine is produced (antidiuresis). Under these conditions, urea recycles between medullary interstitium and tubular fluid and this helps to preserve medullary hyperosmolarity.

Filtration

With a GFR of 125 ml min^{-1}, a plasma Na$^+$ concentration of 150 mmol L^{-1} and a plasma Cl$^-$ of 110 mmol L^{-1}, 180 litres of H_2O are filtered by the kidneys each 24 hours and the filtered load of Na$^+$ is some 19 mmol min^{-1} (27 000 mmol per 24 hour) and of Cl$^-$ some 14 mmol min^{-1} (19 800 mmol per 24 hour). Clearly, changes in either GFR or plasma concentration will change the load presented to the tubules. Figure 18.20 provides a summary of the handling of water and Na$^+$ in the different nephron segments.

Proximal tubules

About two-thirds of the filtered H_2O, Na$^+$ and Cl$^-$ (i.e. 120 litres, and 18 000 mmol Na$^+$ and 13 200 mmol Cl$^-$) is reabsorbed from the proximal tubules per 24 hours. The Na$^+$ enters the cells across the luminal membrane down its electrochemical potential gradient in association with either co-transported solutes or countertransported H$^+$ or NH$_4$$^+$ ions, and is extruded across the basolateral membrane against its electrochemical gradient by the Na$^+$–K$^+$ ATPase (Fig. 18.7). Some Cl$^-$ is reabsorbed through the cells, luminal entry being effected by a Cl$^-$–formate exchanger (Fig. 18.7). In addition, the reabsorption of solutes (glucose, amino acids, HCO$_3$$^-$) with water early in the proximal tubule increases tubular Cl$^-$ concentration. This favours paracellular passive diffusion of Cl$^-$ and contributes to salt absorption in this segment.

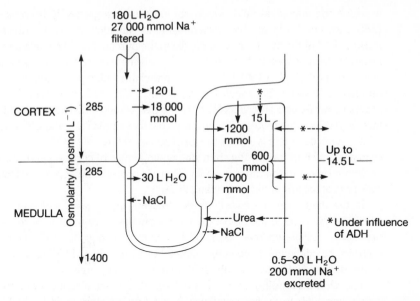

Fig. 18.20 A summary of the contribution of the various segments to the daily handling of Na^+ and water by the human kidney. Note that ADH promotes water reabsorption in the last part of the distal tubule as well as in the collecting ducts. It therefore affects the reabsorption of up to about 30 litres.

The Na^+ accumulated locally in the lateral intercellular spaces (with Cl^- to maintain electroneutrality) provides the force for isosmotic water reabsorption. The hydraulic conductivity of this segment of the nephron is so large than an osmotic imbalance as little as $1–10$ mosmol L^{-1} can account for the observed rate of reabsorption. H_2O follows both cellular and paracellular routes in moving from lumen to interstitium.

The amounts of salt and H_2O reabsorbed from the proximal tubule are also influenced by the balance of hydrostatic and osmotic forces in the peritubular capillaries. An increased colloid osmotic pressure or a decreased hydrostatic pressure in peritubular capillaries favours uptake of interstitial fluid into capillaries and assists reabsorption. Conversely, decreased colloid osmotic pressure or increased hydrostatic pressure hinders reabsorption. For example, if GFR increases with constant RPF, the filtration fraction will increase and the increased colloid osmotic pressure in the peritubular capillaries will assist in reabsorbing the increased volume of filtrate. Proximal tubular fluid reabsorption matches GFR closely over a wide range of GFR and such adjustments of capillary forces help to account for this **glomerulotubular balance**.

Loops of Henle

A volume of about 60 litres containing 9000 mmol Na^+ leaves the proximal tubules and enters the **descending limbs** of the loops of Henle each day as an isosmotic solution. Those loops from the juxtamedullary nephrons which run deep into the medulla (some 20% in the human kidney) pass through a region in which the interstitial fluid osmolarity increases progressively from isosmotic (285 mosmol L^{-1}) at the corticomedullary junction to some 1200–1400 mosmol L^{-1} at the tip of the renal papilla in the human. NaCl and urea contribute in similar proportions to this gradient. The mechanism by which it is generated and sustained is discussed later. The epithelium lining this segment is quite permeable to water, but much less so to Na^+, Cl^- and urea. Therefore, H_2O moves passively and progressively from the descending limb to the interstitium as fluid flows along the tubules. It is also possible that a little Na^+ and Cl^- enters the tubule. Each day there is a net removal of about 30 litres of H_2O from this tubular segment, much of it from the short descending limbs of cortical nephrons which penetrate only into the outer medulla. There is probably no active transepithelial transport of solutes in the descending limbs.

In the medullary descending limbs, fluid has equilibrated with the adjacent interstitial fluid and at the tips of the loops now has an osmolarity of some 1200 mosmol L^{-1}. But, whereas NaCl and urea contribute equally (600 mosmol L^{-1} each) to the interstitial osmolarity, in the tubular fluid about 1150 mosmol L^{-1} is contributed by NaCl and only about 50 mosmol L^{-1} by urea.

The **thin ascending limb** of the loop differs significantly from the thin descending limb in its permeabilities, being impermeable to H_2O, highly permeable to Na^+ and Cl^- and somewhat permeable to urea. Consequently, Na^+ and Cl^-, which are at much higher concentrations in the tubular fluid, diffuse from the tubules to the interstitium as the fluid flows back towards the cortex, and some urea enters the tubules down its concentration gradient. The loss of NaCl greatly exceeds the gain in urea so that, by the time that the thick ascending segment is reached, the tubular fluid has become hypo-osmotic compared with the adjacent interstitial fluid.

Unlike the thin ascending limb, the **thick ascending limb**, and its continuation in the cortex as the first part of the distal tubule, avidly reabsorb NaCl from the tubular fluid by the process illustrated in Fig. 18.21. This involves a combination of co-transport of Cl^- at the luminal membrane with passive diffusion at the basolateral membrane. The positive charge of the luminal relative to the peritubular fluid helps to promote the reabsorption of cations such as K^+, Ca^{2+} and Mg^{2+} at this site. The impermeability to H_2O means that the removal of NaCl dilutes the luminal fluid. By the time this fluid enters the distal tubule in the cortex, it has lost about 7000 mmol of Na^+, much of this from the thick ascending limb, and has become hypo-osmotic to cortical plasma. In total, 60 litres of fluid with 9000 mmol of Na^+ entered all of the loops of Henle and 30 litres with 2000 mmol returned to the distal tubules. The Na^+ concentration is now about 60 mmol L^{-1}, urea concentration is 25 mmol L^{-1} and osmolarity about 150 mosmol L^{-1}.

Distal tubules

It is now appreciated that the distal tubule is not a homogenous segment. The first part is lined by an epithelium of the same type as the adjacent ascending limb of the loop of Henle and the two together comprise the **diluting segment** of the nephron. The last part, the connecting segment, is analogous in its properties to the collecting tubule which it joins. NaCl is reabsorbed throughout the length of the distal tubule, some 1200 mmol of Na^+ (5% of the filtered load) being removed

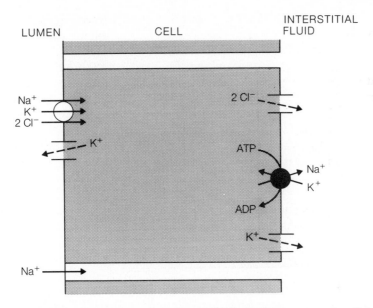

Fig. 18.21 A model of the major transport pathways for ions across the cells in the thick ascending limb of the loop of Henle. Note that the Na^+–K^+ – $2Cl^-$ co-transporter is specifically inhibited by the loop diuretics such as bumetanide and frusemide.

from the tubular fluid each day. Figure 18.22 illustrates one possible model for Na^+ reabsorption in the distal convoluted and connecting tubules.

The permeability of the distal tubule to H_2O reflects its heterogeneity. While the early part is impermeable to H_2O, the permeability of the last part is determined by the level of circulating **ADH**. With normal hydration some 10–20 litres of water per day are absorbed by the distal tubule, leaving up to 20 litres containing some 800 mmol of Na^+ to pass through the collecting ducts. The distal tubule is impermeable to urea. Therefore, urea entering the thin ascending limb of the loop of Henle is retained within the tubule in this portion of the nephron.

Collecting system

Though only about 20% of the loops of Henle run deep into the medulla, all of the fluid remaining at the end of the distal tubule enters the collecting duct system which runs through the hyperosmotic medulla to drain into the renal pelvis and ureters. The cells lining the collecting ducts form a tight epithelium. Active Na^+ reabsorption continues here and Na^+ concentration in the urine may fall to as low as 10 mmol L^{-1}. On a normal diet, about 600 mmol is reabsorbed each day and this reabsorption is largely independent of water handling in this segment.

Water diuresis (the production of a large volume of a dilute urine) is associated with low circulating levels of ADH. In the absence of this hormone (as in the rare disease **diabetes insipidus**), the water permeability of the collecting duct epithelial cells is very low, little water absorption will occur and some patients have been known to excrete as much as 30 litres of urine per day (a flow rate of ~20 ml min^{-1}) with a urine osmolarity lower than 50 mosmol L^{-1}.

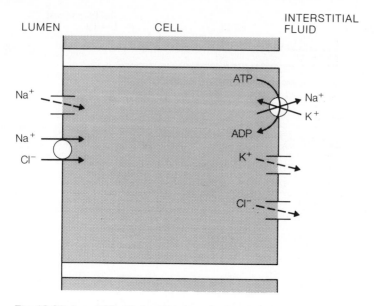

Fig. 18.22 A model for Na$^+$ reabsorption by the distal convoluted and connecting tubules. Note that the Na$^+$–Cl$^-$ co-transporter is specifically inhibited by the thiazide diuretics, whereas the luminal membrane Na$^+$ channels are blocked by the diuretic amiloride.

ADH acts via adenylate cyclase, cyclic adenosine monophosphate and protein kinase and causes cytoplasmic vesicles that contain water channels to fuse with the luminal membrane, thus increasing its water permeability. Water is then free to move through the cell down the osmotic gradient from tubular lumen to interstitial fluid.

ADH increases water permeability of both cortical and medullary portions of the collecting ducts, as well as of the most distal portion of the distal tubules (the **connecting tubule**). Under these conditions much of the water is reabsorbed from the cortical collecting ducts. With maximal plasma concentrations of ADH (**maximal antidiuresis**), as little as 0.5 litre of urine is excreted per day (about 0.3 ml min^{-1}), with a urine osmolarity up to 1200 mosmol L^{-1}. In addition, ADH increases the permeability of the **inner medullary collecting ducts** to urea and promotes its reabsorption in this segment only. Therefore, in antidiuresis, any urea that passes from the medulla into the thin ascending limb of the loop of Henle then diffuses back into the inner medullary interstitium. Its recycling between medullary interstitium and tubular fluid helps to preserve medullary hyperosmolality. During water diuresis, the loss in the urine of much of the urea passing into the ascending limb contributes substantially to the reduction in the medullary osmotic gradient under these conditions.

Generation and maintenance of the medullary hyperosmotic gradient

Generation of the gradient of medullary hyperosmolality requires that about one-quarter of the filtered NaCl is reabsorbed in the medulla while the water associated with it returns to the cortex and is reabsorbed there. Urea is also accumulated within the medulla. The combination of medullary NaCl reabsorption, recycling of urea between medullary interstitium and tubular fluid, and the arrangement of the vasa recta in loops results in the creation and maintenance of an osmotic gradient in the medullary interstitium.

The generation of the gradient of medullary hyperosmolality depends primarily upon the energy-dependent reabsorption of 25% of the filtered load of NaCl in a water-impermeable region of the tubule—the thick ascending limb of the loop of Henle.

The solute so reabsorbed would be washed away from the medulla were it not for the fact that the capillaries within the medulla are arranged in loops—the vasa recta—the descending and ascending limbs of which are close to each other and to the adjacent loops of Henle. Reabsorbed solute diffusing into an ascending capillary loop will tend to increase the concentration of NaCl in the capillary at that point. As it is carried towards the cortex, it flows past plasma in the adjacent descending capillary loop in which the NaCl concentration is lower. At every level there will be a tendency therefore for NaCl to diffuse from the ascending to the descending capillary loop and thus to be retained within the renal medulla rather than be lost to the cortex (Fig. 18.23).

Similarly, at every level, plasma in the descending limb coming from a more dilute region of the medulla will be slightly less hyperosmotic than plasma in the ascending limb which is emerging from the deeper, more hyperosmotic medulla. Thus, in contrast to solute, H_2O will tend at every level to pass from the descending to the ascending limb driven by the osmotic gradient and, thereby, be shunted away from the deeper medulla. Urea, passing from the medullary collecting ducts into the intersitium during antidiuresis, will also be trapped within the medulla by diffusion from ascending to descending vasa recta. These diffusional movements between the two limbs of the vasa recta are often described as **countercurrent exchange**.

The arrangement of the vasa recta allows the medullary gradient to be maintained but, were it not for the sources of water-free solute (from energy-dependent NaCl reabsorption in the thick ascending limb and from urea diffusion from the collecting ducts into the medulla during antidiuresis), the large osmotic gradient could not be created. This process is often referred to as **countercurrent multiplication**. In effect, what happens is that some of the solute filtered at the glomerulus as an isosmotic solution with H_2O has been trapped in the medulla while its associated H_2O is either lost from the distal tubule in the cortex or excreted in the urine.

Diuresis

There are two types of diuresis: water and osmotic. Water diuresis occurs whenever water intake exceeds body needs; it results from suppression of ADH secretion. Osmotic diuresis results when more solute is presented to the tubules than can be reabsorbed.

An increased rate of production of urine (diuresis) can be of two types: water

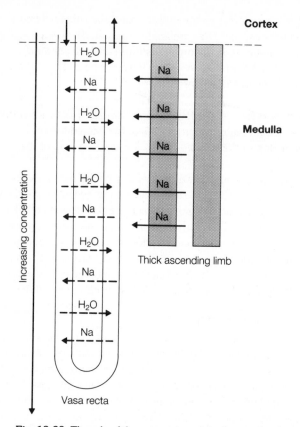

Fig. 18.23 The role of the vasa recta and thick ascending limb of the loop of Henle in generating and maintaining the medullary interstitial concentration gradient. For convenience the only solute shown is Na^+.

diuresis and solute or osmotic diuresis. **Water diuresis** results when water is ingested or administered in excess of the body's requirements. ADH secretion is suppressed (p. 646), the collecting ducts become relatively impermeable to water and the excess water is lost without solute. A typical water diuresis is illustrated in Fig. 18.24. There is an inverse relationship between urine osmolarity and urine

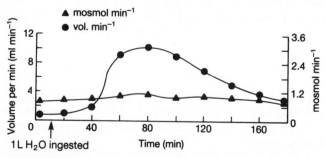

Fig. 18.24 An example of the changes in the rates of excretion of water (volume min^{-1}) and of total solutes (mosmol min^{-1}) in a water diuresis in the human.

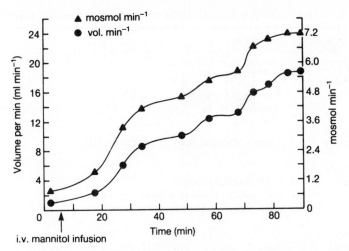

Fig. 18.25 An example of the dependence of the rate of water excretion on solute excretion in an osmotic diuresis induced by the intravenous (i.v.) infusion of mannitol.

flow rate; flow rate × osmolarity (the amount of solute excreted per minute) is relatively constant and independent of flow rate. Thus the kidney can adjust its excretion of water without markedly affecting its handling of solutes.

Osmotic diuresis results when more solute is presented to the tubules than they can reabsorb. For example, an osmotic diuresis occurs if:

1 a non-reabsorbable solute, e.g. mannitol, is filtered;

2 the concentration of glucose in the plasma in diabetes mellitus rises so that the filtered load exceeds the tubular maximum; or

3 tubular function is inhibited (e.g. by drugs which block reabsorption of NaCl in one or more nephron segments).

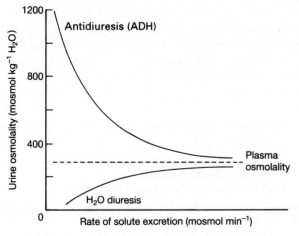

Fig. 18.26 The relationship between the rate of solute excretion and urine osmolality. As solute load increases, urine osmolarity approaches that of plasma. For further explanation see text.

Table 18.2 Summary of the functions of the different segments of mammalian nephrons

Segment of nephron	Functions
Renal corpuscle	Ultrafiltration: 125 ml min^{-1}, approx. 20% RPF
Proximal tubule Reabsorbs	Two-thirds filtered Na$^+$, Cl$^-$ and H$_2$O Most of the HCO$_3^-$, glucose, K$^+$, HPO$_4^{2-}$, amino acids, uric acid 50% urea
Secretes	Organic acids and bases, H$^+$ and some NH$_4^+$
Loop of Henle Reabsorbs	H$_2$O (descending limb) Na$^+$ and Cl$^-$ (ascending limb) (also, some K$^+$, Ca^{2+} and Mg^{2+})
Distal tubule Reabsorbs Secretes	Na$^+$, Cl$^-$ (HCO$_3^-$), (K$^+$), H$_2$O H$^+$, NH$_4^+$, K$^+$
Collecting duct Reabsorbs Secretes	Na$^+$, Cl$^-$, H$_2$O, (K$^+$), urea H$^+$, NH$_4^+$, (K$^+$)

Brackets indicate very small effect under normal conditions. If the body needs to retain water, ADH secreted from the neurohypophysis into the blood increases the permeability of the collecting duct luminal membrane to water, and more water is retained. Urinary volume may be only about 0.5 ml min^{-1}. Therefore, more than 99.5% filtered water has been reabsorbed. Conversely, in the absence of ADH, urinary volume can be as much as 20 ml min^{-1} or about 17% of GFR.

A typical osmotic diuresis is illustrated in Fig. 18.25. In contrast to water diuresis, urinary flow rate depends upon urinary solute content. Moreover, the greater the rate of solute excretion, the lower is the maximal attainable urinary concentration, even with maximal concentrations of ADH (Fig. 18.26), for the larger volume of water reabsorbed from the collecting ducts dilutes the medullary interstitial fluid. Faster flow through the ascending limbs of loops of Henle, especially combined with a decreased concentration of Na$^+$ when glucose or mannitol is present, may also decrease reabsorption so that less Na$^+$ is deposited to maintain the medullary osmotic gradient. This gradient cannot be demonstrated in kidneys removed during osmotic diuresis.

The roles of the different segments of the nephron are summarized in Table 18.2.

18.7 FUNCTION IN DISEASED KIDNEYS

Damaged kidneys may fail to maintain the normal composition and volume of the body fluids because they lose or excessively retain natural constituents of the body or else because they fail to remove waste products.

Disturbances of glomerular function

In the nephrotic syndrome there is damage to the glomerular basement mem-

brane with consequent leakage of protein into the tubular fluid. In chronic renal failure, glomeruli are progressively destroyed. Decreased GFR results in retention of urea and creatinine in the body with increased plasma concentrations of these solutes.

The glomeruli may leak protein or may fail to produce their normal daily quota of 180 litres of protein-free filtrate for the following reasons.

Protein loss from the kidney

In the nephrotic syndrome, an abnormal glomerular basement membrane allows plasma proteins, especially albumin, to escape; the urine may contain from 5 to as much as 30 grams of protein per day. When synthesis in the liver fails to keep up with renal loss, there is less circulating albumin and its concentration in the plasma falls, so that the volume and the colloid osmotic pressure of the plasma tend to be reduced. There is a body-wide disturbance of the Starling equilibrium (p. 428), with gross generalized oedema, caused and sustained by excessive retention of Na^+ by the kidneys.

Failure to produce enough glomerular filtrate

In chronic renal failure, destruction of nephrons gradually reduces GFR (Fig. 18.27).

About three-quarters of a person's nephrons can be destroyed before renal function is obviously impaired. The remaining nephrons hypertrophy and adapt so that total function is surprisingly well-maintained. For substances excreted primarily by the glomeruli, excretion depends upon filtered load, not upon GFR as such. Consequently, for creatinine or urea, the rate of excretion will be maintained if plasma concentration increases in proportion to dwindling GFR. The normal amount of creatinine can be eliminated in one-quarter as much glomerular filtrate if each volume of filtrate contains four times as much creatinine, i.e. if plasma creatinine concentration is four times normal. Similarly, the rising plasma urea

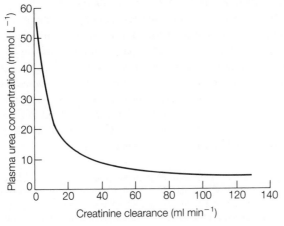

Fig. 18.27 The relationship between plasma urea concentration and the clearance of creatinine, which estimates GFR.

concentration with advancing renal failure is not merely a sign of falling GFR; it is an effective compensating mechanism that allows the excretion of urea to be maintained.

In a steady state, the rate of excretion of urea must equal the rate of production. If GFR suddenly falls to half, excretion will at first lag behind production and plasma urea concentration will increase until it reaches twice normal, when excretion will catch up with production. The body's urea pool will then have twice as much in it, but retention of urea will no longer be increasing. If GFR falls to 12 ml min^{-1} (10% of normal), plasma urea will have to rise to 40 mmol instead of 4 mmol and the body will contain 1700 mmol instead of a normal 170 mmol of urea, but the rates of production and of excretion will again be equal and normal at about 500 mmol or 30 g per day.

Disturbances of tubular function

Because glomerular capillaries are damaged in chronic glomerular disease, peritubular capillaries are deprived of blood, and the associated tubules are therefore also destroyed. Each functional nephron that remains handles a greater load than normal and operates in a state of osmotic diuresis. The ability to concentrate or dilute the urine is lost. Na$^+$ balance is often maintained until late in the disease process, partly at least because, although less Na$^+$ can be filtered, the osmotic diuresis means that less can be reabsorbed distally. This means that a greater fraction of the filtered load is excreted. Loss of functional renal mass results eventually in too little K$^+$ secretion so that the body retains K$^+$. It will also result in a metabolic (renal) acidosis when HCO$_3^-$ synthesis becomes insufficient to replace HCO$_3^-$ consumed in buffering the H$^+$ ions produced by metabolism.

In chronic destructive diseases renal tissue gradually disappears, but the amount of solute to be excreted is not diminished. The hypertrophied and adapted nephrons that are left have to do all the work that was formerly done by a full complement of normal nephrons. If 10% of the nephrons remain, each one of these must on average handle 10 times as much solute as a nephron in a normal kidney. Under such conditions of permanent osmotic diuresis, the urine cannot be concentrated. During water deprivation, normal kidneys can make urine concentrated to four times the osmolarity of the plasma, but only at rates less than ~0.6 ml min^{-1} (Fig. 18.26). The attainable concentration falls as flow rate increases during osmotic diuresis, and at 15 ml min^{-1} urine can be little more than isosmolar. With only 10% of the nephrons remaining, the urine cannot be more concentrated than the plasma if the rate of production exceeds 1.5 ml min^{-1}. This is about 2000 ml per day, the volume of isosmolar urine required to contain the daily output of urinary solutes, which explains the fixed osmolarity and specific gravity of 1010 (isosthenuria) characteristic of renal failure.

Inability to concentrate and to reduce the volume of the urine also accounts for the **nocturia** which may be an early sign of renal damage. Note that a simple test of the power of the kidneys to concentrate the urine (such as the urinary osmolarity or specific gravity after 12 hours overnight without fluid) is both the simplest and the most sensitive test for detecting early impairment of renal function, especially in diseases like pyelonephritis which attack the medulla. Note too that the increased urinary volume that results from the kidney's loss of concentrating power promotes the excretion of urea; the clearance of urea increases with rising flow rate and becomes maximal above about 2 ml min^{-1} (Fig. 18.9).

Handling of sodium

In contrast to urea, retention of Na$^+$ with increasing concentration in the plasma would disturb the balance between cells and extracellular fluid. Retention is avoided by reabsorbing a smaller fraction of filtered Na$^+$ and since normally more than 99% of filtered Na$^+$ is reabsorbed, there is a large margin of safety. If GFR fell to 12 ml min^{-1} (one-tenth of normal), Na$^+$ balance could still be maintained by reabsorbing 97.5% of the filtered load of 2500 mmol per day. Even with GFR reduced to 3 ml min^{-1} (one-fortieth of normal), the filtered load would be 600 mmol per day and balance could be maintained by reabsorption of five-sixths of the filtered Na$^+$. Possible causes of reduced fractional reabsorption include:

1 osmotic diuresis, which decreases contact time for distal reabsorption;
2 reduced availability of NH$_4$$^+$, which means that Na$^+$ must accompany conjugate bases of urinary acids; and
3 possibly the action of natriuretic substances.

Sometimes, and especially when tubular functions have been damaged more than glomerular, too much Na$^+$ may be lost, with depletion of the volume of extracellular fluid which further depresses GFR and aggravates retention of nitrogenous substances.

Handling of potassium

As nephrons are lost, there are fewer cells in total capable of secreting K$^+$. This, together with increased catabolism of tissues in uraemia which releases K$^+$ from broken-down cells, may lead to a dangerous elevation of plasma K$^+$ concentration requiring treatment by dialysis.

Handling of hydrogen ions

Renal acidosis is a metabolic acidosis arising from impaired capacity of the renal tubular cells to secrete H$^+$ ions and add HCO$_3$$^-$ to the plasma, and is accordingly often characterized as **renal tubular acidosis**. Possible causes include:

1 failure to secrete H$^+$ into the urine and add HCO$_3$$^-$ to the plasma at a sufficient rate;
2 failure to bring urine to a sufficiently low pH, so that less H$^+$ can be excreted as titratable acid and ammonium;
3 failure to supply sufficient NH$_4$$^+$ to carry H$^+$ ions into the urine; and
4 shortage of filtered buffer to make into titratable acid.

Only the last of these depends primarily upon GFR. The rest depend upon metabolic failure to transport H$^+$ ions against gradients or to provide NH$_4$$^+$. The biggest single factor is usually the inadequate supply of NH$_4$$^+$ from the dwindling of the mass of metabolizing tubular tissue as nephrons are destroyed.

In summary, patients with failing kidneys have an undiminished need for the services of their kidneys, and so must overload a dwindling population of functioning nephrons. Nitrogenous end-products accumulate in the blood (**azotaemia**) because there are too few glomeruli, and GFR falls. Lack of tubular tissue leads to failure to secrete sufficient K^+, H^+ and NH_4^+ and sometimes to retention of Na^+. The overloaded remaining nephrons, possibly with a disorganized medulla with its countercurrent arrangements wrecked, cannot concentrate the urine; and so the patient must live with an increased turnover of water or risk dehydration. Two well-organized kidneys have been reduced to a motley collection of a few nephrons.

18.8 MICTURITION

The bladder can accommodate up to about 300–400 ml of urine with little increase in tension. As volume increases further and wall tension rises, this is appreciated as a sensation of fullness. If appropriate, an autonomic spinal cord reflex—the micturition reflex—is triggered and urination occurs. This involves a coordinated relaxation of internal and external sphincters, contraction of the bladder muscle and of the abdominal wall and pelvic floor musculature.

Autonomic reflexes, under the control of higher centres, bring about the emptying of the bladder. This hollow organ has stretch receptors in its wall which signal the wall tension and thus the organ volume. The signals are appreciated consciously as a sensation of fullness and they also activate reflex connections in the spinal cord.

The smooth muscle of the bladder wall (the **detrusor muscle**) and neck (the **internal sphincter**) is innervated by sympathetic fibres (inhibitory) from the lumbar segments of the spinal cord and by parasympathetic fibres (excitatory) from sacral segments 2 to 4. The **external sphincter** is a striated muscle with a somatic innervation. When bladder-wall tension reaches a certain level, the detrusor contracts reflexly and the internal sphincter relaxes. It should be noted, however, that the bladder wall can accommodate increasing volumes of urine with very little alteration in tension. Only when the volume reaches 300–400 ml is there normally an appreciable degree of discomfort associated with a steeper rise in tension and triggering of the **micturition reflex**. That this emptying reflex can be facilitated from the brain is shown by the fact that the bladder can be emptied at any volume. More often, however, these spinal reflexes are inhibited by cerebral activity, until either it is appropriate to urinate or the sensations become insistent. The reflex activity is then augmented and the external sphincter relaxed. The flow of urine then begins. This is facilitated by ancillary reflexes from the urethra that are stimulated by flow of urine and whose efferent activity reinforces bladder muscle contraction and sphincter relaxation. Contraction of the abdominal wall and the pelvic floor also aids the complete emptying of the bladder.

Following transection of the spinal cord and the onset of spinal shock, the bladder wall is inert and the sphincter closed, resulting in urinary retention with overflow. As the shock wears off and detrusor muscle tone returns, reflex emptying can be brought about by, for example, cutaneous stimulation, which may at the same time cause defecation and other evidence of widespread autonomic activity (**mass reflex**).

Chapter 19
Fluid and Electrolyte Balance

19.1 WATER

Distribution of water in the body

In a man about 60% of body weight is due to water (p. 3). A 70-kg man therefore has about 42 litres. In a woman, the relatively greater percentage of the body weight contributed by fat cells (which contain little water) means that about 55% of body weight is water. A 50-kg woman therefore has about 27.5 litres of water. Of the total body water, about 55% is in the cells; 45% is extracellular.

This extracellular water in a 70-kg man is further subdivided into plasma (3 litres), interstitial fluid and lymph (15 litres) and transcellular fluids (cerebrospinal fluid, ocular, pleural, peritoneal and synovial fluids; 1 litre). The volumes of these compartments can be estimated from the volumes of distribution of substances thought to equilibrate in different compartments. For example, total body water has been estimated from the volume of distribution of urea and isotopes of water (deuterium oxide, tritiated water). Similarly, inulin, sucrose, mannitol and isotopes of sodium and chloride have been used to estimate extracellular water, and isotopically labelled albumin to estimate plasma water. Interstitial water cannot be measured directly but is (ignoring the transcellular volume) the difference between extracellular volume and plasma volume.

The best guide to changes in body water is alteration in body weight. In estimating body water from body weight, however, it must be remembered that fat contains very little water, so that the greater the body fat as a proportion of body weight, the lower the body water content.

Water balance

Normally, total body water remains constant. Therefore, over a 24-hour period, intake and loss of water must balance exactly.

Water is taken in drinks as well as up to a litre per 24 hours in solid food. It is also formed from the oxidation of metabolites (about 300 ml per 24 hours).

Water is lost by evaporation through the respiratory system (0.5 litre) and skin (0.5 litre) and, depending upon the need to increase heat loss, from the skin in sweat. Faeces contain about 0.1 litre of water per 24 hours. Urinary water loss is variable. About 600 mosmol of solutes, including end-products of metabolism, must be excreted each day in the urine. The maximal achievable urinary osmolarity of about 1200 mosmol L^{-1} therefore demands a minimal urinary volume of 0.4–0.5 litre per day. Normally, water intake is such that about 1.5 litres of urine is excreted each day.

Regulation of total body water

The intake and urinary loss of water are both controlled, by thirst and antidiuretic hormone (ADH), respectively. In health, the primary determinant of both thirst and ADH release is the osmolarity of the plasma perfusing the hypothalamus.

If water loss exceeds gain, there is a reduction in total body water content and the osmolarity of the body fluids increases. This excites **thirst**, so that water may be ingested, and releases **ADH** so that water is retained by the kidneys. Conversely, an excess intake of water expands and dilutes the body fluids; the decrease in osmolarity eliminates thirst and inhibits the release of ADH, so that water diuresis rids the body of the excess. These mechanisms can hold the osmolarities of plasma and other body fluids constant in any one person to within ±2%. However, in any normal population, plasma osmolarities range between about 280 and 300 mosmol L^{-1}.

Both thirst and the release of ADH appear to be triggered by the same mechanisms. Of prime importance, normally, is the osmolarity of the plasma perfusing the hypothalamus. Neurons in the supraoptic nucleus synthesize ADH which is stored with neurophysin in the nerve endings in the neurohypophysis (p. 276). The blood–brain barrier is deficient in bthis region of the hypothalamus, and nearby osmoreceptors which relay to the supraoptic nucleus respond to a local increase in plasma Na^+ concentration. A separate set of osmoreceptors in the same hypothalamic region appears to be involved in signalling thirst.

The receptors that signal thirst have an osmotic threshold about 10 mosmol higher than that of the osmoreceptors involved in ADH release. Thus, normally, thirst is not experienced until ADH release has ensured that the ingested water will be retained by the kidneys.

If plasma osmolarity is increased by solutes which penetrate plasma membranes, e.g. urea or glucose, thirst and ADH release are not initiated. This suggests that changes in the volumes of the osmoreceptor cells link the changes in plasma osmolarity to the response.

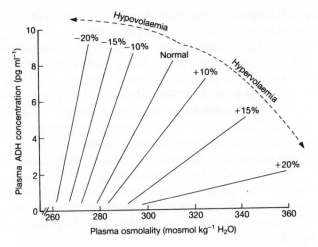

Fig. 19.1 The effect of blood volume and of plasma osmolarity on antidiuretic hormone (ADH) secretion. Note the relative contributions of changes in plasma osmolarity and changes in circulating blood volume to ADH secretion. (Adapted from Robertson, S.L., Shelton, R.L. & Athar, S. (1976) *Kidney Int.* **10**, 25–37.)

Other mechanisms which stimulate thirst and ADH release include:

1 decreased central venous pressure, signalled via the low pressure receptors in atria and great veins;

2 decreased arterial blood pressure, signalled via the carotid and aortic baro-receptors; and

3 increased angiotensin II (ANG II) in the brain (produced locally or delivered via the systemic circulation, or both).

These mechanisms are of particular importance in conditions where circulating blood volume is deficient (p. 441 and 460). Water may even be retained in excess of solute in response to the depleted circulating volume and the control of extracellular osmolarity may be sacrificed. Figure 19.1 illustrates the relationship between plasma osmolarity, circulating ADH and changes in circulating blood volume.

Disturbances of total body water content

Changes in total body water affect the concentrations of solutes in all of the body compartments. Plasma Na^+ concentration is raised (hypernatraemia) when fluid intake is not possible or when the thirst response is blunted. Loss of water is shared between the cells and the extracellular compartment. Plasma Na^+ concentration is decreased (hyponatraemia) by consumption or administration of inappropriately large volumes of water in situations where the renal excretion of water is impaired. Gain of water is shared between the cells and the extracellular compartment.

Hypernatraemia

The effectiveness of the thirst–ADH mechanism ensures that a primary loss of water (**primary dehydration**) *only* occurs when fluid intake is not possible or when the thirst response is blunted (e.g. in the absence of a source of water, in the very young, the aged, the confused, the unconscious or in patients with severe vomiting who cannot drink and retain water). Primary dehydration is most likely to be seen when loss of water is excessive, e.g. with hyperventilation (particularly at high altitude where humidity is low), high environmental temperature, fever or abnormal urinary loss resulting either from a failure to produce or secrete ADH (**neurogenic diabetes insipidus**) or from a decreased responsiveness of the collecting ducts to ADH (**nephrogenic diabetes insipidus**).

The loss of water increases the osmolarity in all fluid compartments and the concentration of Na^+ in the plasma (**hypernatraemia**). The loss is shared between the cells and the extracellular compartment, so that cell volume is decreased but extracellular volume is better preserved than when there is a loss of extracellular Na^+. Nevertheless, the contraction of extracellular volume promotes renal retention of Na^+.

Apart from thirst which may dominate consciousness, symptoms associated with water loss are vague, e.g. weakness and lethargy. The urine is scanty and concentrated, except in diabetes insipidus. Plasma osmolarity and sodium concentration are increased. It is important that the lost water be replaced by mouth or, if necessary, intravenously (as 5% glucose). In neurogenic diabetes insipidus, ADH is also administered.

Hyponatraemia

This occurs with consumption or administration of inappropriately large volumes of water in situations where the renal excretion of water is impaired, e.g. in anuria, with inappropriate secretion of ADH as in trauma, and with neoplasms that produce ADH. **Hyponatraemia** may also follow excessive loss of extracellular fluid (ECF) associated with loss of blood or of Na^+ when the thirst–ADH mechanism is stimulated through the low-pressure central venous stretch receptors, and the maintenance of normal osmolarity is sacrificed in order better to maintain circulating plasma volume.

As the retention of water lowers extracellular osmolarity and plasma Na^+ concentration, water moves from the ECF into the cells, which swell until the osmolarities of the fluid compartments are equalized.

Depending on the degree of water retention, weakness and muscle cramps may develop, followed by confusion (water intoxication), convulsions, coma and death. The plasma hypo-osmolarity and low plasma Na^+ concentration can be reversed by restricting water intake and treating the underlying cause of the condition.

Note that the brain cells, in particular, seem to be able to regulate their volume by adjusting their solute content when exposed to alterations in plasma osmolarity. In hypernatraemia, they shrink initially but gradually increase their solute content over some hours and take up water to restore their volume towards normal. Conversely, in hyponatraemia, they first swell but then lose solute and with it water, again partly restoring their volume over some hours. This ability to adjust volume has obvious value for an organ enclosed in a rigid container (the skull). However, it means that hyper- and hyponatraemia ought never to be reversed too rapidly. For example, in hypernatraemia, the adapted cells contain more solute than normal and a sudden restoration of normal plasma osmolarity will result in pronounced swelling of these cells, raised intracranial pressure and symptoms and signs like those of hyponatraemia.

19.2 SODIUM

Distribution of sodium in the body

Sodium is the major cation of the ECF. A typical 70-kg man, with an extracellular Na^+ concentration of 150 mmol L^{-1} and an extracellular volume of 19 litres, would have 2850 mmol Na^+ in the extracellular compartment. At an average concentration of 10 mmol L^{-1}, 23 litres of cellular water would contain 230 mmol of Na^+. The bones contain a further 2500 mmol of Na^+, less than half of which is readily exchangeable with extracellular Na^+ or other cations; none of this contributes to body fluid osmolarity. Total exchangeable Na^+, measured by isotope dilution, is about 50 mmol kg^{-1} body weight.

Sodium balance

On western-type diets, 100–300 mmol Na^+ are consumed daily. Almost all of this is absorbed from the gut (together with the much larger amount secreted into the gut; p. 595); the faeces normally contain only 5–10 mmol daily. The only additional route for Na^+ loss, other than the renal, is from the skin in sweat. This loss is extremely variable and depends solely upon the need to maintain a relatively constant central body temperature. Each litre of sweat contains 30–50 mmol of Na^+ so that the loss of a few litres can cause a significant Na^+ loss from the extracellular compartment. Renal loss of Na^+ is adjusted to maintain Na^+ balance and can range from a few millimoles up to about 500 mmol per day.

Regulation of body sodium

Day-to-day variations in the amount of body Na^+ result in changes in ECF volume. The major variable that is monitored to provide information about total body Na^+ is the central venous volume through the low-pressure stretch receptors in the central veins and atria. Information received from these and other receptors is processed in a variety of cardiovascular centres in the brain. The efferent limb of the control mechanism involves regulation of the renal handling of Na^+. Both filtered load and tubular reabsorption may be altered. Among factors affecting tubular reabsorption are changes in the balance of the Starling forces between the peritubular capillaries and the luminal fluid; the renin–angiotensin–aldosterone system; prostaglandins; atrial natriuretic peptide; renal nerves; and humoral inhibitors of Na^+–K^+ ATPase.

Most of the 'mobile' Na^+ that can vary in amount from day to day is in solution in the ECF. For every 150 mmol change in Na^+ content, there is a corresponding change of 1 litre in ECF volume. Indeed *the volume of the ECF is regulated by regulating Na^+ content.*

Unlike the regulation of total body water through the thirst–ADH mechanism, which is comparatively well-understood, the mechanisms involved in the regulation of NA^+ and thereby extracellular volume remain controversial. A brief simplified outline is provided here.

Regulation requires the monitoring of some function of the variable to be controlled—central coordination— and the apropriate control of an effector organ, in this case the kidney.

Monitoring of body Na$^+$

Sodium concentration is already fixed through the thirst–ADH mechanism, and neither total extracellular Na$^+$ nor volume is monitored directly. Rather, the volume of a subcompartment—**central venous volume**—is monitored through the low-pressure stretch receptors in the central veins and atria (p. 441).

But how can the volume of this subcompartment adequately reflect total extracellular volume? Most of the ECF is interstitial fluid which forms a weak gel with mucopolysaccharides, largely hyaluronic acid, and thus cannot move as a free fluid under gravitational forces, although it offers no hindrance to diffusion. One hypothesis proposes that this interstitial gel is unsaturated, because pressures as low as 6 mmHg below atmospheric have been measured within it. This imbibition pressure reflects the tendency for the gel to take up and thereby immobilize fluid, and while the gel remains unsaturated the vascular system and extravascular interstitial spaces have similar compliances. Consequently, fluid gained or lost from the extracellular compartment will be shared in proportion between plasma and interstitial fluid, and monitoring the volume of one will monitor the volume of both. However, if sufficient fluid accumulates to saturate the gel (3–4 litres), additional fluid in the interstitial compartment now moves freely under gravity to the most dependent parts of the body (resulting in oedema), the total extracellular volume is no longer monitored by changes in central venous pressure, and regulation of volume breaks down.

The low-pressure receptors in the great veins and atria appear to play the major role in monitoring ECF volume. The vagal nerves carry information from the atrial low-pressure stretch receptors to brainstem centres which also relay to hypothalamic areas involved in cardiovascular homeostasis (p. 446). The carotid sinus and aortic arch baroreceptors, which monitor systemic arterial blood pressure, probably contribute significantly to the regulation of ECF volume only when intravascular volume is so severely depleted that arterial blood pressure is also reduced. In addition, there is evidence that the granular cells in the juxtaglomerular apparatus sense changes in renal perfusion pressure. As well as modifying renin release in response to such changes, information is sent via renal afferent nerves to cardiovascular centres in the brain.

Renal handling of Na$^+$

The efferent limb of the control mechanism involves regulation of the renal handling of Na$^+$. Overall, Na$^+$ excretion = filtered load – tubular reabsorption.

Filtered load

When Na$^+$ concentration in the plasma remains constant, filtered load varies only with glomerular filtration rate (GFR). Factors such as blood loss or loss of extracellular volume, exercise, and adoption of the upright posture activate cardiovascular reflexes mediated by sympathetic nerves to the kidneys and cause catecholamine release from the adrenal medulla. Afferent and/or efferent glomerular arteriolar resistances are altered, thereby changing GFR and filtered load.

Autoregulation maintains GFR relatively constant over a wide range of arterial blood pressure. However, an alteration of only 1 ml min^{-1} in GFR would over 24 hours result in a change in filtered load of Na$^+$ of about 200 mmol—more Na$^+$ than many people consume in a day! Unfortunately, the methods for measuring GFR are not accurate enough to detect such changes and it is not known how important minor changes in GFR are in determining renal Na$^+$ excretion.

Tubular reabsorption

The handling of Na$^+$ by the different parts of the nephron is discussed in Section 18.6 (p. 631). Here factors that influence reabsorption are outlined.

Intrarenal factors

1 Changes in the balance of the **Starling forces** (p. 428) between the peritubular capillaries and the luminal fluid alter net Na$^+$ and water reabsorption, particularly in the proximal tubules.

For example, a decreased plasma protein concentration (as would follow an increase in plasma volume), will favour both increased GFR and decreased tubular reabsorption, and result in an appropriately increased salt and water excretion. Increased efferent arteriolar resistance will increase GFR and decrease peritubular hydrostatic pressure. This will favour greater reabsorption of the filtrate and assist in maintaining glomerulotubular balance.

2 In addition to any extrarenal activation, the **renin–angiotensin system** can be completely activated within the kidneys. **ANG II** is formed through the following sequence (Fig. 19.2). Renin, a specific proteolytic enzyme produced in the juxtaglomerular apparatus in the kidney, splits off a decapeptide angiotensin I (ANG I) from angiotensinogen, an α_2-globulin synthesized in the liver. ANG I is converted in turn to ANG II (an octapeptide) by **converting enzyme**. This is found

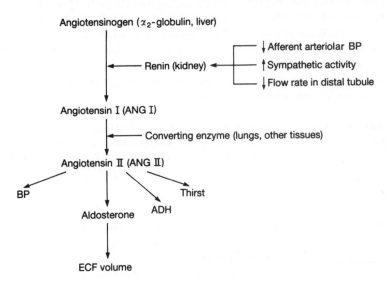

Fig. 19.2 The formation and major physiological roles of angiotensin II. BP, Blood pressure.

within the kidney. Outside of the kidney, converting enzyme is most plentiful in the lungs. The rate-limiting step in this sequence is normally the release of renin from the granular cells in the juxtaglomerular apparatus.

In the kidney, renin release from the granular cells is stimulated by:

(a) decreased arterial blood pressure acting directly on the afferent arterioles;

(b) increased sympathetic activity (including circulating catecholamines) acting through β-adrenoceptors; and

(c) a decrease in the rate of delivery of fluid past the macula densa in the distal tubule.

The ANG II then formed has two important intrarenal effects. It stimulates Na^+ reabsorption in most nephron segments and it constricts the glomerular arterioles. Together these effects favour Na^+ retention and the restoration of ECF volume.

3 Several **prostaglandins** (PG) are synthesized in the kidney. PGE_2 inhibits tubular Na^+ reabsorption and antagonizes the effect of ADH on collecting duct water permeability. Also, PGE_2 and PGI_2 counteract the vasoconstricting effects of ANG II on glomerular arterioles. This may be of importance in maintaining renal blood flow during exercise. The prostaglandins thus favour renal excretion of Na^+ and water and oppose the effects of the renin-angiotensin system.

4 The **kallikrein–kinin system, endothelin, adenosine** and **dopamine** all have effects on renal haemodynamics and/or tubular function and can influence salt and water excretion.

Extrarenal factors

1 As well as its intrarenal function, the **renin–angiotensin system** has important extrarenal actions that affect renal salt and water handling.

Secretion of **aldosterone** (p. 294) from the zona glomerulosa of the adrenal cortex is stimulated principally by circulating ANG II (p. 651). (Of lesser importance are decreased Na^+ and increased K^+ concentrations in the blood perfusing the adrenal gland and also adrenocorticotrophic hormone.)

Aldosterone promotes the reabsorption of Na^+ by a variety of tight epithelia, principally the distal portion of the nephron, but also the epithelium of the colon and the rectum, and the ducts of salivary and of sweat glands. In all of these, it promotes Na^+ absorption and K^+ secretion.

Aldosterone stimulates transepithelial Na^+ transport by first, increasing the availability of luminal Na^+ channels thereby favouring Na^+ entry to the cells from the lumen; second, increasing synthesis and insertion of Na^+-K^+ ATPase into the basolateral membrane, and third, increasing synthesis of some of the enzymes involved in the Krebs cycle, thereby making more energy available to the cells.

The effects of aldosterone on Na^+ reabsorption take about 1 hour to become apparent and up to 24 hours to become fully established. Therefore this hormone is important for long-term rather than minute-to-minute adjustments of Na^+ excretion. It is inactivated in the liver.

As well as promoting the secretion of aldosterone, ANG II is a potent vasoconstrictor in the systemic circulation and can increase total peripheral resistance and therefore renal perfusion pressure. An increase in mean arterial pressure, however, inhibits further release of renin. ANG II also promotes thirst and stimulates the release of ADH by acting directly on circumventricular organs which are outside the blood–brain barrier. Thus it not only stimulates Na^+ retention through aldosterone, but also the acquisition and retention of water, and so plays a major role in the regulation of extracellular volume. It is inactivated by angiotensinases in the plasma.

In summary, in both its intrarenal and extrarenal effects the renin–angiotensin system plays a major role in the conservation of Na^+ and water under both physiological and pathophysiological conditions.

2 Atrial natriuretic peptide (ANP) is released from the cardiac atria in response to stretch. It is also found in the hypothalamus and other regions of the brain where it may have a role in the central control of fluid balance and blood pressure.

ANP increases the excretion of Na^+ by increasing the rate of glomerular filtration, inhibiting Na^+ reabsorption (either directly or indirectly) in proximal tubule and cortical and medullary collecting ducts, and reducing the secretion of renin and aldosterone. It also has systemic effects, reducing venous return (at least in part by promoting fluid shift from capillaries to the interstitial space) and total peripheral resistance, thereby decreasing cardiac output and arterial blood pressure.

Whether ANP plays a role in Na^+ homeostasis under physiological conditions is still unclear. However, it appears to be of importance in protecting against acute overloading of the circulation in pathological states.

3 Renal nerves play a major role in Na^+ homeostasis. *Afferent* fibres from the afferent arterioles send information about renal perfusion pressure to cardiovascular centres in the brain. Low levels of *efferent* sympathetic nerve stimulation release renin from the juxtaglomerular apparatus; higher levels cause afferent arteriolar constriction. In addition, proximal tubules are supplied with sympathetic fibres; stimulation promotes Na^+ reabsorption.

The consequences of efferent nerve stimulation are thus a reduced GFR, increased proximal Na^+ reabsorption and increased distal Na^+ reabsorption (via ANG II and aldosterone). The result is Na^+ retention.

4 A variety of other factors may influence renal handling of Na^+. There is some evidence for a **hypothalamic cardiac glycoside-like inhibitor of Na^+–K^+ ATPase** which would inhibit tubular Na^+ reabsorption throughout the nephron. As well as its established role in increasing the water permeability of the collecting ducts, **ADH** promotes Na^+ reabsorption in the thick ascending limb of the loop of Henle and collecting duct and thereby contributes to the generation of the medullary osmotic gradient that is required for the production of a concentrated urine.

An overview of Na^+ homeostasis

Normally, dietary intake and total losses of Na^+ from the body are in balance. If intake is increased, there is a lag of several days before a new steady state is reached so that initially there is some increase in total body Na^+ and ECF volume. This results in increased central venous pressure with consequent inhibition of the renal mechanisms that favour Na^+ retention and activation of those that favour Na^+ excretion. Thus increased Na^+ intake reduces renin release and therefore

ANG II formation and aldosterone secretion, as well as renal sympathetic nerve activity. At the same time, renal prostaglandin production is enhanced and ANP may also be released from the atria. The net effect is to restore the balance between Na^+ intake and excretion. Normally this is so effective that balance is re-established without excessive volume expansion or hypertension.

Converse changes will occur if Na^+ intake is decreased. Again, balance will be re-established without excessive loss of extracellular volume or hypotension.

Disturbances of body sodium content

Changes in body Na^+ content result in changes in extracellular volume. Retention of Na^+ is accompanied by sufficient retention of water to maintain the osmolarity of the ECF and results in generalized oedema. This is only seen in an adult if more than 3–4 litres of fluid is accumulated. It can occur because the kidneys can no longer excrete sufficient Na^+ to maintain Na^+ balance (as may occur in renal failure) or if there is a widespread disturbance of the capillary Starling forces so that the effective circulating plasma volume is reduced and the kidneys are thereby stimulated to retain Na^+ (as may occur in heart failure, nephrotic syndrome and chronic liver disease). Depletion of Na^+ is caused by absent or diminished Na^+ intake, often combined with excessive loss from skin (sweat), gastrointestinal tract (e.g. vomiting, diarrhoea) or through the kidneys (absence of aldosterone, osmotic diuresis). Loss of Na^+ leads to isosmotic loss of water which is shared between the plasma and the extravascular ECF. The cell water is unaffected or may even increase if Na^+ concentration is depressed. Hence, for a given loss of water, the consequences are much more serious for the circulation than are those of primary loss of water.

Sodium retention and oedema

Oedema is a clinically detectable excess of extravascular ECF. It may be either localized or generalized.

Localized oedema need not imply an increased amount of Na^+ in the body if the affected region is small. It results from local disturbance of the Starling equilibrium (p. 428) by:

1   increased capillary hydrostatic pressure (e.g. with venous stasis, thrombosis or obstruction);

2   increased pericapillary colloid osmotic pressure (e.g. with lymphatic obstruction or inflammatory exudation); or

3   increased capillary permeability, leading to increased loss of protein to the interstitial spaces (e.g. inflammatory and allergic conditions).

Generalized oedema can occur because the kidneys can no longer excrete sufficient Na^+ to balance intake (as can occur in the late stages of chronic renal failure) or there is widespread disturbance of the Starling equilibrium (as in congestive heart failure, liver disease and the nephrotic syndrome), with a shift of fluid from the circulation to the interstitial space, resulting in decreased plasma volume and secondary renal salt and water retention. There is no single and wholly satisfactory explanation for the development of oedema in all circumstances.

Nevertheless, it is argued that the renal salt and water retention is a consequence of a decreased **effective circulating plasma volume** with diminished renal perfusion pressure.

Generalized oedema is associated with increased body weight, the movement of free fluid under gravity, tissue swelling and pitting on pressure. There is typically at least 3–4 litres of excess fluid (enough to saturate the interstitial gel and overflow) and hence there must be an excessive amount of Na^+ in the body—150 mmol for each litre of oedematous fluid or kilogram of excess weight. This disturbance cannot occur by the mere shifting of fluid from the plasma, for there is less than 4 litres of plasma, and so a patient whose whole plasma volume was shifted into the tissue spaces would not be oedematous — and would also not be alive!

Two important causes of widespread loss of fluid from blood vessels are:
1 lack of plasma protein, especially albumin, so that the colloid osmotic pressure is low everywhere. This may occur in gross undernutrition, in diseases of the liver which impair production of albumin, and when the glomerulus becomes leaky to protein so that albumin is lost in the urine faster than the liver can replace it (**nephrotic syndrome**); and
2 a widespread increase in hydrostatic pressure in the capillaries, as in chronic cardiac failure with venous congestion. In **cirrhosis of the liver**, obstruction to the portal vessels results in raised pressure in the portal system and fluid accumulates in the abdominal cavity—a condition known as **ascites**.

Excessive retention of Na^+ by the kidneys is essential for the development of oedema. When plasma albumin is low, and there is a diminished plasma volume, renal blood flow (RBF) and GFR are reduced so that less Na^+ is filtered; secretion of renin is stimulated and secretion of aldosterone is increased, promoting Na^+ reabsorption. In liver disease, breakdown of aldosterone may be slower, increasing the circulating concentration. When plasma albumin is depleted, retained NaCl and accompanying H_2O do not remain in the vessels to restore plasma volume, signal the increase in total ECF volume, and turn off the enhanced retention of Na^+. The kidney acts in a manner appropriate for the correction of a low volume of blood and ECF.

There must, however, be other mechanisms, because the volume of circulating plasma is often not reduced in patients with nephrotic oedema whose total volume of ECF is greatly increased.

When cardiac failure results in decreased cardiac output and increased capillary hydrostatic pressure, reflex vasoconstriction reduces RBF and GFR, so that less Na^+ is filtered. Renin may be released and aldosterone secretion increased, while breakdown of aldosterone in the congested liver may be slowed. Reabsorption of Na^+ is therefore enhanced, Na^+ is retained and ECF volume increases.

Retention of Na^+ would first tend to elevate plasma Na^+ concentration, but this stimulates the thirst–ADH mechanism so water is retained and, in uncomplicated oedema, plasma Na^+ concentration and total plasma osmolarity are within normal limits. (The reduced colloid osmolarity when albumin is deficient amounts to no more than 1 or 2 $mosmol\ L^{-1}$ and this decrease is not detected when total osmolarity is measured.)

However, if effective circulating plasma volume remains low, increased ADH secretion can result, with additional water retention and hyponatraemia. Thus it is possible to have hyponatraemia despite an increased total body Na^+.

Since it is dietary Na^+ that is retained to make oedema fluid, treatment by restriction of dietary salt is logical. It is, however, unpleasant, and often ineffective as it tends to mobilize aldosterone. A variety of diuretic drugs which inhibit renal tubular reabsorption are commonly used to combat retention of Na^+.

Depletion and deficiency of sodium

This is caused by absent or diminished Na^+ intake, often combined with excessive loss. Under extreme environmental conditions, up to 15 litres per day of **sweat** with 30–50 mmol L^{-1} of Na^+ may be lost. Normally, about 8 litres of fluid containing up to 1 mole of Na^+ are secreted and reabsorbed per day in the **gastrointestinal tract**. This Na^+ can be lost through vomiting, aspiration of the gut contents, by diarrhoea or through fistulae. In cholera and other secretory diarrhoeas, the rate of intestinal secretion may be increased enormously. Normally losses through non-renal routes are countered by the virtual disappearance of Na^+ from the urine. But the urine can become a vehicle of loss in **Addison's disease** (absence of aldosterone), during osmotic diuresis, and in diabetes mellitus when Na^+ is lost not only as a consequence of the osmotic diuresis but also with the conjugate bases of ketoacids.

So long as osmoregulation continues, loss of Na^+ leads to loss of water at the rate of 1 litre per 150 mmol Na^+. This loss of water is shared between the plasma and the extravascular ECF. The cell water is unaffected or may even increase if Na^+ concentration is depressed. Hence, *for a given loss of water, the consequences are much more serious for the circulation than are those of primary loss of water*, since the ECF and plasma bear the brunt of the water deficiency. This is shown by haemoconcentration, with increases in red-cell count, plasma protein concentration and viscosity, and evidence of impaired renal function with retention of urea and other waste products as a consequence of the depressed GFR. Moreover, thirst is less prominent than in pure water loss with its increased osmolarity, and the cardinal manifestation of depletion of body Na^+ is peripheral circulatory failure.

There is also the danger of a vicious cycle developing if tissue perfusion falls enough for cells deprived of oxygen to swell, taking up H_2O and Na^+, and further reducing ECF volume,

Table 19.1 Summary of disturbances of body water

	Osmolarity of body fluids	Compartmentrs affected	Manifestations
Water excess			
Primary	↓	ICF↑ECF↑	H_2O intoxication
Secondary to ↑ Na^+	Normal*	only ECF↑	Oedema
Decreased water			
Primary	↑	ICF↓ECF↓	Thirst
Secondary to ↓ Na^+	Normal†	only ECF↓	Circulatory failure

* If the effective circulating volume remains depressed, the thirst–ADH mechanism may result in H_2O intake and retention, which dilutes ECF in an attempt to maintain extracellular volume. Thus hyponatraemia and its pathophysiology may be superimposed on primary Na^+ retention.
† If depletion of extracellular volume is sustained, the thirst–ADH mechanism may result in H_2O intake and retention, which dilutes ECF in an attempt to maintain extracellular volume. Thus hyponatraemia and its pathophysiology may be superimposed on primary Na^+ depletion.

circulating blood volume and perfusion of the tissues. As the loss of ECF continues, the thirst–ADH mechanism is activated as a consequence of decreased venous volume and, eventually, of decreased arterial blood pressure. This may result in retention of H_2O without Na^+ and consequently in decreased body fluid osmolarity and hyponatraemia. Thus it is possible to see hyponatraemia associated with a decreased volume of total body water as well as with pure water retention with overexpansion of total body water.

Note that retention of water alone is relatively ineffective in restoring circulating blood volume in people who are Na^+-deficient. In a 70-kg man with about 23 litres of cell water and 19 litres of extracellular water, less than half of any additional water would be retained within the extracellular compartment, and only one-sixth of this (about 75 ml for each additional litre) would remain in the plasma in the steady state.

In contrast to hypernatraemia, it is Na^+, not water, that needs to be administered either orally (with glucose to promote intestinal absorption) or as an isosmotic intravenous solution.

19.3 POTASSIUM

Distribution of potassium in the body

Potassium is the chief intracellular cation of people and of the animals and plants that they eat. An adult contains about 3500 mmol K^+, of which 300 mmol is in the skeleton, 80 mmol is in the ECF and the remaining 90% is in the cells, some 2700 mmol of it in skeletal muscle, the largest component of the cell mass. Total exchangeable K^+ is about 50 mmol kg^{-1} of body weight in men and about 40 mmol kg^{-1} in women, whose muscles form a lower proportion of body weight. The intracellular concentration ($K_i \sim 150$ mmol L^{-1}) is not critical but a two- or threefold increase or decrease in the extracellular concentration (K_o 4–5 mmol L^{-1}) can paralyse muscles or stop the heart. Rapid loss of 5% of cellular K^+ into the ECF would be lethal. Extracellular K^+ concentrations below 2.5 or above 7 mmol L^{-1} produce weakness of limb and trunk muscles. Below 2 and above 9 mmol L^{-1}, flaccid paralysis occurs, the respiratory musculature is affected and the heart may cease to beat. These changes in cardiac and skeletal muscle activity are consequences of altered K^+ gradients with a normal cellular K^+. Low plasma K^+ concentrations (**hypokalaemia**) increase the diffusion gradient, hyperpolarizing the plasma membrane so that the nerve and muscle cells become less easily depolarized and therefore less excitable. High plasma K^+ concentrations (**hyperkalaemia**) decrease the diffusion gradient, depolarizing the plasma membrane. Initially, as threshold is approached, cells may become more excitable but, once the membrane potential has depolarized below threshold, action potentials can no longer be generated and paralysis results. Hence acute alterations in plasma K^+ concentration, which leave cellular K^+ concentration unaltered, are less well-tolerated than are chronic alterations in which cellular K^+ concentrations change in the same direction and the ratio of cell to plasma $[K^+]$ remains relatively constant.

Potassium balance

Ordinary diets supply about 100 mmol. The faeces remove about 10 mmol and the urine 90 mmol per day.

Functions of potassium

1   In the cell it maintains an optimal environment for cellular enzymes and is important for cell growth (DNA, protein synthesis).

2   It is important in cell-volume regulation because it is the dominant inorganic cation balancing charge on cell anions.

3   It plays a role in acid–base balance by exchanging for H$^+$ when ECF pH alters.

If ECF pH decreases (acidosis), H$^+$ enters cells in exchange for cell K$^+$; if ECF pH increases (alkalosis), H$^+$ leaves the cell in exchange for ECF K$^+$.

3   The resting membrane potential (RMP) is largely a K$^+$ diffusion potential (p. 18) and, as discussed above, the ratio K_i/K_o is important in maintaining RMP and therefore tissue excitability.

Regulation of body potassium

About half of an acute K$^+$ load is lost in the urine over the first 6 hours or so; the remainder is taken up by cells, stimulated by insulin and adrenaline, and is excreted subsequently. Thus, plasma concentration of K$^+$ is little affected by K$^+$ intake. In the distal portion of the nephron, K$^+$ is secreted by the principal cells and is reabsorbed by the intercalated cells. Of less importance to K$^+$ balance is K$^+$ secretion in the colon. Given that the K$^+$ concentration in cells of the distal nephron and colon mirrors that of cells elsewhere, if losses of K$^+$ to the lumen are adjusted to maintain a normal cell K$^+$ concentration, body K$^+$ content will be regulated.

Though losses of K$^+$, principally in the urine, maintain body K$^+$ constant despite variations in intake, the details of this regulation are poorly understood. Following an acute K$^+$ load, about a half is lost in the urine over the next 6 hours or so; the remainder largely disappears from the ECF into cells and is excreted subsequently. Thus plasma concentration of K$^+$ is little affected by K$^+$ intake. Conversely, in patients depleted of K$^+$, cells lose K$^+$ in an attempt to maintain plasma K$^+$ concentration within normal limits.

Several hormones affect the distribution of K$^+$ between cells and the ECF. The most important physiologically is insulin. A raised plasma K$^+$ concentration stimulates release of insulin, and insulin promotes cellular uptake of K$^+$. Insulin may therefore be used clinically in an emergency to lower plasma K$^+$ concentration. Adrenaline appears to have a physiological role in stimulating cellular K$^+$ uptake after meals and in ameliorating the K$^+$ loss from skeletal muscle during severe exercise.

Both insulin and adrenaline appear to increase K$^+$ uptake by stimulating the Na$^+$–K$^+$ ATPase. Adrenaline does this by acting at β-adrenoceptors to increase cyclic adenosine monophosphate (cAMP). The link between insulin and stimulation of the Na$^+$–K$^+$ ATPase is unclear.

In addition to its well-established effects in promoting K$^+$ secretion, together with Na$^+$ reabsorption in the distal nephron, colon and other Na$^+$ absorbing tight epithelia, aldosterone may also promote K^+ uptake into skeletal muscle. In comparison to the very rapid actions of insulin and adrenaline, the full effects of aldosterone take some hours to be established.

The acid–base balance also influences exchange of K^+ between cells and extracellular fluid. During acidosis, H^+ ions enter cells in exchange for K^+; during alkalosis, K^+ enters cells in exchange for H^+ ions. These effects are more marked with metabolic than with respiratory disturbances.

The major excretory route for K^+ is through the kidneys (p. 630). In the late distal tubule and collecting ducts, K^+ ions are both reabsorbed and secreted. Of the two cell types lining the nephrons in this region, the **principal** cells reabsorb Na^+ and secrete K^+ (Fig. 18.19) whereas the **intercalated cells** secrete H^+ ions and reabsorb K^+ (Fig. 18.18). Factors which promote K^+ *secretion* by the *principal* cells include:

1 increased dietary K^+ intake, which results in proliferation of the basolateral membranes and increased Na^+–K^+ ATPase activity;
2 aldosterone;
3 increased rate of luminal Na^+ delivery;
4 increased flow rate; and
5 metabolic alkalosis, which stimulates secretion by increasing the activities of the basolateral Na^+–K^+ ATPase and of the luminal K^+ channels.

Secretion of K^+ by the principal cells is *decreased* by metabolic acidosis which inhibits these pathways. (In contrast, respiratory alkalosis and acidosis have little effect on distal K^+ secretion.)

Factors which promote K^+ *reabsorption* by the *intercalated* cells include decreased dietary K^+ intake, which is associated with changes consistent with the insertion of additional transporters (presumbaly H^+–K^+ ATPase) into the luminal membrane, and acidosis, which stimulates H^+ secretion by these cells and therefore K^+ reabsorption.

The urinary excretion of K^+, therefore, reflects the relative activities of the reabsorptive and secretory processes and the clearance ratio for K^+ can range between 0.1 (reabsorption predominant) and 3.5 (secretion predominant). In a person in K^+ balance, with dietary intake of 100 mmol per day, the clearance ratio ordinarily approximates 0.15.

Potassium is also secreted by epithelial cells lining the colon. Uptake of Na^+ from the lumen appears to stimulate secretion of K^+, and aldosterone promotes both processes. Normally, gut secretion contributes no more than 10% to the total excretion of dietary K^+.

Whereas the control of water balance and, to a lesser extent, of Na^+ balance are fairly well-understood, there is no comparable framework for discussing the regulation of K^+ balance. Are there receptors sensitive to a function of body K^+? Is there central integration of information about body K^+? How is the function of the effector organs (kidney and colon) regulated so that K^+ excretion matches K^+ intake? These questions are still unanswered. Most discussions of K^+ balance imply that the K^+ concentration in cells of the distal nephron and colon mirrors that of cells elsewhere so that factors affecting cellular K^+ generally affect these cells as well. Then if losses of K^+ from principal cells to the luminal fluid and from colonic epithelial cells to the gut lumen are adjusted to maintain the concentration of K^+ in those cells constant, body K^+ content will perforce be regulated. Whether this simple view is adequate remains to be established.

Disturbances of body potassium content

Plasma K^+ concentration does not necessarily parallel body K^+ content. Chronic depletion of body K^+ may be associated with hyperkalaemia (e.g. in a metabolic acidosis) and body K^+ may be greater than normal in the presence of hypokalaemia (e.g. in a metabolic alkalosis). Loss of body K^+ usually reflects a combination of diminished intake with increased loss through the gut or kidneys (e.g. vomiting, diarrhoea, diuretics) and in these situations may result in hypokalaemia. Excessive loss of cellular K^+ combined with diminished renal function (e.g. shock, crush injuries, terminal renal failure) can result in hyperkalaemia.

Potassium is normally excreted by the kidneys so effectively that body K^+ content remains constant, any increase in intake being matched by urinary excretion. However, plasma K^+ concentration *does not* necessarily parallel body K^+ content.

Potassium depletion and hypokalaemia

These must be distinguished from each other. Chronic depletion of body K^+ may be associated with hyperkalaemia, for example, in a metabolic acidosis where cell K^+ is exchanged for the extracellular H^+ ions. Conversely, body K^+ may be greater than normal in the presence of hypokalaemia, for example, in a metabolic alkalosis. Normally, plasma K^+ concentration remains relatively constant, excess or deficit in body K^+ being accommodated by the cells. Thus measurements of plasma K^+ concentration *do not* allow conclusions to be drawn about body K^+ content. Estimates of the latter have been obtained by using isotopes or by extracting K^+ from muscle biopsy specimens. A recent finding that the K^+ content of red blood cells mirrors that of other cells offers a simpler way to estimate total body K^+.

Chronic K^+ depletion is associated with degenerative changes in myocardium and skeletal muscle as well as with functional changes (muscular weakness, paralysis). Vacuolar changes in renal epithelial cells accompany an impaired urinary concentrating ability and a polyuria which is insensitive to exogenous ADH.

Since the kidneys normally conserve K^+, loss of body K^+ usually reflects a combination of diminished intake with increased loss through the gut or kidneys. Vomiting, diarrhoea, aspiration of gastrointestinal contents and fistulae may all lead to hypokalaemia. Aldosterone and a variety of diuretics which facilitate K^+ secretion by the epithelial cells through additional distal delivery of Na^+ promote loss of K^+ in the urine. Typical electrocardiogram (ECG) changes may give early warning of hypokalaemia (Fig. 19.3).

Hyperkalaemia

Clinically hyperkalaemia is usually caused by excessive loss of cellular K^+ combined with diminished renal function as, for example, in:

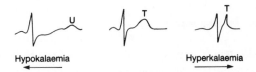

Hypokalaemia   Hyperkalaemia

Fig. 19.3 Characteristic changes in ECG produced by alterations in plasma K^+ concentration. Hypokalaemia results in a reduction in amplitude or reversal of the T wave and the appearance of a U wave; hyperkalaemia results in a reduced P wave, a bizarre QRS complex and a tall narrow T wave.

1 shock, where the fall in mean arterial blood pressure results in hypoperfusion of tissue and a loss of K^+ from hypoxic cells, as well as decreased renal perfusion;
2 crush injuries, where damaged cells leak K^+ into the plasma and myoglobinuria disrupts renal function; and
3 terminal renal failure, where increased tissue catabolism releases K^+ from cells and depressed renal function prevents its excretion.
In all of these conditions, plasma K^+ rises because K^+ is lost from cells faster than the kidneys can excrete it. Total body K^+ may actually be falling, though plasma K^+ is elevated.

The symptoms and signs of hyperkalaemia *per se* are vague; cardiac arrhythmias and cardiac arrest may develop before excitability of nerve and skeletal muscle has been clinically affected. As with hypokalaemia, typical progressive changes in the ECG (Fig. 19.3) give the earliest indications of the severity of hyperkalaemia. In an emergency, plasma K^+ concentration may be decreased abruptly by insulin which drives K^+ (and glucose) into cells. Bicarbonate also lowers plasma K^+ by stimulating cellular uptake of K^+ in exchange for H^+ ions.

Potassium and acid–base balance

Cells exchange K^+ and H^+ ions with plasma. In metabolic acidosis, plasma K^+ concentration increases even though body K^+ may become depleted.

In comparison, respiratory acidosis has little effect on plasma K^+, possibly because the raised Pco_2 generates intracellular HCO_3^- and K^+ remains in the cells to balance the charge on this anion.

In metabolic alkalosis plasma K^+ concentration may decrease. But though cells gain K^+ initially, chronic alkalosis may lead to loss of body K^+ because of increased K^+ secretion by renal principal cells, which is facilitated by increased delivery of Na^+ to this segment of the tubule. This encourages exchange of tubular Na^+ for cell K^+ with retention of K^+ in the lumen to preserve electroneutrality.

Conversely, chronic K^+ depletion can lead to an alkalosis—increased renal secretion of NH_4^+ and H^+, resulting in an increased transfer of cell HCO_3^- to the plasma. This leads to the paradoxical association of an acid urine with an alkaline plasma.

19.4 MAGNESIUM

Distribution of magnesium in the body

Over 60% (~500 mmol) of the total content of Mg^{2+} is found in bone; much of the remainder is in the cells, mostly bound to proteins, not in free solution. The plasma concentration is ~ 1 mmol L^{-1} with about one-third protein-bound, so that the interstitial concentration is about 0.6 mmol L^{-1}.

Magnesium balance

Ordinary diets supply about 20 mmol each day. About 8 mmol (40%) of this is absorbed from the gut, so that 12 mmol is excreted in the faeces. In steady-state conditions, the urine therefeore contains 8 mmol each day. However, in a person with Mg^{2+} deficiency, this can fall to 0.5–1 mmol per day.

Disturbances of body magnesium content

Magnesium depletion

Magnesium depletion has been described after prolonged gastrointestinal aspirations, chronic watery diarrhoea, steatorrhoea and severe chronic alcoholism. It may also be due to sustained renal loss, e.g. following chronic diuretic administration. A pure Mg^{2+} deficiency almost never occurs for a deficiency in any of the three major cellular elements—magnesium, potassium and phosphorus—results in deficiencies in the others. The reason for this is that deficiency of any one of these cellular elements is associated with anorexia, cell atrophy, a negative nitrogen balance and a net loss of the other two elements.

There is thus a close correlation between skeletal muscle and myocardial Mg^{2+} and K^+; in primary K^+ deficiency, for every 10 mmol decrease in K^+ muscle Mg^{2+} content decreases by 0.5 mmol. Conversely, with chronic Mg^{2+} depletion, muscle K^+ also decreases and K^+ will not be restored to normal until Mg^{2+} depletion is corrected.
As with K^+, a deficiency in Mg^{2+} can exist despite a normal plasma concentration. Within weeks on a Mg^{2+}-deficient diet, anorexia, nausea, vomiting and weakness develop. After months, when the plasma Mg^{2+} concentration has fallen (< 0.5 mmol L^{-1}), a syndrome resembling tetany but with emphasis on mental rather than physical irritability is found. Hypocalcaemia contributes to the tetany and related symptoms, for severe hypomagnesaemia inhibits parathyroid hormone release and interferes with the skeletal release of Ca^{2+} in response to that hormone. Table 19.2 summarizes some of the effects of Ca^{2+} and Mg^{2+} on neuromuscular excitability.
The major risk in Mg^{2+} depletion is cardiac arrhythmias and cardiac arrest. The ECG may reveal typical changes in electrical activity.

Magnesium excess

Hypermagnesaemia is rare.

It may be seen in patients in end-stage renal failure or be precipitated by ingestion of large quantities of Mg^{2+} salts. Muscle paralysis and coma can result if plasma concentrations exceed 5–7 mmol L^{-1}.

Calcium metabolism is discussed in Chapter 10 and is not considered here.

Table 19.2 Summary of the effects of ionized Ca^{2+} and Mg^{2+} on neuromuscular excitability

Site of action	Plasma concentration			
	$\uparrow Ca^{2+}$	$\downarrow Ca^{2+}$	$\uparrow Mg^{2+}$	$\downarrow Mg^{2+}$
Nerve axon excitability	$\downarrow$	$\uparrow$	$\downarrow$	$\uparrow$
Neuromuscular junctions				
Release of acetylcholine	$\uparrow$	$\downarrow$	$\downarrow$	$\uparrow$
Response of muscle membrane	$\downarrow$	$\uparrow$	$\downarrow$	$\uparrow$
Overall effect	Block	Tetany	Block	Tetany

19.5 ACID–BASE METABOLISM AND REGULATION OF pH

Cells, especially those of the nervous system, are sensitive to the pH of their surroundings. Too great an alkalinity leads to headache, lassitude, tetany and convulsions. Too great an aciditiy makes respiration deep and sighing, and causes drowsiness progressing to coma. Plasma pH below 7 or above 7.8 is usually lethal, so that H^+ concentration must remain between 16 and 100 nmol L^{-1}.

The blood in humans is normally kept blandly alkaline at pH 7.4 ± 0.04, a H^+ concentration of 36–44 nmol L^{-1}. Cells are less alkaline than plasma, with pH nearer 7.0.

Note how little free H^+ there is in the body. Forty litres of body water at a pH of 7.0 would contain 4000 nmol or 0.004 mmol of H^+; holding this constant within 10% means a tolerance of 0.0004 mmol; and each day the body produces by metabolism 13 000 mmol of potentially acid-forming CO_2 and 50–80 mmol of stronger, non-volatile acids—H_2SO_4 and H_3PO_4. A little arithmetic shows that the precision of control is far better than that of a watch keeping time to 1 s per day!

Mechanisms of control

Stability is achieved by buffers and exchanges of ions with bones and cells which minimize changes of pH pending disposal of acids or alkalis by lungs or kidneys. The important buffer bases that take up H^+ ions in the body are the conjugate bases of three weak acids—carbonic acid, proteins and phosphoric acid. The reactions may be written:

Bicarbonate: $H^+ + HCO_3^- \rightleftharpoons H_2CO_3 \rightleftharpoons H_2O + CO_2$

Phosphate: $H^+ + HPO_4^{2-} \rightleftharpoons H_2PO_4^-$

Proteinate: $H^+ + Prot^{n-} \rightleftharpoons HProt^{(n-1)-}$

or, in general: $H^+ + Buf^- \rightleftharpoons HBuf$

Note that this is reversible. If pH tends to fall, buffer base takes up H^+ while, if pH tends to rise, buffer acid supplies H^+, and so pH is stabilized and changes are buffered. Note also that just as much Buf^- is used up as H^+ is neutralized. For example, when sulphuric acid is added to the plasma:

$SO_4^{2-} + 2H^+ + 2Buf^- + 2Na^+ \rightarrow 2Na^+ + 2HBuf + SO_4^{2-}$

Only the very weak base SO_4^{2-} and the weak buffer acid HBuf remain; *pH is controlled, but buffer base has disappeared.*

Why plasma pH is 7.4

The Henderson–Hasselbalch equation describes the relationship between pH and the concentration of conjugate base and acid. When applied to the HCO_3^-/CO_2 buffer system, it reveals that when the kidneys hold the HCO_3^- concentration constant at 24 mmol L^{-1} and the respiratory system holds P_{CO_2} at 40 mmHg, plasma pH must be 7.40. Use of this equation provides a rational classification and approach to acid–base disturbances.

The Henderson–Hasselbalch equation,

$$pH = pK' + \log_{10} \frac{[\text{conjugate base}]}{[\text{conjugate acid}]}$$

may be applied to any of the plasma's buffer pairs, most conveniently to the predominant one, bicarbonate/carbonic acid. pK', the acid dissociaton constant, for the sequence

$$CO_2 + H_2O \rightarrow H_2CO_3 \rightarrow H^+ + HCO_3^-$$

is 6.1, and the solubility of CO_2 in plasma is 0.03 mmol mmHg. Hence,

$$pH_{\text{plasma}} = 6.1 + \log_{10} \frac{[HCO_3^-]}{0.03\, P_{CO_2}}$$

With $[HCO_3^-]$ maintained at about 24 mmol L^{-1} by the kidney and P_{CO_2} regulated close to 40 mmHg by the respiratory system (p. 537).

$$pH = 6.1 + \log_{10} \frac{24}{0.03 \times 40} = 7.40$$

The Henderson–Hasselbalch equation indicates the important variables: pH varies with $[HCO_3^-]$ and inversely with P_{CO_2}. Hence a change in pH that results from an alteration in *either* $[HCO_3^-]$ or P_{CO_2} can be avoided or corrected by changing the *other variable* to preserve or restore the buffer ratio.

In a solution in which bicarbonate is the only buffer base, doubling or halving $[HCO_3^-]$ or P_{CO_2} shifts the pH by 0.3 unit. In blood, which has other buffers besides bicarbonate, doubling or halving P_{CO_2} produces a smaller final change in pH of 0.2 unit because other buffers release or take up H^+ and $[HCO_3^-]$ changes in the same direction as P_{CO_2}.

The Henderson-Hasselbalch equation also summarizes the basis for physiological regulation of pH: the kidneys control the numerator and the respiratory system the denominator of the terms in the equation. The respiratory chemoreceptors are affected by pH as well as by P_{CO_2} (p. 543). Adding acid to the blood lowers pH as well as $[HCO_3^-]$. The reduction in pH stimulates respiration so that P_{CO_2} is lowered to match the lowered $[HCO_3^-]$, and pH is rapidly returned towards normal.

Finally, the Henderson–Hasselbalch equation offers a rational classification of disturbances of pH (p. 670). The numerator, $[HCO_3^-]$, is affected by ingestion, production or loss of acid or alkali, giving rise to metabolic (including renal) acidosis or alkalosis. The denominator, 0.03 P_{CO_2}, is affected by alterations in pulmonary ventilation or in the composition of inspired air, giving rise to respiratory acidosis or alkalosis. Fundamentally, disturbances of acid–base balance fall into one or other of the following categories; clinically, mixed disturbances are sometimes seen:

1   reduced $[HCO_3^-]$: **metabolic acidosis;**
2   increased $[HCO_3^-]$: **metabolic alkalosis;**
3   increased P_{CO_2}: **respiratory acidosis**; and
4   decreased P_{CO_2}: **respiratory alkalosis**.

Production and initial buffering of acids

Although some 13 000 mmol of CO_2 is generated metabolically each day, the associated H^+ is buffered very efficiently and the CO_2 is excreted rapidly by the lungs. So long as the rates of production and excretion are in balance, plasma pH will remain stable. Approximately 50–80 mmol of non-volatile acids (principally H_2SO_4 and H_3PO_4) is produced each day. The anions are excreted in the urine. Mechanisms which minimize the fall in pH prior to disposal of the associated H^+ ions are dilution in total body water, buffering in blood, interstitial fluid and cells, buffering by carbonate in bone and exchange of H^+ ions for bone cations. These together constitute whole-body buffering.

Carbonic acid

CO_2 is produced at the rate of about 13 000 mmol per day (10 mmol min^{-1}) and represents potential acid that would need 13 000 mmol per day of NaOH to neutralize it to $NaHCO_3$. In fact, CO_2 is excreted by the lungs, and the associated H^+ ions are very effectively buffered in transit from tissues to lungs in the blood, mainly by haemoglobin (p. 536). Removal of O_2 as CO_2 is being taken up, replaces oxygenated haemoglobin (O_2Hb^-) by Hb^- which is a much stronger base. About 0.7 mmol of H^+ for each mmol of O_2 given up (about 25 ml at body temperature) is required to prevent the blood becoming even more alkaline than normal. Hence only about 0.1 mmol of the 0.8 mmol of CO_2 produced from 1 mmol of O_2 (with respiratory quotient of 0.8) yields H^+ ions which need to be handled by buffers other than Hb, and venous blood is only about 0.03 pH more acid than arterial. In summary:

$$CO_2 + H_2O \rightarrow H_2CO_3 \rightarrow H^+ + HCO_3^-$$
(carbonic anhydrase in rbc)

$$H^+ + O_2Hb^- \rightarrow HHb + O_2$$

Summing these two,

$$O_2Hb^- + H_2O + CO_2 \rightarrow O_2 + HHb + HCO_3^-$$

Because red-cell HCO_3^- exchanges readily with plasma Cl^-, CO_2 taken up into the blood is mostly carried as HCO_3^- in the plasma (p. 535). The process is reversed in the pulmonary capillaries.

Non-volatile acids

Each day metabolism yields a total of about 50–80 mmol of H_2SO_4 (from oxidation of sulphur in sulphur-containing amino acids) and H_3PO_4 (derived from phosphoproteins and phospholipids). Organic acids produced in metabolism are normally oxidized to CO_2 and H_2O.

In shock, and in exercise severe enough to be partly anaerobic, oxidation may fail to keep up with production, and lactic and pyruvic acids may also accumulate. In diabetes, as much as 750 mmol of acetoacetic and β-hydroxybutyric acids may be produced daily. The H^+ ions from all these non-volatile acids must be buffered in the body.

Mechanisms which minimize the fall in pH caused by H^+ ions pending final disposal are as follows.

1   Dilution in total body water. Acid produced in cells diffuses into ECF and into other cells; each 10-fold dilution raises pH by 1 unit.

2   Buffering in blood. It would take 28 mmol of H^+ to bring 1 litre of whole blood from pH 7.4 to 7.0. The H^+ ions would be shared between the buffers of the blood as follows: HCO_3^-, 18 mmol; inorganic phosphate, 0.3 mmol; plasma protein, 1.7 mmol; and haemoglobin, 8 mmol. Five litres of blood could take up 140 mmol H^+ (twice the normal daily production) without a lethal decrease in pH; but note that $[HCO_3^-]$ would be reduced from 24 to 6 mmol L^{-1}.

3   Buffering in extravascular ECF. This is mainly achieved by HCO_3^-, for there is no haemoglobin and much less protein than in the blood. The 14 litres of ECF containing 24 mmol L^{-1} of HCO_3^- could take up a further 200 mmol of H^+ before pH was reduced to 7.0.

Processes 1, 2 and 3 are rapid, being completed in minutes, and total ECF including plasma could take about 350 mmol, or some five times the day's production without disaster. This is ample reserve for normal purposes.

4   Buffering in cells. Intracellular buffers, mainly proteinate and organic phosphate, can back up extracellular buffers by coping with H^+ ions that enter cells which exchange them for K^+ or Na^+. Their capacity is probably similar to that of extracellular buffers but the process is much slower (hours) because of limited permeability of cell membranes to H^+.

5   Carbonate in bone. This can contribute some HCO_3^- to the plasma and ECF to replace some of that which is used up in 2 and 3; again this probably takes hours.

6   Exchange of ions with bone mineral. Extracellular H^+ ions can exchange with cations from bone and cells, e.g. $H^+ + M^+$ in bone $\rightarrow H^+$ in bone $+ M^+$, where M^+ may represent Na^+ or K^+ or $\frac{1}{2}Ca^{2+}$ or $\frac{1}{2}Mg^{2+}$. These are slow processes, taking hours to days. Prolonged metabolic acidosis depletes bones of Ca^{2+} and cells of K^+.

These six processes all add up to **whole-body buffering.**

The effectiveness of whole-body buffering was well-demonstrated in 1953 by R.F. Pitts who intravenously titrated a dog with molar HCl. The blood pH decreased from 7.44 to 7.14 after addition of 150 ml of acid (i.e. 150 mmol H^+). By contrast, the addition of the same amount of HCl to 11 kg of water (the estimated water content of the dog) brought the pH to 1.84. Since pH 7.44 corresponds to a $[H^+]$ of 36 nmol L^{-1} and pH 7.14 to a $[H^+]$ of 72 nmol L^{-1}, 36 × 11 or about 400 nmol of the added H^+ remained free in the dog, i.e. of 150 mmol or 150 000 000 nmol of the H^+ added to the dog, 400 remained free and

149 999 600 was buffered in the body. Of this, about a half was buffered extracellularly and a half by cells and bone.

Final disposal of hydrogen ions and regeneration of body buffers

The kidneys manufacture HCO_3^- which they add to the plasma. Much of this replaces the filtered HCO_3^- which reacts with H^+ in the lumen and is converted to CO_2 and H_2O. In addition, to maintain acid–base balance, the kidneys must manufacture sufficient new HCO_3^- to replace all of the HCO_3^- used up in the body in buffering the H^+ ions produced each day from metabolism. This new HCO_3^- represents HCO_3^- formed with H^+ ions that are buffered in the urine and are measured as titratable acid, and HCO_3^- formed as a consequence of the synthesis of ammonium.

It falls to the kidneys to rectify the changes produced by additional acid or alkali in the body. Because of respiratory compensation, pH may be not far from normal but the buffer stores, indicated by the concentration of HCO_3^- in the plasma, may be either depleted (by acids) or augmented (by excess alkali) and these need to be restored to their normal levels.

In terms of the body's principal buffer base, HCO_3^- , what happens when, for example, sulphuric acid is added, may be summarized thus:

$$H_2SO_4 \rightarrow SO_4^{2-} + 2H^+$$

$$2H^+ + 2HCO_3^- \rightarrow H_2CO_3 \rightarrow 2H_2O + 2CO_2(\text{expired})$$

After this, two H^+ ions have become hydrogen atoms in body water. Two CO_2 have been breathed out. The pH has fallen a little, but in the plasma two HCO_3^- have been replaced by one SO_4^{2-}. The kidneys easily dispose of sulphate and the conjugate bases of other unwanted acids that are replacing HCO_3^- in the plasma. These pass into the glomerular filtrate and need only to be left unreabsorbed by the tubules to be excreted in the urine. The H^+ ions no longer have a free existence; they are present as parts of buffer bases or of body water, and are not directly available for excretion.

Eventually, because all the buffers are in equilibrium, as the kidneys manufacture HCO_3^- which they add to the plasma, H^+ ions temporarily buffered by other buffer bases will finally be transferred to HCO_3^- and converted to H_2O. But for every H_2O molecule so formed, a molecule of HCO_3^- has been destroyed. *The role of the kidney is to manufacture HCO_3^- to replace the HCO_3^- used up.* The kidney cells make this HCO_3^-, which they restore to the plasma, together with H^+ which they secrete into the lumen. The H^+ ions excreted in the urine are, therefore, simply a byproduct of the manufacture of HCO_3^-. Because equal numbers of H^+ and HCO_3^- ions are formed in this way, by the time that the HCO_3^- used up in coping with H^+ ions in the body has been replaced, an equivalent amount of H^+ formed in the kidney will have been excreted in the urine.

Bicarbonate production in the kidney

Figure 19.4 illustrates the relationships between HCO_3^-, H^+ and ammonium production.

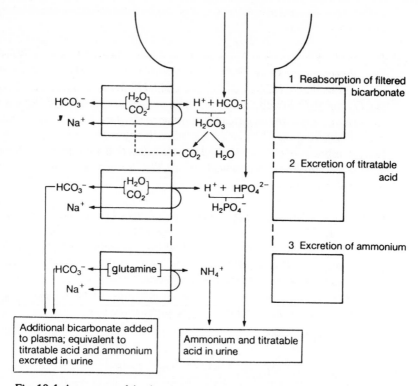

Fig. 19.4 A summary of the three processes by which renal tubular epithelial cells are enabled to synthesize sufficient HCO_3^- (and to secrete H^+ and NH_4^+ ions) to maintain plasma HCO_3^- constant. (Modified from Robinson, J.R. (1972) *Fundamentals of Acid–Base Regulation*, 4th edn, p. 76. Blackwell Scientific Publications, Oxford.)

1 Replacement of filtered bicarbonate with newly formed HCO_3^- from tubular cells; this consumes about 4500 mmol H^+ per day. Within the lumen, carbonic anhydrase attached to the brush borders catalyses the breakdown of H_2CO_3 and allows CO_2 to recirculate. Normally, this process is largely complete by the end of the proximal tubule and simply restores to the body HCO_3^- that would otherwise have been lost in the glomerular filtrate.

2 Production of titratable acid. Buffers in the glomerular filtrate react with secreted H^+ ions. Normally, the predominant filtered buffer is phosphate but, in diabetes mellitus and in shock, weak organic acids may make a major contribution, and as much as 250 mmol per day of titratable acid may be excreted compared with the normal 20–30 mmol per day. This process consumes less H^+ than process 1 but the H^+ is secreted against an increasing gradient as urine is acidified, and an equivalent amount of HCO_3^- is added to plasma over and above the amount rescued from the filtrate.

3 Production of ammonium. Renal ammonium production and secretion are discussed on p. 626. Figure 19.4 summarizes the net result which is the secretion of NH_4^+ and restoration of an equivalent amount of HCO_3^- to the plasma. The

Table 19.3 Production of HCO_3^- and excretion of H^+ and NH_4^+ (mmol per day)

	Metabolic alkalosis	Normal	Metabolic acidosis
[HCO_3^-] plasma	33	25	17
HCO_3^- filtered	6000	4500	3000
HCO_3^- replaced	5600	4500	3000
HCO_3^- excreted	400	0	0
Total H^+ secreted	5600	4580	3400
Titratable acid excreted	0	30	100
NH_4^+ excreted	0	50	300
Total titratable acid and NH_4^+ excreted*	0	80	400

* Additional HCO_3^- synthesized by kidney and added to plasma.

normal range of NH_4^+ excretion is 30–50 mmol per day, but up to 500 mmol per day of NH_4^+ can be secreted.

Note that 1, 2 and 3 are sequential processes with some overlap and are not confined to anatomically separate segments of nephrons. H^+ ions consumed in 1 are not excreted in the urine. H^+ ions involved in 2 are excreted as urinary titratable acid; process 3 allows additional HCO_3^- to be returned to the plasma without H^+ ions being secreted into the lumen and lowering urinary pH (which cannot fall below 4.4 to 4.6).

The amounts of HCO_3^- ions contributed by these three basic processes change under different conditions, as shown by typical round figures in Table 19.3. Note that less H^+ is actually secreted during metabolic acidosis than under normal or alkalotic conditions. But during acidosis, since tubular fluid pH falls to a lower level, more H^+ has to be secreted against a substantial gradient. Also note that, in every situation, most of the H^+ secreted is used up in the replacement of filtered HCO_3^-.

Measurement of urinary titratable acid and ammonium

Because urine contains buffers, its pH cannot provide an estimate of total H^+ excretion in the urine. Urinary titratable acid is measured by adding OH^- ions to a sample of urine until its pH has been restored to the pH of the plasma from which it was formed. Since the pK' of the NH_3/NH_4^+ pair is 9.3, only about one-hundredth of the NH_4^+ has been titrated at the normal plasma pH of 7.4.

The titratable acid does not therefore include NH_4^+, and separate chemical methods are used to measure total urinary ammonium.

The importance of the measurement of titratable acid and total ammonium lies in the fact that their sum is equal to the total acid excreted in the urine, and this is equal to the total amount of additional HCO_3^- added to the plasma after replacing all the filtered HCO_3^- destroyed in the renal tubules. Under steady-state conditions, this additional HCO_3^- is equal to the HCO_3^- consumed in the body in buffering H^+ from non-volatile acids. That is, over a 24-hour period, titratable acid + total ammonium = production of non-volatile acid.

Regeneration of buffer stores

Bicarbonate added to the plasma by the tubules first replaces plasma HCO_3^- that had been lost temporarily in the glomerular filtrate. This conserves existing stores of HCO_3^- but does not add to them. This 'extra' HCO_3^-, corresponding to H^+ ions excreted as titratable acid and ammonium:

1   replaces HCO_3^- that had been used in buffering H^+ ions in the body;
2   replaces bone carbonate that had supplied bicarbonate to the ECF;
3   regenerates other (non-bicarbonate) buffer bases that had been used up:

$$HCO_3^- + HBuf \rightarrow Buf^- + H_2CO_3 \rightarrow H_2O + CO_2; \text{ and}$$

4   reverses ion exchanges with bone and cells:

$$HCO_3^- + M^+ + H^+ \text{ in bone or cell} \rightarrow M^+ \text{ in bone or cell} + H_2CO_3.$$

Regulation of hydrogen ion secretion and urinary acidity and alkalinity

Most filtered HCO_3^- is reabsorbed from the proximal tubules and not more than about 5% ordinarily reaches the distal tubules, but process 1 must give place to process 2 (see Fig. 19.4) when filtered HCO_3^- has all been destroyed within the lumen. Whether the final urine is to be acid or alkaline depends upon the relation between the rate of filtration of HCO_3^- and the rate of secretion of H^+ ions (Table 19.3).

The filtered load of

$$HCO_3^- = [HCO_3^-]_{plasma} \times GFR$$

and is proportional to the plasma $[HCO_3^-]$ so long as GFR remains constant. The rate of secretion of H^+ ions is roughly proportional to P_{CO_2} in the blood supplying the kidneys. Increased P_{CO_2} probably enhances H^+ secretion by making more CO_2 available as a source of carbonic acid and also by lowering pH in the cells so that the gradient from cell to lumen becomes more favourable. Factors which increase Na^+ reabsorption in the distal tubules and collecting ducts (e.g. aldosterone, increased Na^+ delivery) also promote secretion of H^+, as does hypokalaemia.

Disturbances of acid–base balance

Acid–base disturbances are summarized in Table 19.4. Metabolic acidosis results from a reduction in plasma HCO_3^- concentration. The consequent fall in pH stimulates respiration and the lowered P_{CO_2} then provides partial compensation for the pH change. In addition, ammonium synthesis is increased and this generates additional HCO_3^- to be returned to the plasma. Metabolic alkalosis results from a raised plasma HCO_3^- concentration. The consequent increase in pH inhibits respiration and the elevated P_{CO_2} then provides partial compensation for the pH change. Respiratory compensations for metabolic disturbances take minutes to develop. Respiratory acidosis arises from reten-

Table 19.4 Summary of acid–base disturbances

	Uncompensated			Compensated		
	pH	[HCO_3^-]	P_{CO_2}	pH	[HCO_3^-]	P_{CO_2}
Metabolic acidosis	↓	↓↓	—	↓Less	↓↓	↓
Metabolic alkalosis	↑	↑↑	—	↑Less	↑↑	↑
Respiratory acidosis	↓	—	↑	↓Less	↑	↑
Respiratory alkalosis	↑	—	↓	↑Less	↓	↓

Metabolic disturbances affect [HCO_3^-] far more than pH.
Respiratory compensation for metabolic disturbances:
for acidosis: ~1.3 mmHg↓ in P_{CO_2} for each 1 mmol ↓ HCO_3^-
for alkalosis: ~ 0.7 mmHg ↑ in P_{CO_2} for each 1 mmol ↑ in HCO_3^-

Respiratory disturbances affect pH far more than [HCO_3^-].
Renal compensation for chronic respiratory disturbances:
for acidosis: ~4 mmol ↑ in HCO_3^- for each 10 mmHg ↑ in P_{CO_2}
for alkalosis: ~4 mmol ↓ in HCO_3^- for each 10 mmHg ↓ in P_{CO_2}

Compensation is never perfect and the cause of the primary disturbance must be removed if pH, P_{CO_2} and HCO_3^- are to return to normal. But compensation does allow the body to function with greater disturbances in blood chemistry than would otherwise be possible.

tion of CO_2. The elevated P_{CO_2} contributes to increased HCO_3^- synthesis by the kidneys. This provides partial compensation for the pH change. Respiratory alkalosis arises when P_{CO_2} is reduced. This reduction reduces HCO_3^- synthesis by the kidneys, thus providing partial compensation for the pH change. Renal compensation for respiratory disturbances may take days to develop.

Metabolic disturbances

1 Metabolic acidosis. The primary finding is a reduced plasma [HCO_3^-]. This can arise from:

(a) ingestion of HCl or NH_4Cl ($Cl^- + NH_4^+ \rightarrow Cl^- + H^+ + NH_3$, converted by the liver to urea);

(b) production of ketone acids in diabetes or of lactic acid in shock;

(c) loss of alkaline secretions such as bile, pancreatic or intestinal secretions; and

(d) failure to generate adequate HCO_3^- in the kidney (**renal tubular acidosis**).

The fall in pH is largely corrected by increased pulmonary ventilation. Hence the chief effect is a reduction in [Buf^-], especially [HCO_3^-].

2 Metabolic alkalosis. The primary finding is an increased plasma [HCO_3^-]. This can arise from:

(a) ingestion of $NaHCO_3$ or alkaline stomach powders;

(b) loss of acid secretions, notably gastric juice; and

(c) depletion of body potassium.

The rise in pH is partly compensated by reduced pulmonary ventilation with retention of CO_2, and the chief effect is increased [HCO_3^-].

Compensation for metabolic disturbances

Alterations in plasma [HCO_3^-] provide automatic adjustments to minimize metabolic disturbances. In metabolic acidosis, when plasma [HCO_3^-] is reduced, less HCO_3^- is filtered, less of the H^+ secreted is consumed in reabsorption of filtered HCO_3^- and more H^+ is excreted as titratable acid. In addition, over several days ammonium synthesis is increased and this results in additional HCO_3^- being added to the plasma, thus restoring [HCO_3^-] towards normal.

This adaptive increase in ammonium production helps to limit the loss of Na^+ in the urine. Na^+ is inevitably lost to maintain electroneutrality in urines containing excessive amounts of conjugate bases of acids such as H_2SO_4 or HCl (experimentally) or ketone acids (in diabetic ketoacidosis) and there is a threat of Na^+ depletion until sufficient NH_4^+ is available to be excreted instead.

Alterations in P_{CO_2} caused by ventilatory responses to changed plasma pH retard but do not prevent these automatic adjustments. Thus, in **metabolic acidosis**, stimulation of ventilation lowers P_{CO_2} and reduces the rate of secretion of H^+.

In **metabolic akalosis**, when [HCO_3^-] is increased, the filtered load of HCO_3^- requires more H^+ for its reabsorption, and less H^+ goes to form titratable acid. Also, less ammonium is produced and excreted. When filtered HCO_3^- exceeds the total rate of H^+ secretion, process 1 in Fig. 19.4 cannot be completed, and unreabsorbed HCO_3^- is excreted in an alkaline urine, reducing the plasma [HCO_3^-] towards normal.

Retention of CO_2 by reduced ventilation increases P_{CO_2} and the rate of H^+ secretion (see illustrative values in Table 19.3). The body's store of HCO_3^- is normalized more slowly than it would be if the ventilatory response did not occur. In the meantime, however, the ventilatory responses protect the nervous system against deleterious alterations in pH.

Respiratory disturbances

1   Respiratory acidosis. Here the primary finding is an increased P_{CO_2}. This can arise from hypoventilation caused, for example, by respiratory obstruction or paralysis, drugs or toxins; and high inspired P_{CO_2}. Retention of CO_2 increases P_{CO_2} substantially. Since total anions must still equal total cations, [HCO_3^-] can increase no more than other buffer bases are reduced with falling pH; hence initially the main effect is a reduction in plasma pH, later to be corrected by slow replacement of Cl^- by HCO_3^- from the kidneys if the retention of CO_2 persists.

2   Respiratory alkalosis: Here the primary finding is a decreased P_{CO_2}. This results from hyperventilation which may be voluntary (experimental); nervous, with tension and anxiety (hyperventilation syndrome); or hypoxic, at high altitudes, from peripheral chemoreceptor drive (carotid body). P_{CO_2} is substantially reduced, but initially alteration in [HCO_3^-] is minimal, so that pH is increased with no change in [Buf^-]. Later, if hyperventilation persists, the kidneys excrete HCO_3^- in an alkaline urine and slowly restore pH towards normal by reducing plasma [HCO_3^-].

Renal compensation for respiratory disturbances of pH

The dependence of H^+ secretion upon P_{CO_2} provides slow renal compensation for respiratory disturbances of pH. In respiratory acidosis, the increased P_{CO_2} increases H^+ secretion. While the filtered load of HCO_3^- is not increased appreciably, more H^+ is excreted as titratable acid and NH_4^+, and the corresponding HCO_3^- added to the plasma increases plasma $[HCO_3^-]$ and slowly corrects the buffer ratio.

In respiratory alkalosis, the low P_{CO_2} slows H^+ secretion so that the initially undiminished filtered load of HCO_3^- is not completely reabsorbed. Bicarbonate is excreted in an alkaline urine. This lowers plasma $[HCO_3^-]$ so that the excessive alkalinity of the plasma is slowly corrected.

Remember that initially, before renal compensation can be effective (which takes days), *metabolic disturbances affect $[HCO_3^-]$ far more than pH; respiratory disturbances affect pH far more than $[HCO_3^-]$*.

Significance of plasma bicarbonate

After renal compensation for respiratory disturbances, note that respiratory acidosis ends up with a *high* plasma $[HCO_3^-]$ whereas respiratory alkalosis ends up with a *low* plasma $[HCO_3^-]$. Consequently, an *increased* plasma $[HCO_3^-]$ may indicate either metabolic akalosis or respiratory acidosis (chronic and renally compensated). A *decreased* plasma $[HCO_3^-]$ may indicate either metabolic acidosis or respiratory alkalosis (chronic and renally compensated). Laboratory differentiation between these alternatives requires the measurement of plasma pH or P_{CO_2}, as well as HCO_3^-. The conditions can usually be distinguished clinically from observation of the patient and the history of the disturbance.

The role of CO_2 in the body

Although it is a metabolic waste product, CO_2 is very important physiologically. It is the source of renal H^+ and HCO_3^- and is involved in both renal and respiratory control of pH for it affects both denominator (CO_2) directly and numerator (HCO_3^-) indirectly of the Henderson–Hasselbalch equation. In both controls and is controlled by respiration.

Chapter 20
Energy Metabolism and
Temperature Regulation

20.1 ENERGY METABOLISM

Ultimately the body acquires all its energy from outside itself, mostly in the form of food, although thermal energy is also exchanged with the environment. Because energy is neither created nor destroyed: energy of food absorbed = internal heat produced + energy stored + external work performed + energy losses in urine.

Energy is stored in the chemical bonds of organic molecules. The **energy of food absorbed** in the gastrointestinal tract is liberated during metabolism. Most of this energy appears immediately as heat, some of it may be stored as glycogen, fat or protein, and the rest is used to perform work. Internal work is performed during cardiac contraction, gastrointestinal motility and secretion, etc. and, together with the heat liberated during metabolism, contributes to the **internal heat produced**. External work is performed in moving external objects. No more than 20% of the total available chemical energy of food can be used to perform external work (p. 689); the rest is lost as heat. Energy losses in the urine include those in urea and other products.

The **chemical energy of food** can be determined by combusting it with pure oxygen in a bomb calorimeter. Since the final oxidation products of this process are similar to those of human metabolism, at least for carbohydrates and fats, the heat produced by combustion is a measure of the calorific value for the food. The end-products of protein metabolism in humans are urea and ammonia, and not the nitrogen or its oxides which are produced in the bomb calorimeter, but a correction can be made for this.

Metabolic rate

Providing no external work is being performed and body weight is constant, the chemical energy of food is transformed into heat. It is possible to measure the rate of body heat production (**metabolic rate**) in a whole-body calorimeter. However, such equipment is not readily available and it is usual to estimate metabolic rate indirectly from respiratory measurements of steady-state O_2 uptake and CO_2 production and their ratio, the respiratory quotient (RQ; p. 500). For example, the oxidation of 1 mole of glucose requires 6 moles of O_2 and produces 6 moles of CO_2, i.e. the RQ is 1. The total energy produced is 2880 kJ for each mole of glucose utilized. There are thus 480 kJ of energy released for each mole of O_2 involved. For pure carbohydrates, which have an RQ of 1, 21.4 kJ of energy is released for each litre of O_2 (measured at STPD) that is consumed. This is called the **energetic equivalent of oxygen** and it varies for different substrates (Table 20.1). Given the actual RQ and also the urinary nitrogen excretion, the mix of the substrates being metabolized can be estimated and the appropriate energy equivalent of oxygen can be assigned.

For the purposes of comparing individuals, the **basal metabolic rate** (BMR) is determined while the subject is resting in a comfortable environment, some 12 hours after eating. The BMR is calculated as watts ($1 \text{ W} = 1 \text{ J s}^{-1}$) or W m^{-2} of body surface because it correlates better with surface area than with weight and height. The resting metabolic rate of a standard 70-kg man is 100 W (equivalent to 1.433 kcal min^{-1}, i.e. 58 W m^{-2} of surface area, which has been defined as 1 **met**). Other factors that affect BMR include age, sex, race and disease. It is important that the subject be fasting for at least 12 hours because food, particularly protein, increases heat production by its so-called **specific dynamic action** (SDA). Most of this SDA is due to oxidative deamination of amino acids in the liver. Hormones, e.g. adrenaline and thyroxine, also affect the BMR and the determination of BMR was formerly used in the diagnosis of hypothyroidism and hyperthyroidism (p. 286).

The metabolic rate of a person can increase by as much as 15 times during severe exercise. Indeed, one of the major ways of combating a cold environment is by shivering. This is an involuntary contraction of skeletal muscle fibres which does no external work so that all energy appears as heat. There are also non-shivering mechanisms for increasing metabolic rate and these are discussed below.

Energy storage and utilization

For a given individual, weight remains remarkably constant during adult life and it

Table 20.1 Energetic equivalent of oxygen for different substrates

	Substrate		
	Glucose	Palmitic acid	Alanine
Respiratory quotient (RQ)	1.0	0.7	0.8
Energetic equivalent of O_2 (kJ mol^{-1})	480	425	440
Energetic equivalent of O_2 (kJ L^{-1}STPD)	21.4	19.0	19.6

is clear that there is a very precise control mechanism involved. Over a large population weights vary widely, with a normal distribution. Body composition, however, cannot be determined with the precision that weight can. Thus it is not clear whether or not a constant weight implies a constant composition. The statement that 'within every fat person a thin person is striving to get out' illustrates the concept of a **lean body mass** of muscle, skeleton and essential organs embedded in fat. The lean body mass is related to the amount of regular physical activity whereas the fat represents the stored energy. Fat stores in an average person are equivalent to the amount of energy normally consumed over some 40 days. By comparison, carbohydrate stored as glycogen provides for about a day's energy needs and glucose for about an hour's energy needs. The advantage of storing energy in the form of fat is that fat contains about twice as much energy per gram as protein or carbohydrate. Furthermore, fat cells contain relatively little water and so the volume in which energy is stored is minimized, and, as we are all aware, there is no apparent limit to the amount of energy that can be stored in this form!

In the steady state (i.e. no weight gain or loss) all of our daily absorbed nutrients must contribute to the total energy production. Table 20.2 summarizes some aspects of energy storage and utilization. Note that although the energy stored as carbohydrate represents only about 1% of the total, carbohydrate metabolism contributes about half of the basal energy consumed each day. Also, though about 20% of the energy is stored as protein, the central role of proteins in cellular structure and in skeletal muscle activity precludes the use of these stores to any extent until relatively late in starvation. Without mobility, food cannot be sought!

Given the preferred role of glucose as an immediate source of energy, and the fact that glucose is stored in cells as glycogen, it is not surprising that several hormones play key roles in regulating carbohydrate metabolism. Principal amongst these are insulin, glucagon, adrenaline, growth hormone and cortisol (p. 300). Insulin, released from pancreatic β islet cells when blood glucose rises, acts to **lower** blood glucose; it facilitates glucose entry into most cells (other than brain, gut and kidney) and favours carbohydrate storage by stimulating glycogenesis and inhibiting glycogenolysis. In contrast, the other hormones, whose secre-

Table 20.2 Body composition, energy stores and energy consumption of a typical 70-kg man at rest. The energy content of protein and carbohydrate is 17 kJ g^{-1} and of fat 37 kJ g^{-1}

Component	Site of storage	Body weight (%)	Amount (kg)	Total energy (J × 10^6)	Daily consumption (kg)	Daily energy consumption (%)
Protein	Mainly skeletal muscle	12	8.4	143	0.07	17
Fat	Adipose tissue	18	12.6	466	0.06	33
Carbohydrate	As *glycogen* in liver (one-third) and skeletal muscle (two-thirds) and as *glucose* in body fluids	0.5	0.4	7	0.2	50

tion is favoured by a decrease in blood glucose (amongst other stimuli), tend to **increase** plasma glucose by collectively inhibiting glucose uptake and facilitating glycogenolysis and gluconeogenesis.

In addition to glucose, metabolism of free fatty acids also provides a substantial proportion of daily energy requirements. Free fatty acids are derived either directly from the diet, or are released from the fat stored in adipose tissue. When plasma glucose levels are high, uptake of fatty acids by adipose tissue is enhanced, as it is by insulin, whereas glucocorticoids inhibit it. Conversely, the release of fatty acids from adipose tissue is inhibited by insulin but stimulated by adrenaline, glucagon and growth hormone. Thus the hormones involved in regulating blood sugar also regulate fat metabolism.

Finally, a dynamic balance is also maintained between protein and amino-acid metabolism. Amino acids absorbed in the gastrointestinal tract are resynthesized into new proteins to replace those continually being broken down. This process is stimulated by growth hormone and insulin. In addition, some of the amino acids are converted, principally in the liver, to glucose via a variety of intermediates (gluconeogenesis). Once again, glucagon, glucocorticoids and growth hormone favour this process, while insulin opposes it.

There is, therefore, a dynamic balance between anabolism and catabolism that is finely regulated by a variety of hormones. A summary of energy storage and utilization during the absorptive and postabsorptive states is given in Fig. 20.1. When energy is readily available, as after feeding, blood glucose and amino acid

ABSORPTIVE STATE

Energy source for most tissues is glucose.

Energy storage occurs in
    liver and muscle as glycogen and in
    adipose tissue as triglycerides obtained from
    — ingested triglycerides
    — triglycerides synthesized from glucose
      in adipose tissue
    — triglycerides synthesized from glucose
      and amino acids in liver and transported
      to adipose tissue.

POST–ABSORPTIVE STATE

Brain is an obligate user of glucose which is provided
in the short term by glycogenolysis

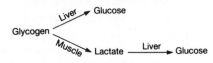

and in the long term by gluconeogenesis

Amino acids ⎤
          ⎬ —Liver→ Glucose
Glycerol ⎦

Other tissues undergo glucose-sparing reactions and
utilize free fatty acids and ketones for their energy needs.

Fig. 20.1 Summary of energy storage and utilization during the absorptive and postabsorptive states.

levels tend to rise, insulin is secreted and cellular uptake of glucose, glycogenesis, protein synthesis and fat storage are favoured. The secretion of glucagon is suppressed and so too may be that of glucocorticoids, growth hormone and adrenaline. Conversely, when an animal is deprived of food, blood glucose falls, insulin secretion is depressed, and glucose uptake by most cells is inhibited, ensuring that what glucose is available is supplied to the cells of the brain. In addition, the hypoglycaemia stimulates secretion of glucagon and perhaps also glucocorticoids, growth hormone and adrenaline. Consequently glycogenolysis and gluconeogenesis are stimulated, increasing the supply of glucose, and fatty acids are released from adipose tissue, providing an alternative source of energy to cells deprived of glucose. Furthermore, the sensation of hunger is aroused and food intake thereby encouraged.

Control of food intake

It is generally held that energy intake rather than energy expenditure is regulated to maintain a steady state. Though such regulation has been described in terms of hypothalamic satiety and hunger centres, a more complex picture is emerging. Between meals, blood glucose remains remarkably constant, reflecting neural and endocrine matching of hepatic glucose production to peripheral uptake. The neural contribution may involve participation of **glucose-sensitive neurons** in the **nucleus of the tractus solitarius**, the **vagal nuclei** and the **hypothalamus** in the regulation of autonomic outflow to the pancreatic islets and adrenal glands.

Feeding is initiated by a complex interaction of factors. Experimentally, a fall in blood glucose can be demonstrated several minutes before feeding begins, and this fall may be detected by neurons in the **lateral hypothalamus**. This area also receives a wealth of olfactory, gustatory and visual inputs and is believed to be responsible for initiating food intake. In contrast, the **ventromedial hypothalamic area** appeas to regulate the utilization and restoration of body fat stores. When this area is activated, sympathetic outflow inhibits pancreatic insulin secretion and stimulates lipolysis in adipose tissue. Destruction of this area, therefore, raises plasma insulin concentration and inhibits lipolysis, leading to increased food intake, increased body fat, weight gain and obesity. It is unclear what normally controls the activity of this area and therefore determines the balance between lipolysis and lipogenesis. However, it is suggested that the availability of glucose to cells in the area may be critical. With glucose availability diminished, insulin secretion would be inhibited and lipolysis favoured; with sufficient glucose, insulin secretion would be stimulated and lipogenesis would predominate.

Though the lateral hypothalamic area initiates feeding, the amount of food eaten is determined by a variety of factors. These include the palatability of the food and the presence of food in the mouth, stomach and intestine, as well as a range of psychological and learned responses. In particular, activation of stretch receptors and chemoreceptors in the stomach and duodenum, as well as gastrointestinal hormones (e.g. cholecystokinin) control food intake. Eating normally ceases *before* much of the ingested food has been absorbed. Other regions of the central nervous system (CNS) also affect food intake (p. 254).

Obesity

People are regarded as being obese if their adipose tissue is more than 30–40% of the total body weight. The causes of obesity are obscure. Genetic factors play a role in certain strains of obese animals. Hormonal disturbances can sometimes cause obesity; for example, it is common in eunuchs, in hypothyroidism and in hyperadrenocorticalism. Most experts, however, consider that psychological factors are the main cause of obesity. Eating habits are a learned phenomenon and cater for a multitude of sensory and emotional needs over and above energy balance. Obesity is a health hazard because it predisposes to cardiovascular disease and diabetes mellitus.

20.2 HEAT AND TEMPERATURE

Heat is a form of energy and is measured in joules (J) or in calories (1 kcal = 4184 J). The same quantity of heat can be distributed over a larger or smaller amount of material. The former will have a lower temperature than the latter. Temperature is a measure of the average kinetic energy per degree of freedom of the constituent molecules. Heat is lost from the body through radiation, evaporation, conduction and convection.

As the thermal energy in an object increases, the temperature rises. The **specific heat capacity** is the amount of energy required to be added per kilogram of material to raise the temperature by one degree. At 37°C, the addition of 4178 joules of energy (4.178 kJ) to 1 kilogram of water will raise its temperature by 1 degree (kelvin or celsius), i.e. the specific heat capacity of water at 37°C is 4.178 kJ kg^{-1}. The specific heat capacity of body tissue is about 85% that of water, i.e. 3.6 kJ kg^{-1}.

The velocity with which molecules move is dictated by their thermal energy in relation to any attractive forces between molecules. Adding energy permits the more energetic molecules to escape the attractive forces, leaving behind the less energetic molecules. In this way, energy is carried away from a liquid by evaporation. Similarly, the melting of ice to make (liquid) water involves a large increase in the velocity of its molecules. Each change requires an addition of thermal energy (**latent heat**), but while the change is occurring the temperature remains constant. To melt ice, 334 kJ kg^{-1} of thermal heat (**specific heat of fusion**) is required. The evaporation of water at 100°C requires 2257 kJ kg^{-1} (**specific heat of vaporization**) at normal atmospheric pressure (at normal human body temperature the heat of vaporization of water is 2410 kJ kg^{-1}).

Heat transfer

Heat is lost from the body through:
1   conduction and evaporation from the skin to air, which may be enhanced by convection due to air movement, and from the respiratory tract aided by the convection of tidal airflow;
2   radiation from the naked skin and to some extent between layers of clothing;
3   urination and defecation.

Although technically urination and defecation result in heat loss, they do not cool the body (except by wetting clothing and evaporating).

It is clearly difficult to say what proportion of heat loss is attributable to the various routes as it will depend on the environment and the clothing. What we can say is that it must balance heat gain and mechanisms exist to increase, say, evaporative heat loss if conductive loss becomes impossible due to high ambient temperatures. At room temperature and in the resting state, about 60% of heat is lost by conduction, 25% by radiation and 15% by evaporation. At temperatures above the thermal comfort zone (see later), about 50% of heat is lost by conduction, 20% by radiation and 30% by evaporation. Heat lost by respiration accounts for about 5% of the total.

Conduction

Within a volume of material the process of heat conduction allows heat energy to pass from areas of high temperature to those of lower. Materials differ in their capacity for heat conduction and they can be compared under standard conditions.

Thermal conductivity is the flux of energy per second through material per square metre of area per degree celsius per metre temperature gradient. So the units of thermal conductivity are $J\,s^{-1}\,m^{-2}\,{}^{\circ}C^{-1}\,m^{-1}$, which reduces to $W\,m^{-1}\,{}^{\circ}C^{-1}$, since $1\,J\,s^{-1}$ is $1\,W$. As shown in Table 20.3, the thermal conductivity of a slab of fat or muscle is about the same as that of cork.

Conduction through clothing is thus proportional to the area for conduction and the temperature gradient across it. The conductance is primarily that of air, which occupies between 0.7 and 0.9 of the volume of the fabric (the index ϕ) and is taken as $0.035\,W\,m^{-1}\,{}^{\circ}C^{-1}$ for all fabrics.

Convection

External transfers

A warm object surrounded by cooler fluid transfers heat to the adjacent fluid by conduction, as explained above. In addition, as the fluid warms, it expands, lowering its density, and an ascending flow, termed **natural convection**, results. Fluid flow resulting from wind or movement of the object is **forced convection**. Fluid moving in to replace that convected away will have the characteristics of the environment at large, rather than that of the microclimate surrounding the body.

Table 20.3 The specific thermal conductivities of different materials

Material	Specific thermal conductivity ($W\,m^{-1}\,{}^{\circ}C^{-1}$)
Felt	0.004
Air (static)	0.025
Cork	0.054
Fat (unperfused)	0.06
Fat (perfused)	0.60
Silver	418.6

So the temperature immediately adjacent to the skin will be closer to that of the environment than in the absence of convection.

However produced, flow over the skin surface will modify heat transfer between the body and the environment. The direction of heat flux depends on whether the temperature of the external fluid is higher or lower than the body, and the heat loss will depend on the thermal conductivity and specific heat of the fluid. The specific heat of water is 1000 times greater and its heat conductance 25 times greater than air, so heat loss into water, particularly stirred water, is much faster than into air.

Internal transfers

As body tissues have a very low thermal conducitivty the core would become very hot if heat were not redistributed by blood flow. Blood is warmed as it passes through an actively metabolizing region. The amount of heat in an organ, and thus its temperature, is a balance between production and removal. Heat transfer depends critically on blood flow per mass of tissue.

The fatty subcutaneous tissue has variable thermal properties. Unperfused it has the thermal conductivity of cork (Table 20.3). However, there are loops of blood vessels perpendicular to the skin and blood flow can vary from between 0.2 and 2.5 L min^{-1} at normal temperatures up to about 8 L min^{-1} at elevated body temperatures. Such enhanced body flow can increase the thermal conductivity of subcutaneous tissue 10-fold (Table 20.3). By this means, the effectiveness of the insulating layer between core and environment and also the amount of heat delivered to the skin can be varied.

Radiation

All molecules, unless at absolute zero temperature, continuously emit energy in an electromagnetic form. They simultaneously absorb such energy. In this way, energy is transferred at a distance without collision between molecules. The quantity of the radiated energy depends on the fourth power of the absolute temperature of the object and also on the ability of the surface to emit radiation, i.e. the emissivity. This also applies to radiant energy reaching the body from hotter objects. The net energy exchange by radiation is thus related to the temperature difference between the emitting surface and the receiving surfaces. The quality of the surface determines the fraction of the incident energy which is absorbed and the fraction of the maximum possible radiant energy which is actually emitted. An efficient emitter is also an efficient absorber. The **emissivity** is a measure of this efficiency and is rated as 1 for a surface absorbing all incident radiation (a black body), whereas a perfect reflector absorbs no energy and emits no energy, having an emissivity of 0. All real surfaces fall between these. Human skin emits radiation in the infrared region, with a peak at 10 μm, regardless of skin colour. The emissivity of skin in this region is approximately 0.98, which is typical of organic materials. Clothing will also lose heat by radiation but it is the surface temperature of the clothing, not the skin temperature, which will determine this heat loss.

Not all of the surface area of the naked human body is available to radiate energy. The **effective** surface area varies as body posture changes. The net radiant energy loss is 34 W from a 70-kg person, whose effective surface area is 1.5 m^2 with a skin temperature of 33°C in an environment at 29°C, and this is about a third of the resting heat production (85–100 W).

20.3 BODY TEMPERATURE

The body does not have the same temperature throughout. The core tempera-ture is the temperature of the thoracic and abdominal contents and of the brain, which must be kept at optimal temperature. The rest of the body can be allowed to cool considerably.

There is a range of temperatures from the body core to the hands and feet. (Fig. 20.2). Nevertheless, over a wide range of environmental conditions there is remarkably little variation in core temperature. This constancy implies a very precise control of heat balance and has given rise to the concept of a set-point for temperature regulation. Deviation of core temperature from the set-point pro-duces an 'error signal' that drives heat-loss or heat-retention mechanisms.

Core temperature

There are only very small temperature differences between the main internal organs and it is their temperature that is referred to as deep-body temperature or **core temperature**, often measured in the rectum. Body temperatures obtained from the mouth and axilla are about 0.5°C lower than rectal temperature and show greater variation. There is normally a diurnal variation in core temperature,

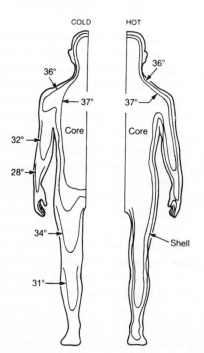

Fig. 20.2 Isotherms in the body in a cold (left) and hot (right) environment. The core, which is surrounded by 37°C isotherm, contracts in a cold and expands in a hot environment. (Adapted from Aschoff, J. & Weaver, R. (1958) *Naturwissenschaften* **45**, 477.)

the temperature being the highest in the evening (37.3°C) and lowest in the early morning (35.8°C). In females, core temperature varies during the menstrual cycle, being about 0.5°C higher in the latter half of the cycle.

Peripheral (shell) temperature

Skin temperature is somewhat less than core temperature, being the surface from which heat is lost to the environment. Between the core and the skin there is a temperature gradient and the tissues involved, the **shell**, thus provide a thermal buffer and a variable degree of thermal insulation. The extremities have a large surface area relative to their mass and the muscles within them provide heat when active. Apart from this, their only endogenous heat source is the blood flowing from the trunk. They would be regarded as part of the shell when cold environmental conditions necessitate heat conservation, but in other circumstances considerable portions would be at very close to core temperature (Fig. 20.2). This enlarged core would be associated with an increased heat content in the body.

20.4 THERMAL COMFORT AND THERMONEUTRALITY

A thermostat-like centre in the hypothalamus signals any discrepancy between the set-point and the actual body temperature. Skin receptors also have an input which gives a feeling of comfort or discomfort. Thermoregulatory responses depend on these inputs.

Brain temperature is typical of the core and is monitored by a thermoregulatory centre in the hypothalamus. It compares the hypothalamic temperature with a set-point. If these are similar and inputs received here from receptors in the skin indicate preferred temperatures, the individual feels comfortable. The ambient temperature chosen by a subject as comfortable is the **preferred comfort zone**, which covers a very narrow span. It is that combination of air temperature, air movement, humidity and radiation intensity that leads to a skin temperature of about 33°C, but showing some individual variation. The elderly and infirm may have reduced thermal responses and blunted thermal perception, and so may require a higher background temperature to prevent inadvertent hypothermia.

Around the preferred comfort zone is a range of ambient temperature in which it is possible to maintain a normal core temperature by adjusting heat loss without the expenditure of energy and without sweating. This is known as the **thermoneutral zone**. Below this at a certain **critical temperature**, the metabolic rate rises linearly as temperature is lowered further (Fig. 20.3). This rise, reflecting the activity of skeletal muscle and other mechanisms (see later), is sufficient to maintain deep-body temperature until the heat loss becomes too great. Then core temperature begins to fall as the **zone of hypothermia** is entered, regulation is lost and metabolic rate falls exponentially with the core temperature, accelerating the temperature fall. With increasing environmental temperatures above the thermoneutral zone, conductive heat loss diminishes and evaporative heat loss is enhanced by sweating. The metabolic rate remains fairly steady and minimal until at an upper limit the rate rises again. Here the **zone of hyperthermia** is reached as body temperature rises. As in the hypothermic zone, the resultant change in core

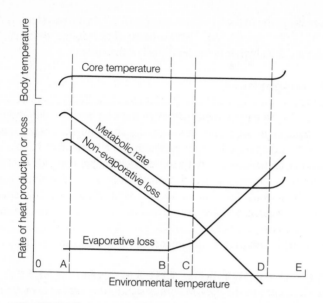

Fig. 20.3 Effects of environmental temperature on core temperature, metabolic rate, and rates of evaporative and non-evaporative heat loss. O–A, zone of hypothermia; B, lower critical temperature; B–C, thermoneutral zone; D–E, zone of hyperthermia. (From Mount, L.E. (1974) In: Monteith, J.L. & Mount, L.E. (eds) *Heat Loss from Animals and Man*, pp. 425–439. Butterworth, London.)

temperature alters the metabolic rate (up in this case), causing an accelerating loss of thermal control. It must be emphasized that these zones are of ambient temperature, not body temperature.

Thermoreception

The skin, mouth and pharynx contain thermal receptors (**peripheral thermoreceptors**), the signals of which reach consciousness. There are both warm and cold thermoreceptors whose properties have been described on p. 154. As well as the superficial receptors, there is evidence for **deep-body thermoreceptors** located within the oesophagus, stomach and perhaps duodenum and possibly in the intra-abdominal veins.

Thermoregulatory responses in a variety of animal species can be elicited by local thermal stimulation of various areas in the CNS—the spinal cord, medulla oblongata and midbrain reticular formations and, most importantly, the **preoptic–anterior hypothalamic** region. This region in the hypothalamus not only monitors brain (core) temperature; it also acts as a thermostat for determining the required core temperature, causing cooling or heat retention as necessary.

The thermal drives from various parts of the body may interact synergistically or antagonistically. For instance, during heavy exercise in cold weather, skin temperature can be low at the same time as heat production is high and core temperature is above resting levels. Thus information from temperature-sensitive neurons in the hypothalamus is combined with information from other parts of the body. These combinations can alter the threshold temperature of the hypothalamus at which thermoregulatory responses occur. As well as this, the combined information can modify the sensitivity of the hypothalamus to the input and thus the briskness of the reflex response. However, if core temperature moves significantly away from normal, the hypothalamic responses increasingly dominate.

20.5 THERMOREGULATORY EFFECTORS

The responses of the body to thermal stimuli include behavioural avoidance of excessive heat input or output, increasing metabolic heat production, alteration of thermal conductivity by changing blood-flow patterns and enhancing evaporative heat loss by sweating.

The most primitive and obvious thermoregulatory responses are behavioural. In animals this involves seeking shade or sunlight, avoiding wind and rain or revelling in a cool breeze. Alteration of posture to present more or less of the surface to the environment by curling up or stretching out is present in sleep, and in decorticate individuals. In humans selection of clothing and bedding is almost automatic, and artificial control of the environment, e.g. by opening and shutting windows, is extremely accurate even in the absence of appropriate gauges.

Short-term control of physiological change is initiated via efferent nervous pathways, both somatic and autonomic.

Shivering thermogenesis

This is an involuntary contraction of skeletal muscle fibres which can raise the metabolic rate threefold and is controlled from the hypothalamus. The muscle activity is greatest in the upper half of the torso and may reach 16% of maximum voluntary activity. Limb muscles have much less ability to shiver, and are not very effective when they do, because of their distance from the core. Purposeful movements inhibit shivering.

Non-shivering thermogenesis

In the newborn human (and adult rats) there are highly vascular tissues known as **brown fat** situated between the scapulae and around the intra-abdominal vessels. Brown fat is capable of metabolizing free fatty acids, directly generating heat which is distributed to the rest of the body via the blood. In the newborn, this, rather than shivering, is the principal means of increasing heat production and non-shivering thermogenesis can increase metabolic rate to about 2.5 times the resting value.

Although cooling the preoptic region of the hypothalamus in animals has been shown to increase the rate of secretion of thyroid hormone, the effect in adult humans is uncertain. Variations in thyroid function do not seem to play a part in long-term adaptation to cold environments but pathological alterations in thyroid function certainly alter heat production.

Cutaneous heat transfer

For heat to be lost from the body, it must be conveyed by the blood to the skin. If ambient temperature is below the skin temperature, heat can be lost by conduction and radiation. If it is above skin temperature, heat can be lost only by evaporation of sweat. A brisk blood flow to the skin is important to provide the heat necessary to evaporate the sweat, and also the water for sweat formation. This is equally true of the upper respiratory tract in panting animals.

The control of blood flow through the skin in response to changes in temperature is discussed on p. 454. The adjustments of the cardiovascular system to the strains imposed by thermal stress are discussed on p. 446. During exercise, cutaneous vasodilatation is confined to the profusely sweating areas of head, trunk and proximal limbs, while skin blood flow in distal portions of limbs does not alter much.

Cutaneous veins have a specialized function in thermoregulation. Their calibre is under noradrenergic sympathetic control and varies directly with heat load. The pathway for blood returning from the limbs can be either via superficial or deep veins. In cold conditions, it is predominantly in the deep veins and this permits transfer of heat directly from artery to vein, thus preventing heat loss from body surfaces. This is a thermal countercurrent system for heat conservation. Hot arterial blood is cooled so that the limb is perfused without transferring heat, and returning venous blood is warmed, preventing cooling of the core. In hot conditions, however, the returning venous blood passes close to the skin, and so remains cool, and returns to the core without contacting the arterial outflow, thus cooling the core.

Sweating

Sweating is an active secretory process from eccrine sweat glands which are widely distributed over the surface of the body. They are innervated by sympathetic postganglionic cholinergic fibres and produce a hypo-osmotic secretion. Their activity is blocked by atropine, the administration of which can lead to elevated body temperature.

Sweating can result in a fluid loss of as much as 1 litre per hour. This can rapidly diminish body water and electrolyte content and if these are severely depleted the rate of sweating diminishes and body temperature can rise to fatal levels. For effective heat loss, complete evaporation is needed. Therefore, sweating becomes more obvious but is less effective as environmental humidity increases.

20.6 ALTERED TEMPERATURES

The body temperature may be altered by fever as a result of infection or other illness, by exercise, or by climatic extremes. Survival probability is highly temperature-dependent.

Fever

A variety of substances resulting from infections (pyrogens) can elevate the set-point. The subject, even though body temperature is still normal, feels cold.

Peripheral vasoconstriction and shivering then ensure that heat balance is re-established at a higher temperature. This is called fever or **pyrexia**. Aspirin can lower the temperature associated with fever. Since aspirin also inhibits prostaglandin synthetase, prostaglandins appear to be involved in the mechanism of temperature regulation.

Hyperthermia can also arise following damage to the CNS, from excessive heat production related to hyperthyroidism or to malignant hyperthermia (abnormal muscle response to anaesthetics), or from failure of the heat-loss mechanism, such as in dehydration.

Exercise

During continuous exercise the core temperature varies directly with the work rate, and it can be elevated by as much as 3°C. At a given work rate the increase in core temperature remains constant over a wide range of environmental temperatures. This can be interpreted to mean that the set-point is raised; but it is also possible that the increased core temperature simply provides the error signal needed to drive the mechanisms that dissipate the heat generated during exercise.

Survival and temperature

Life is a low-temperature phenomenon existing in a universe in which temperatures range from near absolute zero to greater than 10^6 K. The rate of enzymatic chemical reaction varies with temperature, increasing about 2.5-fold for each 10°C rise. However, at temperatures above 45°C, the enzymes that catalyse living reactions lose their structural integrity and are said to be denatured. Unlike lower animals, which do not regulate their body temperature (**poikilotherms**), mammals and birds derive an advantage by maintaining their body temperature constant (**homoiotherms**) and close to denaturing temperatures. Enzyme reactions of the body function optimally within a narrow range. Conscious intelligence requires an even narrower range of approximately 35–40°C.

Body temperatures above 40°C cause heat stroke, mental confusion and unconsciousness; temperatures below 34°C give rise to amnesia and also to unconsciousness. Lowering cardiac temperature below about 30°C causes arrhythmia and then cardiac arrest. However, if the circulation is externally supported, the heart can survive for a period and functions well on rewarming. In fact, the whole body can be cooled to 10°C with consequently no circulation and yet with a very high probability of unimpaired survival on rewarming. Cooling of individual organs and tissues is deliberately used to permit survival in the absence of circulation.

Chapter 21
Exercise

Exercise consists of voluntary activation of skeletal muscle. This chapter deals first with the accompanying increased metabolic activity in the muscle, then the changes that occur in the circulatory and respiratory systems to meet the needs for supply of more O_2 and for removal of CO_2, heat and metabolites. Finally, some aspects of performance and, training are dealt with in a section on applied aspects of exercise physiology.

21.1 ENERGETICS

Muscles convert about 20% of the energy of adenosine triphosphate (ATP) to external work, the remaining energy appearing as heat. The demand for ATP is met by anaerobic and aerobic metabolism. Consumption of oxygen by the aerobic system takes time to develop and continues for some time after exercise stops.

The energy for muscular work is derived from the breakdown of ATP and creatine phosphate (p. 122). As stores of ATP are limited, exercise lasting for more than a few seconds demands an increased supply from **aerobic** and **anaerobic metabolism**. In mild exerise ATP is produced by the aerobic pathway by oxidation of fatty acids and ketones, while in more vigorous exercise glycogen is oxidized. At higher levels of exercise the additional demand for ATP is met by the anaerobic pathway, which produces ATP more rapidly by converting glycogen to lactic acid. The relative contributions from the aerobic and anaerobic pathway depend on the duration of the exercise. Aerobic metabolism contributes very little of the total energy utilized during brief explosive exercise. The contribution increases to about 50% for events such as a 400-metre race lasting under a minute, and exceeds 90% in events lasting 10 minutes or more.

    Fatigue is the sensation associated with increasing difficulty or inability to sustain a given level of work output or exercise. Fatigue occurring in the first few seconds of very intense exercise is caused by depletion of ATP and creatine phosphate stores in the active muscles. For slightly less intense exercise, fatigue occurs within a few minutes as the build-up of lactic acid and other metabolites in the muscles inhibits glycolysis or the contractile mechanism. Fatigue also results from depletion of intramuscular glycogen stores after long periods of exercise ('hitting the wall' in the marathon).

The **efficiency** of exercise is a measure of the conversion of chemical energy into external work. In estimating efficiency, the exercise physiologist is usually content to measure the steady-state O_2 consumption ($\dot{V}_{O_2}$) of a subject in conditions where the rate of external work can be estimated reasonably accurately. It is assumed either that aerobic metabolism is the sole source of power, or that any products of anaerobic metabolism are oxidized. The efficiency can then be calculated after converting the O_2 cost of exercise ($\dot{V}_{O_2}$ in exercise − $\dot{V}_{O_2}$ at rest) measured, for example, in millilitres per second, into units of power. On a mixed diet, $1 \, \text{ml s}^{-1}$ of oxygen (STPD) yields 20 W of total power (determined using whole-body calorimetry), and therefore:

$$\% \text{ efficiency} = \frac{\text{rate of external work}}{(\dot{V}_{O_2} \text{ in exercise} - \dot{V}_{O_2} \text{ at rest}) \times 20} \times 100$$

The external work of exercise may be very difficult to estimate, as for example in running on the level, where work is done to accelerate and decelerate the limbs and to raise and lower the centre of gravity with each step. On the other hand, the external work of cycling is easily determined with a cycle ergometer, and the efficiency of cycling is found to be 20–25%. This means that someone doing 200 W of external work on a cycle ergometer is also generating about 800 W of heat, which must be transferred to the skin from where it can be dissipated (p. 686).

If exercise is started abruptly and a constant moderate work output is maintained, the O_2 uptake takes several minutes to reach a steady state (Fig. 21.1). The missing O_2 that would have been consumed, if the body's response to the O_2 demand of exercise was immediate, is known as the **oxygen deficit**. For mild to moderate exercise the oxygen deficit is accounted for by:

1 the lower content of ATP and creatine phosphate in the active muscles;
2 the reduction in O_2 content of myoglobin in the active muscles;
3 the reduction in O_2 content of the venous blood leaving the muscles.

The **oxygen debt** or **excess postexercise oxygen consumption** represents repayment of the O_2 deficit (Fig. 21.1). When the exercise is vigorous or prolonged the debt is considerably greater than the deficit. Following such exercise the ATP and creatine phosphate stores and the O_2 content of myoglobin and venous blood return to normal within a minute or two, but there is a residual

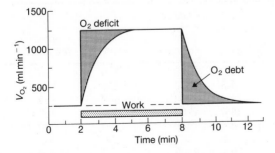

Fig. 21.1 Time-course of O_2 consumption ($\dot{V}_{O_2}$) during moderate exercise, illustrating the O_2 deficit and debt.

elevated O_2 consumption that can last for minutes to hours depending on the intensity and duration of the exercise. This extra O_2 consumption is attributed to the energy needed to dissipate heat and restore intracellular electrolytes to normal concentrations, and to the higher metabolic rate caused by the increase in body temperature and in circulating levels of catecholamines and thyroxine. Regeneration of intramuscular glycogen stores also produces a very small increase in O_2 consumption over a period of a day or two following the exercise. Lactate produced by the anaerobic pathway is metabolized within 30 minutes after exercise but, contrary to popular belief, metabolism of lactate following anaerobic exercise does not contribute significantly to the O_2 debt.

21.2 CIRCULATORY RESPONSES IN EXERCISE

Blood flow to muscles increases during exercise as a result of vasodilatation by local factors. Activity in the sympathetic nervous system also increases cardiac output and reduces flow to other tissues to meet the demand for increased muscle blood flow.

Rhythmic exercise produces an increase in blood flow in the active muscles, which at maximum can reach 20 times the resting blood flow. The vasodilatation that gives rise to the increase in blood flow may be due, in part, to an increase in activity of the cholinergic sympathetic nerves to blood vessels in muscles, but most of it is caused by local factors such as the decrease in P_{O_2} and pH, the increase in P_{CO_2}, extracellular $[K^+]$ and temperature, and release of other metabolites such as adenosine (p. 423). The exercise-induced vasodilatation results in a large increase in the number of capillaries carrying blood through the muscle; this shortens the mean path length between capillaries and mitochondria in the muscle fibres, allowing a greater extraction of O_2 from the haemoglobin. Off-loading of O_2 from haemoglobin is also assisted by the increased P_{CO_2}, temperature and acidity in the working muscles (p. 530), resulting in the extraction of about 90% of the O_2 from blood perfusing muscles at maximum exercise.

Changes in blood flow to the rest of the body during exercise are shown in Table 21.1. Blood flow to heart muscle increases, meeting the extra O_2 demand of

Table 21.1 Changes in blood flow distribution during exercise

| | Rest | | Exercise | | | |
	(ml min^{-1})	(%)	Light (ml min^{-1})	Medium (ml min^{-1})	Heavy (ml min^{-1})	(%)
Cerebral	750	13.0	750	750	750	3.0
Coronary	250	4.5	350	650	1000	4.0
Renal	1100	19.0	900	600	250	1.0
Splanchnic	1400	24.0	1100	600	300	1.2
Skin	500	8.5	1500	1800	600	2.4
Others	600	10.5	400	300	100	0.4
Skeletal muscle	1200	20.5	4500	10 800	22 000	88.0
Cardiac output	5800	100.0	9500	15 500	25 000	100.0

the increased cardiac work; as in skeletal muscle this flow is under local control, although circulating adrenaline may also contribute to the dilatation of the coronary vessels via activation of β_2-receptors. Blood flow to the gut and kidney is reduced by sympathetic activity, making more blood available for the exercising muscles. Blood flow to the skin increases as the need to dissipate heat increases, but at maximum exercise skin flow decreases, giving the subject a pale or ashen pallor just before exhaustion; this indicates that heat dissipation is compromised when cardiac output cannot be increased further to meet the demands of exercise. The constancy of blood flow to the brain reflects the unchanged metabolic rate of the brain whatever the level of exercise, and there appears to be no reserve here that can be drawn upon for the exercising muscles.

The effects of exercise on circulatory and respiratory variables are summarized in Table 21.2 and Fig. 21.2. The increase in O_2 consumption (about 15-fold at maximum in a young adult) is accommodated by an increase in the arteriovenous O_2 concentration difference (up to threefold) and an increase in cardiac output (up to fivefold). The increase in cardiac output results from venoconstriction, vasodilatation and increased myocardial contractility, as well as from the repeated squeezing of the veins by the active muscles (the muscle pump). The increase in heart rate limits the increase in stroke volume to a factor of 1.5. Most of this increase in stroke volume occurs in light to moderate levels of exercise and results from an increase in end-diastolic volume (increased filling due to higher central venous pressure) and from a decrease in end-systolic volume (increase emptying due to sympathetically increased contractility). At high levels of exercise the end-diastolic volume is no greater than at rest, and all the increase in stroke volume is due to a greater emptying (p. 403). Systolic pressure increases and diastolic pressure decreases during exercise, resulting in a pronounced increase in pulse pressure; however, in healthy young people mean blood pressure increases only slightly during moderate dynamic exercise.

The nature of the primary signal activating the sympathetic nervous system during exercise is still not entirely clear, but it is probably activity in sensory nerves

Table 21.2 Cardiorespiratory variables and indices at rest and during exercise

	Rest	Exercise		
		Light	Medium	Heavy
Oxygen consumption(ml min^{-1})	250	1500	2800	3500
Oxygen consumption of muscles (ml min^{-1})	50	1250	2200	3150
Cardiac output (L min^{-1})	5	13	16	21
Arterio venous oxygen difference (ml L^{-1})	50	100	127	130
$P_{\bar{v}}O_2$(mmHg)	40	26	23	20
Heart rate (beats min^{-1})	72	144	160	190
Diastolic filling time (ms)	500	200	170	150
Stroke volume (ml)	70	90	100	110
Mean blood pressure (mmHg)	85	90	95	100
Pulse pressure (mmHg)	40	60	90	120
Systolic/diastolic pressure (mmHg)	110/70	130/70	155/65	180/60

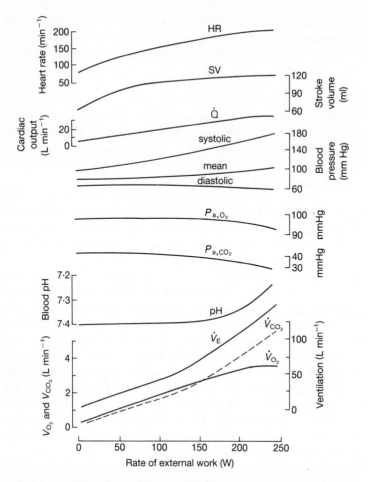

Fig. 21.2 Changes in metabolism, ventilation, arterial pH, arterial gases and cardiovascular variables of a healthy male subject as a function of steady-state work on a cycle ergometer. HR, Heart rate; Q, cardiac output; SV, stroke volume.

that detect the factors inducing vasodilatation in the active muscles (p. 444). This means that control of the circulatory system via the sympathetic nervous system is by negative feedback, because the circulatory response (increased cardiac output or blood pressure) tends to remove the initiating stimulus (metabolites released in the muscles) by increasing perfusion of the muscles. There may be additional stimulation of the sympathetic system from neural activity in the motor cortex or in sensory pathways carrying information about movement, and also from neural activity in the respiratory system. The baroreceptor reflex (p. 440) helps to buffer changes in blood pressure resulting from mismatch of cardiac output and peripheral resistance in dynamic exercise. Note, however, that in exercise where perfusion of muscles is absent or limited (e.g. sustained static contractions or arm work with arms raised above the head), vasodilator factors accumulate in the muscles, stimulate sensory nerves and produce dramatic increases in blood pressure. The time-course of the changes in cardiac output during a bout of moderate exercise is shown in Fig. 21.3. There is a sudden increase in cardiac output at the start of exercise, followed by a gradual exponential rise to the steady state; when exercise is stopped, there is a sudden decrease in cardiac output, then an exponential fall. The sudden changes at the onset and cessation of exercise are attributed to the effect of the muscle pump on venous return, and also to stimulation of the cardiovascular system from the neural activity associated with movement. The slower changes reflect the time-course of vasodilatation in active muscles and of stimulation of the cardiovascular system via the sensory nerves that detect the vasodilator factors in the muscles.

21.3 RESPIRATORY RESPONSES IN EXERCISE

The increase in pulmonary ventilation that occurs during exercise clears CO_2 from the venous blood and maintains arterial P_{O_2}, but the main stimulus for the increase in ventilation remains a mystery.

Pulmonary ventilation ($\dot{V}_E$) increases with light to moderate exercise in proportion to the increased demand for the supply of O_2 and the excretion of CO_2 (Fig. 21.2). This involves increases in both the frequency of breathing and tidal volume. The transient response of $\dot{V}_E$ during a bout of exercise (Fig. 21.3) shows rapid and gradual changes similar to those for cardiac output.

The nature of the stimuli that produce the increase in ventilation with exercise is still uncertain. The sudden changes in $\dot{V}_E$ at the beginning and end of exercise appear to result from neuronal input to the respiratory centre from neural activity in the motor cortex and possibly also from movement receptors in the exercising limbs (p. 540). However, this neural drive appears to play only a minor role in the steady-state response, which has been shown to be coupled very closely to the delivery of CO_2 to the lungs. Detection of CO_2 by a chemoreceptor is therefore implicated, but a venous or pulmonary CO_2 receptor driving ventilation cannot be demonstrated. Arterial chemoreception also seems to be ruled out, because an increase in mean arterial P_{CO_2} or a decrease in P_{O_2} would be necessary to provide

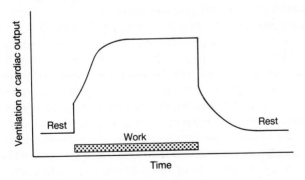

Fig. 21.3 Time-course of abrupt and slow changes in ventilation and cardiac output during exercise.

the stimulus for the increased $\dot{V}_E$, and these are not observed (Fig. 21.2). This has led to the suggestion that the sensitivity of chemoreception by the carotid bodies is altered during exercise; another suggestion is that the increase in venous CO_2 load increases the rate of change of arterial pH associated with the breath-to-breath oscillations in arterial P_{CO_2}, and that this is sensed by the carotid bodies and provides the primary stimulus to ventilation. Neither of these hypotheses has firm experimental support.

At higher levels of exercise there are disproportionate increases in $\dot{V}_E$ and $\dot{V}_{CO_2}$ compared with increases in $\dot{V}_{O_2}$ (Fig. 21.2). This hyperventilation appears to arise from additional stimulation of the carotid bodies, because it is virtually absent when the carotid bodies are denervated. The stimulus is usually taken to be the fall in arterial pH that occurs when lactic acid is released by the muscles and starts to accumulate in the blood; the hyperventilation is viewed as an additional buffering mechanism that lowers arterial P_{CO_2} and thereby prevents arterial blood from becoming too acidic. However, other substances released by active muscles, for example K^+, may also stimulate the carotid chemoreceptors and contribute to the hyperventilation response.

At the most intense levels of exercise, O_2 consumption begins to reach a plateau that defines its maximum value, the $\dot{V}_{O_2max}$ (Fig. 21.2). Utilization of O_2 by muscles (oxidative capacity) is not considered to be a limiting factor for $\dot{V}_{O_2max}$, and in most people the limitation appears to be the supply of O_2 to muscles by cardiac output. However, in some highly trained athletes, significant arterial hypoxaemia (P_{O_2} <70mmHg) develops as $\dot{V}_{O_2max}$ is approached, indicating that for these individuals the limitation may be in the lungs. The hypoxaemia results from a combination of a reduced hyperventilation response and a reduced transit time for blood in the pulmonary capillaries.

21.4 PHYSIOLOGICAL INDICES OF PERFORMANCE

The only useful physiological indices of physical performance are maximum oxygen consumption and the anaerobic threshold.

Some measurements made by the physiologist in the laboratory under defined conditions can be useful indices of the current performing ability of athletes. However, in sprint or power events the relationship between physiological measures and physical performance is not good. For example, oxygen deficit, oxygen debt and blood lactate concentration in short-term maximal exercise tests might be expected to correlate with performance in sprint or power events, but the relationship turns out to be poor. On the other hand, the performance of endurance athletes correlates well with several physiological measures. One of these is $\dot{V}_{O_2max}$, which is usually determined by exercising the subject gradually to maximum on a treadmill or cycle ergometer over a 5–15 minute period. An even better measure is the so-called **anaerobic threshold**. This is the workload above which the concentration of lactic acid in the blood builds up rapidly. The high correlation between the anaerobic threshold and endurance performance arises most probably because the anaerobic threshold represents the highest level of exercise that can be performed without fatigue occurring from the accumulation of acid in the muscles. Muscle cells can produce ATP and lactic acid via anaerobic metabolism, even when aerobic metabolism is not limited by the supply of O_2, so the anaerobic threshold is not necessarily a threshold for the onset of hypoxia.

Training can be for strength or short-term power (anaerobic training) or for endurance (aerobic training). The various changes that occur can be regarded as adaptations to the stress of exercise, but the mechanisms that bring about the changes are poorly understood. Anaerobic training, which consists of many repetitions of brief, intense exercise, results in increases in muscle-fibre diameter and activities of enzymes of anaerobic metabolism. Increases in blood pressure during anaerobic training may also produce thickening of the left ventricular wall of the heart. Aerobic training is achieved by exercising at moderate to high intensity for at least 20 minutes several times per week. The most well-known adaptation to such training is an increase in the $\dot{V}_{O_{2max}}$, which is attributed to an increase in maximum cardiac output and maximum blood flow to the muscles. The increase in cardiac output is itself a consequence of an increased stroke volume with no change in maximum heart rate (Table 13.2, p. 403). The stroke volume is increased at all workloads, including rest, as a result of an increase in circulating blood volume and in the volume of the ventricles. Cardiac output at rest does not change, so the increased stroke volume entails a reduced resting heart rate. Endurance-trained muscles have significant increases in the concentration of oxidative enzymes and myoglobin, and in the density of mitochondria and of capillaries (p. 132); these changes probably contribute to the increase in the $\dot{V}_{O_2}$ max and anaerobic threshold.

Appendix: Units of Measurement

Throughout this text the units are those commonly used by physiologists. Unfortunately, physiologists and their clinical colleagues have been slow to adopt the now accepted SI units familiar to chemists and physicists. To assist readers trained in this system (and those who believe that physiology should use a logical and consistent system), a summary of SI units and conversion for common units used by physiologists follows.

Basic SI units

	Unit
Length	metre (m)
Mass	kilogram (kg)
Time	second (s)
Electric current	ampere (A)
Thermodynamic temperature	kelvin (K)
Amount of substance	mole (mol)
Luminous intensity	candela (cd)

Derived SI units and relationships to other units in common use

	Unit	Definition	Other units
Area		m^2	
Volume		m^3	1 litre (L) = 1 dm^3
Velocity		$m\,s^{-1}$	
Concentration		$mol\,dm^{-3}$ ($mol\,L^{-1}$)	
Density		$kg\,m^{-3}$	
Force	newton (N)	$m\,kg\,s^{-2}$	1 dyne (dyn) = 10^{-5} N
Pressure	pascal (Pa)	$N\,m^{-2}$ ($m^{-1}\,kg\,s^{-2}$)	1 atmosphere (atm)
			= 101.3 kPa
			= 760 mmHg
			1 mmHg = 1 torr
			= 133.3 Pa
			1 cmH_2O = 98.07 Pa
Energy	joule (J)	$N\,m$ ($m^2\,kg\,s^{-2}$)	1 calorie (cal) = 4.19 J
Power	watt (W)	$J\,s^{-1}$ ($m^2\,kg\,s^{-3}$)	
Dynamic viscosity		$Pa\,s$ ($m^{-1}\,kg\,s^{-1}$)	1 poise (P) = 10^{-1} Pa s
Surface tension		$N\,m^{-1}$ ($kg\,s^{-2}$)	1 dyne cm^{-1} = 10^{-5} N cm^{-1}
Electric charge	couloumb (C)	$A\,s$	
Electric potential difference	volt (V)	$J\,C^{-1}$ ($m^2\,kg\,A^{-1}\,s^{-3}$)	

	Unit	Definition	Other units
Electric resistance	ohm (Ω)	$V A^{-1} (m^2 kg A^{-2} s^{-3})$	
Electric conductance	siemens (S)	$\Omega^{-1} (A^2 s^3 m^{-2} kg^{-1})$	
Electric capacitance	farad (F)	$C V^{-1} (m^{-2} kg^{-1} A^2 s^4)$	
Frequency	hertz (Hz)	s^{-1}	

Physical constants

Name	Symbol	Numerical value
Gravitational constant	G	$6.67 \times 10^{-11} N m^2 kg^{-2}$
Universal gas constant	R	$8.314 J K^{-1} mol^{-1}$
		(at atmospheric pressure and
		$0°C$, i.e. $P_o = 101.3$ kPa, $T_o = 273.15$ K)
Molar volume of ideal gas	V_o	$2.24 \times 10^{-2} m^3$ (22.4 L)
		(at $P_o = 101.3$ kPa, $T_o = 273.15$ K)
Avogadro constant	N_A	$6.02 \times 10^{23} mol^{-1}$
Faraday constant	F	$9.65 \times 10^4 C mol^{-1}$

Note: R is also sometimes expressed as 8.206×10^{-2} L atm $K^{-1} mol^{-1}$

Decimal multiples and submultiples of units

kilo	(k)	for 10^3
hecto	(h)	for 10^2
deca	(da)	for 10^1
deci	(d)	for 10^{-1}
centi	(c)	for 10^{-2}
milli	(m)	for 10^{-3}
micro	(μ)	for 10^{-6}
nano	(n)	for 10^{-9}
pico	(p)	for 10^{-12}
femto	(f)	for 10^{-15}
atto	(a)	for 10^{-18}

Note: 1 Angstrom (Å) = 10^{-10} metre

Index

699